GW01388306

European ABS Antilock Brake Systems TechBook

Charles White

Systems covered

Bendix Addonix	Clayton Dewandre Wabco	Teves II
Bendix Integrated	Kelsey Hayes 415	Teves Mk.IV (ITT 04)
Bendix Mecatronic II/III	Lucas/Girling 2/2	Teves IV-GI (ITT 04-GI)
Bosch 2	Lucas 4/4F	Teves 20-I
Bosch 5.0	Lucas EBC 430	Teves 20GI (ITT 20GI)
Bosch 5.3	Lucas SCS	Teves 20IE (ITT 20IE)

Printed by **J H Haynes & Co Ltd, Sparkford, Nr Yeovil, Somerset BA22 7JJ, England**

Haynes Publishing
Sparkford, Nr Yeovil, Somerset BA22 7JJ, England

Haynes North America, Inc
861 Lawrence Drive, Newbury Park, California 91320, USA

Editions Haynes
Tour Aurore – IBC, 18 Place des Reflets
92975 Paris La Defense 2 Cedex, France

Haynes Publishing Nordiska AB
Box 1504, 751 45 UPPSALA, Sweden

30130 119040508

Contents

GENERAL INFORMATION

SYSTEM SPECIFICS

REFERENCE

Introduction

This book is devoted to unravelling the mysteries of the Antilock Brake Systems (ABS) used on mainly European, or European derived vehicles. The book first describes the ABS principles of operation, followed by details of test equipment, general test routines and fault diagnosis. Further chapters describe system operation, and test procedures for each specific system. Even if the reader has no intention of actually attempting to investigate faults on their own vehicle, the book provides valuable insight into ABS operation, testing and fault diagnosis.

On the other hand, if you relish the task of electronic fault diagnosis, this book will provide you with much of the background knowledge necessary to test the ABS components and circuits on your car. Generally we describe how to diagnose faults using test equipment available from most good auto retailers. We also mention where the use of specialised equipment is necessary, and describe some of the common routines used by the professional garage trade.

This book does delve into electrical or electronic theory, but there are many other excellent publications available for that purpose. See *'Automotive Electrical and Electronic Systems Manual'*, available from the publishers of this book.

The Vehicle Manufacturer (VM) may not specifically endorse a number of our tests and routines. Mainly this is because VM test procedures are becoming more focused upon specialised test equipment that is not readily available to the 'aftermarket'. In almost all instances, our own tests follow well defined test methods taught in independent training schools and used by many modern vehicle technical specialists. We mainly describe simple methods of testing that are possible with the aid of the ubiquitous Digital Multi-Meter (DMM). Our test procedures are necessarily generic. However, by following the procedures in conjunction with a good wiring diagram you will be able to diagnose the reason for most faults.

The routines and test methods that we describe are perfectly safe to carry out on electronic systems so long as certain simple rules are observed. These rules are actually the observation of good electrical practice. Be aware that damage resulting in the replacement of a very expensive Electronic Control Module (ECM) may be the result of not following the rules. Please refer to the Warnings section in Reference, and heed them implicitly.

We have described the operation and test procedures of most of the modern antilock brake systems fitted to mainly European, or European derived vehicles. However you must be aware of the generic nature of our illustrations and wiring diagrams. Each system may have many variations according to the specific vehicle to which it is actually fitted. Differences which are often encountered are variations in the type and number of relays used in the system, and the allocation of ECM pins (particularly in the area of wheel speed sensors). In this instance, particular care must be taken during system testing and fault diagnosis. Although we have tried to define most of the system variations, we have only provided an example of the wiring diagrams for any particular system. It is essential that a specific wiring diagram for the vehicle under test is obtained where serious tracing of system faults is required.

Throughout the world, the various VMs tend to use their own particular terms to describe a particular component. Of course, all of these terms tend to be different and the problem is exacerbated by translation from the various languages. This often leads to confusion when several terms are used to describe the same component. There have been attempts to bring all the VMs into line with a common standard for all, and one does now exist, (J1930). It seems unlikely that all VMs will adopt this particular standard and we are unsure that the terms used are particularly meaningful anyway. So, the terms used in this book will tend to follow terms that are commonly used in the UK. To avoid confusion we will tend to use similar terms over the whole range of systems, and alternatives will be listed in the Glossary.

Acknowledgements

We would like to thank all of those people at Sparkford and elsewhere that helped us in the production of this manual. In particular we would like to thank Equiptech for permission to use illustrations from the 'CAPS' ABS fault diagnosis database and for providing much of the technical information that was used in authoring this book. We also thank Desmond Broomfield and Ross Forest who drew many of the illustrations, and John Mead and Simon Ashby for the provision of additional technical information, specific system knowledge and general guidance in the preparation of this manual.

We take great pride in the accuracy of information given in this manual, but vehicle manufacturers make alterations and design changes during the production run of a particular vehicle of which they do not inform us. No liability can be accepted by the authors or publishers for loss, damage or injury caused by errors in, or omissions from, the information given.

Safety First!

Working on your car can be dangerous. This page shows just some of the potential risks and hazards, with the aim of creating a safety-conscious attitude.

General hazards

Scalding

• Don't remove the radiator or expansion tank cap while the engine is hot.
• Engine oil, automatic transmission fluid or power steering fluid may also be dangerously hot if the engine has recently been running.

Burning

• Beware of burns from the exhaust system and from any part of the engine. Brake discs and drums can also be extremely hot immediately after use.

Crushing

• When working under or near a raised vehicle, always supplement the jack with axle stands, or use drive-on ramps. *Never venture under a car which is only supported by a jack.*
• Take care if loosening or tightening high-torque nuts when the vehicle is on stands. Initial loosening and final tightening should be done with the wheels on the ground.

Fire

• Fuel is highly flammable; fuel vapour is explosive.
• Don't let fuel spill onto a hot engine.
• Do not smoke or allow naked lights (including pilot lights) anywhere near a vehicle being worked on. Also beware of creating sparks (electrically or by use of tools).
• Fuel vapour is heavier than air, so don't work on the fuel system with the vehicle over an inspection pit.
• Another cause of fire is an electrical overload or short-circuit. Take care when repairing or modifying the vehicle wiring.
• Keep a fire extinguisher handy, of a type suitable for use on fuel and electrical fires.

Electric shock

• Ignition HT voltage can be dangerous, especially to people with heart problems or a pacemaker. Don't work on or near the ignition system with the engine running or the ignition switched on.

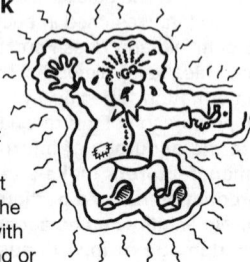

• Mains voltage is also dangerous. Make sure that any mains-operated equipment is correctly earthed. Mains power points should be protected by a residual current device (RCD) circuit breaker.

Fume or gas intoxication

• Exhaust fumes are poisonous; they often contain carbon monoxide, which is rapidly fatal if inhaled. Never run the engine in a confined space such as a garage with the doors shut.
• Fuel vapour is also poisonous, as are the vapours from some cleaning solvents and paint thinners.

Poisonous or irritant substances

• Avoid skin contact with battery acid and with any fuel, fluid or lubricant, especially antifreeze, brake hydraulic fluid and Diesel fuel. Don't syphon them by mouth. If such a substance is swallowed or gets into the eyes, seek medical advice.
• Prolonged contact with used engine oil can cause skin cancer. Wear gloves or use a barrier cream if necessary. Change out of oil-soaked clothes and do not keep oily rags in your pocket.
• Air conditioning refrigerant forms a poisonous gas if exposed to a naked flame (including a cigarette). It can also cause skin burns on contact.

Asbestos

• Asbestos dust can cause cancer if inhaled or swallowed. Asbestos may be found in gaskets and in brake and clutch linings. When dealing with such components it is safest to assume that they contain asbestos.

Special hazards

Hydrofluoric acid

• This extremely corrosive acid is formed when certain types of synthetic rubber, found in some O-rings, oil seals, fuel hoses etc, are exposed to temperatures above 400°C. The rubber changes into a charred or sticky substance containing the acid. *Once formed, the acid remains dangerous for years. If it gets onto the skin, it may be necessary to amputate the limb concerned.*
• When dealing with a vehicle which has suffered a fire, or with components salvaged from such a vehicle, wear protective gloves and discard them after use.

The battery

• Batteries contain sulphuric acid, which attacks clothing, eyes and skin. Take care when topping-up or carrying the battery.
• The hydrogen gas given off by the battery is highly explosive. Never cause a spark or allow a naked light nearby. Be careful when connecting and disconnecting battery chargers or jump leads.

Air bags

• Air bags can cause injury if they go off accidentally. Take care when removing the steering wheel and/or facia. Special storage instructions may apply.

Diesel injection equipment

• Diesel injection pumps supply fuel at very high pressure. Take care when working on the fuel injectors and fuel pipes.

⚠ *Warning: Never expose the hands, face or any other part of the body to injector spray; the fuel can penetrate the skin with potentially fatal results.*

Remember...

DO

• Do use eye protection when using power tools, and when working under the vehicle.

• Do wear gloves or use barrier cream to protect your hands when necessary.

• Do get someone to check periodically that all is well when working alone on the vehicle.

• Do keep loose clothing and long hair well out of the way of moving mechanical parts.

• Do remove rings, wristwatch etc, before working on the vehicle – especially the electrical system.

• Do ensure that any lifting or jacking equipment has a safe working load rating adequate for the job.

DON'T

• Don't attempt to lift a heavy component which may be beyond your capability – get assistance.

• Don't rush to finish a job, or take unverified short cuts.

• Don't use ill-fitting tools which may slip and cause injury.

• Don't leave tools or parts lying around where someone can trip over them. Mop up oil and fuel spills at once.

• Don't allow children or pets to play in or near a vehicle being worked on.

Index of vehicles covered

Model	Year	System
Alfa Romeo		
145/146	1994-1996	Bosch 2EH
145/146	1996-2000	Bosch 5.3
155	1992-1996	Bosch 2EH
155	1996-1998	Bosch 5.3
156	1996-2000	Bosch 5.3
164	1990-1992	Bosch 2S
164	1992-1998	Bosch 2EH
166	1999-2000	Bosch 5.3
Audi		
80	1988-1991	Bosch 2S
80	1992-1995	Bosch 2E
90	1988-1991	Bosch 2S
100	1988-1992	Bosch 2S
100	1992-1995	Bosch 2E
200	1986-1991	Bosch 2S
A3	1996-2000	Teves 20IE (ITT 20IE)
A4	1995-1996	Bosch 5.0
A4	1997-2000	Bosch 5.3
A6	1994-1998	Bosch 5.0
A6	1998-2000	Bosch 5.3
A8	1994-1997	Bosch 5.0
A8	1997-2000	Bosch 5.3
V8	1990-1992	Bosch 2S
V8	1992-1994	Bosch 2E
Coupe	1988-1991	Bosch 2S
Coupe	1992-1996	Bosch 2EH
TT	1999-2000	Teves 20IE (ITT 20IE)
BMW		
3-Series (E30)	1982-1993	Bosch 2S
3-Series (E36)	1990-1996	Teves Mk.IV (ITT 04) - 55-pin
3-Series (E36)	1996-1997	Teves Mk.IV (ITT 04) - 42-pin
5-Series (E28)	1982-1987	Bosch 2S
5-Series (E34)	1988-1992	Bosch 2S
5-Series (E34)	1992-1996	Bosch 2E
5-Series (E39)	1996-1999	Bosch 5.0/TC
7-Series (E32)	1987-1992	Bosch 2S
7-Series (E32)	1992-1994	Bosch 2E
7-Series (E38)	1994-2000	Bosch 5.0

Citroën	Year	System
AX	1991-1997	Bendix Addonix
Berlingo	1996-2000	Bosch 5.0
BX	1987-1992	Teves II
CX	1988-1991	Bosch 2S
Dispatch	1995-2000	Bendix Addonix
Evasion	1995-1999	Bendix Addonix
Relay	1994-1997	Bendix Addonix
Relay	1997-2000	Bosch 5.3
Saxo	1996-1998	Teves IV-GI (ITT 04-GI)
Saxo	1998-1999	Teves 20IE (ITT 20IE)
Synergie	1995-1997	Bendix Addonix
Synergie	1997-1999	Bosch 5.3
Xantia	1993-1998	Teves Mk.IV (ITT 04) - 37-pin
Xantia	1998-1999	Teves 20IE (ITT 20IE)
XM	1989-1998	Bendix Addonix
XM (2.0 and Diesel)	1996-2000	Teves Mk.IV (ITT 04) - 37-pin
Xsara	1997-2000	Bosch 5.3
ZX	1991-1996	Bosch 2EH
ZX	1996-1997	Bosch 5.0

Daewoo		
Matiz	1998-2000	Lucas EBC 430

Fiat		
Barchetta	1995-1997	Bosch 2SH
Barchetta	1997-2000	Bosch 5.3
Bravo/Brava	1995-2000	Teves 20GI (ITT 20GI)
Croma	1986-1991	Bosch 2S
Croma	1991-1993	Bosch 2EH
Fiat Coupe	1995-1998	Bosch 2EH
Fiat Coupe	1998-2000	Bosch 5.3
Fiorino	1997-2000	Bosch 5.3
Marea/Weekend	1996-2000	Lucas EBC 430
Punto	1994-1996	Bosch 2SH
Punto	1996-2000	Bosch 5.3
Seicento	1998-2000	Bosch 5.3
Tempra (1.6/1.8)	1988-1993	Lucas/Girling 2/2
Tempra (2.0/Diesel)	1988-1997	Bosch 2S
Tipo (1.4/1.6)	1988-1993	Lucas/Girling 2/2
Tipo (2.0/Diesel)	1988-1995	Bosch 2S
Ulysse	1995-1997	Bendix Addonix
Ulysse	1997-2000	Bosch 5.3

Ford		
Cougar	1998-2000	Bosch 5.3
Escort/Orion	1985-1990	Lucas SCS
Escort/Orion	1990-1996	Teves Mk.IV (ITT 04) - 55-pin
Escort	1996-2000	Teves 20-I
Escort Cosworth	1992-1996	Teves II
Fiesta/Fiesta RS	1989-1996	Lucas SCS
Fiesta	1996-2000	Teves 20-I
Focus	1998-2000	Teves 20-1
Galaxy	1995-1997	Teves IV-GI (ITT 04-GI)
Galaxy	1997-2000	Teves 20IE (ITT 20IE)
Granada/Scorpio	1985-1991	Teves II
Granada/Scorpio	1992-1994	Teves Mk. IV (ITT 04)
Granada/Scorpio 2.5 D	1992-1994	Teves II (55-pin)

Ford (continued)	Year	System
Ka	1996-1999	Teves 20-I
Mondeo	1993-1996	Bendix Mecatronic II
Mondeo	1996-1998	Bendix Mecatronic III
Mondeo	1998-2000	Bosch 5.3
Puma	1997-2000	Teves 20-I
Scorpio	1995-1997	Teves IV-GI (ITT 04-GI)
Scorpio	1997-1999	Teves 20-I
Sierra/Sapphire	1985-1992	Teves II
Transit	1991-1997	Teves Mk. IV (ITT 04) - 55-pin
Transit	1997-1999	Bosch 5.3

Jaguar/Daimler

	Year	System
XJ6/Sovereign	1986-1991	Bosch 2S
XJ12 Sovereign HE	1983-1992	Bosch 2S
XJ6/XJ12	1993-1994	Teves II
XJS	1992-1996	Teves II

Lancia

	Year	System
Dedra	1990-1992	Bosch 2S
Dedra	1992-1994	Lucas/Girling 2/2
Thema	1986-1994	Bosch 2S

Land Rover

	Year	System
Discovery	1994-2000	Clayton Dewandre Wabco Type II
Range Rover	1989-2000	Clayton Dewandre Wabco Type I

Mercedes-Benz

	Year	System
190 (201)	1985-1993	Bosch 2S
C-Class (202)	1993-2000	Bosch 2E
E-Class (124)	1985-1992	Bosch 2S
E-Class (124)	1992-1995	Bosch 2E
S-Class (126)	1985-1991	Bosch 2S
SL (107)	1985-1990	Bosch 2S

Peugeot

	Year	System
106	1991-1996	Bendix Addonix
106	1996-2000	Teves IV-GI (ITT 04-GI)
Partner	1996-2000	Bosch 5.0
205	1990-1994	Bendix Addonix
206	1998-2000	Teves 20IE (ITT 20IE)
306	1993-1997	Bendix Mecatronic II
306	1993-1997	Bosch 2EH
306	1996-1997	Bosch 5.0
306	1997-2000	Bosch 5.3
309	1991-1993	Bendix Addonix
405	1988-1991	Bendix Addonix
405	1992-1997	Bosch 2EH
405 Mi-16	1988-1991	Bendix integrated
405 Mi-16	1992-1996	Bosch 2EH
406	1996-1997	Bosch 5.0
406	1997-2000	Bosch 5.3
505	1983-1990	Teves II
605	1990-1992	Bendix integrated
605	1993-1996	Bosch 2SE
605	1996-1997	Bosch 5.0
605	1997-1999	Bosch 5.3
806	1995-1997	Bendix Addonix
806	1997-1999	Bosch 5.3
Boxer	1995-2000	Bendix Addonix

Renault	Year	System
19	1990-1993	Bendix integrated
19	1993-1996	Bosch 2EH
21	1988-1992	Teves II
25	1985-1992	Bosch 2S
Clio	1991-1997	Bosch 2EH
Clio	1998-2000	Bosch 5.3
Espace	1990-1994	Bosch 2SE
Espace	1995-2000	Bosch 5.0
Kangoo	1997-2000	Bosch 5.3
Laguna	1995-1997	Teves Mk.IV (ITT 04) - 37-pin
Laguna	1998-2000	Teves 20IE (ITT 20IE)
Master	1997-2000	Teves 20IE (ITT 20IE)
Megane	1995-1998	Teves IV-GI (ITT 04-GI)
Megane	1998-2000	Bosch 5.3
Safrane	1992-1997	Bosch 2EH
Safrane	1997-2000	Bosch 5.3

Rover/MG		
200	1995-1997	Bosch 5.0
200	1998-1999	Bosch 5.3
400	1992-1995	Bosch 2EH
400	1995-1996	Bosch 5.0
600	1996-1998	Bosch 5.0
800	1986-1991	Bosch 2S
800	1992-1996	Bosch 2EH
800	1996-1999	Bosch 5.0
MGF	1995-2000	Bosch 5.0

Saab		
900	1989-1993	Teves II
900	1996-1998	Bosch 5.3
9-3	1998-2000	Bosch 5.3
9-5	1997-2000	Bosch 5.3
9000	1987-1993	Teves II
9000	1993-1999	Teves Mk.IV (ITT 04) - 55-pin

Seat		
Alhambra	1996-2000	Teves IV-GI (ITT 04-GI)
Arosa	1996-2000	Teves 20IE (ITT 20IE)
Cordoba	1993-1995	Teves Mk.IV (ITT 04) - 55-pin
Cordoba	1995-2000	Teves 20GI (ITT 20GI)
Ibiza	1993-1995	Teves Mk.IV (ITT 04) - 55-pin
Ibiza	1995-2000	Teves 20GI (ITT 20GI)
Inca	1995-1997	Teves 20GI (ITT 20GI)
Inca	1997-2000	Teves 20IE (ITT 20IE)
Toledo	1991-1995	Teves Mk.IV (ITT 04) - 55-pin
Toledo	1995-1999	Teves 20GI (ITT 20GI)

Skoda		
Felicia	1995-2000	Teves 20GI (ITT 20GI)
Octavia	1998-2000	Teves 20IE (ITT 20IE)

Vauxhall/Opel		
Astra-E/Belmont	1989-1991	Bosch 2E
Astra-F	1992-1998	Bosch 2EH
Astra-G	1998-2000	Bosch 5.3
Calibra (2WD)	1990-1991	Bosch 2E
Calibra (2WD)	1992-1998	Bosch 2EH
Carlton	1986-1994	Bosch 2S
Cavalier	1989-1992	Bosch 2S

Vauxhall/Opel	Year	System
Cavalier	1992-1996	Bosch 2EH
Corsa-B	1993-1996	Lucas 4/4F
Corsa-B	1996-2000	Bosch 5.3
Frontera	1995-2000	Kelsey Hayes 415
Kadett-E	1989-1991	Bosch 2E
Monza	1986-1987	Bosch 2S
Omega-A	1986-1994	Bosch 2S
Omega-B	1994-1998	Bosch 2SH
Omega-B	1998-2000	Bosch 5.3
Senator	1986-1994	Bosch 2S
Tigra	1994-1996	Lucas 4/4F
Tigra	1996-2000	Bosch 5.3
Vectra-A	1989-1992	Bosch 2S
Vectra-A	1992-1996	Bosch 2EH
Vectra-B	1996-2000	Kelsey Hayes 415
Vectra-B	1994-1997	Bosch 5.0/TC

Volkswagen	Year	System
Bora	1999-2000	Teves 20IE (ITT 20IE)
Caddy	1995-1997	Teves 20GI (ITT 20GI)
Caddy	1997-2000	Teves 20IE (ITT 20IE)
Corrado	1988-1992	Teves II
Corrado	1992-1996	Teves Mk.IV (ITT 04) - 55-pin
Golf	1989-1992	Teves II (55-pin)
Golf	1992-1994	Teves Mk.IV (ITT 04) - 55-pin
Golf	1995-1998	Teves 20GI (ITT 20GI)
Golf	1998-2000	Teves 20IE (ITT 20IE)
Jetta	1989-1992	Teves II (55-pin)
LT	1997-1998	Bosch 5.0
LT	1998-2000	Bosch 5.3
Lupo	1999-2000	Teves 20IE (ITT 20IE)
Passat	1990-1992	Teves II
Passat	1992-1995	Teves Mk.IV (ITT 04) - 55-pin
Passat	1995-1997	Teves 20GI (ITT 20GI)
Passat	1997-2000	Bosch 5.3
Polo	1994-1997	Teves IV-GI (ITT 04-GI)
Polo	1994-1997	Teves 20GI (ITT 20GI)
Polo	1997-2000	Teves 20IE (ITT 20IE)
Sharan	1995-1997	Teves IV-GI (ITT 04-GI)
Sharan	1997-2000	Teves 20IE (ITT 20IE)
Transporter	1992-1995	Bosch 2E
Transporter	1996-1997	Bosch 5.0
Transporter	1998-2000	Bosch 5.3
Vento	1992-1994	Teves Mk.IV (ITT 04) - 55-pin
Vento	1995-1998	Teves 20GI (ITT 20GI)

Volvo	Year	System
240	1990-1992	Bosch 2S
440/460/480	1987-1992	Teves II
440/460/480	1993-1997	Teves Mk.IV (ITT 04) - 37-pin
740/760	1987-1991	Bosch 2S
850	1992-1995	Teves Mk.IV (ITT 04) - 55-pin
850	1996	Teves 20GI (ITT 20GI)
940/960	1990-1992	Bosch 2S
940/960	1994-1997	Bosch 2E
S40/V40	1996-1998	Bosch 5.0
S40/V40	1998-2000	Bosch 5.3
S70/V70/C70	1996-1998	Teves 20GI (ITT 20GI)
S70/V70/C70	1999-2000	Teves 20IE (ITT 20IE)
S90/V90	1997-1999	Bosch 2E

Chapter 1
Antilock Brake System principles of operation

Contents

1 Introduction

Optimum braking force and vehicle directional stability is dependent on the grip between the tyres and the road surface, which can only be maintained when the wheels are rotating. Once a wheel stops rotating, the brakes no longer have any effect because the wheel is already locked, and steering control will be impaired or lost completely. Prior to the advent of ABS, only skilful driving and the use of 'cadence braking' could prevent total loss of control in such situations.

Cadence braking is an advanced driving technique used by professional and competition drivers for many years. During heavy braking, or even moderate braking on loose or slippery surfaces, the wheels are inclined to lock. The problem is exacerbated if braking is necessary while the vehicle is cornering. If the driver 'pumps' or rapidly depresses and releases the brake pedal many times during the braking manoeuvre, the vehicle will come to a safe and controlled halt **(see illustration 1.1)**. In addition, if the front wheels are allowed to momentarily lock and the steering is turned to the left or right, the vehicle will continue to slide in the direction of travel. If the vehicle is sliding forward with the wheels turned and the brakes are then released, the vehicle will rapidly turn to travel in the direction to which the front wheels are pointing.

However, cadence braking is slow and crude and requires some skilled input from the driver to obtain maximum benefit. Advances in electronic technology made possible an electronically controlled braking system that emulated and improved upon the cadence method. The Antilock Brake System (ABS) was the result. Under the cadence method, pumping the brake pedal causes each wheel to lock and release simultaneously. However, during ABS operation, the wheels can be individually regulated which allows for even greater control.

ABS does not replace the conventional braking system, but is a supplementary feature that allows powerful and controlled braking when driving situations and road conditions dictate.

ABS does not come into operation during normal braking situations, and in some vehicles may never come into play during the whole life of the vehicle. In the average vehicle life and under average driving conditions, ABS may operate no more than six to ten times.

The main safety advantage of ABS is obvious in that a vehicle braking to a controlled stop will decelerate more rapidly than a vehicle where one or more wheels are locked and skidding. Less well known is a number of other major ABS safety features. Vehicles braking under ABS control are much more stable. When braking sharply without ABS the vehicle may veer and lose control as wheel lock is not uniform on all four wheels. Sharp braking during a cornering manoeuvre can be very dangerous without ABS, and ABS control will usually result in a safe stop for this situation. In addition, when the vehicle is braking under ABS control, the wheels do not lock and steering control is retained, which allows for a rapid change of direction in order to miss an object in the vehicle's path.

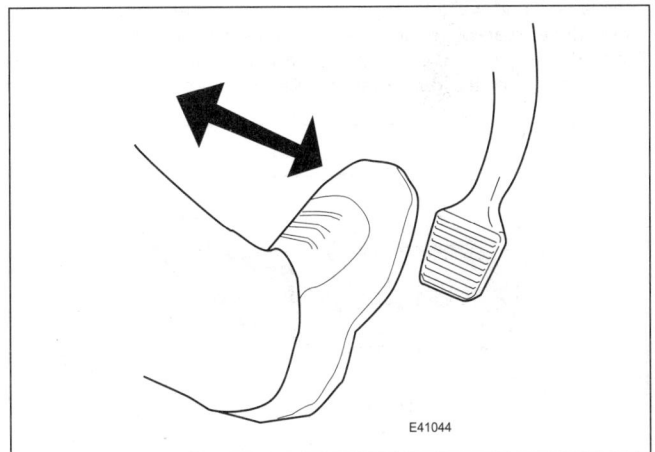

E41044

1.1 The technique of cadence braking where the brake pedal is rapidly pumped and released

2 System types and layouts

System types

With the exception of the Lucas/Girling SCS (Stop Control System) all the antilock brake systems covered in this book are electronically controlled by means of a sophisticated computer known as an electronic control module (ECM) **(see illustration 1.2)**.

Early attempts at devising a system to control vehicle-braking application resulted in the limited development of experimental mechanical systems. These systems were adapted from those used in aircraft applications but compared to the modern electronic system, were bulky, slow to respond, and expensive to manufacture.

With the rapid advances in electronic circuitry and component reliability that occurred in the 1970s it became practical to develop an electro-mechanical system for braking control. In this context, interpretation of braking conditions (by means of detecting whether a roadwheel was rotating or locked) and then transmitting that data in the form of control signals was carried out electronically, with the brake hydraulic pressure required to actually apply the brake being generated mechanically. Although now far more advanced, this basic principle still applies to all antilock brake systems currently in use.

Essentially there are three distinct antilock brake system types; semi-integrated, integrated and additional.

Semi-integrated systems use a conventional master cylinder to provide brake hydraulic pressure to part of the brake system and a hydraulic booster unit to operate the remaining part. When braking under ABS control, a pump operated high-pressure system is activated and the master cylinder is bypassed.

On integrated systems, the components used for ABS operation are also required during the operation of the conventional braking system. Typically, a pressure cylinder, electric pump unit and a pressure accumulator replace the conventional master cylinder. The pump draws hydraulic fluid from a fluid reservoir and feeds it under high pressure to the accumulator. The pressurised fluid is then used for operation of the braking system.

Additional systems (more often referred to as 'add-on' systems) retain the conventional braking system components and use a hydraulic control unit operating in parallel with the conventional system, for ABS operation. The majority of the antilock brake systems currently in use are of the add-on type.

System layouts

Variations in system layouts are numerous and are governed by a number of factors. Essentially, the layout will be determined by the vehicle size and weight, vehicle type, drivetrain arrangement (front-wheel drive, rear-wheel drive or four-wheel drive) and system cost. From these (and other) considerations an appropriate system configuration for the particular vehicle will be established. This will

result in two, three, or four controlled hydraulic channels, split front-to-rear or diagonally, and two, three or four wheel speed sensor circuits. The various combinations of the hydraulic circuits and sensor circuits are as follows; starting with those most commonly used.

Four-channel, diagonal split with four wheel speed sensors

One wheel speed sensor monitoring each roadwheel and a separate ABS hydraulic channel controlling each brake. One front brake and one diagonally opposite rear brake on each hydraulic circuit.

Four-channel, front-to-rear split with four wheel speed sensors

One wheel speed sensor monitoring each roadwheel and a separate ABS hydraulic channel controlling each brake. Both front brakes and both rear brakes on separate hydraulic circuits.

Three-channel, front-to-rear split with three wheel speed sensors

One wheel speed sensor monitoring each front roadwheel and one monitoring the rear roadwheels as a pair. One separate ABS hydraulic channel controlling each front brake with the third channel controlling the rear brakes jointly. Both front brakes and both rear brakes on separate hydraulic circuits.

Two-channel, front-to-rear split with three wheel speed sensors

One wheel speed sensor monitoring each front roadwheel and one monitoring the rear roadwheels as a pair. One ABS hydraulic channel controlling the front brakes and one controlling the rear brakes. Both front brakes and both rear brakes on separate hydraulic circuits. ABS modulation of both front brakes determined by the roadwheel with the highest adhesion.

Two-channel, front-to-rear split with two wheel speed sensors

One wheel speed sensor monitoring one front roadwheel and one monitoring one rear roadwheel. One ABS hydraulic channel controlling the front brakes and one controlling the rear brakes. Both front brakes and both rear brakes on separate hydraulic circuits. ABS modulation of both front brakes and both rear brakes determined by the roadwheel in that channel with the highest adhesion.

Two-channel, diagonal split with two wheel speed sensor

One wheel speed sensor monitoring each front roadwheel and a separate ABS hydraulic channel controlling each front brake. Rear brakes controlled through pressure regulating valves. One front brake and one diagonally opposite rear brake on each hydraulic circuit.

Generally speaking, during straight line braking with all four tyres on a similar road surface, each of the above ABS layouts operate equally effectively. However, when braking with the tyres on one side of the vehicle on a road surface with a different co-efficient of friction than the tyres on the other side, some are more effective than others. This can affect the stability of the vehicle, causing considerable input from the driver to correct a pull, or swing to one side under certain conditions. This may be particularly noticeable when the tyres on one side of the car are initially on surfaces with different friction co-efficients and then come onto a uniform surface, or when the vehicle is being steered during braking.

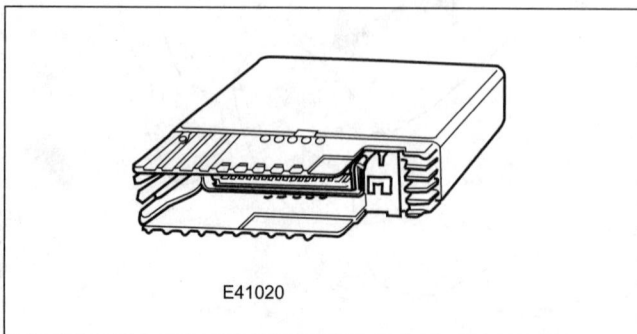

E41020

1.2 Typical ABS electronic control module (ECM)

3 System operation

All electronically controlled antilock brake systems consist of three main components or combinations of these components:

a) *Electronic control module (ECM).*
b) *Wheel speed sensors.*
c) *Hydraulic control unit.*

The electronic control module is a powerful microprocessor which controls the operation of the entire system based on inputs from the system sensors. In terms of ABS control, these inputs inform the ECM of the rotational speed of the roadwheels (from the wheel speed sensors) and that the brakes have been applied (usually by means of a switch or sensor on the brake pedal) **(see illustration 1.3)**. From these inputs the ECM is able to determine the onset of wheel lock at a given wheel, or pair of wheels, and instigate a sequence of hydraulic pressure regulations until wheel rotation is restored. Additionally, the ECM has complex circuitry to continually evaluate the functionality of the system and to disable ABS operation in the event of a system fault. Later ECMs also have the ability to recognise individual system faults and store these as codes for fault diagnosis.

To enable the ECM to determine the onset of wheel lock, the rotational speed of each wheel or pair of wheels, and any changes in that rotational speed are recorded by the wheel speed sensors. Two types of sensor are currently used - inductive and active. Inductive sensors work on the same principle as a pulse generator used in engine management systems to determine engine RPM. A permanent magnet contained in the sensor body is mounted adjacent to a toothed ring that rotates at roadwheel speed. As the roadwheel (and toothed ring) rotate, an AC voltage signal is produced which varies according to the speed of the roadwheel **(see illustration 1.4)**. With active sensors, the toothed ring itself is magnetised, with elements

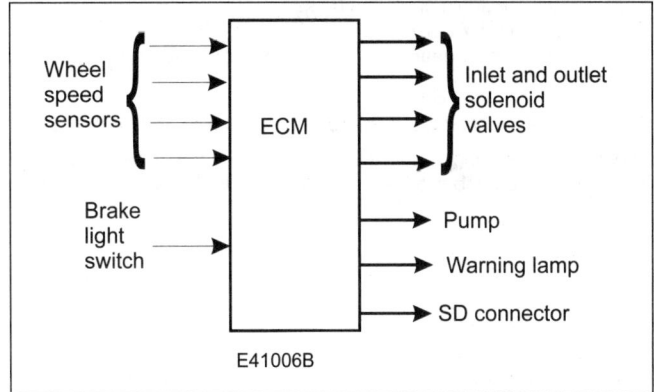

1.3 ECM sensor inputs and control signal outputs in a typical system

containing alternating north/south polarities. As the ring rotates, the changes in magnetic flux alter the internal resistance of the sensor to produce a DC voltage signal, varying with roadwheel speed.

The ECM compares the signals from the wheel speed sensors, and the speed of the fastest wheel is used as a reference value. If the onset of lock at any wheel is detected (a received speed signal being less than the reference value) regulation of the brake hydraulic pressure for the relevant wheel(s) will begin.

Regulation of the brake hydraulic pressure is carried out by the hydraulic control unit, under ECM control. The type and operation of the hydraulic control unit varies considerably according to system, but essentially the same principle applies to all **(see illustration 1.5)**. ABS pressure regulation is typically carried out in phases - pressure holding, pressure reduction and pressure build-up. When the onset of wheel lock is initially detected, the ECM signals the hydraulic control unit to maintain (or hold) the pressure in the hydraulic channel supplying the brake of the affected wheel. This is done by closing off the hydraulic fluid supply to that channel, typically by the activation of solenoid valves within the hydraulic control unit. The pressure cannot now be increased in the controlled channel by any further application of the brake pedal.

If the roadwheel is still exhibiting the tendency to lock, the hydraulic pressure in the controlled channel will now be reduced (again, typically by the activation of solenoid valves) thus releasing the brake completely. The hydraulic fluid is then returned to the master cylinder

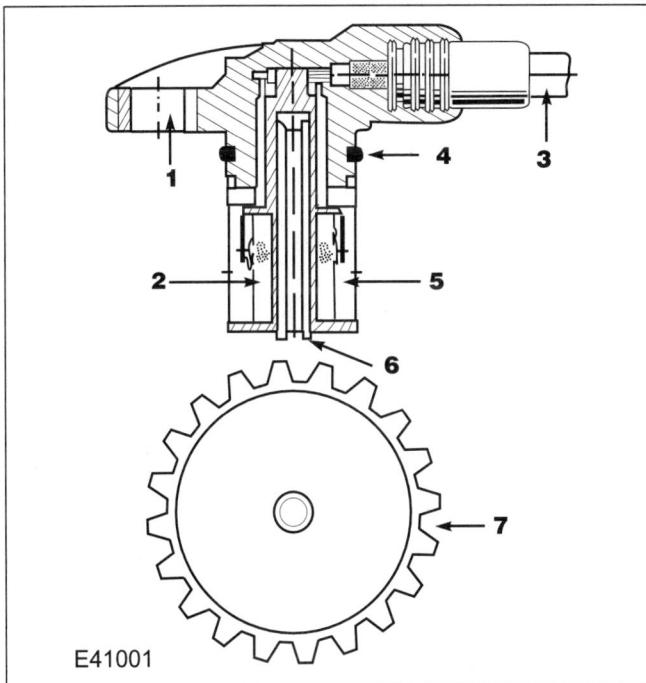

1.4 Cut-away view of an inductive wheel speed sensor

1 *Mounting hole*
2 *Permanent magnet*
3 *Wiring harness*
4 *O-ring*
5 *Coil*
6 *Sensor tip*
7 *Sensor ring*

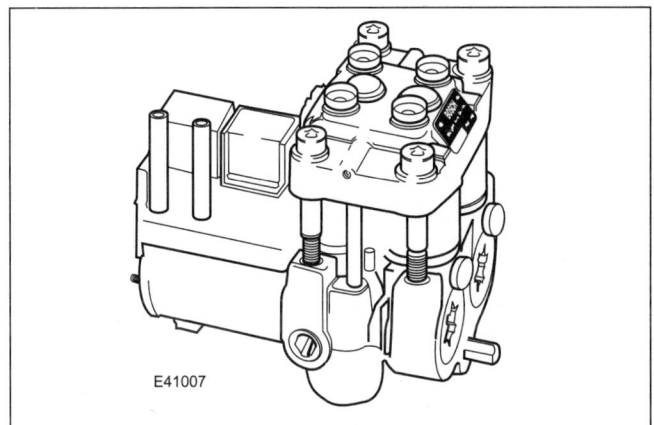

1.5 Hydraulic control unit as used on the Bosch 2S system

or hydraulic control unit usually by means of an electrically driven pump.

Once the roadwheel rotation has stabilised, the ECM will initiate a further pressure holding phase or a pressure build-up phase, whereby hydraulic fluid pressure in the controlled channel will be allowed to increase to once again apply the brake.

This controlled ABS cycle occurs many times a second until the vehicle comes nearly to a stop, or until the brake pedal is released.

Traction control and electronic stability programs

With the increasing sophistication of the software within the ECM, it has become possible to enhance the antilock brake function to include traction control (TC) and electronic stability programs (ESP). In its simplest form, traction control can be compared to ABS in reverse. Instead of the brake of a locking wheel being released in ABS mode, the brake of a spinning wheel is applied in traction control mode. In conjunction with additional forms of control such as electronic differential lock (EDL) engine torque can be applied effectively to the driven wheel with the greatest adhesion.

Enhanced versions of traction control and electronic stability programs can allow communication between the ABS ECM and the engine management and automatic transmission ECMs. This allows engine throttle opening, ignition timing and gear selection to be electronically regulated so that maximum vehicle driveability is maintained under all road conditions.

Chapter 2

Test equipment, training and technical data

Contents

1 Introduction

Testing systems on the modern automobile is a serious business. To be good at it, you need to seriously invest in three areas. We can liken the three areas to the good old three-legged stool. In our automotive stool, the legs are equipment, training and information. Kick one leg away, and the others are left a little shaky. Those with serious diagnostic intentions will make appropriate investments in all three areas.

That is not to say that those without the best equipment, or the necessary know-how, or the information, are completely stuck. It will just require a little more time and patience, that's all.

However, we do have to say at the outset that although quite a lot can be accomplished with a digital multimeter (DMM), a number of tests and checks do require specialist equipment. For the sake of completeness, we will detail test procedures that do require the use of sophisticated equipment.

Fault diagnosis then, and your method of diagnosis, will largely depend upon the equipment available and your expertise. There is a definite trade-off in time against cost. The greater the level of investment in equipment and training, the speedier the diagnosis. The less investment, the longer it will take. Obvious, really!

2 Equipment

Within the motor trade, there are several different approaches to testing vehicles and diagnosing faults. Let us take a look at the various options.

Multi-meter

This is the equipment required for the most basic approach. These days, the meter will probably be digital, and must be designed for use with electronic circuits **(see illustration 2.1)**. An analogue meter or even a test lamp may be used, so long as it meets the same requirements as the digital meter. Depending on the sophistication of the meter, the DMM can be used to test for basic voltage (AC and DC), resistance, frequency, rpm, duty cycle, temperature etc. A selection of thin probes and banana plugs for connecting to a breakout box (BOB) will also be useful. Use a splitpin or hairgrip for probing into square or oblong multi-plugs. The round DMM probes can 'spread' the connection and turn a good connection into a bad one.

If the fault is a straightforward electrical fault, the meter will often be adequate. However, the drawback is that it cannot analyse the complex electrical waveforms produced by many electronic sensors and actuators, and test results can sometimes be misleading.

Programmed test equipment

This kind of proprietary equipment will interface between the ECM and the ECM multi-plug. This equipment checks the input and output signals moving between the ECM and its sensors and actuators. If one or more of the signals are outside of pre-programmed parameters, the equipment will display the erroneous signals as faults. Once again, other test equipment may be required to pinpoint the actual fault.

Oscilloscope

An oscilloscope is essentially a graphic voltmeter. Voltage is rarely still, and tends to rise and fall over a period of time. The oscilloscope (or 'scope) measures voltage against time, and displays it as a waveform. Even when the voltage change is very rapid, the scope can usually capture the changes. Circuit faults can often be spotted much faster than when using other types of test instrument. Traditionally, the 'scope has been used for many years to diagnose faults in the primary and secondary ignition systems of conventional non-electronic vehicles. With the advent of electronics, the 'scope has become even more important, and when a labscope function is available, analysis of complex waveforms is possible. This equipment is often used in conjunction with other equipment, for speedy diagnosis of a wide range of vehicle problems. Within the context of the test procedures contained in this book, the oscilloscope is particularly suited to evaluating wheel speed sensor operation.

Battery saver

Actually, 'battery saver' is a misnomer, since the function of this device is to hold power to permanently live circuits whilst the battery is removed or changed. The live circuits may provide power to the radio security and station memory, and to the ECM adaptive memory, etc.

Jump leads with surge protection

It is possible to destroy an ECM if unprotected jump leads are used to provide emergency power to the battery. Rather than use jump leads, it is far safer to charge the battery before attempting to start the vehicle. A poor engine or chassis earth, flat battery or tired starter motor and unprotected jump leads are a recipe for total disaster.

ECM testing equipment

Usually the province of those companies that specialise in the repair of the ECM, and not available for purchase by the garage or workshop. One company (ATP) offers an ECM test via a modem over the telephone network if the ECM is taken to one of their agents. Other ECM testing companies require that the ECM be sent to them by post for evaluation.

Jumper wires

Useful for checking out circuits, and bridging or 'by-passing' a relay **(see illustration 2.2)**.

Break-out box (BOB)

The BOB is a box containing a number of connectors that allows easy access to the ECM input and output signals, without directly probing the ECM pins. The BOB loom terminates in a universal connector. A multi-plug harness of similar construction to the ECM harness is interfaced between the ECM and its multi-plug, and the other end is connected to the BOB loom. The BOB will now intercept all signals that go to and from the ECM. If a DMM or an oscilloscope

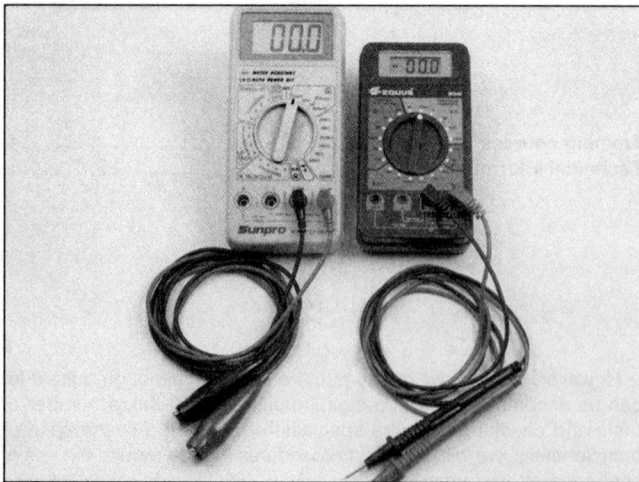

2.1 Two typical high impedance DMMs with similar performance but different sets of leads and probes. The left unit is equipped with alligator clips, and the right unit with spiked probes. Using the alligator clips frees your hands for other tasks, whilst the probes are useful for backprobing multi-plug connectors

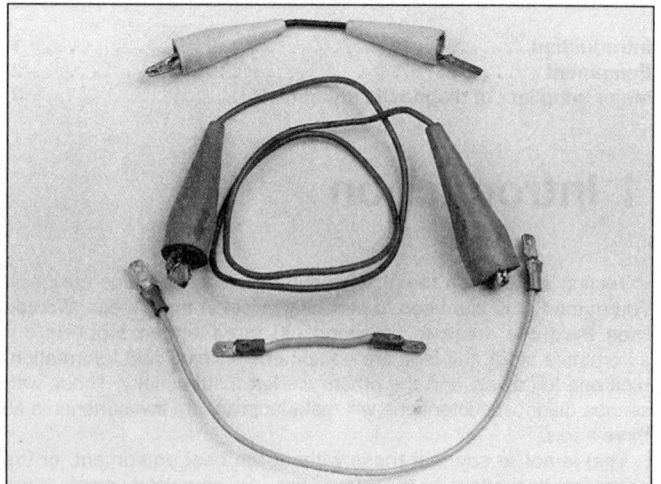

2.2 A selection of temporary jumper wires

2

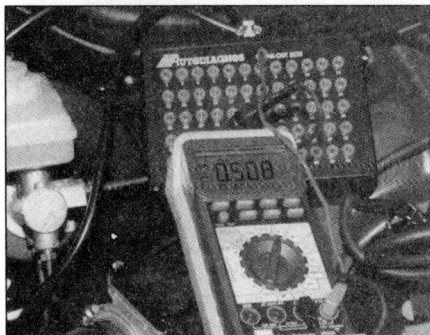

2.3 Using a BOB to obtain voltages at the ECM pins

2.4 A typical FCR or 'scanner'

2.5 Typical computer based dedicated test equipment that contains a very sophisticated and interactive test program

or any other suitable kind of test equipment is connected to the relevant BOB connectors, the ECM signals can be easily measured. The main drawback is the number of different ECM multi-plug connectors required for a good coverage of electronic systems **(see illustration 2.3)**.

There are three main reasons why use of a BOB is desirable in order to access the signals:

1) *Ideally, the connection point for measuring data values from sensors and actuators is at the ECM multi-plug (with the ECM multi-plug connected). The ECM multi-plug is the point through which all incoming and outgoing signals will pass, and dynamically testing at this point is considered to give more accurate results.*

2) *In modern vehicles, the multi-plug is becoming more heavily insulated, and removing the insulation or dismantling the ECM multi-plug so that back-probing is possible, is becoming more difficult, and in some instances, almost impossible. To a certain extent, the same is true of components, with certain components becoming increasingly difficult to backprobe or even probe at all.*

3) *ECM multi-plug terminals (pins) are at best fragile, and frequent probing or backprobing can cause damage. Some pins are gold-plated, and will lose their conductivity if the plating is scraped off. Using a BOB protects the pins from such damage.*

Fault code reader (FCR) - sometimes called a 'scanner'

A number of manufacturers market test equipment for connecting to the ECM diagnostic serial port. These general-purpose FCR allow data on a wide range of vehicles and systems to be obtained. The FCR could be used to obtain and clear fault codes, display datastream information on the state of the various sensors and actuators, and also "fire" the system actuators. The FCR is very useful for pointing the engineer in the direction of a specific fault. However, other test equipment may be required to pinpoint the actual fault, and the faults detected may be limited by the self-diagnostics designed into the vehicle ECM **(see illustration 2.4)**.

Franchised vehicle dealer

Will often use dedicated test equipment that relies on programmed test methods. The equipment will interface with the ECM, usually through the serial port, and lead the engineer through a programmed test procedure. Depending on its sophistication, the test equipment may be able to test most circuits, or may refer the engineer to test procedures using additional equipment. This equipment is dedicated to one vehicle manufacturer, and may not be available to other workshops outside of the franchised network **(see illustration 2.5)**.

3 Major suppliers of diagnostic equipment

Note: *The details below are correct at the time of writing (summer 2000).*

ASNU (UK) Ltd
27 Bournehall Avenue
Bushey, Herts
WD2 3AU
Tel: 020 8420 4494

ATP Electronic Developments Ltd
Victoria St
Hednesford, Staffordshire
WS12 5BU
Tel: 05438 79788

AutoDiagnos (UK) Ltd
Preston Technology Centre
Marsh Lane
Preston, Lancashire
PR1 8UD
Tel: 01772 887774

Crypton Ltd
Bristol Road
Bridgwater, Somerset
TA6 4BX
Tel: 01278 436210

Fluke (UK) Ltd
Colonial Way
Watford, Herts, WD2 4WD
Tel: 01923 240511

Gunson Ltd
Pudding Mill Lane
London, E15 2PJ
Tel: 020 8555 7421

Intermotor
Occupational Road
Hucknall, Nottingham
NG15 6DZ
Tel: 0602 528000

J.H. Haynes & Co. Ltd.
Sparkford
Yeovil, Somerset
BA22 7JJ
Tel: 01963 440635

Lucas Test
International Training Centre
Unit 7, Mica Close
Tamworth, Staffs
B77 4QH
Tel: 0827 63503

Omitech Instrumentation Ltd
Hopton Industrial Estate
London Road
Devizes, Wiltshire
SN10 2EU
Tel: 01380 729256

3 Major suppliers of diagnostic equipment (continued)

OTC Europe Ltd
VL Churchill Ltd
PO Box 3
London Road
Daventry, Northants
NN11 4NF
Tel: 01327 704461

Robert Bosch Ltd
PO Box 98
Broadwater Park
Denham
Uxbridge, Middx
UB9 5HJ
Tel: 01895 838353

Sun Electric (UK) Ltd
Oldmedow Road
Kings Lynn, Norfolk, PE30 4JW
Tel: 01553 692422

Sykes-Pickavant Ltd
Kilnhouse Lane
Lytham St. Annes, Lancs., FY8 3DU
Tel: 01253 784800

4 Training courses

There are a number of companies that specialise in training for the motor industry. The same training courses are usually available to the general public. Please contact the various bodies listed at the end of this chapter to if you wish to learn more about training for the automotive industry.

Note: *The details below are correct at the time of writing (summer 2000).*

AA External Training Courses
Widmerpool Hall
Keyworth, Notts
NG12 5QB
Tel: 021 501 7357/7389

Crypton Ltd
Bristol Road
Bridgwater, Somerset
TA6 4BX
Tel: 01278 436210

Lucas Test
International Training Centre
Unit 7, Mica Close
Tamworth, Staffs
B77 4QH
Tel: 0827 63503

MasterTech
Freepost RM1109
Wickford, Essex
SS11 8BR
Tel: 01268 570100

OTC Europe Ltd
VL Churchill Ltd
PO Box 3
London Road
Daventry, Northants
NN11 4NF
Tel: 01327 704461

Sun Electric (UK) Ltd
Oldmedow Road
Kings Lynn, Norfolk
PE30 4JW
Tel: 01553 692422

Sykes-Pickavant Ltd
Kilnhouse Lane
Lytham St. Annes, Lancs.
FY8 3DU
Tel: 01253 784800

5 Technical information

Specific information on the various systems is essential if effective diagnosis and repairs are to be completed. Companies that specialise in automotive technical information are listed below.

Note: *The details below are correct at the time of writing (summer 2000).*

Autologic Data Systems Ltd
Arnewood Bridge Road
Sway, Lymington, Hants
SO41 6DA
Tel: 01590 683868

Equiptech
Yawl House, Main Road
Marchwood, Southampton
SO40 4UZ
Tel: 023 8086 2240

Glass's Information Services Ltd
Elgin House
St George's Avenue
Weybridge, Surrey
KT13 OBX

Haynes Publishing
Sparkford
Yeovil, Somerset
BA22 7JJ
Tel: 01963 440635

3

Chapter 3

General test procedures

Contents

1 Introduction

Note: *Before carrying out any tests on the antilock brake system, refer to the warnings in the Reference Section of this manual.*

Generally speaking, test results obtained using a voltmeter or oscilloscope (particularly recommended) are more reliable and may reveal more faults than the ohmmeter. Voltage tests are much more dynamic and are obtained with voltage applied to the circuit, which is far more likely to reveal a problem than if the circuit is broken and the component measured for resistance. In some instances, disconnecting a multi-plug may break the actual connection that is at fault, and the circuit test may then reveal 'no fault found'.

In addition, the oscilloscope may reveal some faults that the voltmeter fails to find; the 'scope is particularly useful for analysing and displaying the signals and waveforms from components such as wheel speed sensors. With the proliferation of small, portable handheld oscilloscopes at a cost of less than £2500, the 'scope is not quite in the province of the home mechanic, but every workshop that is serious about fault diagnosis should certainly have one.

For the purposes of this book, we will generally test the majority of components with reference to the voltmeter. Resistance tests or continuity tests using an ohmmeter will be described where applicable.

Ideally, the connection point for measuring data values from sensors and actuators is at the ECM multi-plug (with the ECM multi-plug connected). The ECM multi-plug is the point through which all incoming and outgoing signals will pass, and dynamically testing at this point is considered to give more accurate results. However, for a variety of reasons it is not always possible to test at the ECM multi-plug, and other points of testing may also give satisfactory results.

2 Wheel speed sensor signals

Inductive wheel speed sensors

1 Inductive wheel speed sensors consisting of a permanent magnet pulse generator and adjacent toothed sensor ring are used on the majority of vehicles covered in this book. The sensors generate an AC voltage signal which varies with the rotational speed of the wheel and are the prime components used by the ECM for control and operation of the ABS. Any problems with the signal generated will be detected as a fault condition, causing the ECM to disable ABS operation.

2 When testing the wheel speed sensors, bear in mind that the voltage figures given in the test procedures of the system specific Chapters should be treated as typical voltage readings that may be expected from sensors of this type.

3 The actual voltage generated by the sensors is not the critical factor when evaluating sensor operation, provided that the voltage is of sufficient magnitude to be detected by the ECM. What is critical, however, is that the voltage generated is virtually the same for all the sensors on the vehicle, for a given roadwheel speed. The difference between the voltage readings from all the sensors should be no more than 15% to 20% and the waveform (when viewed on an oscilloscope) should be even and regular.

4 A number of factors may cause a variation in the output voltage of the sensors and these include:

a) *Wheel speed sensor air gap - the size of the air gap between the tip of the sensor and the adjacent toothed sensor ring will significantly affect the voltage output. The gap is not usually adjustable and is initially dependent on correct fitting of the sensor and sensor ring. Any variations in the air gap in service are usually caused by damage to the sensor ring teeth, or by excessive free play in the wheel bearings or components to which the sensor rings are fitted.*

b) *Dirt and swarf build-up - The wheel speed sensors and sensor rings typically operate in a hostile environment and are particularly susceptible to an accumulation of dirt and swarf between the sensor ring teeth and around the tip of the sensor. This can cause a variation not only in the air gap between the sensor tip and sensor ring, but also in the gap between the teeth on the sensor ring. Both of these conditions can affect the voltage output of the sensor.*

c) *Wiring and connectors - Again, considering the environment in which the sensors operate, they are very susceptible to chafing of the wiring on adjacent suspension components, and corrosion of the wiring connectors due to dirt or water ingress. This will lead to a variation in the resistance in the wiring between the sensor and ECM, resulting in a variation in the received voltage signals, or a total loss of signal altogether.*

'Active' wheel speed sensors

5 Active (or switching) wheel speed sensors have certain advantages over inductive sensors and are beginning to appear on an increasing number of vehicles. The active wheel speed sensor assembly consists of a sensor ring containing a series of magnetic elements with alternating poles located around its circumference, and an adjacent sensor. The sensor is supplied with a voltage from the ECM, and as the sensor ring rotates with roadwheel speed, the internal electrical resistance of the sensor varies according to the changes in intensity and direction of the lines of force of the sensor ring magnetic field. This causes the sensor signal to switch between the supply voltage and a lower value as the wheel rotates.

6 The main advantage of the active wheel speed sensor is that a DC signal is generated obviating the need for an analogue-to-digital signal converter in the ABS ECM. Also, as the sensor is supplied with a voltage from the ECM, it is possible for the ECM to detect any problems with the sensor circuit during the initial self-test.

7 When testing the wheel speed sensors, bear in mind that the voltage figures given in the test procedures of the system specific Chapters should be treated as typical voltage readings that may be expected from sensors of this type.

The points mentioned in paragraphs 3 and 4 above are equally applicable to active wheel speed sensors, but due to the nature of their operation they are less susceptible to problems caused by incorrect air gap, and dirt and swarf build-up.

3 ABS warning lamp circuit

1 On all the electronically controlled antilock brake systems covered in this book, a fault in the system is indicated by the illumination of the ABS warning lamp on the instrument panel. The lamp receives a fused voltage supply and is then illuminated by the ECM, or system relay providing the earth path when necessary.

2 On the current generation of antilock brake systems the ECM, relays and hydraulic control unit are all combined in one compact unit with all sensor signals and power supplies arriving via a single multi-plug connector.

3 On these systems, if the ECM multi-plug is disconnected, the ABS will obviously be disabled, but the warning lamp will not be illuminated because the relays and the ECM are also disabled.

4 To ensure that the warning lamp will illuminate under these conditions, many systems incorporate a mechanical switch in the multi-plug itself which provides an earth path for the warning lamp circuit if the multi-plug is disconnected.

4 Voltage testing

Connecting equipment probes

1 Connect the voltmeter negative probe to a sound vehicle earth.

2 Ensure that the voltmeter is set to the correct voltage range, and observe the stated conditions for the test (ie, ignition on or off).

3 Connect the positive probe to the actual terminals of the component under test.

Note: *This procedure will give acceptable results in most instances, and is the one we would recommend to non-professionals.*

4 If possible, connect a break-out-box (BOB) between the ECM and its multi-plug. This is the preferred method, and will avoid any possibility of damage to the ECM terminals.

5 If a BOB is not available, remove the protective cover over the ECM

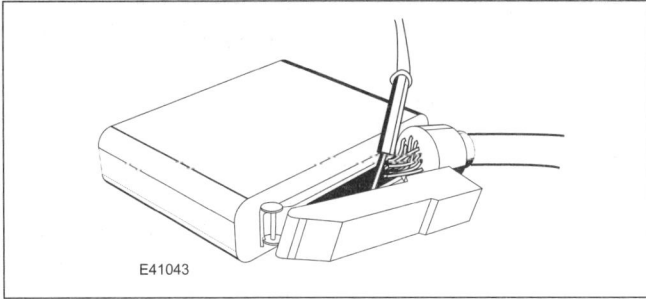

3.1 Remove the protective cover over the ECM multi-plug and backprobe the terminals using the equipment probes

multi-plug and backprobe the terminals using the equipment probes **(see illustration 3.1)**.

6 Alternatively, the ECM multi-plug could be disconnected and the ECM multi-plug terminals probed for voltages **(see illustration 3.2)**. **Note:** *This procedure is mainly used for checking voltage supplies to the ECM and the integrity of the earth connections. Great care must be taken when using this method, as the ECM terminal pins can be easily damaged.*

7 Unless otherwise stated, attach the voltmeter negative test lead to a sound earth on the vehicle, and probe or backprobe the component terminal under test with the voltmeter positive test lead. **Note:** *DO NOT push round tester probes into square or oblong terminal connectors. This leads to terminal deformation and poor connections. A split pin is the correct shape for inserting into square or oblong terminals.*

8 In this book, the multi-plug diagram usually shows the terminals of the harness connector. When backprobing the multi-plug (or viewing the sensor connector terminals) the terminal positions will be reversed.

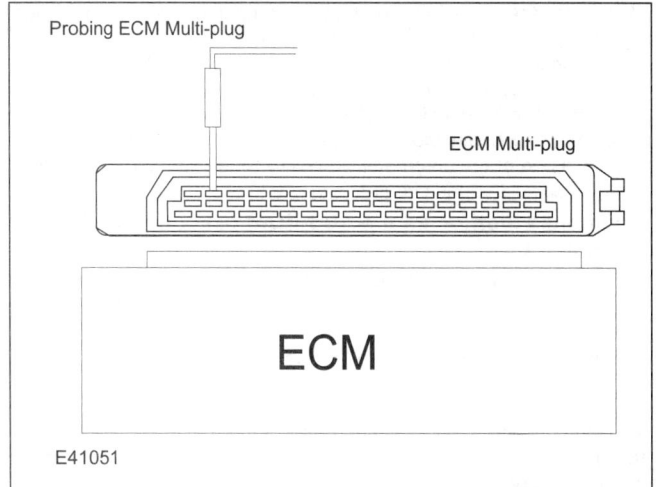

3.2 Probing the disconnected ECM multi-plug for voltages

Probing for supply voltage

9 With the component multi-plug connected or disconnected and the ignition switched on or off, as stated in the relevant test procedure, probe or backprobe for nominal battery voltage.

Earth or return

10 Method one: With the ignition switched on and the component multi-plug connected, backprobe for 0.25 volts maximum.

11 Method two: With the component multi-plug connected or disconnected, attach the voltmeter positive test lead to the supply terminal and the negative lead to the earth or return terminal. The voltmeter should indicate supply voltage if the earth is satisfactory.

5 Resistance testing

1 Ensure that the ignition is switched off and the component or circuit under test is isolated from a voltage supply.

2 DO NOT push round tester probes into square or oblong terminal connectors. This leads to terminal deformation and poor connections.

3 Circuits that begin and end at the ECM are best tested for resistance (and continuity) at the ECM multi-plug (after it has been disconnected).

4 The use of a BOB is also recommended for resistance tests, but the BOB must be connected to the ECM multi-plug, and NOT to the ECM itself.

5 If the resistance test for a sensor circuit is made at the ECM multi-plug pins, and the sensor has a common connection to the ECM, the multi-plug connectors for the remaining components must be disconnected. If this procedure is not followed the results may be inaccurate.

6 When checking continuity of a circuit or continuity to earth, the maximum resistance should be less than 1.0 ohm.

7 When checking the resistance of a component against specifications, care should be taken in evaluating the condition of that component as the result of a good or bad test result. A component with a resistance outside of its operating parameters may not necessarily be faulty. Conversely, a circuit that measures within its operating parameters may still be faulty. However, an open circuit or a comparatively high resistance will almost certainly indicate a fault. The ohmmeter is more useful for checking circuit continuity than it is for indicating faulty components.

Checking circuit continuity

Note: *These tests can be used to quickly check for continuity of a circuit between most components (sensors and actuators) and the ECM.*

Resistance method 1 (two-pin component)

8 Use this method when the component earth and supply wires are both connected to the ECM.

9 With the ignition switched off disconnect the ECM multi-plug, and the component multi-plug.

10 Connect a temporary bridging wire between terminals 1 and 2 at the component multi-plug **(see illustration 3.3)**.

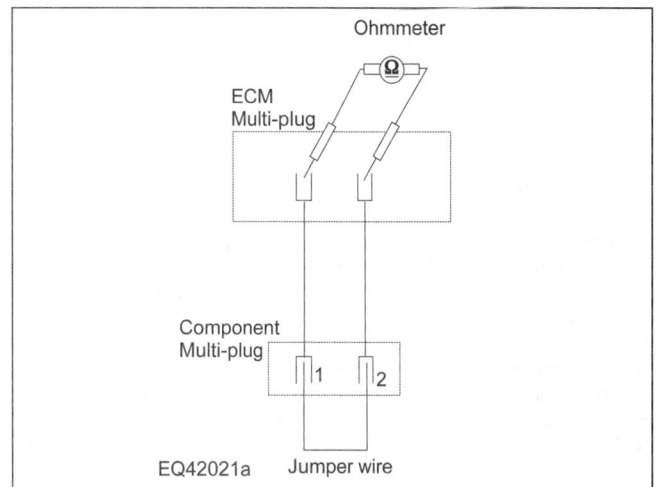

3.3 Checking circuit continuity using resistance method 1

11 Identify the two ECM pins that are connected to the component under test.

12 Connect an ohmmeter between the two pins at the ECM multi-plug. The meter should display continuity of the circuit.

13 If no continuity check for a break in the wiring or a bad connection between the ECM pin and its corresponding terminal at the multi-plug.

14 Move one of the ohmmeter probes to touch earth. The ohmmeter should display an open circuit.

Resistance method 2 (one ABS-ECM pin connected to a component)

15 Use this method when the component contains one wire from the ECM to a component, and the component is some distance from the ECM.

16 With the ignition switched off disconnect the ECM multi-plug, and the component multi-plug.

17 Connect a temporary jumper wire between the component multi-plug terminal and earth **(see illustration 3.4)**.

18 Identify an ECM earth pin and the ECM pin that is connected to the component under test.

19 Connect an ohmmeter between the two pins at the ECM multi-plug. The meter should display continuity of the circuit.

20 If no continuity check for a break in the wiring or a bad connection between the ECM pin and its corresponding terminal at the multi-plug.

21 Disconnect the temporary jumper wire and the ohmmeter should display an open circuit.

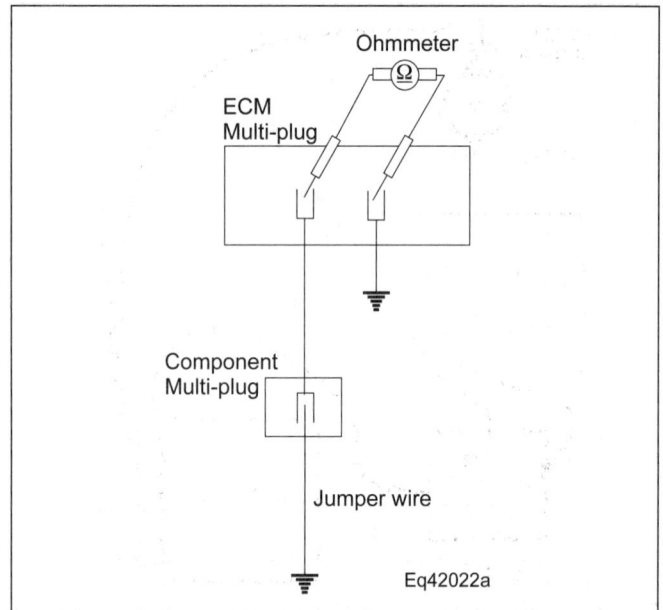

3.4 Checking circuit continuity using resistance method 2

6 Testing with an oscilloscope

1 Where in-depth fault diagnosis and system testing is to be undertaken, an oscilloscope is strongly recommend for all voltage tests, and various illustrations in this book will often show the instrument in use. Although a voltmeter may also be used for voltage testing, be warned that it may not reveal all faults in voltage circuits.

2 An oscilloscope is essentially a graphic voltmeter. Voltage is rarely still, and tends to rise and fall over a period of time. The oscilloscope (or 'scope) measures voltage against time, and displays it as a waveform. Even when the voltage change is very rapid, the scope can usually capture the changes. Circuit faults can often be spotted much faster than when using other types of test instrument. Within the context of the test procedures contained in this book, the oscilloscope is particularly suited to evaluate wheel speed sensor operation.

3 The signal received by the oscilloscope from the component under test will be displayed as a graph showing voltage (displayed vertically) and time (displayed horizontally). From the display, it is also possible to observe the shape of the waveform (sine wave, square wave, saw tooth etc), the pulse width (the period of time in which the signal is on) and the repeated waveform pattern.

Wheel speed sensor oscilloscope tests

4 Testing of the wheel speed sensors using an oscilloscope is extremely effective and with inductive wheel speed sensors, can be carried out either at the ECM multi-plug or at the 2-pin wheel speed sensor-wiring plug. Whichever method is being used, the ignition should be switched off and the multi-plug disconnected.

5 Active wheel speed sensors require a voltage supply from the ECM, and oscilloscope testing of these sensors should be carried out with a BOB connected between the ECM and ECM multi-plug.

6 Refer to the wiring diagrams and test procedures in the relevant system specific Chapter to identify the ECM pins for the sensor under test. Where inductive wheel speed sensors are being tested, connect the oscilloscope to the appropriate ECM multi-plug pins (preferably using a BOB connected between the ECM and multi-plug) or the wheel speed sensor wiring plug terminals **(see illustration 3.5)**. Where active wheel speed sensors are being tested, connect the

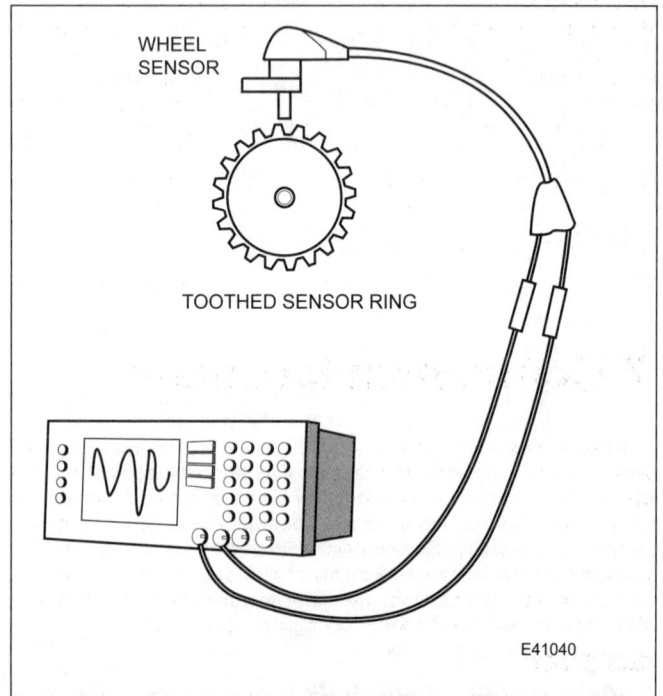

3.5 Checking inductive wheel speed sensor output with an oscilloscope connected to the wheel speed sensor wiring plug

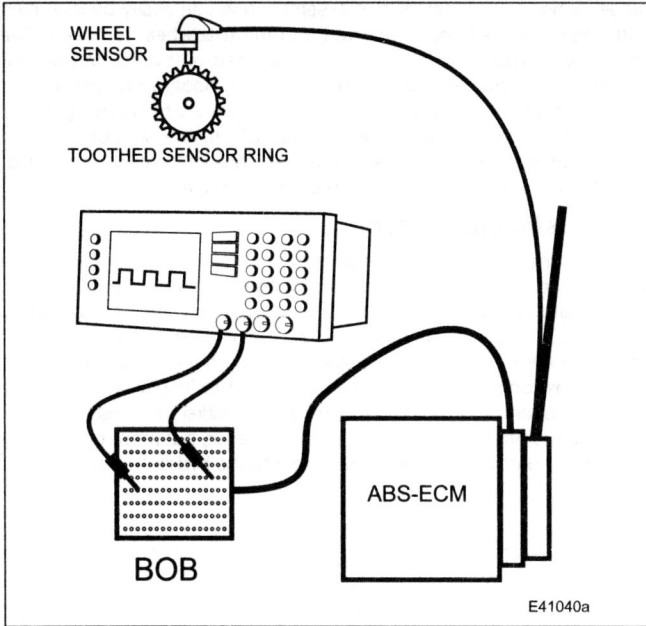

3.6 Checking active wheel speed sensor output with an oscilloscope connected to the relevant BOB terminals

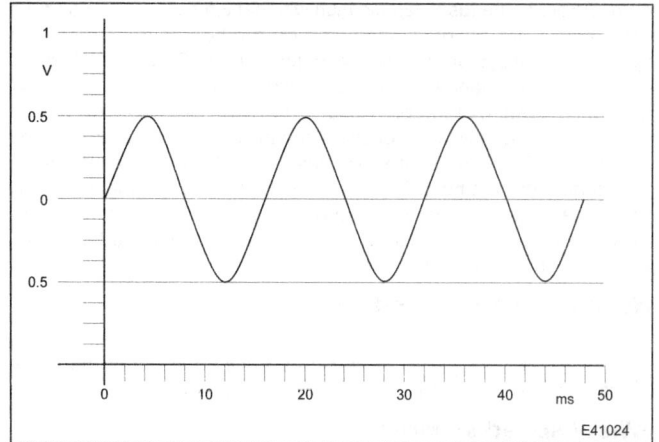

3.7 Typical inductive wheel speed sensor sine wave as displayed on an oscilloscope

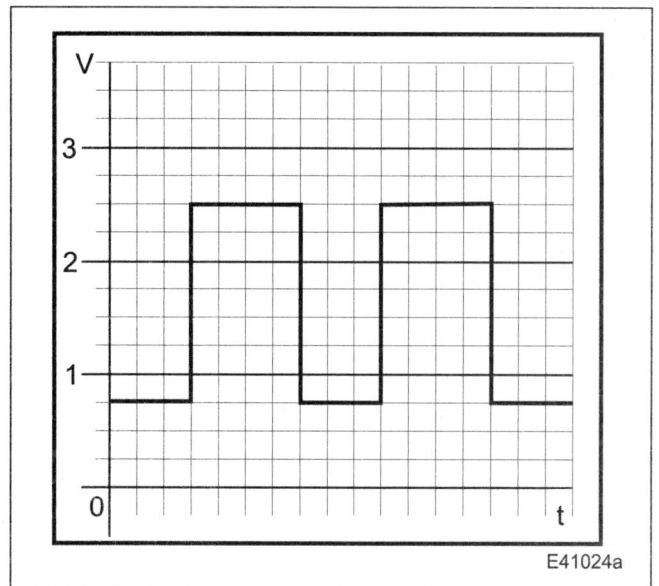

3.8 Typical active wheel speed sensor square wave as displayed on an oscilloscope

oscilloscope to the appropriate terminals on the BOB for the sensor under test (see illustration 3.6).

7 Raise and suitably support the vehicle so that the roadwheel can be rotated. If active wheel speed sensors are being tested, switch the ignition on.

8 Rotate the roadwheel at approximately 60-rpm (one revolution per second) and observe the oscilloscope display.

9 For inductive sensors, a sinusoidal AC voltage waveform should be displayed on the 'scope (see illustration 3.7). The amplitude of the waveform is related to the speed of rotation of the roadwheel, the faster the roadwheel is rotating the higher the output amplitude.

10 For active sensors, a DC voltage square wave should be displayed on the oscilloscope, switching between approximately 0.5 and 2.5 volts as the roadwheel is rotated (see illustration 3.8).

11 Check the waveform for even peaks. One or more peaks much smaller than the others would indicate a missing or damaged toothed sensor ring.

12 A low signal amplitude could indicate an excessive air gap between the sensor and the toothed wheel (this is not adjustable and is probably as a result of a damaged sensor).

13 In addition to determining that the wheel speed sensors are actually producing a voltage output, the 'scope can be used to compare the output being produced by each sensor. This is a very important feature as it is essential that the output from each sensor is the same for any given roadwheel speed.

7 Component locations

Because of the wide range of models in most vehicle manufacturers' range, and the number of different antilock brake systems that may be fitted to each model, it is impossible to specifically identify the location of the various ABS components on all vehicles. Also, information on component location, wiring connectors etc, is in many instances not available from the VM or even from the ABS manufacturer. The following information is therefore provided as a general guide to the likely location of the more important ABS components on most vehicles.

ABS ECM

On the current generation of antilock brake systems the ECM, relays and hydraulic control unit are all combined in one compact unit and its location is usually very obvious within the engine compartment. In some instances the unit may be located under a panel near to the front left-hand or right-hand side of the vehicle. Tracing the run of the brake hydraulic pipes from the master cylinder onwards will lead to the location of the unit. On earlier systems, the ECM was almost always located within the passenger compartment, or on some vehicles, within the luggage compartment. When in the passenger compartment, the location is typically under the facia or behind a trim panel in the footwell. When in the luggage compartment, the unit is usually located under a panel near to the front of the compartment.

ABS fuses and relays

The ABS fuses are located in the vehicle main or auxiliary fuse/relay box, which may be in the engine compartment or passenger compartment. The fuse identification within the fuse/relay box and the fuse ratings are generally indicated on the box cover. On current systems the relays are located together with the ECM and hydraulic control unit. This can be located as described previously for the ECM. On earlier systems (notably Bosch) the relays are located under a cover on the hydraulic control unit. Alternatively, the relays are located in the vehicle main or auxiliary fuse/relay box in the engine or passenger compartment. The relay identification within the fuse/relay box is generally indicated on the box cover. On systems where an overvoltage protection relay is used, this is often situated remotely within the engine compartment.

Hydraulic control unit

The location of this unit is usually always obvious, and as mentioned previously, can be found by tracing the run of the brake hydraulic pipes from the master cylinder onwards.

Wheel speed sensors

The front wheel speed sensors are always located on the front wheel hubs or suspension strut assembly and can easily be identified with the car raised and supported and the roadwheel removed. The rear wheel speed sensors are either located outboard on the rear wheel hubs or rear suspension assembly, or inboard on, or adjacent to, the differential housing. To locate the relevant wheel speed sensor wiring connector, trace the wiring back from the sensor until the connector is found. In most instances this will entail removal of underbody wheel arch panels or internal trim panels.

Diagnostic connector

Due to the variations in wiring harness layouts used by the various VM, the locations of the diagnostic connector are numerous. Some attempt at standardisation has emerged on later vehicles and usually (but not always) the connector is part of a 16-terminal socket that is used for interrogation of all the system ECM on the particular vehicle. This is now commonly located adjacent to the fuse/relay box in the vehicle interior, or in the centre console. On earlier vehicles, again the connector is often located in the vehicle interior, but more often in the engine compartment, generally adjacent to the hydraulic control unit.

Chapter 4

Fault diagnosis

Contents

1 Introduction

Diagnosis of automotive faults is sometimes a time-consuming process. Unless you are very lucky and stumble across the fault immediately, the best and ultimately quickest method is to follow a logical test pattern that checks, tests and evaluates all possibilities.

Interrogate the customer (even if you are the customer)

Use an incident checklist to log details of the incident and record under what conditions the problem occurs. This is an important pre-requisite to determining test procedures, and also prevents misunderstandings between workshop and customer.

Basic inspection

Make a basic inspection, and follow a sequence of visual checks and the problem area can often be quickly diagnosed.

Fault code reader (FCR) diagnosis

Also refer to the appropriate Chapter for the system under test
On systems incorporating self-diagnostics, if possible, connect a FCR to the diagnostic connector and interrogate the system ECM for fault codes. If a FCR is not available, flash codes can sometimes be obtained by following the procedures detailed for the particular system under test. If a fault code is present, test the relevant circuit by following the test procedures contained in the appropriate Chapter for the system under test.

Symptom-related fault diagnosis

Also refer to the appropriate Chapter for the system under test
If a fault code is not present, or if working on systems without a self-diagnostic capability, follow the symptom-related fault diagnosis charts, and systematically test the circuits and components that might be responsible for the fault condition.

2 Incident check list

Acquisition of information relating to the incident and the conditions under which the incident occurs is vital in any fault diagnosis procedure. The following checklist is based around a typical preliminary diagnosis sheet used in professional workshops.

Incident check list

Customer name . . . .   Date . . . .

Vehicle make . . . .   Model and year . . . .

Speedometer mileage reading . . . .   Registration number . . . .

VIN . . . .   ABS system type . . . .

Precise nature of customer complaint

. . . .

. . . .

Conditions under which fault occurs

Vehicle conditions
Stationary   ☐
Below approx 12 mph   ☐
Above approximately 12 mph   ☐
Intermittent   ☐
Light braking   ☐
Heavy braking   ☐
ABS in operation   ☐

Warning lamp conditions
Warning lamp not illuminated   ☐
Warning lamp illuminated with engine running, vehicle stationary ☐
Warning lamp illuminated below approx 12 mph   ☐
Warning lamp illuminated above approx 12 mph   ☐

Faults found during basic inspection

. . . .

. . . .

Fault codes logged (where applicable)

. . . .

. . . .

Any additional information

. . . .

. . . .

3 Basic inspection

No matter what the problem is, the following checks are an essential pre-requisite to the use of diagnostic equipment. In many instances, the fault will be revealed during these procedures. Make a careful visual check of the following items and rectify any problems found as necessary. This basic inspection can save a great deal of valuable diagnostic time.

Item	Check
Battery	Check the condition and security of the battery cables and the battery case. If the battery charge condition is suspect check the battery voltage, electrolyte level, and alternator drivebelt condition and tension.
Hydraulic fluid level	Check the fluid level in the brake master cylinder and top-up as necessary.
Hydraulic fluid leaks	Inspect all hydraulic pipes, hoses and unions for signs of leakage, corrosion or deterioration and rectify as necessary.
Wiring harness and connectors	Check the condition and security of the ECM multi-plug and check all other ABS wiring multi-plugs for loose or corroded connections.
Roadwheels and tyres	Check that the tyres are inflated to the correct pressure. Check that the roadwheel and tyre size are as specified by the vehicle manufacturer.
Brake pad/shoe condition	Check the amount of friction material remaining on the brake pads and/or brake shoes and renew if necessary.
Brake caliper/wheel cylinder condition	Check for any signs of fluid leaks at the brake calipers and/or wheel cylinders and rectify as necessary. Check that the calipers and wheel cylinders are securely mounted, and check the condition of the brake discs or drums.
Pressure regulating valves	Where fitted, check for any signs of fluid leaks from the pressure regulating/load sensing valves and check the condition of any mechanical operating linkages. Where adjustable linkages are used, ensure that the adjustment is correctly set.
ABS wheel speed sensors	Check the security of the wheel speed sensors and the condition of the wiring and multi-plugs. Check the condition of the teeth on the adjacent sensor rings.
Wheel bearings	Check for excess free play at the wheel bearings.
Brake light switch	Check for correct operation of the brake lights when the brake pedal is depressed and released.
Brake pedal	Check the operation and feel of the brake pedal. Long pedal travel and 'sponginess' may indicate air in the system. If so, the hydraulic system should be bleed according to the manufacturer's specific procedure.

4 Fault code reader (FCR) diagnosis

General

1 Before connecting the FCR, it is generally advisable to carry out the checks listed under "Basic inspection" first, primarily for the following reasons.

  a) *Certain electrical faults or battery voltage problems may adversely affect ECM operation, giving incorrect or spurious test results.*

  b) *For the purposes of fault diagnosis, the ABS and the conventional braking system should be treated as a single system consisting of both electrical and mechanical components. Incorrect operation of certain mechanical components may cause an electrical component or system fault to be incorrectly indicated. For example, excessively worn wheel bearings may cause the relevant wheel speed sensor to produce an erratic signal. This may log a fault code indicating problems with the sensor, when in reality the sensor may not be the problem.*

  c) *The sophistication of the ECM self-diagnostic software varies considerably from system to system. Problems relating to hydraulic fluid level, air in the hydraulic system or worn mechanical components in the braking system may be the specific reason for a fault, but may not cause a fault code to be generated.*

Testing self-diagnostic systems

2 If the ABS warning lamp remains illuminated with the engine running or illuminates at any time when the vehicle is driven, this is indicative of a system fault. **Note:** *Be aware that the warning lamp on some systems does not illuminate for faults that are designated as minor faults.*

3 Connect a FCR to the diagnostic connector, and interrogate the ECM for fault codes, or trigger the flash codes on systems where this is possible.

4 If fault codes are logged, use the test procedures for the specific system to check the relevant circuits and components.

5 If fault codes are not logged, use the FCR to view the datastream (live data on system sensors and actuators, and not available for all systems) or follow the symptom-driven fault diagnosis charts.

Limitations of self-diagnostic systems

6 Some may see the FCR as a panacea for solving all electronic problems with the car, but reading the fault code is only the beginning. To a large degree, the information decoded by the FCR is provided by the software designed into the ECM. The FCR makes the most of this information, but if certain facilities or data are not designed to be output to the diagnostic serial port, then these facilities will not be available to the FCR.

7 In many instances, the FCR can provide the answer to a puzzling fault very quickly. However it will not provide all the answers, because some faults (including actual ECM faults) may not even generate a fault code.

8 There are a number of distinct limitations to self-diagnostic systems:

a) *The basic data extracted from the ECM by the FCR is laid down by the manufacturer, and the self-diagnostic system and FCR must work within those limitations.*

b) *A code will not be logged if the ECM is not programmed to recognise that a particular component is faulty.*

c) *Spurious codes can be triggered by electrical faults or electrical interference.*

d) *One or more spurious codes can be triggered by a faulty component that may or may not trigger a code by itself.*

e) *In many instances, the fault code indicates a faulty circuit and not necessarily a faulty component. For example, a code indicating a wheel speed sensor fault may be caused by a faulty sensor, wiring fault, or corroded connector. Always check the wiring and connectors, and apply proper tests to the component before judging it to be faulty.*

f) *Some systems are capable of logging faults that occur intermittently, and others are not.*

g) *Older vehicles with earlier systems may not support self-diagnosis.*

Using a FCR

9 The FCR can be used for the following purposes. Some more sophisticated FCRs may interact with the ECM and allow a diagnostic trail to be followed.

a) *Obtaining fault codes.*

b) *Clearing fault codes.*

c) *Datastream testing (not all systems).*

d) *Testing the system actuators (ie relays and solenoid valves).*

Dynamic test procedures

10 Use the FCR to interrogate the ECM via the diagnostic connector.

11 Once the FCR has diagnosed one or more faults, further tests are usually required, and the technician may use the FCR (where possible), or it may be necessary to use a DMM or an oscilloscope to complete the diagnosis. Refer to the test procedures in the specific Chapter for the system under test.

12 Once the FCR has logged a fault, a datastream enquiry (where possible) is a quick method of determining where the fault might lie. This data may take various forms, but it is essentially electrical data on voltage, frequency, pulse duration etc, provided by the sensors and actuators. Unfortunately, such data is not available in all systems and datastream is not an option if you are working with flash codes. Since the data is in real time, various tests can be made, and the response of the sensor or actuator evaluated.

13 Use an oscilloscope or DMM to check voltages at the faulty component. Compare the results with the data given in the relevant system Chapter.

14 Use an ohmmeter to check the faulty circuit for continuity of the wiring and component resistance. Compare the results with the data given in the relevant system Chapter.

15 A faulty circuit should be tested, and any faults that are discovered should be repaired. The FCR should then be used to clear the fault codes and the ECM should be interrogated once again to see if other fault codes are still present.

16 An important point to bear in mind is that the ECM will only log faults in the ABS electronic circuits; faults in the conventional braking system or other mechanical faults will still require diagnosis using time-honoured methods.

17 Check that the faults remain cleared and do not return after road test.

Intermittent faults

18 Wiggle the component wiring, apply heat from a hairdryer, or freeze with a cold spray.

19 Intermittent faults can be extremely difficult to find, and on-road testing is often desirable, with codes or datastream information being generated as the fault occurs.

5 Fault diagnosis

Note: *The following information is presented as a guide to locating faults that may be of a mechanical nature, and for general fault diagnosis on systems without self-diagnostic capabilities. On systems with self-diagnosis, use the following information together with any fault codes obtained after system interrogation, to pinpoint the area under investigation. The test procedures in the system specific Chapters can then be used to verify and rectify the fault.*

Vehicle pulls to one side under normal braking

☐ Worn, defective, damaged or contaminated brake pads/shoes on one side.

☐ Seized or partially seized front brake caliper or rear wheel cylinder/caliper piston.

☐ A mixture of brake pad/shoe lining materials fitted between sides.

☐ Brake caliper or backplate mounting bolts loose.

☐ Worn or damaged steering or suspension components.

Noise (grinding or high-pitched squeal) when brakes applied

☐ Brake pad or shoe friction-lining material worn down to metal backing.

☐ Excessive corrosion of brake disc or drum. (May be apparent after the vehicle has been standing for some time.

☐ Foreign object (stone chipping, etc) trapped between brake disc and caliper or caliper shield.

Excessive brake pedal travel

☐ Inoperative rear brake self-adjust mechanism (vehicles with drum brakes).

☐ Faulty master cylinder.

☐ Air in hydraulic system.

☐ Faulty vacuum servo unit.

☐ Low hydraulic system pressure (where applicable).

Brake pedal feels spongy when depressed

☐ Air in hydraulic system.

☐ Deteriorated brake hydraulic hose(s).

☐ Master cylinder mountings loose.

☐ Faulty master cylinder.

☐ Low hydraulic system pressure (where applicable).

Excessive brake pedal effort required to stop vehicle

- [ ] Faulty vacuum servo unit.
- [ ] Disconnected, damaged or insecure brake servo vacuum hose.
- [ ] Primary or secondary hydraulic circuit failure.
- [ ] Seized brake caliper or wheel cylinder piston(s).
- [ ] Brake pads or brake shoes incorrectly fitted.
- [ ] Incorrect grade of brake pads or brake shoes fitted.
- [ ] Brake pads or brake shoe linings contaminated.

Judder felt through brake pedal or steering wheel when braking

Note: *During ABS braking, a slight pulsation will be felt through the brake pedal indicating that ABS is in operation. This is a normal condition and should not be confused with pedal judder, which can usually be felt under all braking conditions.*

- [ ] Excessive run-out or distortion of discs/drums.
- [ ] Brake pad or brake shoe linings worn.
- [ ] Brake caliper or brake backplate-mounting bolts loose.
- [ ] Wear in suspension or steering components or mountings.

Brakes binding

- [ ] Seized brake caliper or wheel cylinder piston(s).
- [ ] Incorrectly adjusted handbrake mechanism.
- [ ] Faulty master cylinder.

Rear wheel(s) locking under normal braking

- [ ] Rear brake pad/shoe linings contaminated.
- [ ] Faulty, or incorrectly adjusted pressure regulating/load sensing valves.

Front/rear wheel(s) locking under ABS braking

- [ ] Worn, defective, damaged or contaminated brake pads/shoes.
- [ ] Seized or partially seized front brake caliper or rear wheel cylinder/caliper piston.
- [ ] A mixture of brake pad/shoe lining materials fitted between sides.
- [ ] Brake caliper or backplate mounting bolts loose.
- [ ] Seized brake caliper or wheel cylinder piston(s).
- [ ] Faulty master cylinder.
- [ ] Faulty solenoid valve in hydraulic control unit.
- [ ] ECM faulty.

ABS warning lamp remains on when engine is started

- [ ] Blown fuse.
- [ ] Faulty alternator or alternator wiring connection to ECM.
- [ ] ECM voltage supply too low or too high.
- [ ] Faulty relay.
- [ ] Poor wiring or wiring connections.
- [ ] Incorrect relay supply voltage.
- [ ] Hydraulic pump motor faulty.
- [ ] Faulty solenoid valve(s) in hydraulic control unit.
- [ ] ECM faulty.
- [ ] Faulty wheel speed sensor or sensor wiring.

ABS warning lamp remains off when ignition is switched on

- [ ] Blown fuse.
- [ ] Warning lamp bulb blown.
- [ ] Poor wiring or wiring connections in warning lamp circuit.
- [ ] Faulty warning lamp relay (where applicable).
- [ ] ECM faulty.

ABS warning lamp comes on during driving

- [ ] Blown fuse.
- [ ] Faulty alternator or alternator wiring connection to ECM.
- [ ] ECM voltage supply too low or too high.
- [ ] Faulty relay.
- [ ] Poor wiring or wiring connections.
- [ ] Incorrect relay supply voltage.
- [ ] Hydraulic pump motor faulty.
- [ ] Faulty solenoid valve(s) in hydraulic control unit.
- [ ] Faulty wheel speed sensor or sensor wiring.
- [ ] Incorrect wheel speed sensor air gap.
- [ ] Excessive wheel bearing free play.
- [ ] Damaged toothed sensor ring teeth.
- [ ] Incorrect roadwheel/tyre size.
- [ ] Low hydraulic system pressure (where applicable).
- [ ] Faulty hydraulic system pressure switch (where applicable).
- [ ] Faulty brake light switch or pedal position sensor.
- [ ] ECM faulty.

4

Chapter 5
Bendix Addonix ABS

Contents

Vehicle coverage

Model	Year
Citroën	
AX	1991-1997
Dispatch	1995-2000
Evasion	1995-1999
Relay	1994-1997
Synergie	1995-1997
XM	1989-1998
Fiat	
Ulysse	1995-1997
Peugeot	
106	1991-1996
205	1990-1994
309	1991-1993
405	1988-1991
806	1995-1997
Boxer	1995-2000

Overview of system operation

1 Basic principles and system identification

The Bendix Addonix (often referred to as Bendix additional) Antilock Brake System has been fitted to a number of mainly Peugeot/Citroën derived vehicles since its introduction in the late 1980's. The system is of the additional or 'add-on' type operating in conjunction with the conventional braking system components.

The purpose of the system is to apply the vehicle brakes at maximum efficiency without wheel lock or loss of directional stability. Inductive sensors (wheel speed sensors) monitor the speed of the wheels by generating an electrical signal as the wheel is rotated. This information is passed to the ABS Electronic Control Module (ECM) which is then able to determine wheel speed, wheel acceleration and wheel deceleration. The ECM compares the signals received from each wheel with a calculated reference value and if the onset of lock at any wheel is detected, a signal is sent to the ABS hydraulic control unit which regulates the brake pressure for the relevant wheel(s).

Typically, Bendix Addonix ABS is comprised of the following components (see illustration 5.1):
a) ABS-ECM.
b) Hydraulic control unit.
c) Two or four inductive wheel speed sensors and associated sensor rings.
d) ABS electrical wiring harness and relays.
e) ABS warning lamp.
f) Diagnostic connector.

In addition, the conventional braking system is comprised of the following components:
a) Tandem brake master cylinder.
b) Vacuum servo unit.
c) Brake calipers and hydraulic hoses and pipes.
d) Pressure regulating/load-sensing valve(s) depending on application.

There are a number of differences in the construction of the Addonix system, depending on vehicle application. For the purposes of the information contained in this Chapter, there are three distinct hydraulic arrangements - two-, three- and four-channel. The three- and four-channel systems can be further sub-divided according to ECM type. Early versions generally have an ECM with a single 35-pin multi-plug, whereas on later versions the ECM is equipped with two multi-plugs, one with 15-pins, and one with 22-pins. A breakdown of the vehicles covered and system application is as follows:

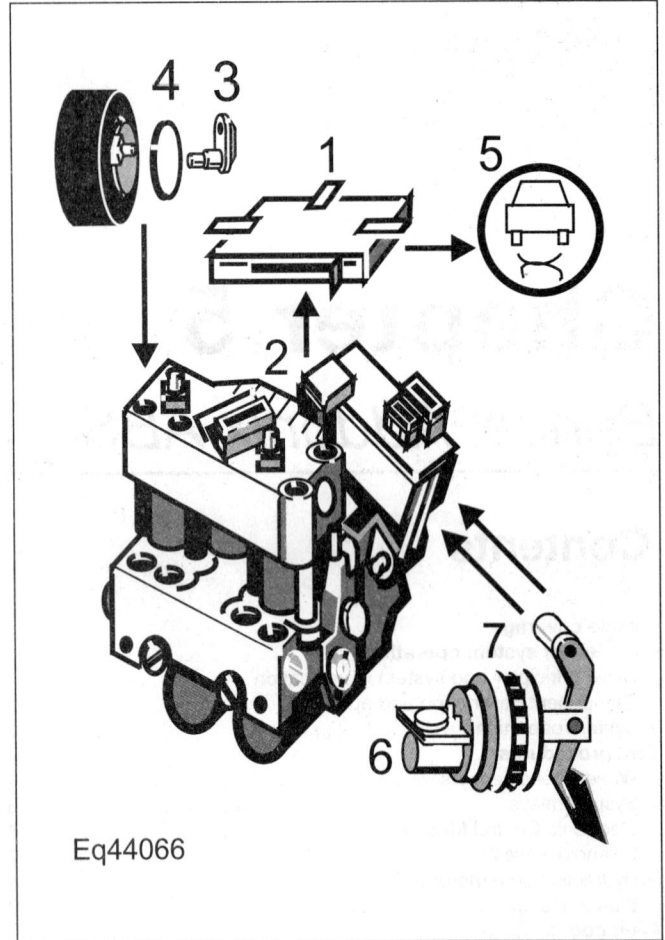

Eq44066

5.1 Bendix Addonix components

1	ABS-ECM	5	ABS warning lamp
2	Hydraulic control unit	6	Tandem master cylinder
3	Wheel speed sensors	7	Vacuum servo unit
4	Sensor rings		

Model	System
Citroën	
AX	Two-channel, 35-pin ECM multi-plug
Dispatch	Four-channel, 15-pin/22-pin ECM multi-plug
Evasion	Four-channel, 15-pin/22-pin ECM multi-plug
Relay	Four-channel, 15-pin/22-pin ECM multi-plug
Synergie	Four-channel, 15-pin/22-pin ECM multi-plug
XM	Three-channel, 35-pin multi-plug
Fiat	
Ulysse	Four-channel, 15-pin/22-pin ECM multi-plug
Peugeot	
106	Two-channel, 35-pin ECM multi-plug
205	Two-channel, 35-pin ECM multi-plug
309	Two-channel, 35-pin ECM multi-plug
405	Four-channel, 35-pin ECM multi-plug
806	Four-channel, 15-pin/22-pin ECM multi-plug
Boxer	Four-channel, 15-pin/22-pin ECM multi-plug

2 Component description and operation

ABS ECM

General

The Bendix Addonix Electronic Control Module (ECM) continually monitors wheel speed from the signals provided by the wheel speed sensors. If the ECM detects the incidence of wheel lock on one or more wheels, a signal is sent to the hydraulic control unit to modulate the hydraulic pressure to the brake of the locking wheel **(see illustration 5.2)**.

The ECM contains two microprocessors and uses digital technology to complete this function and other functions such as fault code memory and power modules for valve and relay activity **(see illustration 5.3)**.

Self-test

The Bendix Addonix ECM is equipped with a self-test capability that initially examines the ABS system when the ignition is switched on, and then examines the wheel speed sensor signals after a wheel speed of approximately 4 mph is reached from all wheels. The ABS self-test program continues to examine the signals from the various components as long as the ignition is switched on. If self-test determines that faults are not present, the ABS is ready for operation once a specified vehicle speed has been achieved.

If the ECM detects that a fault is present, all ABS functions are switched off and the warning lamp is turned on. The conventional braking system continues to operate as normal without ABS assistance.

Self-diagnostics

If the ECM detects a fault during the self-test routine, an internal fault code is stored in the ECM memory. Stored fault codes can be retrieved from the SD connector with the aid of a suitable fault code reader. If the fault clears, the code will remain stored until cleared with an FCR.

Hydraulic control unit

The hydraulic control unit consists of a valve block containing the inlet/outlet and restricting solenoid valves, pressure accumulator, restrictor, brake release valve, pulsation damper, non-return valve, high-frequency chamber, electric motor and return pump. The hydraulic control unit controls the pressure applied to the brake caliper for each individual wheel, or pair of wheels, during the various phases of ABS operation. During a controlled ABS cycle, the pump returns brake fluid, drained off during the pressure reduction phase, back into the brake circuit.

On certain Citroën vehicles an integrated hydraulic system is used

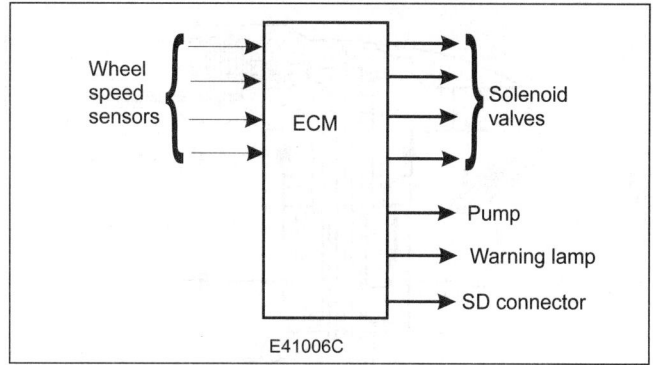

E41006C

5.2 ECM sensor inputs and control signal outputs

whereby hydraulic pressure generated by a mechanical pump is used for both the braking system and the vehicle suspension system. On these vehicles, the ABS hydraulic control unit consists of the solenoid valve circuits only - the electric motor and return pump assembly, and the conventional brake master cylinder being replaced by a compensator/brake control valve which is part of the vehicle main hydraulic system and is unique to Citroën.

According to vehicle application, Bendix Addonix may be either a two-, three- or four-channel system according to the number of solenoid valves used in the hydraulic control unit **(see illustration 5.4)**. On a two-channel system, only the front wheels are under ABS control and there are four solenoid valves, two for each front brake (one inlet/outlet solenoid valve and one restricting solenoid valve). On a three-channel system there are five solenoid valves, two for each front brake (one inlet/outlet solenoid valve and one restricting solenoid valve) and one inlet/outlet solenoid valve controlling both the rear brakes as a pair (the restricting solenoid valves are not used in the rear brake hydraulic circuit). The four-channel system is the same as the three-channel version except that the rear brakes are individually controlled resulting in an additional inlet/outlet solenoid valve.

On certain installations, control of hydraulic pressure to the rear brakes is based on the 'select-low' principle. With the 'select-low' principle, the rear wheel with the lowest adhesion determines the amount of hydraulic pressure to both rear brakes during ABS operation.

Wheel speed sensors

The rotational speed of the roadwheels and any changes in the rotational speed are recorded by inductive wheel speed sensors located at the roadwheels. Depending on system type, there may be a separate sensor for each front wheel only or a separate sensor for all four wheels **(see illustration 5.5)**.

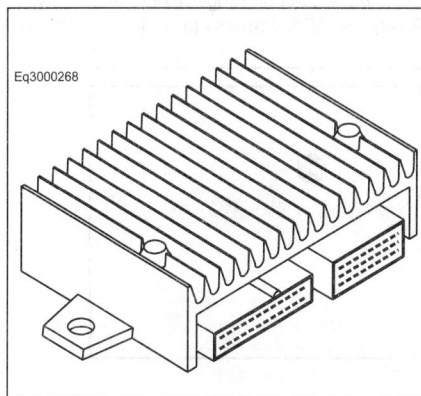

Eq3000268

5.3 Typical Bendix Addonix ECM

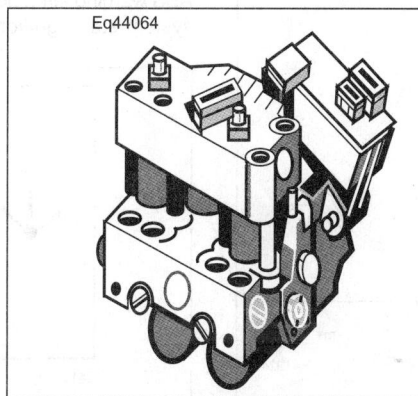

Eq44064

5.4 Bendix Addonix 2-channel hydraulic control unit

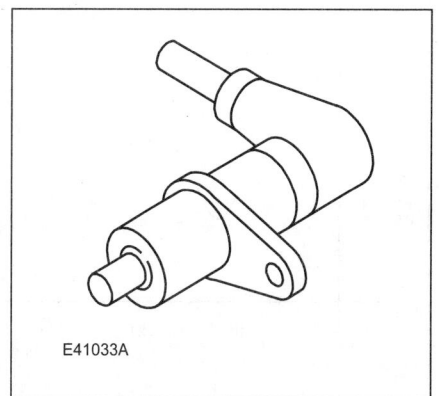

E41033A

5.5 Typical wheel speed sensor

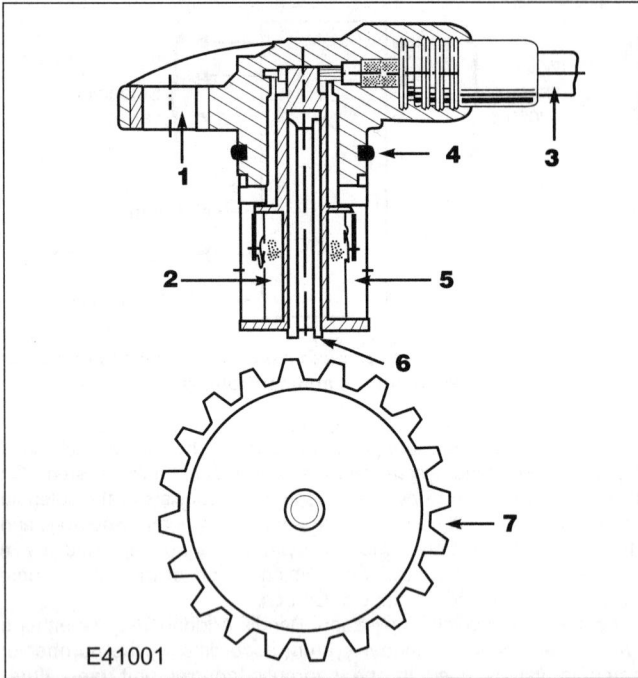

5.6 Sectional view of a wheel speed sensor

1   Mounting bolt location
2   Permanent magnet
3   Wiring harness
4   O-ring
5   Coil
6   Sensor tip
7   Toothed sensor ring

Each wheel speed sensor assembly comprises a toothed sensor ring, which rotates at roadwheel speed, and an adjacent sensor mounted a set distance from the sensor ring **(see illustration 5.6)**.

The sensors are permanent magnet pulse generator types producing an AC voltage sine wave as the sensor ring teeth pass through the magnetic field of the sensor **(see illustration 5.7)**.

The frequency of the waveform produced by the wheel speed sensor is proportional to the road speed. This AC voltage signal is continually being delivered to the ABS-ECM for processing.

The peak to peak voltage of the speed signal (when viewed upon an oscilloscope) can vary considerably according to wheel speed. An analogue to digital converter (ADC) in the ECM transforms the AC pulse into a digital signal **(see illustration 5.8)**.

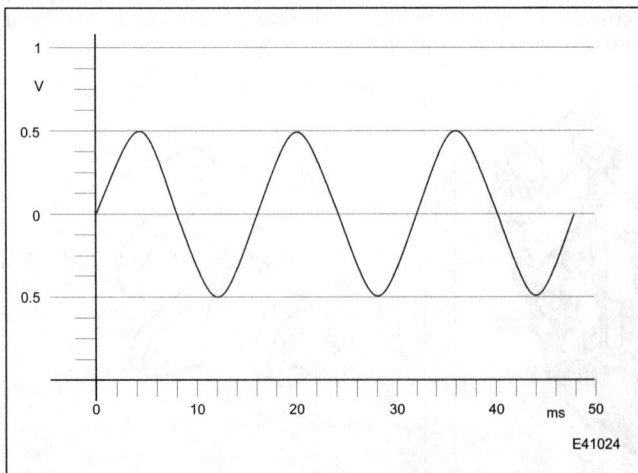

5.8 Wheel speed sensor waveform as viewed on an oscilloscope

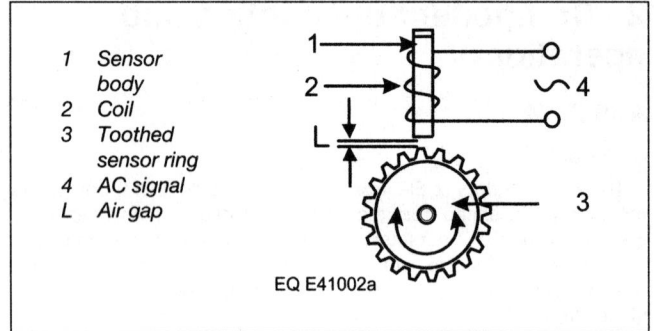

1   Sensor body
2   Coil
3   Toothed sensor ring
4   AC signal
L   Air gap

EQ E41002a

5.7 Wheel speed sensor operation

ABS electrical wiring harness and relays

An integrated main wiring harness is used to connect power and earth to various electrical components. This enables sensor signals to reach the ECM, and the ECM in turn to send command signals to the relays, return pump and hydraulic control unit. The ABS main and pump relays are located in a relay box mounted on top of the hydraulic control unit.

Main relay

Nominal battery voltage is supplied to the relay from the battery. On completion of the ECM self-test phase, the relay is activated by the ECM, allowing battery voltage to be applied to the solenoid valves. The earth for the solenoid valves is controlled by the ECM.

Pump relay

Nominal battery voltage is supplied to one main terminal of the pump relay via the pump motor and a fuse. The other relay main terminal is connected to earth. The ECM controls the pump motor by actuating the relay to provide the earth for the motor.

ABS warning lamp

After the ignition is switched on, the ABS warning lamp on the instrument panel is illuminated for approximately 2-4 seconds as the system executes a self-test routine. If satisfactory operation of the system is detected by the ECM, the lamp is extinguished. During vehicle operation above a pre-determined wheel speed, the ABS-ECM implements a further self-test cycle whereby ABS operation and wheel speed sensor signals are continually monitored. If a fault is detected, the relevant ECM pin is earthed to illuminate the warning lamp on the instrument panel, and the ABS function is disabled. The warning lamp will remain illuminated until the fault is no longer present **(see illustration 5.9)**.

When the ABS-ECM detects a fault, the fault code is stored and the ABS warning lamp activated. If the fault no longer exists after the next system start (ignition on/off) the ABS warning lamp is extinguished

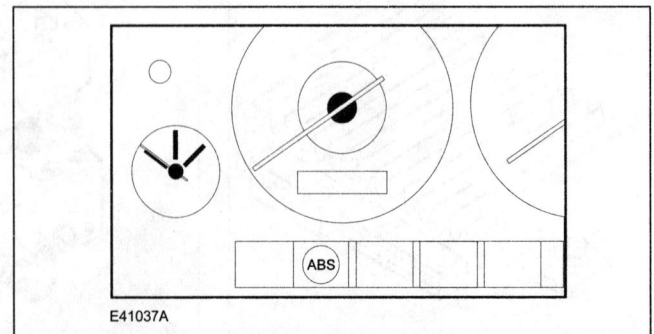

E41037A

5.9 ABS warning lamp

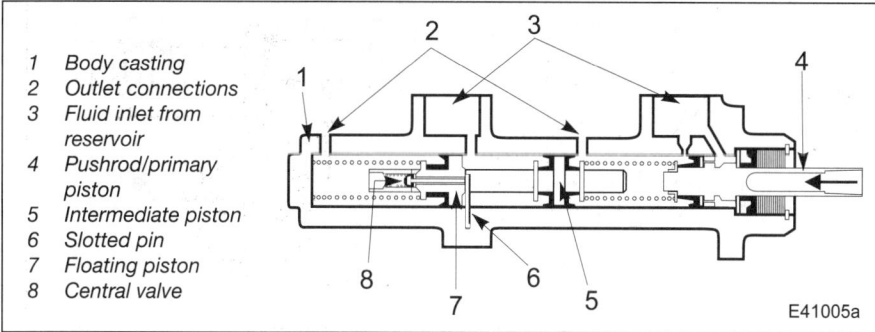

1 Body casting
2 Outlet connections
3 Fluid inlet from reservoir
4 Pushrod/primary piston
5 Intermediate piston
6 Slotted pin
7 Floating piston
8 Central valve

5.10 Sectional view of a typical tandem master cylinder

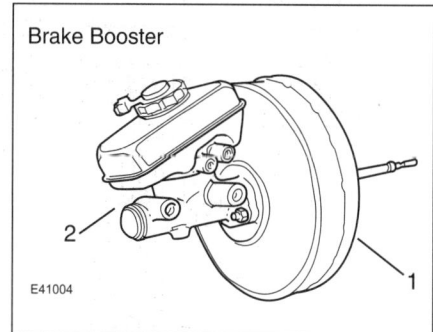

Brake Booster

5.11 Vacuum servo unit (1) and tandem master cylinder (2)

5

after the self-test cycle, however the fault code remains stored in the ECM memory.

Tandem master cylinder

Typically, the tandem master cylinder comprises a body casting incorporating primary and secondary pressure chambers, primary piston, intermediate piston, floating piston, slotted pin and central valve. The cylinder operates as a conventional master cylinder using vacuum assistance from the vacuum servo unit **(see illustration 5.10)**.

When the brake system is at rest, the central valve in the floating piston rests against the slotted pin. In this condition, the central valve is open and brake fluid can discharge out of the pressure chamber back into the brake fluid reservoir. When the brake pedal is depressed, the build-up of hydraulic pressure in the primary pressure chamber acts on the intermediate piston and floating piston, moving them down the cylinder bore. The floating piston contacts the seal on the central valve, closing the connection between the intermediate and secondary pressure chambers. Brake hydraulic pressure can now also increase in the secondary pressure chamber.

Vacuum servo unit

The vacuum servo unit is located between the brake pedal and tandem master cylinder. When the brake pedal is depressed, the servo unit increases the force applied by the pedal, reducing the effort required to operate the brakes **(see illustration 5.11)**.

The unit is operated by vacuum created in the engine inlet manifold (or from a separate vacuum pump on diesel engines) which is applied to a diaphragm contained within the unit casing. A pushrod connected to the centre of the diaphragm acts directly on the primary piston in the master cylinder.

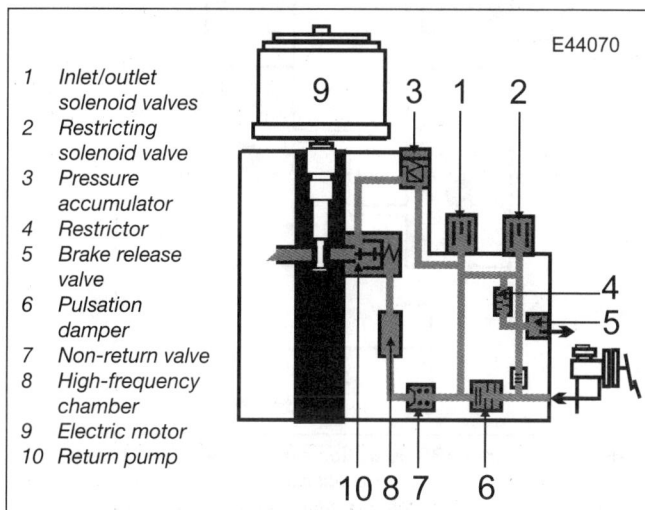

1 Inlet/outlet solenoid valves
2 Restricting solenoid valve
3 Pressure accumulator
4 Restrictor
5 Brake release valve
6 Pulsation damper
7 Non-return valve
8 High-frequency chamber
9 Electric motor
10 Return pump

5.12 Bendix Addonix hydraulic control unit components

When the brake pedal is released, vacuum is applied to both sides of the diaphragm. When the pedal is depressed, one side of the diaphragm is opened to atmosphere and the vacuum acting on the other side deflects the diaphragm which in turn operates the master cylinder primary piston. The resulting force applied to the master cylinder piston is therefore significantly greater than the initial force applied to the brake pedal by the driver.

Pressure regulating/load-sensing valve(s)

Depending on vehicle application, pressure regulating valves or load sensing valves may be incorporated to restrict the hydraulic fluid pressure to the rear brakes. The valves may be pressure conscious whereby the hydraulic fluid supply is restricted once a pre-determined pressure is reached, or load conscious whereby the hydraulic pressure is reduced according to vehicle loading.

3 System operation

The operation of the braking system both conventionally and when under ABS control varies according to the number of channels in the Bendix Addonix system. On a two-channel system, only the front brakes are regulated during ABS operation, the rear brakes being operated conventionally directly from the master cylinder, and typically via pressure regulating/load sensing valves. On three- and four-channel systems, all four brakes are regulated during ABS operation.

Two-channel ABS operation

Brake system at rest

When the system is at rest all the brake components are inoperative. Pressure is non-existent in the hydraulic pipes between the tandem master cylinder and the brake calipers.

Brake system operating under conventional control without ABS

When the brake pedal is activated, the pedal force is applied to the tandem master cylinder by the vacuum servo unit pushrod. The servo unit pushrod acts directly on the pressure piston in the master cylinder, which pressurises the hydraulic fluid in the brake pipes to the hydraulic control unit.

Both the inlet/outlet and restricting solenoid valves in the controlled circuit are 'at rest' allowing an unrestricted hydraulic passage from the master cylinder through the hydraulic control unit to the brake caliper, thus operating the brake **(see illustration 5.12)**.

Brake system operating in conjunction with ABS control

The ABS-ECM continually monitors wheel speed from the signals provided by the wheel speed sensors. If the ECM detects the incidence of wheel lock on either of the front wheels, ABS is automatically initiated in four phases. As Bendix Addonix operates individually on each front wheel, either of the controlled wheels could be in any one of the following phases at any particular moment.

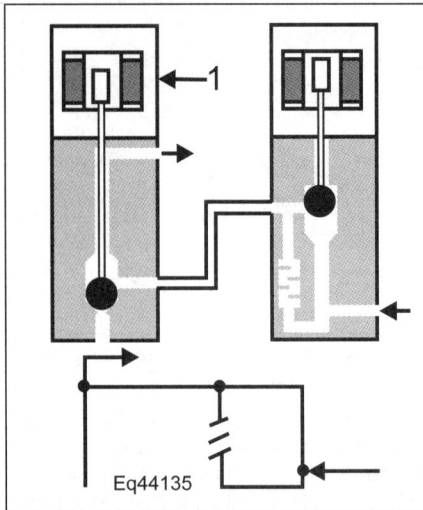

5.13 Two-channel ABS operation - phase one, rapid pressure reduction

1 Inlet/outlet solenoid valve

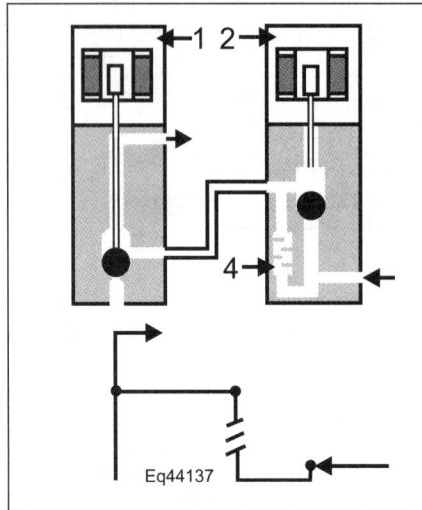

5.14 Two-channel ABS operation - phase two, slow pressure reduction

1 Inlet/outlet solenoid valve
2 Restricting solenoid valve
4 Restrictor

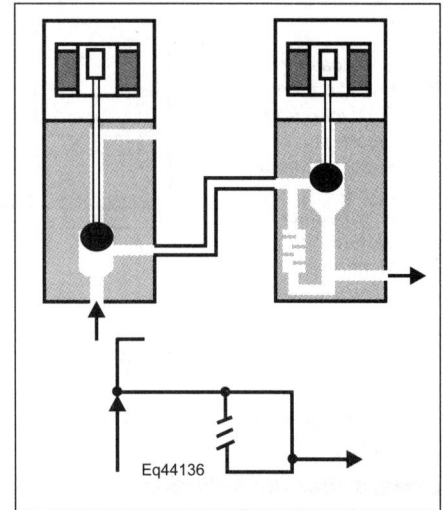

5.15 Two-channel ABS operation - phase three, rapid pressure increase

Phase one, rapid pressure reduction

The ECM energises the inlet/outlet solenoid valve, closing off the direct hydraulic connection between the master cylinder and the front brake caliper in the controlled circuit. Pressure in the hydraulic circuit falls rapidly as fluid from the brake caliper is returned to the pressure accumulator **(see illustration 5.13)**.
At the same time, the ECM actuates the return pump, which transfers the returned fluid to the high-frequency chamber.

Phase two, slow pressure reduction

The rapid pressure reduction phase is always followed by a slow pressure reduction phase. The inlet/outlet solenoid valve remains energised and the restricting solenoid valve is now also energised. Hydraulic fluid returned from the brake caliper is forced to flow through the restrictor as it passes back to the pressure accumulator, thus resulting in a slow pressure reduction in the controlled circuit. When the controlled wheel is once again rotating, phase three is initiated **(see illustration 5.14)**.

Phase three, rapid pressure increase

Both solenoid valves are de-energised by the ECM allowing them to return to the 'at rest' position. The hydraulic circuit from the master cylinder directly to the front brake caliper is re-instated and the hydraulic fluid returned during the two pressure reduction phases is pumped back into the circuit via the non-return valve. This results in a rapid increase in hydraulic pressure in the controlled circuit and the brake is once again applied **(see illustration 5.15)**.

Phase four, slow pressure increase

The rapid pressure increase phase is always followed by a slow pressure increase phase. The inlet/outlet solenoid valve remains 'at rest', but the restricting solenoid valve is once again energised by the ECM. Hydraulic fluid from the master cylinder is forced to flow through the restrictor before passing to the brake caliper, resulting in a slow increase in hydraulic pressure in the controlled circuit **(see illustration 5.16)**.
This pressure increase continues until the wheel speed sensor detects that the brake is once again about to lock at which point phase one is re-instigated by the ECM. The cycle then continues until the vehicle comes to a halt or the brake pedal is released.
The whole ABS control cycle takes place many times per second for each affected wheel and this ensures maximum braking effect and control during ABS operation.

Three- and four-channel ABS operation

Where a three- or four-channel system is used, the operation of the front brakes is the same as a two-channel system. The rear brakes operate in a slightly different manner, as there are no restricting solenoid valves for the rear brake hydraulic circuits. Instead, the inlet/outlet solenoid valve(s) incorporate a restrictor and flow valve and operate as follows.

Brake system at rest

When the system is at rest all the brake components are inoperative. Pressure is non-existent in the hydraulic pipes between the tandem master cylinder and the brake calipers.

Brake system operating under conventional control without ABS

When the brake pedal is activated, the pedal force is applied to the tandem master cylinder by the vacuum servo unit pushrod. The servo

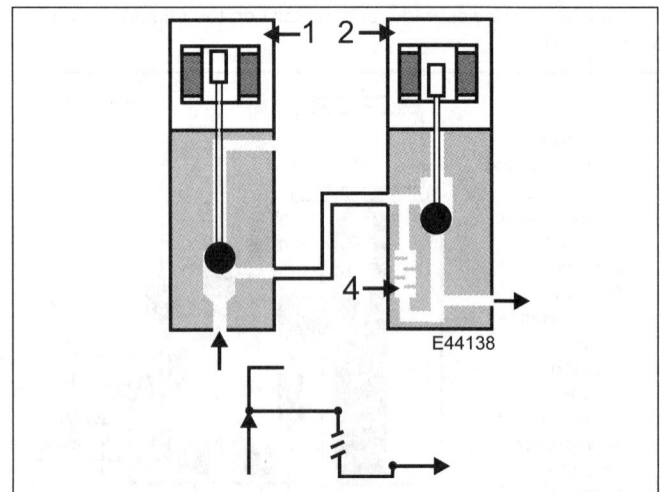

5.16 Two-channel ABS operation - phase four, slow pressure increase

1 Inlet/outlet solenoid valve   2 Restricting solenoid valve
4 Restrictor

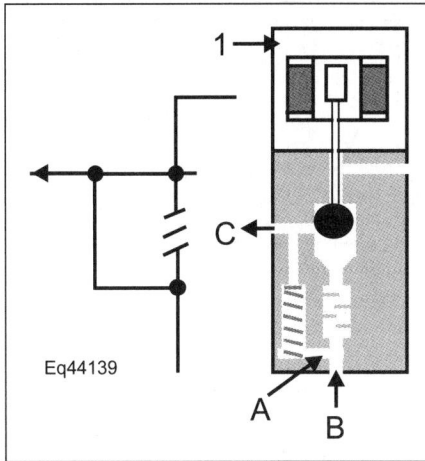

5.17 Three and four-channel ABS operation - conventional braking without ABS

1 Inlet/outlet solenoid valve
a Flow valve
b Hydraulic fluid input from master cylinder
c Hydraulic fluid output to rear brakes

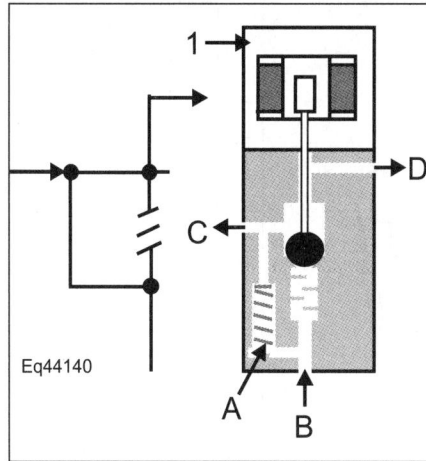

5.18 Three and four-channel ABS operation - phase one, rapid pressure reduction

1 Inlet/outlet solenoid valve
a Flow valve
b Hydraulic fluid input from master cylinder
c Hydraulic fluid output/input to/from rear brakes
d Hydraulic fluid return to pressure accumulator

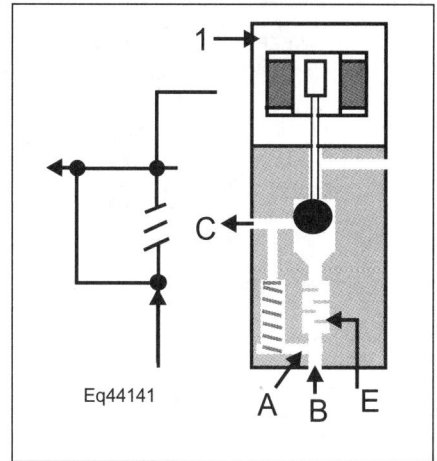

5.19 Three and four-channel ABS operation - phase two, slow pressure increase

1 Inlet/outlet solenoid valve
a Flow valve
b Hydraulic fluid input from master cylinder
c Hydraulic fluid output/input to/from rear brakes
e Restrictor

unit pushrod acts directly on the pressure piston in the master cylinder, which pressurises the hydraulic fluid in the brake pipes to the hydraulic control unit.

The inlet/outlet solenoid valve in the controlled circuit is 'at rest' and the flow valve is open allowing an unrestricted hydraulic passage from the master cylinder through the hydraulic control unit to the individual rear brake caliper, or pair of calipers, thus operating the brakes **(see illustration 5.17)**.

Brake system operating in conjunction with ABS control

The ABS-ECM continually monitors wheel speed from the signals provided by the wheel speed sensors. If the ECM detects the incidence of wheel lock on any of the controlled wheels, ABS is automatically initiated. On four-channel systems where ABS operates individually on each wheel, any of the controlled wheels could be in any one of the following phases at any particular moment.

Phase one, rapid pressure reduction

The ECM energises the inlet/outlet solenoid valve, closing off the direct hydraulic connection between the master cylinder and the rear brake caliper(s) in the controlled circuit. This creates a difference in hydraulic pressure on either side of the flow valve, the greater pressure being on the supply side from the master cylinder. The flow valve therefore closes, shutting the direct

hydraulic connection between the master cylinder and the rear brake caliper(s) in the controlled circuit. Pressure in the hydraulic circuit falls rapidly as fluid from the brake caliper(s) is returned to the pressure accumulator **(see illustration 5.18)**.

At the same time, the ECM actuates the return pump, which transfers the returned fluid to the high-frequency chamber.

Phase two, slow pressure increase

The inlet/outlet solenoid valve is de-energised by the ECM allowing it to return to the 'at rest' position. Hydraulic fluid pressure is still greater on the supply side from the master cylinder causing the flow valve to remain closed. Hydraulic fluid from the master cylinder is forced to flow through the restrictor before passing to the brake caliper(s), resulting in a slow increase in hydraulic pressure in the controlled circuit. Hydraulic fluid returned during the pressure reduction phase is pumped back into the circuit via the non-return valve **(see illustration 5.19)**.

This pressure increase continues until the wheel speed sensor detects that the brake is once again about to lock at which point phase one is re-instigated by the ECM. The cycle then continues until the vehicle comes to a halt or the brake pedal is released.

The whole ABS control cycle takes place many times per second for each affected wheel and this ensures maximum braking effect and control during ABS operation.

Test procedures

Important note: The test procedures, pintables and wiring diagrams contained in this Chapter are necessarily representative of the system depicted. Because of the variations in wiring and other data that often occurs, even between similar vehicles in any particular VM's range, the reader should take great care in identification of ECM pins, and satisfy himself that he has gathered the correct data before failing a particular component.

4 Wheel speed sensors

Checking the wheel speed sensor (general)

1 Inspect the wheel speed sensor for corrosion or damage and check that the sensor is tightly mounted.
2 Check the toothed sensor ring for damage, eccentricity and for broken or missing teeth.

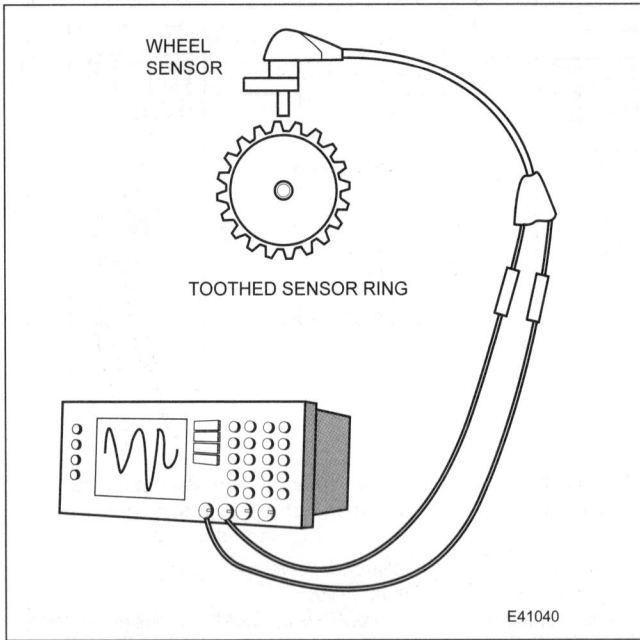

5.20 Checking wheel speed sensor output with an oscilloscope connected to the sensor-wiring plug

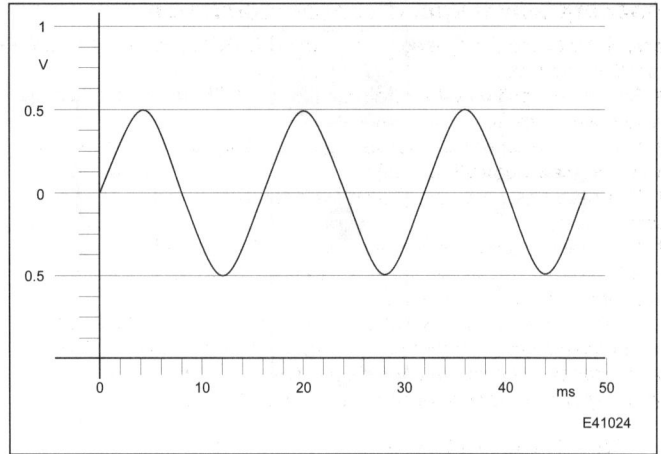

5.21 Typical inductive wheel speed sensor sine wave as displayed on an oscilloscope

3 Inspect the wheel speed sensor-wiring plug for corrosion and damage. One plug for each sensor.

4 Check that the connector terminal pins are pushed fully home and making good contact with the sensor-wiring plug.

5 Check the clearance between the sensor and the toothed sensor ring. The clearance is not normally adjustable but is nominally 0.2 to 1.2 mm. If the clearance is excessive, expect a worn sensor tip or problems with the wheel bearings/hub or sensor ring.

6 When carrying out voltage checks with an oscilloscope or voltmeter, the voltage obtained will be proportional to the speed at which the wheel is rotating. In addition to determining that the wheel speed sensors are actually producing a voltage output, it is essential that the output from the sensors on a particular axle is the same for any given wheel speed.

Checking wheel speed sensor output with an oscilloscope

Note: *Refer to the wiring diagrams for specific ECM pin identification according to model.*

7 Switch the ignition off and disconnect the ECM multi-plug or the relevant wheel speed sensor-wiring plug.

8 Connect an oscilloscope between the terminal pins for the sensor under test **(see illustration 5.20)**.

9 Select a range to cover 80 Hz on the oscilloscope and a free run time base.

10 Raise the wheel and rotate it by hand at approximately one revolution per second.

11 A sinusoidal wave form should be obtained, with amplitude and duration changing with rotational speed **(see illustration 5.21)**.

12 If there is no signal, or a very weak or intermittent signal at the ECM, repeat the test at the sensor-wiring plug. If there is no change in signal status, the sensor is suspect.

13 If the signal is now satisfactory this indicates a fault in the wiring harness, which should be checked for continuity.

Checking wheel speed sensor output with an AC voltmeter

Note: *Refer to the wiring diagrams for specific ECM pin identification according to model.*

14 Switch the ignition off and disconnect the ECM multi-plug or the relevant wheel speed sensor-wiring plug.

15 Connect an AC voltmeter between the terminal pins for the sensor under test **(see illustration 5.22)**.

16 Raise the wheel and rotate it by hand at approximately one revolution per second.

17 A voltage of approximately 0.1 to 3.0 volts (AC RMS) should be obtained. If there is no signal, or a very weak or intermittent signal at the ECM, repeat the test at the sensor-wiring plug. If there is no change in the signal, the sensor is suspect.

18 If the signal is now satisfactory, this indicates a fault in the wiring harness which should be checked for continuity. **Note:** *This test at least proves that a signal is being generated by the sensor. However, the voltage produced is an average voltage and does not clearly indicate damage to the sensor ring or that the sinewave is regular in formation.*

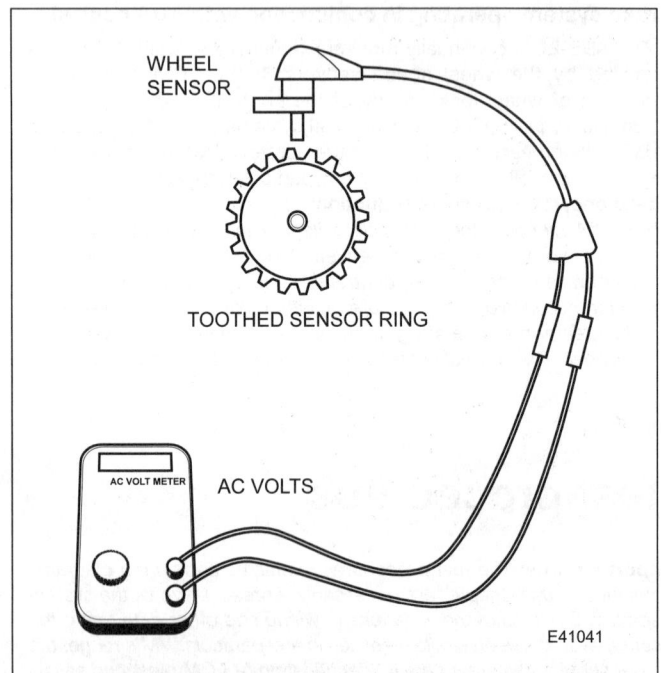

5.22 Checking wheel speed sensor output with a voltmeter connected to the sensor-wiring plug

Checking wheel speed sensor resistance

Note: *Refer to the wiring diagrams for specific ECM pin identification according to model.*

19 Switch the ignition off and disconnect the ECM multi-plug or the relevant wheel speed sensor-wiring plug.

20 Connect an ohmmeter between the terminal pins for the sensor under test **(see illustration 5.23)**.

21 The readings obtained should be between 0.7 and 1.5 kohms approximately.

22 If the resistance is excessively high, or open circuit at the ECM, repeat the test at the sensor multi-plug. If there is no change in resistance, the sensor is suspect.

23 If the resistance is now satisfactory, this indicates a fault in the wiring harness which should be checked for continuity. **Note:** *Even if the resistance is within the quoted specifications, this does not prove that the speed sensor can generate an acceptable signal.*

5 System relays

Relay resistance and continuity tests

Main relay

1 Switch the ignition off.

2 Disconnect the 5-pin multi-plug (D) from the relay box located on the hydraulic control unit **(see illustration 5.24)**.

3 Connect an ohmmeter between terminals 1 and 2 in the relay box (D). A reading of between 50 and 70 ohms should be obtained.

Hydraulic pump relay

4 Switch the ignition off.

5 Disconnect the 5-pin multi-plug (D) and 3-pin multi-plug (C) from the relay box located on the hydraulic control unit **(see illustration 5.24)**.

6 Connect an ohmmeter between terminal 3 of the 5-pin multi-plug base (D), and terminal 2 of the 3-pin multi-plug base (C). A reading of between 30 and 70 ohms should be obtained.

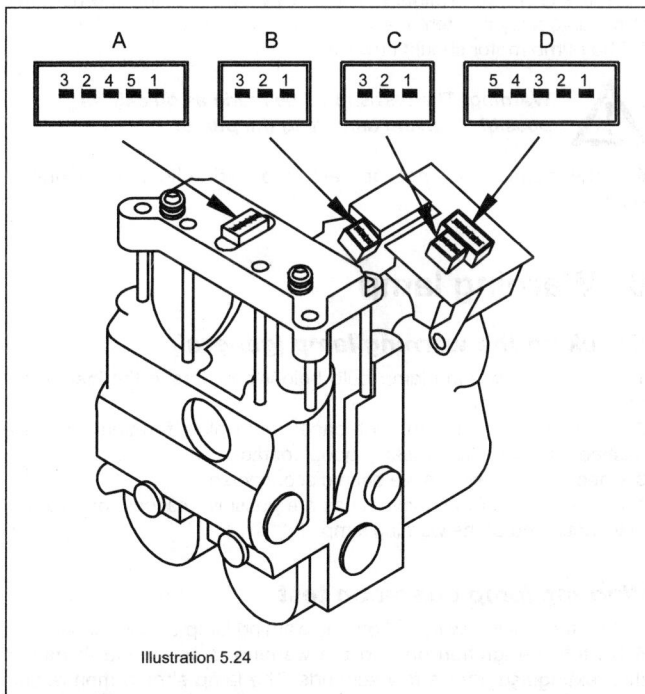

Illustration 5.24

5.24 Hydraulic control unit multi-plug identification

A and D   Five-pin multi-plug connections
B and C   Three-pin multi-plug connections

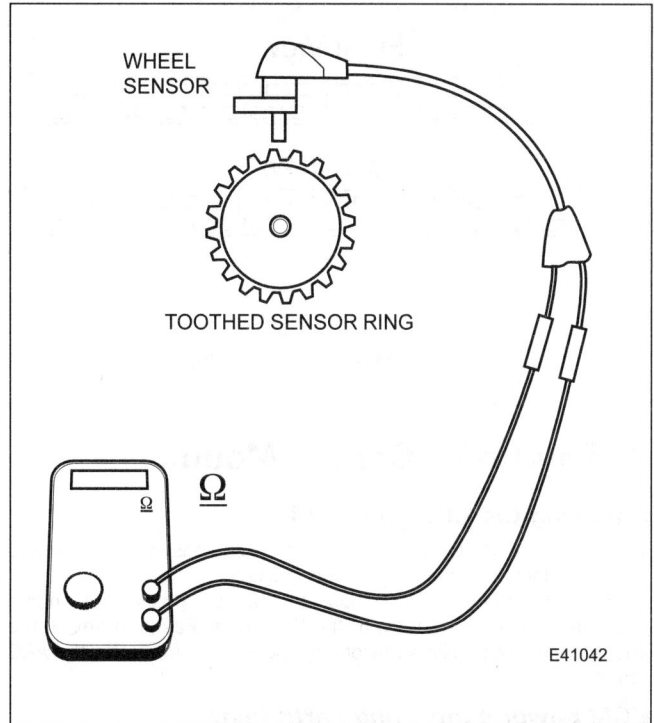

E41042

5.23 Checking wheel speed sensor resistance with an ohmmeter connected to the sensor-wiring plug

7 Connect an ohmmeter between terminal 5 of the 5-pin multi-plug base (D), and terminal 2 of the 3-pin multi-plug base (C). The ohmmeter should indicate an open circuit.

8 If any of the readings are not as specified, the relay box is suspect.

Relay operation tests

Main relay

9 Switch the ignition off.

10 Disconnect the 5-pin multi-plug (D) from the relay box located on the hydraulic control unit **(see illustration 5.24)**.

11 Connect a voltmeter between terminal 3 in the relay box (D) and a vehicle earth. A reading of zero volts should be indicated.

12 Remove the voltmeter.

13 Connect the positive terminal of a 12-volt supply to terminal 2 in the relay box (D), and the negative terminal to relay box terminal 1. There should be an unmistakable click from the relay.

14 If the relay does not operate or if the readings were not as specified, the relay box is suspect.

Relay power supply tests

Main and hydraulic pump relays

15 Switch the ignition off and disconnect the ECM multi-plug.

16 Disconnect the 3-pin multi-plug (C) from the relay box located on the hydraulic control unit **(see illustration 5.24)**.

17 Connect a voltmeter between a vehicle earth and terminals 1 and 3 in turn in the 3-pin multi-plug. The voltmeter should indicate nbv in each case.

18 Disconnect the 5-pin multi-plug (D) from the relay box located on the hydraulic control unit.

19 Connect a voltmeter between a vehicle earth and terminal 4 in the 5-pin multi-plug.

20 Switch the ignition on. The voltmeter should indicate nbv.

21 If no voltage is found, check the relevant fuse and the supply wiring back to the ignition switch.

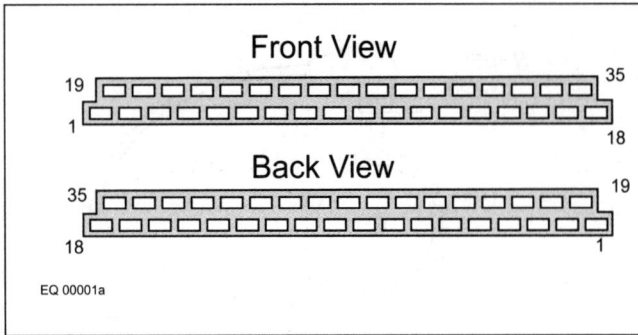

5.25 ECM 35-pin multi-plug

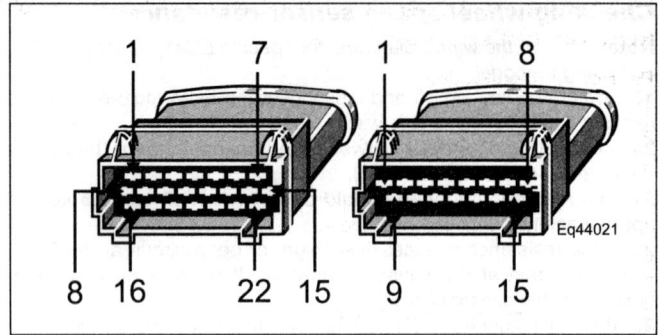

5.26 ECM 22-pin and 15-pin multi-plugs

6 Electronic Control Module

Checking the ECM (general)

1 Inspect the ECM for corrosion or damage and ensure that the unit is securely attached to the hydraulic control unit.
2 Check that the ECM multi-plug terminals are pushed fully home and making good contact with the ECM pins. Faults in any of the above areas are possible reasons for poor performance in the ABS system.

ECM power supply and earth tests

35-pin multi-plug systems

3 Switch the ignition off and disconnect the ECM multi-plug.
4 Connect a voltmeter between terminal 9 in the ECM multi-plug and a vehicle earth (see illustration 5.25). The voltmeter should indicate nbv.
5 Connect a voltmeter between terminal 2 in the ECM multi-plug and a vehicle earth.
6 Switch the ignition on. The voltmeter should indicate nbv.
7 If no voltage is found, check the relevant fuse and the supply wiring back to the battery positive terminal.
8 Switch the ignition off.
9 Connect an ohmmeter between a vehicle earth and ECM earth pins 3, 20, 27 and 28, in turn. The ohmmeter should indicate continuity. If not, check the ECM main earth connection and wiring.

15-pin/22-pin multi-plug systems

10 Switch the ignition off and disconnect the ECM multi-plugs.
11 Connect a voltmeter between terminal 13 in the ECM 15-pin multi-plug and a vehicle earth (see illustration 5.26). The voltmeter should indicate nbv.
12 Connect a voltmeter between terminal 21 in the ECM 22-pin multi-plug and a vehicle earth.
13 Switch the ignition on. The voltmeter should indicate nbv.
14 If no voltage is found, check the relevant fuse and the supply wiring back to the battery positive terminal.
15 Switch the ignition off.
16 Connect an ohmmeter between a vehicle earth and ECM earth pins 8, 14 and 15, in the ECM 15-pin multi-plug in turn. The ohmmeter should indicate continuity. If not, check the ECM main earth connection and wiring.

7 Solenoid valves

Solenoid valve resistance tests

1 Disconnect the 5-pin multi-plug (A) from the hydraulic control unit (see illustration 5.24).
2 Disconnect the 5-pin multi-plug (D) and 3-pin multi-plug (C) from the relay box located on the hydraulic control unit.

3 Connect an ohmmeter between terminal 2 of the relay box 3-pin multi-plug base (C), and terminals 1, 2, 3 and 4 in turn, of the hydraulic control unit 5-pin multi-plug base (A). A reading of between 0.5 and 3.5 ohms should be obtained in each case.
4 If any of the readings are not as specified, the relay box is suspect.

8 Hydraulic pump motor

Pump operation test

1 Switch the ignition off and disconnect the ECM multi-plug.
2 Disconnect the 3-pin multi-plug (B) from the relay box located on the hydraulic control unit (see illustration 5.24).
3 Connect an ohmmeter between terminal 2 and 3 of the relay box 3-pin multi-plug base (B). A reading of between 0.3 and 5.0 ohms should be obtained.
4 Remove the ohmmeter and connect the positive terminal of a 12 volt supply to terminal 2 in the relay box (B), and the negative terminal to relay box terminal 3.
5 The pump motor should now run.

⚠️ Warning: The test should be made as quickly as possible to avoid damaging the pump.

6 If the pump does not operate, the hydraulic control unit is suspect.

9 Warning lamp

Checking the warning lamp (general)

1 Inspect the warning lamp bulb holder contacts in the instrument panel.
2 Check that the instrument panel multi-plug terminal pins are pushed fully home and making good contact.
3 Check that no wires have been disconnected.
4 Faults in any of the above areas are possible reasons for failure or malfunctioning of the warning lamp.

Warning lamp operation test

5 With the ignition switched off, the warning lamp should remain off.
6 Switch the ignition on and the warning lamp should illuminate then extinguish after a few seconds. The lamp should then remain off.
7 If the warning lamp comes on and remains on at any time during vehicle operation, carry out the previously described test procedures on the system components.

Pin table - typical 35-pin (two-channel system)

Note: *Refer to illustration 5.25*

Pin No.	Connection	Test condition	Voltage
1	-	-	-
2	Supply from ignition switch	Ignition on	Nbv
3	ECM earth	Ignition on	0.25 volts (max)
4	Right restricting solenoid valve	Ignition on	Nbv
5	Right inlet/outlet solenoid valve	Ignition on	Nbv
6	-	-	-
7	-	-	-
8	ABS main relay driver	Ignition on/inactive	Nbv
		Actuated	1.0 volt (max)
9	Supply from battery (via pump circuit)	Ignition off/on	Nbv
		Pump actuated	1.0 volts (max)
10	-	-	-
11	-	-	-
12	SD connector	-	-
13	-	-	-
14	-	-	-
15	-	-	-
16	Right wheel speed sensor earth	Roadwheel rotating	0.25 volts (max)
17	-	-	-
18	Left wheel speed sensor earth	Roadwheel rotating	0.25 volts (max)
19	Relay supply	Ignition on	Nbv
20	ECM earth	Ignition on	0.25 volts (max)
21	Left restricting solenoid valve	Ignition on	Nbv
22	Left inlet/outlet solenoid valve	Ignition on	Nbv
23	-	-	-
24	-	-	-
25	Hydraulic pump relay driver	Ignition on/inactive	Nbv
		Actuated	1.0 volt (max)
26	ABS warning lamp	Ignition on:	
		Lamp off	Nbv
		Lamp on	0.25 volts (max)
27	ECM earth	Ignition on	0.25 volts (max)
28	ECM earth	Ignition on	0.25 volts (max)
29	-	-	-
30	-	-	-
31	-	-	-
32	-	-	-
33	Right wheel speed sensor signal	Roadwheel rotating	0.1 to 3.0 volts AC (approx)
34	-	-	-
35	Left wheel speed sensor signal	Roadwheel rotating	0.1 to 3.0 volts AC (approx)

Pin table - typical 35-pin (early three/four-channel system)

Note: *Refer to illustration 5.25*

Pin No.	Connection	Test condition	Voltage
1	-	-	-
2	Supply from ignition switch	Ignition on	Nbv
3	ECM earth	Ignition on	0.25 V (max)
4	Front right restricting solenoid valve	Ignition on	Nbv
5	Front right inlet/outlet solenoid valve	Ignition on	Nbv
6	-	-	-
7	-	-	-
8	Relay supply	Ignition on	Nbv
9**	Supply from battery	Ignition off/on	Nbv
10	-	-	-

Pin table - typical 35-pin (early three/four-channel system) (continued)

Note: *Refer to illustration 5.25*

Pin No.	Connection	Test condition	Voltage
11	-	-	-
12	SD connector	-	-
13	-	-	-
14	-	-	-
15	Rear left wheel speed sensor earth	Roadwheel rotating	0.25 volts (max)
16	Front right wheel speed sensor earth	Roadwheel rotating	0.25 volts (max)
17	Rear right wheel speed sensor earth	Roadwheel rotating	0.25 volts (max)
18	Front left wheel speed sensor earth	Roadwheel rotating	0.25 volts (max)
19	ABS main relay driver	Ignition on/inactive	Nbv
		Actuated	1.0 volt (max)
20	ECM earth	Ignition on	0.25 volts (max)
21	Front left restricting solenoid valve	Ignition on	Nbv
22	Front left inlet/outlet solenoid valve	Ignition on	Nbv
23**	Rear left inlet/outlet solenoid valve	Ignition on	nbv
24*	Rear solenoid valve	Ignition on	Nbv
24**	Rear right inlet/outlet solenoid valve		
25**	Hydraulic pump relay driver	Ignition on/inactive	Nbv
		Actuated	1.0 volt (max)
26	ABS warning lamp	Ignition on:	
		Lamp off	Nbv
		Lamp on	0.25 volts (max)
27	ECM earth	Ignition on	0.25 volts (max)
28	ECM earth	Ignition on	0.25 volts (max)
29	-	-	-
30	-	-	-
31	-	-	-
32	Rear left wheel speed sensor signal	Roadwheel rotating	0.1 to 3.0 volts AC (approx)
33	Front right wheel speed sensor signal	Roadwheel rotating	0.1 to 3.0 volts AC (approx)
34	Rear right wheel speed sensor signal	Roadwheel rotating	0.1 to 3.0 volts AC (approx)
35	Front left wheel speed sensor signal	Roadwheel rotating	0.1 to 3.0 volts AC (approx)

*Three-channel systems
**Four-channel systems

Pin table - typical 15-pin/22-pin (later four-channel system)

Note: *Refer to illustration 5.26*

Pin No.	Connection	Test condition	Voltage
15-pin multi-plug			
1	Front right restricting solenoid valve	Ignition on	Nbv
2	Front left restricting solenoid valve	Ignition on	Nbv
3	Front right inlet/outlet solenoid valve	Ignition on	Nbv
4	Front left inlet/outlet solenoid valve	Ignition on	Nbv
5	Rear right solenoid valve	Ignition on	Nbv
6	Rear left solenoid valve	Ignition on	Nbv
7	Hydraulic pump relay driver	Ignition on/inactive	Nbv
		Actuated	1.0 volt (max)
8	ECM earth	Ignition on	0.25 volts (max)
9	ABS main relay driver	Ignition on/inactive	Nbv
		Actuated	1.0 volt (max)
10	-	-	-
11	Relay supply	Ignition on	Nbv
12	-	-	-
13	Supply from battery	Ignition off/on	Nbv
14	ECM earth	Ignition on	0.25 volts (max)
15	ECM earth	Ignition on	0.25 volts (max)

Pin table - typical 15-pin/22-pin (later four-channel system) (continued)

Note: *Refer to illustration 5.26*

Pin No.	Connection	Test condition	Voltage
22-pin multi-plug			
1	Front right wheel speed sensor signal	Roadwheel rotating	0.1 to 3.0 volts AC (approx)
2	Rear right wheel speed sensor signal	Roadwheel rotating	0.1 to 3.0 volts AC (approx)
3	Front left wheel speed sensor signal	Roadwheel rotating	0.1 to 3.0 volts AC (approx)
4	-	-	-
5*	ECM earth	Ignition on	0.25 volts (max)
6	-	-	-
7	-	-	-
8	Rear left wheel speed sensor signal	Roadwheel rotating	0.1 to 3.0 volts AC (approx)
9	Front right wheel speed sensor earth	Roadwheel rotating	0.25 volts (max)
10	Front left wheel speed sensor earth	Roadwheel rotating	0.25 volts (max)
11	-	-	-
12	-	-	-
13	-	-	-
14	-	-	-
15	SD connector	-	-
16	Rear left wheel speed sensor earth	Roadwheel rotating	0.25 volts (max)
17	Rear right wheel speed sensor earth	Roadwheel rotating	0.25 volts (max)
18	-	-	-
19	-	-	-
20	-	-	-
21	Supply from ignition switch	Ignition on	Nbv
22	ABS warning lamp	Ignition on:	
		Lamp off	Nbv
		Lamp on	0.25 volts (max)

Certain versions only

Fault codes

1 The Bendix Addonix system requires the use of a FCR for obtaining fault codes. Flash codes are not available for output from this system.
2 If a FCR is available, it should be connected to the SD serial connector and used in accordance with the maker's instructions.
3 The FCR can be used for the following purposes:
 a) *Obtaining fault codes.*
 b) *Clearing fault codes.*
 c) *Obtaining datastream information.*
 d) *Testing the system actuators (solenoid valve relay, pump relay and solenoid valves).*

Fault code table
(35-pin ECM - two-channel system)

Code	Item
13	Solenoid valves - wiring and connections
14	Hydraulic pump relay - faulty operation
15	Main relay - open circuit
21	Main relay - short circuit
22	Main relay - faulty operation
23	ABS warning lamp circuit
25	Front right wheel speed sensor - incorrect resistance
32	Front left wheel speed sensor - incorrect resistance
34	Front right wheel speed sensor signal variation
41	Front left wheel speed sensor signal variation
42	Front right inlet/outlet solenoid valve faulty
43	Front right restricting solenoid valve faulty
44	Front left inlet/outlet solenoid valve faulty
45	Front left restricting solenoid valve faulty
53	Hydraulic pump motor wiring, connections
54	Hydraulic system fault
55	ECM disconnected

Fault code table
(35-pin ECM - early three/four-channel system)

Code	Item
12	Start of test sequence
11	End of test sequence
13	Solenoid valves - relay, wiring and connections
14	Hydraulic pump relay - faulty operation
15	Main relay - open circuit
21	Main relay - short circuit
22	Main relay - faulty operation
24	Rear left wheel speed sensor - incorrect resistance
25	Front right wheel speed sensor - incorrect resistance
31	Rear right wheel speed sensor - incorrect resistance
32	Front left wheel speed sensor - incorrect resistance
33	Rear left wheel speed sensor signal variation
34	Front right wheel speed sensor signal variation
35	Rear right wheel speed sensor signal variation
41	Front left wheel speed sensor signal variation
42	Front right inlet/outlet solenoid valve faulty
43	Front right restricting solenoid valve faulty
44	Front left inlet/outlet solenoid valve faulty
45	Front left restricting solenoid valve faulty
51*	Rear inlet/outlet solenoid valve faulty
51**	Rear right inlet/outlet solenoid valve faulty
52**	Rear left inlet/outlet solenoid valve faulty
53	Hydraulic pump motor relay, wiring, connections
54	Hydraulic system fault
55	ECM memory fault

*Three-channel systems
**Four-channel systems

Fault code table (15-pin/22-pin ECM - later four-channel system)

Code	Item
13	Solenoid valves supply voltage
14	Hydraulic pump relay - faulty operation
15	Main relay - short circuit
18	Wheel speed sensors - incoherent wheel speeds
19	Front or rear wheel speed sensors - incoherent signal
21	Main relay - open circuit
22	Main relay - faulty operation
24	Rear left wheel speed sensor - incorrect resistance
25	Front right wheel speed sensor - incorrect resistance
31	Rear right wheel speed sensor - incorrect resistance
32	Front left wheel speed sensor - incorrect resistance
33-36	Rear left wheel speed sensor signal variation
34-37	Front right wheel speed sensor signal variation
35-38	Rear right wheel speed sensor signal variation
39-41	Front left wheel speed sensor signal variation
42	Front right inlet/outlet solenoid valve faulty
43	Front right restricting solenoid valve faulty
44	Front left inlet/outlet solenoid valve faulty
45	Front left restricting solenoid valve faulty
51	Rear right solenoid valve faulty
52	Rear left solenoid valve faulty
53	Hydraulic pump motor wiring, connections
55	ECM disconnected
56	ECM disconnected
57	ECM supply voltage incorrect
66-67	Rear left and front right wheels locking
66-68	Rear wheels locking
67	Front right wheel locking
68-69	Rear right and front left wheels locking
69	Front left wheel locking

Wiring diagrams

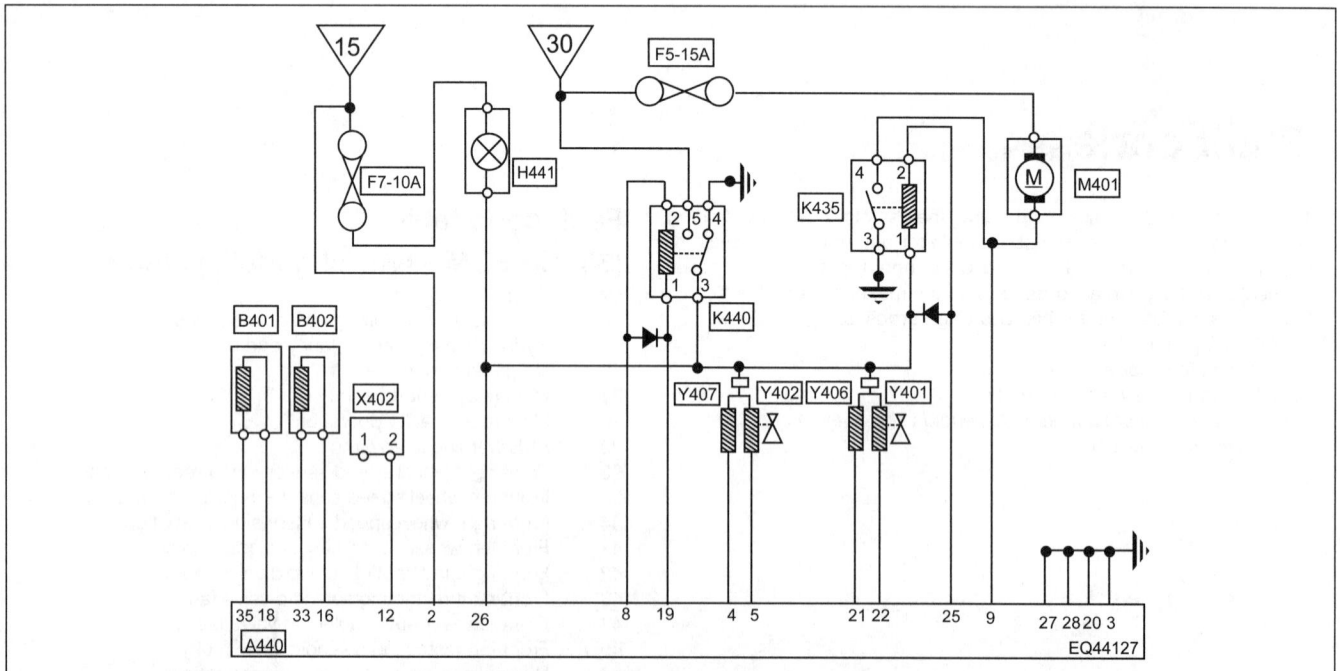

5.27 35-pin wiring diagram, 2-channel, Citroën AX

5

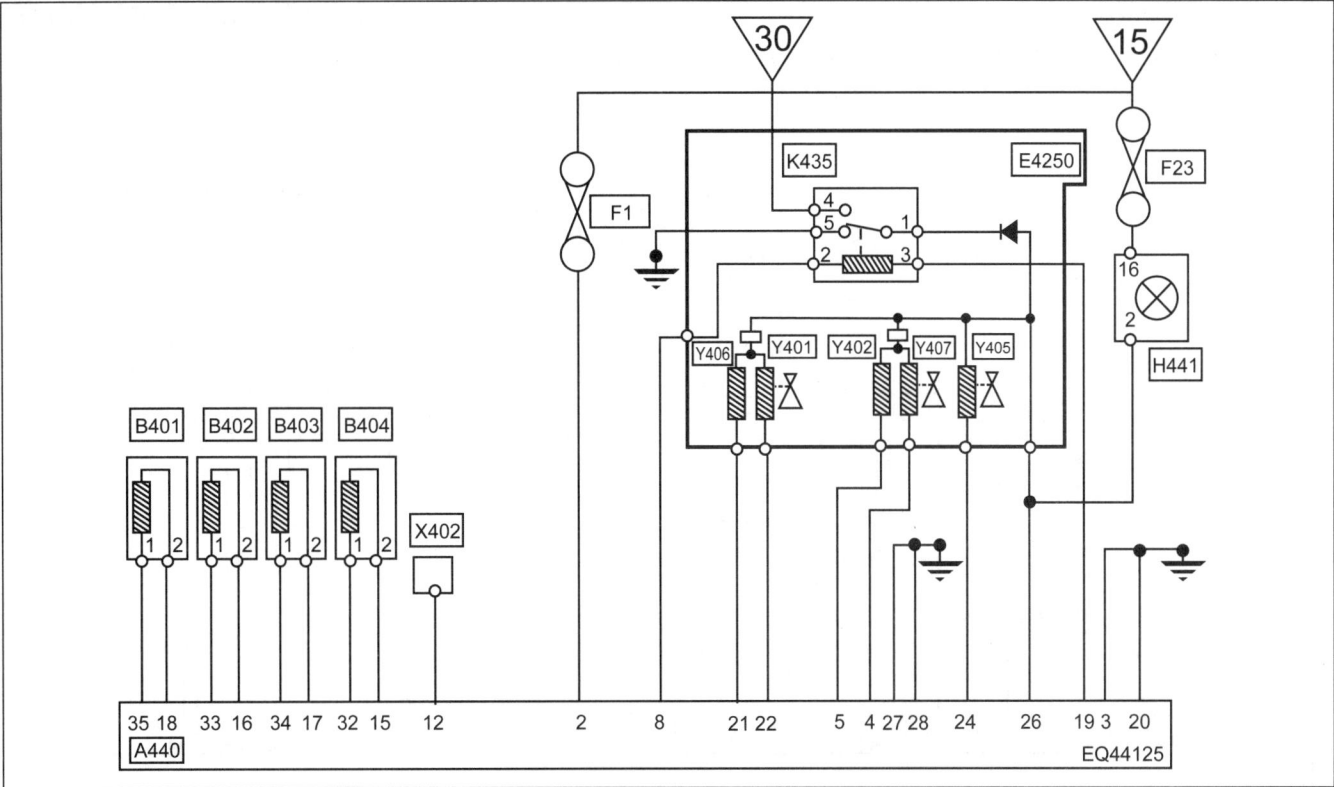

5.28 35-pin wiring diagram, 3-channel, Citroën XM

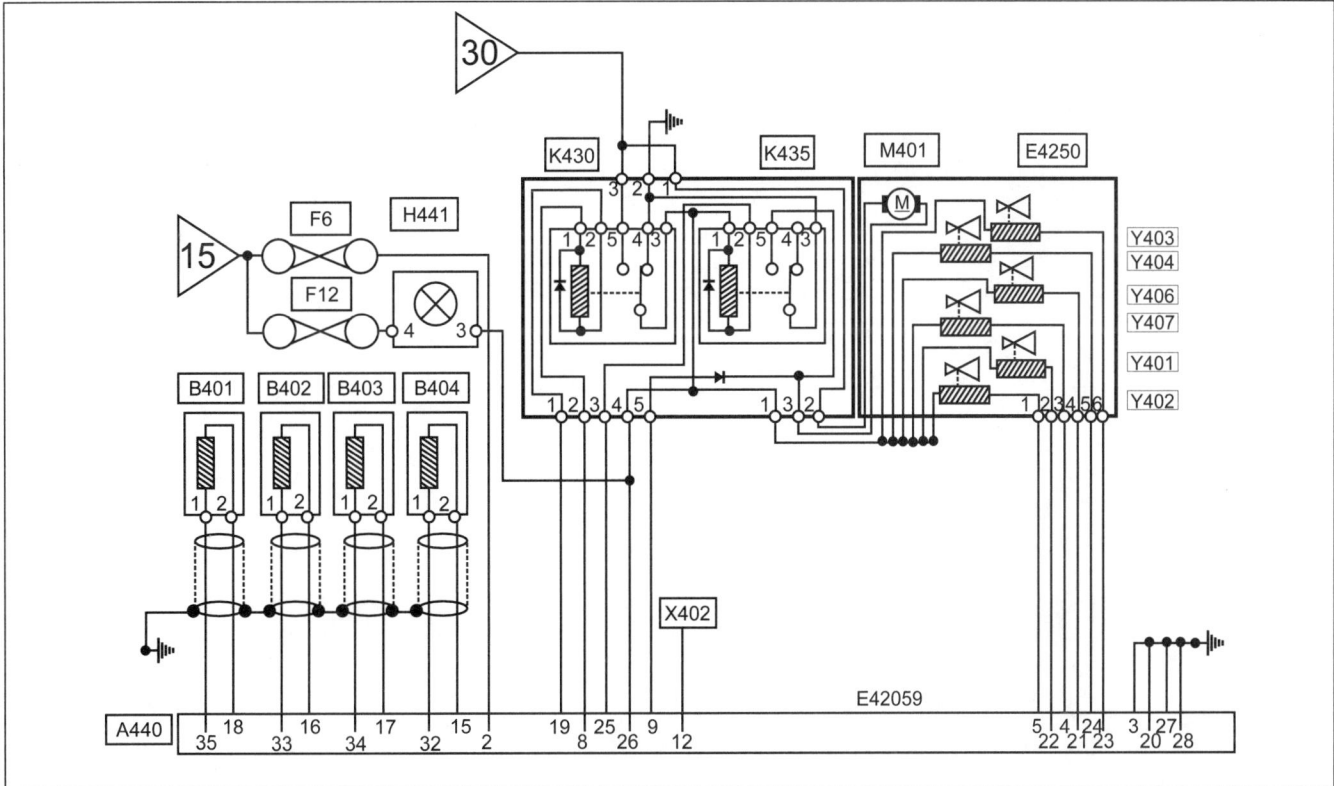

5.29 35-pin wiring diagram, 4-channel, Peugeot 405

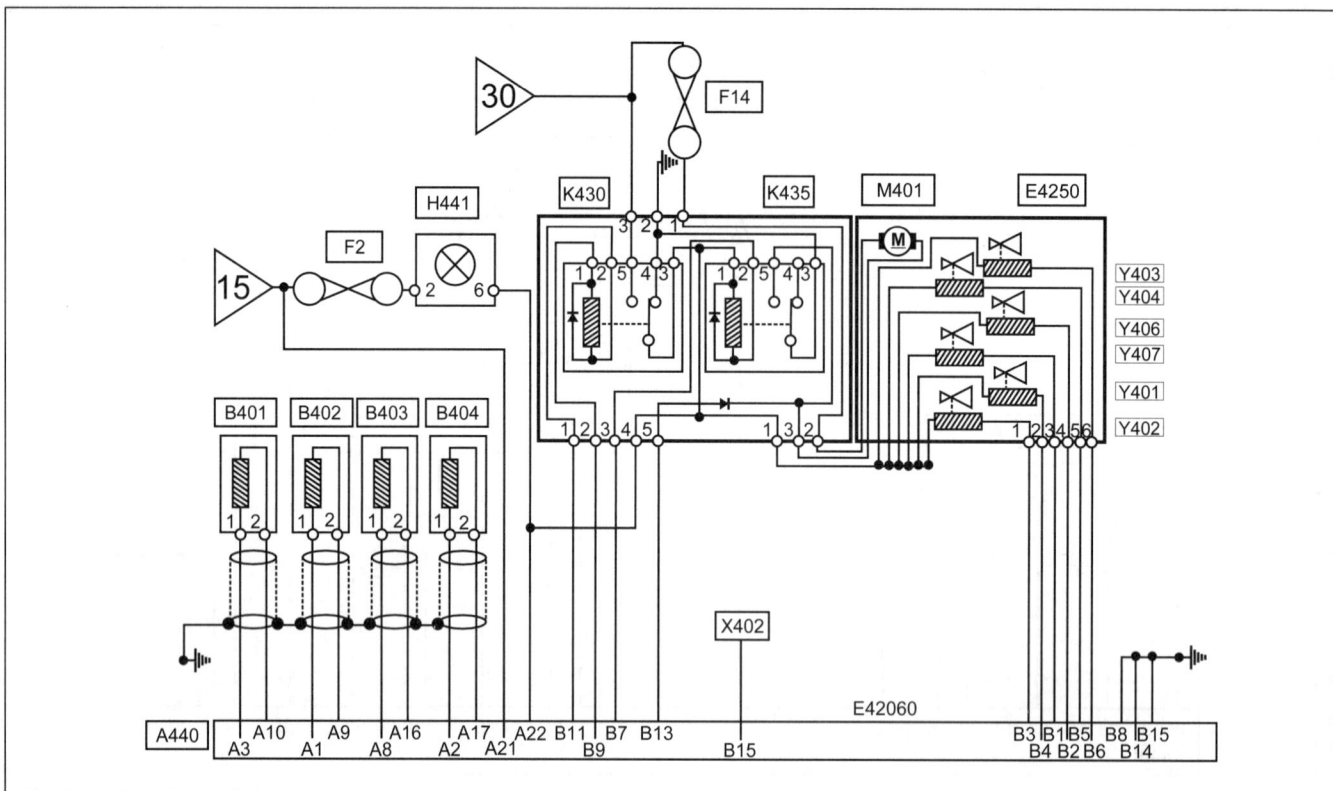

5.30 15/22 pin wiring diagram, 4-channel, Peugeot 806

Chapter 6

Bendix Integrated ABS

Contents

Vehicle coverage

Overview of system operation

1 Basic principles and system identification

The Bendix Integrated Antilock Brake System has been fitted to a limited number of mainly Peugeot passenger vehicles since its introduction in the late 1980s. The system is often referred to as an integrated type whereby the components of the ABS hydraulic control unit are also required during the operation of the conventional braking system.

The purpose of the system is to apply the vehicle brakes at maximum efficiency without wheel lock or loss of directional stability. Inductive sensors (wheel speed sensors) monitor the speed of the wheels by generating an electrical signal as the wheel is rotated. This information is passed to the ABS Electronic Control Module (ECM) which is then able to determine wheel speed, wheel acceleration and wheel deceleration. The ECM compares the signals received from each wheel and if the onset of lock at any wheel is detected, a signal is sent to the ABS hydraulic control unit, which regulates the brake pressure for the relevant wheel(s).

Bendix Integrated ABS is comprised of the following components **(see illustration 6.1)**:

a) *ABS-ECM.*
b) *Electric pump unit.*
c) *Hydraulic control unit.*
d) *Four inductive wheel speed sensors and associated sensor rings.*
e) *ABS electrical wiring harness and relays.*
f) *ABS warning lamps.*
g) *Diagnostic connector.*
h) *Brake calipers and hydraulic hoses and pipes.*
i) *Pressure regulating/load-sensing valve(s) depending on application.*

2 Component description and operation

ABS ECM

General

The Bendix Integrated ABS Electronic Control Module (ECM) continually monitors wheel speed from the signals provided by the wheel speed sensors. If the ECM detects the incidence of wheel lock on one or more wheels, a signal is sent to the hydraulic control unit to modulate the hydraulic pressure to the brake of the locking wheel **(see illustration 6.2)**.

The ECM contains two microprocessors and uses digital technology to complete this function and other functions such as fault code memory and power modules for valve and relay activity **(see illustration 6.3)**.

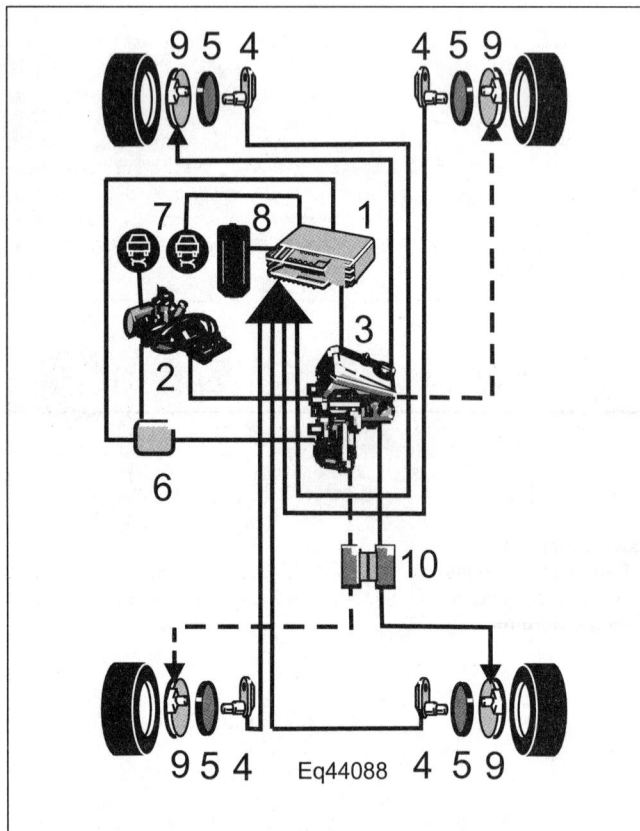

6.1 **Bendix Integrated components**

1 *ABS-ECM.*
2 *Electric pump unit.*
3 *Hydraulic control unit.*
4 *Inductive wheel speed sensors*
5 *Sensor rings.*
6 *ABS electrical wiring harness and relays.*
7 *ABS warning lamps.*
8 *Diagnostic connector.*
9 *Brake calipers and hydraulic hoses and pipes.*
10 *Pressure regulating/load-sensing valve(s) depending on application.*

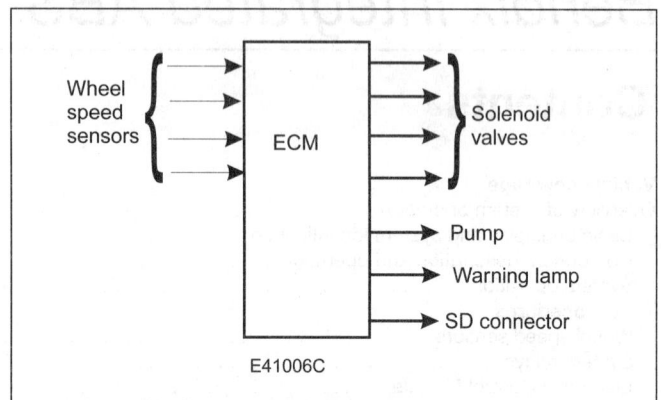

6.2 **ECM sensor inputs and control signal outputs**

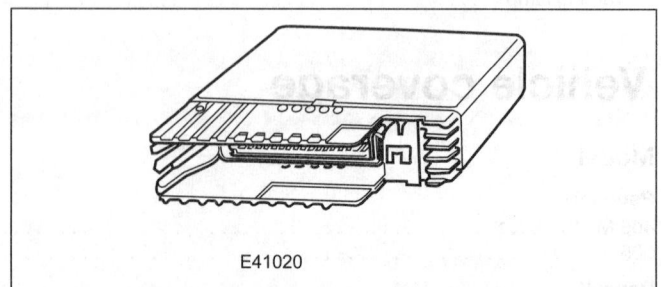

6.3 **Typical Bendix Integrated ECM**

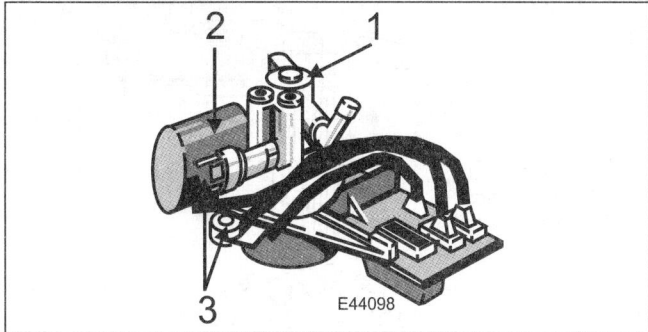

6.4 Bendix Integrated electric pump unit

1   Electric pump   3   Pressure switches
2   Pressure accumulator

Self-test

The Bendix ECM is equipped with a self-test capability that initially examines the ABS system when the ignition is switched on, and then examines the wheel speed sensor signals after a pre-determined wheel speed is reached from all wheels (engine running). The ABS self-test program continues to examine the signals from the various components as long as the ignition is switched on. If self-test determines that faults are not present, the ABS is ready for operation once a specified vehicle speed has been achieved.

If the ECM detects that a fault is present, the ABS functions are restricted or disabled altogether, depending on the nature of the fault. One or both of the warning lamps will also be illuminated indicating the presence of a fault. Conventional braking will still be available on all wheels, however, if the fault is associated with low system pressure, brake pedal travel will be excessive.

Self-diagnostics

If the ECM detects a fault during the self-test routine, an internal fault code is stored in the ECM memory. Stored fault codes can be retrieved from the SD connector with the aid of a suitable fault code reader. If the fault clears, the code will remain stored until cleared with the FCR.

Electric pump unit

Instead of using a conventional vacuum servo unit for braking assistance, the Bendix Integrated system uses an electric pump, pressure accumulator and pressure switches for hydraulic operation. The pump draws hydraulic fluid from the reservoir and feeds it under high pressure to the accumulator (see illustration 6.4).

The accumulator contains two chambers separated by a flexible diaphragm. The upper chamber is filled with high-pressure nitrogen gas and the lower chamber contains the brake hydraulic fluid supplied under pressure from the pump. The hydraulic fluid compresses the nitrogen gas further, thus providing a supply of high-pressure hydraulic fluid for the brake system. Pump operation is controlled by two pressure switches located below the accumulator. The switches actuate the pump relay to operate the electric pump if the accumulator pressure drops below 160 bar. When the pressure reaches 180 bar, the switches deactivate the relay to switch off the pump. If the accumulator pressure falls below 90 bar, a signal is sent to the ECM to switch off the ABS operation. On earlier versions, a third pressure switch is used to provide the low-pressure signal to the ECM.

The pressurised hydraulic fluid from the accumulator is supplied to the hydraulic control unit for brake system operation either with or without ABS control.

Hydraulic control unit

The hydraulic control unit consists of two master cylinder assemblies, one for each brake circuit, a hydraulic valve block

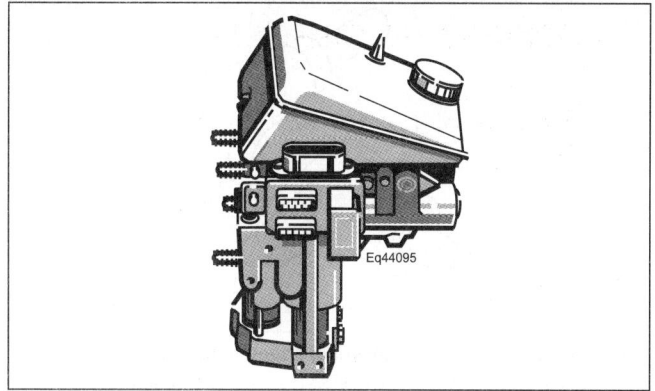

6.5 Bendix Integrated hydraulic control unit

containing the ABS solenoid valves, and a fluid reservoir with integral low fluid level switch (see illustration 6.5).

The unit is mounted on the engine compartment bulkhead in place of the conventional master cylinder, and connected to the electric pump unit by a series of interconnecting hydraulic pipes and pressure hoses.

The hydraulic valve block contains the inlet/outlet and restricting solenoid valves for ABS operation. The valve block modulates the pressure applied to the brake caliper for each individual wheel during the various phases of ABS operation.

Bendix Integrated ABS is a four-channel system incorporating six solenoid valves in the hydraulic valve block - two for each front brake (one inlet/outlet solenoid valve and one restricting solenoid valve) and one inlet/outlet solenoid valve for each rear brake (the restricting solenoid valves are not used in the rear brake hydraulic circuit) (see illustration 6.6).

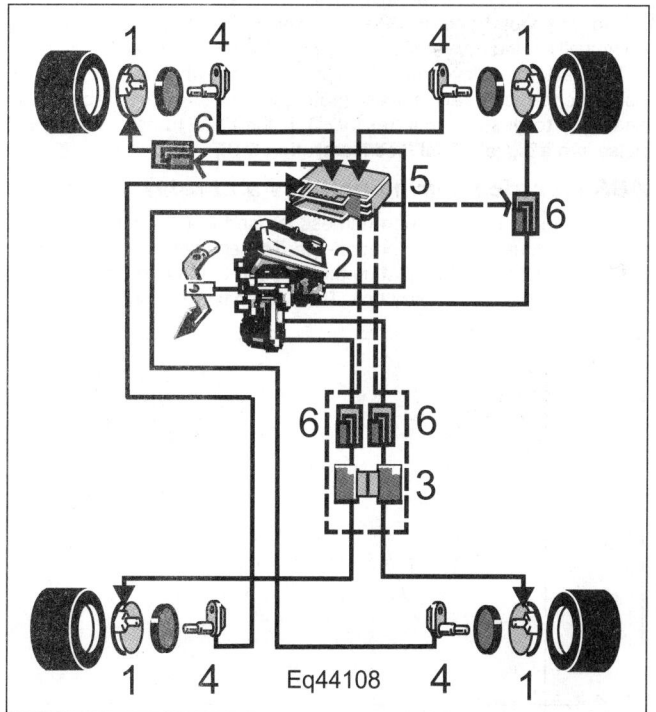

6.6 Bendix Integrated hydraulic circuit layout

1   Brake calipers   4   Wheel speed sensors
2   Hydraulic control unit   5   ABS-ECM
3   Rear brake pressure   6   Solenoid valves
    regulating valve

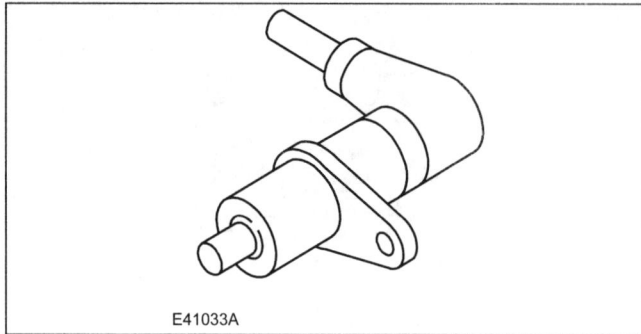

E41033A

6.7 Typical wheel speed sensor

Typically, control of hydraulic pressure to the rear brakes is based on the 'select-low' principle. With the 'select-low' principle, the rear wheel with the lowest adhesion determines the amount of hydraulic pressure to both rear brakes during ABS operation.

Wheel speed sensors

The rotational speed of the roadwheels and any changes in the rotational speed are recorded by inductive wheel speed sensors located at the roadwheels. Depending on system type, there may be a separate sensor for each front wheel only or a separate sensor for all four wheels **(see illustration 6.7)**.

Each wheel speed sensor assembly comprises a toothed sensor ring, which rotates at roadwheel speed, and an adjacent sensor mounted a set distance from the sensor ring **(see illustration 6.8)**.

The sensors are permanent magnet pulse generator types producing an AC voltage sine wave as the sensor ring teeth pass through the magnetic field of the sensor **(see illustration 6.9)**.

The frequency of the waveform produced by the wheel speed sensor is proportional to the road speed. This AC voltage signal is continually being delivered to the ABS-ECM for processing.

The peak to peak voltage of the speed signal (when viewed upon an oscilloscope) can vary considerably according to wheel speed. An analogue to digital converter (ADC) in the ECM transforms the AC pulse into a digital signal **(see illustration 6.10)**.

ABS electrical wiring harness and relays

An integrated main wiring harness is used to connect power and earth to various electrical components. This enables sensor signals to reach the ECM and the ECM, in turn to send command signals to the relays and hydraulic control unit. The ABS main relay and pump relay

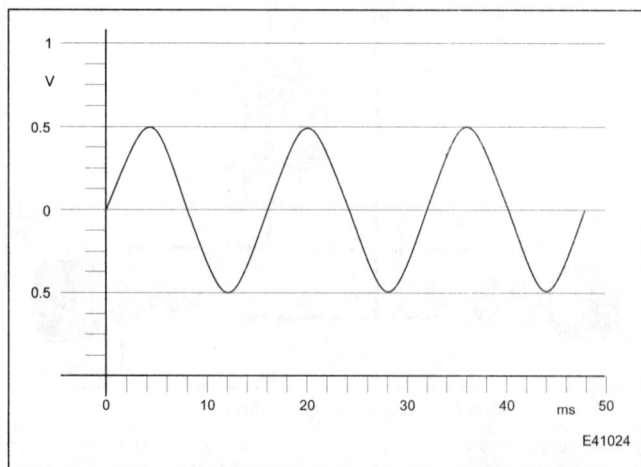

6.10 Wheel speed sensor waveform as viewed on an oscilloscope

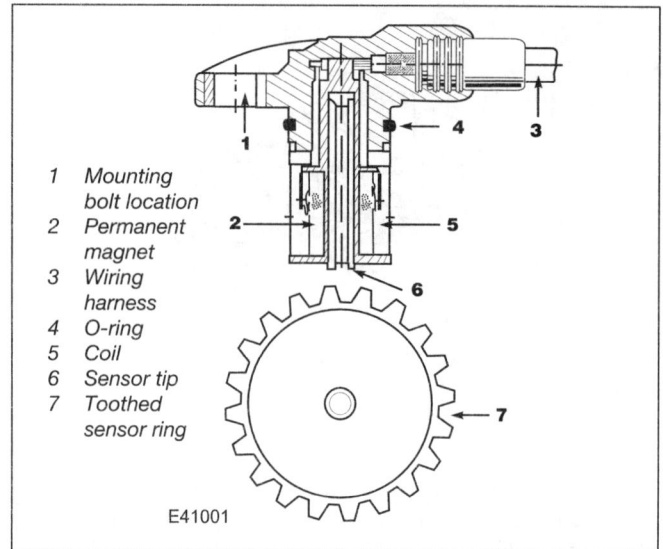

1 Mounting bolt location
2 Permanent magnet
3 Wiring harness
4 O-ring
5 Coil
6 Sensor tip
7 Toothed sensor ring

E41001

6.8 Sectional view of a wheel speed sensor

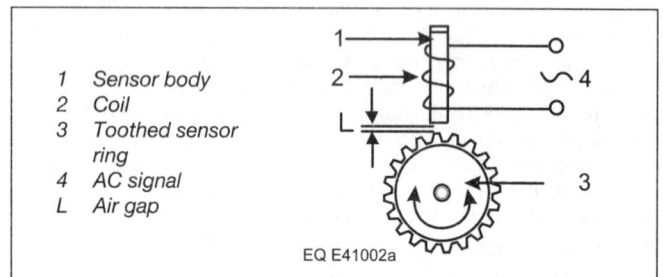

1 Sensor body
2 Coil
3 Toothed sensor ring
4 AC signal
L Air gap

EQ E41002a

6.9 Wheel speed sensor operation

are located either on the hydraulic control unit and electric pump unit respectively, or on a separate relay interconnection plate within the engine compartment. According to application, an additional pressure warning relay and overvoltage relay may be used with the system.

Main relay

Nominal battery voltage is supplied from the battery to the main relay. When the relay is actuated by the ECM, battery voltage is supplied to the solenoid valves, the earth for the solenoid valves being controlled by the ECM.

Pump relay

Nominal battery voltage is supplied to one main terminal of the pump relay via the pump motor and a fuse. The other relay main terminal is connected to earth. The relay is activated by the pressure switches thus operating the pump motor according to system pressure requirements.

Pressure warning relay

When fitted, the pressure-warning relay is used to illuminate the warning lamps on the instrument panel in case of low system pressure.

Overvoltage relay

The overvoltage relay is only fitted to certain systems and is used to protect the ECM from voltage surges during engine starting. When the starter is in operation, the relay cuts off the supply from the ignition switch to the ECM.

ABS warning lamps

After the ignition is switched on, the ABS warning lamp on the instrument panel is illuminated for approximately 3 seconds as the

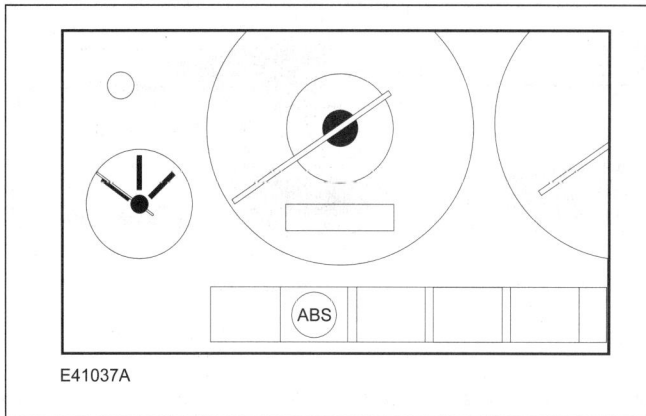

E41037A

6.11 ABS warning lamp

system executes three self-test routines, each of one second duration. If satisfactory operation of the system is detected by the ECM, and if the system hydraulic pressure is greater than 90 bar, the lamp is extinguished. During vehicle operation above a pre-determined wheel speed, the ABS-ECM implements a further self-test cycle whereby ABS operation, hydraulic pressure and wheel speed sensor signals are monitored. If a fault is detected, the relevant ECM pin is earthed to illuminate the warning lamp on the instrument panel and the ABS function is restricted, or completely disabled, depending on the nature of the fault. The warning lamp will remain illuminated until the fault is no longer present **(see illustration 6.11)**.

Depending on application, one or two additional brake system warning lamps are used to indicate low system pressure or low hydraulic fluid level. These warning lamps are illuminated by the pressure switches to indicate low system pressure, or by the level sensor in the hydraulic fluid reservoir to indicate low fluid level.

The fluid level sensor is a two position switch which will illuminate the brake system warning lamp(s) on the instrument panel if the level in the reservoir falls, and also signal the ECM. The ECM will then disable the ABS operation and illuminate the ABS warning lamp. If the level falls still further, the switch will isolate the pump relay to inhibit the operation of the pump so the remaining hydraulic fluid can be retained for normal braking without ABS control.

When the ABS-ECM detects a fault, the fault code is stored and the ABS warning lamp activated. If the fault no longer exists after the next system start (ignition on/off) the ABS warning lamp is extinguished after the self-test cycle, however the fault code remains stored in the ECM memory.

Pressure regulating/load-sensing valve(s)

Depending on vehicle application, pressure regulating valves or load sensing valves may be incorporated to restrict the hydraulic fluid pressure to the rear brakes. The valves may be pressure conscious whereby the hydraulic fluid supply is restricted once a pre-determined pressure is reached, or load conscious whereby the hydraulic pressure is reduced according to vehicle loading.

3 System operation

Brake system at rest

When the system is at rest prior to operation of the brake pedal, hydraulic fluid under pressure from the accumulator is present at the return valve in the master cylinder, keeping the valve closed. The pressurised fluid is also present in the pressure chamber, which, together with the action of the return spring, move the master cylinder piston assembly to the left. In this condition, valve is open and there is a direct connection between the fluid reservoir and the solenoid valves in the hydraulic valve block via the passage. Fluid in the reservoir, and in the remainder of the master cylinder passages and hydraulic valve block is at atmospheric pressure and the brakes are not applied **(see illustration 6.12)**.

Brake system operating under conventional control without ABS

When the brake pedal is activated, the pedal force moves the piston to the right under action of the spring. This movement closes the valve and moves the spool to begin opening the passage from the pressure chamber to the solenoid valves in the hydraulic valve block via the passage **(see illustration 6.13)**. Fluid from the pressure chamber also passes to the spool chamber. Pressure in this chamber acts upon the piston and when the pressure overcomes the tension of the spring, the spool moves to the left, maintaining the existing pressure, but preventing any further high pressure fluid passage to the hydraulic valve block. The hydraulic pressure in the system from the accumulator is therefore always proportional to the effort applied at the brake pedal.

In the hydraulic valve block, both the front brake inlet/outlet and restricting solenoid valves, and rear brake inlet/outlet solenoid valves in the controlled circuit are 'at rest' allowing an unrestricted hydraulic passage from the master cylinder through the valve block to the brake caliper, thus operating the brake. The passage of fluid in the rear

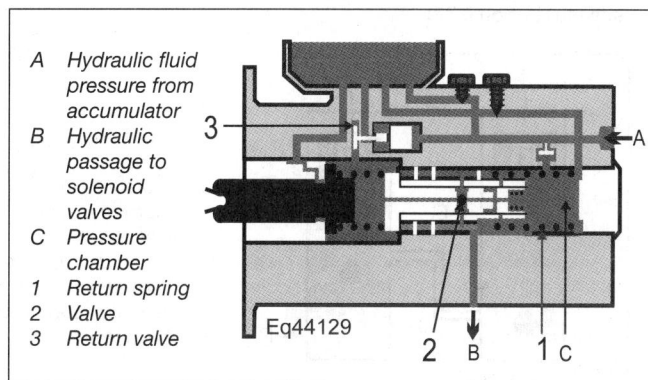

A Hydraulic fluid pressure from accumulator
B Hydraulic passage to solenoid valves
C Pressure chamber
1 Return spring
2 Valve
3 Return valve

Eq44129

6.12 System operation 1 - brake system at rest

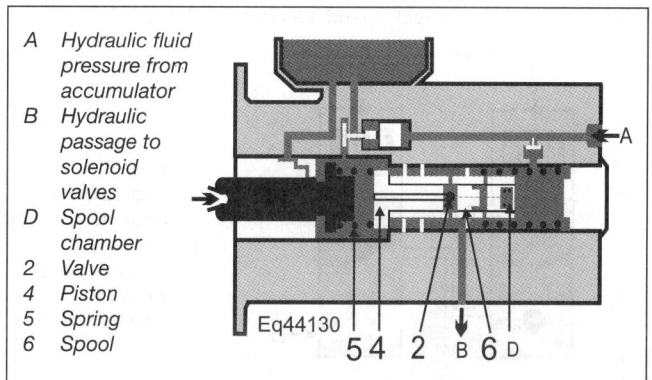

A Hydraulic fluid pressure from accumulator
B Hydraulic passage to solenoid valves
D Spool chamber
2 Valve
4 Piston
5 Spring
6 Spool

Eq44130

6.13 System operation 2 - conventional braking without ABS

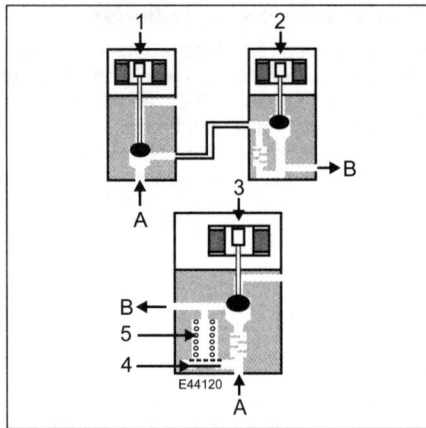

6.14 System operation 3 - conventional braking without ABS

A   Hydraulic fluid pressure from master
    cylinder
B   Hydraulic fluid pressure to brake caliper
1   Front inlet/outlet solenoid valve
2   Front restricting solenoid valve
3   Rear inlet/outlet solenoid valve
4   Flow valve
5   Spring

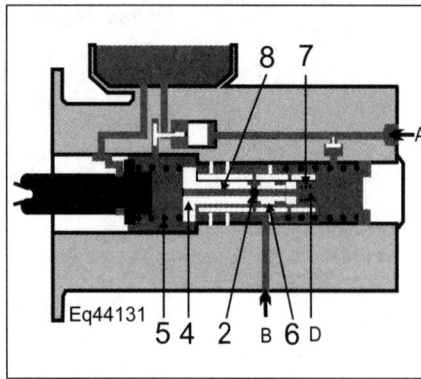

6.15 System operation 4 - conventional braking without ABS

A   Hydraulic fluid pressure from master
    cylinder
B   Hydraulic fluid return from brake caliper
D   Spool chamber
2   Valve
4   Piston
5   Spring
6   Spool
7   Spring
8   Fluid return passage to reservoir

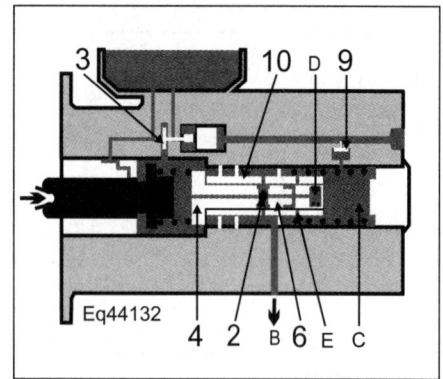

6.16 System operation 5 - conventional braking without ABS

B   Hydraulic fluid pressure to brake caliper
C   Pressure chamber
D   Pressure chamber
E   Delivery passage
2   Valve
3   Shut-off valve
4   Piston
6   Spool
9   Shut-off valve
10 Sleeve

brake inlet/outlet valve is through the flow valve, which is held open by the spring **(see illustration 6.14)**.

When the brake pedal is released, the pressurised hydraulic fluid in the spool chamber, and the action of the spring, force the spool and piston to move to the left. The fluid passage from the pressure chamber is closed by the movement of the spool, and the valve is re-opened. Hydraulic fluid returned from the brake calipers, via the valve block enters through a passage and is returned through a channel to the fluid reservoir **(see illustration 6.15)**.

In the event of failure of the high-pressure side of the system, the unit operates in the same way as a conventional master cylinder. As there is no high-pressure supply from the accumulator, the two shut-off valves are closed under the action of their return springs. The force applied to the brake pedal moves the piston to the right, which in turn moves the sleeve. The valve closes and the spool also moves to the right which opens the delivery passage. This movement increases the hydraulic pressure in the pressure chambers, which is transmitted through a passage, and through the valve block to operate the brakes **(see illustration 6.16)**.

Brake system operating in conjunction with ABS control

The ABS-ECM continually monitors wheel speed from the signals provided by the wheel speed sensors. If the ECM detects the

incidence of wheel lock on any wheel, ABS is automatically initiated. As ABS operates individually on each wheel, any of the wheels could be in any one of the following phases at any particular moment. In addition, the arrangement of the solenoid valves in the hydraulic valve block dictates that a different principle of ABS modulation is used for the front and rear brake hydraulic circuits.

Front brakes - phase one, rapid pressure reduction

The ECM energises the inlet/outlet solenoid valve in the hydraulic valve block, closing off the direct hydraulic connection between the master cylinder and the front brake caliper in the controlled circuit. The restricting solenoid valve remains 'at rest' and pressure in the hydraulic circuit falls rapidly as fluid from the brake caliper is returned to the fluid reservoir **(see illustration 6.17)**.

Front brakes - phase two, slow pressure reduction

The rapid pressure reduction phase is always followed by a slow pressure reduction phase. The inlet/outlet solenoid valve remains energised and the restricting solenoid valve is now also energised. Hydraulic fluid returned from the brake caliper is forced to flow through the restrictor as it passes back to the fluid reservoir, thus resulting in a slow pressure reduction in the controlled circuit. When the controlled wheel is once again rotating, phase three is initiated **(see illustration 6.18)**.

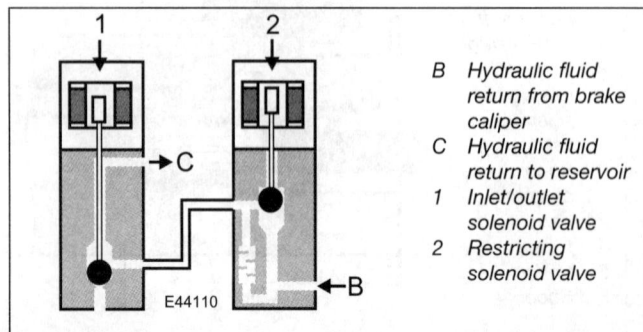

B   Hydraulic fluid
    return from brake
    caliper
C   Hydraulic fluid
    return to reservoir
1   Inlet/outlet
    solenoid valve
2   Restricting
    solenoid valve

6.17 ABS operation - front brakes phase one, rapid pressure reduction

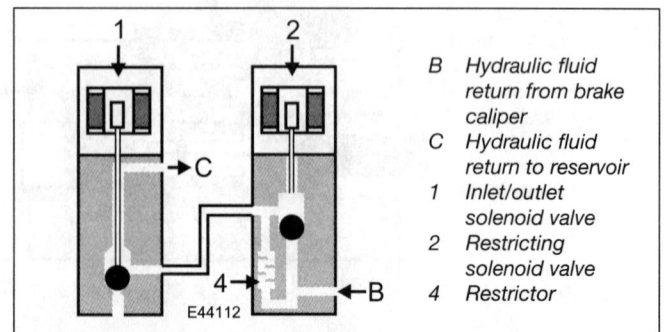

B   Hydraulic fluid
    return from brake
    caliper
C   Hydraulic fluid
    return to reservoir
1   Inlet/outlet
    solenoid valve
2   Restricting
    solenoid valve
4   Restrictor

6.18 ABS operation - front brakes phase two, slow pressure reduction

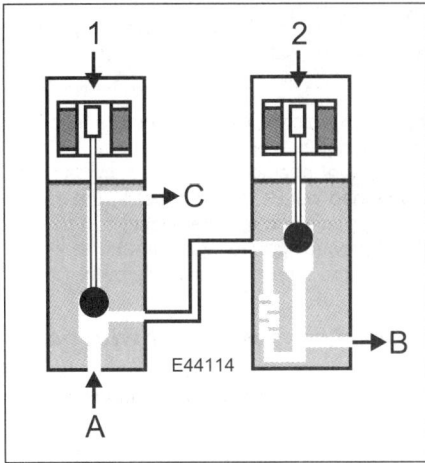

6.19 ABS operation - front brakes phase three, rapid pressure increase

A   Hydraulic fluid pressure from master cylinder
B   Hydraulic fluid pressure to brake caliper
C   Hydraulic fluid return to reservoir
1   Inlet/outlet solenoid valve
2   Restricting solenoid valve

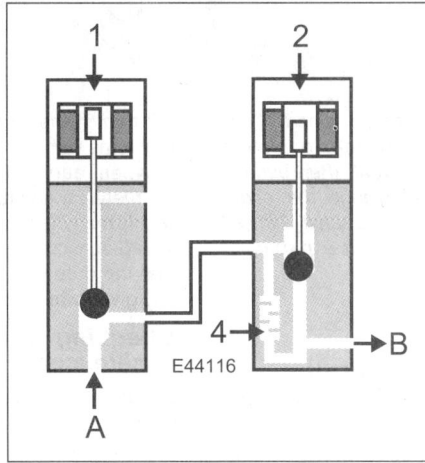

6.20 ABS operation - front brakes phase four, slow pressure increase

A   Hydraulic fluid pressure from master cylinder
B   Hydraulic fluid pressure to brake caliper
1   Inlet/outlet solenoid valve
2   Restricting solenoid valve
4   Restrictor

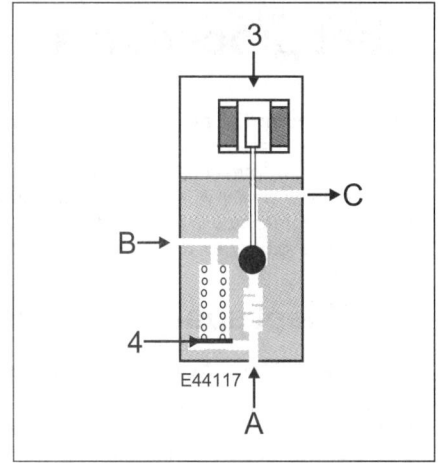

6.21 ABS operation - rear brakes phase one, rapid pressure reduction

A   Hydraulic fluid pressure from master cylinder
B   Hydraulic fluid return from brake caliper
C   Hydraulic fluid return to reservoir
3   Inlet/outlet solenoid valve
4   Flow valve

Front brakes - phase three, rapid pressure increase

Both solenoid valves are de-energised by the ECM allowing them to return to the 'at rest' position. The hydraulic circuit from the master cylinder directly to the front brake caliper is re-instated allowing an unrestricted hydraulic passage from the master cylinder through the valve block to the brake caliper. This results in a rapid increase in hydraulic pressure in the controlled circuit and the brake is once again applied **(see illustration 6.19)**.

Front brakes - phase four, slow pressure increase

The rapid pressure increase phase is always followed by a slow pressure increase phase. The inlet/outlet solenoid valve remains 'at rest', but the restricting solenoid valve is once again energised by the ECM. Hydraulic fluid from the master cylinder is forced to flow through the restrictor before passing to the brake caliper, resulting in a slow increase in hydraulic pressure in the controlled circuit **(see illustration 6.20)**.

This pressure increase continues until the wheel speed sensor detects that the brake is once again about to lock at which point phase one is re-instigated by the ECM. The cycle then continues until the vehicle comes to a halt or the brake pedal is released.

The whole ABS control cycle takes place many times per second for each affected wheel and this ensures maximum braking effect and control during ABS operation.

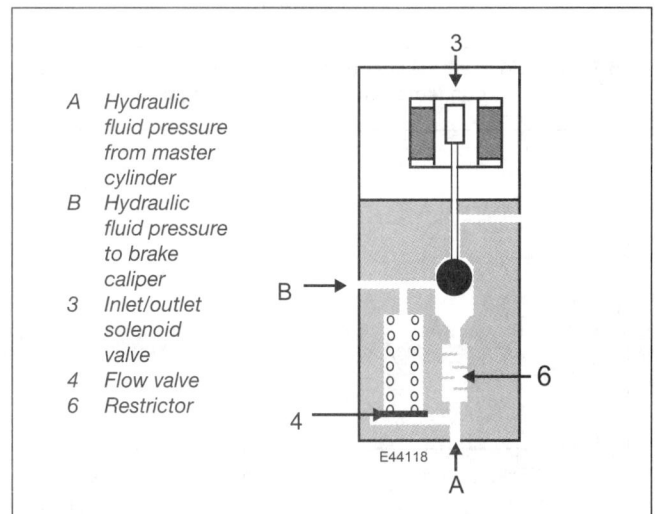

Rear brakes - phase one, rapid pressure reduction

The rear brakes operate in a slightly different manner, as there are no restricting solenoid valves for the rear brake hydraulic circuits. Instead, the inlet/outlet solenoid valves incorporate a restrictor and flow valve.

The ECM energises the inlet/outlet solenoid valve closing off the direct hydraulic connection between the master cylinder and the rear brake caliper in the controlled circuit. This creates a difference in hydraulic pressure on either side of the flow valve, the greater pressure being on the supply side from the master cylinder. The flow valve therefore closes, shutting the direct hydraulic connection between the master cylinder and the rear brake caliper in the controlled circuit. Pressure in the hydraulic circuit falls rapidly as fluid from the brake caliper is returned to the fluid reservoir **(see illustration 6.21)**.

Rear brakes - phase two, slow pressure increase

The inlet/outlet solenoid valve is de-energised by the ECM allowing it to return to the 'at rest' position. Hydraulic fluid pressure is still greater on the supply side from the master cylinder causing the flow valve to remain closed. Hydraulic fluid from the master cylinder is forced to flow through a restrictor before passing to the brake caliper, resulting in a slow increase in hydraulic pressure in the controlled circuit **(see illustration 6.22)**.

This pressure increase continues until the wheel speed sensor detects that the brake is once again about to lock at which point phase one is re-instigated by the ECM. The cycle then continues until the vehicle comes to a halt or the brake pedal is released.

The whole ABS control cycle takes place many times per second for each affected wheel and this ensures maximum braking effect and control during ABS operation.

A   Hydraulic fluid pressure from master cylinder
B   Hydraulic fluid pressure to brake caliper
3   Inlet/outlet solenoid valve
4   Flow valve
6   Restrictor

6.22 ABS operation - rear brakes phase two, slow pressure increase

Test procedures

Important note: *The test procedures and pintables contained in this Chapter are necessarily representative of the system depicted. Because of the variations in wiring and other data that often occurs, even between similar vehicles in any particular VM's range, the reader should take great care in identification of ECM pins, and satisfy himself that he has gathered the correct data before failing a particular component.*

4 Wheel speed sensors

Checking the wheel speed sensor (general)

1 Inspect the wheel speed sensor for corrosion or damage and check that the sensor is tightly mounted.

2 Check the toothed sensor ring for damage, eccentricity and for broken or missing teeth.

3 Inspect the wheel speed sensor-wiring plug for corrosion and damage. One plug for each sensor.

4 Check that the connector terminal pins are pushed fully home and making good contact with the sensor-wiring plug.

5 Check the clearance between the sensor and the toothed sensor ring. The clearance is not normally adjustable but is nominally 0.2 to 1.2 mm. If the clearance is excessive, expect a worn sensor tip or problems with the wheel bearings/hub or sensor ring.

6 When carrying out voltage checks with an oscilloscope or voltmeter, the voltage obtained will be proportional to the speed at which the wheel is rotating. In addition to determining that the wheel speed sensors are actually producing a voltage output, it is essential that the output from the sensors on a particular axle is the same for any given wheel speed.

Checking wheel speed sensor output with an oscilloscope

Note: *Refer to a suitable wiring diagram for specific ECM pin identification according to model.*

7 Switch the ignition off and disconnect the ECM multi-plug or the relevant wheel speed sensor-wiring plug.

8 Connect an oscilloscope between the terminal pins for the sensor under test **(see illustration 6.23)**.

9 Select a range to cover 80 Hz on the oscilloscope and a free run time base.

10 Raise the wheel and rotate it by hand at approximately one revolution per second.

11 A sinusoidal wave form should be obtained, with amplitude and duration changing with rotational speed **(see illustration 6.24)**.

12 If there is no signal, or a very weak or intermittent signal at the ECM, repeat the test at the sensor-wiring plug. If there is no change in signal status, the sensor is suspect.

13 If the signal is now satisfactory this indicates a fault in the wiring harness, which should be checked for continuity.

Checking wheel speed sensor output with an AC voltmeter

Note: *Refer to a suitable wiring diagram for specific ECM pin identification according to model.*

14 Switch the ignition off and disconnect the ECM multi-plug or the relevant wheel speed sensor-wiring plug.

15 Connect an AC voltmeter between the terminal pins for the sensor under test **(see illustration 6.25)**.

16 Raise the wheel and rotate it by hand at approximately one revolution per second.

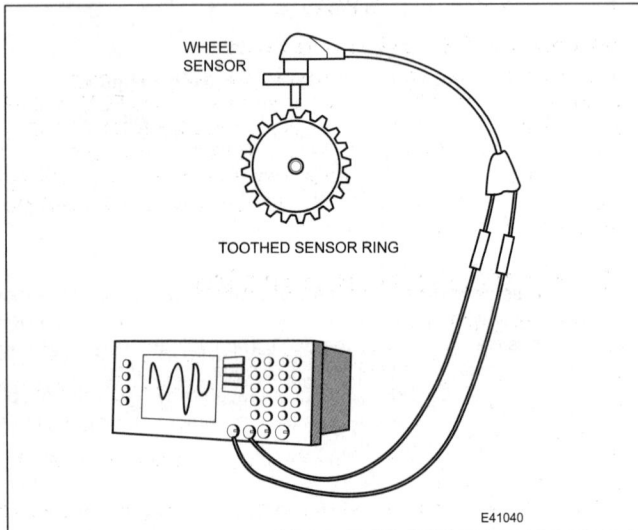

6.23 Checking wheel speed sensor output with an oscilloscope connected to the sensor-wiring plug

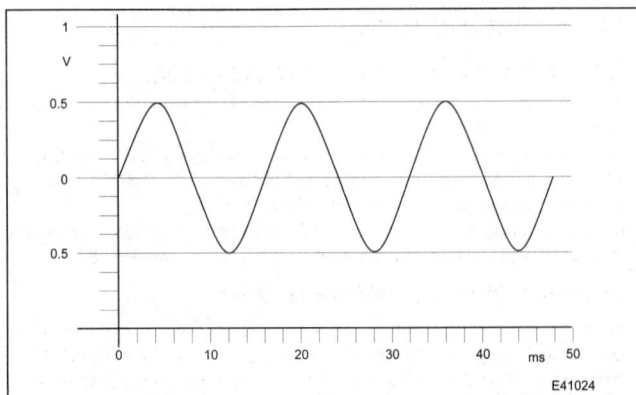

6.24 Typical inductive wheel speed sensor sine wave as displayed on an oscilloscope

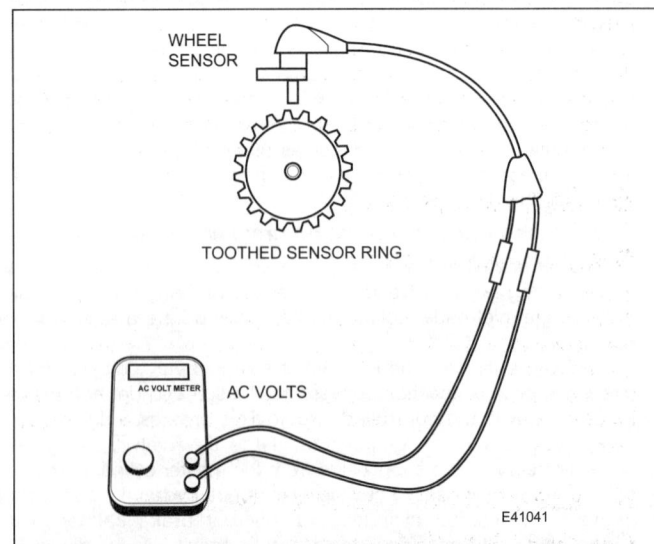

6.25 Checking wheel speed sensor output with a voltmeter connected to the sensor-wiring plug

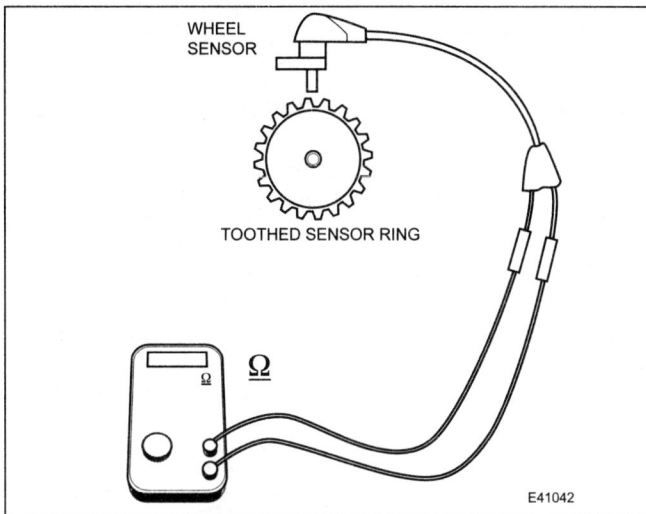

6.26 Checking wheel speed sensor resistance with an ohmmeter connected to the sensor-wiring plug

17 A voltage of approximately 0.5 to 1.5 volts (AC RMS) should be obtained. If there is no signal, or a very weak or intermittent signal at the ECM, repeat the test at the sensor-wiring plug. If there is no change in the signal, the sensor is suspect.

18 If the signal is now satisfactory, this indicates a fault in the wiring harness which should be checked for continuity. **Note:** *This test at least proves that a signal is being generated by the sensor. However, the voltage produced is an average voltage and does not clearly indicate damage to the sensor ring or that the sinewave is regular in formation.*

Checking wheel speed sensor resistance

Note: *Refer to a suitable wiring diagram for specific ECM pin identification according to model.*

19 Switch the ignition off and disconnect the ECM multi-plug or the relevant wheel speed sensor-wiring plug.

20 Connect an ohmmeter between the terminal pins for the sensor under test **(see illustration 6.26)**.

21 The readings obtained should be between 0.7 and 2.2 kohms approximately.

22 If the resistance is excessively high or open circuit at the ECM, repeat the test at the sensor multi-plug. If there is no change in resistance, the sensor is suspect.

23 If the resistance is now satisfactory, this indicates a fault in the wiring harness which should be checked for continuity. **Note:** *Even if the resistance is within the quoted specifications, this does not prove that the speed sensor can generate an acceptable signal.*

5 System relays

Relay resistance tests

1 Switch the ignition off and disconnect the ECM multi-plug.

2 Connect an ohmmeter between ECM multi-plug terminals 8 and 19 **(see illustration 6.27)**. A reading of between 50 and 60 ohms should be obtained.

3 If the reading is not as specified, check the continuity of the wiring between the ECM and hydraulic control unit. If satisfactory, suspect a faulty relay.

6 Electronic Control Module

Checking the ECM (general)

1 Inspect the ECM for corrosion or damage and ensure that the unit is securely attached to the hydraulic control unit.

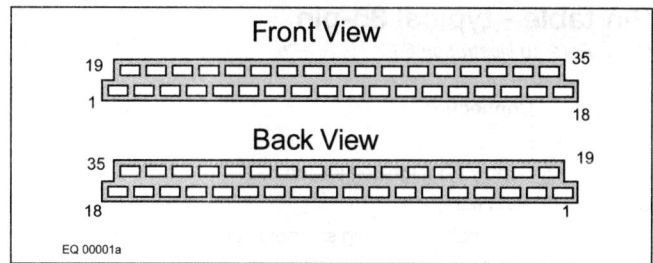

6.27 ECM 35-pin multi-plug

2 Check that the ECM multi-plug terminals are pushed fully home and making good contact with the ECM pins. Faults in any of the above areas are possible reasons for poor performance in the ABS system.

ECM power supply and earth tests

3 Switch the ignition off and disconnect the ECM multi-plug.

4 Connect a voltmeter between terminal 2 in the ECM multi-plug and a vehicle earth **(see illustration 6.27)**.

5 Switch the ignition on. The voltmeter should indicate nbv.

6 If no voltage is found, check the relevant fuse and the ECM main earth connection and wiring.

7 Solenoid valves

Solenoid valve resistance tests

1 Switch the ignition off and disconnect the ECM multi-plug.

2 Connect an ohmmeter between a vehicle earth and ECM multi-plug pins 4, 5, 21, 22, 23 and 24 in turn **(see illustration 6.27)**. A reading of between 2.0 and 4.0 ohms should be obtained in each case.

3 If any of the readings are not as specified check the continuity of the wiring between the ECM and hydraulic control unit. If satisfactory, suspect the hydraulic control unit.

8 Hydraulic pump motor

Power supply test

Note: *The multi-plug arrangement on the hydraulic control unit varies according to vehicle application. Refer to a suitable wiring diagram for specific details.*

1 Switch the ignition off.

2 Disconnect the 7-pin or 4-pin multi-plug (according to model) from the hydraulic control unit.

3 Connect a voltmeter between a vehicle earth and pin 4 of the multi-plug. The voltmeter should indicate nbv.

4 If not, check the relevant fuse and the ECM main earth connection and wiring.

9 Warning lamp

Checking the warning lamp (general)

1 Inspect the warning lamp bulb holder contacts in the instrument panel.

2 Check that the instrument panel multi-plug terminal pins are pushed fully home and making good contact.

3 Check that no wires have been disconnected.

4 Faults in any of the above areas are possible reasons for failure or malfunctioning of the warning lamp.

Warning lamp operation test

5 With the ignition switched off, the warning lamp should remain off.

6 Switch the ignition on and the warning lamp should illuminate then extinguish after a few seconds. The lamp should then remain off.

7 If the warning lamp comes on and remains on at any time during vehicle operation, carry out the previously described test procedures on the system components.

Pin table - typical 35-pin

Note: *Refer to illustration 6.27*

Pin No.	Connection	Test condition	Voltage
1	-	-	-
2	Supply from ignition switch	Ignition on	Nbv
3	ECM earth	Ignition on	0.25 V (max)
4	Front right restricting solenoid valve	Ignition on	Nbv
5	Front right inlet/outlet solenoid valve	Ignition on	Nbv
6	-	-	-
7	-	-	-
8	Relay supply	Ignition on	Nbv
9	Supply from battery	Ignition off/on	Nbv
10	-	-	-
11	-	-	-
12	SD connector	-	-
13	-	-	-
14	-	-	-
15	Rear left wheel speed sensor earth	Roadwheel rotating	0.25 volts (max)
16	Front right wheel speed sensor earth	Roadwheel rotating	0.25 volts (max)
17	Rear right wheel speed sensor earth	Roadwheel rotating	0.25 volts (max)
18	Front left wheel speed sensor earth	Roadwheel rotating	0.25 volts (max)
19	ABS main relay driver	Ignition on/inactive	Nbv
		Actuated	1.0 volt (max)
20	ECM earth	Ignition on	0.25 volts (max)
21	Front left restricting solenoid valve	Ignition on	Nbv
22	Front left inlet/outlet solenoid valve	Ignition on	Nbv
23	Rear left inlet/outlet solenoid valve	Ignition on	Nbv
24	Rear right inlet/outlet solenoid valve	Ignition on	Nbv
25*	Hydraulic fluid level warning lamp	-	-
26	ABS warning lamp	Ignition on:	
		Lamp off	Nbv
		Lamp on	0.25 volts (max)
27	ECM earth	Ignition on	0.25 volts (max)
28	ECM earth	Ignition on	0.25 volts (max)
29	-	-	-
30	-	-	-
31	-	-	-
32	Rear left wheel speed sensor signal	Roadwheel rotating	0.1 to 3.0 volts AC (approx)
33	Front right wheel speed sensor signal	Roadwheel rotating	0.1 to 3.0 volts AC (approx)
34	Rear right wheel speed sensor signal	Roadwheel rotating	0.1 to 3.0 volts AC (approx)
35	Front left wheel speed sensor signal	Roadwheel rotating	0.1 to 3.0 volts AC (approx)

*Certain versions only

Fault codes

10 Peugeot fault codes

1 The Bendix Integrated system requires the use of a FCR for obtaining fault codes. Flash codes are not available for output from this system.

2 If a FCR is available, it should be connected to the SD serial connector and used in accordance with the maker's instructions.

3 The FCR can be used for the following purposes:
   a) *Obtaining fault codes.*
   b) *Clearing fault codes.*
   c) *Obtaining datastream information.*
   d) *Testing the system actuators (relays and solenoid valves).*

11 Renault fault codes

1 On Renault models, internal fault codes are used by the ECM to designate faults in the system components and circuits. A proprietary fault code reader (FCR) or system tester (such as the Renault XR25) is required to interrogate the system. No actual fault code numbers are available although the component circuits checked by the ECM are similar to those shown in the following table.

Fault code table

Code	Item
12	Start of test sequence
11	End of test sequence
13	Solenoid valves - relay, wiring and connections
14	Hydraulic pressure switch - faulty operation
15	Main relay - open circuit
21	Main relay - short circuit
22	Main relay - faulty operation
24	Rear left wheel speed sensor - incorrect resistance
25	Front right wheel speed sensor - incorrect resistance
31	Rear right wheel speed sensor - incorrect resistance
32	Front left wheel speed sensor - incorrect resistance
33	Rear left wheel speed sensor signal variation
34	Front right wheel speed sensor signal variation
35	Rear right wheel speed sensor signal variation
41	Front left wheel speed sensor signal variation
42	Front right inlet/outlet solenoid valve faulty
43	Front right restricting solenoid valve faulty
44	Front left inlet/outlet solenoid valve faulty
45	Front left restricting solenoid valve faulty
51	Rear right inlet/outlet solenoid valve faulty
52	Rear left inlet/outlet solenoid valve faulty
55	ECM memory fault

6

Wiring diagrams

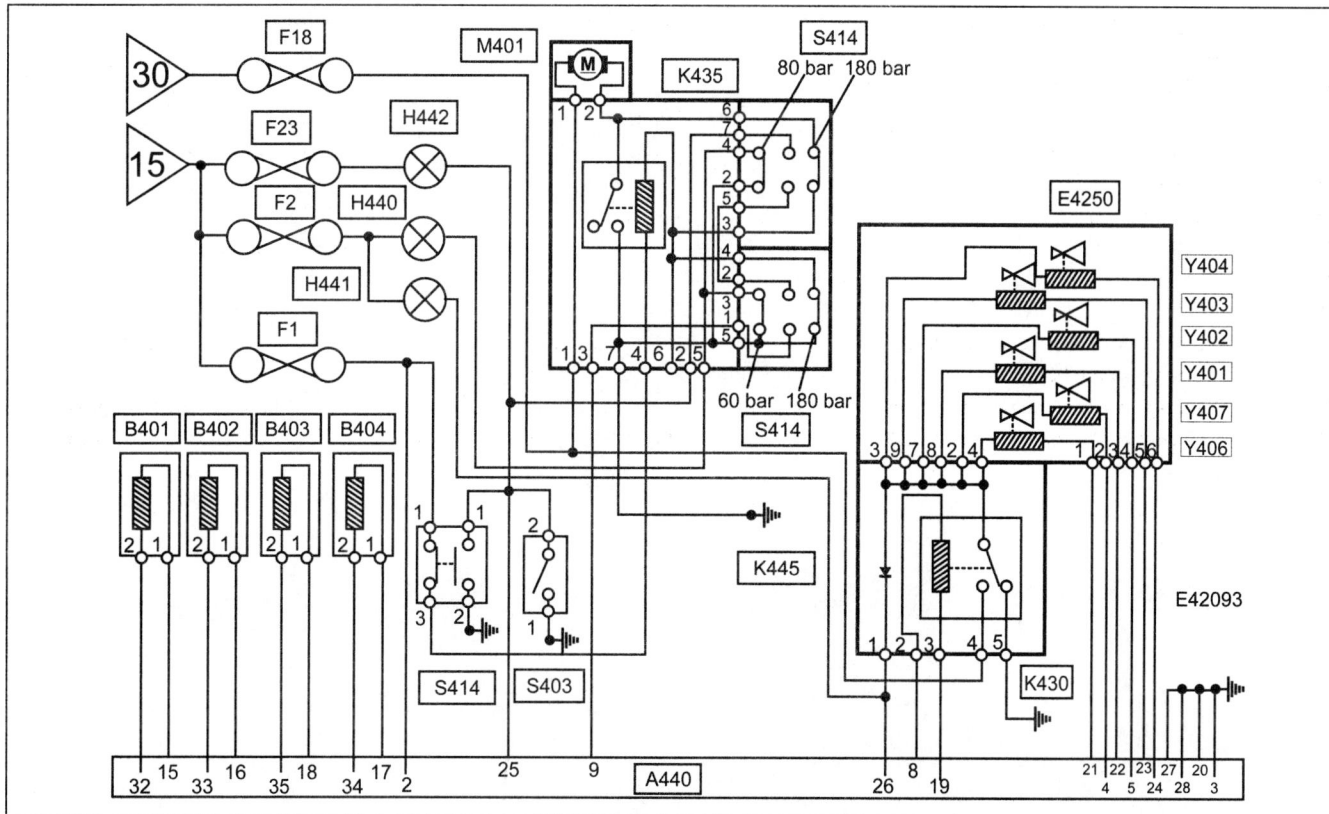

6.28 Wiring diagram, Peugeot 605

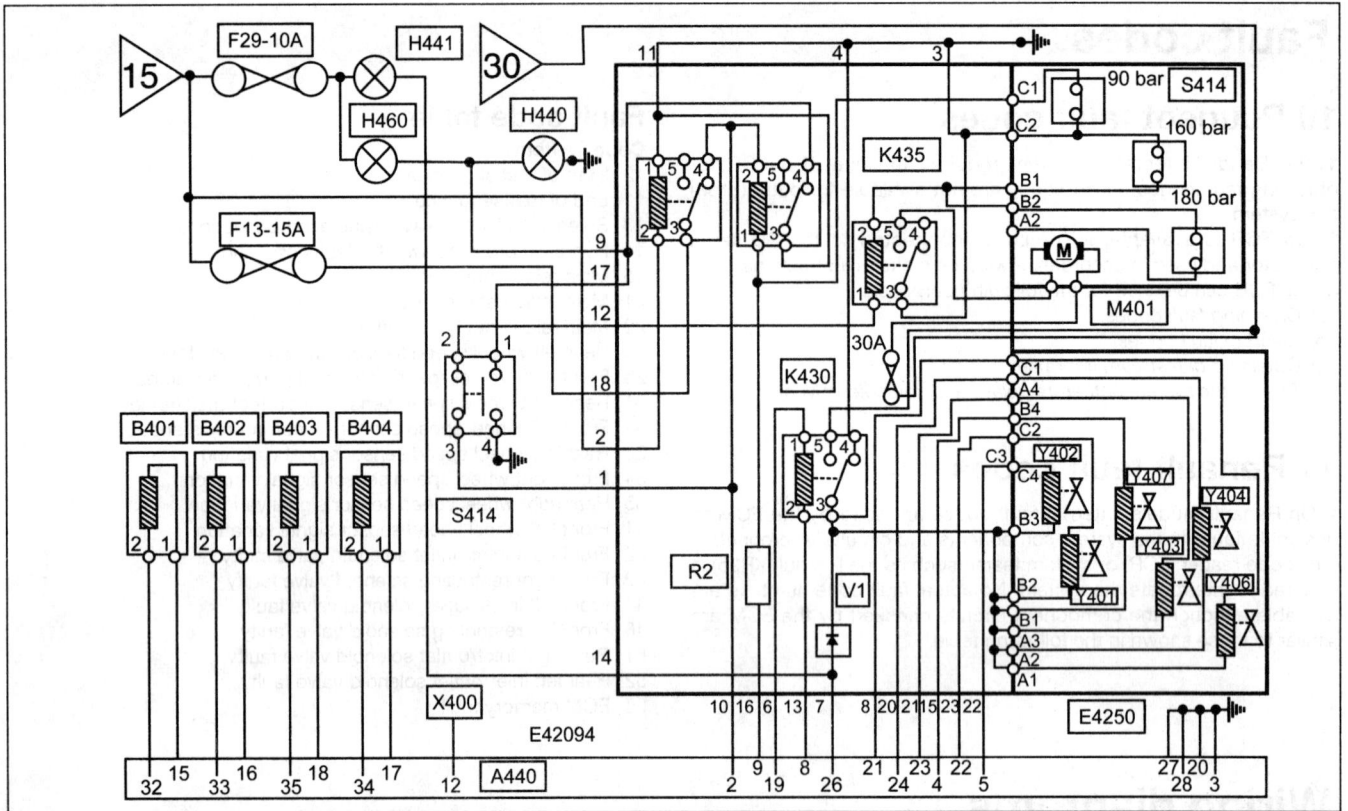

6.29 Wiring diagram, Renault 19

Chapter 7

Bendix Mecatronic II/III ABS

Contents

Vehicle coverage

Model	Year
Ford	
Mondeo	1993-1996
Mondeo	1996-1998
Peugeot	
306	1993-1997

Overview of system operation

1 Basic principles and system identification

The Bendix Mecatronic Antilock Brake System is a development of the Bendix Addonix series ABS, with revisions to the hydraulic control unit, and enhancement of the Electronic Control Module (ECM) software. The system has been fitted to a number of mainly Ford derived vehicles since its introduction in the mid 1990s. Bendix Mecatronic ABS is of the additional or 'add-on' type operating in conjunction with the conventional braking system components.

The purpose of the system is to apply the vehicle brakes at maximum efficiency without wheel lock or loss of directional stability. Inductive sensors (wheel speed sensors) monitor the speed of the wheels by generating an electrical signal as the wheel is rotated. This information is passed to the ABS Electronic Control Module (ECM) which is then able to determine wheel speed, wheel acceleration and wheel deceleration. The ECM compares the signals received from each wheel with a calculated reference value and if the onset of lock at any wheel is detected, a signal is sent to the ABS hydraulic control unit which regulates the brake pressure for the relevant wheel(s).

In four-channel configuration, Bendix Mecatronic ABS is typically comprised of the following components (see illustration 7.1):

a) Hydraulic control unit.
b) ABS-ECM (integral with hydraulic control unit).
c) Inductive wheel speed sensors and associated sensor rings.
d) ABS electrical wiring harness and relays.
e) ABS warning lamp.
f) Diagnostic connector.

In addition, the conventional braking system is comprised of the following components:

a) Tandem brake master cylinder.
b) Vacuum servo unit.
c) Brake calipers and hydraulic hoses and pipes.
d) Pressure regulating/load-sensing valve(s) depending on application.

For the purposes of the information contained in this Chapter, the two Mecatronic systems can be identified by the ECM wiring harness connection. Mecatronic II may be a two- or four-channel system and the ECM is connected to the main wiring harness by means of two multi-plugs; one with 8-pins and one with 22-pins. Mecatronic III is a four-channel system and the ECM connection is by a single, 24-pin multi-plug. A breakdown of the vehicles covered and system application is as follows:

Model	System
Ford	
Mondeo (1993-1996)	Mecatronic II - four-channel
Mondeo (1996-1998)	Mecatronic III - four-channel
Peugeot	
306	Mecatronic II - two-channel

2 Component description and operation

ABS ECM

General

The Bendix Mecatronic Electronic Control Module (ECM) continually monitors wheel speed from the signals provided by the wheel speed sensors. If the ECM detects the incidence of wheel lock on one or more wheels, a signal is sent to the hydraulic control unit to modulate the hydraulic pressure to the brake of the locking wheel (see illustration 7.2).

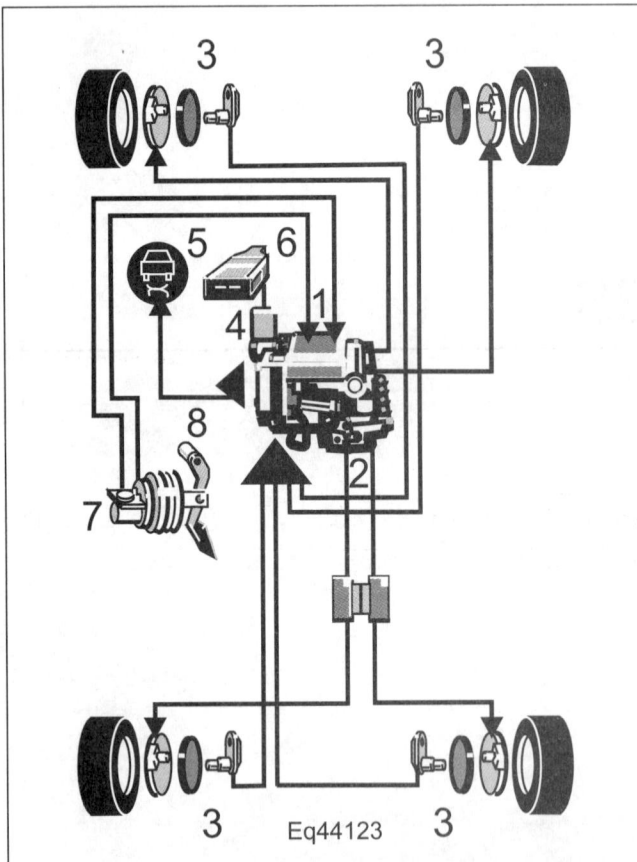

7.1 Bendix Mecatronic components (four-channel system shown)

1 Hydraulic control unit
2 ABS-ECM (integral with hydraulic control unit)
3 Inductive wheel speed sensors and sensor rings
4 ABS electrical wiring harness and relays
5 ABS warning lamp
6 Diagnostic connector.
7 Tandem brake master cylinder
8 Vacuum servo unit

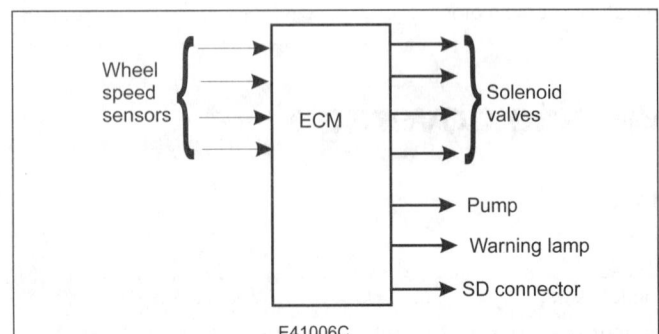

7.2 ECM sensor inputs and control signal outputs

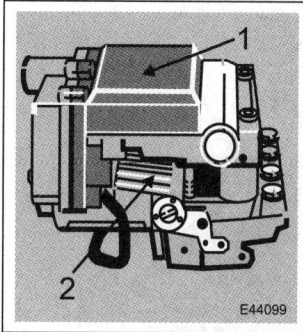

7.3 Typical Mecatronic ECM (1) and hydraulic control unit (2) arrangement

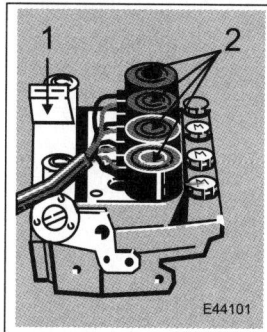

7.4 Mecatronic hydraulic control unit

1 Valve block
2 Inlet/outlet solenoid valves

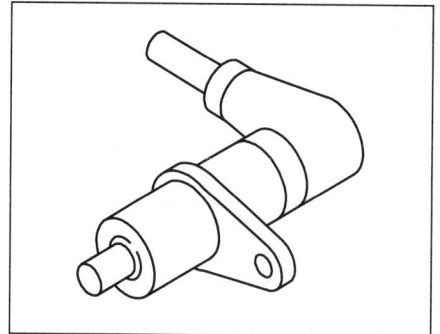

7.5 Typical wheel speed sensor

The ECM contains two microprocessors and uses digital technology to complete this function and other functions such as fault code memory and power modules for valve and relay activity.

To reduce external electrical connections to a minimum and improve reliability, the ECM is integral with the hydraulic control unit **(see illustration 7.3)**. In most applications the ECM and hydraulic control unit cannot be separated and are only available as a complete sealed unit.

Self-test

The Mecatronic ECM is equipped with a self-test capability that initially examines the ABS system when the ignition is switched on, and then examines the wheel speed sensor signals after a wheel speed of approximately 4 mph is reached from all wheels (engine running). The ABS self-test program continues to examine the signals from the various components as long as the ignition is switched on. If self-test determines that faults are not present, the ABS is ready for operation once a specified vehicle speed has been achieved.

If the ECM detects that a fault is present, all ABS functions are switched off and the warning lamp is turned on. The conventional braking system continues to operate as normal without ABS assistance.

Self-diagnostics

If the ECM detects a fault during the self-test routine, an internal fault code is stored in the ECM memory. Stored fault codes can be retrieved from the SD connector with the aid of a suitable fault code reader. If the fault clears, the code will remain stored until cleared with the FCR.

Hydraulic control unit

The hydraulic control unit consists of a valve block containing the inlet/outlet solenoid valves, restricting solenoid valves (two-channel system), pressure accumulator, restrictor, brake release valve, pulsation damper, non-return valve, high-frequency chamber, electric motor and return pump. The hydraulic control unit controls the pressure applied to the brake caliper for each individual wheel during the various phases of ABS operation. During a controlled ABS cycle, the pump returns brake fluid, drained off during the pressure reduction phase, back into the brake circuit **(see illustration 7.4)**.

According to vehicle application, Bendix Mecatronic may be either a two-, or four-channel system and the construction of the hydraulic control unit varies accordingly. On a two-channel system, only the front wheels are under ABS control and there are four solenoid valves, two for each front brake (one inlet/outlet solenoid valve, and one restricting solenoid valve). On a four-channel system, all four wheels are under ABS control and there are also four solenoid valves, but only one solenoid valve (inlet/outlet) is required for each brake.

Wheel speed sensors

The rotational speed of the roadwheels and any changes in the rotational speed are recorded by inductive wheel speed sensors located at the roadwheels. Depending on system type, there may be a separate sensor for each front wheel only or a separate sensor for all four wheels **(see illustration 7.5)**.

Each wheel speed sensor assembly comprises a toothed sensor ring, which rotates at roadwheel speed, and an adjacent sensor mounted a set distance from the sensor ring **(see illustration 7.6)**.

The sensors are permanent magnet pulse generator types producing an AC voltage sine wave as the sensor ring teeth pass through the magnetic field of the sensor **(see illustration 7.7)**.

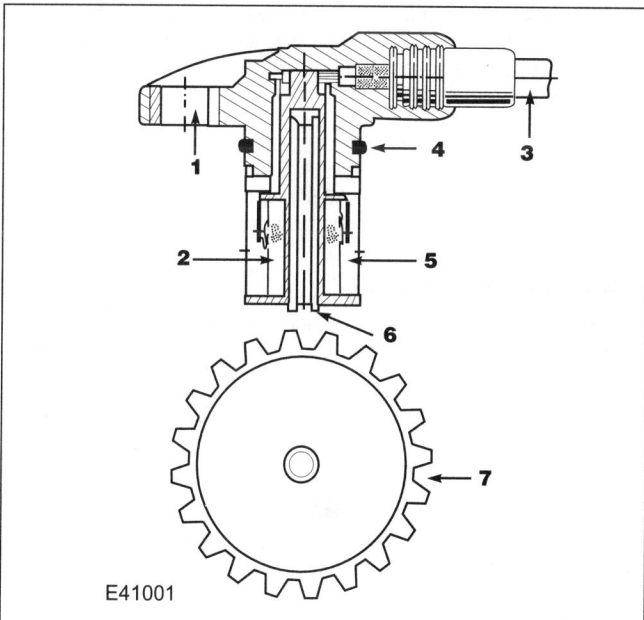

7.6 Sectional view of a wheel speed sensor

1	Mounting bolt location	5	Coil
2	Permanent magnet	6	Sensor tip
3	Wiring harness	7	Toothed sensor ring
4	O-ring		

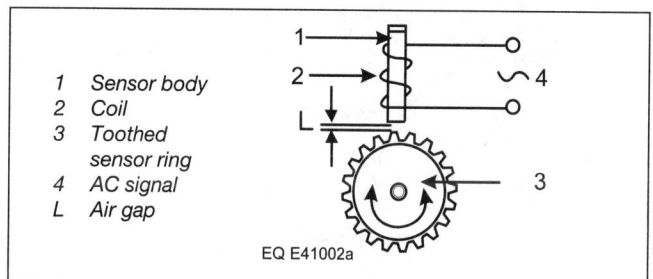

1 Sensor body
2 Coil
3 Toothed sensor ring
4 AC signal
L Air gap

7.7 Wheel speed sensor operation

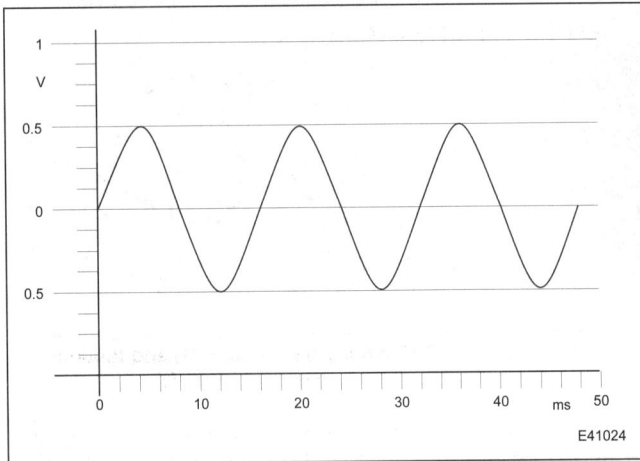

7.8 Wheel speed sensor waveform as viewed on an oscilloscope

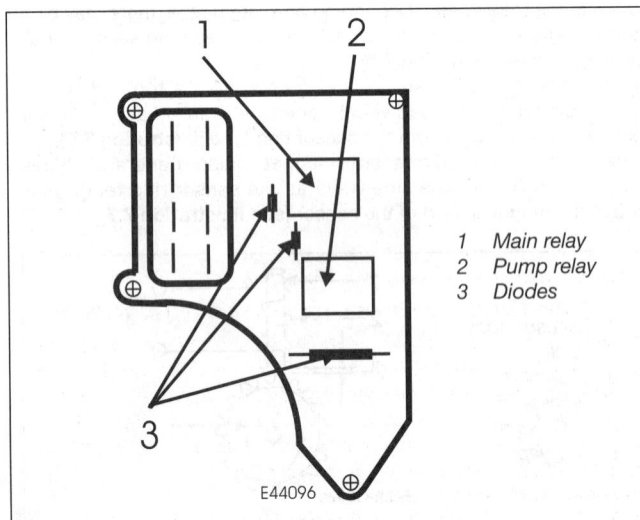

7.9 Mecatronic relay box circuit board (1) and ECM (2)

The frequency of the waveform produced by the wheel speed sensor is proportional to the road speed. This AC voltage signal is continually being delivered to the ABS-ECM for processing.

The peak to peak voltage of the speed signal (when viewed upon an oscilloscope) can vary considerably according to wheel speed. An analogue to digital converter (ADC) in the ECM transforms the AC pulse into a digital signal **(see illustration 7.8)**.

ABS electrical wiring harness and relays

An integrated main wiring harness is used for ABS-ECM power supply and earth connections, and enables sensor signals to reach the ECM and the ECM, in turn, to send output signals to the ABS warning lamp and diagnostic connector. The ABS main relay, pump relay and surge protection diodes are soldered to the relay box circuit board attached to the ECM **(see illustration 7.9)**. Internal connections between the ECM and hydraulic control unit are used to activate the return pump and solenoid valves.

Main relay

Nominal battery voltage is supplied to the relay from the battery. On completion of the ECM self-test phase, the relay is activated by the ECM, allowing battery voltage to be applied to the solenoid valves. The earth for the solenoid valves is controlled by the ECM **(see illustration 7.10)**.

Pump relay

Nominal battery voltage is supplied to the pump relay via a fuse. The ECM controls the pump motor by actuating the relay to supply nominal battery voltage for the motor.

Diodes

Three diodes are used in the electrical circuit to protect against voltage surges and polarity reversal.

ABS warning lamp

After the ignition is switched on, the ABS warning lamp on the instrument panel is illuminated for approximately 3 seconds as the system executes a self-test routine. If satisfactory operation of the system is detected by the ECM, the light is extinguished. During vehicle operation above a pre-determined wheel speed, the ABS-ECM implements a further self-test cycle whereby ABS operation and wheel speed sensor signals are continually monitored. If a fault is detected, the relevant ECM pin is earthed to illuminate the warning lamp on the instrument panel and the ABS function is disabled. The warning lamp will remain illuminated until the fault is no longer present **(see illustration 7.11)**.

When the ABS-ECM detects a fault, the fault code is stored and the ABS warning lamp activated. If the fault no longer exists after the next system start (ignition on/off) the ABS warning lamp is extinguished after the self-test cycle, however the fault code remains stored in the ECM memory.

Tandem master cylinder

Typically, the tandem master cylinder comprises a body casting incorporating primary and secondary pressure chambers, primary

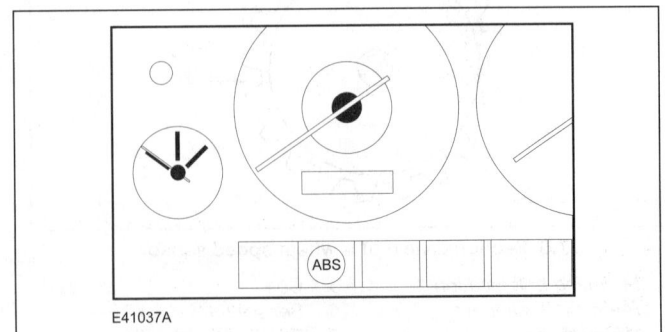

1 Main relay
2 Pump relay
3 Diodes

7.10 Relay and diode locations on the relay box circuit board

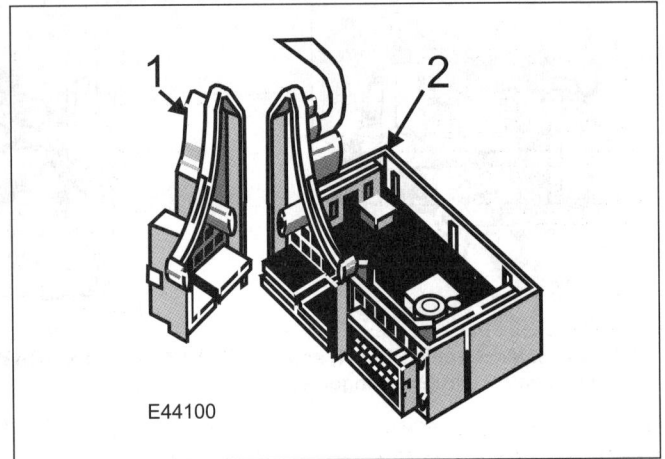

7.11 ABS warning lamp

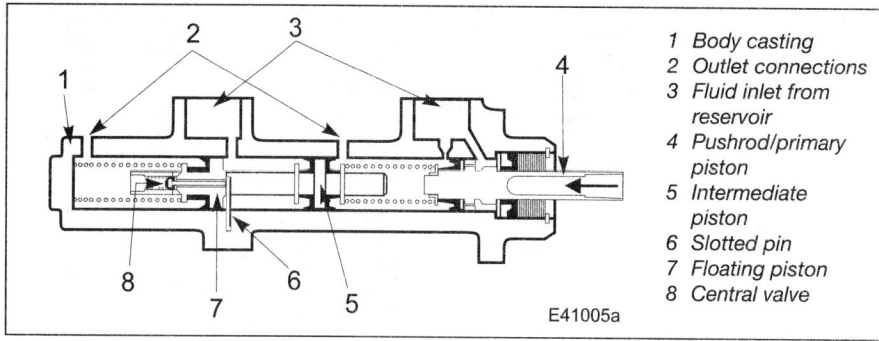

1 Body casting
2 Outlet connections
3 Fluid inlet from reservoir
4 Pushrod/primary piston
5 Intermediate piston
6 Slotted pin
7 Floating piston
8 Central valve

E41005a

7.12 Sectional view of a typical tandem master cylinder

Brake Booster

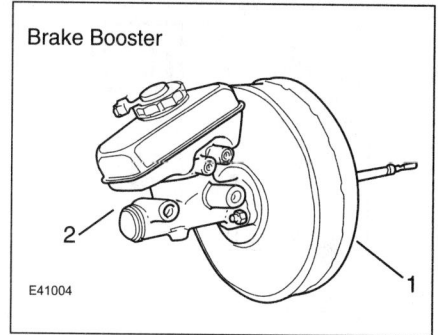

E41004

7.13 Vacuum servo unit (1) and tandem master cylinder (2)

piston, intermediate piston, floating piston, slotted pin and central valve. The cylinder operates as a conventional master cylinder using vacuum assistance from the vacuum servo unit **(see illustration 7.12)**.

When the brake system is at rest, the central valve in the floating piston rests against the slotted pin. In this condition, the central valve is open and brake fluid can discharge out of the pressure chamber back into the brake fluid reservoir. When the brake pedal is depressed, the build-up of hydraulic pressure in the primary pressure chamber acts on the intermediate piston and floating piston, moving them down the cylinder bore. The floating piston contacts the seal on the central valve, closing the connection between the intermediate and secondary pressure chambers. Brake hydraulic pressure can now also increase in the secondary pressure chamber.

Vacuum servo unit

The vacuum servo unit is located between the brake pedal and tandem master cylinder. When the brake pedal is depressed, the servo unit increases the force applied by the pedal, reducing the effort required to operate the brakes **(see illustration 7.13)**.

The unit operates by means of engine inlet manifold vacuum applied to a diaphragm contained within the unit casing. A pushrod connected to the centre of the diaphragm acts directly on the primary piston in the master cylinder.

When the brake pedal is released, vacuum is applied to both sides of the diaphragm. When the pedal is depressed, one side of the

diaphragm is opened to atmosphere and the vacuum acting on the other side deflects the diaphragm which in turn operates the master cylinder primary piston. The resulting force applied to the master cylinder piston is therefore significantly greater than the initial force applied to the brake pedal by the driver.

Pressure regulating/load-sensing valve(s)

Depending on vehicle application, pressure regulating valves or load sensing valves may be incorporated to restrict the hydraulic fluid pressure to the rear brakes. The valves may be pressure conscious whereby the hydraulic fluid supply is restricted once a pre-determined pressure is reached, or load conscious whereby the hydraulic pressure is reduced according to vehicle loading.

3 System operation

The operation of the braking system both conventionally and when under ABS control varies according to the number of channels in the Mecatronic system. On a two-channel system, only the front brakes are regulated during ABS operation, the rear brakes being operated conventionally directly from the master cylinder, via pressure regulating/load sensing valves. On a four-channel system, all four brakes are regulated during ABS operation.

Two-channel ABS system operation

Brake system at rest

When the ABS system is at rest all the brake components are inoperative. Pressure is non-existent in the hydraulic pipes between the tandem master cylinder and the brake calipers.

Brake system operating under conventional control without ABS

When the brake pedal is activated, the pedal force is applied to the tandem master cylinder by the vacuum servo unit pushrod. The servo unit pushrod acts directly on the pressure piston in the master cylinder, which pressurises the hydraulic fluid in the brake pipes to the hydraulic control unit.

Both the inlet/outlet and restricting solenoid valves in the controlled circuit are 'at rest' allowing an unrestricted hydraulic passage from the master cylinder through the hydraulic control unit to the brake caliper, thus operating the brake **(see illustration 7.14)**.

Brake system operating in conjunction with ABS control

The ABS-ECM continually monitors wheel speed from the signals provided by the wheel speed sensors. If the ECM detects the incidence of wheel lock on one or both front wheels, ABS is automatically initiated in four phases. As ABS operates individually on each front wheel, either of the controlled wheels could be in any one of the following phases at any particular moment.

Phase one, rapid pressure reduction

The ECM energises the inlet/outlet solenoid valve closing off the direct hydraulic connection between the master cylinder and the

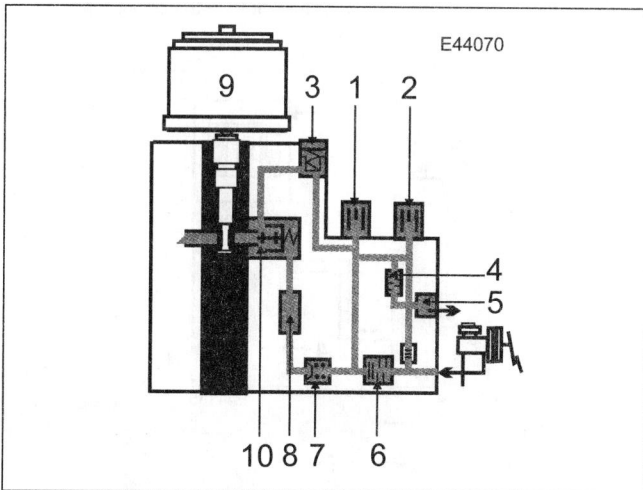

E44070

7.14 Mecatronic hydraulic control unit components

1 Inlet/outlet solenoid valves
2 Restricting solenoid valve
3 Pressure accumulator
4 Restrictor
5 Brake release valve
6 Pulsation damper
7 Non-return valve
8 High-frequency chamber
9 Electric motor
10 Return pump

7

7.15 Two-channel ABS operation - phase one, rapid pressure reduction

1   Inlet/outlet solenoid valve

7.16 Two-channel ABS operation - phase two, slow pressure reduction

1   Inlet/outlet solenoid valve
2   Restricting solenoid valve
4   Restrictor

7.17 Two-channel ABS operation - phase three, rapid pressure increase

front brake caliper in the controlled circuit. Pressure in the hydraulic circuit falls rapidly as fluid from the brake caliper is returned to the pressure accumulator **(see illustration 7.15)**.
At the same time, the ECM actuates the return pump, which transfers the returned fluid to the high-frequency chamber.

Phase two, slow pressure reduction

The rapid pressure reduction phase is always followed by a slow pressure reduction phase. The inlet/outlet solenoid valve remains energised and the restricting solenoid valve is now also energised. Hydraulic fluid returned from the brake caliper is forced to flow through the restrictor as it passes back to the pressure accumulator, thus resulting in a slow pressure reduction in the controlled circuit. When the controlled wheel is once again rotating, phase three is initiated **(see illustration 7.16)**.

Phase three, rapid pressure increase

Both solenoid valves are de-energised by the ECM allowing them to return to the 'at rest' position. The hydraulic circuit from the master cylinder directly to the front brake caliper is re-instated and the hydraulic fluid returned during the two pressure reduction phases is pumped back into the circuit via the non-return valve. This results in a rapid increase in hydraulic pressure in the controlled circuit and the brake is once again applied **(see illustration 7.17)**.

Phase four, slow pressure increase

The rapid pressure increase phase is always followed by a slow pressure increase phase. The inlet/outlet solenoid valve remains 'at rest', but the restricting solenoid valve is once again energised by the ECM. Hydraulic fluid from the master cylinder is forced to flow through the restrictor before passing to the brake caliper, resulting in a slow increase in hydraulic pressure in the controlled circuit **(see illustration 7.18)**.
This pressure increase continues until the wheel speed sensor detects that the brake is once again about to lock at which point phase one is re-instigated by the ECM. The cycle then continues until the vehicle comes to a halt or the brake pedal is released.
The whole ABS control cycle takes place many times per second for each affected wheel and this ensures maximum braking effect and control during ABS operation.

Four-channel ABS system operation

Where a four-channel system is used, all four brakes are individually regulated during ABS operation, however the brakes operate in a slightly different manner, as there are no restricting solenoid valves for

the brake hydraulic circuits. Instead, the inlet/outlet solenoid valves incorporate a restrictor and flow valve to emulate the function of the restricting solenoid valve.
The arrangement and layout of the solenoid valves and control passages in the valve block varies according to application, but typically the system operates on the following principles.

Brake system at rest

As with a two-channel system, when the ABS system is at rest all the brake components are inoperative. Pressure is non-existent in the hydraulic pipes between the tandem master cylinder and the brake calipers.

Brake system operating under conventional control without ABS

When the brake pedal is activated, the pedal force is applied to the tandem master cylinder by the vacuum servo unit pushrod. The servo unit pushrod acts directly on the pressure piston in the master cylinder, which pressurises the hydraulic fluid in the brake pipes to the hydraulic control unit.

7.18 Two-channel ABS operation - phase four, slow pressure increase

1   Inlet/outlet solenoid valve   4   Restrictor
2   Restricting solenoid valve

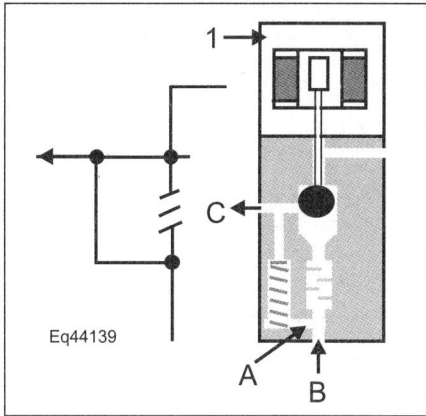

7.19 Four-channel ABS operation - conventional braking without ABS

1   Inlet/outlet solenoid valve
a   Flow valve
b   Hydraulic fluid input from master cylinder
c   Hydraulic fluid output to rear brakes

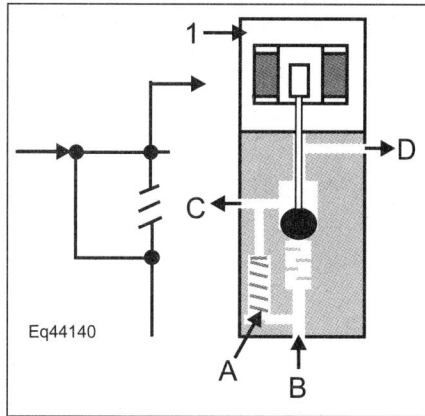

7.20 Four-channel ABS operation - phase one, rapid pressure reduction

1   Inlet/outlet solenoid valve
a   Flow valve
b   Hydraulic fluid input from master cylinder
c   Hydraulic fluid output/input to/from rear brakes
d   Hydraulic fluid return to pressure accumulator

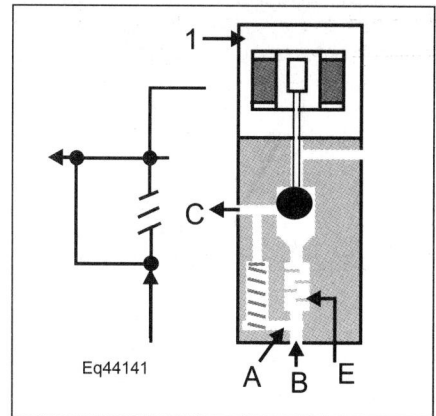

7.21 Four-channel ABS operation - phase two, slow pressure increase

1   Inlet/outlet solenoid valve
a   Flow valve
b   Hydraulic fluid input from master cylinder
c   Hydraulic fluid output/input to/from rear brakes
e   Restrictor

The inlet/outlet solenoid valve in the controlled circuit is 'at rest' and the flow valve is open allowing an unrestricted hydraulic passage from the master cylinder through the hydraulic control unit to the relevant brake caliper, thus operating the brake **(see illustration 7.19)**.

Brake system operating in conjunction with ABS control

The ABS-ECM continually monitors wheel speed from the signals provided by the wheel speed sensors. If the ECM detects the incidence of wheel lock on any of the wheels, ABS is automatically initiated. As ABS operates individually on each wheel, any of the controlled wheels could be in any one of the following phases at any particular moment.

Phase one, rapid pressure reduction

The ECM energises the inlet/outlet solenoid valve closing off the direct hydraulic connection between the master cylinder and the brake caliper in the controlled circuit. This creates a difference in hydraulic pressure on either side of the flow valve, the greater pressure being on the supply side from the master cylinder. The flow valve therefore closes, shutting the direct hydraulic connection between the master cylinder and the brake caliper in the controlled circuit. Pressure in the hydraulic circuit falls rapidly

as fluid from the brake caliper is returned to the pressure accumulator **(see illustration 7.20)**.
At the same time, the ECM actuates the return pump, which transfers the returned fluid to the high-frequency chamber.

Phase two, slow pressure increase

The inlet/outlet solenoid valve is de-energised by the ECM allowing it to return to the 'at rest' position. Hydraulic fluid pressure is still greater on the supply side from the master cylinder causing the flow valve to remain closed. Hydraulic fluid from the master cylinder is forced to flow through the restrictor before passing to the brake caliper, resulting in a slow increase in hydraulic pressure in the controlled circuit. Hydraulic fluid returned during the pressure reduction phase is pumped back into the circuit via the non-return valve **(see illustration 7.21)**.
This pressure increase continues until the wheel speed sensor detects that the brake is once again about to lock at which point phase one is re-instigated by the ECM. The cycle then continues until the vehicle comes to a halt or the brake pedal is released.
The whole ABS control cycle takes place many times per second for each affected wheel and this ensures maximum braking effect and control during ABS operation.

Test procedures

Important note: The test procedures, pintables and wiring diagrams contained in this Chapter are necessarily representative of the system depicted. Because of the variations in wiring and other data that often occurs, even between similar vehicles in any particular VM's range, the reader should take great care in identification of ECM pins, and satisfy himself that he has gathered the correct data before failing a particular component.

4   Wheel speed sensors

Checking the wheel speed sensor (general)

1 Inspect the wheel speed sensor for corrosion or damage and check that the sensor is tightly mounted.
2 Check the toothed sensor ring for damage, eccentricity and for broken or missing teeth.

3 Inspect the wheel speed sensor-wiring plug for corrosion and damage. One plug for each sensor.
4 Check that the connector terminal pins are pushed fully home and making good contact with the sensor-wiring plug.
5 Check the clearance between the sensor and the toothed sensor ring. The clearance is not normally adjustable but is nominally 0.2 to 1.2 mm. If the clearance is excessive, expect a worn sensor tip or problems with the wheel bearings/hub or sensor ring.
6 When carrying out voltage checks with an oscilloscope or voltmeter, the voltage obtained will be proportional to the speed at which the wheel is rotating. In addition to determining that the wheel speed sensors are actually producing a voltage output, it is essential that the output from the sensors on a particular axle is the same for any given wheel speed.

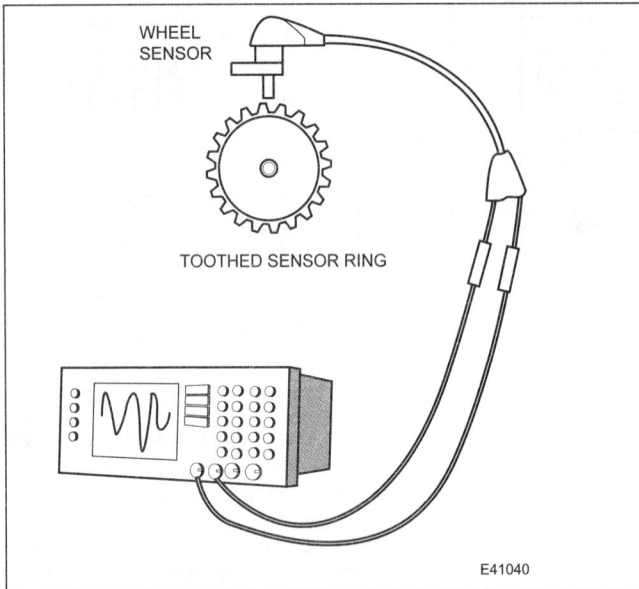

7.22 Checking wheel speed sensor output with an oscilloscope connected to the sensor-wiring plug

Checking wheel speed sensor output with an oscilloscope

Note: *Refer to the wiring diagrams for specific ECM pin identification according to model.*

7 Switch the ignition off and disconnect the ECM multi-plug or the relevant wheel speed sensor-wiring plug.

8 Connect an oscilloscope between the terminal pins for the sensor under test **(see illustration 7.22)**.

9 Select a range to cover 80 Hz on the oscilloscope and a free run time base.

10 Raise the wheel and rotate it by hand at approximately one revolution per second.

11 A sinusoidal wave form should be obtained, with amplitude and duration changing with rotational speed **(see illustration 7.23)**.

12 If there is no signal, or a very weak or intermittent signal at the

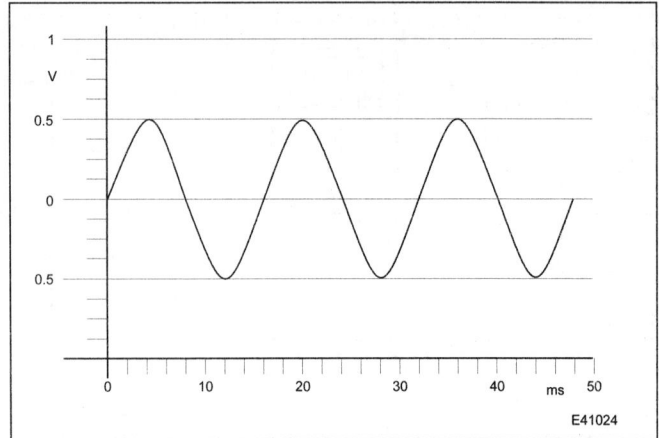

7.23 Typical inductive wheel speed sensor sine wave as displayed on an oscilloscope

ECM, repeat the test at the sensor-wiring plug. If there is no change in signal status, the sensor is suspect.

13 If the signal is now satisfactory this indicates a fault in the wiring harness, which should be checked for continuity.

Checking wheel speed sensor output with an AC voltmeter

Note: *Refer to the wiring diagrams for specific ECM pin identification according to model.*

14 Switch the ignition off and disconnect the ECM multi-plug or the relevant wheel speed sensor-wiring plug.

15 Connect an AC voltmeter between the terminal pins for the sensor under test **(see illustration 7.24)**.

16 Raise the wheel and rotate it by hand at approximately one revolution per second.

17 A voltage of approximately 0.5 to 1.5 volts (AC RMS) should be obtained. If there is no signal, or a very weak or intermittent signal at the ECM, repeat the test at the sensor-wiring plug. If there is no change in the signal, the sensor is suspect.

18 If the signal is now satisfactory, this indicates a fault in the wiring harness which should be checked for continuity. **Note:** *This test at least proves that a signal is being generated by the sensor. However,*

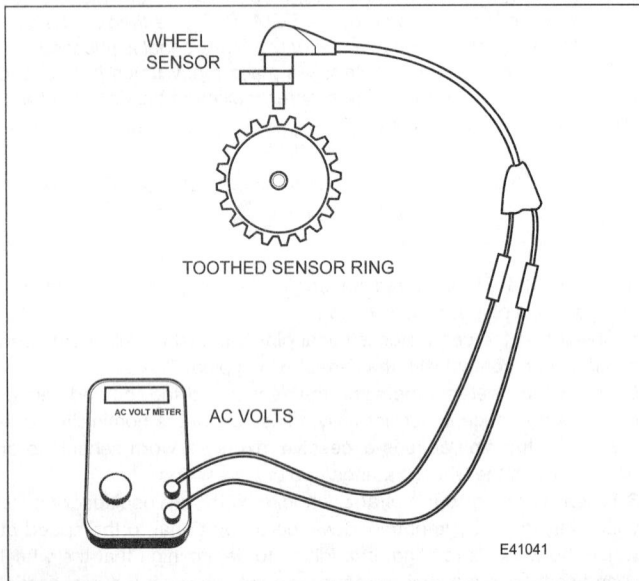

7.24 Checking wheel speed sensor output with a voltmeter connected to the sensor-wiring plug

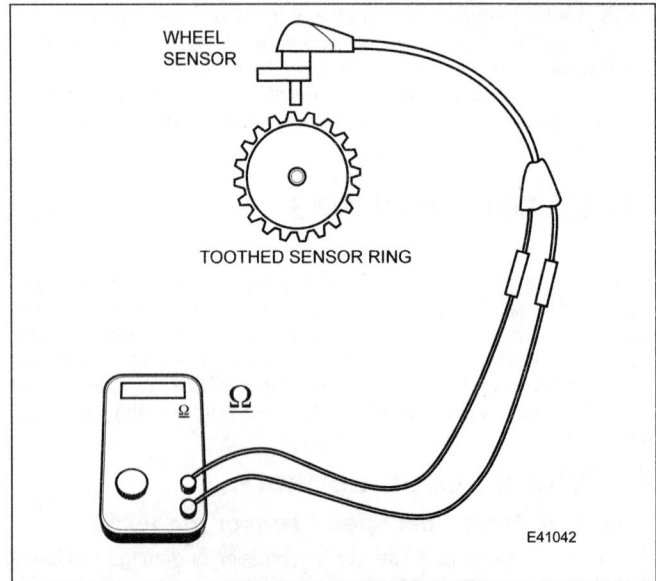

7.25 Checking wheel speed sensor resistance with an ohmmeter connected to the sensor-wiring plug

the voltage produced is an average voltage and does not clearly indicate damage to the sensor ring or that the sinewave is regular in formation.

Checking wheel speed sensor resistance

Note: Refer to the wiring diagrams for specific ECM pin identification according to model.

19 Switch the ignition off and disconnect the ECM multi-plug or the relevant wheel speed sensor-wiring plug.

20 Connect an ohmmeter between the terminal pins for the sensor under test **(see illustration 7.25)**.

21 The readings obtained should be between 0.7 and 2.2 kohms approximately.

22 If the resistance is excessively high or open circuit at the ECM, repeat the test at the sensor multi-plug. If there is no change in resistance, the sensor is suspect.

23 If the resistance is now satisfactory, this indicates a fault in the wiring harness which should be checked for continuity. **Note:** Even if the resistance is within the quoted specifications, this does not prove that the speed sensor can generate an acceptable signal.

5 System relays

Relay operation and power supply tests

The relays are integral with the relay circuit board attached to the ECM and can only be checked using suitable ABS diagnostic test equipment.

6 Electronic Control Module

Checking the ECM (general)

1 Inspect the ECM for corrosion or damage and ensure that the unit is securely attached to the hydraulic control unit.

2 Check that the ECM multi-plug terminals are pushed fully home and making good contact with the ECM pins. Faults in any of the above areas are possible reasons for poor performance in the ABS system.

ECM power supply test

Mecatronic II

3 Switch the ignition off and disconnect the ECM 22-pin multi-plug.

4 Connect an ohmmeter between a vehicle earth and ECM multi-plug pin 5 **(see illustration 7.26)**. The ohmmeter should indicate continuity.

5 Remove the ohmmeter.

6 Attach a negative voltmeter probe to a vehicle earth.

7 Attach a positive voltmeter probe to ECM multi-plug pin 21.

8 Switch the ignition on. The voltmeter should indicate nbv.

9 If no voltage is found, check the relevant fuse and the supply wiring back to the ignition switch.

Mecatronic III

10 Switch the ignition off and disconnect the ECM multi-plug.

11 Attach a negative voltmeter probe to a vehicle earth.

12 Attach a positive voltmeter probe to ECM multi-plug pins 3 and 4 in turn **(see illustration 7.27)**. The voltmeter should indicate nbv in each case.

13 Attach the positive voltmeter probe to ECM multi-plug pin 5.

14 Switch the ignition on. The voltmeter should indicate nbv.

15 If no voltage is found, check the relevant fuse and the supply wiring back to the battery or ignition switch.

7 Solenoid valves

Solenoid valve operation tests

The solenoid valves are integral with the hydraulic control unit and can only be checked using suitable ABS diagnostic test equipment.

8 Hydraulic pump motor

Pump operation test

The hydraulic pump motor is integral with the hydraulic control unit and can only be checked using suitable ABS diagnostic test equipment.

9 Warning lamp

Checking the warning lamp (general)

1 Inspect the warning lamp bulb holder contacts in the instrument panel.

2 Check that the instrument panel multi-plug terminal pins are pushed fully home and making good contact.

3 Check that no wires have been disconnected.

4 Faults in any of the above areas are possible reasons for failure or malfunctioning of the warning lamp.

Warning lamp operation test

5 With the ignition switched off, the warning lamp should remain off.

6 Switch the ignition on and the warning lamp should illuminate then extinguish after a few seconds. The lamp should then remain off.

7 If the warning lamp comes on and remains on at any time during vehicle operation, carry out the previously described test procedures on the system components.

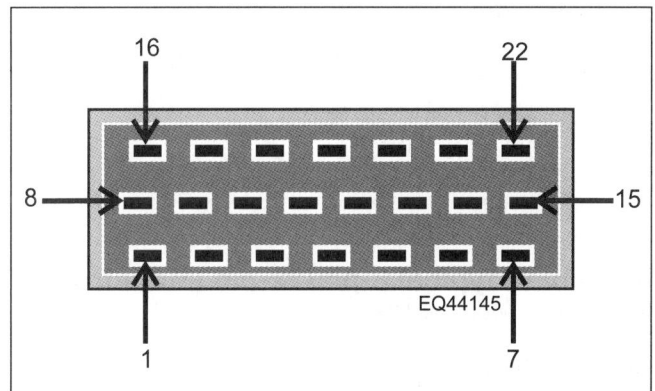

7.26 ECM 22-pin multi-plug

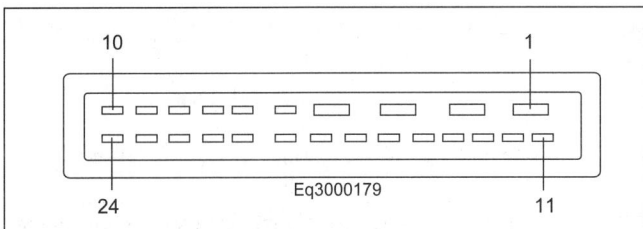

7.27 ECM 24-pin multi-plug

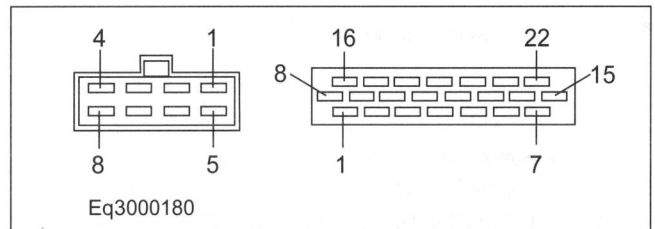

7.28 ECM 8-pin and 22-pin multi-plugs

Pin table - typical 8-pin/22-pin (Mecatronic II, two-channel)

Note: *Refer to illustration 7.28*

Pin No.	Connection	Test condition	Voltage
8-pin multi-plug			
1	ECM earth	Ignition on	0.25 volts (max)
2	-	-	-
3	ABS warning lamp	Ignition on:	
		Lamp off	Nbv
		Lamp on	0.25 volts (max)
4	-	-	-
5	Supply from battery	Ignition off/on	Nbv
6	Supply from battery	Ignition off/on	Nbv
7	ECM earth	Ignition on	0.25 volts (max)
8	-	-	-
22-pin multi-plug			
1	Right wheel speed sensor signal	Roadwheel rotating	0.5 to 1.5 volts AC (approx)
2	-	-	-
3	Left wheel speed sensor signal	Roadwheel rotating	0.5 to 1.5 volts AC (approx)
4	-	-	-
5	ECM earth	Ignition on	0.25 volts (max)
6	-	-	-
7	-	-	-
8	-	-	-
9	Right wheel speed sensor earth	Roadwheel rotating	0.25 volts (max)
10	Left wheel speed sensor earth	Roadwheel rotating	0.25 volts (max)
11	-	-	-
12	-	-	-
13	-	-	-
14	-	-	-
15	SD connector	-	-
16	-	-	-
17	-	-	-
18	-	-	-
19	-	-	-
20	-	-	-
21	Supply from ignition switch	Ignition on	Nbv
22	ABS warning lamp	Ignition on:	
		lamp off	nbv
		lamp on	0.25 volts (max)

Pin table - typical 8-pin/22-pin (Mecatronic II, four-channel)

Note: *Refer to illustration 7.28*

Pin No.	Connection	Test condition	Voltage
8-pin multi-plug			
1	ECM earth	Ignition on	0.25 volts (max)
2	-	-	-
3	ABS warning lamp	Ignition on:	
		Lamp off	Nbv
		Lamp on	0.25 volts (max)
4	-	-	-
5	Supply from battery	Ignition off/on	Nbv
6	Supply from battery	Ignition off/on	Nbv
7	ECM earth	Ignition on	0.25 volts (max)
8	-	-	-

22-pin multi-plug

1	Front right wheel speed sensor signal	Roadwheel rotating	0.5 to 1.5 volts AC (approx)
2	Rear right wheel speed sensor signal	Roadwheel rotating	0.5 to 1.5 volts AC (approx)
3	Front left wheel speed sensor signal	Roadwheel rotating	0.5 to 1.5 volts AC (approx)
4	-	-	-
5	ECM earth	Ignition on	0.25 volts (max)
6	-	-	-
7	-	-	-
8	Rear left wheel speed sensor signal	Roadwheel rotating	0.5 to 1.5 volts AC (approx)
9	Front right wheel speed sensor earth	Roadwheel rotating	0.25 volts (max)
10	Front left wheel speed sensor earth	Roadwheel rotating	0.25 volts (max)
11	-	-	-
12	-	-	-
13	-	-	-
14	-	-	-
15	SD connector	-	-
16	Rear left wheel speed sensor earth	Roadwheel rotating	0.25 volts (max)
17	Rear right wheel speed sensor earth	Roadwheel rotating	0.25 volts (max)
18	-	-	-
19	-	-	-
20	-	-	-
21	Supply from ignition switch	Ignition on	Nbv
22	ABS warning lamp	Ignition on: lamp off lamp on	 Nbv 1.0 volt (max)

Pin table - typical 24-pin (Mecatronic III)

Note: *Refer to illustration 7.27*

Pin No.	Connection	Test condition	Voltage
1	ECM earth	Ignition on	0.25 volts (max)
2	ECM earth	Ignition on	0.25 volts (max)
3	Supply from battery	Ignition on/off	Nbv
4	Supply from battery	Ignition on/off	Nbv
5	Supply from ignition switch	Ignition on	Nbv
6	-	-	-
7	Front right wheel speed sensor signal	Roadwheel rotating	0.5 to 1.5 volts AC (approx)
8	Front right wheel speed sensor earth	Roadwheel rotating	0.25 volts (max)
9	Rear right wheel speed sensor signal	Roadwheel rotating	0.5 to 1.5 volts AC (approx)
10	Rear right wheel speed sensor earth	Roadwheel rotating	0.25 volts (max)
11	SD connector	-	-
12	-	-	-
13	-	-	-
14	-	-	-
15	-	-	-
16	-	-	-
17	-	-	-
18	-	-	-
19	ECM earth	Ignition on	0.25 volts (max)
20	ABS warning lamp	Ignition on: Lamp on Lamp off	 1.0 volt (max) Nbv
21	Front left wheel speed sensor signal	Roadwheel rotating	0.5 to 1.5 volts AC (approx)
22	Front left wheel speed sensor earth	Roadwheel rotating	0.25 volts (max)
23	Rear left wheel speed sensor signal	Roadwheel rotating	0.5 to 1.5 volts AC (approx)
24	Rear left wheel speed sensor earth	Roadwheel rotating	0.25 volts (max)

7

Fault codes

10 Mecatronic II fault codes

1 The Bendix Mecatronic II system requires the use of a FCR for obtaining fault codes. Flash codes are not available for output from this system.

2 If a FCR is available, it should be connected to the SD serial connector and used in accordance with the maker's instructions.

3 The FCR can be used for the following purposes:

  a) Obtaining fault codes.

  b) Clearing fault codes.

  c) Obtaining datastream information.

  d) Testing the system actuators (relays and solenoid valves).

11 Mecatronic III fault codes

1 On the Mecatronic III system, internal fault codes are used by the ECM to designate faults in the system components and circuits. A proprietary fault code reader (FCR) or system tester (such as the Ford FDS 2000) is required to interrogate the system. No actual fault code numbers are available although the component circuits checked by the ECM are similar to those shown in the following tables.

Fault code table (Mecatronic II, two-channel)

Code	Item	Fault
12	Start of diagnosis	-
13	Solenoid valve supply	Incorrect voltage
14	Pump motor	Defective (non-operation of pump after actuation of relay)
15	ABS main relay	Short circuit
19	Wheel speed sensor(s)	Inconsistent speed signal
21	ABS main relay	Open circuit
22	ABS main relay supply	Incorrect voltage
25	Right wheel speed sensor	Short circuit
32	Left wheel speed sensor	Short circuit
34	Right wheel speed sensor signal	Incorrect air gap, damaged rotor teeth
37	Right wheel speed sensor signal	Sensor missing, incorrect air gap, damaged rotor teeth
39	Left wheel speed sensor signal	Sensor missing, incorrect air gap, damaged rotor teeth
41	Left wheel speed sensor signal	Incorrect air gap, damaged rotor teeth
42	Right inlet/outlet solenoid valve	Defective (signal interruption, short/open circuit)
43	Right restricting solenoid valve	Defective (signal interruption, short/open circuit)
44	Left inlet/outlet solenoid valve	Defective (signal interruption, short/open circuit)
45	Left restricting solenoid valve	Defective (signal interruption, short/open circuit)
53	Pump relay	Defective (signal interruption, short/open circuit, incorrect voltage)
55	ECM	Incorrect voltage, ECM defective
56	ECM	Defective
57	ECM supply	Incorrect voltage
67	Front right wheel locked	Hydraulic system fault, incorrect wheel speed sensor signal
69	Front left wheel locked	Hydraulic system fault, incorrect wheel speed sensor signal
11	End of diagnosis	-

Fault code table (Mecatronic II, four-channel)

Code	Item	Fault
37	ECM	Incorrect coding, 22-pin multi-plug wiring fault
38	Front right wheel speed sensor	Incorrect resistance
42	Wheel speed sensors (all)	Signal variation
44	Front right wheel speed sensor signal	Incorrect air gap, damaged rotor teeth
48	Front left wheel speed sensor	Incorrect resistance
52	Wheel speed sensors (all)	Signal variation
54	Front left wheel speed sensor signal	Incorrect air gap, damaged rotor teeth
58	Rear right wheel speed sensor	Incorrect resistance
62	Wheel speed sensors (all)	Signal variation
64	Rear right wheel speed sensor signal	Incorrect air gap, damaged rotor teeth
68	Rear left wheel speed sensor	Incorrect resistance
72	Wheel speed sensors (all)	Signal variation
74	Rear left wheel speed sensor signal	Incorrect air gap, damaged rotor teeth
78	Rear wheel speed sensors	Signal variation
79	Wheel speed sensors (all)	Signal variation
81	Relay supply	Incorrect voltage, wiring, connections
89	Relay supply	Incorrect voltage, wiring, connections
91	ABS relays	Faulty operation, wiring, connections
92	ABS relays	Faulty operation, wiring, connections
93	ABS relays	Faulty operation, wiring, connections
94	Hydraulic pump motor supply	Incorrect voltage, fuses, wiring, connections
95	ABS relay supply and earth	Wiring, connections
96	ECM supply and earth	Incorrect voltage, fuses, wiring, connections
97	8-pin multi-plug	Wiring, connections

7

Wiring diagrams

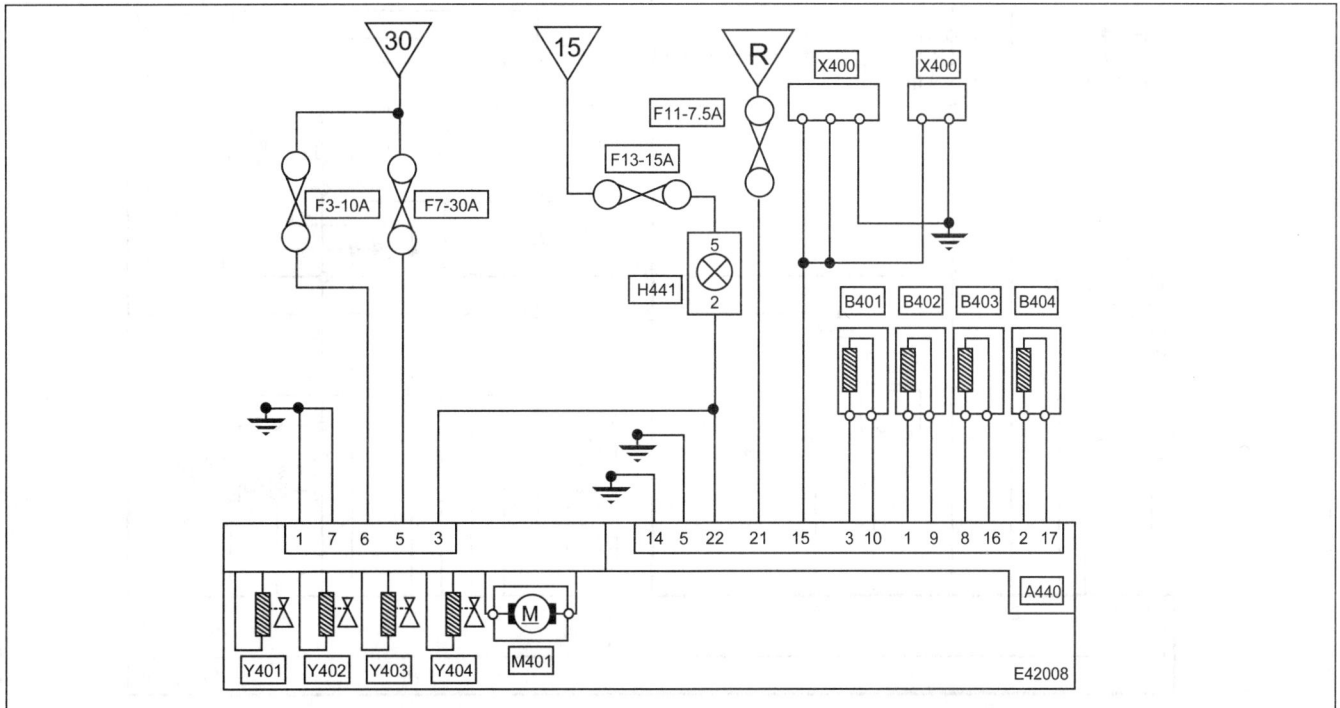

7.29 8-pin and 22-pin wiring diagram, Ford Mondeo (Mecatronic II)

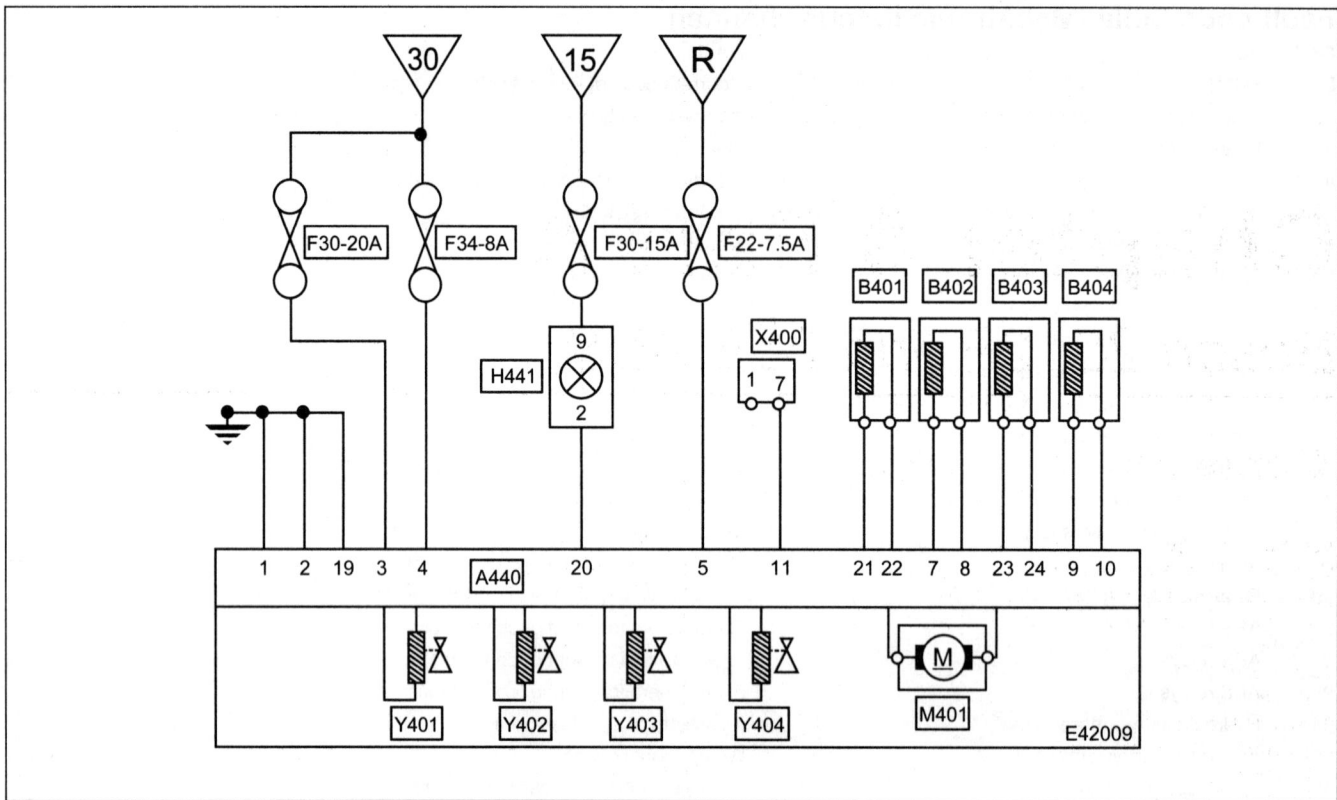

7.30 8-pin and 22-pin wiring diagram, Ford Mondeo (Mecatronic III)

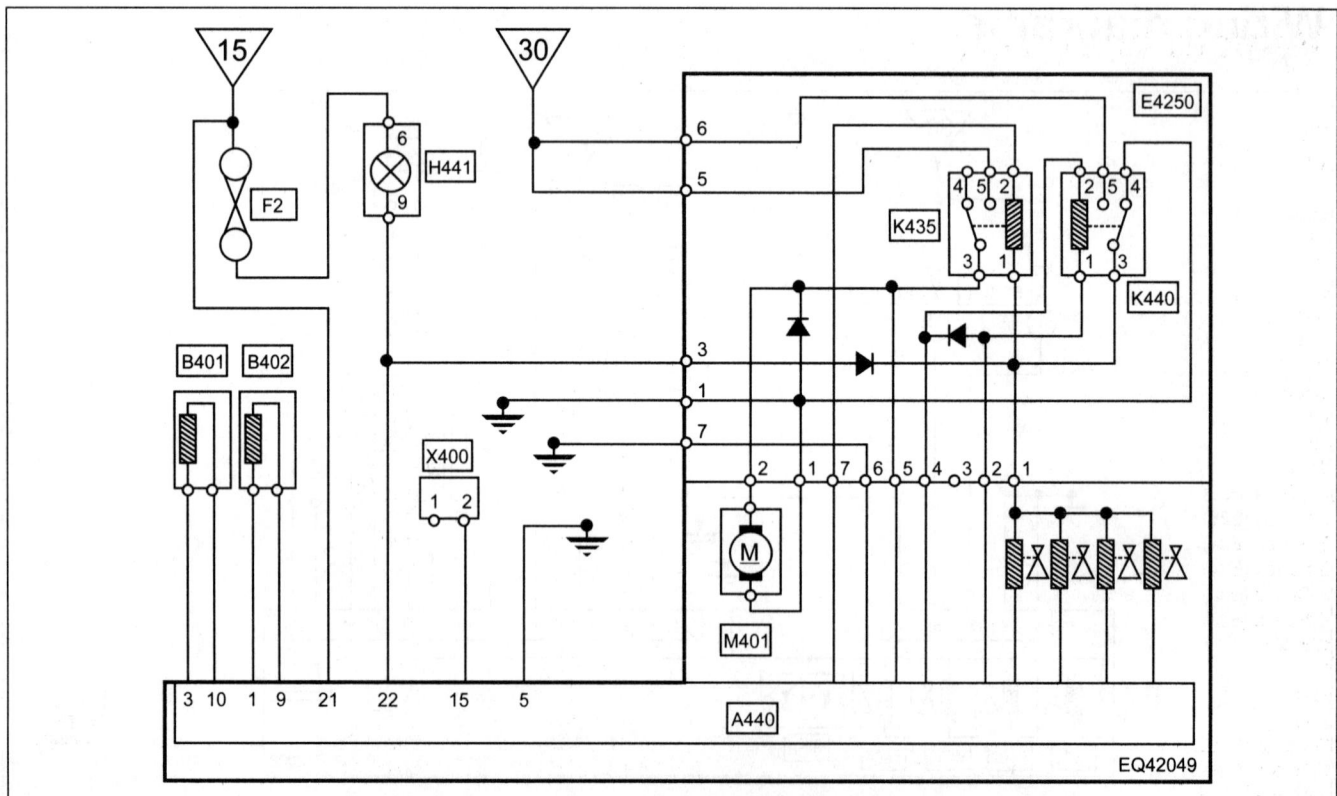

7.31 24-pin wiring diagram, Peugeot 306

Chapter 8 Part A
Bosch 2S, 2E and 2SE ABS

Contents

Vehicle coverage

Model	Year	System
Alfa Romeo		
164	1990-1992	Bosch 2S
Audi		
80	1988-1991	Bosch 2E
80	1992-1995	Bosch 2E
90	1988-1991	Bosch 2S
100	1988-1992	Bosch 2S
100	1992-1995	Bosch 2E
200	1986-1991	Bosch 2S
Coupe	1988-1991	Bosch 2S
V8	1990-1992	Bosch 2S
V8	1992-1994	Bosch 2E
BMW		
3-Series (E30)	1982-1993	Bosch 2S
5-Series (E28)	1982-1987	Bosch 2S
5-Series (E34)	1988-1992	Bosch 2S
5-Series (E34)	1992-1996	Bosch 2E
7-Series (E32)	1987-1992	Bosch 2S
7-Series (E32)	1992-1994	Bosch 2E
Citroën		
CX	1988-1991	Bosch 2S
Fiat		
Croma	1986-1991	
Tempra (2.0/Diesel)	1988-1997	Bosch 2S
Tipo (2.0/Diesel)	1988-1995	Bosch 2S
Jaguar/Daimler		
XJ6/Sovereign	1986-1991	Bosch 2S
XJ12 Sovereign HE	1983-1992	Bosch 2S

Vehicle coverage (continued)

Model	Year	System
Lancia		
Dedra	1990-1992	
Thema	1986-1994	Bosch 2S
Mercedes-Benz		
190 (201)	1985-1993	Bosch 2S
C-Class (202)	1993-2000	Bosch 2E
E-Class (124)	1985-1992	Bosch 2S
E-Class (124)	1992-1995	Bosch 2E
S-Class (126)	1985-1991	Bosch 2S
SL (107)	1985-1990	Bosch 2S
Peugeot		
605	1993-1996	Bosch 2SE
Renault		
25	1985-1992	Bosch 2S
Espace	1990-1994	Bosch 2SE
Rover		
800	1986-1991	Bosch 2S
Vauxhall/Opel		
Astra-E/Belmont	1989-1991	Bosch 2E
Calibra	1990-1991	Bosch 2E
Carlton	1986-1994	Bosch 2S
Cavalier	1989-1992	Bosch 2S
Kadett-E	1989-1991	Bosch 2E
Monza	1986-1987	Bosch 2S
Omega-A	1986-1994	Bosch 2S
Senator	1986-1994	Bosch 2S
Vectra-A	1989-1992	Bosch 2S
Volkswagen		
Transporter	1992-1995	Bosch 2E
Volvo		
240	1990-1992	Bosch 2S
740/760	1987-1991	Bosch 2S
940/960	1990-1992	Bosch 2S
940/960	1994-1997	Bosch 2E
S90/V90	1997-1999	Bosch 2E

Overview of system operation

1 Basic principles and system identification

The Bosch 2 Antilock Brake System has been fitted to a vast range of passenger vehicles since its introduction in the late 1970's. The system is of the additional or 'add-on' type operating in conjunction with the conventional braking system components.

The purpose of the system is to apply the vehicle brakes at maximum efficiency without wheel lock or loss of directional stability. Inductive sensors (wheel speed sensors) monitor the speed of the wheels by generating an electrical signal as the wheel is rotated. This information is passed to the ABS Electronic Control Module (ECM) which is then able to determine wheel speed, wheel acceleration and wheel deceleration. The ECM compares the signals received from each wheel and if the onset of lock at any wheel is detected, a signal is sent to the ABS hydraulic control unit, which regulates the brake pressure for the relevant wheel(s).

The Bosch 2 system has been refined and uprated over the years in accordance with advances in electrical technology and demands from vehicle manufacturers. Essentially there are three distinct versions of the system, with the main differences being in the location of the ECM

(either remotely mounted within the vehicle, or attached to the hydraulic control unit) the ECM software, and the ECM multi-plug which may be a 35-pin, 15-pin or 25-pin type.

For clarity, 35-pin ECM multi-plug systems, with remotely mounted ECM are covered in this Part of Chapter 8, with 15-pin, and 25-pin systems with integral ECM and hydraulic control unit being covered in Part B .

In general, the various versions can be classified as follows, according to certain distinct features:

35-pin multi-plug systems covered in this Part of Chapter 8.

Bosch 2S

Original version with separate ECM and hydraulic control unit. ECM without self-diagnostics.

Bosch 2E

Uprated version of Bosch 2S. ECM now incorporates self-diagnostics.

Bosch 2SE

A development of Bosch 2S but can have additional ECM circuitry for 4WD vehicle applications. With/without self-diagnostics.

15-pin/25-pin multi-plug systems covered in Part B of Chapter 8.

Bosch 2EH

Similar to Bosch 2E, but with ECM integral with hydraulic control unit.

Bosch 2SH

Similar to Bosch 2EH but can have additional ECM circuitry for traction control.

All versions of Bosch 2 ABS are similar in operation and are comprised of the following main components **(see illustration 8A.1)**.

a) *ABS-ECM.*
b) *Hydraulic control unit.*
c) *Inductive wheel speed sensors and associated sensor rings.*
d) *ABS electrical wiring harness and relays.*
e) *Brake light switch.*
f) *ABS warning lamp.*
g) *G-force sensor (certain 4WD applications only).*
h) *Diagnostic connector (where applicable).*
i) *Copy valve assembly (where applicable).*

In addition, the conventional brake system is comprised of the following components:

a) *Tandem brake master cylinder.*
b) *Vacuum servo unit.*
c) *Brake calipers/wheel cylinders and hydraulic hoses and pipes.*
d) *Pressure regulating/load-sensing valve(s) depending on application.*

Certain Bosch 2E systems covered in this Part of Chapter 8 may be installed as an antilock brake system only, or as an antilock brake system incorporating traction control.

In ABS mode, the purpose of the system is to apply the vehicle brakes at maximum efficiency without wheel lock or loss of directional stability. Inductive sensors (wheel speed sensors) monitor the speed of the wheels by generating an electrical signal as the wheel is rotated. This information is passed to the ABS-ECM which compares the signals received from each wheel and uses the speed of the fastest wheel as a reference value. The ECM continually monitors the speed of each wheel and if the onset of lock at any wheel is detected (a received speed signal being less than the reference value) a signal is sent to the ABS hydraulic control unit which regulates the brake pressure for the relevant wheel(s).

Where the system incorporates traction control, essentially the reverse principle is applied. When the ECM detects that one or more wheels are rotating faster than the reference value, the brake is actually applied on the relevant wheel(s) to reduce the rotational speed. Additionally, when wheel spin is detected, various signals may be sent to the engine management ECM to control engine torque and rpm.

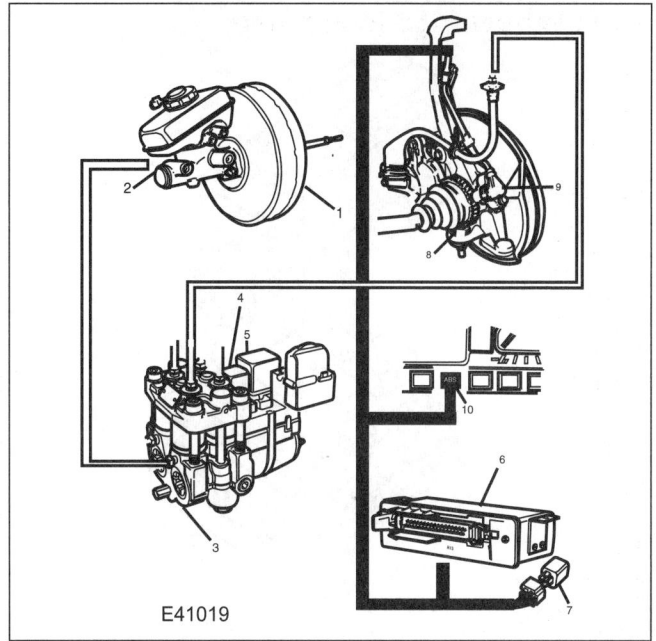

E41019

8A.1 Typical Bosch 2S main components

1 Vacuum servo unit
2 Tandem master cylinder
3 Hydraulic control unit
4 Solenoid valve relay
5 Pump relay
6 ABS-ECM
7 Overvoltage protection relay
8 Sensor rings
9 Wheel speed sensors
10 ABS warning lamp

8A

2 Component description and operation

ABS ECM

General

The Bosch Electronic Control Module (ECM) continually monitors wheel speed from the signals provided by the wheel speed sensors, and brake application from the brake light switch signal. If the ECM detects the incidence of wheel lock (or wheel spin, if traction control is incorporated) on one or more wheels, a signal is sent to the hydraulic control unit to modulate the hydraulic pressure to the brake of the locking or spinning wheel(s) **(see illustration 8A.2)**.

The ECM contains two microprocessors and uses digital technology to complete this function and other functions such as, fault code memory (self-diagnostic versions) and power modules for valve and relay activity.

On Bosch 2S, 2E and 2SE systems, the ECM is located remotely from the hydraulic control unit, generally within the passenger compartment **(see illustration 8A.3)**.

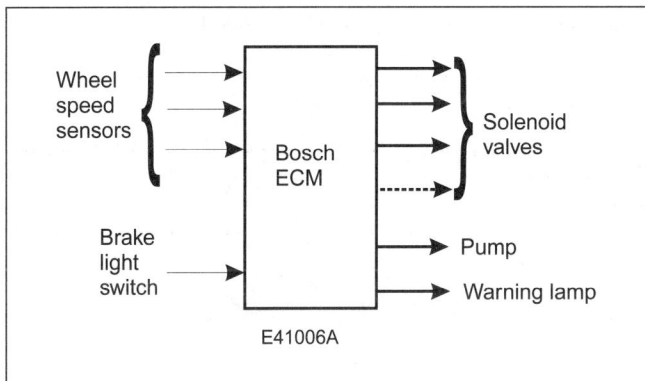

E41006A

8A.2 ECM sensor inputs and control signal outputs

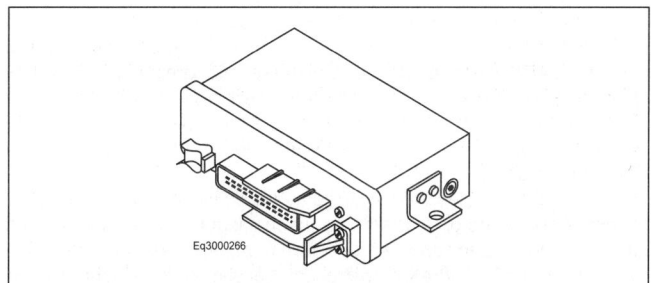

Eq3000266

8A.3 Typical Bosch 2 ECM

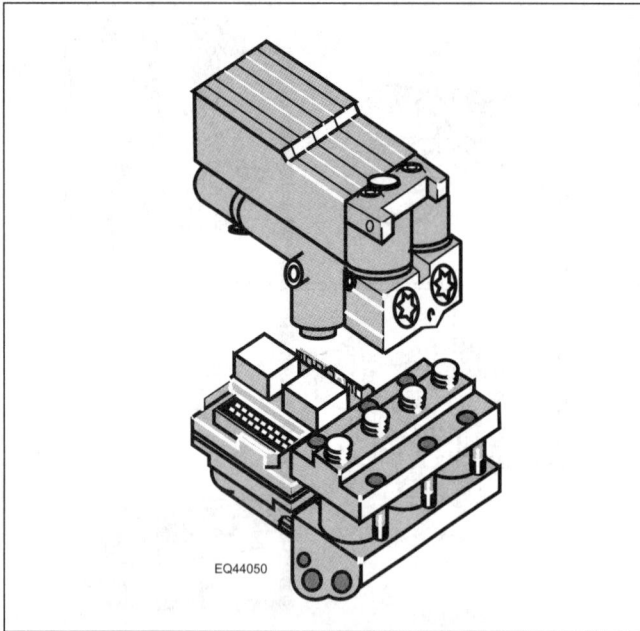

EQ44050

8A.4 Bosch 2E hydraulic control unit

Self-test

The Bosch 2 ECM is equipped with a self-test capability that initially examines the ABS system when the ignition is switched on, and then examines the wheel speed sensor signals after a wheel speed of approximately 4 mph is reached from all wheels (engine running). The ABS self-test program continues to examine the signals from the various components as long as the ignition is switched on. If self-test determines that faults are not present, the ABS is ready for operation once a specified vehicle speed has been achieved.

If the ECM detects that a fault is present, all ABS functions are switched off and the warning lamp is turned on. The conventional braking system continues to operate as normal without ABS assistance.

Self-diagnostics

On Bosch 2E and some 2SE systems with self-diagnostics (SD), if the ECM detects a fault during the self-test routine, an internal fault code is stored in the ECM memory. Stored fault codes can be retrieved from the SD connector with the aid of a suitable fault code reader or, on certain models, can be output as flash codes by means of the ABS warning lamp, or on-board diagnostic unit. If the fault clears, the code will remain stored until cleared, either manually or with an FCR.

Hydraulic control unit

The hydraulic control unit consists of an electric motor and radial piston return pump, solenoid valves and pressure accumulator. The unit controls the hydraulic pressure applied to the brake for each individual front wheel and each individual rear wheel, or pair of wheels, according to application. The return pump is switched on when the ABS is activated. It returns brake fluid, drained off during the pressure reduction phase, back into the brake circuit. The pump motor relay, solenoid valve relay and the connection plug for the power supply can be located under the cover on top of the unit (see illustration 8A.4).

On Citroën vehicles an integrated hydraulic system is used whereby hydraulic pressure generated by a mechanical pump is used for both the braking system and the vehicle suspension system. On these vehicles, the ABS hydraulic control unit consists of the solenoid valves only - the return pump assembly, pressure accumulator and conventional brake master cylinder being replaced by a compensator/brake control valve which is part of the vehicle main hydraulic system and is unique to Citroën.

According to vehicle application, Bosch 2S, 2E and 2SE may be either a three- or four-channel system according to the number of solenoid valves used in the hydraulic control unit. On a four-channel system, there is one solenoid valve for each brake. On a three-channel system, there is a solenoid valve for each front brake, with the remaining valve controlling the rear brakes as a pair.

On certain Bosch 2E three-channel systems, the rear solenoid valve operates in conjunction with a copy valve to control the hydraulic pressure to the rear brakes during normal braking and when braking under ABS control.

The copy valve is an additional mechanical valve assembly, which can be either located remotely on the vehicle, or integrated within the hydraulic control unit. The unit comprises two hydraulic chambers linked by a dual piston assembly. Hydraulic pressure from the rear solenoid valve in the hydraulic control unit is supplied to the first chamber, and hydraulic pressure from the master cylinder is supplied to the second chamber. During normal braking without ABS control, hydraulic pressure flows through the open rear solenoid valve to one of the rear brakes, and also to the first pressure chamber in the copy valve. The other rear brake receives hydraulic pressure from the second pressure chamber in the copy valve. Pressure differences within the copy valve chambers act on the piston, which moves to open or close hydraulic passages within the valve. The action of the piston assembly in the copy valve balances the hydraulic pressure applied to both the rear brakes.

When ABS is in operation, the hydraulic pressure in the circuit to one of the rear brakes is modulated by the rear solenoid valve in the hydraulic control unit. The other rear brake receives exactly the same hydraulic pressure via the operation of the piston assembly in the copy valve.

Wheel speed sensors

The rotational speed of the roadwheels and any changes in the rotational speed are recorded by inductive wheel speed sensors, located at the roadwheels, or in the differential casing. Depending on system type and vehicle application, there may be a separate sensor for each wheel, or each pair of wheels (see illustration 8A.5).

Each wheel speed sensor assembly comprises a toothed sensor ring, which rotates at roadwheel speed, and an adjacent sensor mounted a set distance from the sensor ring (see illustration 8A.6).

The sensors are permanent magnet pulse generator types producing an AC voltage sine wave as the sensor ring teeth pass through the magnetic field of the sensor (see illustration 8A.7).

The frequency of the waveform produced by the wheel speed sensor is proportional to the road speed. This AC voltage signal is continually being delivered to the ABS-ECM for processing.

The peak to peak voltage of the speed signal (when viewed upon an oscilloscope) can vary considerably according to wheel speed. An analogue to digital converter (ADC) in the ECM transforms the AC pulse into a digital signal (see illustration 8A.8).

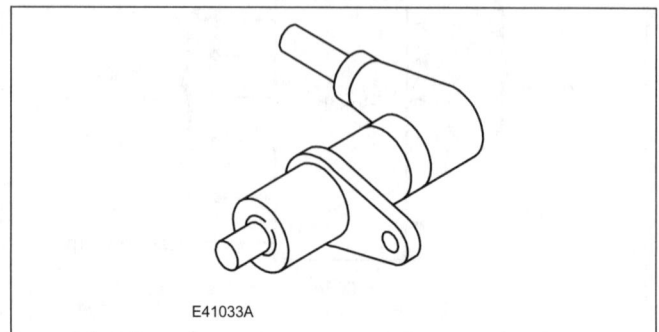

E41033A

8A.5 Typical wheel speed sensor

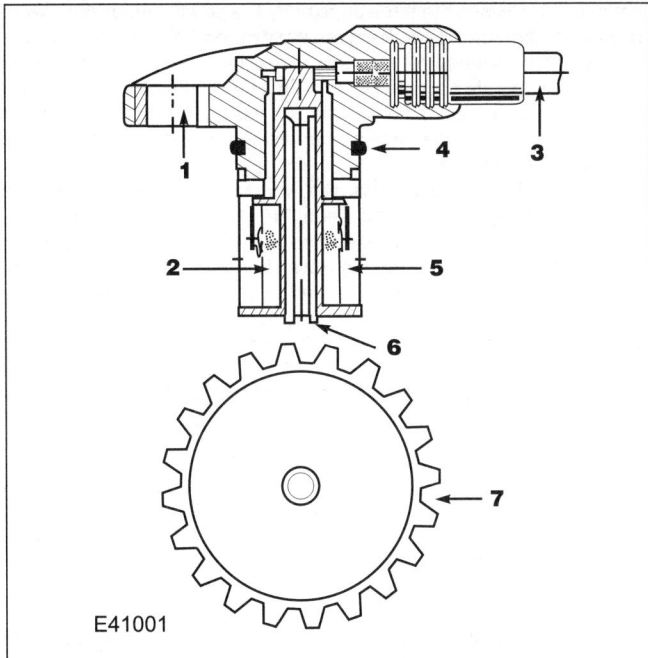

8A.6 Sectional view of a wheel speed sensor

1 Mounting bolt location
2 Permanent magnet
3 Wiring harness
4 O-ring
5 Coil
6 Sensor tip
7 Toothed sensor ring

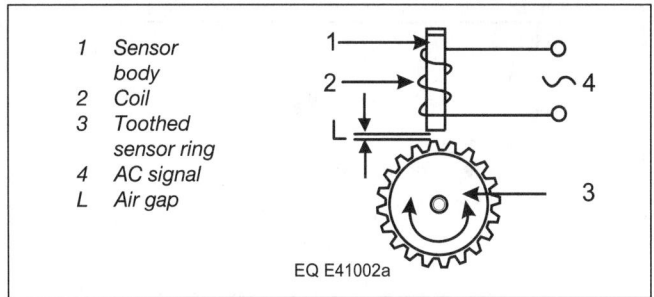

1 Sensor body
2 Coil
3 Toothed sensor ring
4 AC signal
L Air gap

8A.7 Wheel speed sensor operation

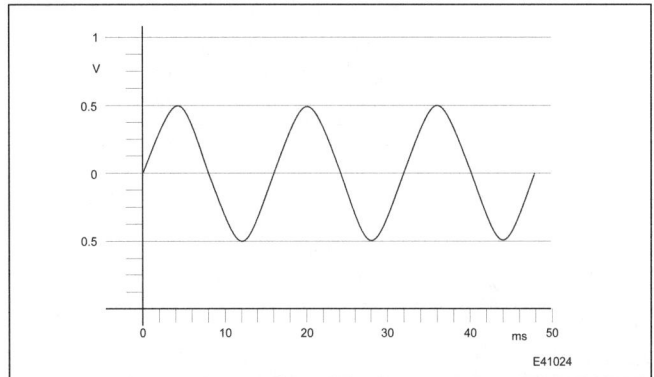

8A.8 Wheel speed sensor waveform as viewed on an oscilloscope

ABS electrical wiring harness and relays

An integrated 35-pin main wiring harness is used to connect power and earth to various electrical components. This enables sensor signals to reach the ECM and the ECM, in turn, to send command signals to the relays, return pump and hydraulic control unit. The hydraulic control unit harness connector, pump relay and main (solenoid valve) relay are located under a cover on the top of the unit **(see illustration 8A.9)**. Where an overvoltage protection relay is used, this is typically located remotely within the engine compartment.

Pump relay

The pump relay commonly has a black cover, and is furthest from the solenoid valves. A permanent voltage is supplied to the relay contacts from the battery or fuel injection main relay, with the relay coil typically being supplied from the overvoltage protection relay or from the ECM. The pump relay is controlled by the ECM applying an earth to the relay coil, which actuates the relay allowing battery voltage to be applied to the pump. The hydraulic control unit provides the earth for the pump motor.

Main (solenoid valve) relay

The main (solenoid valve) relay commonly has a transparent cover, and is closest to the solenoid valves. A permanent voltage is supplied to the relay contacts from the battery or fuel injection main relay, with the relay coil typically being supplied from the overvoltage protection relay or from the ECM. The valve relay is controlled by the ECM applying an earth to the relay coil, which actuates the relay allowing battery voltage to be applied to the solenoid valves, with the ECM controlling the earth.

Overvoltage protection relay

Where an overvoltage protection relay is used, nominal battery voltage is supplied from the battery to the relay contacts via a fuse. One terminal of the relay coil receives battery voltage from the ignition switch, with the other terminal being connected to earth. When the ignition is switched on, the relay is activated allowing battery voltage to be supplied to the ECM, and in some applications, to the pump relay and solenoid valve relay coils.

Brake light switch

The brake light switch comprises a switch body and contact pin and is located above the brake pedal **(see illustration 8A.10)**.

When the brake pedal is depressed, closing the brake light switch, a signal is sent to the ECM indicating that the brakes are being applied. Once this signal is received, the ECM will begin monitoring the wheel speed via the wheel speed sensors and activate the ABS if necessary.

ABS warning lamp

After the ignition is switched on, the ABS warning lamp on the instrument panel is illuminated, and remains illuminated until the engine is started and the ECM receives a voltage signal, typically from

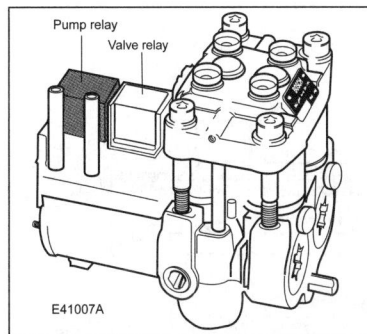

8A.9 Pump and solenoid valve relay locations on the hydraulic control unit (Bosch 2S unit shown)

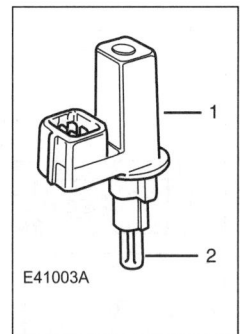

8A.10 Typical brake light switch body (1) and contact pin (2)

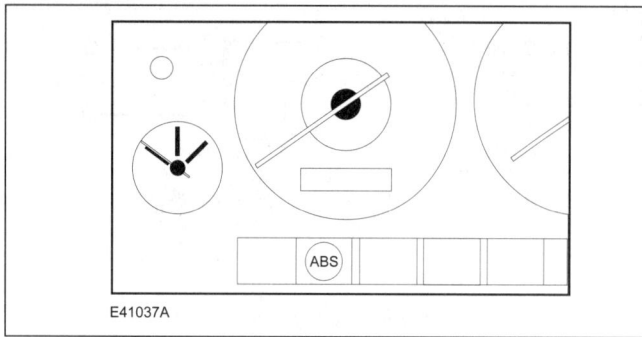

8A.11 ABS warning lamp

the alternator or, on some systems, from the engine oil pressure switch. If satisfactory operation of the system is detected by the ECM, the light is extinguished. During vehicle operation above a pre-determined wheel speed, the ABS-ECM implements a further self-test cycle whereby ABS operation and wheel speed sensor signals are continually monitored. If a fault is detected, the relevant ECM pin is earthed to illuminate the warning lamp on the instrument panel, and the ABS function is disabled. The warning lamp will remain illuminated until the fault is no longer present (see illustration 8A.11).

When the ABS-ECM detects a fault, the ABS warning lamp activated and, on systems with self-diagnostics, a fault code is stored in the ECM memory. If the fault no longer exists after the next system start (ignition on/off) the ABS warning lamp is extinguished after the self-test cycle, however the fault code (where applicable) remains stored in the ECM memory.

Tandem master cylinder

Typically, the tandem master cylinder comprises a body casting incorporating primary and secondary pressure chambers, primary piston, intermediate piston, floating piston, slotted pin and central valve. The cylinder operates as a conventional master cylinder using vacuum assistance from the vacuum servo unit (see illustration 8A.12).

When the brake system is at rest, the central valve in the floating piston rests against the slotted pin. In this condition, the central valve is open and brake fluid can discharge out of the pressure chamber back into the brake fluid reservoir. When the brake pedal is depressed, the build-up of hydraulic pressure in the primary pressure chamber acts on the intermediate piston and floating piston, moving them down the cylinder bore. The floating piston contacts the seal on the central valve, closing the connection between the intermediate and secondary pressure chambers. Brake hydraulic pressure can now also increase in the secondary pressure chamber.

Vacuum servo unit

The vacuum servo unit is located between the brake pedal and tandem master cylinder. When the brake pedal is depressed, the servo unit increases the force applied by the pedal, reducing the effort required to operate the brakes (see illustration 8A.13).

The unit is operated by vacuum created in the engine inlet manifold (or from a separate vacuum pump on diesel engines) which is applied to a diaphragm contained within the unit casing. A pushrod connected to the centre of the diaphragm acts directly on the primary piston in the master cylinder.

When the brake pedal is released, vacuum is applied to both sides of the diaphragm. When the pedal is depressed, one side of the diaphragm is opened to atmosphere and the vacuum acting on the other side deflects the diaphragm which in turn operates the master cylinder primary piston. The resulting force applied to the master cylinder piston is therefore significantly greater than the initial force applied to the brake pedal by the driver.

Pressure regulating/load-sensing valve(s)

Depending on vehicle application, pressure regulating valves or load sensing valves may be incorporated to restrict the hydraulic fluid pressure to the rear brakes. The valves may be pressure conscious whereby the hydraulic fluid supply is restricted once a pre-determined pressure is reached, or load conscious whereby the hydraulic pressure is reduced according to vehicle loading.

G-force sensor

On 4WD vehicles with all four roadwheels linked through the transmission, there is only minimal difference in wheel speed when 4WD is engaged. Additional sensitivity is necessary for the ABS-ECM to detect the onset of wheel lock in these conditions. On certain applications, a G-force sensor is used to signal the ABS-ECM once vehicle deceleration reaches a pre-determined rate. When the ABS-ECM receives this signal, ABS sensitivity is significantly increased, typically by 50%.

3 System operation

Brake system at rest

When the ABS system is at rest all the brake components are inoperative. Pressure is non-existent in the hydraulic pipes between the tandem master cylinder and the brake calipers or wheel cylinders.

Brake system operating under conventional control without ABS

When the brake pedal is activated, the pedal force is applied to the tandem master cylinder by the vacuum servo unit pushrod. The servo unit pushrod acts directly on the pressure piston in the master cylinder, which pressurises the hydraulic fluid in the brake pipes to the hydraulic control unit. Pressure acting upon the valves in the hydraulic control unit causes them to open and hydraulic pressure is

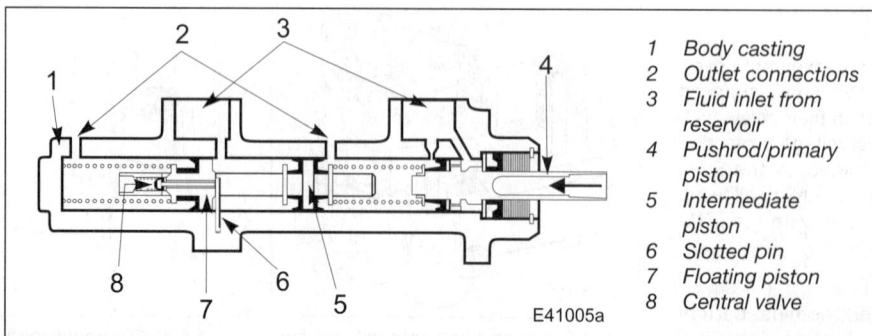

1 Body casting
2 Outlet connections
3 Fluid inlet from reservoir
4 Pushrod/primary piston
5 Intermediate piston
6 Slotted pin
7 Floating piston
8 Central valve

8A.12 Sectional view of a typical tandem master cylinder

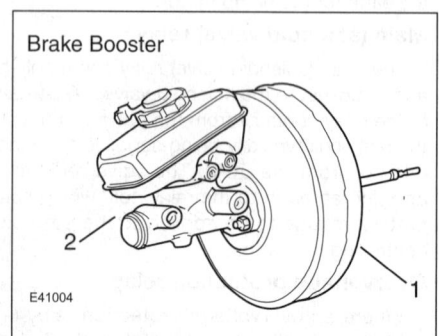

Brake Booster

8A.13 Vacuum servo unit (1) and tandem master cylinder (2)

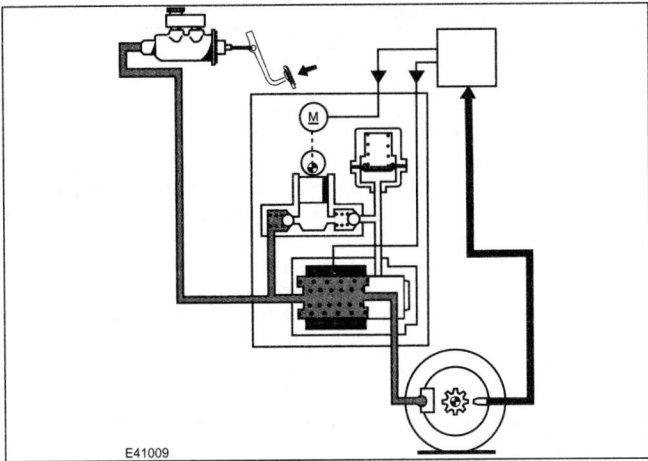

8A.14 Brake system operating under conventional control without ABS

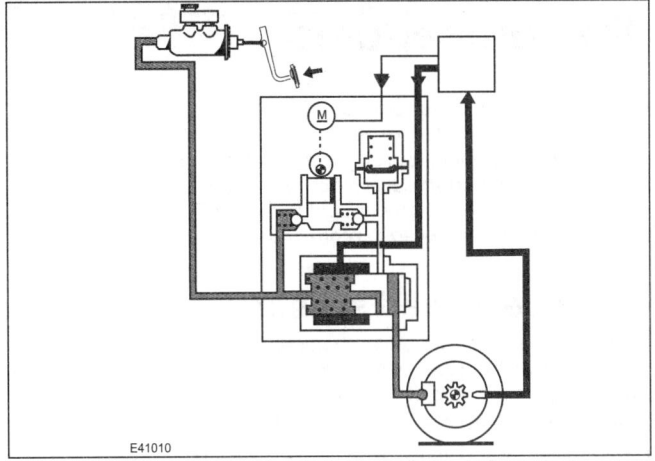

8A.15 ABS operation - first phase, pressure holding

transmitted to each brake caliper or wheel cylinder, thus operating the brakes **(see illustration 8A.14)**.

Brake system operating in conjunction with ABS control

The ABS-ECM continually monitors wheel speed from the signals provided by the wheel speed sensors. If the ECM detects the incidence of wheel lock on one or more wheels, ABS is automatically initiated in three phases. Typically, ABS operates individually on each front wheel and either individually or jointly on the rear wheels. In which case, all or any of the wheels could be in any one of the following phases at any particular moment.

First ABS phase, pressure holding

The ECM earths the appropriate solenoid valve driver and a current of 2.0 amps flows in the solenoid valve circuit. The solenoid valve actuates to move the corresponding piston upward against spring force so that the hydraulic fluid line from the tandem master cylinder to the brake caliper or wheel cylinder is closed. The hydraulic fluid in the controlled circuit is maintained at a constant pressure and this effectively removes the braking force from the controlled circuit. The pressure cannot now be increased in the controlled circuit by any further application of the brake pedal **(see illustration 8A.15)**.

Second ABS phase, pressure reduction

If the ECM detects wheel instability, a pressure reduction phase is initiated. The current flow in the solenoid valve circuit is increased to

5.0 amps, which causes the solenoid valve to move the corresponding piston further upward, so that a channel to the return pump is opened. The result is that hydraulic fluid flows quickly out of the brake caliper or wheel cylinder into the pressure accumulator in the hydraulic control unit. The return pump is actuated by the ECM so that hydraulic fluid is pumped back into the master cylinder. This process creates a pulsation, which can be felt in the brake pedal action **(see illustration 8A.16)**.

After pressure reduction, the ECM returns to the pressure holding phase and wheel speed is closely monitored.

Third ABS phase, pressure build-up

If the wheel speed accelerates too quickly during the second pressure holding phase, the solenoid valve driver is released which switches off the solenoid valve circuit. Spring action returns the piston to the lowest position, which reopens the hydraulic fluid line from the tandem master cylinder to the brake caliper or wheel cylinder. Hydraulic pressure is reinstated in the controlled circuit thus reintroducing operation of the brake. After a brief period, a short pressure holding phase is re-introduced and the ECM continually shifts between pressure build-up and pressure holding until the wheel has decelerated to a sufficient degree where pressure reduction is once more required **(see illustration 8A.17)**.

The whole ABS control cycle takes place 4 to 10 times per second for each affected wheel and this ensures maximum braking effect and control during ABS operation.

8A

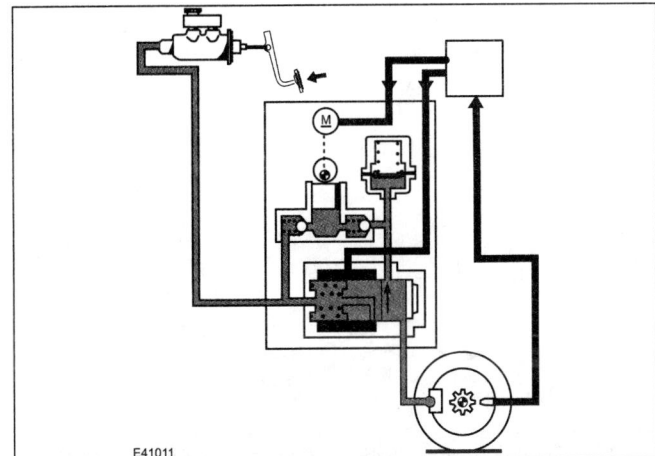

8A.16 ABS operation - second phase, pressure reduction

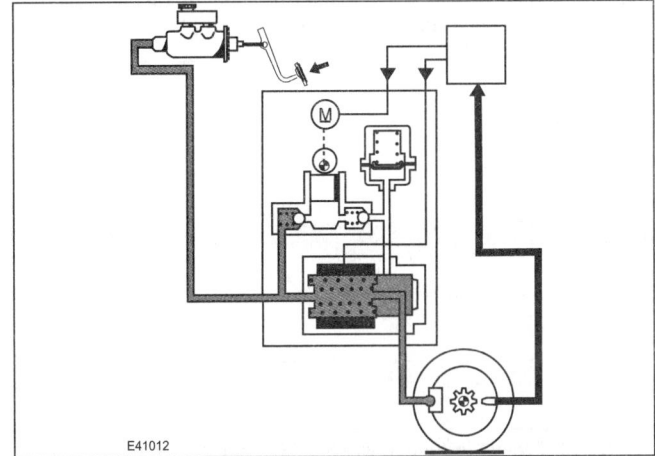

8A.17 ABS operation - third phase, pressure build-up

Test procedures

Important note: *The test procedures, pintables and wiring diagrams contained in this Chapter are necessarily representative of the system depicted. Because of the variations in wiring and other data that often occurs, even between similar vehicles in any particular VM's range, the reader should take great care in identification of ECM pins, and satisfy himself that he has gathered the correct data before failing a particular component.*

4 Wheel speed sensors

Note: *There are considerable variations in wheel speed sensor pin connections at the ECM, depending on vehicle type and model. The wiring diagrams in this Chapter show typical pin connections, but some variations are likely to be encountered.*

Checking the wheel speed sensor (general)

1 Inspect the wheel speed sensor for corrosion or damage and check that the sensor is tightly mounted.
2 Check the toothed sensor ring for damage, eccentricity and for broken or missing teeth.
3 Inspect the wheel speed sensor-wiring plug for corrosion and damage. One plug for each sensor.

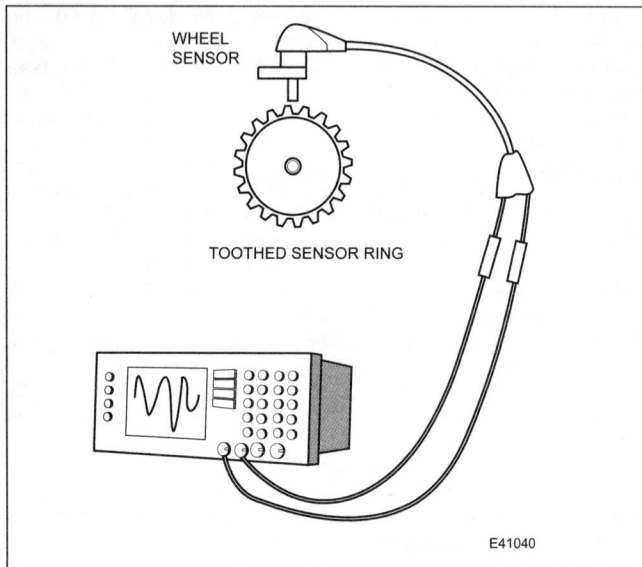

8A.18 Checking wheel speed sensor output with an oscilloscope connected to the sensor-wiring plug

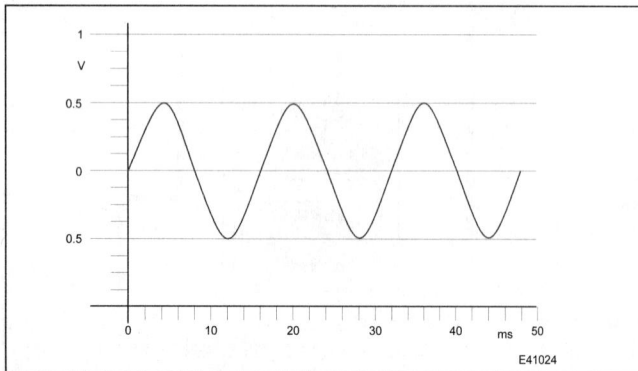

8A.19 Typical inductive wheel speed sensor sine wave as displayed on an oscilloscope

4 Check that the connector terminal pins are pushed fully home and making good contact with the sensor-wiring plug.
5 Check the clearance between the sensor and the toothed sensor ring. The clearance is not normally adjustable but is nominally 0.2 to 1.2 mm. If the clearance is excessive, expect a worn sensor tip or problems with the wheel bearings/hub or sensor ring.
6 When carrying out voltage checks with an oscilloscope or voltmeter, the voltage obtained will be proportional to the speed at which the wheel is rotating. In addition to determining that the wheel speed sensors are actually producing a voltage output, it is essential that the output from the sensors on a particular axle is the same for any given wheel speed.

Checking wheel speed sensor output with an oscilloscope

Note: *Refer to the note at the beginning of this Section concerning wheel speed sensor pin connections before proceeding.*
7 Switch the ignition off and disconnect the ECM multi-plug or the relevant wheel speed sensor-wiring plug.
8 Connect an oscilloscope between the terminal pins for the sensor under test **(see illustration 8A.18)**.
9 Select a range to cover 80 Hz on the oscilloscope and a free run time base.
10 Raise the wheel and rotate it by hand at approximately one revolution per second.
11 A sinusoidal wave form should be obtained, with amplitude and duration changing with rotational speed **(see illustration 8A.19)**.
12 If there is no signal, or a very weak or intermittent signal at the ECM, repeat the test at the sensor-wiring plug. If there is no change in signal status, the sensor is suspect.
13 If the signal is now satisfactory this indicates a fault in the wiring harness, which should be checked for continuity.

Checking wheel speed sensor output with an AC voltmeter

Note: *Refer to the note at the beginning of this Section concerning wheel speed sensor pin connections before proceeding.*
14 Switch the ignition off and disconnect the ECM multi-plug or the relevant wheel speed sensor-wiring plug.
15 Connect an AC voltmeter between the terminal pins for the sensor under test **(see illustration 8A.20)**.

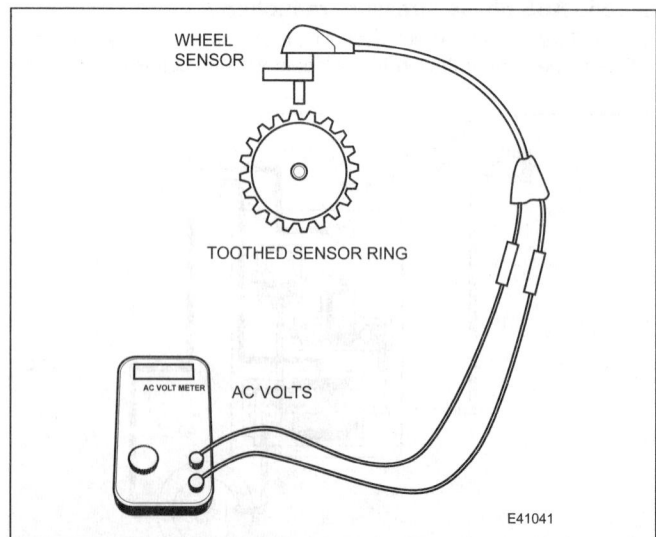

8A.20 Checking wheel speed sensor output with a voltmeter connected to the sensor-wiring plug

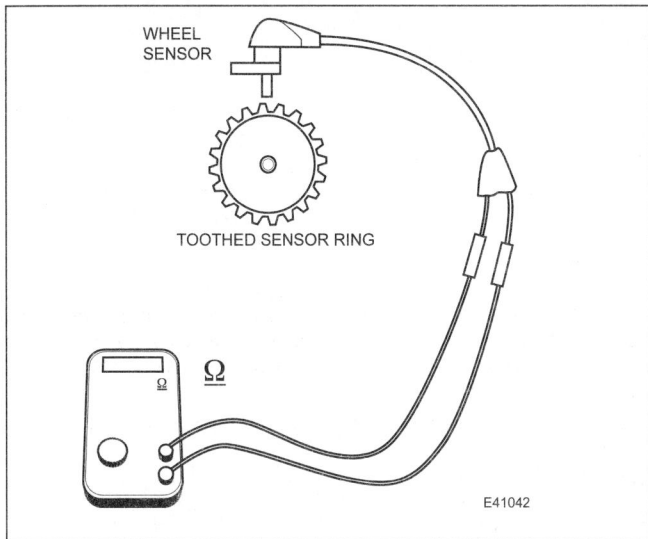

8A.21 Checking wheel speed sensor resistance with an ohmmeter connected to the sensor-wiring plug

16 Raise the wheel and rotate it by hand at approximately one revolution per second.

17 A voltage of approximately 0.5 to 2.0 volts (AC RMS) should be obtained. If there is no signal, or a very weak or intermittent signal at the ECM, repeat the test at the sensor-wiring plug. If there is no change in the signal, the sensor is suspect.

18 If the signal is now satisfactory, this indicates a fault in the wiring harness which should be checked for continuity. **Note:** *This test at least proves that a signal is being generated by the sensor. However, the voltage produced is an average voltage and does not clearly indicate damage to the sensor ring or that the sinewave is regular in formation.*

Checking wheel speed sensor resistance

Note: *Refer to the note at the beginning of this Section concerning wheel speed sensor pin connections before proceeding.*

19 Switch the ignition off and disconnect the ECM multi-plug or the relevant wheel speed sensor-wiring plug.

20 Connect an ohmmeter between the terminal pins for the sensor under test **(see illustration 8A.21)**.

21 The readings obtained should be between 0.7 and 2.2 kohms approximately.

22 If the resistance is excessively high or open circuit at the ECM, repeat the test at the sensor multi-plug. If there is no change in resistance, the sensor is suspect.

23 If the resistance is now satisfactory, this indicates a fault in the wiring harness which should be checked for continuity. **Note:** *Even if the resistance is within the quoted specifications, this does not prove that the speed sensor can generate an acceptable signal.*

5 System relays

Checking relays (general)

1 The following tests are applicable to the ABS main (solenoid valve) relay, hydraulic pump relay and overvoltage protection relay.

2 The main (solenoid valve) and pump relays are located on the hydraulic control unit, usually under a detachable cover, whereas the overvoltage protection relay is commonly located in the engine compartment fuse/relay box.

3 Before carrying out specific tests on the relays or power supplies inspect the relays and relay base for corrosion and damage.

4 Check that the relays are pushed fully home and making good contact with the base.

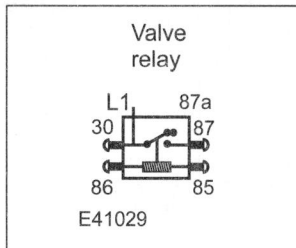

8A.22 Typical main (solenoid valve) relay

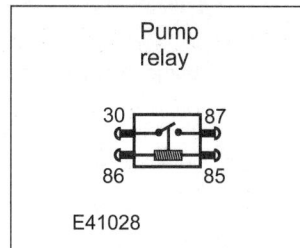

8A.23 Typical pump relay

5 A fault in any of the above areas are possible reasons for failure or malfunctioning of the relay.

Relay operation tests

Main (solenoid valve) relay

Note: *On some models, relay terminals 87 and 30 are transposed from the arrangement shown in the illustration and described in the text. Where applicable, refer to the wiring diagrams for specific relay terminal numbers according to model.*

6 Switch the ignition off and remove the relay from the relay base.

7 Connect an ohmmeter between the relay output terminal (typically 30) and earth terminal (typically 87a) **(see illustration 8A.22)**. The ohmmeter should indicate continuity.

8 Connect an ohmmeter between the relay main voltage supply terminal (typically 87) and the output terminal (typically 30). The ohmmeter should indicate an open circuit.

9 Attach the positive lead of a 12 volt supply to the relay coil supply terminal (typically 86) and the negative lead to the relay coil earth terminal (typically 85). There should be an unmistakable click from the relay and the ohmmeter should indicate continuity.

10 Where applicable, connect an ohmmeter or a diode tester between the relay output terminal (typically 30) and ABS warning lamp terminal L1.

11 Measure the circuit and then reverse the connections. The instrument should indicate continuity in one direction (diode conducting), but not the other (diode blocking).

12 Renew the relay if it fails these tests. If the relay cover is transparent, inspect the relay for burns. A faulty relay usually has a burnt appearance.

Hydraulic pump relay

Note: *On some models, relay terminals 87 and 30 are transposed from the arrangement shown in the illustration and described in the text. Where applicable, refer to the wiring diagrams for specific relay terminal numbers according to model.*

13 Switch the ignition off and remove the relay from the relay base.

14 Connect an ohmmeter between the relay main voltage supply terminal (typically 87) and the output terminal (typically 30) **(see illustration 8A.23)**. The ohmmeter should indicate an open circuit.

15 Attach the positive lead of a 12 volt supply to the relay coil supply terminal (typically 86) and the negative lead to the relay coil earth terminal (typically 85). There should be an unmistakable click from the relay and the ohmmeter should indicate continuity.

16 Renew the relay if it fails these tests. If the relay cover is transparent, inspect the relay for burns. A faulty relay usually has a burnt appearance.

Overvoltage protection relay

Note: *In the following test, the relay terminal numbers shown in brackets refer to a typical overvoltage protection relay as shown in the illustration. Where applicable, refer to the wiring diagrams for specific relay terminal numbers according to model.*

17 Switch the ignition off and remove the relay from the relay base.

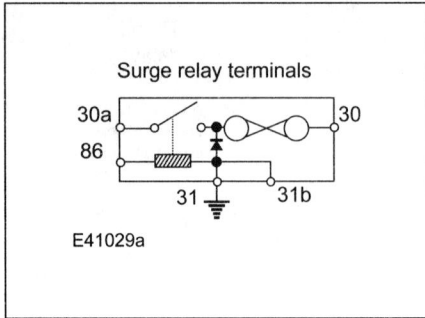

8A.24 Typical overvoltage (surge) protection relay

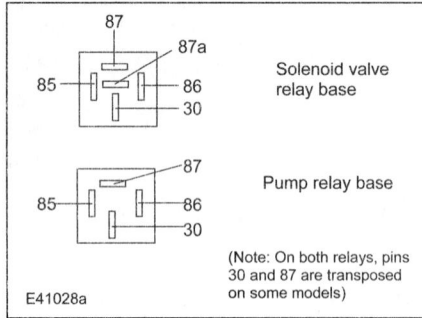

8A.25 Typical main (solenoid valve) and pump relay base terminal identification

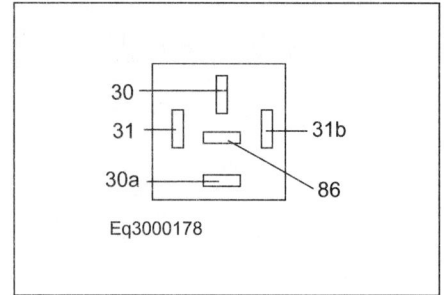

8A.26 Typical overvoltage (surge) protection relay base terminal identification

18 Connect an ohmmeter between the relay main voltage supply terminal (30) and output terminal (30a) **(see illustration 8A.24)**.

19 Attach the positive lead of a 12 volt supply to the relay coil supply terminal (typically 86) and the negative lead to the relay coil earth terminal (typically 31). There should be an unmistakable click from the relay and the ohmmeter should indicate continuity.

20 Connect an ohmmeter or a diode tester between the two-diode terminals (typically 30 and 31).

21 Measure the circuit and then reverse the connections. The instrument should indicate continuity in one direction (diode conducting), but not the other (diode blocking).

22 Renew the relay if it fails these tests. If the relay cover is transparent, inspect the relay for burns. A faulty relay usually has a burnt appearance.

Relay power supply tests

Main (solenoid valve) and hydraulic pump relays

Note: *On some models, relay terminals 87 and 30 are transposed from the arrangement shown in the illustration and described in the text. Where applicable, refer to the wiring diagrams for specific relay terminal numbers according to model.*

23 Switch the ignition off and remove both relays leaving the bases exposed **(see illustration 8A.25)**.

24 Using a voltmeter, in turn probe between the main voltage supply terminal 87 or 30, as applicable, of each relay base and earth. The reading should indicate nbv.

25 If no voltage is found, check the wiring back to the voltage supply.

26 Switch the ignition on.

27 Using a voltmeter, in turn probe between the relay coil supply terminal (typically 86) of each relay base and earth. The reading should indicate nbv.

28 If no voltage is found, check the supply wiring back to the battery positive terminal, fuel injection relay, or overvoltage protection relay as applicable.

Overvoltage protection relay

Note: *In the following test, the relay base terminal numbers shown in brackets refer to a typical overvoltage protection relay as shown in the illustration. Where applicable, refer to the wiring diagrams for specific terminal numbers according to model.*

29 Switch the ignition off and remove the relay from the relay base.

30 Using a voltmeter probe between the main voltage supply terminal (30) in the relay base and earth. The reading should indicate nbv **(see illustration 8A.26)**.

31 If no voltage is found, check the wiring back to the voltage supply.

32 Switch the ignition on.

33 Using a voltmeter probe between the relay coil supply terminal (86) in the relay base and earth. The reading should indicate nbv.

34 If no voltage is found, check the supply wiring back to the battery positive terminal or ignition switch as applicable.

6 Electronic Control Module

Checking the ECM (general)

1 Inspect the ECM for corrosion or damage and ensure that the unit is mounted securely.

2 Check that the ECM multi-plug terminals are pushed fully home and making good contact with the ECM pins. A fault in any of the above areas are possible reasons for poor performance in the ABS system.

ECM power supply and earth tests

Note: *Where applicable, refer to the wiring diagrams for specific terminal numbers according to model.*

3 Switch the ignition off and disconnect the ECM multi-plug.

4 Switch the ignition on.

5 Attach a negative voltmeter probe to a vehicle earth.

6 Attach a positive voltmeter probe to the ECM power supply pin 1. The voltmeter should indicate nbv **(see illustration 8A.27)**.

7 If no voltage is found, check the ECM main earth connection, and the supply wiring back to the battery positive terminal, overvoltage protection relay, or fuel injection relay (according to model).

8 Switch the ignition off.

9 Connect an ohmmeter between a vehicle earth and ECM earth pins 10, 20 and 34, in turn **(see illustration 8A.28)**. The ohmmeter should indicate continuity. If not, check the ECM main earth connection and wiring.

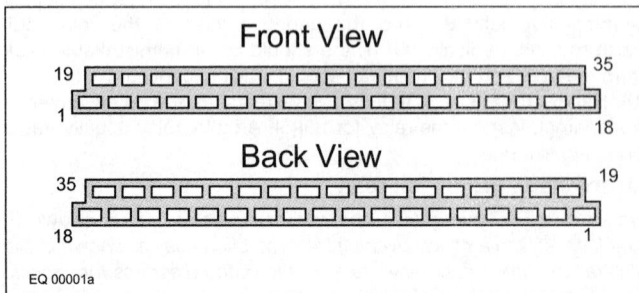

8A.27 ECM 35-pin multi-plug

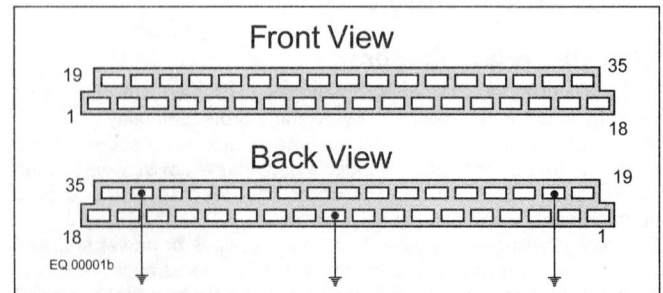

8A.28 ECM multi-plug earth connections

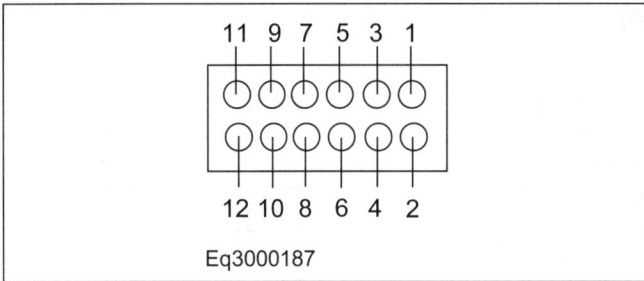

11 9 7 5 3 1

12 10 8 6 4 2

Eq3000187

**8A.29 Hydraulic control unit 12-pin multi-plug terminal
identification**

7 Solenoid valves

Solenoid valve resistance tests

Note: *Where applicable, refer to the wiring diagrams for specific terminal numbers according to model.*
1 Switch the ignition off and disconnect the ECM multi-plug.
2 Using an ohmmeter measure the resistance between pin 32 in the ECM multi-plug and each of the following multi-plug pins in turn - 2, 35, 18, 19. The resistance should be in the order of 1.0 ohm for each solenoid valve **(see illustration 8A.27)**. **Note:** *Vehicles with a front/rear split hydraulic circuit have only three solenoid valves, and ECM pin 19 is not used.*
3 If the resistance is not as specified, disconnect the hydraulic control unit 12-pin multi-plug.
4 Using an ohmmeter, measure the resistance between terminal 4 and terminals 7, 5, 3 and 1 in turn in the hydraulic control unit **(see illustration 8A.29)**. **Note:** *Vehicles with a front/rear split hydraulic circuit have only three solenoid valves, and terminal 7 is not used.*
5 The resistance should be in the order of 1.0 ohm for each solenoid valve. If correct, check the wiring between the hydraulic control unit and the ECM. If the resistance is not as specified, the hydraulic control unit is suspect.

8 Hydraulic pump motor

Pump operation test

1 Switch the ignition off and disconnect the hydraulic control unit 12-pin multi-plug.
2 Connect a 12-volt positive supply to terminal 9 of the hydraulic control unit **(see illustration 8A.29)**. The pump motor should now run.

⚠ *Warning: The test should be made as quickly as possible to avoid damaging the pump.*

3 If the pump does not operate as described, renew the hydraulic control unit.

9 Brake light switch

Checking the brake light switch (general)

1 Check that the brake light switch is correctly and securely mounted and that the plunger moves smoothly with no trace of binding.
2 Check that the wiring multi-plug is pushed fully home and making good contact.
3 Check that no wires have been disconnected.
4 A fault in any of the above areas are possible reasons for failure or malfunctioning of the switch.

Brake light switch voltage and continuity tests

Voltage test

Note: *Where applicable, refer to the wiring diagrams for specific terminal numbers according to model.*
5 Switch the ignition off and disconnect the ECM multi-plug from the ECM.
6 Connect a voltmeter between a vehicle earth and the ECM multi-plug brake light switch input pin, typically 25 **(see illustration 8A.27)**.
7 Switch the ignition on and depress the brake pedal. The voltmeter should indicate nbv.
8 If no voltage is found, the fuse, the brake light switch and the wiring are suspect.
9 Release the brake pedal. The voltage should drop to zero as the switch opens.

Continuity test

10 Switch the ignition off and disconnect the brake light switch multi-plug.
11 Connect an ohmmeter between the terminal pins of the brake light switch.
12 Operate the brake light switch and check for continuity. If the test fails, renew the switch.

10 Warning lamp

Checking the warning lamp (general)

1 Inspect the warning lamp bulb holder contacts in the instrument panel.
2 Check that the instrument panel multi-plug terminal pins are pushed fully home and making good contact.
3 Check that no wires have been disconnected.
4 A fault in any of the above areas are possible reasons for failure or malfunctioning of the warning lamp.

Warning lamp operation test

5 With the ignition switched off, the warning lamp should remain off.
6 Switch the ignition on and the warning lamp should illuminate.
7 Start the engine and allow it to idle, the lamp should then extinguish after a few seconds.
8 If the warning lamp comes on and remains on at any time during vehicle operation, carry out the previously described test procedures on the system components.

8A

Pin table - typical Bosch 2S, 35-pin (Alfa Romeo)

Note: *Refer to illustration 8A.27*

Pin No.	Connection	Test condition	Voltage
1	Supply from relay	Ignition on	Nbv
2	Front left solenoid valve	Engine running:	
		Inactive	Nbv
		Actuated	1.0 volt (max)
3	-	-	-
4	Front left wheel speed sensor signal	Roadwheel rotating	0.5 to 1.5 volts AC (approx)
5	Front left wheel speed sensor earth	Roadwheel rotating	0.25 volts (max)
6	-	-	-
7	Rear right wheel speed sensor signal	Roadwheel rotating	0.5 to 1.5 volts AC (approx)
8	-	-	-
9	Rear left wheel speed sensor signal	Roadwheel rotating	0.5 to 1.5 volts AC (approx)
10	ECM earth	Ignition on	0.25 volts (max)
11	-	-	-
12	-	-	-
13	-	-	-
14	Hydraulic pump relay display input	Ignition on	0 volts
		Pump relay actuated	Nbv
15	Alternator voltage input	Engine running	13.5 to 15.0 volts
16	-	-	-
17	-	-	-
18	Rear left solenoid valve	Engine running:	
		Inactive	Nbv
		Actuated	1.0 volt (max)
19	Rear right solenoid valve	Engine running:	
		Inactive	Nbv
		Actuated	1.0 volt (max)
20	ECM earth	Ignition on	0.25 volts (max)
21	Front right wheel speed sensor signal	Roadwheel rotating	0.5 to 1.5 volts AC (approx)
22	-	-	-
23	Front right wheel speed sensor earth	Roadwheel rotating	0.25 volts (max)
24	Rear right wheel speed sensor earth	Roadwheel rotating	0.25 volts (max)
25	Brake light switch input	Ignition on:	
		Brake pedal released	0 volts
		Brake pedal depressed	Nbv
26	Rear left wheel speed sensor earth	Roadwheel rotating	0.25 volts (max)
27	Solenoid valve relay driver	Ignition on/inactive	Nbv
		Actuated	1.0 volt (max)
28	Hydraulic pump relay driver	Ignition on/inactive	Nbv
		Actuated	1.0 volt (max)
29	ABS warning lamp	Ignition on	1.0 volt (max)
		Engine running:	
		Lamp on, faults present	1.0 volt (max)
		Lamp off, no faults present	Nbv
30	-	-	-
31	-	-	-
32	Solenoid valve relay display input	Ignition on/inactive	0 volts
		Actuated	Nbv
33	-	-	-
34	ECM earth	Ignition on	0.25 volts (max)
35	Front right solenoid valve	Engine running:	
		Inactive	Nbv
		Actuated	1.0 volt (max)

Pin table - typical Bosch 2S, 35-pin (Audi)

Note: *Refer to illustration 8A.27*

Pin No.	Connection	Test condition	Voltage
1	Supply from relay	Ignition on	Nbv
2	Front left solenoid valve	Engine running: Inactive Actuated	 Nbv 1.0 volt (max)
3	-	-	-
4	Front left wheel speed sensor earth	Roadwheel rotating	0.25 volts (max)
5	Front left wheel speed sensor signal	Roadwheel rotating	0.5 to 1.5 volts AC (approx)
6	-	-	-
7	Rear left wheel speed sensor earth	Roadwheel rotating	0.25 volts (max)
8	-	-	-
9	Rear left wheel speed sensor signal	Roadwheel rotating	0.5 to 1.5 volts AC (approx)
10	ECM earth	Ignition on	0.25 volts (max)
11*	Front right wheel speed sensor signal	Roadwheel rotating	0.5 to 1.5 volts AC (approx)
12	-	-	-
13	-	-	-
14	Hydraulic pump relay display input	Ignition on Pump relay actuated	0 volts Nbv
15	Alternator voltage input	Engine running	13.5 to 15.0 volts
16	-	-	-
17	-	-	-
18	Rear left solenoid valve	Engine running: Inactive Actuated	 Nbv 1.0 volt (max)
19	Rear right solenoid valve	Engine running: Inactive Actuated	 Nbv 1.0 volt (max)
20	ECM earth	Ignition on	0.25 volts (max)
21	Front right wheel speed sensor signal	Roadwheel rotating	0.5 to 1.5 volts AC (approx)
22	-	-	-
23	Front right wheel speed sensor earth	Roadwheel rotating	0.25 volts (max)
24	Rear right wheel speed sensor earth	Roadwheel rotating	0.25 volts (max)
25	Brake light switch input	Ignition on: Brake pedal released Brake pedal depressed	 0 volts Nbv
26	Rear right wheel speed sensor signal	Roadwheel rotating	0.5 to 1.5 volts AC (approx)
27	Solenoid valve relay driver	Ignition on/inactive Actuated	Nbv 1.0 volt (max)
28	Hydraulic pump relay driver	Ignition on/inactive Actuated	Nbv 1.0 volt (max)
29	ABS warning lamp	Ignition on Engine running: Lamp on, faults present Lamp off, no faults present	1.0 volt (max) 1.0 volt (max) Nbv
30	-	-	-
31	-	-	-
32	Solenoid valve relay display input	Ignition on/inactive Actuated	0 volts Nbv
33	-	-	-
34	ECM earth	Ignition on	0.25 volts (max)
35	Front right solenoid valve	Engine running: Inactive Actuated	 Nbv 1.0 volt (max)

Alternative

Pin table - typical Bosch 2S, 35-pin (BMW 3/5-Series)
Note: *Refer to illustration 8A.27*

Pin No.	Connection	Test condition	Voltage
1	Supply to ECM	Ignition on/engine running	Nbv
2	Front left solenoid valve	Engine running:	
		Inactive	Nbv
		Actuated	1.0 volt (max)
3	-	-	-
4	Front left wheel speed sensor earth	Roadwheel rotating	0.25 volts (max)
5	-	-	-
6	Front left wheel speed sensor signal	Roadwheel rotating	0.5 to 1.5 volts AC (approx)
7	-	-	-
7*	Rear left wheel speed sensor signal	Roadwheel rotating	0.5 to 1.5 volts AC (approx)
8	Rear left wheel speed sensor signal	Roadwheel rotating	0.5 to 1.5 volts AC (approx)
8*	-		
9	Rear left wheel speed sensor earth	Roadwheel rotating	0.25 volts (max)
10	ECM earth	Ignition on	0.25 volts (max)
11	Front right wheel speed sensor signal	Roadwheel rotating	0.5 to 1.5 volts AC (approx)
12	-	-	-
13	-	-	-
14	Hydraulic pump relay display input	Ignition on	0 volts
		Pump relay actuated	Nbv
15	Alternator voltage input	Engine running	13.5 to 15.0 volts
16	-	-	-
17	-	-	-
18	Rear/rear left solenoid valve	Engine running:	
		Inactive	Nbv
		Actuated	1.0 volt (max)
19*	Rear right solenoid valve	Engine running:	
		Inactive	Nbv
		Actuated	1.0 volt (max)
20	ECM earth	Ignition on	0.25 volts (max)
21	Front right wheel speed sensor earth	Roadwheel rotating	0.25 volts (max)
22	-	-	-
23	-	-	-
24	Rear right wheel speed sensor signal	Roadwheel rotating	0.5 to 1.5 volts AC (approx)
25	Brake light switch input	Ignition on:	
		Brake pedal released	0 volts
		Brake pedal depressed	Nbv
26	Rear right wheel speed sensor earth	Roadwheel rotating	0.25 volts (max)
27	Solenoid valve relay driver	Ignition on/inactive	Nbv
		Actuated	1.0 volt (max)
28	Hydraulic pump relay driver	Ignition on/inactive	Nbv
		Actuated	1.0 volt (max)
29	ABS warning lamp	Ignition on	1.0 volt (max)
		Engine running:	
		Lamp on, faults present	1.0 volt (max)
		Lamp off, no faults present	Nbv
30	-	-	-
31	-	-	-
32	Solenoid valve relay display input	Ignition on/inactive	0 volts
		Actuated	Nbv
33	-	-	-
34	ECM earth	Ignition on	0.25 volts (max)
35	Front right solenoid valve	Engine running:	
		Inactive	Nbv
		Actuated	1.0 volt (max)

*5-Series only

Pin table - typical Bosch 2S, 35-pin (Citroën)

Note: *Refer to illustration 8A.27*

Pin No.	Connection	Test condition	Voltage
1	Supply to ECM	Ignition on/engine running	Nbv
2	Front left solenoid valve	Engine running:	
		Inactive	Nbv
		Actuated	1.0 volt (max)
3	-	-	-
4	Front left wheel speed sensor signal	Roadwheel rotating	0.5 to 1.5 volts AC (approx)
5	Front left wheel speed sensor earth	Roadwheel rotating	0.25 volts (max)
6	-	-	-
7	-	-	-
8	Rear left wheel speed sensor signal	Roadwheel rotating	0.5 to 1.5 volts AC (approx)
9	Rear left wheel speed sensor earth	Roadwheel rotating	0.25 volts (max)
10	ECM earth	Ignition on	0.25 volts (max)
11	Front right wheel speed sensor signal	Roadwheel rotating	0.5 to 1.5 volts AC (approx)
12	-	-	-
13	-	-	-
14	-	-	-
15	Engine oil pressure switch input	Engine running	Nbv
16	-	-	-
17	-	-	-
18	Rear solenoid valve	Engine running:	
		Inactive	Nbv
		Actuated	1.0 volt (max)
20	ECM earth	Ignition on	0.25 volts (max)
21	Front right wheel speed sensor earth	Roadwheel rotating	0.25 volts (max)
22	-	-	-
23	-	-	-
24	Rear right wheel speed sensor signal	Roadwheel rotating	0.5 to 1.5 volts AC (approx)
25	Brake light switch input	Ignition on:	
		Brake pedal released	0 volts
		Brake pedal depressed	Nbv
26	Rear right wheel speed sensor earth	Roadwheel rotating	0.25 volts (max)
27	Solenoid valve relay driver	Ignition on/inactive	Nbv
		Actuated	1.0 volt (max)
29	ABS warning lamp	Ignition on	1.0 volt (max)
		Engine running:	
		Lamp on, faults present	1.0 volt (max)
		Lamp off, no faults present	Nbv
30	-	-	-
31	-	-	-
32	Solenoid valve relay display input	Ignition on/inactive	0 volts
		Actuated	Nbv
33	-	-	-
34	ECM earth	Ignition on	0.25 volts (max)
35	Front right solenoid valve	Engine running:	
		Inactive	Nbv
		Actuated	1.0 volt (max)

Pin table - typical Bosch 2S, 35-pin (Fiat, Lancia, Renault, Rover)

Note: *Refer to illustration 8A.27*

Pin No.	Connection	Test condition	Voltage
1	Supply from relay	Ignition on	Nbv
2	Front left solenoid valve	Engine running:	
		Inactive	Nbv
		Actuated	1.0 volt (max)
3	-	-	-
4	Front left wheel speed sensor signal	Roadwheel rotating	0.5 to 1.5 volts AC (approx)
5	Front left wheel speed sensor earth	Roadwheel rotating	0.25 volts (max)
6	-	-	-
7	Rear left wheel speed sensor signal	Roadwheel rotating	0.5 to 1.5 volts AC (approx)
8	-	-	-
9	Rear left wheel speed sensor earth	Roadwheel rotating	0.25 volts (max)
10	ECM earth	Ignition on	0.25 volts (max)
11*	-	-	-
12	-	-	-
13	-	-	-
14	Hydraulic pump relay display input	Ignition on	0 volts
		Pump relay actuated	Nbv
15	Alternator voltage input	Engine running	13.5 to 15.0 volts
16	-	-	-
17	-	-	-
18	Rear left solenoid valve	Engine running:	
		Inactive	Nbv
		Actuated	1.0 volt (max)
19	Rear right solenoid valve	Engine running:	
		Inactive	Nbv
		Actuated	1.0 volt (max)
20	ECM earth	Ignition on	0.25 volts (max)
21*	Front right wheel speed sensor signal	Roadwheel rotating	0.5 to 1.5 volts AC (approx)
22	-	-	-
23*	Front right wheel speed sensor earth	Roadwheel rotating	0.25 volts (max)
24	Rear right wheel speed sensor signal	Roadwheel rotating	0.5 to 1.5 volts AC (approx)
25	Brake light switch input	Ignition on:	
		Brake pedal released	0 volts
		Brake pedal depressed	Nbv
26	Rear right wheel speed sensor earth	Roadwheel rotating	0.25 volts (max)
27	Solenoid valve relay driver	Ignition on/inactive	Nbv
		Actuated	1.0 volt (max)
28	Hydraulic pump relay driver	Ignition on/inactive	Nbv
		Actuated	1.0 volt (max)
29	ABS warning lamp	Ignition on	1.0 volt (max)
		Engine running:	
		Lamp on, faults present	1.0 volt (max)
		Lamp off, no faults present	Nbv
30	-	-	-
31	-	-	-
32	Solenoid valve relay display input	Ignition on/inactive	0 volts
		Actuated	Nbv
33	-	-	-
34	ECM earth	Ignition on	0.25 volts (max)
35	Front right solenoid valve	Engine running:	
		Inactive	Nbv
		Actuated	1.0 volt (max)

*Pin connections interchanged on certain models

Pin table - typical Bosch 2S, 35-pin (Jaguar/Daimler)

Note: *Refer to illustration 8A.27*

Pin No.	Connection	Test condition	Voltage
1	Supply to ECM	Ignition on/engine running	Nbv
2	Front left solenoid valve	Engine running:	
		Inactive	Nbv
		Actuated	1.0 volt (max)
3	-	-	-
4	Front left wheel speed sensor earth	Roadwheel rotating	0.25 volts (max)
5	-	-	-
6	Front left wheel speed sensor signal	Roadwheel rotating	0.5 to 1.5 volts AC (approx)
7	-	-	-
8	Rear left wheel speed sensor signal	Roadwheel rotating	0.5 to 1.5 volts AC (approx)
9	Rear left wheel speed sensor earth	Roadwheel rotating	0.25 volts (max)
10	ECM earth	Ignition on	0.25 volts (max)
11	Front right wheel speed sensor signal	Roadwheel rotating	0.5 to 1.5 volts AC (approx)
12	-	-	-
13	-	-	-
14	Hydraulic pump relay display input	Ignition on	0 volts
		Pump relay actuated	Nbv
15	Alternator voltage input	Engine running	13.5 to 15.0 volts
16	-	-	-
17	-	-	-
18	Rear/rear left solenoid valve	Engine running:	
		Inactive	Nbv
		Actuated	1.0 volt (max)
19	-		
20	ECM earth	Ignition on	0.25 volts (max)
21	Front right wheel speed sensor earth	Roadwheel rotating	0.25 volts (max)
22	-	-	-
23	-	-	-
24	Rear right wheel speed sensor signal	Roadwheel rotating	0.5 to 1.5 volts AC (approx)
25	Brake light switch input	Ignition on:	
		Brake pedal released	0 volts
		Brake pedal depressed	Nbv
26	Rear right wheel speed sensor earth	Roadwheel rotating	0.25 volts (max)
27	Solenoid valve relay driver	Ignition on/inactive	Nbv
		Actuated	1.0 volt (max)
28	Hydraulic pump relay driver	Ignition on/inactive	Nbv
		Actuated	1.0 volt (max)
29	ABS warning lamp	Ignition on	1.0 volt (max)
		Engine running:	
		Lamp on, faults present	1.0 volt (max)
		Lamp off, no faults present	Nbv
30	-	-	-
31	-	-	-
32	Solenoid valve relay display input	Ignition on/inactive	0 volts
		Actuated	Nbv
33	-	-	-
34	ECM earth	Ignition on	0.25 volts (max)
35	Front right solenoid valve	Engine running:	
		Inactive	Nbv
		Actuated	1.0 volt (max)

8A

Pin table - typical Bosch 2S, 35-pin (Mercedes-Benz)

Note: *Refer to illustration 8A.27*

Pin No.	Connection	Test condition	Voltage
1	Supply from relay	Ignition on/engine running	Nbv
2	Front left solenoid valve	Engine running:	
		Inactive	Nbv
		Actuated	1.0 volt (max)
3	-	-	-
4	Front left wheel speed sensor signal	Roadwheel rotating	0.5 to 1.5 volts AC (approx)
5	-		
6	Front left wheel speed sensor earth	Roadwheel rotating	0.25 volts (max)
7	Rear wheel speed sensor signal	Roadwheel rotating	0.5 to 1.5 volts AC (approx)
8	-	-	-
9	Rear wheel speed sensor earth	Roadwheel rotating	0.25 volts (max)
10	ECM earth	Ignition on	0.25 volts (max)
11	-	-	-
12	-	-	-
13	-	-	-
14	Hydraulic pump relay display input	Ignition on	0 volts
		Pump relay actuated	Nbv
15	Alternator voltage input	Engine running	13.5 to 15.0 volts
16	-	-	-
17	-	-	-
18	Rear left solenoid valve	Engine running:	
		Inactive	Nbv
		Actuated	1.0 volt (max)
19	-	-	-
20	ECM earth	Ignition on	0.25 volts (max)
21	Front right wheel speed sensor signal	Roadwheel rotating	0.5 to 1.5 volts AC (approx)
22	-	-	-
23	Front right wheel speed sensor earth	Roadwheel rotating	0.25 volts (max)
24	-	-	-
25	Brake light switch input	Ignition on:	
		Brake pedal released	0 volts
		Brake pedal depressed	Nbv
26	-	-	-
27	Solenoid valve relay driver	Ignition on/inactive	Nbv
		Actuated	1.0 volt (max)
28	Hydraulic pump relay driver	Ignition on/inactive	Nbv
		Actuated	1.0 volt (max)
29	ABS warning lamp	Ignition on	1.0 volt (max)
		Engine running:	
		Lamp on, faults present	1.0 volt (max)
		Lamp off, no faults present	Nbv
30	-	-	-
31	-	-	-
32	Solenoid valve relay display input	Ignition on/inactive	0 volts
		Actuated	Nbv
33	-	-	-
34	ECM earth	Ignition on	0.25 volts (max)
35	Front right solenoid valve	Engine running:	
		Inactive	Nbv
		Actuated	1.0 volt (max)

Pin table - typical Bosch 2S, 35-pin (Vauxhall/Opel)

Note: *Refer to illustration 8A.27*

Pin No.	Connection	Test condition	Voltage
1	Supply from relay	Ignition on	Nbv
2	Front left solenoid valve	Engine running:	
		Inactive	Nbv
		Actuated	1.0 volt (max)
3	-	-	-
4	Front left wheel speed sensor signal	Roadwheel rotating	0.5 to 1.5 volts AC (approx)
5	Front left wheel speed sensor earth	Roadwheel rotating	0.25 volts (max)
6	-	-	-
7	Rear left wheel speed sensor signal	Roadwheel rotating	0.5 to 1.5 volts AC (approx)
8	-	-	-
9	Rear left wheel speed sensor earth	Roadwheel rotating	0.25 volts (max)
10	ECM earth	Ignition on	0.25 volts (max)
11	Front right wheel speed sensor signal	Roadwheel rotating	0.5 to 1.5 volts AC (approx)
12	-	-	-
13	-	-	-
14	Hydraulic pump relay display input	Ignition on	0 volts
		Pump relay actuated	Nbv
15	Alternator voltage input	Engine running	13.5 to 15.0 volts
16	-	-	-
17	-	-	-
18	Rear left solenoid valve	Engine running:	
		Inactive	Nbv
		Actuated	1.0 volt (max)
19	Rear right solenoid valve	Engine running:	
		Inactive	Nbv
		Actuated	1.0 volt (max)
20	ECM earth	Ignition on	0.25 volts (max)
21	Front right wheel speed sensor earth	Roadwheel rotating	0.25 volts (max)
22	-	-	-
23	-	-	-
24	Rear right wheel speed sensor signal	Roadwheel rotating	0.5 to 1.5 volts AC (approx)
25	Brake light switch input	Ignition on:	
		Brake pedal released	0 volts
		Brake pedal depressed	Nbv
26	Rear right wheel speed sensor earth	Roadwheel rotating	0.25 volts (max)
27	Solenoid valve relay driver	Ignition on/inactive	Nbv
		Actuated	1.0 volt (max)
28	Hydraulic pump relay driver	Ignition on/inactive	Nbv
		Actuated	1.0 volt (max)
29	ABS warning lamp	Ignition on	1.0 volt (max)
		Engine running:	
		Lamp on, faults present	1.0 volt (max)
		Lamp off, no faults present	Nbv
30	-	-	-
31	-	-	-
32	Solenoid valve relay display input	Ignition on/inactive	0 volts
		Actuated	Nbv
33	-	-	-
34	ECM earth	Ignition on	0.25 volts (max)
35	Front right solenoid valve	Engine running:	
		Inactive	Nbv
		Actuated	1.0 volt (max)

8A

Pin table - typical Bosch 2S, 35-pin (Volvo)

Note: *Refer to illustration 8A.27*

Pin No.	Connection	Test condition	Voltage
1	Supply to ECM	Ignition on/engine running	Nbv
2	Front left solenoid valve	Engine running:	
		Inactive	Nbv
		Actuated	1.0 volt (max)
3	-	-	-
4	Front left wheel speed sensor signal	Roadwheel rotating	0.5 to 1.5 volts AC (approx)
5	-	-	-
6	Front left wheel speed sensor earth	Roadwheel rotating	0.25 volts (max)
7	Rear wheel speed sensor earth	Roadwheel rotating	0.25 volts (max)
8	-	-	-
9	Rear wheel speed sensor signal	Roadwheel rotating	0.5 to 1.5 volts AC (approx)
10	ECM earth	Ignition on	0.25 volts (max)
11*	Front right wheel speed sensor signal	Roadwheel rotating	0.5 to 1.5 volts AC (approx)
12	-	-	-
13	-	-	-
14	Hydraulic pump relay display input	Ignition on	0 volts
		Pump relay actuated	Nbv
15	Alternator voltage input	Engine running	13.5 to 15.0 volts
16	-	-	-
17	-	-	-
18	Rear left solenoid valve	Engine running:	
		Inactive	Nbv
		Actuated	1.0 volt (max)
19	-	-	-
20	ECM earth	Ignition on	0.25 volts (max)
21	Front right wheel speed sensor earth	Roadwheel rotating	0.25 volts (max)
22	-	-	-
23**	Front right wheel speed sensor signal	Roadwheel rotating	0.5 to 1.5 volts AC (approx)
24	-	-	-
25	Brake light switch input	Ignition on:	
		Brake pedal released	0 volts
		Brake pedal depressed	Nbv
26	-	-	-
27	Solenoid valve relay driver	Ignition on/inactive	Nbv
		Actuated	1.0 volt (max)
28	Hydraulic pump relay driver	Ignition on/inactive	Nbv
		Actuated	1.0 volt (max)
29	ABS warning lamp	Ignition on	1.0 volt (max)
		Engine running:	
		Lamp on, faults present	1.0 volt (max)
		Lamp off, no faults present	Nbv
30	-	-	-
31	-	-	-
32	Solenoid valve relay display input	Ignition on/inactive	0 volts
		Actuated	Nbv
33	-	-	-
34	ECM earth	Ignition on	0.25 volts (max)
35	Front right solenoid valve	Engine running:	
		Inactive	Nbv
		Actuated	1.0 volt (max)

*Volvo 240 only
**Not Volvo 240

Pin table - typical Bosch 2E, 35-pin (Audi)

Note: *Refer to illustration 8A.27*

Pin No.	Connection	Test condition	Voltage
1	Supply to ECM from ignition switch	Ignition on	Nbv
2	Front left solenoid valve	Engine running:	
		Inactive	Nbv
		Actuated	1.0 volt (max)
3	-	-	-
4	Front left wheel speed sensor signal	Roadwheel rotating	0.5 to 1.5 volts AC (approx)
5	Front left wheel speed sensor earth	Roadwheel rotating	0.25 volts (max)
6	-	-	-
7	Rear left wheel speed sensor earth	Roadwheel rotating	0.25 volts (max)
8	-	-	-
9	Rear left wheel speed sensor signal	Roadwheel rotating	0.5 to 1.5 volts AC (approx)
10	-	-	-
11	Front right wheel speed sensor earth	Roadwheel rotating	0.25 volts (max)
12	-	-	-
13	-	-	-
14	Hydraulic pump relay display input	Ignition on	0 volts
		Pump relay actuated	Nbv
15	Alternator voltage input	Engine running	13.5 to 15.0 volts
16	-	-	-
17	Hydraulic pump relay supply	Ignition on	Nbv
18	Rear left solenoid valve	Engine running:	
		Inactive	Nbv
		Actuated	1.0 volt (max)
19	Rear right solenoid valve	Engine running:	
		Inactive	Nbv
		Actuated	1.0 volt (max)
20	ECM earth	Ignition on	0.25 volts (max)
21	Front right wheel speed sensor signal	Roadwheel rotating	0.5 to 1.5 volts AC (approx)
22	-	-	-
23	-	-	-
24	Rear right wheel speed sensor earth	Roadwheel rotating	0.25 volts (max)
25	Brake light switch input	Ignition on:	
		Brake pedal released	0 volts
		Brake pedal depressed	Nbv
26	Rear right wheel speed sensor signal	Roadwheel rotating	0.5 to 1.5 volts AC (approx)
27	Solenoid valve relay driver	Ignition on/inactive	Nbv
		Actuated	1.0 volt (max)
28	Hydraulic pump relay driver	Ignition on/inactive	Nbv
		Actuated	1.0 volt (max)
29	ABS warning lamp	Ignition on	1.0 volt (max)
		Engine running:	
		Lamp on, faults present	1.0 volt (max)
		Lamp off, no faults present	Nbv
30	-	-	-
31	SD connector	-	-
32	Solenoid valve relay display input	Ignition on/inactive	0 volts
		Actuated	Nbv
33	-	-	-
34	ECM earth	Ignition on	0.25 volts (max)
35	Front right solenoid valve	Engine running:	
		Inactive	Nbv
		Actuated	1.0 volt (max)

8A

Pin table - typical Bosch 2E, 35-pin (BMW 5/7-Series)

Note: *Refer to illustration 8A.27*

Pin No.	Connection	Test condition	Voltage
1	Supply to ECM from relay	Ignition on/engine running	Nbv
2	Front left solenoid valve	Engine running:	
		Inactive	Nbv
		Actuated	1.0 volt (max)
3	-	-	-
4	Front left wheel speed sensor earth	Roadwheel rotating	0.25 volts (max)
5	-	-	-
6	Front left wheel speed sensor signal	Roadwheel rotating	0.5 to 1.5 volts AC (approx)
7	-	-	-
8*	Rear left wheel speed sensor signal	Roadwheel rotating	0.5 to 1.5 volts AC (approx)
9	Rear left wheel speed sensor earth	Roadwheel rotating	0.25 volts (max)
10	-	-	-
11	Front right wheel speed sensor signal	Roadwheel rotating	0.5 to 1.5 volts AC (approx)
12	-	-	-
13	-	-	-
14	Hydraulic pump relay display input	Ignition on	0 volts
		Pump relay actuated	Nbv
15	Alternator voltage input	Engine running	13.5 to 15.0 volts
16	-	-	-
17	Hydraulic pump relay supply	Ignition on	Nbv
18	Rear/rear left solenoid valve	Engine running:	
		Inactive	Nbv
		Actuated	1.0 volt (max)
19**	Rear right solenoid valve	Engine running:	
		Inactive	Nbv
		Actuated	1.0 volt (max)
20	ECM earth	Ignition on	0.25 volts (max)
21	Front right wheel speed sensor earth	Roadwheel rotating	0.25 volts (max)
22	-	-	-
23	-	-	-
24	Rear right wheel speed sensor signal	Roadwheel rotating	0.5 to 1.5 volts AC (approx)
25	Brake light switch input	Ignition on:	
		Brake pedal released	0 volts
		Brake pedal depressed	Nbv
26	Rear right wheel speed sensor earth	Roadwheel rotating	0.25 volts (max)
27	Solenoid valve relay driver	Ignition on/inactive	Nbv
		Actuated	1.0 volt (max)
28	Hydraulic pump relay driver	Ignition on/inactive	Nbv
		Actuated	1.0 volt (max)
29	ABS warning lamp	Ignition on	1.0 volt (max)
		Engine running:	
		Lamp on, faults present	1.0 volt (max)
		Lamp off, no faults present	Nbv
30	SD connector	-	-
31	SD connector	-	-
32	Solenoid valve relay display input	Ignition on/inactive	0 volts
		Actuated	Nbv
33	-	-	-
34	ECM earth	Ignition on	0.25 volts (max)
35	Front right solenoid valve	Engine running:	
		Inactive	Nbv
		Actuated	1.0 volt (max)

*Pin 7 on 7-Series

**7-Series only

Pin table - typical Bosch 2E, 35-pin (Mercedes-Benz)

Note: *Refer to illustration 8A.27*

Pin No.	Connection	Test condition	Voltage
1	Supply to ECM from relay	Ignition on/engine running	Nbv
2	Front left solenoid valve	Engine running:	
		Inactive	Nbv
		Actuated	1.0 volt (max)
3	-	-	-
4	Front left wheel speed sensor earth	Roadwheel rotating	0.25 volts (max)
5	Instrument panel connection	-	-
6	Front left wheel speed sensor signal	Roadwheel rotating	0.5 to 1.5 volts AC (approx)
7	Rear wheel speed sensor signal	Roadwheel rotating	0.5 to 1.5 volts AC (approx)
8	-	-	-
9	Rear wheel speed sensor earth	Roadwheel rotating	0.25 volts (max)
10	-	-	-
11	-	-	-
12	-	-	-
13	-	-	-
14	Hydraulic pump relay display input	Ignition on	0 volts
		Pump relay actuated	Nbv
15	Alternator voltage input	Engine running	13.5 to 15.0 volts
16	-	-	-
17	Hydraulic pump relay supply	Ignition on	Nbv
18	Rear solenoid valve	Engine running:	
		Inactive	Nbv
		Actuated	1.0 volt (max)
19	-	-	-
20	ECM earth	Ignition on	0.25 volts (max)
21	Front right wheel speed sensor earth	Roadwheel rotating	0.25 volts (max)
22	-	-	-
23	Front right wheel speed sensor signal	Roadwheel rotating	0.5 to 1.5 volts AC (approx)
24	-	-	-
25	Brake light switch input	Ignition on:	
		Brake pedal released	0 volts
		Brake pedal depressed	Nbv
26	-	-	-
27	Solenoid valve relay driver	Ignition on/inactive	Nbv
		Actuated	1.0 volt (max)
28	Hydraulic pump relay driver	Ignition on/inactive	Nbv
		Actuated	1.0 volt (max)
29	ABS warning lamp	Ignition on	1.0 volt (max)
		Engine running:	
		Lamp on, faults present	1.0 volt (max)
		Lamp off, no faults present	Nbv
30	SD connector	-	-
31	-	-	-
32	Solenoid valve relay display input	Ignition on/inactive	0 volts
		Actuated	Nbv
33	-	-	-
34	ECM earth	Ignition on	0.25 volts (max)
35	Front right solenoid valve	Engine running:	
		Inactive	Nbv
		Actuated	1.0 volt (max)

8A

Pin table - typical Bosch 2E, 35-pin (Vauxhall/Opel)

Note: *Refer to illustration 8A.27*

Pin No.	Connection	Test condition	Voltage
1	Supply to ECM from relay	Ignition on	Nbv
2	Front left solenoid valve	Engine running:	
		Inactive	Nbv
		Actuated	1.0 volt (max)
3	-	-	-
4	Front left wheel speed sensor signal	Roadwheel rotating	0.5 to 1.5 volts AC (approx)
5	Front left wheel speed sensor earth	Roadwheel rotating	0.25 volts (max)
6	-	-	-
7	-	-	-
8	Rear left wheel speed sensor signal	Roadwheel rotating	0.5 to 1.5 volts AC (approx)
9	Rear left wheel speed sensor earth	Roadwheel rotating	0.25 volts (max)
10	ECM earth	Ignition on	0.25 volts (max)
11	Front right wheel speed sensor signal	Roadwheel rotating	0.5 to 1.5 volts AC (approx)
12	-	-	-
13	-	-	-
14	Hydraulic pump relay display input	Ignition on	0 volts
		Pump relay actuated	Nbv
15	-	-	-
16	-	-	-
17	Relay supply	Ignition on	Nbv
18	Rear solenoid valve	Engine running:	
		Inactive	Nbv
		Actuated	1.0 volt (max)
19	-	-	-
20	ECM earth	Ignition on	0.25 volts (max)
21	-	-	-
22	-	-	-
23	-	-	-
24	Rear right wheel speed sensor signal	Roadwheel rotating	0.5 to 1.5 volts AC (approx)
25	Brake light switch input	Ignition on:	
		Brake pedal released	0 volts
		Brake pedal depressed	Nbv
26	Rear right wheel speed sensor earth	Roadwheel rotating	0.25 volts (max)
27	Solenoid valve relay driver	Ignition on/inactive	Nbv
		Actuated	1.0 volt (max)
28	Hydraulic pump relay driver	Ignition on/inactive	Nbv
		Actuated	1.0 volt (max)
29	ABS warning lamp	Ignition on	1.0 volt (max)
		Engine running:	
		Lamp on, faults present	1.0 volt (max)
		Lamp off, no faults present	Nbv
30	SD connector	-	-
31	SD connector	-	-
32	Solenoid valve relay display input	Ignition on/inactive	0 volts
		Actuated	Nbv
33	-	-	-
34	ECM earth	Ignition on	0.25 volts (max)
35	Front right solenoid valve	Engine running:	
		Inactive	Nbv
		Actuated	1.0 volt (max)

Pin table - typical Bosch 2E, 35-pin (Volkswagen)

Note: *Refer to illustration 8A.27*

Pin No.	Connection	Test condition	Voltage
1	Supply to ECM from ignition switch	Ignition on	Nbv
2	Front left solenoid valve	Engine running:	
		Inactive	Nbv
		Actuated	1.0 volt (max)
3	-	-	-
4	Front left wheel speed sensor earth	Roadwheel rotating	0.25 volts (max)
5	Front left wheel speed sensor signal	Roadwheel rotating	0.5 to 1.5 volts AC (approx)
6	-	-	-
7	Rear left wheel speed sensor signal	Roadwheel rotating	0.5 to 1.5 volts AC (approx)
8	-	-	-
9	Rear left wheel speed sensor earth	Roadwheel rotating	0.25 volts (max)
10	-	-	-
11	Front right wheel speed sensor signal	Roadwheel rotating	0.5 to 1.5 volts AC (approx)
12	-	-	-
13	-	-	-
14	Hydraulic pump relay display input	Ignition on	0 volts
		Pump relay actuated	Nbv
15	-	-	-
16	-	-	-
17	Relay supply	Ignition on	Nbv
18	Rear left solenoid valve	Engine running:	
		Inactive	Nbv
		Actuated	1.0 volt (max)
19	Rear right solenoid valve	Engine running:	
		Inactive	Nbv
		Actuated	1.0 volt (max)
20	ECM earth	Ignition on	0.25 volts (max)
21	Front right wheel speed sensor earth	Roadwheel rotating	0.25 volts (max)
22	-	-	-
23	-	-	-
24	Rear right wheel speed sensor signal	Roadwheel rotating	0.5 to 1.5 volts AC (approx)
25	Brake light switch input	Ignition on:	
		Brake pedal released	0 volts
		Brake pedal depressed	Nbv
26	Rear right wheel speed sensor earth	Roadwheel rotating	0.25 volts (max)
27	Solenoid valve relay driver	Ignition on/inactive	Nbv
		Actuated	1.0 volt (max)
28	Hydraulic pump relay driver	Ignition on/inactive	Nbv
		Actuated	1.0 volt (max)
29	ABS warning lamp	Ignition on	1.0 volt (max)
		Engine running:	
		Lamp on, faults present	1.0 volt (max)
		Lamp off, no faults present	Nbv
30	SD connector	-	-
31	SD connector	-	-
32	Solenoid valve relay display input	Ignition on/inactive	0 volts
		Actuated	Nbv
33	-	-	-
34	ECM earth	Ignition on	0.25 volts (max)
35	Front right solenoid valve	Engine running:	
		Inactive	Nbv
		Actuated	1.0 volt (max)

8A

Pin table - typical Bosch 2E, 35-pin (Volvo)

Note: *Refer to illustration 8A.27*

Pin No.	Connection	Test condition	Voltage
1	Supply to ECM from ignition switch	Ignition on	Nbv
2	Front left solenoid valve	Engine running:	
		Inactive	Nbv
		Actuated	1.0 volt (max)
3	-	-	-
4	Front left wheel speed sensor earth	Roadwheel rotating	0.25 volts (max)
5	-	-	-
6	Front left wheel speed sensor signal	Roadwheel rotating	0.5 to 1.5 volts AC (approx)
7	Rear wheel speed sensor signal	Roadwheel rotating	0.5 to 1.5 volts AC (approx)
8	-	-	-
9	Rear wheel speed sensor earth	Roadwheel rotating	0.25 volts (max)
10	-	-	-
11	Front right wheel speed sensor signal	Roadwheel rotating	0.5 to 1.5 volts AC (approx)
12	-	-	-
13	-	-	-
14	Hydraulic pump relay display input	Ignition on	0 volts
		Pump relay actuated	Nbv
15	Alternator voltage input	Engine running	13.5 to 15.0 volts
16	-	-	-
17	Relay supply	Ignition on	Nbv
18	Rear solenoid valve	Engine running:	
		Inactive	Nbv
		Actuated	1.0 volt (max)
19	-	-	-
20	ECM earth	Ignition on	0.25 volts (max)
21	Front right wheel speed sensor earth	Roadwheel rotating	0.25 volts (max)
22	-	-	-
23	-	-	-
24	-	-	-
25	Brake light switch input	Ignition on:	
		Brake pedal released	0 volts
		Brake pedal depressed	Nbv
26	-	-	-
27	Solenoid valve relay driver	Ignition on/inactive	Nbv
		Actuated	1.0 volt (max)
28	Hydraulic pump relay driver	Ignition on/inactive	Nbv
		Actuated	1.0 volt (max)
29	ABS warning lamp	Ignition on	1.0 volt (max)
		Engine running:	
		Lamp on, faults present	1.0 volt (max)
		Lamp off, no faults present	Nbv
30	-		
31	SD connector	-	-
32	Solenoid valve relay display input	Ignition on/inactive	0 volts
		Actuated	Nbv
33	-	-	-
34	ECM earth	Ignition on	0.25 volts (max)
35	Front right solenoid valve	Engine running:	
		Inactive	Nbv
		Actuated	1.0 volt (max)

Pin table - typical Bosch 2SE, 35-pin (Peugeot, Renault)

Note: *Refer to illustration 8A.27*

Pin No.	Connection	Test condition	Voltage
1	Supply to ECM from relay	Ignition on	Nbv
2	Front left solenoid valve	Engine running:	
		Inactive	Nbv
		Actuated	1.0 volt (max)
3	-	-	-
4	Front left wheel speed sensor signal	Roadwheel rotating	0.5 to 1.5 volts AC (approx)
5	Front left wheel speed sensor earth	Roadwheel rotating	0.25 volts (max)
6	-	-	-
7*	Rear left wheel speed sensor signal	Roadwheel rotating	0.5 to 1.5 volts AC (approx)
8**	Rear left wheel speed sensor signal	Roadwheel rotating	0.5 to 1.5 volts AC (approx)
9	Rear left wheel speed sensor earth	Roadwheel rotating	0.25 volts (max)
10	ECM earth	Ignition on	0.25 volts (max)
11	Front right wheel speed sensor signal	Roadwheel rotating	0.5 to 1.5 volts AC (approx)
12	-	-	-
13	-	-	-
14	Hydraulic pump relay display input	Ignition on	0 volts
		Pump relay actuated	Nbv
15*	Engine oil pressure switch	-	-
16	-	-	-
17	Relay supply	Ignition on	Nbv
18	Rear left solenoid valve	Engine running:	
		Inactive	Nbv
		Actuated	1.0 volt (max)
19	Rear right solenoid valve	Engine running:	
		Inactive	Nbv
		Actuated	1.0 volt (max)
20	ECM earth	Ignition on	0.25 volts (max)
21	Front right wheel speed sensor earth	Roadwheel rotating	0.5 to 1.5 volts AC (approx)
22	-	-	-
23	-	-	-
24	Rear right wheel speed sensor signal	Roadwheel rotating	0.5 to 1.5 volts AC (approx)
25	Brake light switch input	Ignition on:	
		Brake pedal released	0 volts
		Brake pedal depressed	Nbv
26	Rear right wheel speed sensor earth	Roadwheel rotating	0.25 volts (max)
27	Solenoid valve relay driver	Ignition on/inactive	Nbv
		Actuated	1.0 volt (max)
28	Hydraulic pump relay driver	Ignition on/inactive	Nbv
		Actuated	1.0 volt (max)
29	ABS warning lamp	Ignition on	1.0 volt (max)
		Engine running:	
		Lamp on, faults present	1.0 volt (max)
		Lamp off, no faults present	Nbv
30	-	-	-
31	SD connector	-	-
32	Solenoid valve relay display input	Ignition on/inactive	0 volts
		Actuated	Nbv
33	-	-	-
34	ECM earth	Ignition on	0.25 volts (max)
35	Front right solenoid valve	Engine running:	
		Inactive	Nbv
		Actuated	1.0 volt (max)

*Renault only
**Peugeot only

8A

Fault codes

11 General fault codes

1 With the exception of Vauxhall and Volvo vehicles, the Bosch 2E and 2SE systems that incorporate self-diagnostics require the use of a proprietary fault code reader (FCR) or system tester to obtain fault codes. Flash codes are not available for output from this system in these applications.

2 If a FCR is available, it should be connected to the SD serial connector and used in accordance with the maker's instructions.

3 The FCR can be used for the following purposes:
  a) Obtaining fault codes.
  b) Clearing fault codes.
  c) Obtaining datastream information.
  d) Testing the system actuators (relays and solenoid valves).

12 Vauxhall fault codes

1 If a FCR is available, it can be connected to the SD serial connector (the Vauxhall/Opel term for SD connector is ALDL) and used as described above to obtain fault codes. If a FCR is not available, it is still possible to obtain fault codes, which will be displayed as a series of flashes of the ABS warning lamp.

Obtaining codes without a FCR

2 Locate the 10-pin diagnostic connector, which is typically situated on the left-hand side of the engine compartment.

3 Use a jumper lead to bridge terminals A and K in the diagnostic connector **(see illustration 8A.30)**.

4 Switch the ignition on and the ECM will enter diagnostic mode.

5 The fault codes will now be output in numerical sequence as a series of flashes of the ABS warning lamp on the instrument panel. By counting the flashes and referring to the fault code table, faults can thus be determined.

Clearing fault codes from the ECM memory

6 When the fault code output is complete the codes must be erased from the ECM memory. To do this, switch the ignition off and remove the bridging wire from the diagnostic connector. Switch the ignition on and off for a minimum of twenty times to erase the memory.

13 Volvo fault codes

1 On Volvo vehicles, it is possible to obtain fault codes by means of the on-board diagnostic unit. The codes are displayed as a series of flashes of the red LED on the diagnostic unit.

Obtaining codes using the on-board diagnostic unit

2 Ensure that the ignition is switched off, then remove the cover from the diagnostic unit located on the left-hand side of the engine compartment.

3 Unclip the flylead from the holder on the side of the unit and insert it into socket 3 of the unit **(see illustration 8A.31)**. The three-digit codes will be displayed as a series of blinks of the red LED (located on the top face of the unit, next to the test button) with a slight pause between each digit.

4 With the flylead inserted switch on the ignition. Press the test button on the top of the unit once, for about one second, then release it and wait for the LED to flash. As the LED flashes, copy down the fault code. Press the button again and copy down the next fault code, if there is one. Continue until the first fault code is displayed again indicating that all the stored codes have been accessed then switch off the ignition.

5 If code 1-1-1 is obtained, this indicates that there are no fault codes stored in the ECM memory and the system is operating correctly.

Clearing fault codes from the ECM memory

6 Once all the fault codes have been recorded they should be deleted from the ECM memory, using the diagnostic unit. Note that the codes cannot be deleted until all of them have been displayed at least once, and the first one is displayed again.

7 With the flylead still inserted in position 3 of the diagnostic unit, switch on the ignition, press the test button and hold it down for approximately five seconds. Release the test button and after three seconds the LED will light. When the LED lights, press and hold the test button down for a further five seconds then release it - the LED will go out.

8 Switch off the ignition and check that all the fault codes have been deleted by switching the ignition on again and pressing the test button for one second - code 1-1-1 should appear. If a code other than 1-1-1 appears, record the code then repeat the deleting procedure. When all the codes have been deleted, switch off the ignition, locate the flylead in its holder and refit the unit cover.

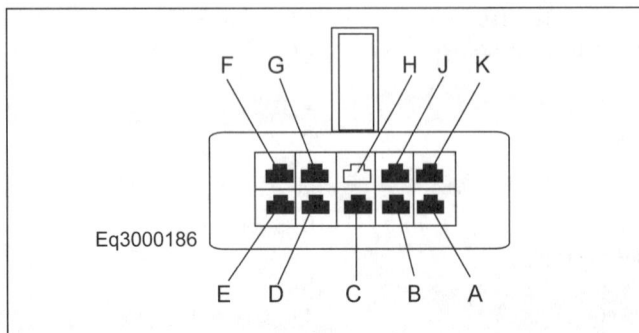

Eq3000186

8A.30 Vauxhall/Opel 10-pin diagnostic connector terminal identification

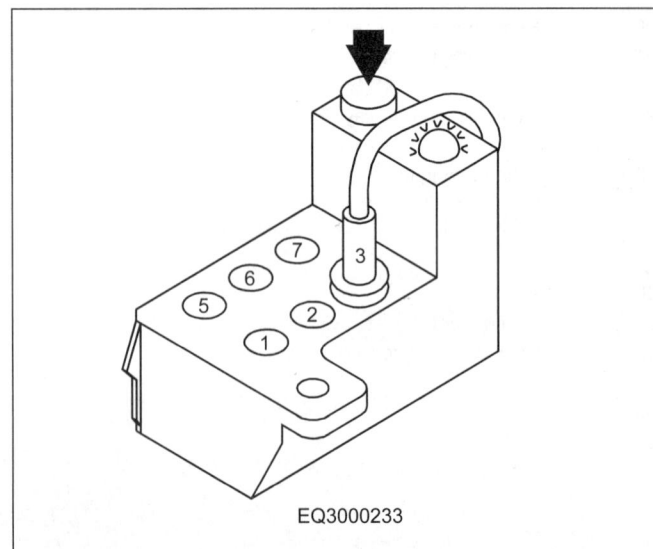

EQ3000233

8A.31 Volvo on-board diagnostic unit showing test button (arrowed) and flylead sockets

Fault code table (Audi, Volkswagen)

Code	Item	Fault
00000	No faults recognised	
00277	Front left solenoid valve	Break, short circuit or poor wiring connection; solenoid valve defective
00283	Front left wheel speed sensor	Break, short circuit or poor wiring connection; incorrect air gap; sensor or sensor ring dirty or damaged
00284	Front right solenoid valve	Break, short circuit or poor wiring connection; solenoid valve defective
00285	Front right wheel speed sensor	Break, short circuit or poor wiring connection; incorrect air gap; sensor or sensor ring dirty or damaged
00286	Rear left solenoid valve	Break, short circuit or poor wiring connection; solenoid valve defective
00287	Rear right wheel speed sensor	Break, short circuit or poor wiring connection; incorrect air gap; sensor or sensor ring dirty or damaged
00289	Rear right solenoid valve	Break, short circuit or poor wiring connection; solenoid valve defective
00290	Rear left wheel speed sensor	Break, short circuit or poor wiring connection; incorrect air gap; sensor or sensor ring dirty or damaged
00301	Hydraulic pump	Pump defective
00302	Solenoid valve relay	Defective (signal interruption, short to earth, supply voltage too low)
00526	Brake light switch	Poor wiring connection, switch defective or incorrectly adjusted
00532	Supply voltage	Outside expected values
00597	Wheel speed sensor signal variation	Incorrect wheel/tyre sizes, incorrect sensor air gap, sensor or sensor ring dirty or damaged
65535	ECM	ECM defective

8A

Fault code table (BMW)

Code	Item	Fault
4	Rear left wheel speed sensor	Wiring open or short circuit
5	Rear right wheel speed sensor	Wiring open or short circuit
6	Front right wheel speed sensor	Wiring open or short circuit
7	Front left wheel speed sensor	Wiring open or short circuit
8	Rear left solenoid valve	Wiring open or short circuit, valve inoperative
9	Rear right solenoid valve	Wiring open or short circuit, valve inoperative
10	Front right solenoid valve	Wiring open or short circuit, valve inoperative
11	Front left solenoid valve	Wiring open or short circuit, valve inoperative
14	Solenoid valve relay	Wiring open or short circuit, relay inoperative, ECM defective
15	Hydraulic pump motor	Relay inoperative, open or short circuit, faulty motor
21	ECM	Defective
24	Wheel speed sensors	Incorrect number of teeth on sensor ring
25	Brake light switch	Wiring open or short circuit

Fault code table (Mercedes-Benz)

Code	Item	Fault
002	Front left wheel speed sensor	Wiring open or short circuit, incorrect sensor resistance
003	Front right wheel speed sensor	Wiring open or short circuit, incorrect sensor resistance
004	Rear wheel speed sensor	Wiring open or short circuit, incorrect sensor resistance
006	Front left solenoid valve	Wiring open or short circuit, valve inoperative
007	Front right solenoid valve	Wiring open or short circuit, valve inoperative
008	Rear solenoid valve	Wiring open or short circuit, valve inoperative
010	Hydraulic pump motor	Relay inoperative, wiring open or short circuit, faulty motor
011	Solenoid valve relay	Wiring open or short circuit, relay inoperative, ECM defective
015	ECM	Incorrect supply voltage, ECM defective
016	Wheel speed sensors (all)	Signal variation, incorrect air gap, damaged sensor ring
017	Supply voltage	Voltage too low
025	Front left wheel speed sensor	Incorrect air gap or number of teeth on sensor ring
026	Front right wheel speed sensor	Incorrect air gap or number of teeth on sensor ring
027	Rear wheel speed sensor	Incorrect air gap or number of teeth on sensor ring

Fault code table (Peugeot)

Code	Item	Fault
12	Start of diagnosis	-
15	Solenoid valve relay	Defective (signal interruption, short to earth, supply voltage too low)
16	Brake light switch	Open circuit / poor connections
18	Toothed sensor ring	Wrong number of teeth
24	Rear left wheel speed sensor	Open circuit / incorrect resistance
25	Front right wheel speed sensor	Open circuit / incorrect resistance
31	Rear right wheel speed sensor	Open circuit / incorrect resistance
32	Front left wheel speed sensor	Open circuit / incorrect resistance
33	Rear left wheel speed sensor	No signal / incorrect signal
34	Front right wheel speed sensor	No signal / incorrect signal
35	Rear right wheel speed sensor	No signal / incorrect signal
41	Front left wheel speed sensor	No signal / incorrect signal
42	Front right solenoid valve	Defective (signal interruption, short to earth)
44	Front left solenoid valve	Defective (signal interruption, short to earth)
51	Rear right solenoid valve	Defective (signal interruption, short to earth)
52	Rear left solenoid valve	Defective (signal interruption, short to earth)
53	Pump relay / pump motor	Defective / poor connections
55	ECM	Defective
57	Battery voltage	Voltage too low
11	End of diagnosis	-

Fault code table (Vauxhall/Opel)

Code	Item	Fault
12	Start of diagnosis	
16	Front left solenoid valve	Defective (signal interruption, short to earth)
17	Front right solenoid valve	Defective (signal interruption, short to earth)
18	Rear solenoid valve	Defective (signal interruption, short to earth)
19	Solenoid valve relay	Defective (signal interruption, short to earth, supply voltage too low)
25	Wheel speed sensor toothed sensor ring	Wrong number of teeth
35	Pump motor	Defective (non-operation of pump after actuation of relay)
37	Brake light switch	Defective (signal interruption)
39	Front left wheel speed sensor	No/poor signal
41	Front left wheel speed sensor	Signal interrupted
42	Front right wheel speed sensor	No/poor signal
43	Front right wheel speed sensor	Signal interrupted
44	Rear left wheel speed sensor	No/poor signal
45	Rear left wheel speed sensor	Signal interrupted
46	Rear right wheel speed sensor	No/poor signal
47	Rear right wheel speed sensor	Signal interrupted
48	Battery voltage	Voltage too low
55	ECM	Defective

Fault code table (Volvo)

Code	Item	Fault
1-1-1	No fault detected	
1-2-5	Wheel speed sensors	Permanent signal loss
1-3-5	ECM	Internal circuit fault
1-4-2	Brake light switch	Open or short circuit
1-4-3	ECM	Memory fault
1-5-1	Front left wheel speed sensor	Open or short circuit
1-5-2	Front right wheel speed sensor	Open or short circuit
1-5-5	Rear wheel speed sensor	Open or short circuit
2-1-5	Solenoid valve relay	Open or short circuit
2-3-1	Front left wheel speed sensor	No signal
2-3-2	Front right wheel speed sensor	No signal
2-3-5	Rear wheel speed sensor	No signal
4-1-1	Front left inlet solenoid valve	Open or short circuit
4-1-3	Front right inlet solenoid valve	Open or short circuit
4-1-5	Rear solenoid valve	Open or short circuit
4-4-3	Hydraulic pump motor	Electrical or mechanical fault, relay fault

Wiring diagrams

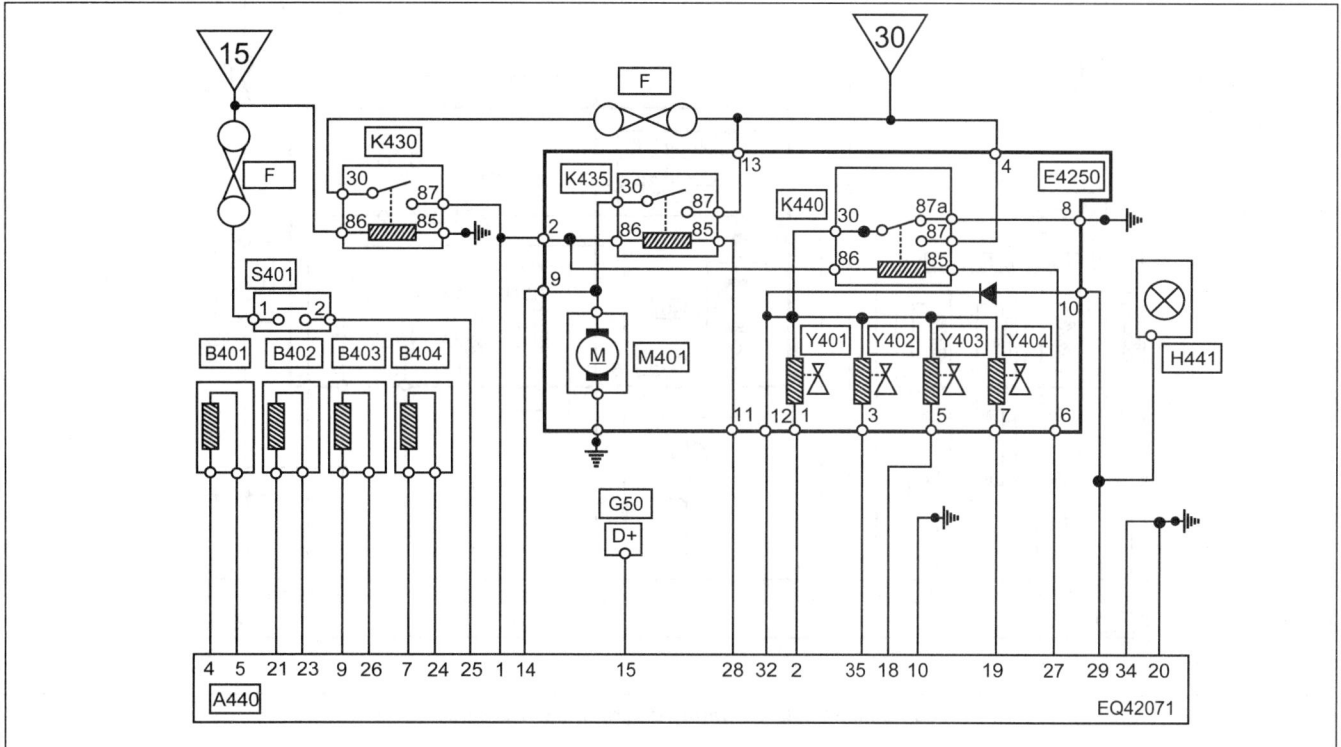

8A.32 Bosch 2S wiring diagram, Alfa Romeo 164

8A

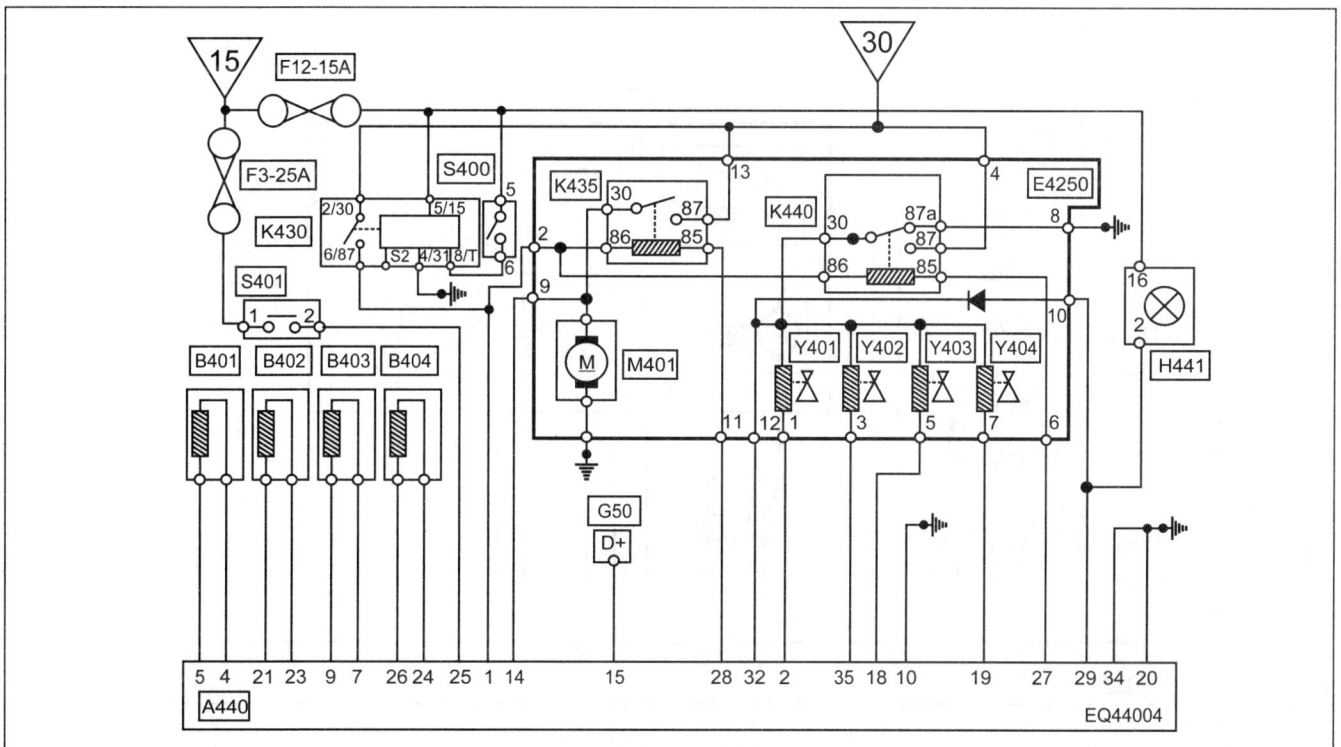

8A.33 Bosch 2S 35-pin wiring diagram, Audi 80/90/100/200/V8/Coupe

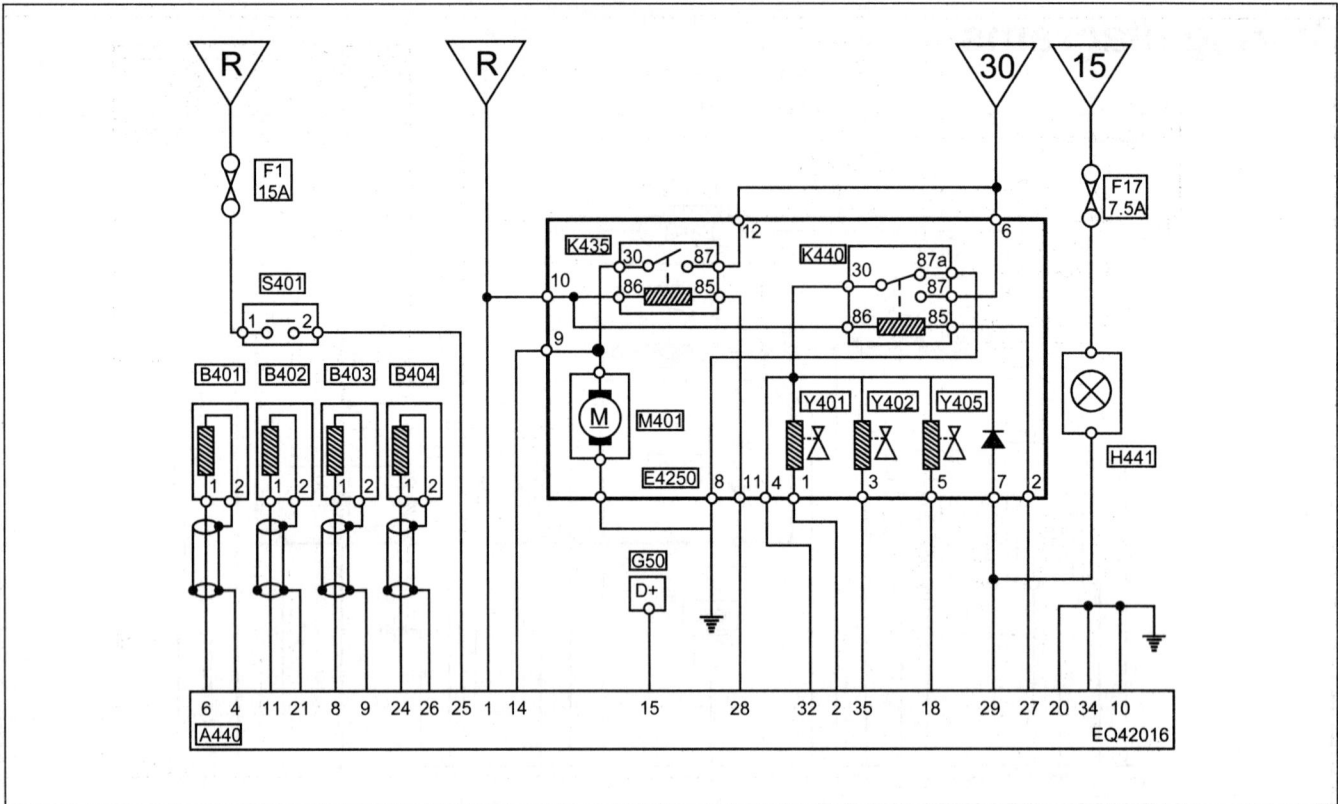

8A.34 Bosch 2S 35-pin wiring diagram, BMW 3-Series (E30)

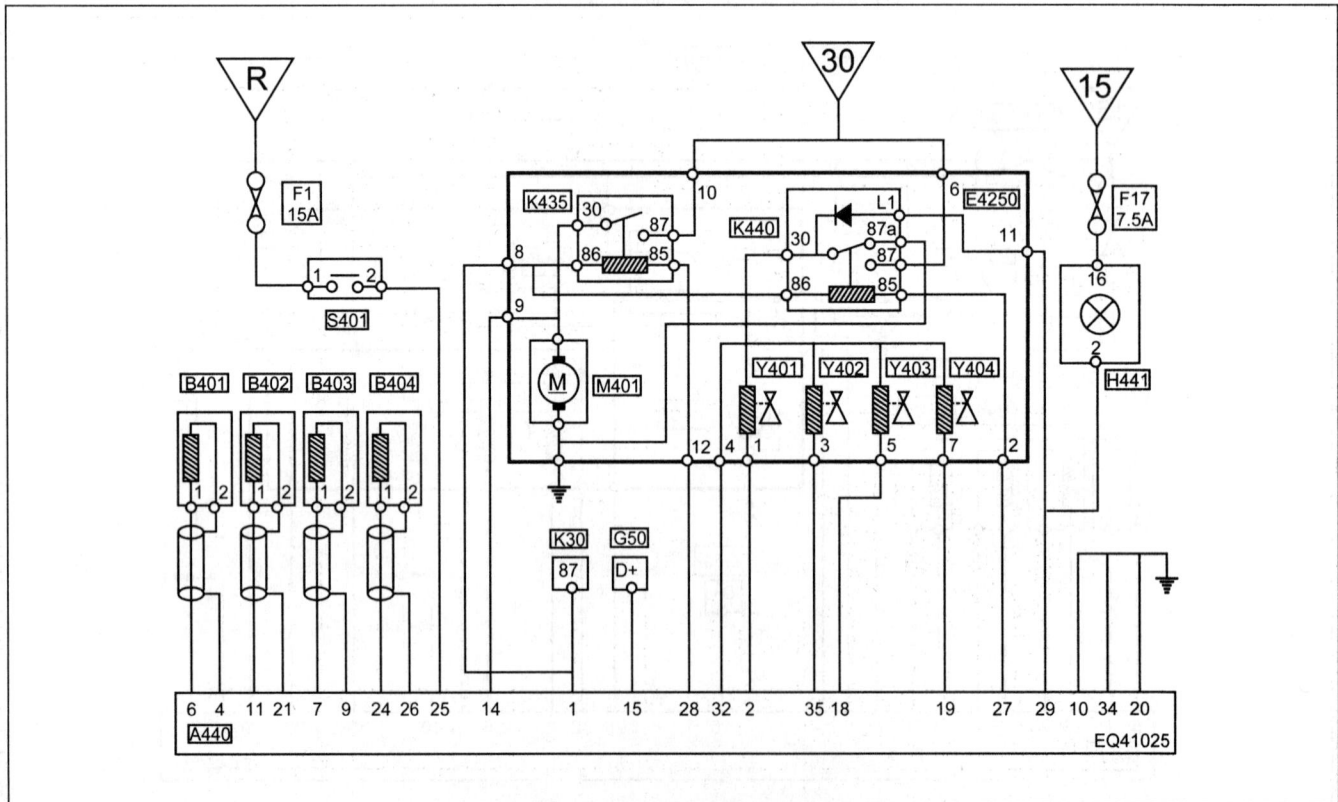

8A.35 Bosch 2S 35-pin wiring diagram, BMW 5-Series (E34)

8A

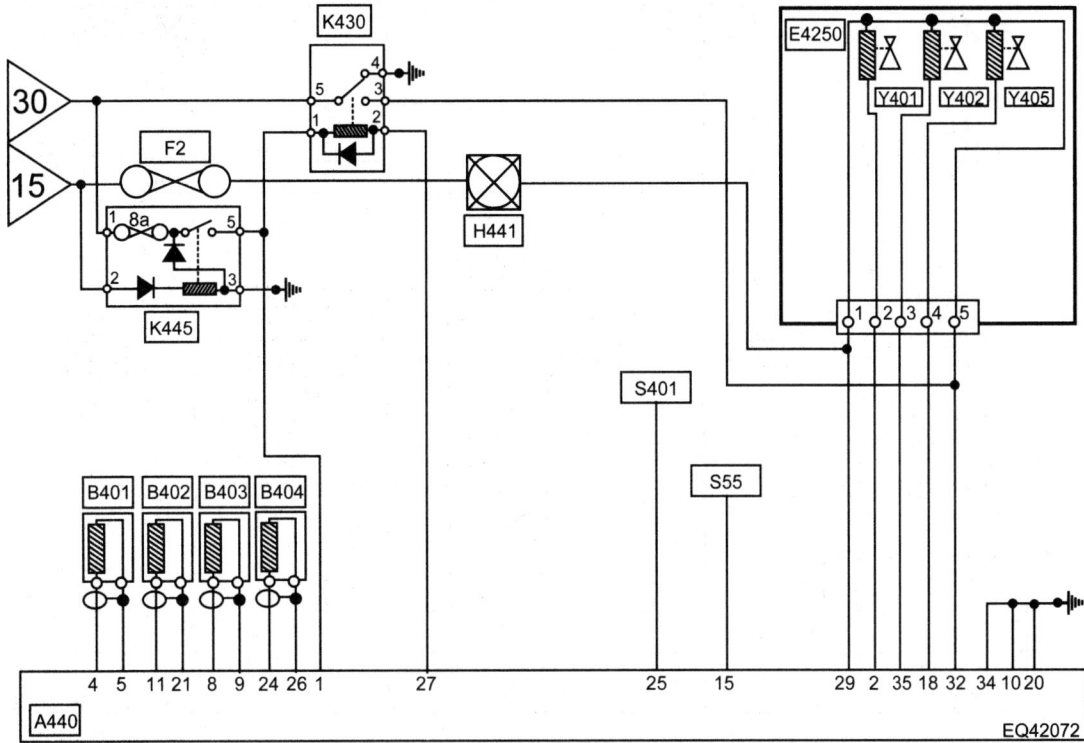

8A.36 Bosch 2S wiring diagram, Citroën CX

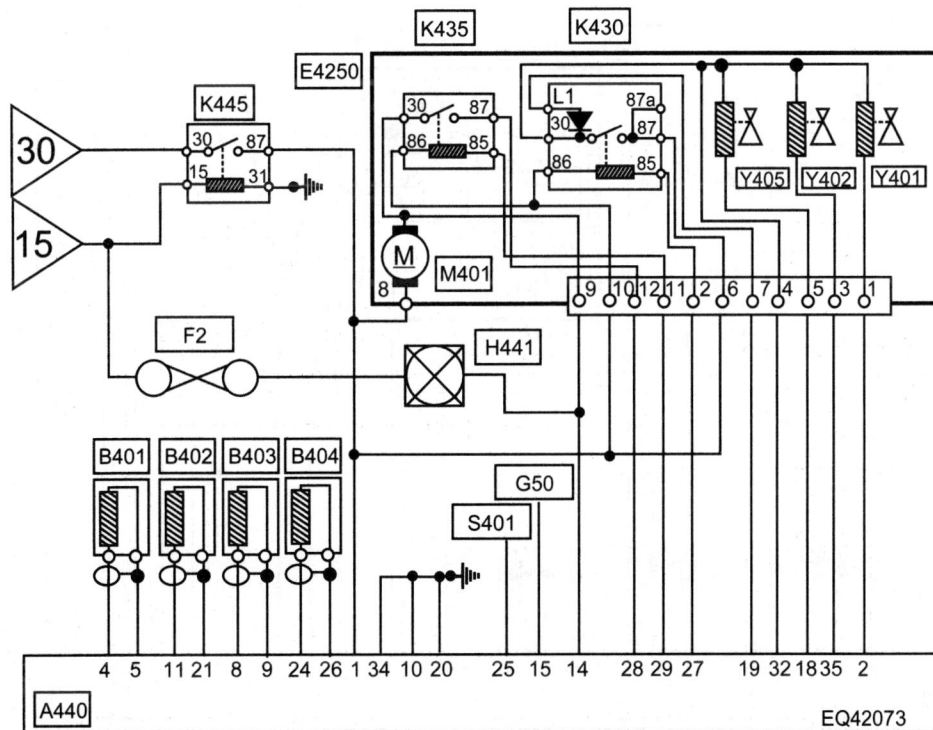

8A.37 Bosch 2S wiring diagram, Fiat Croma/Tempra (2.0/Diesel)/Tipo (2.0/Diesel)/Lancia Dedra/Thema

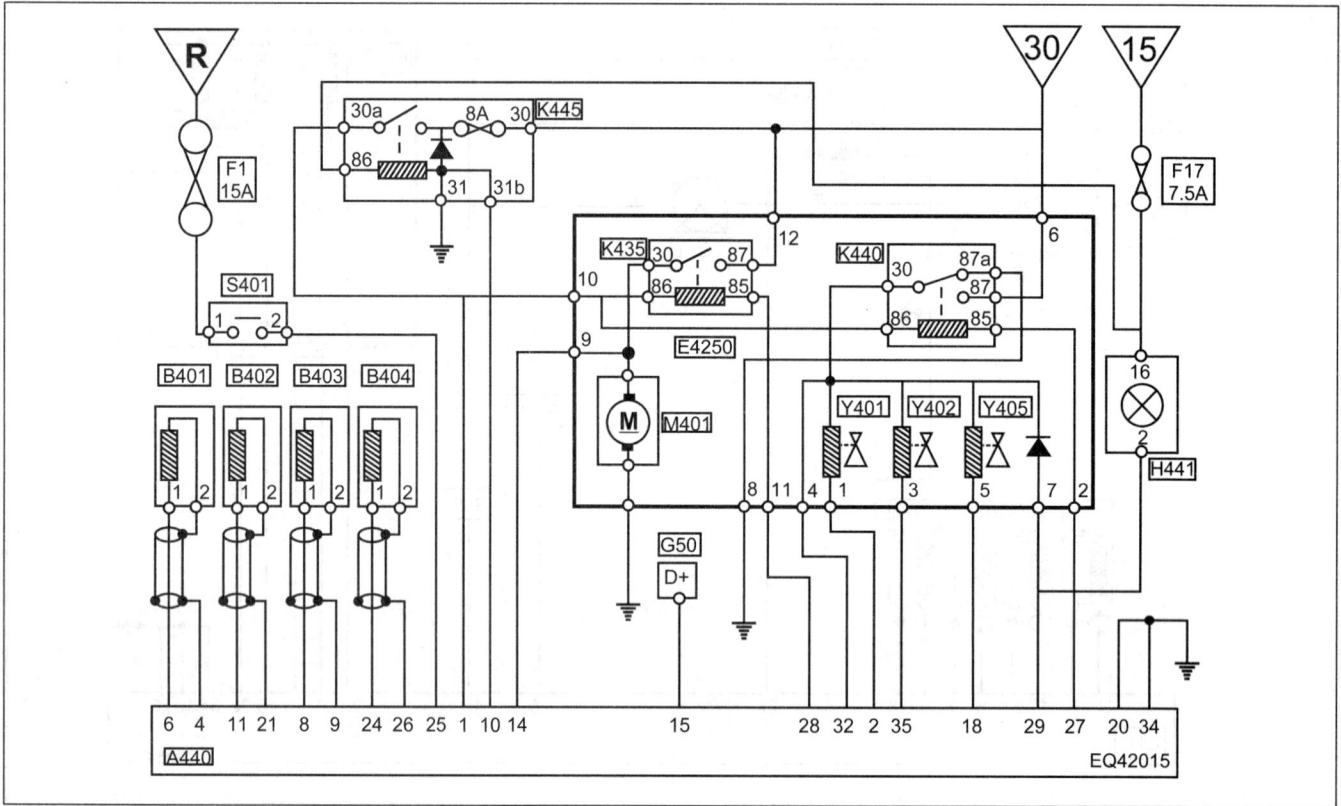

8A.38 Bosch 2S 35-pin wiring diagram, Jaguar/Daimler XJ6/Sovereign/XJ12 Sovereign HE

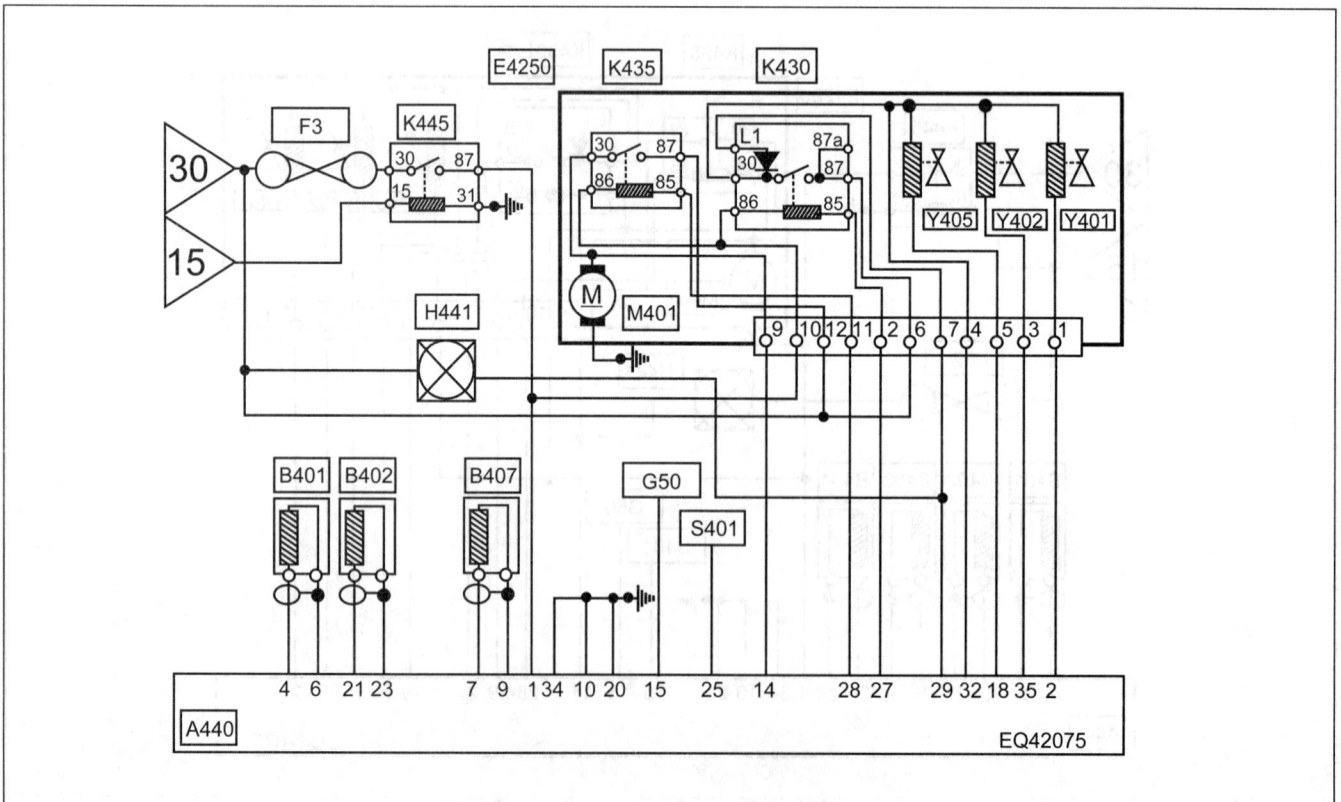

8A.39 Bosch 2S wiring diagram, Mercedes-Benz 190 (201)/E-Class (124)/S-Class (126)/SL (107)

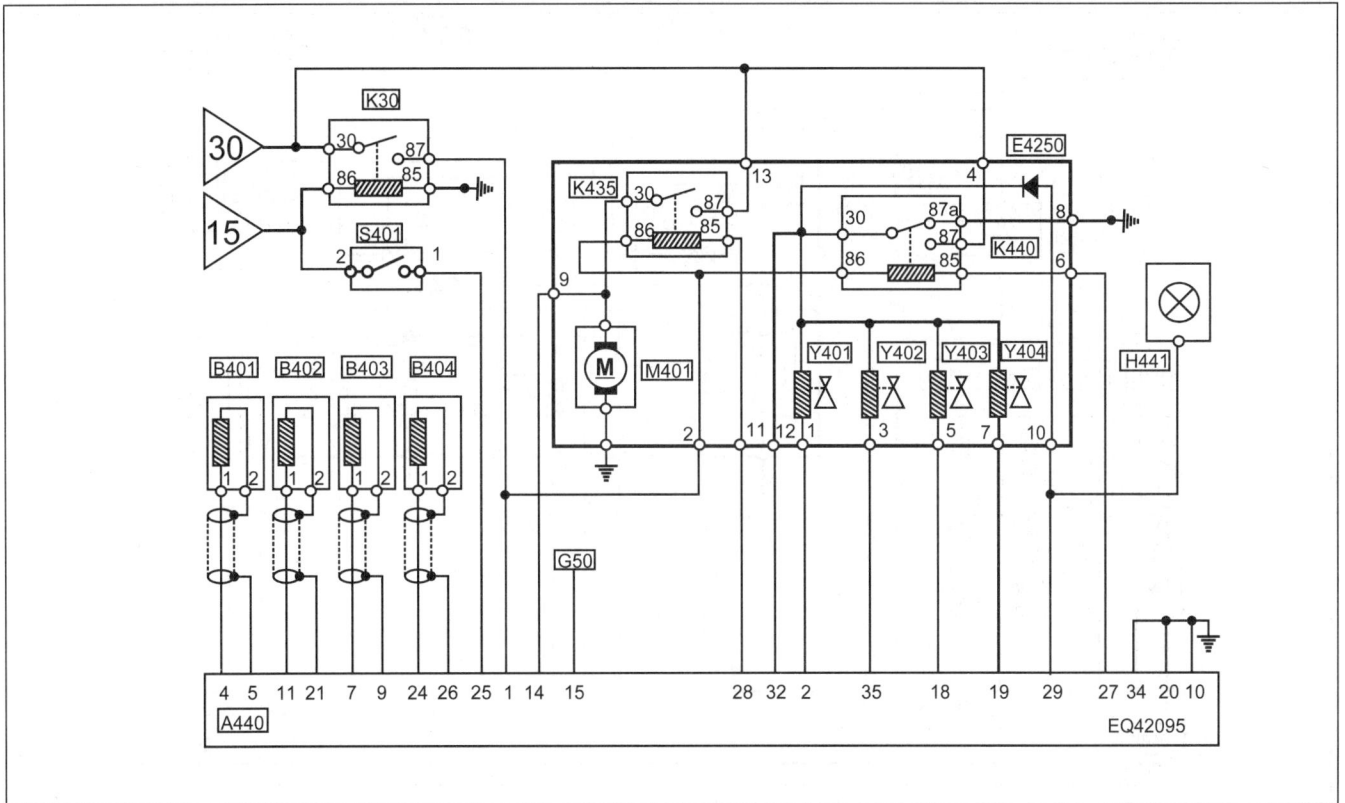

8A.40 Bosch 2S 35-pin wiring diagram, Vauxhall/Opel Carlton/Cavalier/Monza/Omega-A/Senator/Vectra-A

8A

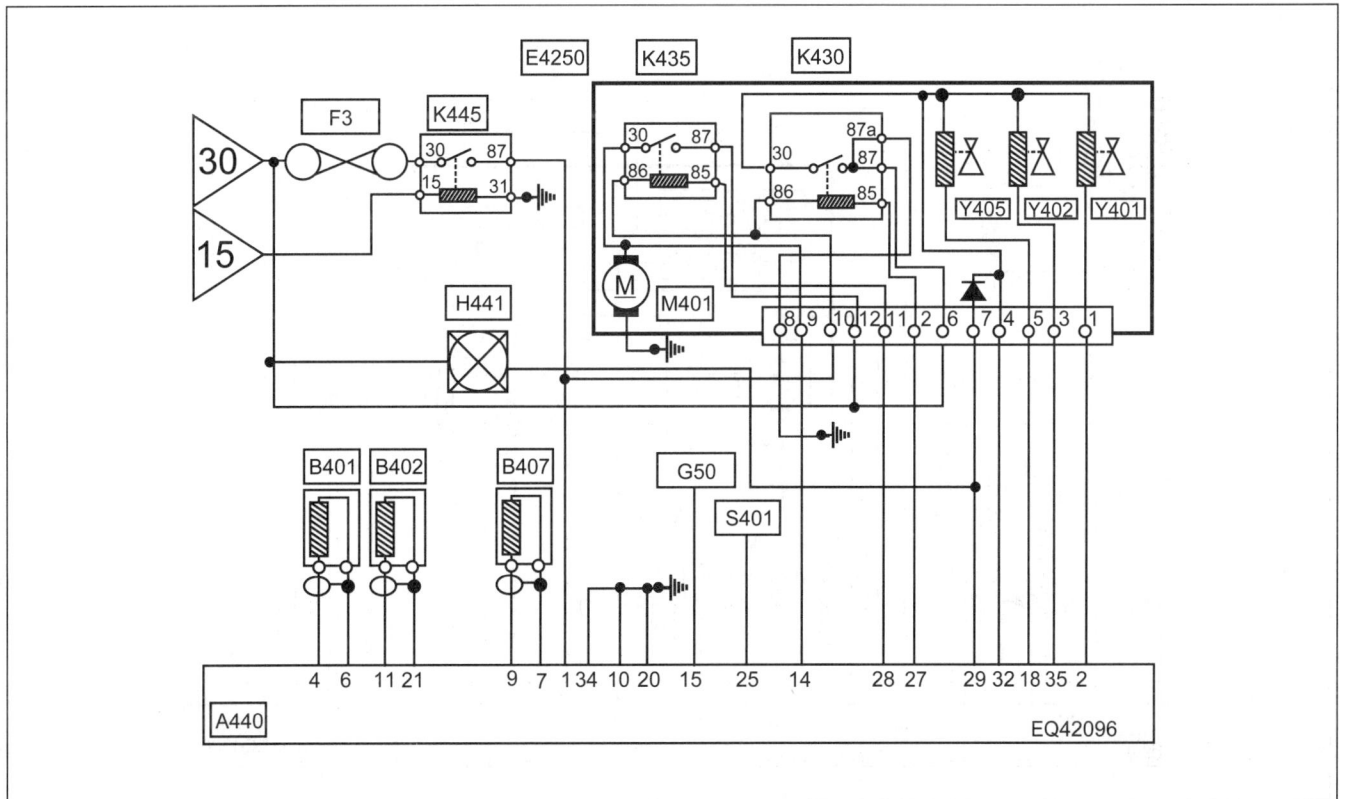

8A.41 Bosch 2S 35-pin wiring diagram, Volvo 240/740/760/940/960

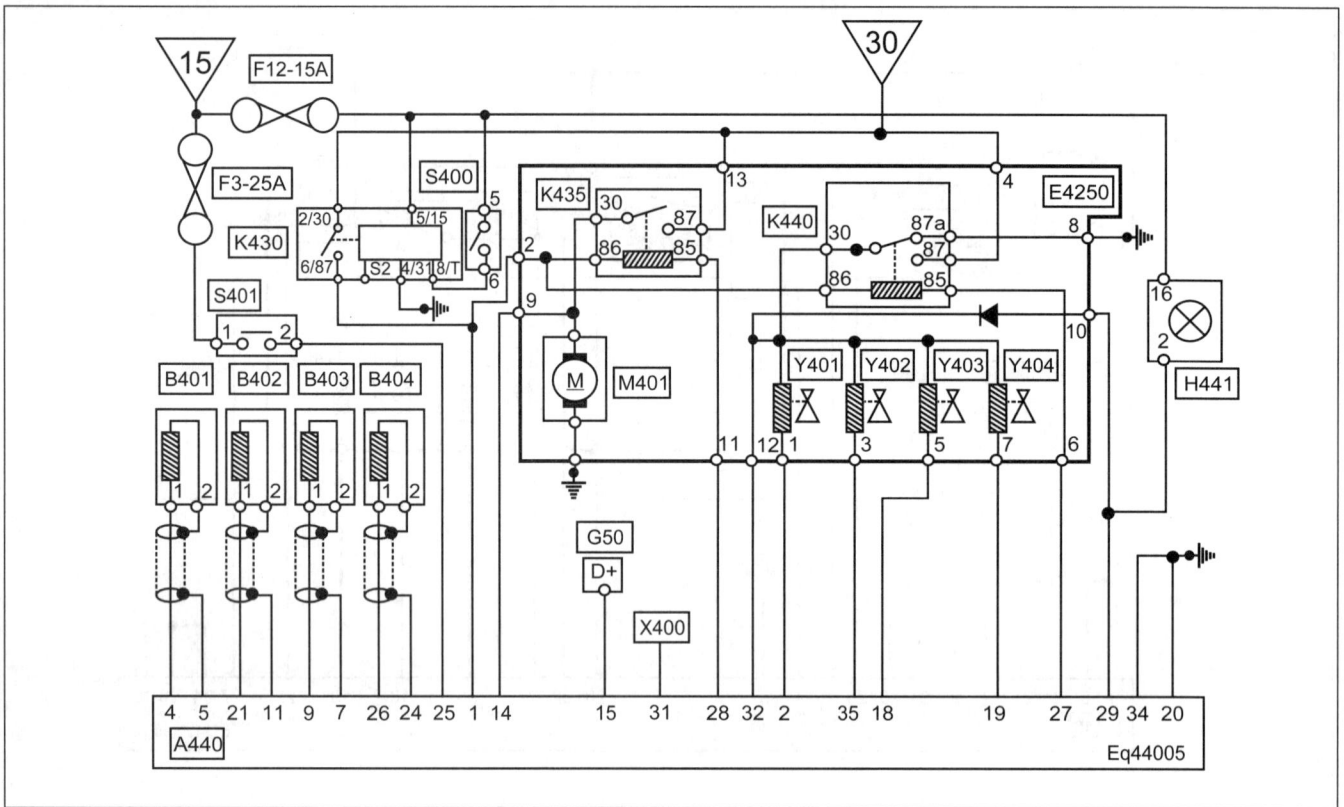

8A.42 Bosch 2E wiring diagram, Audi 80/100/V8

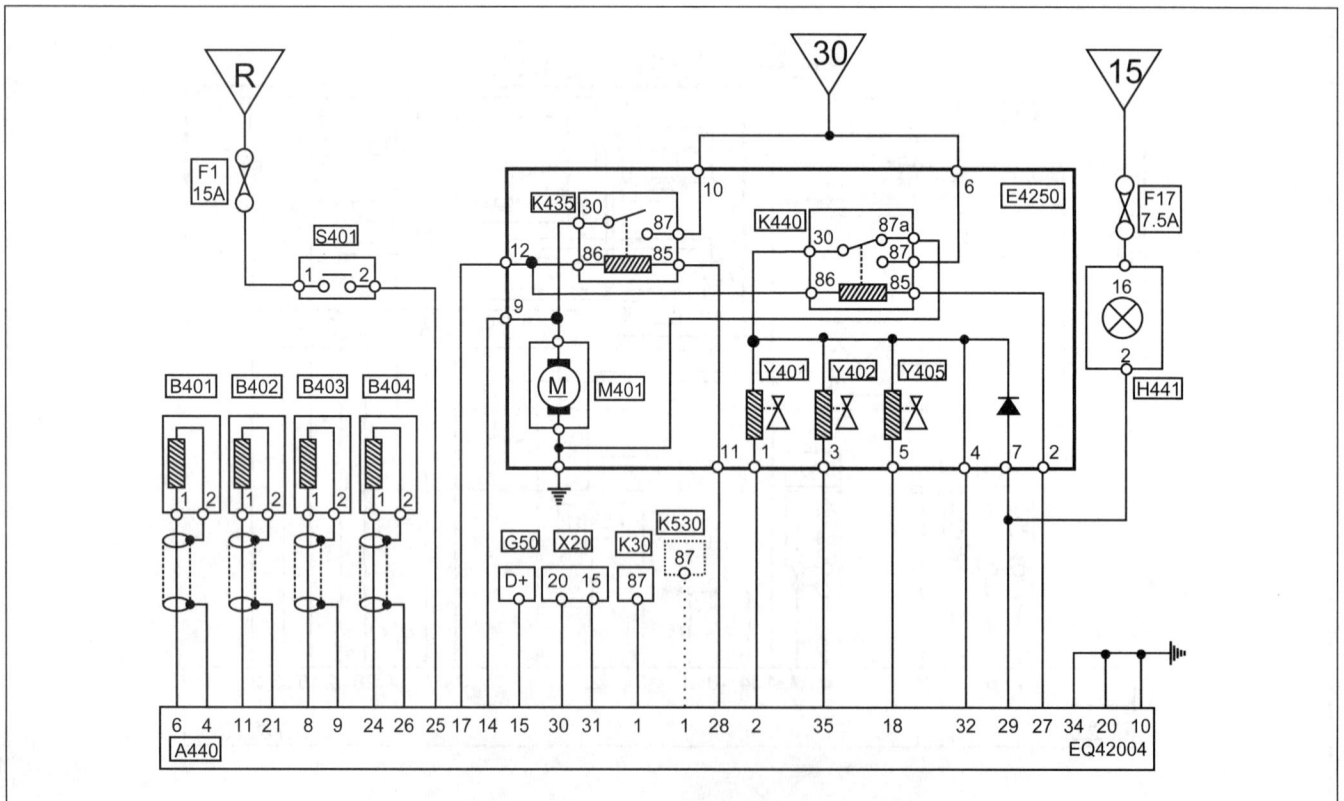

8A.43 Bosch 2E 35-pin wiring diagram, BMW 5-Series (E34)

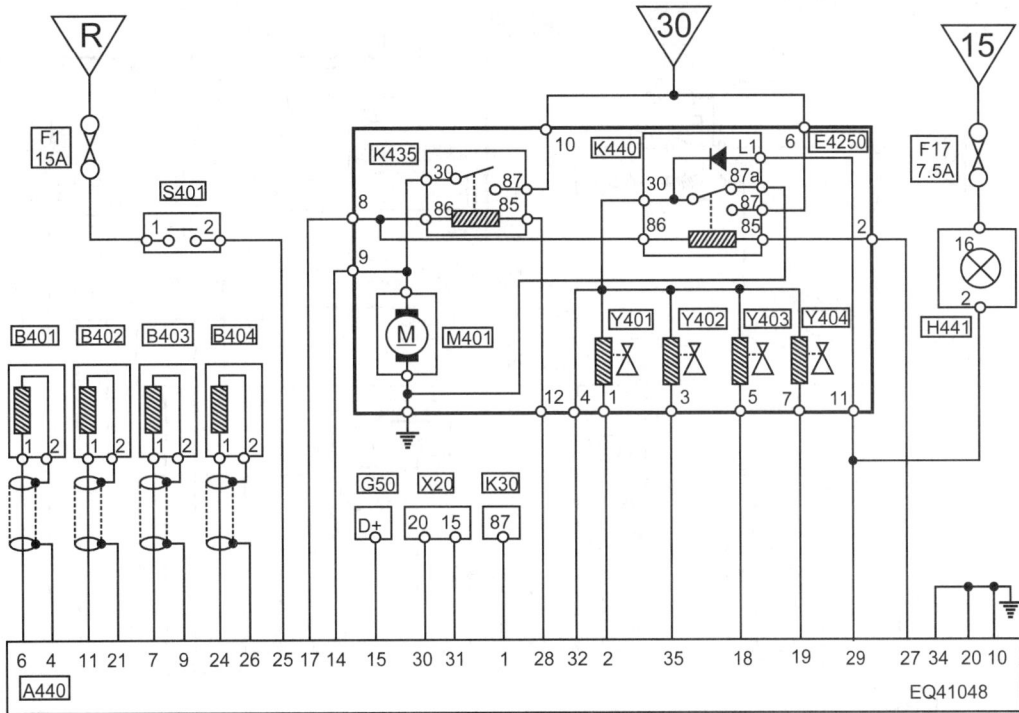

8A.44 Bosch 2E 35-pin wiring diagram, BMW 7-Series (E32)

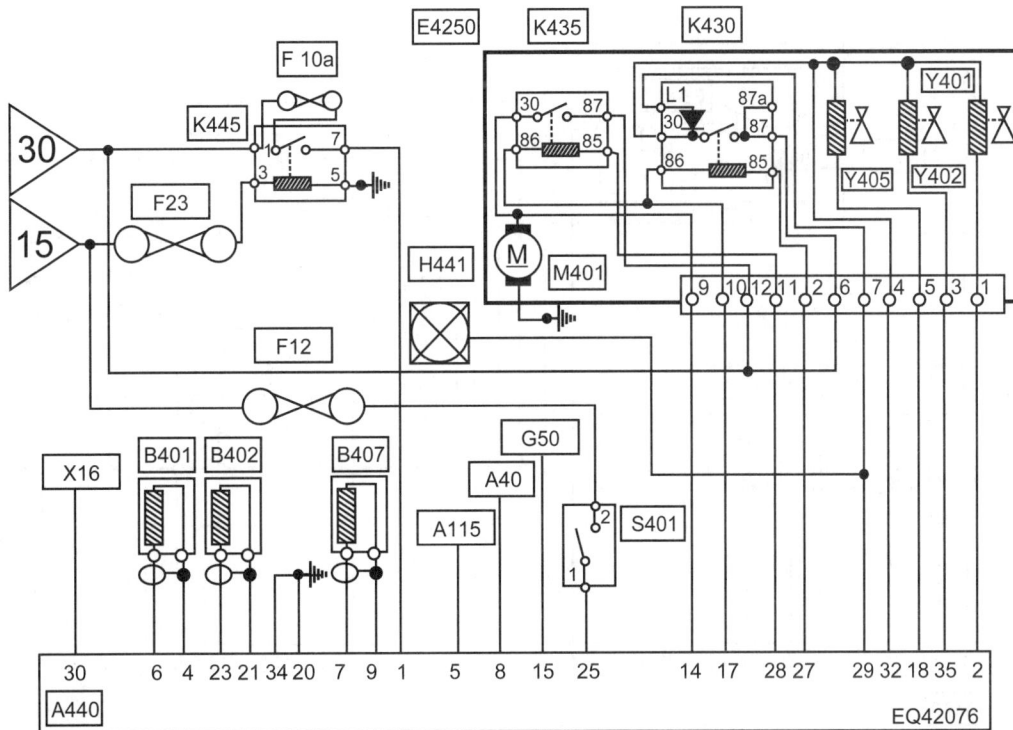

8A.45 Bosch 2E wiring diagram, Mercedes-Benz C-Class (202)

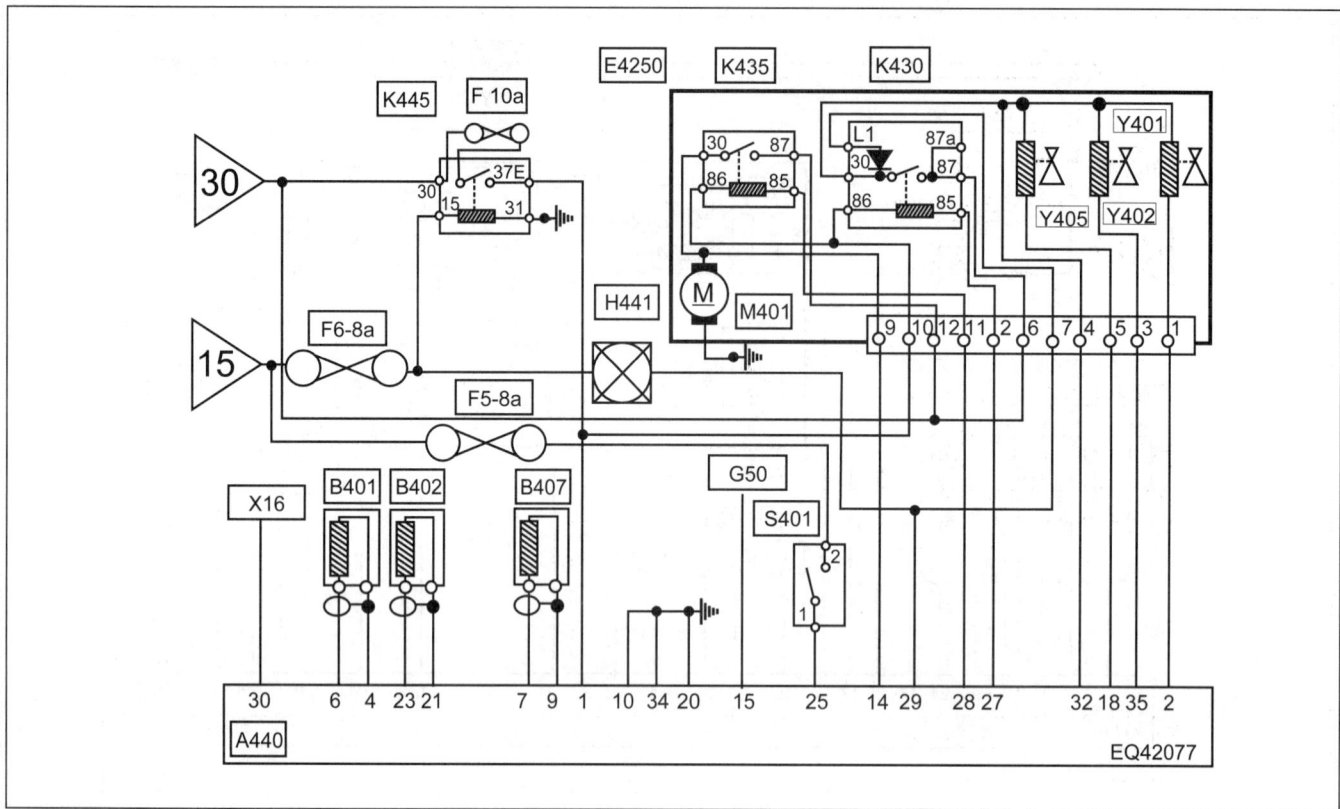

8A.46 Bosch 2E wiring diagram, Mercedes-Benz E-Class (124)

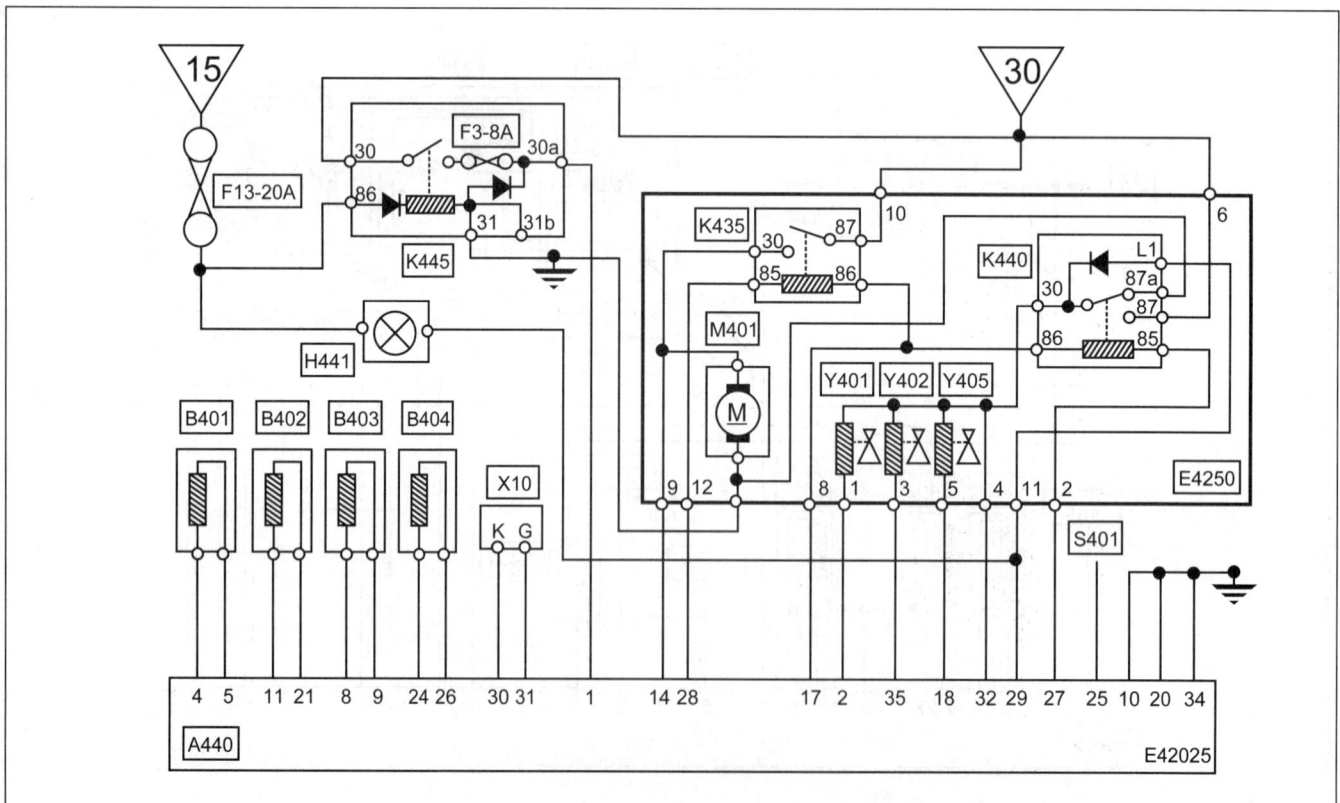

8A.47 Bosch 2E 35-pin wiring diagram, Vauxhall/Opel Astra-E/Belmont/Calibra/Kadett-E

8A

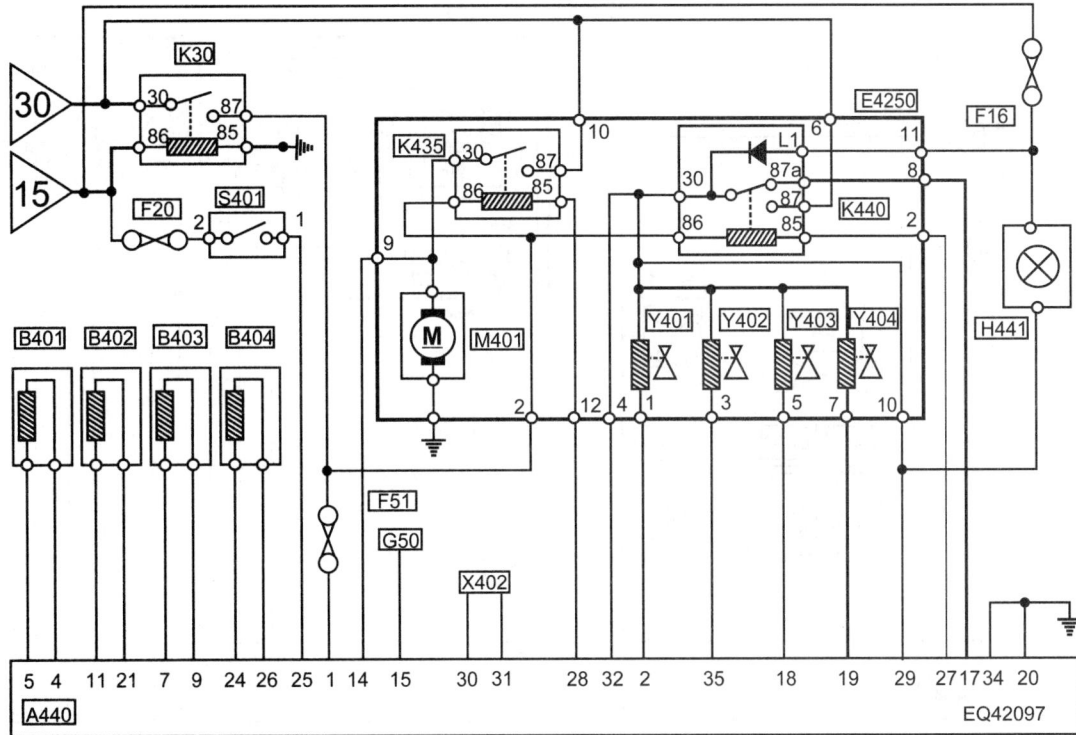

8A.48 Bosch 2E wiring diagram, Volkswagen Transporter

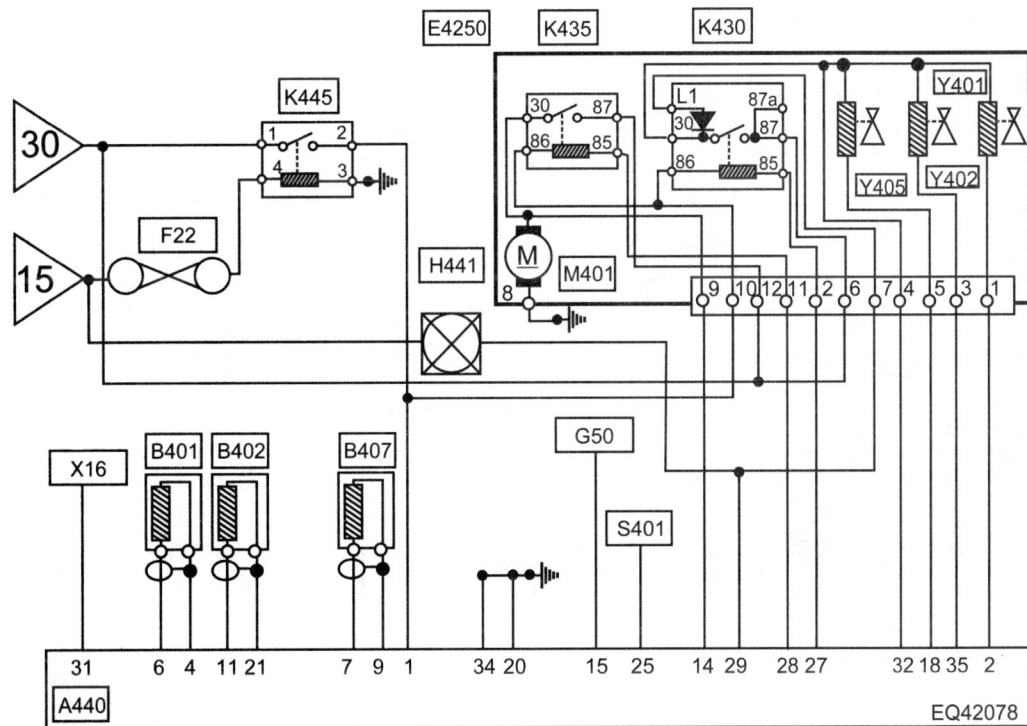

8A.49 Bosch 2E wiring diagram, Volvo 940/960

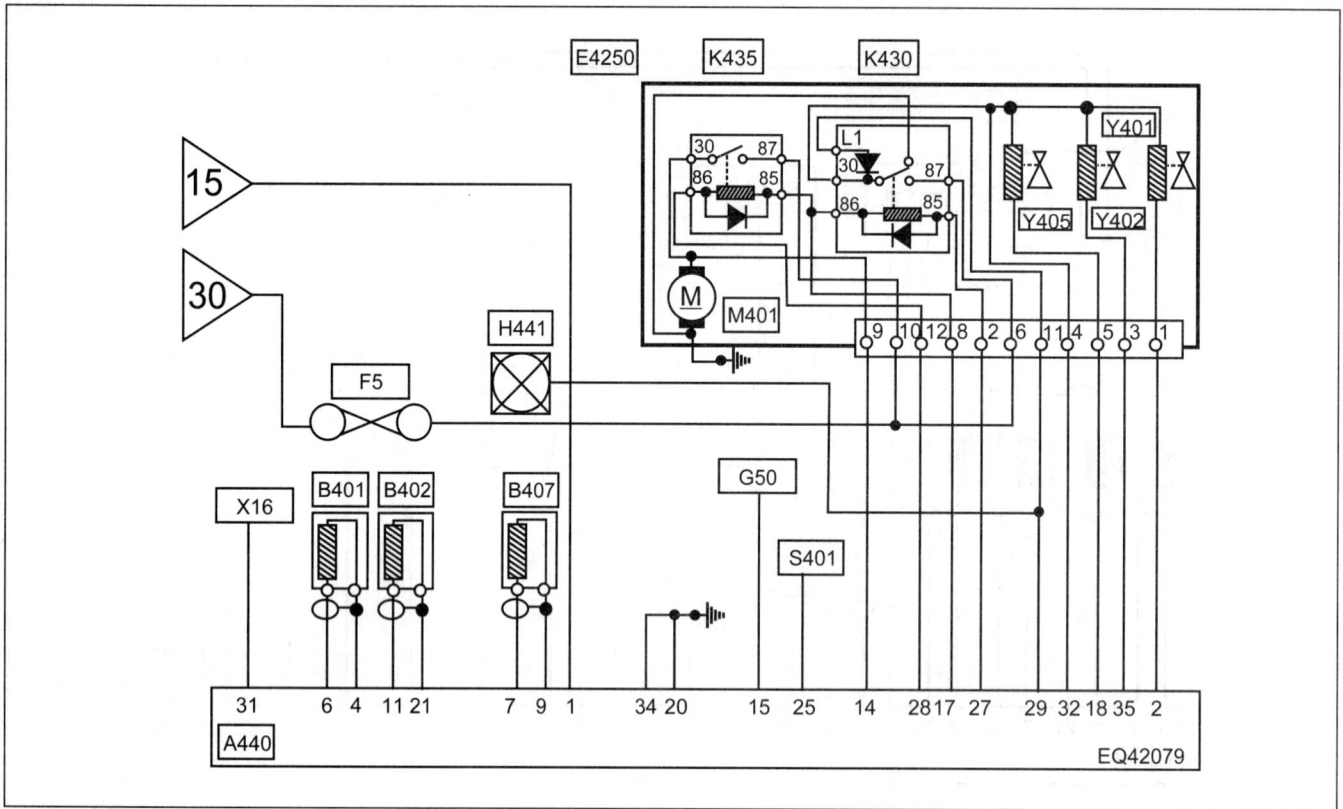

8A.50 Bosch 2E wiring diagram, Volvo S90/V90

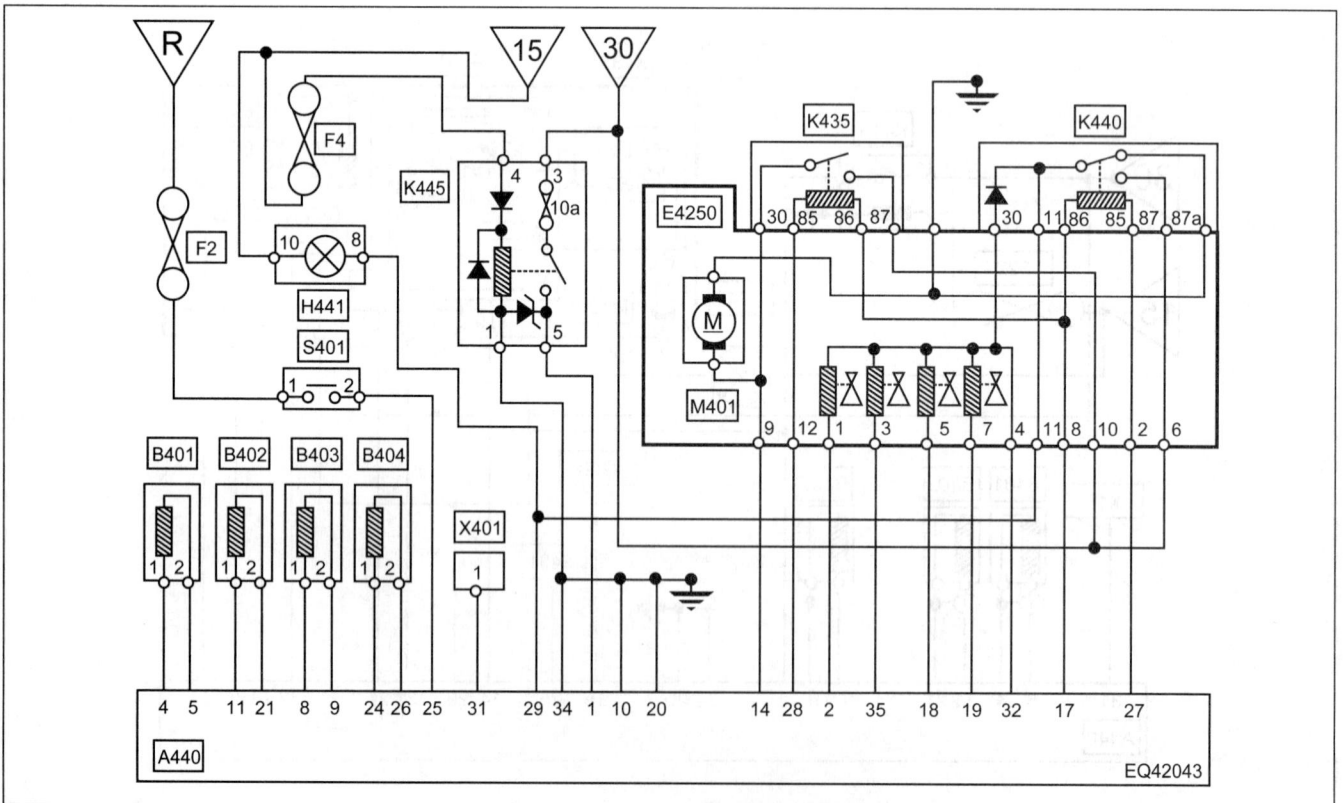

8A.51 Bosch 2SE 35-pin wiring diagram, Peugeot 605 (SD)

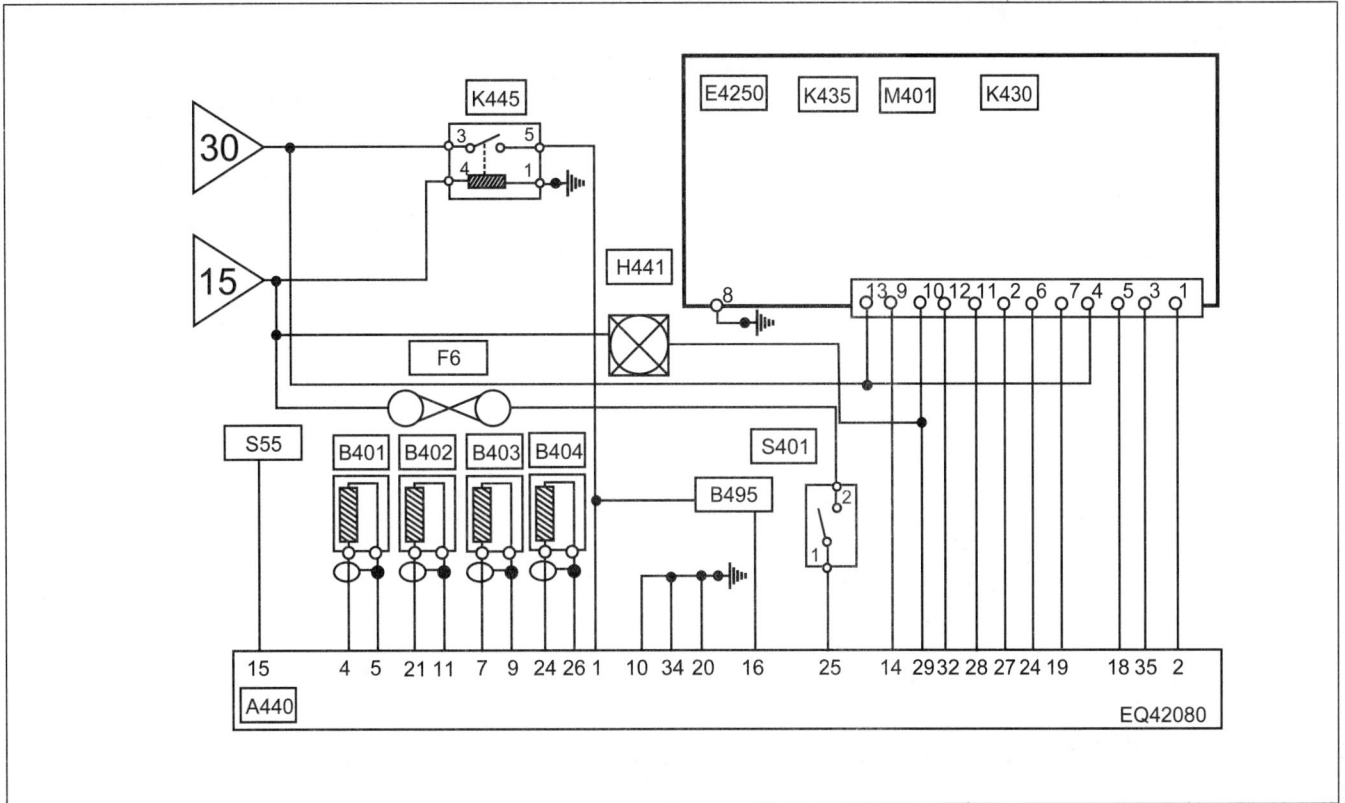

8A.52 Bosch 2SE wiring diagram, Renault Espace (non-SD)

8A

Chapter 8 Part B

Bosch 2EH and 2SH ABS

Contents

Vehicle coverage

Model	Year	System
Alfa Romeo		
145/146	1994-1996	Bosch 2EH
155	1992-1996	Bosch 2EH
164	1992-1998	Bosch 2EH
Audi		
Coupe	1992-1996	Bosch 2EH
Citroën		
ZX	1991-1996	Bosch 2EH
Fiat		
Barchetta	1995-1997	Bosch 2SH
Coupe	1995-1998	Bosch 2EH
Croma	1991-1993	Bosch 2EH
Punto	1994-1996	Bosch 2SH
Peugeot		
306	1993-1997	Bosch 2EH
405	1992-1997	Bosch 2EH
405 Mi-16	1992-1996	Bosch 2EH
Renault		
19	1993-1996	Bosch 2EH
Clio	1991-1997	Bosch 2EH
Safrane	1992-1997	Bosch 2EH
Rover		
400	1992-1995	Bosch 2EH
800	1992-1996	Bosch 2EH
Vauxhall/Opel		
Astra-F	1992-1998	Bosch 2EH
Calibra (2WD)	1992-1998	Bosch 2EH
Cavalier	1992-1996	Bosch 2EH
Omega-B	1994-1998	Bosch 2SH
Vectra-A	1992-1996	Bosch 2EH

Overview of system operation

1 Basic principles and system identification

The Bosch 2 Antilock Brake System has been fitted to a vast range of passenger vehicles since its introduction in the late 1970s. The system is of the additional or 'add-on' type operating in conjunction with the conventional braking system components.

The Bosch 2 system has been refined and uprated over the years in accordance with advances in electrical technology and demands from vehicle manufacturers. Essentially there are three distinct versions of the system, with the main differences being in the location of the ECM (either remotely mounted within the vehicle, or attached to the hydraulic control unit), the ECM software and the ECM multi-plug which may be a 35-pin, 15-pin or 25-pin type.

For clarity, 15-pin, and 25-pin systems with integral ECM and hydraulic control unit are covered in this Part of Chapter 8, and 35-pin ECM multi-plug systems, with remotely mounted ECM are covered in Part A.

In general, the various versions can be classified as follows, according to certain distinct features:

35-pin multi-plug systems covered in Part A of Chapter 8.

Bosch 2S

Original version with separate ECM and hydraulic control unit. ECM without self-diagnostics.

Bosch 2E

Uprated version of Bosch 2S. ECM now incorporates self-diagnostics.

Bosch 2SE

A development of Bosch 2S but can have additional ECM circuitry for 4WD vehicle applications. With/without self-diagnostics.

15-pin/25-pin multi-plug systems covered in this Part of Chapter 8.

Bosch 2EH

Similar to Bosch 2E, but with ECM integral with hydraulic control unit.

Bosch 2SH

Similar to Bosch 2EH but can have additional ECM circuitry for traction control.

All versions of Bosch 2EH and 2SH ABS are similar in operation and are comprised of the following main components.
a) Hydraulic control unit with integral ABS-ECM.
b) Inductive wheel speed sensors and associated sensor rings.
c) Diagnostic connector.
d) ABS electrical wiring harness and relays.
e) Brake light switch.
f) ABS warning lamp.
g) Copy valve assembly.

In addition, the conventional brake system is comprised of the following components.
a) Tandem brake master cylinder.
b) Vacuum servo unit.
c) Brake calipers/wheel cylinders and hydraulic hoses and pipes.
d) Pressure regulating/load-sensing valves (depending on application).

The systems covered in this Part of Chapter 8 may be installed as an antilock brake system only, or as an antilock brake system incorporating traction control.

In ABS mode, the purpose of the system is to apply the vehicle brakes at maximum efficiency without wheel lock or loss of directional stability. Inductive sensors (wheel speed sensors) monitor the speed of the wheels by generating an electrical signal as the wheel is rotated. This information is passed to the ABS-ECM, which compares the signals received from each wheel, and uses the speed of the fastest wheel as a reference value. The ECM continually monitors the speed of each wheel and if the onset of lock at any wheel is detected (a received speed signal being less than the reference value) a signal is sent to the ABS hydraulic control unit which regulates the brake pressure for the relevant wheel(s).

Where the system incorporates traction control, essentially the reverse principle is applied. When the ECM detects that one or more wheels are rotating faster than the reference value, the brake is actually applied on the relevant wheel(s) to reduce the rotational speed. Additionally, when wheel spin is detected, various signals are sent to the engine management ECM to control engine torque and rpm.

2 Component description and operation

ABS ECM

General

The Bosch Electronic Control Module (ECM) continually monitors wheel speed from the signals provided by the wheel speed sensors, and brake application from the brake light switch signal. If the ECM detects the incidence of wheel lock (or wheel spin, if traction control is incorporated) on one or more wheels, a signal is sent to the hydraulic control unit to modulate the hydraulic pressure to the brake of the locking or spinning wheel(s) **(see illustration 8B.1)**. The ECM contains two microprocessors and uses digital technology to complete this function and other functions such as, fault code memory and power modules for valve and relay activity.

To reduce external electrical connections to a minimum and improve reliability, the ECM is integral with the hydraulic control unit. In some applications the ECM and hydraulic control unit cannot be separated and are only available as a complete sealed unit.

Self-test

The Bosch 2 ECM is equipped with a self-test capability that initially examines the ABS system when the ignition is switched on, and then examines the wheel speed sensor signals after a wheel speed of approximately 4 mph is reached from all wheels (engine running). The

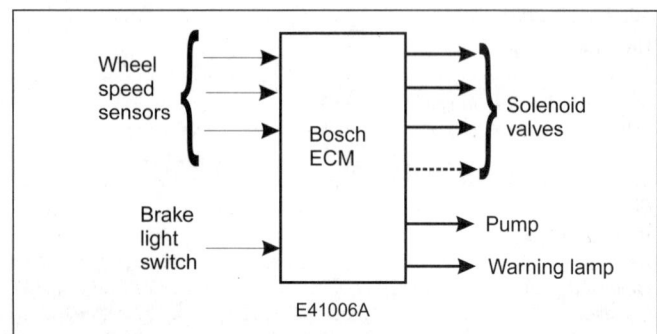

8B.1 ECM sensor inputs and control signal outputs

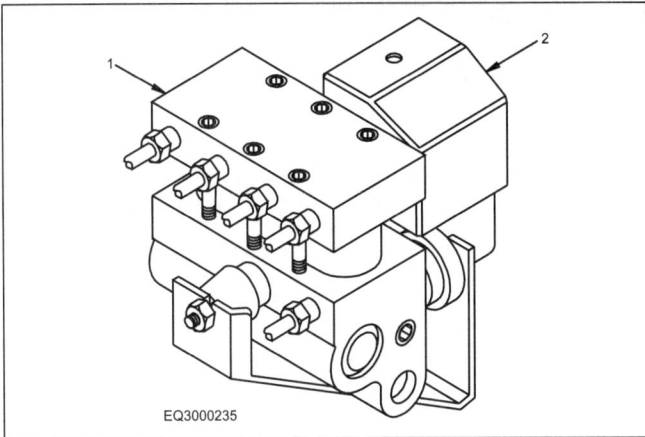

8B.2 Bosch 2EH hydraulic control unit assembly

1 Hydraulic control unit
2 Cover over ECM, relays and wiring connectors

ABS self-test program continues to examine the signals from the various components as long as the ignition is switched on. If self-test determines that faults are not present, the ABS is ready for operation once a specified vehicle speed has been achieved.

If the ECM detects that a fault is present, all ABS functions are switched off and the warning lamp is turned on. The conventional braking system continues to operate as normal without ABS assistance.

Self-diagnostics

If the ECM detects a fault during the self-test routine, an internal fault code is stored in the ECM memory. Stored fault codes can be retrieved from the SD connector with the aid of a suitable fault code reader or, on certain models, can be output as flash codes by means of the ABS warning lamp. If the fault clears, the code will remain stored until cleared, either manually or with a FCR.

Hydraulic control unit

The hydraulic control unit consists of an electric motor and radial piston return pump, solenoid valves and pressure accumulator. The unit controls the hydraulic pressure applied to the brake for each individual front wheel and for the rear wheels as a pair. The return pump is switched on when the ABS is activated and returns brake fluid, drained off during the pressure reduction phase, back into the brake circuit. The pump motor relay, solenoid valve relay and the connection plugs for the power supply and ECM can be located under the cover on top of the unit **(see illustration 8B.2)**.

Bosch 2EH and 2SH are both three-channel hydraulic systems incorporating three solenoid valves in the hydraulic control unit. One solenoid valve is used for each front brake with the remaining (rear) solenoid valve operating in conjunction with a copy valve to control the rear brakes as a pair.

The copy valve is an additional mechanical valve assembly, which can be either located remotely on the vehicle, or integrated within the hydraulic control unit. The unit comprises two hydraulic chambers linked by a dual piston assembly. Hydraulic pressure from the rear solenoid valve in the hydraulic control unit is supplied to the first chamber, and hydraulic pressure from the master cylinder is supplied to the second chamber. During normal braking without ABS control, hydraulic pressure flows through the open rear solenoid valve to one of the rear brakes, and to the first pressure chamber in the copy valve. The other rear brake receives hydraulic pressure from the second pressure chamber in the copy valve. Pressure differences within the copy valve chambers act on the piston, which moves to open or close hydraulic passages within the valve. The action of the piston assembly in the copy valve balances the hydraulic pressure applied to both the rear brakes.

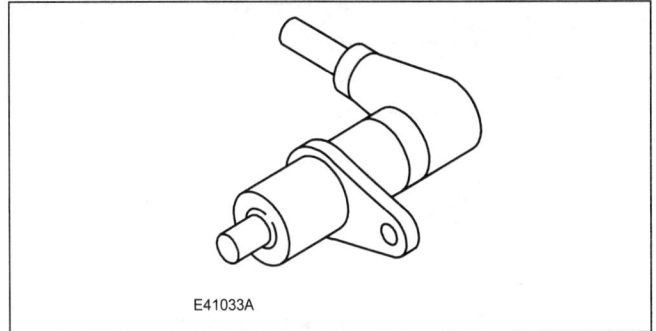

8B.3 Typical wheel speed sensor

When ABS is in operation, the hydraulic pressure in the circuit to one of the rear brakes is modulated by the rear solenoid valve in the hydraulic control unit. The other rear brake receives exactly the same hydraulic pressure via the operation of the piston assembly in the copy valve.

Wheel speed sensors

The rotational speed of the roadwheels and any changes in the rotational speed are recorded by inductive wheel speed sensors, one located at each front roadwheel, and one located at each rear roadwheel or on either side of the differential casing **(see illustration 8B.3)**.

Each wheel speed sensor assembly comprises a toothed sensor ring, which rotates at roadwheel speed, and an adjacent sensor mounted a set distance from the sensor ring **(see illustration 8B.4)**.

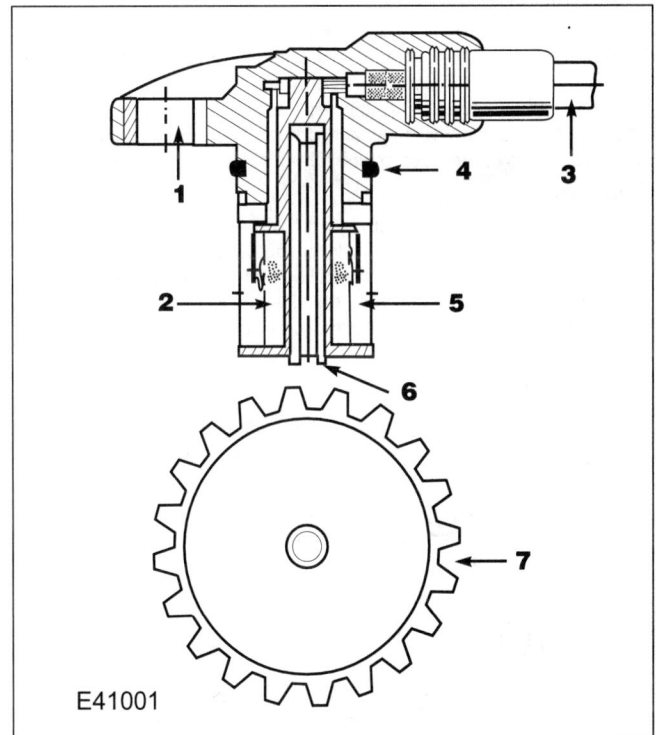

8B.4 Sectional view of a wheel speed sensor

1	Mounting bolt location	5	Coil
2	Permanent magnet	6	Sensor tip
3	Wiring harness	7	Toothed sensor ring
4	O-ring		

8B

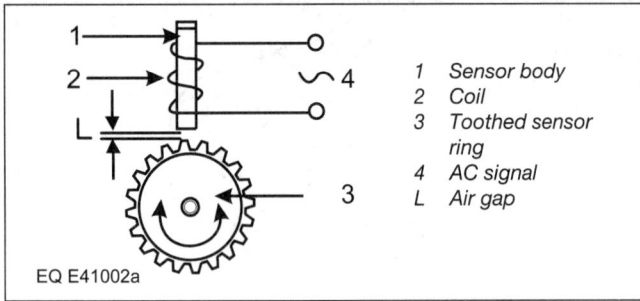

1 Sensor body
2 Coil
3 Toothed sensor ring
4 AC signal
L Air gap

EQ E41002a

8B.5 Wheel speed sensor operation

The sensors are permanent magnet pulse generator types producing an AC voltage sine wave as the sensor ring teeth pass through the magnetic field of the sensor **(see illustration 8B.5)**.

The frequency of the waveform produced by the wheel speed sensor is proportional to the road speed. This AC voltage signal is continually being delivered to the ABS-ECM for processing.

The peak to peak voltage of the speed signal (when viewed upon an oscilloscope) can vary considerably according to wheel speed. An analogue to digital converter (ADC) in the ECM transforms the AC pulse into a digital signal **(see illustration 8B.6)**.

ABS electrical wiring harness and relays

An integrated 15-pin or 25-pin ECM wiring harness is used to enable sensor input signals to reach the ECM. Additionally 4-pin and 6-pin wiring harnesses are used to connect power and earth to the relays and solenoid valves in the hydraulic control unit. The harness connectors, pump relay and main (solenoid valve) relay are located under a cover on the top of the unit.

Solenoid valve relay

A permanent voltage is supplied to the relay contacts from the battery, with the relay coil typically being supplied from the ignition switch. The valve relay is controlled by the ECM applying an earth to the relay coil which actuates the relay allowing battery voltage to be applied to the solenoid valves and pump relay coil. The earth for the solenoid valves is provided by the ECM.

Pump relay

A permanent voltage is supplied to the relay contacts from the battery, with the relay coil typically being supplied from the solenoid valve relay. The pump relay is controlled by the ECM applying an earth to the relay coil, which actuates the relay allowing battery voltage to be applied to the pump. The hydraulic control unit provides the earth for the pump motor.

Brake light switch

The brake light switch comprises a switch body and contact pin and is located above the brake pedal **(see illustration 8B.7)**.

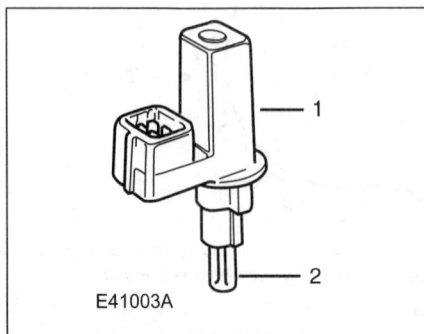

E41024

8B.6 Wheel speed sensor waveform as viewed on an oscilloscope

When the brake pedal is depressed, closing the brake light switch, a signal is sent to the ECM indicating that the brakes are being applied. Once this signal is received, the ECM will begin monitoring the wheel speed via the wheel speed sensors and activate the ABS if necessary.

ABS warning lamp

After the ignition is switched on, the ABS warning lamp on the instrument panel is illuminated for approximately 2 to 4 seconds during the initial self-test routine, then extinguished if satisfactory operation of the system is detected by the ECM. During vehicle operation above a pre-determined wheel speed, the ABS-ECM implements a further series of self-test cycles whereby the system components are continually monitored. If a fault is detected at any time during the self-test cycles, the ECM will illuminate the warning lamp on the instrument panel and disable ABS operation. However, the conventional braking system will continue to operate as normal. The warning lamp will then remain illuminated until the fault is no longer present **(see illustration 8B.8)**.

When the ABS-ECM detects a fault, the fault code is stored and the ABS warning lamp activated. If the fault no longer exists after the next system start (ignition on/off) the ABS warning lamp is extinguished after the initial self-test cycle, however the fault code remains stored in the ECM memory.

Tandem master cylinder

Typically, the tandem master cylinder comprises a body casting incorporating primary and secondary pressure chambers, primary piston, intermediate piston, floating piston, slotted pin and central valve. The cylinder operates as a conventional master cylinder using

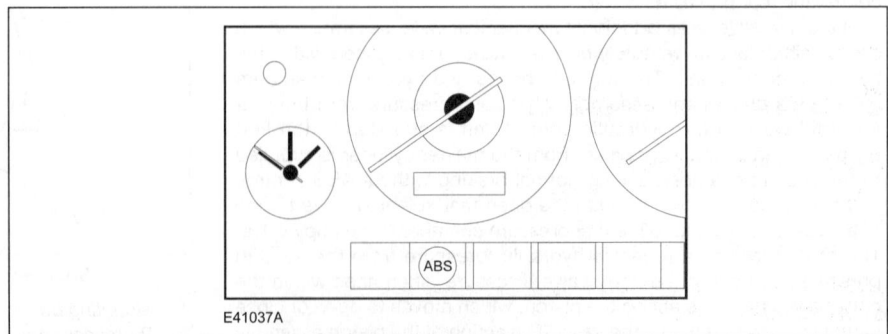

E41003A

8B.7 Typical brake light switch body (1) and contact pin (2)

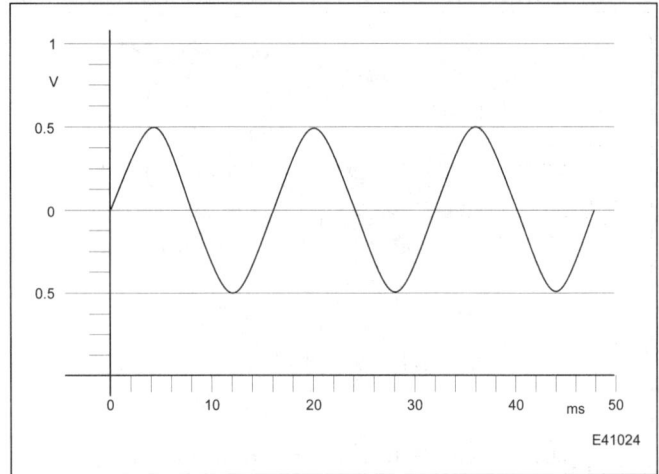

E41037A

8B.8 ABS warning lamp

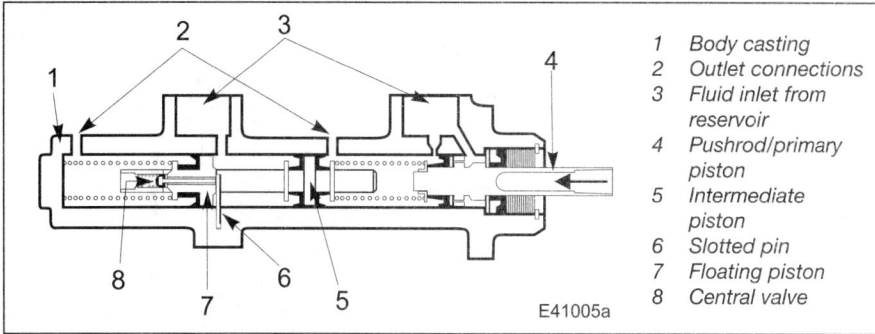

1 Body casting
2 Outlet connections
3 Fluid inlet from reservoir
4 Pushrod/primary piston
5 Intermediate piston
6 Slotted pin
7 Floating piston
8 Central valve

E41005a

8B.9 Sectional view of a typical tandem master cylinder

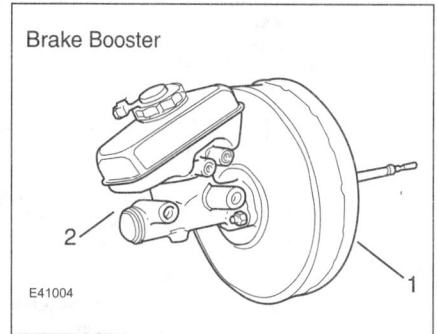

Brake Booster

E41004

8B.10 Vacuum servo unit (1) and tandem master cylinder (2)

8B

vacuum assistance from the vacuum servo unit **(see illustration 8B.9).**

When the brake system is at rest, the central valve in the floating piston rests against the slotted pin. In this condition, the central valve is open and brake fluid can discharge out of the pressure chamber back into the brake fluid reservoir. When the brake pedal is depressed, the build-up of hydraulic pressure in the primary pressure chamber acts on the intermediate piston and floating piston, moving them down the cylinder bore. The floating piston contacts the seal on the central valve, closing the connection between the intermediate and secondary pressure chambers. Brake hydraulic pressure can now also increase in the secondary pressure chamber.

Vacuum servo unit

The vacuum servo unit is located between the brake pedal and tandem master cylinder. When the brake pedal is depressed, the servo unit increases the force applied by the pedal, reducing the effort required to operate the brakes **(see illustration 8B.10).**

The unit is operated by vacuum created in the engine inlet manifold (or from a separate vacuum pump on diesel engines) which is applied to a diaphragm contained within the unit casing. A pushrod connected to the centre of the diaphragm acts directly on the primary piston in the master cylinder.

When the brake pedal is released, vacuum is applied to both sides of the diaphragm. When the pedal is depressed, one side of the diaphragm is opened to atmosphere and the vacuum acting on the other side deflects the diaphragm. This in turn operates the master cylinder primary position. The resulting force applied to the master cylinder piston is therefore significantly greater than the initial force applied to the brake pedal by the driver.

Pressure regulating/load-sensing valve(s)

Depending on vehicle application, pressure regulating valves or load sensing valves may be incorporated to restrict the hydraulic fluid pressure to the rear brakes. The valves may be pressure conscious whereby the hydraulic fluid supply is restricted once a pre-determined pressure is reached, or load conscious whereby the hydraulic pressure is reduced according to vehicle loading.

3 System operation

Brake system at rest

When the ABS system is at rest all the brake components are inoperative. Pressure is non-existent in the hydraulic pipes between the tandem master cylinder and the brake calipers or wheel cylinders.

Brake system operating under conventional control without ABS

When the brake pedal is activated, the pedal force is applied to the tandem master cylinder by the vacuum servo unit pushrod. The servo unit pushrod acts directly on the pressure piston in the master

cylinder, which pressurises the hydraulic fluid in the brake pipes to the hydraulic control unit. Pressure acting upon the valves in the hydraulic control unit causes them to open and hydraulic pressure is transmitted to each brake caliper or wheel cylinder, thus operating the brakes **(see illustration 8B.11).**

Brake system operating in conjunction with ABS control

The ABS-ECM continually monitors wheel speed from the signals provided by the wheel speed sensors. If the ECM detects the incidence of wheel lock on one or more wheels, ABS is automatically initiated in three phases. During ABS operation, all or any of the wheels could be in any one of the following phases at any particular moment.

First ABS phase, pressure holding

The ECM earths the appropriate solenoid valve driver and a current of 2.0 amps flows in the solenoid valve circuit. The solenoid valve actuates to move the corresponding piston upward against spring force so that the hydraulic fluid line from the tandem master cylinder to the brake caliper, wheel cylinder or copy valve is closed. The hydraulic fluid in the controlled circuit is maintained at a constant pressure and this effectively removes the braking force from the controlled circuit. The pressure cannot now be increased in the

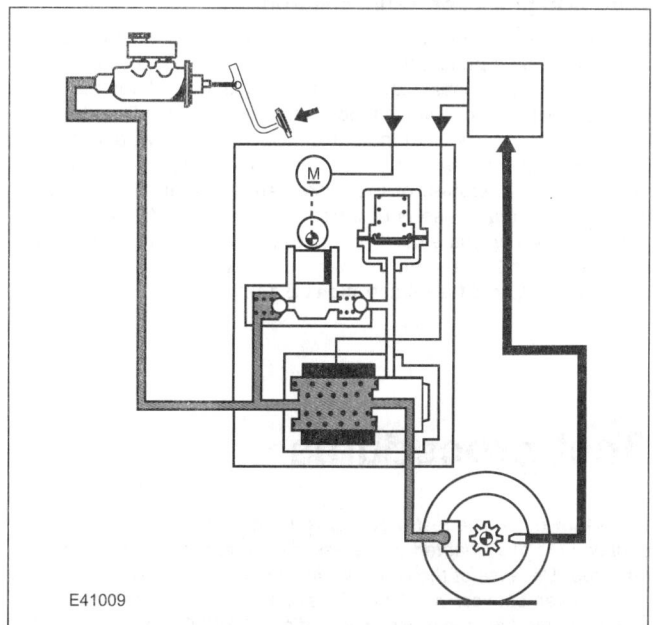

E41009

8B.11 Brake system operating under conventional control without ABS

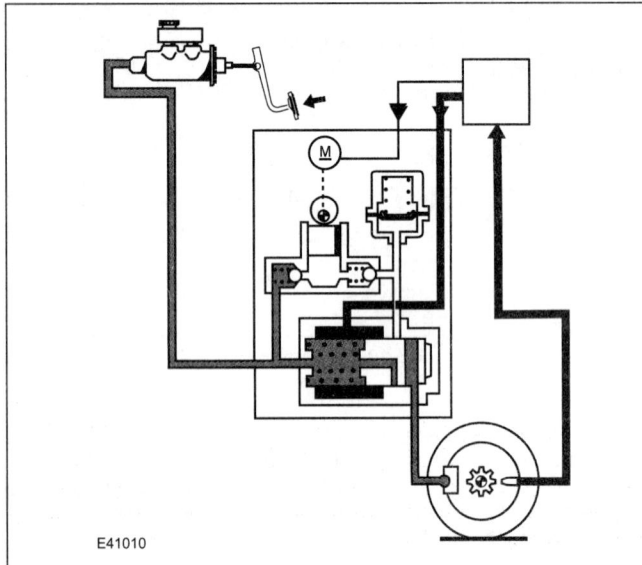

8B.12 ABS operation - first phase, pressure holding

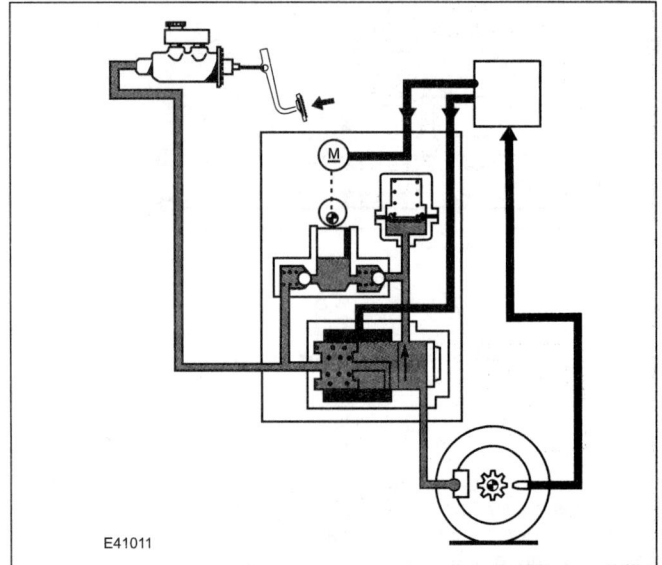

8B.13 ABS operation - second phase, pressure reduction

controlled circuit by any further application of the brake pedal **(see illustration 8B.12)**.

Second ABS phase, pressure reduction

If the ECM detects wheel instability, a pressure reduction phase is initiated. The current flow in the solenoid valve circuit is increased to 5.0 amps, which causes the solenoid valve to move the corresponding piston further upward, so that a channel to the return pump is opened. The result is that hydraulic fluid flows quickly out of the brake caliper or wheel cylinder into the pressure accumulator in the hydraulic control unit. The return pump is actuated by the ECM so that hydraulic fluid is pumped back into the master cylinder. This process creates a pulsation, which can be felt in the brake pedal action **(see illustration 8B.13)**.

After pressure reduction, the ECM returns to the pressure holding phase and wheel speed is closely monitored.

Third ABS phase, pressure build-up

If the wheel speed accelerates too quickly during the second pressure holding phase, the solenoid valve driver is released which switches off the solenoid valve circuit. Spring action returns the piston to the lowest position, which reopens the hydraulic fluid line from the tandem master cylinder to the brake caliper, wheel cylinder, or copy valve. Hydraulic pressure is reinstated in the controlled circuit thus reintroducing operation of the brake. After a brief period, a short pressure holding phase is re-introduced and the ECM continually shifts between pressure build-up and pressure holding until the wheel has decelerated to a sufficient degree where pressure reduction is once more required **(see illustration 8B.14)**.

The whole ABS control cycle takes place 4 to 10 times per second for each affected wheel and this ensures maximum braking effect and control during ABS operation.

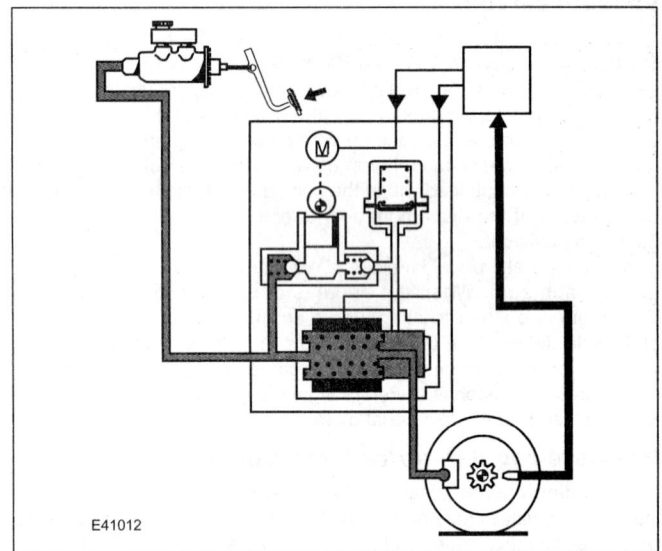

8B.14 ABS operation - third phase, pressure build-up

Test procedures

Important note: *The test procedures, pin-tables and wiring diagrams contained in this Chapter are necessarily representative of the system depicted. Because of the variations in wiring and other data that often occurs, even between similar vehicles in any particular VM's range, the reader should take great care in identification of ECM pins, and satisfy himself that he has gathered the correct data before failing a particular component.*

4 Wheel speed sensors

Note: *There are considerable variations in wheel speed sensor pin connections at the ECM, depending on vehicle type and model. The wiring diagrams in this Chapter show typical pin connections, but some variations are likely to be encountered.*

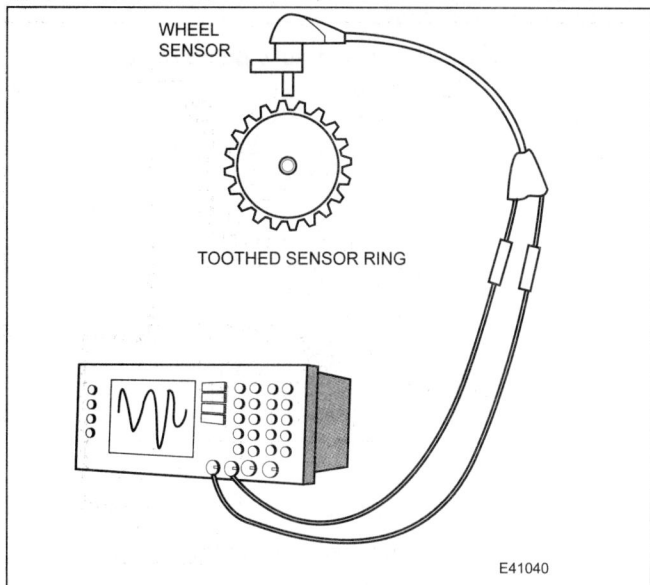

8B.15 Checking wheel speed sensor output with an oscilloscope connected to the sensor-wiring plug

Checking the wheel speed sensor (general)

1 Inspect the wheel speed sensor for corrosion or damage and check that the sensor is tightly mounted.

2 Check the toothed sensor ring for damage, eccentricity and for broken or missing teeth.

3 Inspect the wheel speed sensor-wiring plug for corrosion and damage. One plug for each sensor.

4 Check that the connector terminal pins are pushed fully home and making good contact with the sensor-wiring plug.

5 Check the clearance between the sensor and the toothed sensor ring. The clearance is not normally adjustable but is nominally 0.2 to 1.2 mm. If the clearance is excessive, expect a worn sensor tip or problems with the wheel bearings/hub or sensor ring.

6 When carrying out voltage checks with an oscilloscope or voltmeter, the voltage obtained will be proportional to the speed at which the wheel is rotating. In addition to determining that the wheel

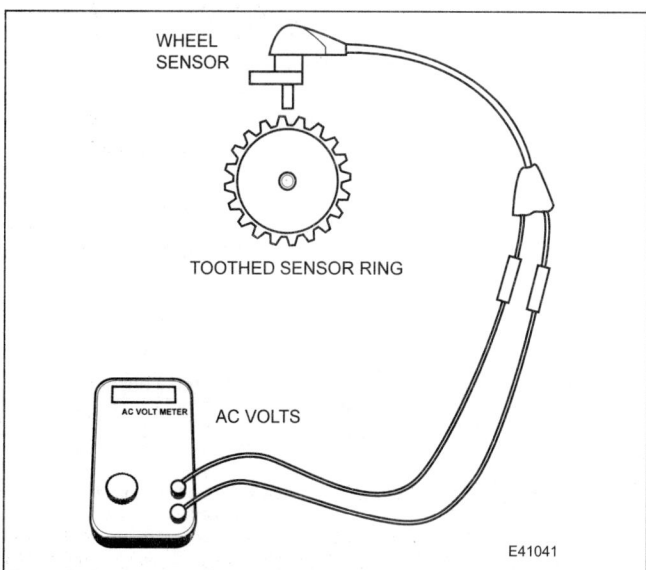

8B.17 Checking wheel speed sensor output with a voltmeter connected to the sensor-wiring plug

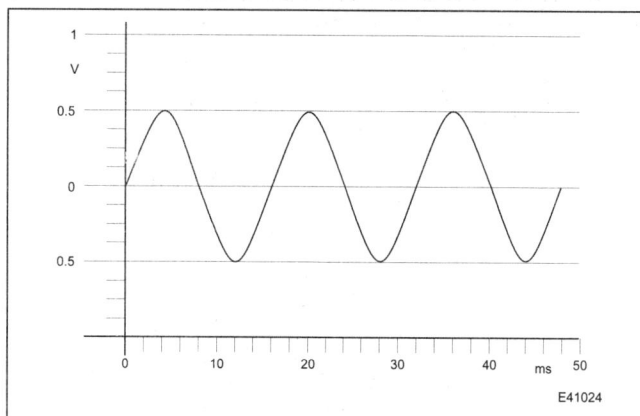

8B.16 Typical inductive wheel speed sensor sine wave as displayed on an oscilloscope

speed sensors are actually producing a voltage output, it is essential that the output from the sensors on a particular axle is the same for any given wheel speed.

Checking wheel speed sensor output with an oscilloscope

Note: *Refer to the note at the beginning of this Section concerning wheel speed sensor pin connections before proceeding.*

7 Switch the ignition off and disconnect the ECM multi-plug or the relevant wheel speed sensor-wiring plug.

8 Connect an oscilloscope between the terminal pins for the sensor under test **(see illustration 8B.15)**.

9 Select a range to cover 80 Hz on the oscilloscope and a free run time base.

10 Raise the wheel and rotate it by hand at approximately one revolution per second.

11 A sinusoidal wave form should be obtained, with amplitude and duration changing with rotational speeds **(see illustration 8B.16)**.

12 If there is no signal, or a very weak or intermittent signal at the ECM, repeat the test at the sensor-wiring plug. If there is no change in signal status, the sensor is suspect.

13 If the signal is now satisfactory this indicates a fault in the wiring harness, which should be checked for continuity.

Checking wheel speed sensor output with an AC voltmeter

Note: *Refer to the note at the beginning of this Section concerning wheel speed sensor pin connections before proceeding.*

14 Switch the ignition off and disconnect the ECM multi-plug or the relevant wheel speed sensor-wiring plug.

15 Connect an AC voltmeter between the terminal pins for the sensor under test **(see illustration 8B.17)**.

16 Raise the wheel and rotate it by hand at approximately one revolution per second.

17 A voltage of approximately 0.5 to 1.5 volts (AC RMS) should be obtained. If there is no signal, or a very weak or intermittent signal at the ECM, repeat the test at the sensor-wiring plug. If there is no change in the signal, the sensor is suspect.

18 If the signal is now satisfactory, this indicates a fault in the wiring harness which should be checked for continuity. **Note:** *This test at least proves that a signal is being generated by the sensor. However, the voltage produced is an average voltage and does not clearly indicate damage to the sensor ring or that the sinewave is regular in formation.*

Checking wheel speed sensor resistance

Note: *Refer to the note at the beginning of this Section concerning wheel speed sensor pin connections before proceeding.*

19 Switch the ignition off and disconnect the ECM multi-plug or the relevant wheel speed sensor-wiring plug.

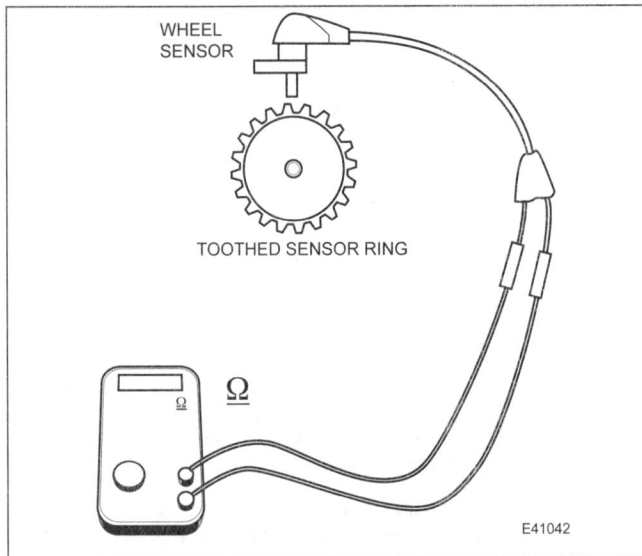

8B.18 Checking wheel speed sensor resistance with an ohmmeter connected to the sensor-wiring plug

20 Connect an ohmmeter between the terminal pins for the sensor under test **(see illustration 8B.18)**.

21 The readings obtained should be between 0.7 and 2.2 kohms approximately.

22 If the resistance is excessively high or open circuit at the ECM, repeat the test at the sensor multi-plug. If there is no change in resistance, the sensor is suspect.

23 If the resistance is now satisfactory, this indicates a fault in the wiring harness which should be checked for continuity. **Note:** *Even if the resistance is within the quoted specifications, this does not prove that the speed sensor can generate an acceptable signal.*

5 System relays

Checking relays (general)

1 The following tests are applicable to the ABS main (solenoid valve) relay and hydraulic pump relay which are located on the hydraulic control unit, under a detachable cover.

2 Before carrying out specific tests on the relays or power supplies inspect the relays and relay base for corrosion and damage.

3 Check that the relays are pushed fully home and making good contact with the base.

4 A fault in any of the above areas are possible reasons for failure or malfunctioning of the relay.

Caution: On certain Bosch 2EH systems, the solenoid valve and pump relays are integral with the ECM cannot be removed from their bases. Where integral relays are encountered, check the relay operation by testing the components supplied by the specific relay.

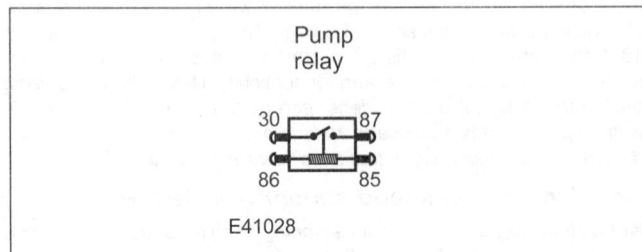

8B.19 Typical pump relay terminal identification

Relay operation tests

Main (solenoid valve) and hydraulic pump relays

Note: *On some models, relay terminals 87 and 30 are transposed from the arrangement shown in the illustration and described in the text. Where applicable, refer to the wiring diagrams for specific relay terminal numbers according to model.*

5 Switch the ignition off and remove the relevant relay from the relay base.

6 Connect an ohmmeter between the relay main voltage supply terminal (typically 87) and the output terminal (typically 30) **(see illustration 8B.19)**. The ohmmeter should indicate an open circuit.

7 Attach the positive lead of a 12 volt supply to the relay coil supply terminal (typically 86) and the negative lead to the relay coil earth terminal (typically 85). There should be an unmistakable click from the relay and the ohmmeter should indicate continuity.

8 Renew the relay if it fails these tests. If the relay cover is transparent, inspect the relay for burns. A faulty relay usually has a burnt appearance.

Relay power supply tests

Main (solenoid valve) and hydraulic pump relays

Note: *On some models, relay terminals 87 and 30 are transposed from the arrangement shown in the illustration and described in the text. Where applicable, refer to the wiring diagrams for specific relay terminal numbers according to model.*

9 Switch the ignition off and remove both relays leaving the bases exposed **(see illustration 8B.20)**.

10 Using a voltmeter, in turn probe between the main voltage supply terminal 87 or 30, as applicable, of each relay base and earth. The reading should indicate nbv.

11 If no voltage is found, check the wiring back to the voltage supply.

12 Switch the ignition on.

13 Using a voltmeter, in turn probe between the relay coil supply terminal (typically 86) of each relay base and earth. The reading should indicate nbv.

14 If no voltage is found, check the supply wiring or fuse as applicable.

6 Electronic Control Module

Checking the ECM (general)

1 Inspect the ECM for corrosion or damage and ensure that the unit is mounted securely.

2 Check that the ECM multi-plug terminals are pushed fully home and making good contact with the ECM pins. A fault in any of the above areas is possible reason for poor performance in the ABS system.

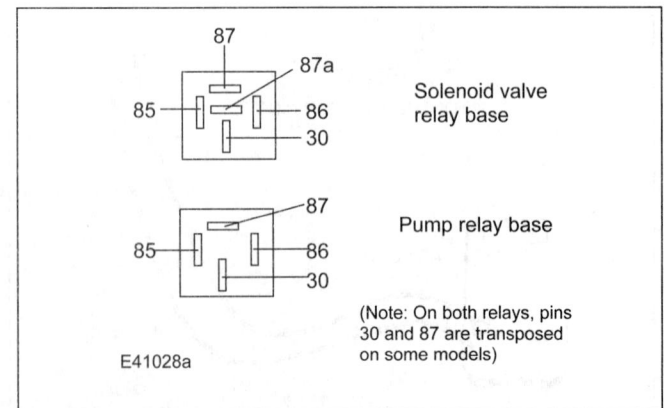

8B.20 Typical main (solenoid valve) and pump relay base terminal identification

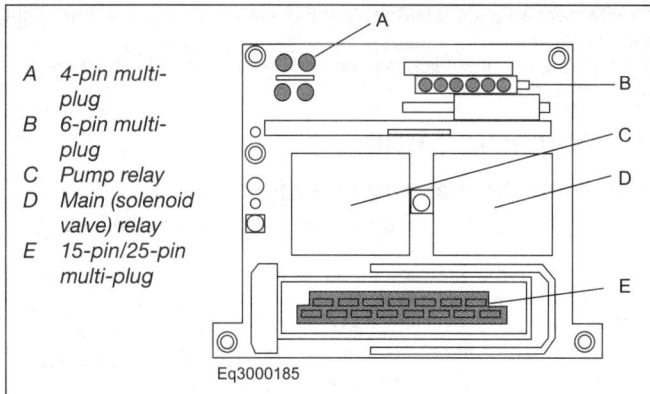

A 4-pin multi-plug
B 6-pin multi-plug
C Pump relay
D Main (solenoid valve) relay
E 15-pin/25-pin multi-plug

8B.21 Hydraulic control unit connections

ECM power supply and earth tests

Note: *Where applicable, refer to the wiring diagrams for specific terminal numbers according to model.*

1 Switch the ignition off and disconnect the hydraulic control unit 4-pin multi-plug **(see illustration 8B.21)**.
2 Attach a negative voltmeter probe to a vehicle earth.
3 Attach a positive voltmeter probe to pin 2 of the multi-plug. The voltmeter should indicate nbv **(see illustration 8B.22)**.
4 If no voltage is found, check the supply wiring back to the battery positive terminal.
5 Attach a negative voltmeter probe to a vehicle earth.
6 Attach a positive voltmeter probe to pin 1 of the multi-plug.
7 Switch the ignition on. The voltmeter should indicate nbv.
8 If no voltage is found, check the relevant fuse and the supply wiring back to the ignition switch.
9 Switch the ignition off.
10 Connect an ohmmeter between pin 3 of the multi-plug and a vehicle earth.
11 The ohmmeter should indicate continuity. If not, check the main earth connection and wiring.
12 Disconnect the ECM 15-pin or 25-pin multi-plug according to system.
13 Attach a negative voltmeter probe to a vehicle earth.
14 Attach a positive voltmeter probe to pin 1 of the multi-plug **(see illustration 8B.23)**.
15 Switch the ignition on. The voltmeter should indicate nbv.
16 If no voltage is found, check the relevant fuse and the supply wiring back to the ignition switch.

7 Solenoid valves

Solenoid valve resistance tests

1 Switch the ignition off and disconnect the 6-pin multi-plug at the hydraulic control unit **(see illustration 8B.21)**.
2 Using an ohmmeter measure the resistance between terminals 1 and 2, 3 and 4, and 5 and 6, in turn, in the 6-pin multi-plug **(see illustration 8B.24)**.
3 Resistance should be in the order of 1.5 ohms between each pair of terminals.
4 If the resistance is not as specified, the hydraulic control unit is suspect.

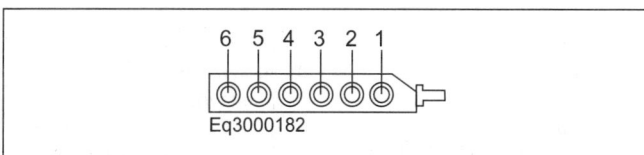

8B.24 6-pin multi-plug terminal identification

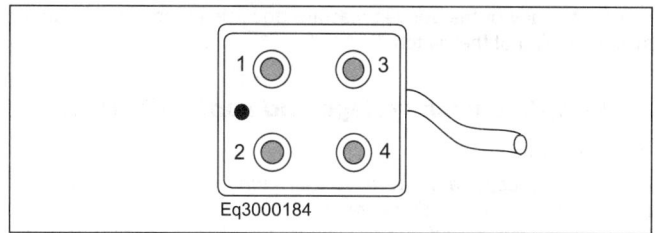

8B.22 4-pin multi-plug terminal identification

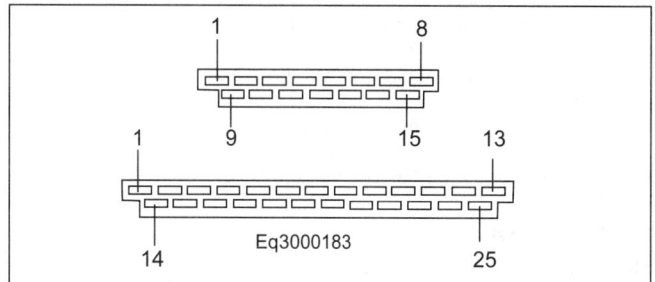

8B.23 ECM 15-pin and 25-pin multi-plugs

8 Hydraulic pump motor

Pump operation test

1 Switch the ignition off and remove the pump motors relay from the relay base.
2 Using a fused jumper lead, bridge relay base terminals 30 and 87 in the hydraulic control unit **(see illustration 8B.25)**. The pump motor should now run.

⚠️ **Warning: The test should be made as quickly as possible to avoid damaging the pump.**

3 If the pump does not operate as described, check the hydraulic control unit earth connection. If satisfactory, renew the hydraulic control unit.

9 Brake light switch

Checking the brake light switch (general)

1 Check that the brake light switch is correctly and securely mounted and that the plunger moves smoothly with no trace of binding.
2 Check that the wiring multi-plug is pushed fully home and making good contact.
3 Check that no wires have been disconnected.

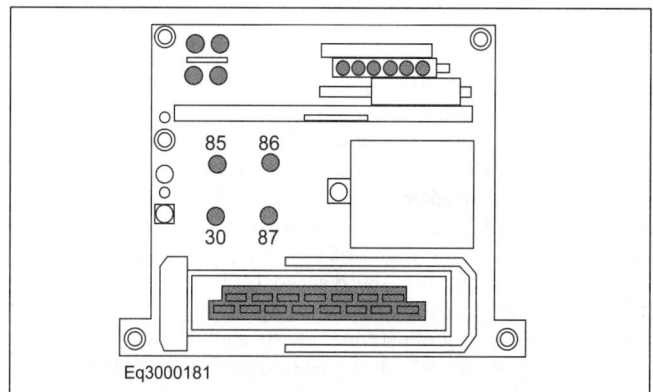

8B.25 Hydraulic pump relay base terminal identification

4 A fault in any of the above areas are possible reasons for failure or malfunctioning of the switch.

Brake light switch voltage and continuity tests

Voltage test

Note: *Where applicable, refer to the wiring diagrams for specific terminal numbers according to model.*

5 Switch the ignition off and disconnect the ECM multi-plug from the ECM.

6 Connect a voltmeter between a vehicle earth and the ECM multi-plug brake light switch input pin, (typically pin 9 for 15-pin multi-plugs, and pin 23 for 25-pin multi-plugs).

7 Switch the ignition on and depress the brake pedal. The voltmeter should indicate nbv.

8 If no voltage is found, the fuse, the brake light switch and the wiring are suspect.

9 Release the brake pedal. The voltage should drop to zero as the switch opens.

Continuity test

10 Switch the ignition off and disconnect the brake light switch multi-plug.

11 Connect an ohmmeter between the terminal pins of the brake light switch.

12 Operate the brake light switch and check for continuity. If the test fails, renew the switch.

10 Warning lamp

Checking the warning lamp (general)

1 Inspect the warning lamp bulb holder contacts in the instrument panel.

2 Check that the instrument panel multi-plug terminal pins are pushed fully home and making good contact.

3 Check that no wires have been disconnected.

4 A fault in any of the above areas are possible reasons for failure or malfunctioning of the warning lamp.

Warning lamp operation test

5 With the ignition switched off, the warning lamp should remain off.

6 Switch the ignition on and the warning lamp should illuminate and then extinguish after a few seconds.

7 If the warning lamp comes on and remains on at any time during vehicle operation, carry out the previously described test procedures on the system components.

Pin table - typical 15-pin

Note: *Refer to illustration 8B.23*

Pin No.	Connection	Test condition	Voltage
1	ABS warning lamp	Ignition on: Lamp on Lamp off	 1.0 volt (max) Nbv
2	Rear left wheel speed sensor signal	Roadwheel rotating	0.5 to 1.5 volts AC (approx)
2*	Rear left wheel speed sensor earth	Roadwheel rotating	0.25 volts (max)
2*	Rear right wheel speed sensor earth	Roadwheel rotating	0.25 volts (max)
3*	-	-	-
3*	Brake light switch input	Ignition on: Brake pedal released Brake pedal depressed	 0 volts Nbv
4*	Rear left wheel speed sensor signal	Roadwheel rotating	0.5 to 1.5 volts AC (approx)
4	Rear left wheel speed sensor earth	Roadwheel rotating	0.25 volts (max)
5	Front right wheel speed sensor signal	Roadwheel rotating	0.5 to 1.5 volts AC (approx)
5*	Front right wheel speed sensor earth	Roadwheel rotating	0.25 volts (max)
6	Rear right wheel speed sensor signal	Roadwheel rotating	0.5 to 1.5 volts AC (approx)
6*	Rear right wheel speed sensor earth	Roadwheel rotating	0.25 volts (max)
7	Front left wheel speed sensor signal	Roadwheel rotating	0.5 to 1.5 volts AC (approx)
7*	Front left wheel speed sensor earth	Roadwheel rotating	0.25 volts (max)
8	-	-	-
9	Brake light switch input	Ignition on: Brake pedal released Brake pedal depressed	 0 volts Nbv
10*	Front right wheel speed sensor earth	Roadwheel rotating	0.25 volts (max)
10*	Front right wheel speed sensor earth	Roadwheel rotating	0.25 volts (max)
11*	Front right wheel speed sensor signal	Roadwheel rotating	0.5 to 1.5 volts AC (approx)
11	Front right wheel speed sensor earth	Roadwheel rotating	0.25 volts (max)
12	SD connector	-	-
13	Front left wheel speed sensor earth	Roadwheel rotating	0.25 volts (max)
13*	Front left wheel speed sensor signal	Roadwheel rotating	0.5 to 1.5 volts AC (approx)
14	Rear left wheel speed sensor earth	Roadwheel rotating	0.25 volts (max)
14*	Rear right wheel speed sensor signal	Roadwheel rotating	0.5 to 1.5 volts AC (approx)
14*	Rear right wheel speed sensor earth	Roadwheel rotating	0.25 volts (max)
14*	Rear right/left wheel speed sensor common earth	Roadwheel rotating	0.25 volts (max)
15	SD connector	-	-

*Alternative pin connection on certain models

Pin table - typical 25-pin

Note: *Refer to illustration 8B.23*

Pin No.	Connection	Test condition	Voltage
1	ABS warning lamp	Ignition on: Lamp on Lamp off	 1.0 volt (max) Nbv
2	Brake light switch input	Ignition on: Brake pedal released Brake pedal depressed	 0 volts Nbv
3*	Output to engine/transmission control ECM		
4	-	-	-
5*	Input from engine/transmission control ECM		
6*	Output to engine/transmission control ECM		
7	-	-	-
8	Front left wheel speed sensor signal	Roadwheel rotating	0.5 to 1.5 volts AC (approx)
9	Rear left wheel speed sensor earth	Roadwheel rotating	0.25 volts (max)
10	Front right wheel speed sensor signal	Roadwheel rotating	0.5 to 1.5 volts AC (approx)
11	-	-	-
12	Front right wheel speed sensor earth	Roadwheel rotating	0.25 volts (max)
13	-	-	-
14*	Output to engine/transmission control ECM		
15	SD connector	-	-
16	-	-	-
17*	Output to transmission control ECM		
18	-	-	-
19	-	-	-
20	Front left wheel speed sensor earth	Roadwheel rotating	0.25 volts (max)
21	Rear left wheel speed sensor signal	Roadwheel rotating	0.5 to 1.5 volts AC (approx)
22	Rear right wheel speed sensor earth	Roadwheel rotating	0.25 volts (max)
23	Rear right wheel speed sensor signal	Roadwheel rotating	0.5 to 1.5 volts AC (approx)
24	-	-	-
25	-	-	-

*Versions with traction control only

8B

Fault codes

11 General fault codes

1 If a FCR is available, it can be connected to the SD serial connector (the Vauxhall/Opel term for SD connector is ALDL) and used as described in paragraphs 2 and 3 for Fiat and Renault fault codes. If a FCR is not available, it is still possible to obtain fault codes, which will be displayed as a series of flashes of the ABS warning lamp on the instrument panel.

Obtaining codes without a FCR

2 Locate the diagnostic connector and use a jumper lead to bridge the relevant terminals in the connector as follows **(see illustration 8B.26):**

Vehicle	Diagnostic connector location	Terminals to bridge
Alfa Romeo	Centre of engine compartment bulkhead	A and B
Citroën/Peugeot	Engine compartment fuse/relay box (grey plug)	A to earth
Rover	Under the facia on the driver's side	A and C*
Vauxhall/Opel	Under the facia on the left-hand side	A and K

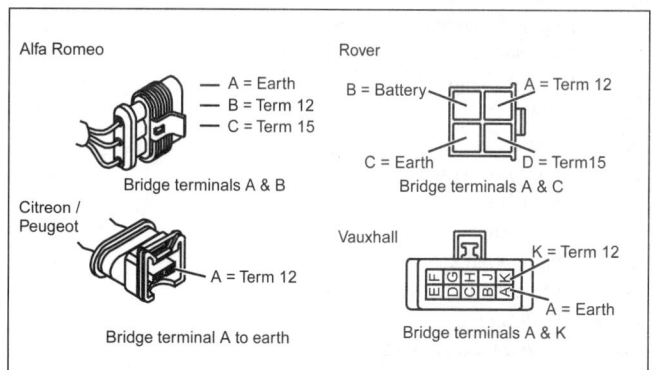

8B.26 Method of obtaining fault codes by bridging diagnostic connector terminals as shown

3 With the bridging wire in place switch the ignition on and the ECM will enter diagnostic mode.

* Remove the link wire connecting terminals C and D in the diagnostic connector before connecting the bridging wire. Reconnect the link wire before clearing the fault codes from the ECM memory.

4 The fault codes will now be output in numerical sequence as a series of flashes of the ABS warning lamp. By counting the flashes and referring to the fault code table, faults can thus be determined.

5 The first code to be output will be code 12 (start of diagnosis) which will be displayed as one flash of the warning light, then a pause, then two flashes. Code 12 will be displayed three times followed by a three second pause and then the next code.

Clearing fault codes from the ECM memory

6 When the fault code output is complete, and any faults repaired, the existing codes must be erased from the ECM memory. To do this, switch the ignition off and remove the bridging wire from the diagnostic connector (refit the link wire on Rover vehicles).

7 Switch the ignition on, wait until the warning lamp extinguishes, then switch the ignition off. Repeat this procedure for a minimum of twenty times to erase the memory.

12 Fiat and Renault fault codes

1 On Fiat and Renault vehicles, internal fault codes are used by the ECM to designate faults in the system components and circuits. Flash codes are not available for output on these vehicles and a proprietary fault code reader (FCR) or system tester (such as the Renault XR25) is required to interrogate the ECM memory. No actual fault code numbers are available, although the component circuits checked by the ECM are similar to those shown for the other vehicles listed.

2 If a FCR is available, it should be connected to the SD serial connector and used in accordance with the maker's instructions.

3 The FCR can be used for the following purposes:

a) Obtaining fault codes.
b) Clearing fault codes.
c) Obtaining datastream information.
d) Testing the system actuators (relays and solenoid valves).

Fault code table

Code	Item	Fault
11*	End of diagnosis	
12	Start of diagnosis	
16	Front left solenoid valve	Defective (signal interruption, short to earth)
17	Front right solenoid valve	Defective (signal interruption, short to earth)
18	Rear solenoid valve	Defective (signal interruption, short to earth)
19	Solenoid valve relay	Defective (signal interruption, short to earth, supply voltage too low)
25	Wheel speed sensor toothed sensor ring	Wrong number of teeth
31**	Engine management ECM	No signal
35	Pump motor	Defective (non-operation of pump after actuation of relay)
37	Brake light switch	Defective (signal interruption)
39	Front left wheel speed sensor	No/poor signal
41	Front left wheel speed sensor	Signal interrupted
42	Front right wheel speed sensor	No/poor signal
43	Front right wheel speed sensor	Signal interrupted
44	Rear left wheel speed sensor	No/poor signal
45	Rear left wheel speed sensor	Signal interrupted
46	Rear right wheel speed sensor	No/poor signal
47	Rear right wheel speed sensor	Signal interrupted
48	Battery voltage	Voltage too low
55	ECM	Defective
56	Diagnosis error	
57*	Battery voltage	Voltage too low
65**	EEPROM	Not programmed
66**	Engine management ECM	No throttle valve angle signal
67**	Engine management ECM	No engine torque reduction signal
71	Automatic transmission ECM	No torque control signal

*Certain models only

**Models with traction control only

Wiring diagrams

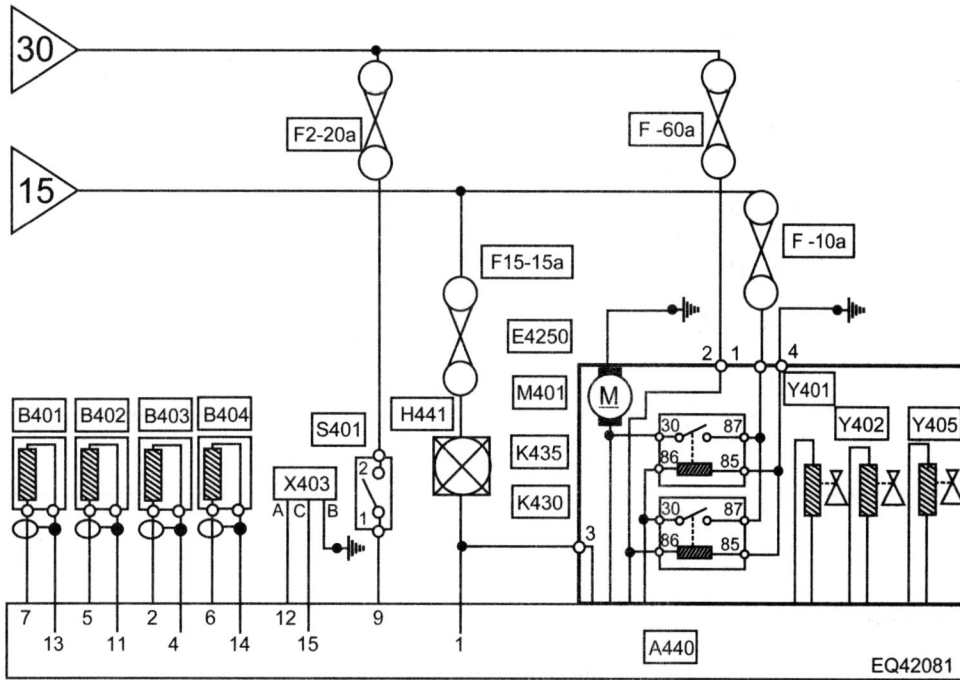

8B.27 Bosch 2EH wiring diagram, Alfa Romeo 145/146

8B

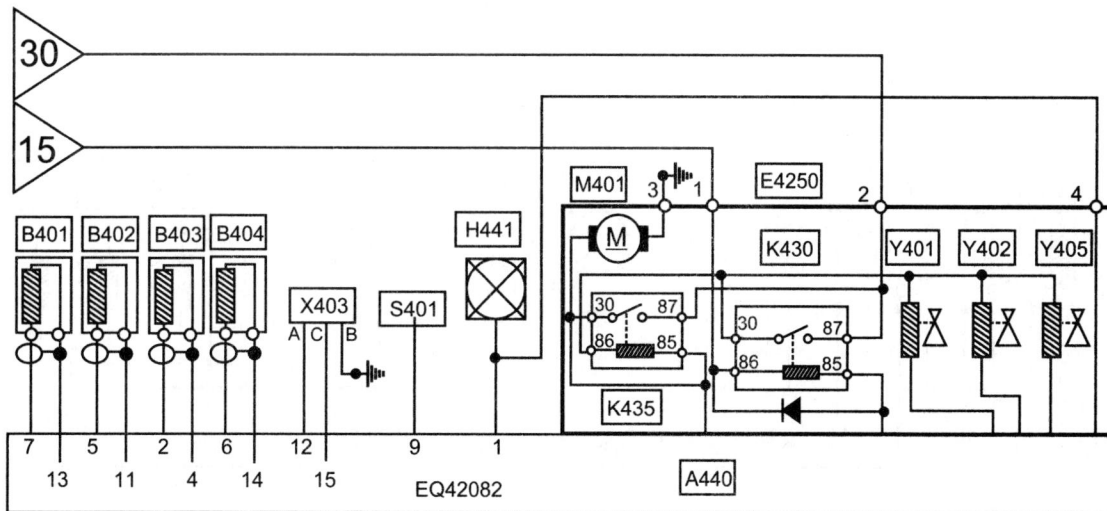

8B.28 Bosch 2EH wiring diagram, Alfa Romeo 155

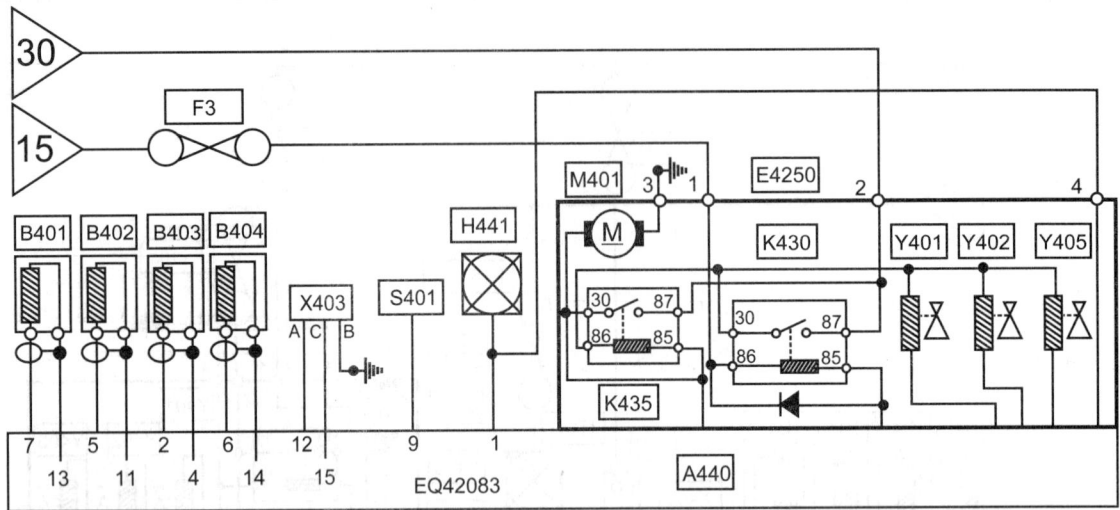

8B.29 Bosch 2EH wiring diagram, Alfa Romeo 164

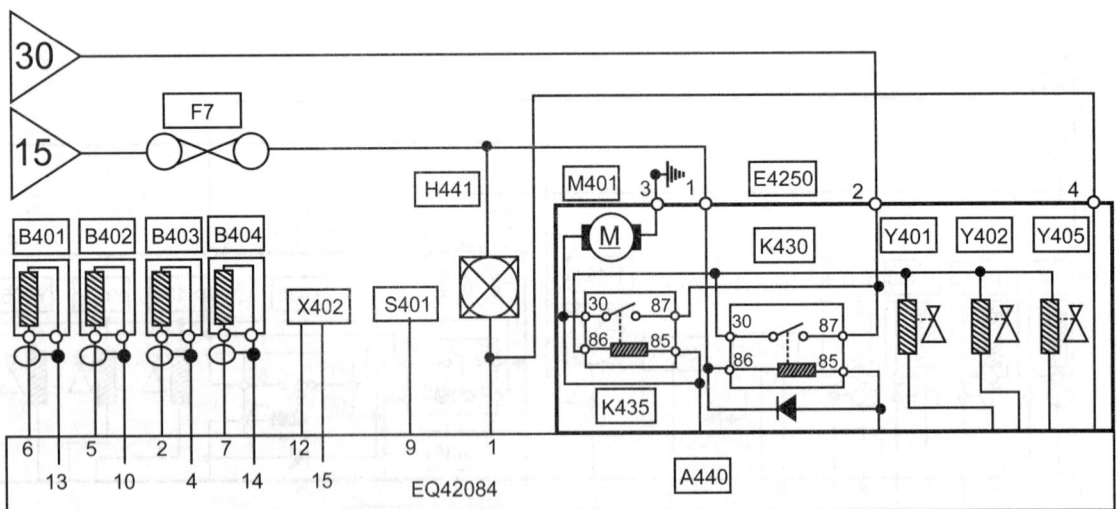

8B.30 Bosch 2EH wiring diagram, Citroën ZX

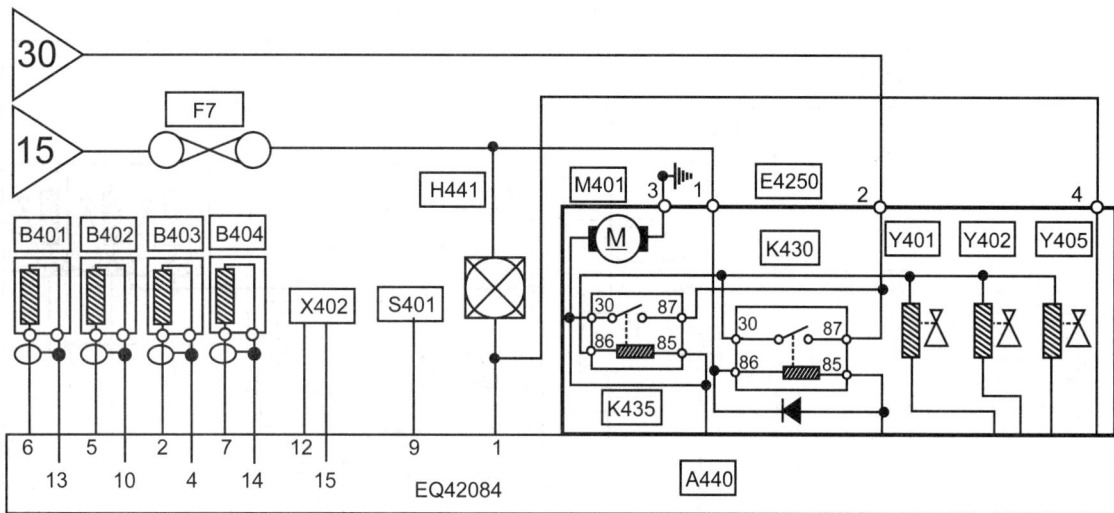

8B.31 Bosch 2EH wiring diagram, Fiat Croma

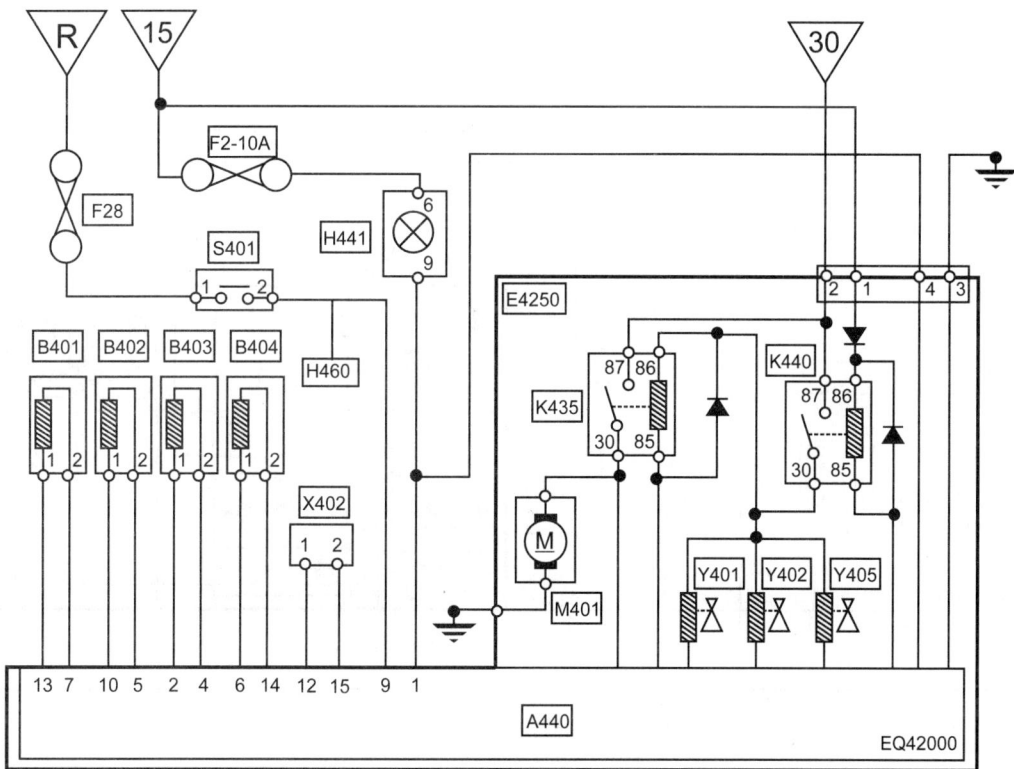

8B.32 Bosch 2EH/2SH, 15-pin wiring diagram, Peugeot 306

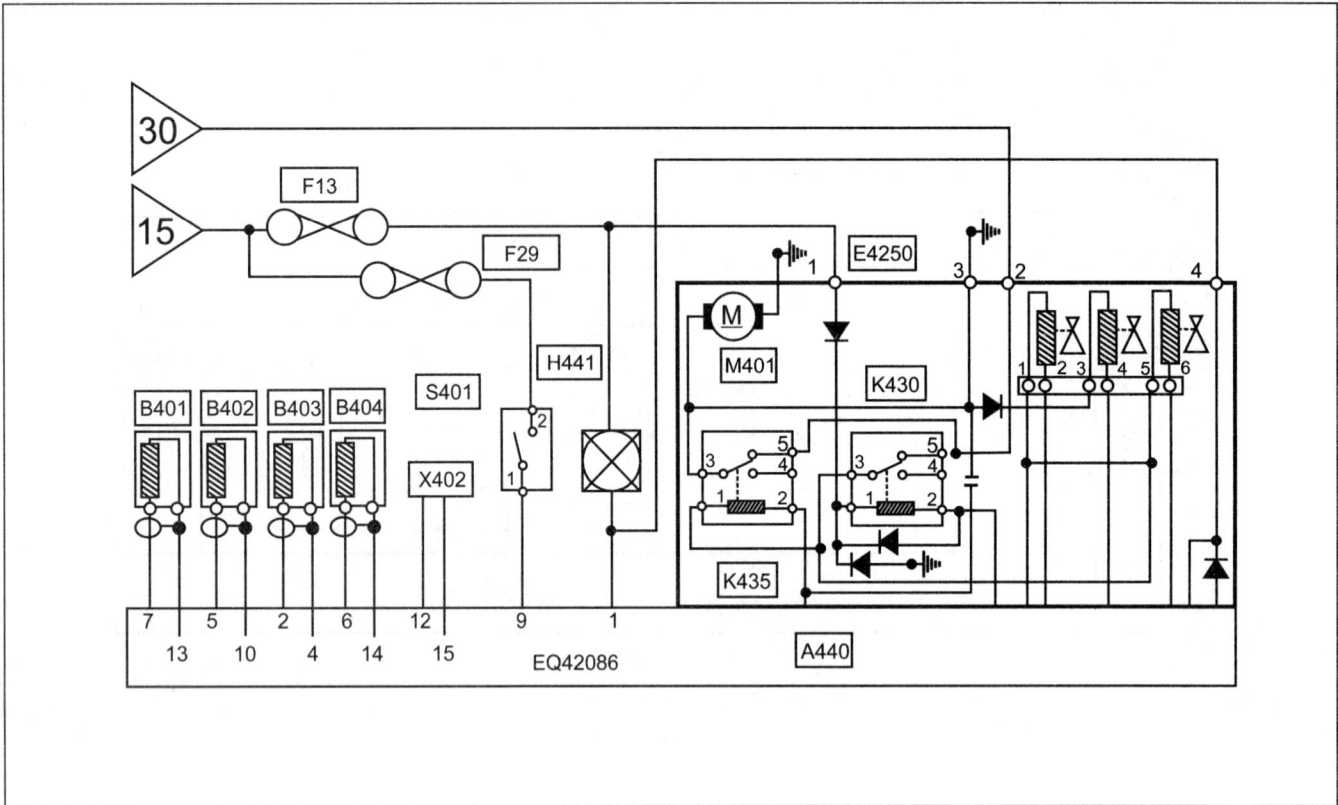

8B.33 Bosch 2EH wiring diagram, Renault 19

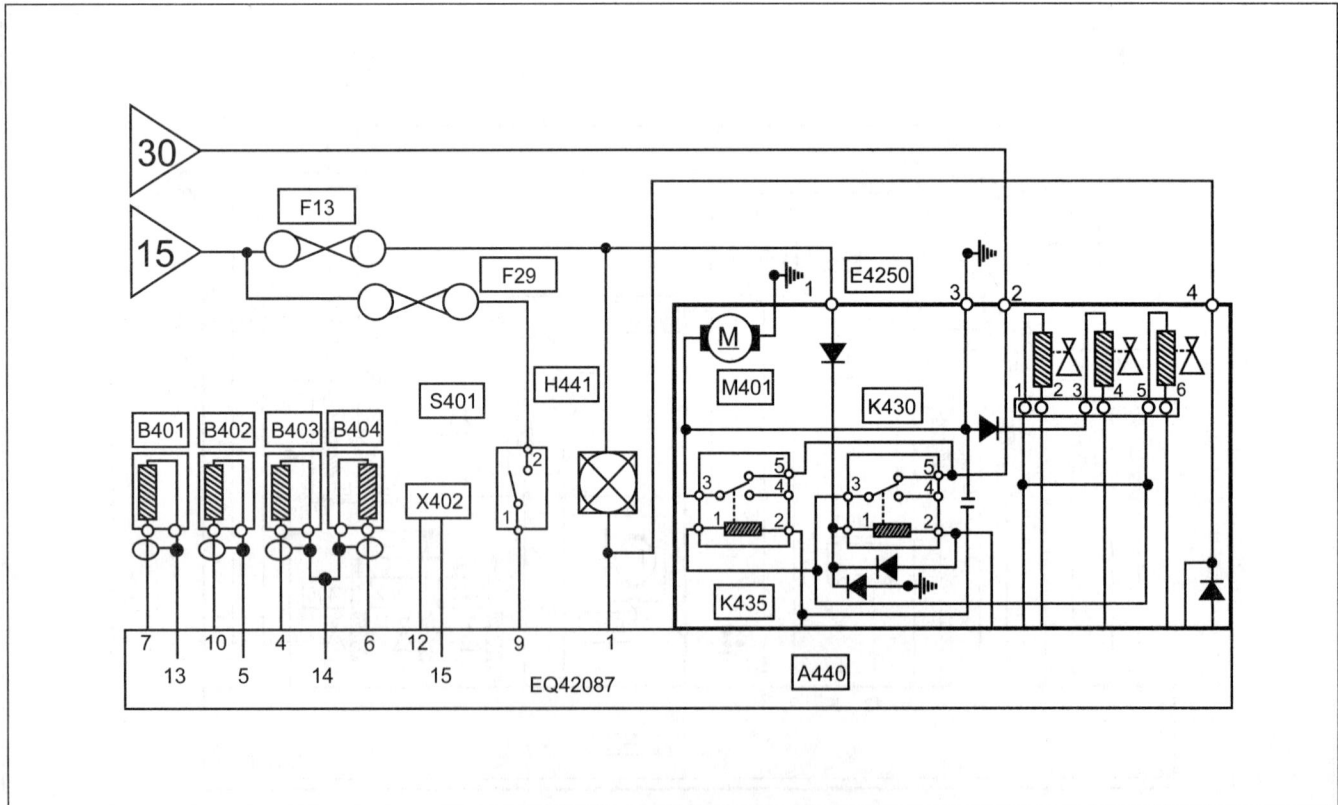

8B.34 Bosch 2EH wiring diagram, Renault Clio

8B

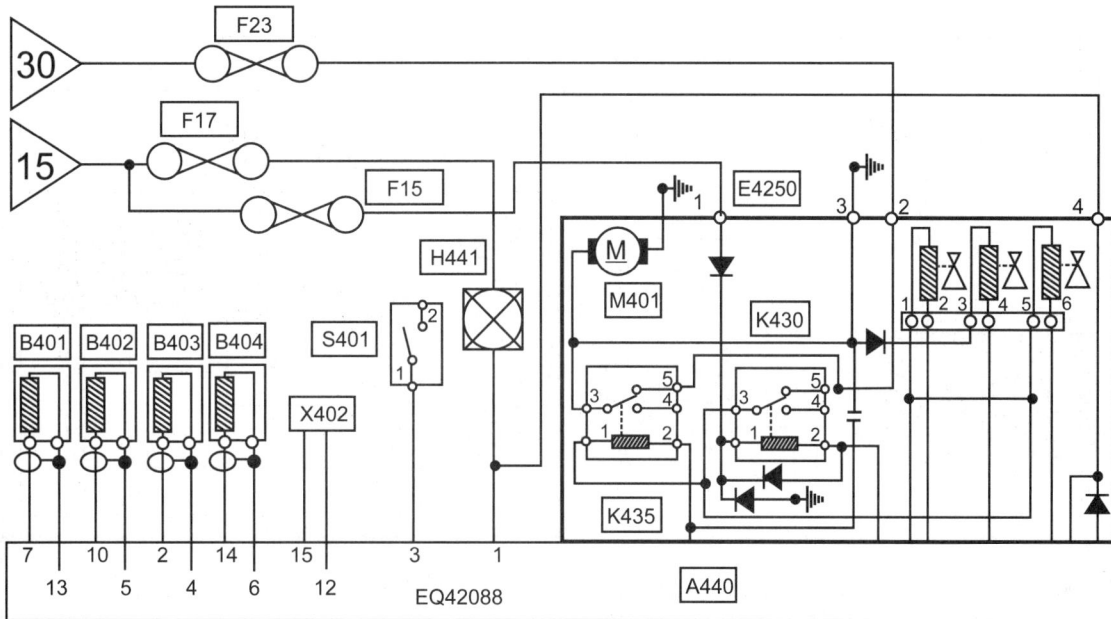

8B.35 Bosch 2EH wiring diagram, Renault Safrane

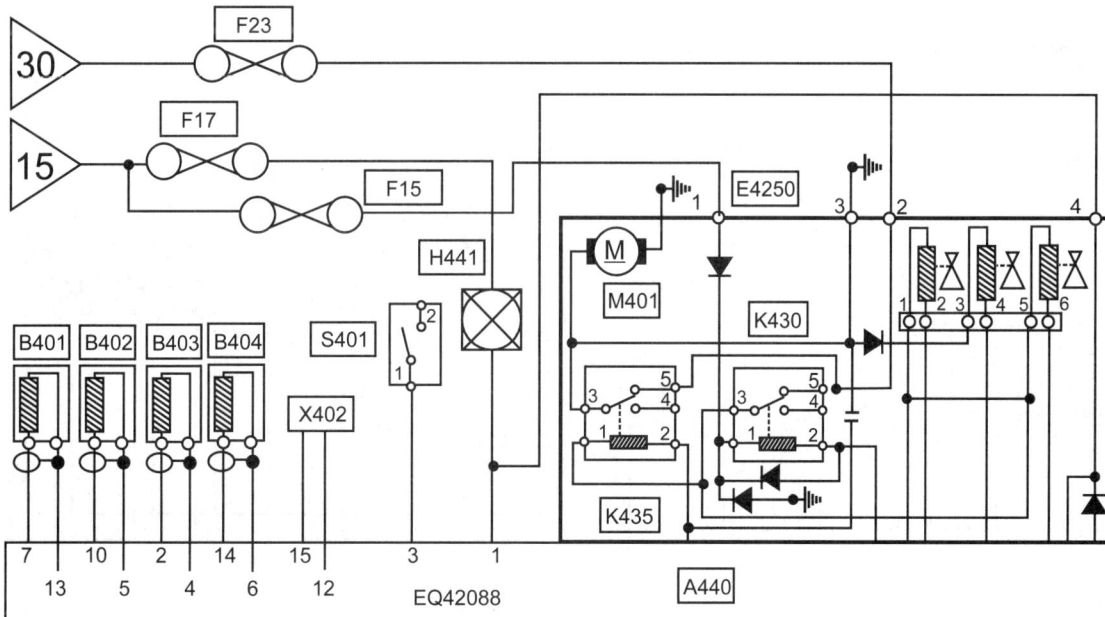

8B.36 Bosch 2EH wiring diagram, Vauxhall/Opel Astra-F

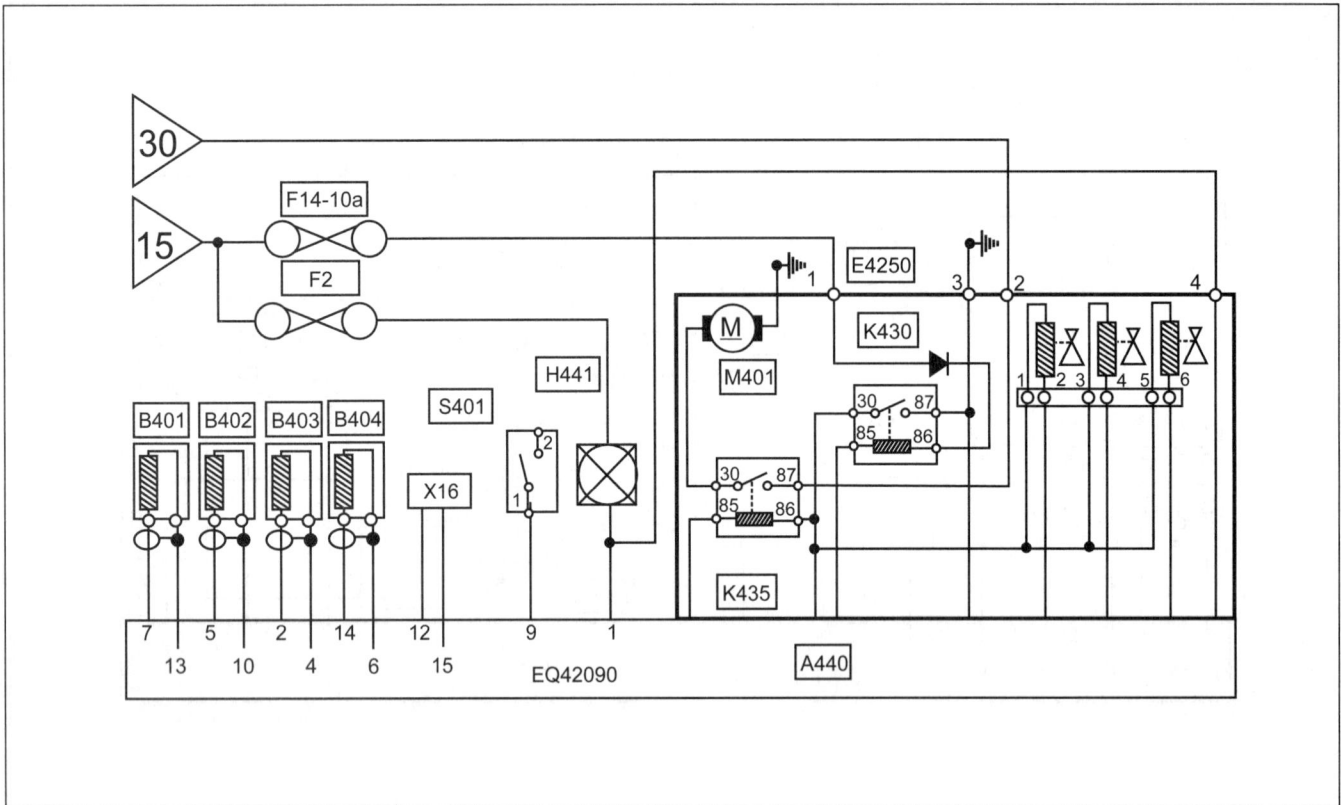

8B.37 Bosch 2EH wiring diagram, Vauxhall/Opel Calibra (2WD)/Cavalier

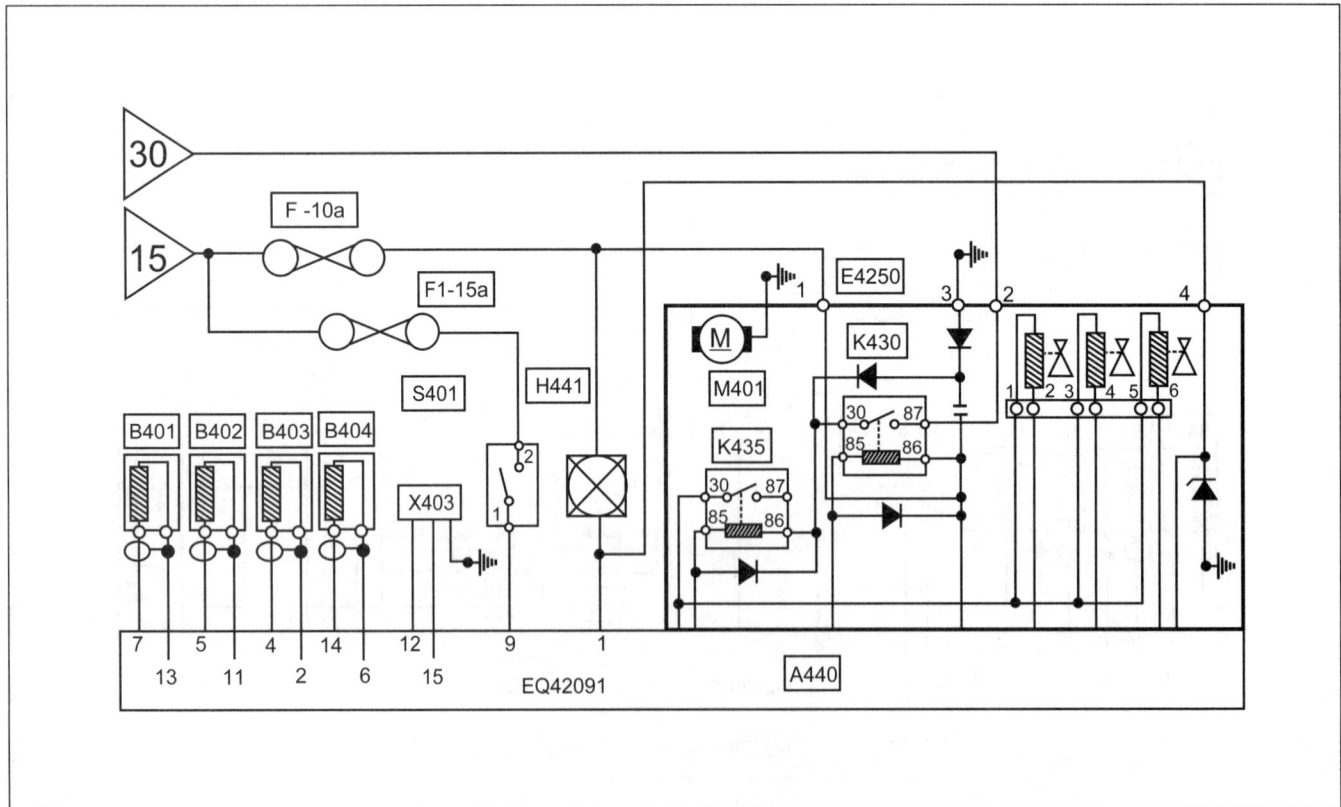

8B.38 Bosch 2SH wiring diagram, Fiat Punto

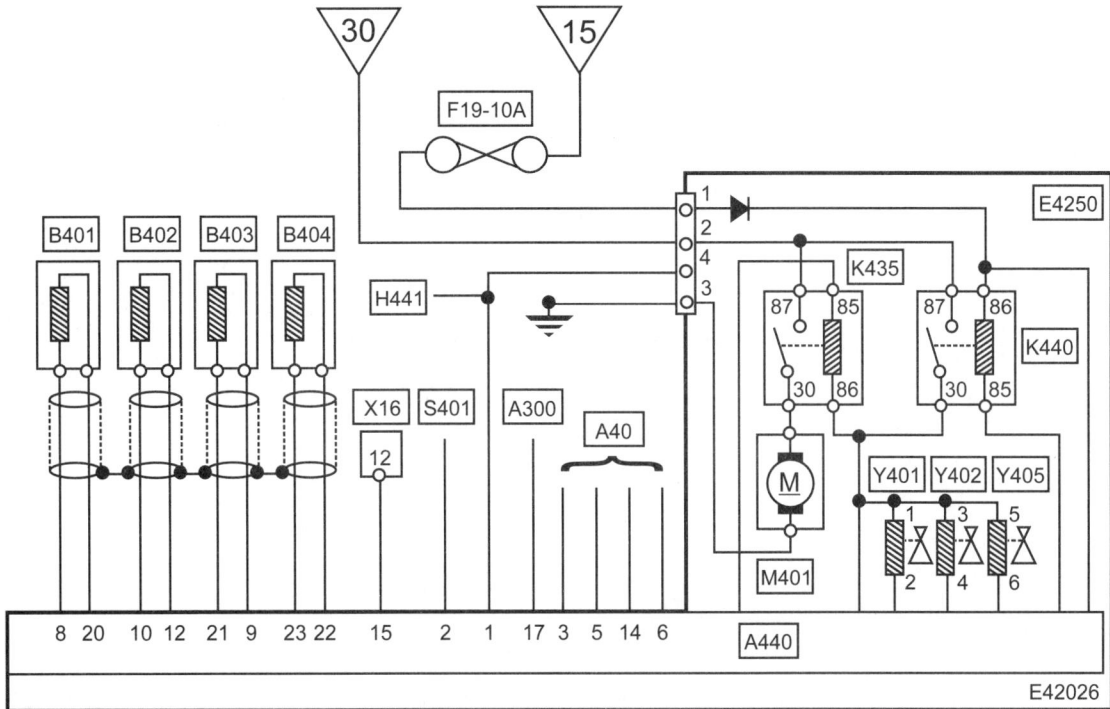

8B.39 Bosch 2SH, 25-pin wiring diagram, Vauxhall/Opel Omega-B

8B

Chapter 9
Bosch 5.0 ABS

Contents

9

Vehicle coverage

Model	Year	System
Audi		
A4	1995-1996	88-pin ECM
A6	1994-1998	88-pin ECM
A8	1994-1997	88-pin ECM
BMW		
5-Series (E39)	1996-1999	88-pin ECM
7-Series (E38)	1996-2000	88-pin ECM
Citroën		
Berlingo	1996-2000	40-pin ECM
ZX	1996-1997	40-pin ECM
Peugeot		
Partner	1996-2000	40-pin ECM
306	1996-1997	40-pin ECM
406	1996-1997	40-pin ECM
605	1996-1997	40-pin ECM
Renault		
Espace	1995-2000	40-pin ECM
Rover/MG		
200	1995-1997	40-pin ECM
400	1995-1996	40-pin ECM
600	1996-1998	40-pin ECM
800	1996-1999	40-pin ECM
MGF	1995-2000	40-pin ECM
Vauxhall/Opel		
Vectra-B	1994-1997	26-pin ECM
Volkswagen		
LT	1997-1998	40-pin ECM
Transporter	1996-1997	40-pin ECM
Volvo		
S40/V40	1996-1998	40-pin ECM

Overview of system operation

1 Basic principles and system identification

The fifth generation of Bosch Antilock Brake Systems (Bosch 5.0) is a development of the earlier Bosch 2 series ABS, with significant improvements to the hydraulic control unit, and the Electronic Control Module (ECM) software. Bosch 5.0 has been fitted to a wide range of passenger vehicles since its introduction in the mid 1990's. The system is of the additional or 'add-on' type operating in conjunction with the conventional braking system components.

Depending on application, Bosch 5.0 may be installed as an antilock braking system only, or as an antilock braking system incorporating traction control (Bosch 5.0/TC).

In ABS mode, the purpose of the system is to apply the vehicle brakes at maximum efficiency without wheel lock or loss of directional stability. Inductive, or 'active' sensors (wheel speed sensors) monitor the speed of the wheels by generating an electrical signal as the wheel is rotated. This information is passed to the ABS-ECM which compares the signals received from each wheel and uses the speed of the fastest wheel as a reference value. The ECM continually monitors the speed of each wheel and if the onset of lock at any wheel is detected (a received speed signal being less than the reference value) a signal is sent to the ABS hydraulic control unit which regulates the brake pressure for the relevant wheel(s).

Where the system incorporates traction control, essentially the reverse principle is applied. When the ECM detects that one or more wheels are rotating faster than the reference value, the brake is actually applied on the relevant wheel(s) to reduce the rotational speed. Additionally, when wheel spin is detected, various signals are sent to the engine management ECM to control engine torque and rpm.

Bosch 5.0 ABS is comprised of the following main components **(see illustration 9.1)**:

a) *Hydraulic control unit with integral ABS-ECM.*
b) *Inductive wheel speed sensors and associated sensor rings.*
c) *Brake light switch.*
d) *ABS warning lamp.*
e) *Diagnostic connector.*

In addition, the conventional brake system is comprised of the following components:

a) *Tandem brake master cylinder.*
b) *Vacuum servo unit.*
c) *Brake calipers/wheel cylinders and hydraulic hoses and pipes.*
d) *Pressure regulating/load-sensing valve(s) depending on application.*

Although the hydraulic operation of the Bosch 5.0 system is essentially the same for all models, there are considerable differences in the ECM software and pin connections, and in the construction of the hydraulic control unit. For the purposes of the test procedures described later in this Chapter, Bosch 5.0 can be divided into three distinct groups; 26-pin ECM types, 40-pin ECM types, and 88-pin ECM types.

2 Component description and operation

ABS ECM

General

The Bosch 5.0 Electronic Control Module (ECM) continually monitors wheel speed from the signals provided by the wheel speed sensors, and brake application from the brake light switch signal. If the ECM detects the incidence of wheel lock (or wheel spin, if traction control is incorporated) on one or more wheels, a signal is sent to the hydraulic control unit to modulate the hydraulic pressure to the brake of the locking or spinning wheel(s). The ECM contains two microprocessors and uses digital technology to complete this function and other functions such as, fault code memory and power modules for valve and relay activity **(see illustration 9.2)**.

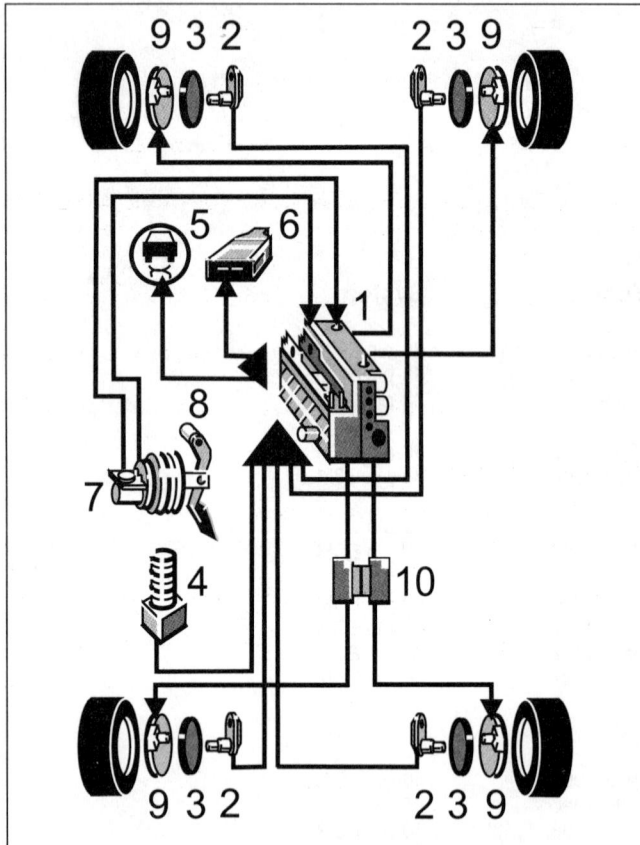

9.1 Typical Bosch 5.0 main components

1. Hydraulic control unit with integral ECM
2. Wheel speed sensors
3. Sensor rings
4. Brake light switch
5. ABS warning lamp
6. Diagnostic connector
7. Tandem master cylinder
8. Vacuum servo unit
9. Brake calipers/wheel cylinders and hydraulic hoses and pipes
10. Pressure regulating/load-sensing valve(s)

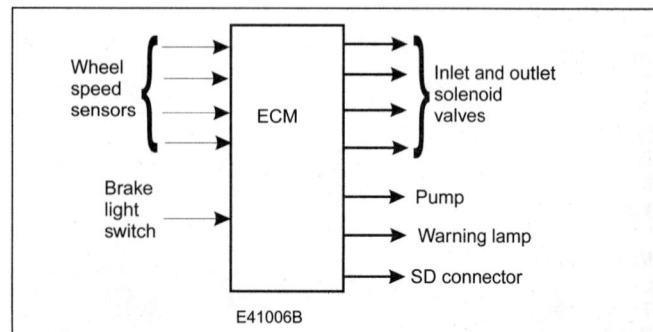

E41006B

9.2 ECM sensor inputs and control signal outputs

9.3 Bosch 5.0 ECM (1) and integral hydraulic control unit (2)

To reduce external electrical connections to a minimum and improve reliability, the ECM is generally integral with the hydraulic control unit **(see illustration 9.3)**. In certain applications (notably VW) the ECM and hydraulic control unit cannot be separated and are only available as a complete sealed unit.

Self-test

The Bosch 5.0 ECM is equipped with a self-test capability that initially examines the ABS system when the ignition is switched on, and then examines the wheel speed sensor signals after a wheel speed of approximately 4 mph is reached from all wheels. The ABS self-test program continues to examine the signals from the various components as long as the ignition is switched on. If self-test determines that faults are not present, the ABS is ready for operation once a specified vehicle speed has been achieved.

If the ECM detects that a fault is present, all ABS functions are switched off and the warning lamp is turned on. The conventional braking system continues to operate as normal without ABS assistance.

Self-diagnostics

If the ECM detects a fault during the self-test routine, an internal fault code is stored in the ECM memory. Stored fault codes can be retrieved from the SD connector with the aid of a suitable fault code reader. If the fault clears, the code will remain stored until cleared with an FCR.

Hydraulic control unit

Bosch 5.0 is typically a four-channel system with a separate hydraulic circuit for each brake. On BMW models, however, the system is installed in three-channel configuration with a separate hydraulic circuit for each front brake, but with the rear brakes controlled as a pair. The hydraulic control unit consists of an electric motor and radial piston return pump, inlet and outlet solenoid valves, pressure accumulators and pulsation dampers **(see illustration 9.4)**. The unit controls the hydraulic pressure applied to the brake for each individual front wheel and each individual rear wheel, or pair of wheels. The return pump is switched on when the ABS is activated and returns hydraulic fluid, drained off during the pressure reduction phase, back into the brake circuit.

In most installations, the 'select-low' principle is employed for control of the rear brakes during ABS operation. With the 'select-low'

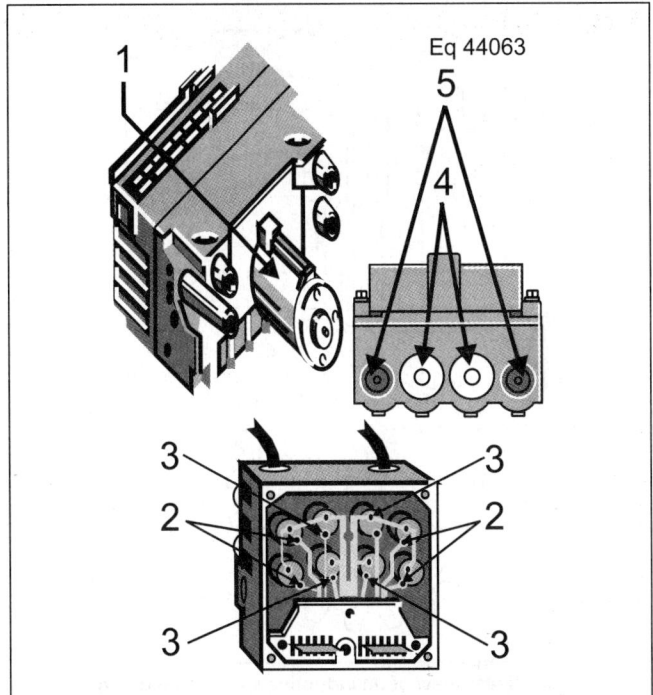

9.4 Bosch 5.0 hydraulic control unit components

1	Pump motor	4	Pressure accumulators
2 and 3	Solenoid valves	5	Pulsation dampers

principle, the wheel with the lowest adhesion determines the amount of hydraulic pressure to be supplied to both rear brakes during a controlled ABS cycle.

In certain installations, the ABS-ECM contains additional software for rear brake hydraulic fluid pressure regulation when ABS is not in operation. This is generally known as 'Electronic Brake Force Distribution' and in these applications, mechanical pressure regulating/load sensing valves are not required as the rear brake hydraulic pressure is controlled by the ABS hydraulic control unit.

Wheel speed sensors

Inductive type wheel speed sensors

The rotational speed of the roadwheels and any changes in the rotational speed are recorded either by inductive, or 'active' wheel speed sensors, one located at each roadwheel **(see illustration 9.5)**.

Where inductive sensors are used, each wheel speed sensor assembly comprises a toothed sensor ring, which rotates at

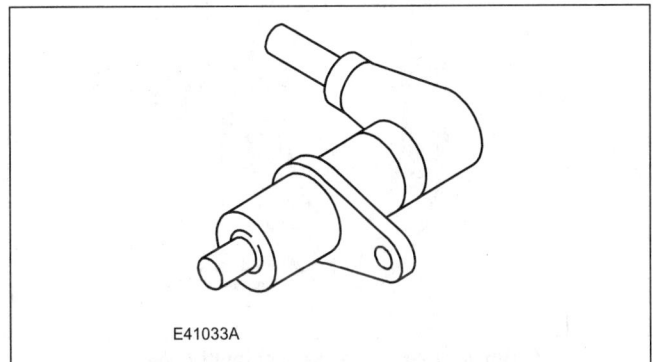

9.5 Typical inductive wheel speed sensor

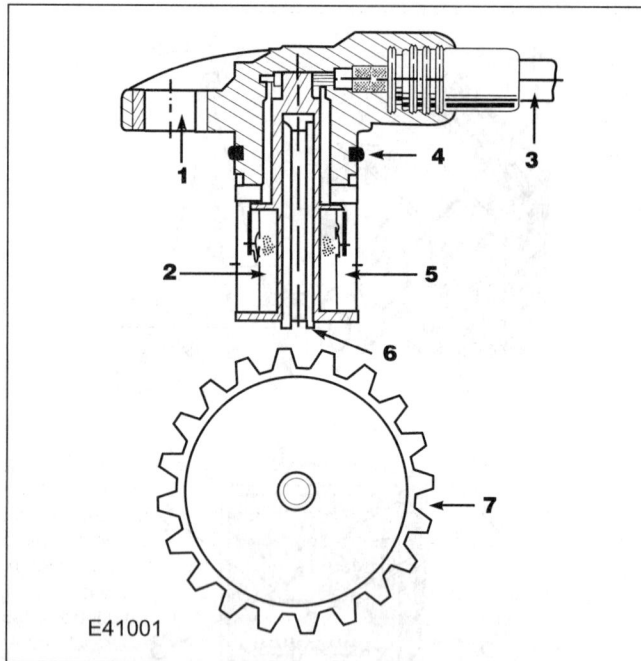

9.6 Sectional view of an inductive wheel speed sensor

1	Mounting bolt location	5	Coil
2	Permanent magnet	6	Sensor tip
3	Wiring harness	7	Toothed sensor ring
4	O-ring		

roadwheel speed, and an adjacent sensor mounted a set distance from the sensor ring **(see illustration 9.6)**.

The sensors are permanent magnet pulse generator types producing an AC voltage sine wave as the sensor ring teeth pass through the magnetic field of the sensor **(see illustration 9.7)**.

The frequency of the waveform produced by the wheel speed sensor is proportional to the road speed. This AC voltage signal is continually being delivered to the ABS-ECM for processing.

The peak to peak voltage of the speed signal (when viewed upon an oscilloscope) can vary considerably according to wheel speed. An analogue to digital converter (ADC) in the ECM transforms the AC pulse into a digital signal **(see illustration 9.8)**.

'Active' type wheel speed sensors

On the Bosch 5.0 system fitted to BMW 5-Series models, the

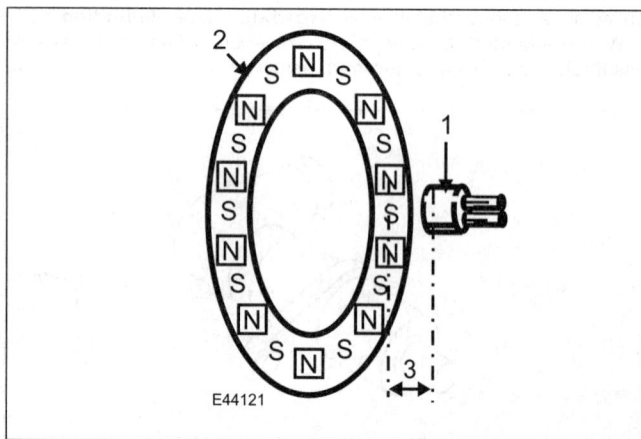

9.9 Active wheel speed sensor multi-polar ring

1 Wheel speed sensor   2 Multi-Polar ring   3 air gap

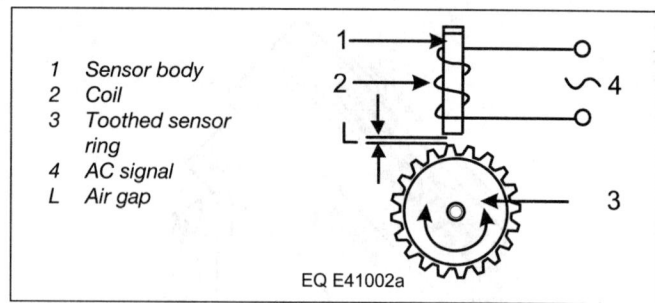

1	Sensor body
2	Coil
3	Toothed sensor ring
4	AC signal
L	Air gap

EQ E41002a

9.7 Inductive wheel speed sensor operation

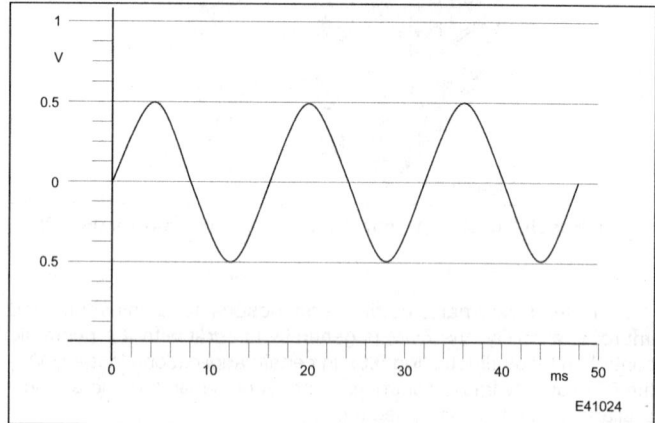

9.8 Inductive wheel speed sensor waveform as viewed on an oscilloscope

rotational speed of the roadwheels and any changes in the rotational speed are recorded by 'active' wheel speed sensors.

On the majority of antilock brake systems, the wheel speed sensors are of the permanent magnet pulse generator type, as described previously. On the Bosch 5.0 system with 'active' sensors, the internal electrical resistance of the sensor is altered by changes in intensity and direction of the lines of force of an external magnetic field.

This external magnetic field is created by a unique type of sensor ring known as a 'multi-polar ring'. The multi-polar ring consists of a series of magnetic elements with alternating north/south polarities, located around the circumference of the ring **(see illustration 9.9)**.

The multi-polar ring is attached to the driveshaft or wheel hub, with the adjacent sensor mounted a set distances from the ring. As the ring rotates at roadwheel speed, a DC voltage square wave, constant amplitude signal is produced; the frequency of which is proportional to the road speed. This signal is continually being delivered by the sensor to the ABS-ECM for processing **(see illustration 9.10)**.

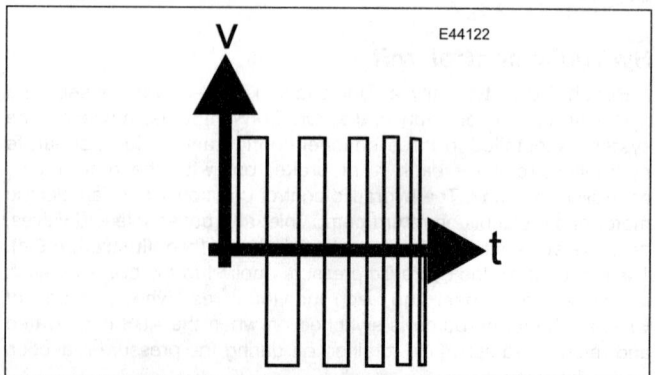

9.10 DC voltage square wave signal produced by the active wheel speed sensors

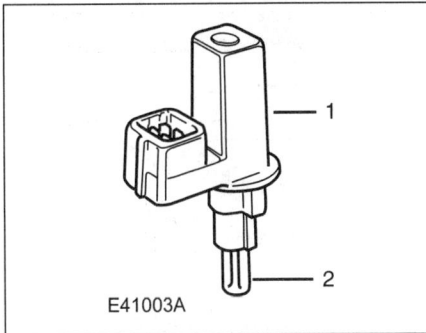

**9.11 Typical brake light switch body (1)
and contact pin (2)**

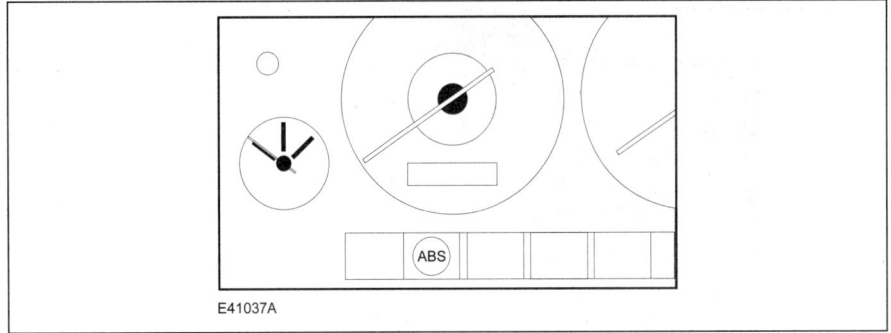

9.12 ABS warning lamp

ABS electrical wiring harness and relays

An integrated main wiring harness is used for ABS-ECM power supply and earth connections, and enables sensor signals to reach the ECM and the ECM, in turn, to send output signals to the ABS warning lamp and diagnostic connector. On most versions, the main (solenoid valve) relay and return pump relay are an integral part of the ECM and cannot be separately removed. Internal connections between the ECM and hydraulic control unit are used to activate the return pump motor.

Brake light switch

The brake light switch comprises a switch body and contact pin and is located above the brake pedal **(see illustration 9.11)**.

When the brake pedal is depressed, closing the brake light switch, a signal is sent to the ECM indicating that the brakes are being applied. Once this signal is received, the ECM will begin monitoring the wheel speed via the wheel speed sensors and activate the ABS if necessary.

ABS warning lamp

After the ignition is switched on, the ABS warning lamp on the instrument panel is illuminated for approximately 4 seconds, then extinguished. During vehicle operation above a pre-determined wheel speed, the ABS-ECM implements a self-test cycle whereby ABS operation is monitored. If a fault is detected, the ECM illuminates the warning lamp on the instrument panel. The ECM switches off the ABS, however the conventional braking system continues to operate as normal. The warning lamp will remain illuminated until the fault is no longer present **(see illustration 9.12)**.

When the ABS-ECM detects a fault, the fault code is stored and the ABS warning lamp activated. If the fault no longer exists after the next system start (ignition on/off) the ABS warning lamp is extinguished after the self-test cycle, however the fault code remains stored in the ECM memory.

Tandem master cylinder

Typically, the tandem master cylinder comprises a body casting incorporating primary and secondary pressure chambers, primary piston, intermediate piston, floating piston, slotted pin and central valve. The cylinder operates as a conventional master cylinder using vacuum assistance from the vacuum servo unit **(see illustration 9.13)**.

When the brake system is at rest, the central valve in the floating piston rests against the slotted pin. In this condition, the central valve is open and brake fluid can discharge out of the pressure chamber back into the brake fluid reservoir. When the brake pedal is depressed, the build-up of hydraulic pressure in the primary pressure chamber acts on the intermediate piston and floating piston, moving them down the cylinder bore. The floating piston contacts the seal on the central valve, closing the connection between the intermediate and secondary pressure chambers. Brake hydraulic pressure can now also increase in the secondary pressure chamber.

Vacuum servo unit

The vacuum servo unit is located between the brake pedal and tandem master cylinder. When the brake pedal is depressed, the servo unit increases the force applied by the pedal, reducing the effort required to operate the brakes **(see illustration 9.14)**.

The unit is operated by vacuum created in the engine inlet manifold (or from a separate vacuum pump on diesel engines) which is applied to a diaphragm contained within the unit casing. A pushrod connected to the centre of the diaphragm acts directly on the primary piston in the master cylinder.

When the brake pedal is released, vacuum is applied to both sides of the diaphragm. When the pedal is depressed, one side of the diaphragm is opened to atmosphere and the vacuum acting on the other side deflects the diaphragm which in turn operates the master cylinder primary piston. The resulting force applied to the master cylinder piston is therefore significantly greater than the initial force applied to the brake pedal by the driver.

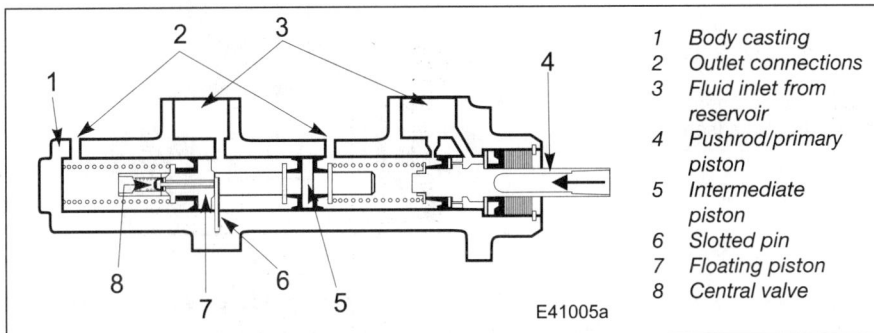

1 Body casting
2 Outlet connections
3 Fluid inlet from reservoir
4 Pushrod/primary piston
5 Intermediate piston
6 Slotted pin
7 Floating piston
8 Central valve

9.13 Sectional view of a typical tandem master cylinder

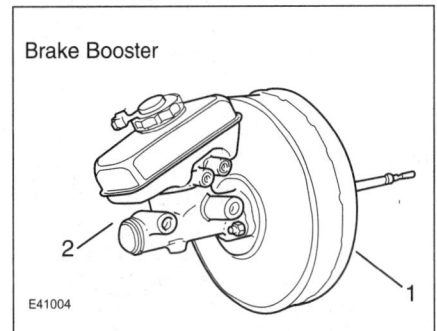

Brake Booster

9.14 Vacuum servo unit (1) and tandem master cylinder (2)

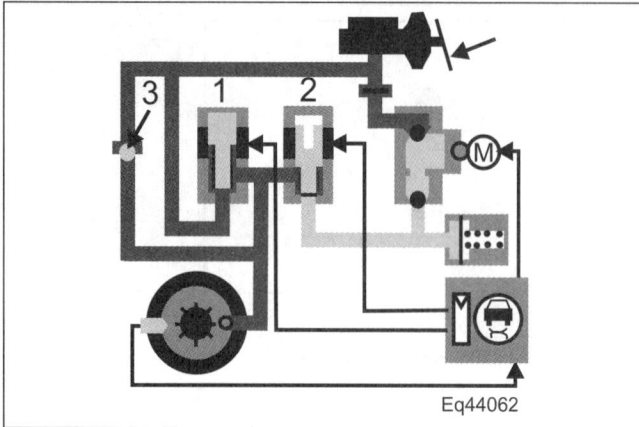

9.15 Brake system operating under conventional control without ABS

1   Inlet solenoid valve
2   Outlet solenoid valve
3   One-way valve

Pressure regulating/load-sensing valve(s)

Depending on vehicle application, pressure regulating valves or load sensing valves may be incorporated to restrict the hydraulic fluid pressure to the rear brakes. The valves may be pressure conscious whereby the hydraulic fluid supply is restricted once a pre-determined pressure is reached, or load conscious whereby the hydraulic pressure is reduced according to vehicle loading.

On certain versions, the ABS-ECM contains additional software for rear brake hydraulic fluid pressure regulation during normal braking, generally known as 'Electronic Brake Force Distribution'. In these applications, mechanical pressure regulating/load sensing valves are not required as the rear brake hydraulic pressure is controlled by the ABS hydraulic control unit.

3   System operation

Brake system at rest

When the system is at rest all the brake components are inoperative. Pressure is non-existent in the hydraulic pipes between the tandem master cylinder and the brake calipers. The inlet solenoid valves in the hydraulic control unit valve block are open and the outlet solenoid valves are closed.

Brake system operating under conventional control without ABS

When the brake pedal is activated, the pedal force is applied to the tandem master cylinder by the vacuum servo unit pushrod. The servo unit pushrod acts directly on the pressure piston in the master cylinder, which pressurises the hydraulic fluid in the brake pipes to the hydraulic control unit. The inlet solenoid valve and outlet solenoid valve both remain in the 'at rest' position (inlet solenoid valve open and outlet solenoid valve closed). Hydraulic pressure is transmitted to each brake caliper, thus operating the brakes.

When the brake pedal is released, the one-way valve opens allowing the hydraulic pressure in the circuit to rapidly decrease (see illustration 9.15).

Brake system operating in conjunction with ABS control

The ABS-ECM continually monitors wheel speed from the signals provided by the wheel speed sensors. If the ECM detects the incidence of wheel lock on one or more wheels, ABS is automatically initiated in three phases. As Bosch 5.0 ABS typically operates individually on each wheel, all or any of the wheels could be in any one of the following phases at any particular moment.

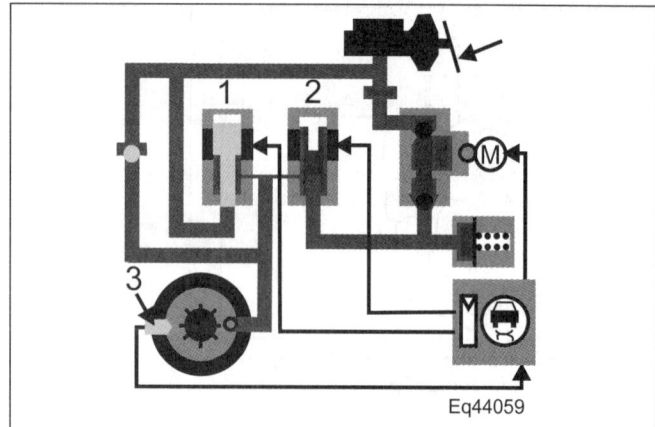

9.16 ABS operation - first phase, pressure holding

1   Inlet solenoid valve
2   Outlet solenoid valve
3   Wheel speed sensor

First ABS phase, pressure holding

To prevent any further build-up of hydraulic pressure in the circuit being controlled, the ECM closes the inlet solenoid valve and allows the outlet solenoid valve to remain closed. The hydraulic fluid line from the tandem master cylinder to the brake caliper or wheel cylinder is closed, and the hydraulic fluid in the controlled circuit is maintained at a constant pressure. This effectively removes the braking force from the controlled circuit. The pressure cannot now be increased in that circuit by any further application of the brake pedal (see illustration 9.16).

If the wheel speed sensor signals indicate that wheel rotation has now stabilised, the ECM will instigate the pressure build-up phase, allowing braking to continue. If wheel lock is still detected after the pressure holding phase, the ECM instigates the pressure reduction phase.

Second ABS phase, pressure reduction

If the ECM detects wheel instability, a pressure reduction phase is initiated. The inlet solenoid valve remains closed and the outlet solenoid valve is opened by means of a series of short activation pulses. The pressure in the controlled circuit decreases rapidly as the fluid flows from the brake caliper or wheel cylinder into the pressure accumulator. At the same time, the ECM actuates the electric motor to operate the return pump. The hydraulic fluid is then pumped back into the pressure side of the master cylinder. This process creates a pulsation which can be felt in the brake pedal action, but which is softened by the pulsation damper (see illustration 9.17).

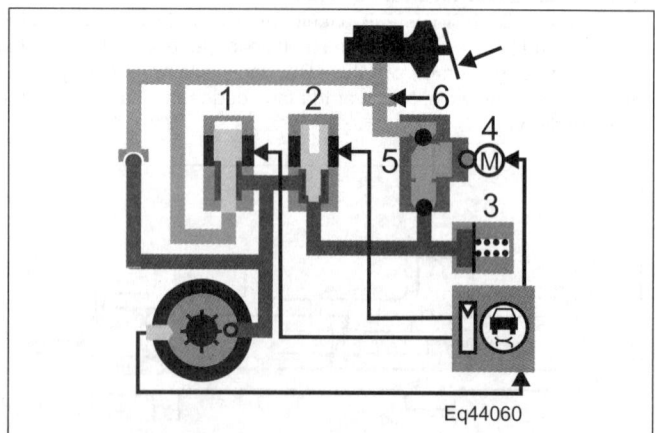

9.17 ABS operation - second phase, pressure reduction

1   Inlet solenoid valve
2   Outlet solenoid valve
3   Pressure accumulator
4   Pump motor
5   Return pump
6   Pulsation damper

Third ABS phase, pressure build-up

The pressure build-up phase is instigated after the wheel rotation has stabilised. The inlet and outlet solenoid valves are returned to the at rest position (inlet solenoid valve open and outlet solenoid valve closed) which re-opens the hydraulic fluid line from the tandem master cylinder to the brake caliper or wheel cylinder. Hydraulic pressure is reinstated, thus re-introducing operation of the brake. After a brief period, a short pressure holding phase is re-introduced and the ECM continually shifts between pressure build-up and pressure holding until the wheel has decelerated to a sufficient degree where pressure reduction is once more required (see illustration 9.18).

The whole ABS control cycle takes place 4 to 10 times per second for each affected wheel and this ensures maximum braking effect and control during ABS operation.

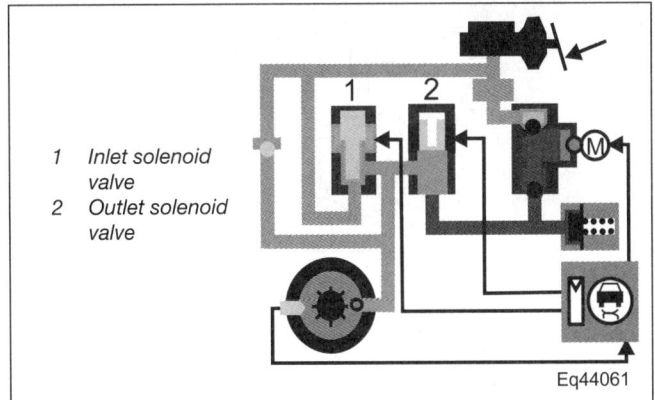

1 Inlet solenoid valve
2 Outlet solenoid valve

Eq44061

9.18 ABS operation - third phase, pressure build-up

Test procedures

Important note: *The test procedures, pintables and wiring diagrams contained in this Chapter are necessarily representative of the system depicted. Because of the variations in wiring and other data that often occurs, even between similar vehicles in any particular VM's range, the reader should take great care in identification of ECM pins, and satisfy himself that he has gathered the correct data before failing a particular component.*

4 Wheel speed sensors

Note: *'Active' type wheel speed sensors are fitted to BMW 5-Series models. All other models utilise inductive wheel speed sensors.*

Inductive type wheel speed sensors

Checking the wheel speed sensor (general)

1 Inspect the wheel speed sensor for corrosion or damage and check that the sensor is tightly mounted.
2 Check the toothed sensor ring for damage, eccentricity and for broken or missing teeth.
3 Inspect the wheel speed sensor-wiring plug for corrosion and damage. One plug for each sensor.

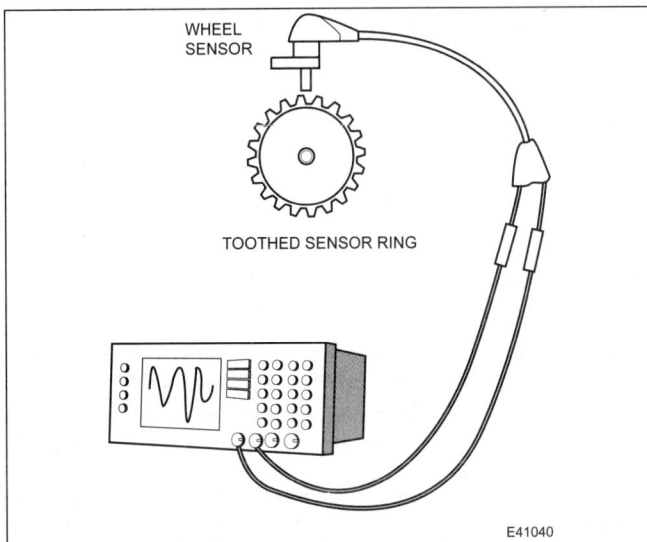

4 Check that the connector terminal pins are pushed fully home and making good contact with the sensor-wiring plug.
5 Check the clearance between the sensor and the toothed sensor ring. The clearance is not normally adjustable but is nominally 0.3 to 0.5 mm. If the clearance is excessive, expect a worn sensor tip or problems with the wheel bearings/hub or sensor ring.
6 When carrying out voltage checks with an oscilloscope or voltmeter, the voltage obtained will be proportional to the speed at which the wheel is rotating. In addition to determining that the wheel speed sensors are actually producing a voltage output, it is essential that the output from the sensors on a particular axle is the same for any given wheel speed.

Checking wheel speed sensor output with an oscilloscope

Note: *Refer to the wiring diagrams for specific ECM pin identification according to model.*
7 Switch the ignition off and disconnect the ECM multi-plug or the relevant wheel speed sensor-wiring plug.
8 Connect an oscilloscope between the terminal pins for the sensor under test (see illustration 9.19).
9 Select a range to cover 80 Hz on the oscilloscope and a free run time base.
10 Raise the wheel and rotate it by hand at approximately one revolution per second.
11 A sinusoidal wave form should be obtained, with amplitude and duration changing with rotational speed (see illustration 9.20).

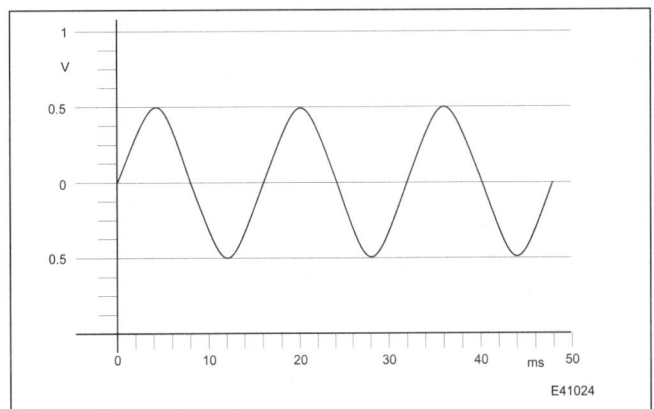

WHEEL SENSOR

TOOTHED SENSOR RING

E41040

9.19 Checking inductive wheel speed sensor output with an oscilloscope connected to the sensor-wiring plug

E41024

9.20 Typical inductive wheel speed sensor sine wave as displayed on an oscilloscope

9

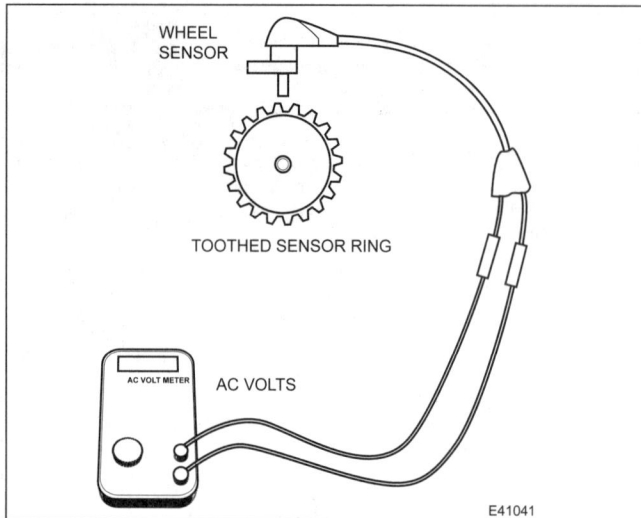

9.21 Checking inductive wheel speed sensor output with a voltmeter connected to the sensor-wiring plug

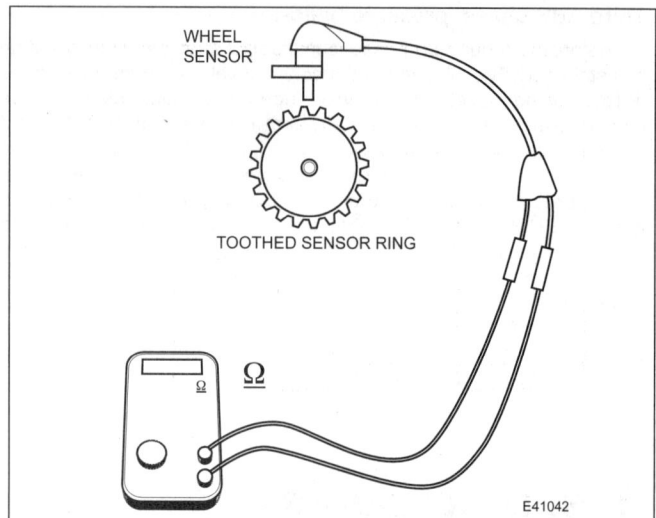

9.22 Checking inductive wheel speed sensor resistance with an ohmmeter connected to the sensor-wiring plug

12 If there is no signal, or a very weak or intermittent signal at the ECM, repeat the test at the sensor-wiring plug. If there is no change in signal status, the sensor is suspect.

13 If the signal is now satisfactory this indicates a fault in the wiring harness, which should be checked for continuity.

Checking wheel speed sensor output with an AC voltmeter

Note: *Refer to the wiring diagrams for specific ECM pin identification according to model.*

14 Switch the ignition off and disconnect the ECM multi-plug or the relevant wheel speed sensor-wiring plug.

15 Connect an AC voltmeter between the terminal pins for the sensor under test **(see illustration 9.21)**.

16 Raise the wheel and rotate it by hand at approximately one revolution per second.

17 A voltage of approximately 0.1 to 0.5 volts (AC RMS) should be obtained. If there is no signal, or a very weak or intermittent signal at the ECM, repeat the test at the sensor-wiring plug. If there is no change in the signal, the sensor is suspect.

18 If the signal is now satisfactory, this indicates a fault in the wiring harness which should be checked for continuity. **Note:** *This test at*

least proves that a signal is being generated by the sensor. However, the voltage produced is an average voltage and does not clearly indicate damage to the sensor ring or that the sinewave is regular in formation.

Checking wheel speed sensor resistance

Note: *Refer to the wiring diagrams for specific ECM pin identification according to model.*

19 Switch the ignition off and disconnect the ECM multi-plug or the relevant wheel speed sensor-wiring plug.

20 Connect an ohmmeter between the terminal pins for the sensor under test **(see illustration 9.22)**.

21 The readings obtained should be between 0.5 and 2.3 kohms approximately.

22 If the resistance is excessively high or open circuit at the ECM, repeat the test at the sensor multi-plug. If there is no change in resistance, the sensor is suspect.

23 If the resistance is now satisfactory, this indicates a fault in the wiring harness which should be checked for continuity. **Note:** *Even if the resistance is within the quoted specifications, this does not prove that the speed sensor can generate an acceptable signal.*

'Active' type wheel speed sensors

Checking the wheel speed sensor (general)

24 Inspect the wheel speed sensor for corrosion or damage and check that the sensor is tightly mounted.

25 Check the toothed sensor ring for damage, eccentricity and for broken or missing teeth.

26 Inspect the wheel speed sensor-wiring plug for corrosion and damage. One plug for each sensor.

27 Check that the connector terminal pins are pushed fully home and making good contact with the sensor-wiring plug.

28 Check the clearance between the sensor and the toothed sensor ring. The clearance is not normally adjustable but is nominally 0.2 to 2.0 mm. If the clearance is excessive, expect a worn sensor tip or problems with the wheel bearings/hub or sensor ring.

Checking wheel speed sensor signal with an oscilloscope

Note: *Refer to the wiring diagrams for specific ECM pin identification according to model.*

29 Switch the ignition off and connect a BOB between the ECM and the harness multi-plug.

30 Connect an oscilloscope to the relevant BOB terminals for the sensor under test, and set the oscilloscope to measure DC voltage **(see illustration 9.23)**.

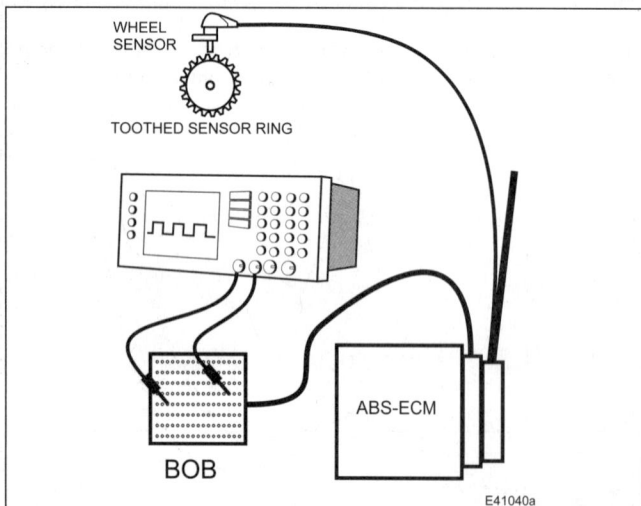

9.23 Checking active wheel speed sensor output with a break-out-box connected between the oscilloscope and ECM multi-plug

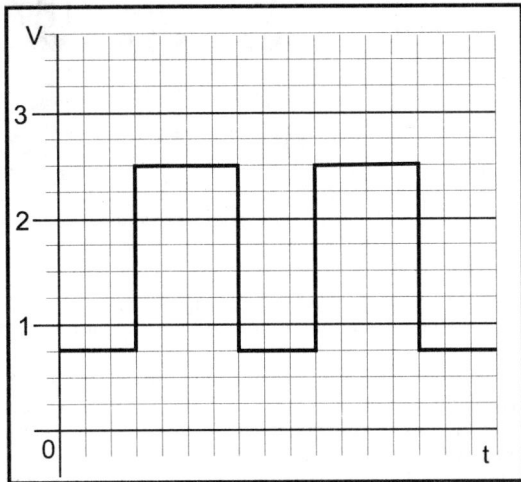

9.24 Active wheel speed sensor square wave output as viewed on an oscilloscope

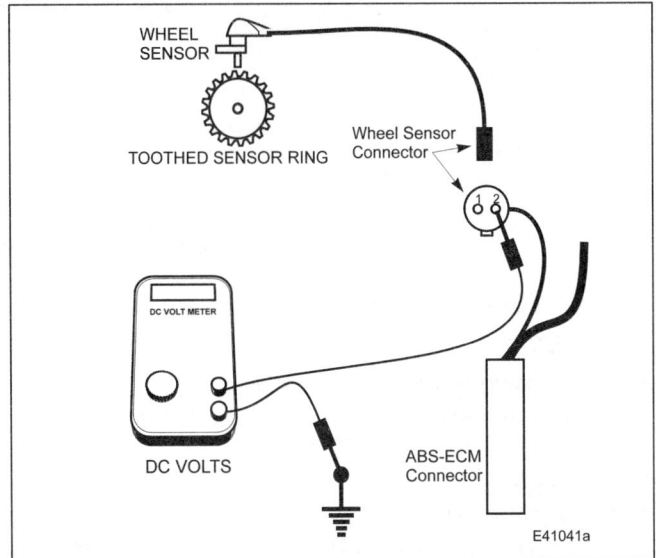

9.25 Checking active wheel speed sensor output using a voltmeter

31 Raise and support the vehicle.
32 Switch the ignition on and rotate the wheel by hand at approximately one revolution per second.
33 A square waveform should be obtained, switching between approximately 0.5 and 2.5 volts as the wheel is rotated **(see illustration 9.24)**.
34 If the output is not as specified, check the supply voltage to the sensor and the wiring harness for continuity.
35 If the supply voltage and wiring are satisfactory, the sensor is suspect.

Checking wheel speed sensor signal with a DC voltmeter

Note: *Refer to the wiring diagrams for specific ECM pin identification according to model.*
36 Switch the ignition off and connect a BOB between the ECM and the harness multi-plug.
37 Connect a DC voltmeter between the relevant BOB terminal for the sensor under test, and earth.
38 Raise and support the vehicle.
39 Switch the ignition on and rotate the wheel very slowly.
40 A voltage switching between approximately 0.5 and 2.5 volts should be obtained as the wheel is rotated.
41 If the voltage is not as specified, check the supply voltage to the sensor and the wiring harness for continuity.
42 If the supply voltage and wiring are satisfactory, the sensor is suspect.

Checking wheel speed sensor supply voltage

43 Switch the ignition off and disconnect the wheel speed sensor wiring harness 2-pin multi-plug, of the sensor under test.
44 Connect a DC voltmeter between a vehicle earth and pin 2 of the multi-plug **(see illustration 9.25)**.
45 Switch the ignition on. A voltage of approximately 8.0 volts should be obtained.
46 If the voltage is not as specified, repeat the test at the ECM multi-plug using the BOB as described previously. If there is no change in the reading, the ABS-ECM is suspect. If the voltage is now satisfactory, this indicates a fault in the wiring harness, which should be checked for continuity. **Note:** *This test at least proves that a supply voltage is available for the sensor. However, the integral electronics in the sensor must be active to produce an output.*

Checking wheel speed sensor resistance

Note: *Refer to the wiring diagrams for specific ECM pin identification according to model.*
47 Switch the ignition off and disconnect the ECM multi-plug or the relevant wheel speed sensor 2-pin multi-plug.
48 Connect an ohmmeter between the terminal pins for the sensor under test.
49 The readings obtained should be between 2.5 and 4.5 kohms **(see illustration 9.26)**.
50 If the resistance is excessively high or open circuit at the ECM, repeat the test at the sensor multi-plug. If there is no change in resistance, the sensor is suspect.
51 If the resistance is now satisfactory, this indicates a fault in the wiring harness which should be checked for continuity. **Note:** *Even if the resistance is within the quoted specifications, this does not prove that the speed sensor can generate an acceptable signal.*

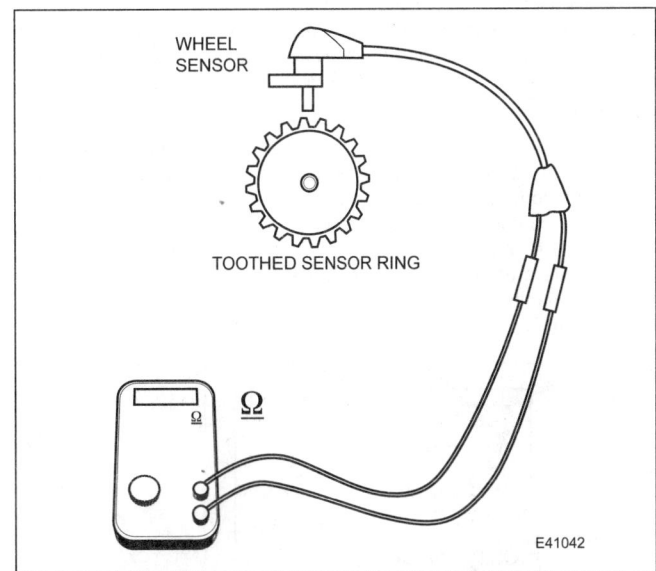

9.26 Checking active wheel speed sensor resistance with an ohmmeter connected to the sensor-wiring plug

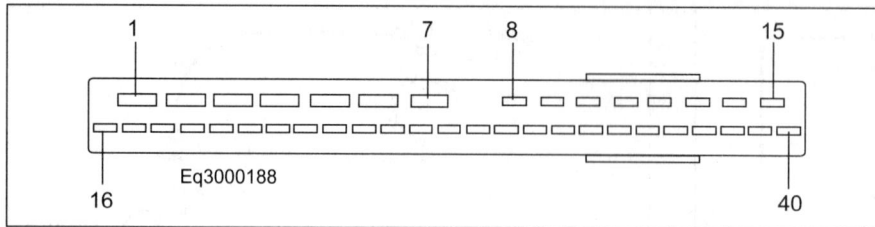

9.27 ECM 40-pin multi-plug terminal identification

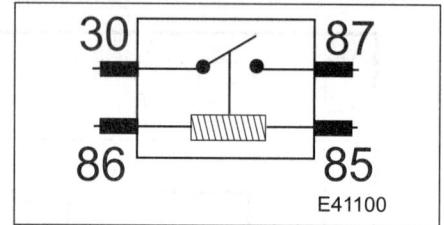

9.28 Typical relay terminal identification

5 System relays

26-pin ECM

Relay operation and power supply tests

1 The relays are integral with the ECM and can only be checked using suitable ABS diagnostic test equipment.

40-pin ECM

Relay operation tests

2 The relays are integral with the ECM and can only be checked using suitable ABS diagnostic test equipment.

Relay power supply tests

Note: *Refer to the wiring diagrams for specific ECM pin identification according to model.*

Main (solenoid valve) relay

3 Switch the ignition off and disconnect the ECM multi-plug.

4 Switch the ignition on.

5 Attach a negative voltmeter probe to a vehicle earth.

6 Attach a positive voltmeter probe to ECM multi-plug pin 3. The voltmeter should indicate nbv (see illustration 9.27). If no voltage is found, check the supply wiring back to the battery positive terminal.

Pump relay

7 Switch the ignition off and disconnect the ECM multi-plug.

8 Switch the ignition on.

9 Attach a negative voltmeter probe to a vehicle earth.

10 Attach a positive voltmeter probe to ECM multi-plug pin 2 (see illustration 9.27). The voltmeter should indicate nbv.

11 If no voltage is found, check the relevant fuse and the supply wiring back to the battery positive terminal.

88-pin ECM (BMW 5/7-Series)

Note: *The following tests are applicable to BMW 5/7-series models only. On Audi models, the relays are integral with the ECM and can only be checked using suitable ABS diagnostic test equipment.*

Checking relays (general)

12 The following tests are applicable to the ABS main (or solenoid valve) relay and hydraulic pump relay, which are commonly located in the engine compartment power distribution box.

13 Before carrying out specific tests on the relays or power supplies inspect the relays and relay base for corrosion and damage. Check that the relays are pushed fully home and making good contact with the base. Faults in any of the above areas are possible reasons for failure or malfunctioning of the relay.

Relay operation tests

Main (solenoid valve) and hydraulic pump relays

14 Switch the ignition off, then remove the relevant relay from the relay base.

15 Connect an ohmmeter between relay terminals 30 and 87 (see illustration 9.28). The ohmmeter should indicate an open circuit.

16 With the ohmmeter still connected attach a wire between the relay driver supply terminal 86 and a 12-volt supply. Attach a second wire between the relay driver earth terminal 85 and earth. There should be an unmistakable click from the relay and the ohmmeter should indicate continuity.

17 Renew the relay if it fails these tests. If the relay cover is transparent, inspect the relay for burns. A faulty relay usually has a burnt appearance.

Relay power supply tests

Main (solenoid valve) and hydraulic pump relays

18 Switch the ignition off then remove both relays leaving the bases exposed.

19 Using a voltmeter, in turn probe between the main voltage supply terminal 30, of each relay base and earth (see illustration 9.29). The reading should indicate nbv. If no voltage is found, check the wiring back to the voltage supply.

20 Switch the ignition on. Using a voltmeter, in turn probe between the relay driver supply terminal 86, of each relay base and earth. The reading should indicate nbv. If no voltage is found, check the supply wiring back to ABS-ECM pin 2.

6 Electronic Control Module

Checking the ECM (general)

1 Inspect the ECM for corrosion or damage and ensure that the unit is securely attached to the hydraulic control unit.

2 Check that the ECM multi-plug terminals are pushed fully home and making good contact with the ECM pins. Faults in any of the above areas are possible reasons for poor performance in the ABS system.

ECM power supply and earth tests

26-pin and 40-pin ECM

3 Switch the ignition off and disconnect the ECM multi-plug.

4 Attach a negative voltmeter probe to a vehicle earth.

5 Attach a positive voltmeter probe to ECM multi-plug pin 2, and pin 3 in turn. The voltmeter should indicate nbv in each case (see illustrations 9.27 and 9.30). If no voltage is found, check the relevant fuse and the supply wiring back to the battery.

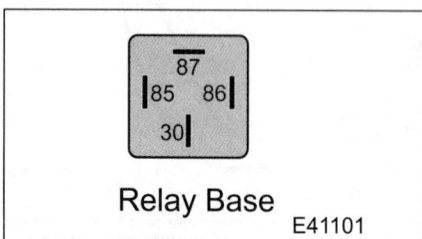

Relay Base

9.29 Relay base terminal identification

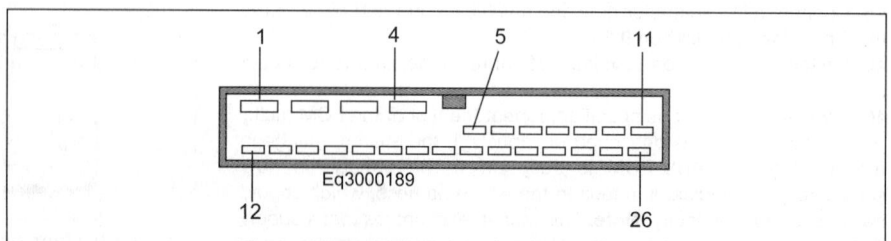

9.30 ECM 26-pin multi-plug terminal identification

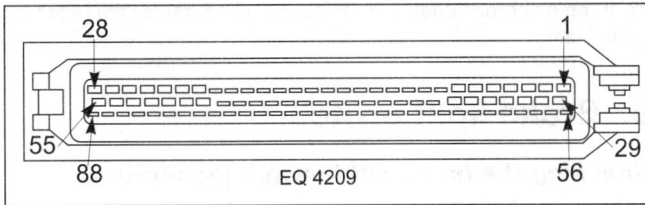

9.31 ECM 88-pin multi-plug terminal identification

6 Switch the ignition on.

7 Attach a negative voltmeter probe to a vehicle earth.

8 Attach a positive voltmeter probe to ECM multi-plug pin 4. The voltmeter should indicate nbv. If no voltage is found, check the relevant fuse and the supply wiring back to the ignition switch.

9 Switch the ignition off.

10 Connect an ohmmeter between a vehicle earth and ECM multi-plug pin 1. The ohmmeter should indicate continuity. If not, check the ECM main earth connection and wiring.

88-pin ECM

11 Switch the ignition off and connect a BOB between the ECM and the harness multi-plug.

12 Switch the ignition on and connect a voltmeter between terminal 1 of the BOB and earth **(see illustration 9.31)**. The voltmeter should indicate nbv. If no voltage is found, check the relevant fuse, supply wiring and relay.

13 Switch the ignition off.

14 Connect an ohmmeter between a vehicle earth and terminals 28 and 29 of the BOB in turn. The ohmmeter should indicate continuity in each case. If not, check the ECM main earth connection and wiring.

7 Solenoid valves

Solenoid valve continuity and resistance tests

26-pin ECM

1 The solenoid valves are integral with the ECM and can only be checked using suitable ABS diagnostic test equipment.

40-pin ECM

2 Switch the ignition off and disconnect the ECM multi-plug.

3 Remove the ECM from the hydraulic control unit, for access to the hydraulic control unit terminals. **Note:** *On some versions of Bosch 5.0, the hydraulic control unit and ECM are a sealed unit and cannot be separated. On these systems, testing of the solenoid valves can only be carried out using suitable ABS diagnostic test equipment.*

4 Connect an ohmmeter between terminal 7 in the hydraulic control unit and each of the following hydraulic control unit terminals in turn **(see illustration 9.32)**:

Terminal 14 - Front left-hand inlet solenoid
Terminal 1 - Front right-hand inlet solenoid
Terminal 2 - Rear left-hand inlet solenoid
Terminal 13 - Rear right-hand inlet solenoid
Terminal 9 - Front left-hand outlet solenoid
Terminal 6 - Front right-hand outlet solenoid
Terminal 5 - Rear left-hand outlet solenoid
Terminal 10 - Rear right-hand outlet solenoid

5 The resistance should be in the order of 7.5 to 12.0 Ohms for each inlet solenoid valve, and 4.0 to 6.5 Ohms for each outlet solenoid valve

6 If there is no reading, or one reading is excessively high, possible solenoid valve failure is indicated.

7 On completion refit the ECM to the hydraulic control unit and reconnect the multi-plug.

88-pin ECM (Audi)

8 Switch the ignition off and connect a BOB between the ECM and the harness multi-plug.

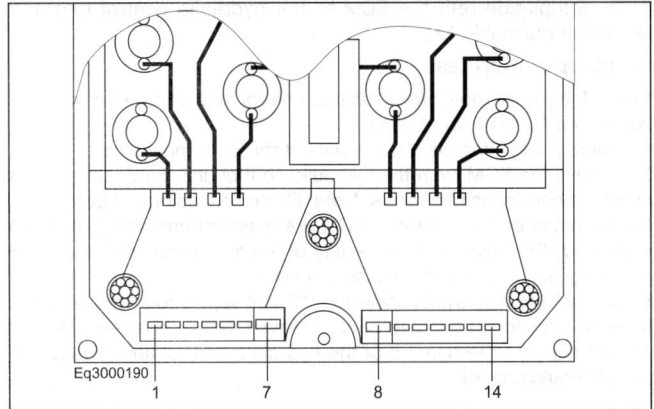

9.32 Hydraulic control unit terminal identification

9 Connect an ohmmeter between each off the following BOB terminals in turn **(see illustration 9.31)**.

Terminals 5 and 33 - Front left-hand inlet and outlet solenoids
Terminals 26 and 54 - Front right-hand inlet and outlet solenoids
Terminals 25 and 53 - Rear left-hand inlet and outlet solenoids
Terminals 6 and 34 - Rear right-hand inlet and outlet solenoids

10 The resistance should be in the order of 9.0 to 22.0 Ohms for each pair of solenoid valves.

11 If there is no reading, or one reading is excessively high, possible solenoid valve failure is indicated.

88-pin ECM (BMW)

12 Switch the ignition off and disconnect the ECM multi-plug.

13 Remove the solenoid valve relay from the relay base.

14 Using an ohmmeter, check for continuity between terminal 87 of the relay base and each of the following ECM multi-plug pins in turn **(see illustrations 9.29 and 9.31)**:

Pin five - Front left-hand inlet solenoid
Pin 33 - Front left-hand outlet solenoid
Pin 54 - Front right-hand inlet solenoid
Pin 26 - Front right-hand outlet solenoid
Pin 53 - Rear axle inlet solenoid
Pin 25 - Rear axle outlet solenoid

15 Now measure the resistance between terminal 87 of the relay base and each of the inlet valves in turn (ECM pins 5, 54 and 53). The resistance should be in the order of 7.4 to 13.2 Ohms for each solenoid valve.

16 Repeat the test on each of the outlet valves (ECM pins 33, 26 and 25). The resistance should be in the order of 3.8 to 6.9 Ohms for each solenoid valve.

17 If there is no reading, or one reading is excessively high, possible solenoid valve failure is indicated.

8 Hydraulic pump motor

26-pin ECM

Pump resistance and operation tests

1 The pump motor is integral with the ECM and hydraulic control unit and can only be checked using suitable ABS diagnostic test equipment.

40-pin ECM

Pump resistance test

2 Switch the ignition off and disconnect the ECM multi-plug.

3 Remove the ECM from the hydraulic control unit, for access to the hydraulic control unit terminals. **Note:** *On some versions of Bosch 5.0, the hydraulic control unit and ECM are a sealed unit and cannot be separated. On these systems, testing of the pump can only be carried out using suitable ABS diagnostic test equipment.*

4 Connect an ohmmeter between terminal 8 in the hydraulic control unit and a vehicle earth **(see illustration 9.32)**. A reading of approximately 1.5 Ohms should be obtained.

9

5 On completion refit the ECM to the hydraulic control unit and reconnect the multi-plug.

Pump operation test

Note: *The pump motor (or hydraulic control unit) earth lead must be connected for the following test.*

6 Switch the ignition off and disconnect the ECM multi-plug.

7 Remove the ECM from the hydraulic control unit, for access to the hydraulic control unit terminals. **Note:** *On some versions of Bosch 5.0, the hydraulic control unit and ECM are a sealed unit and cannot be separated. On these systems, testing of the pump can only be carried out using suitable ABS diagnostic test equipment.*

8 Connect the negative terminal of a 12-volt supply to a vehicle earth.

9 Using a fused jumper lead briefly connect the positive terminal of a 12-volt supply to terminal 8 of the hydraulic control unit. The pump motor should now run.

> ⚠ **Warning: The test should be made as quickly as possible to avoid damaging the pump.**

10 If the pump does not operate as described, renew the hydraulic control unit.

11 On completion refit the ECM to the hydraulic control unit and reconnect the multi-plug.

88-pin ECM (Audi)

Pump resistance test

12 Switch the ignition off and connect a BOB between the ECM and the harness multi-plug.

13 Connect an ohmmeter between terminal 19 in the BOB and a vehicle earth **(see illustration 9.31)**. A reading of approximately 0.1 to 1.0 Ohm should be obtained.

Pump operation test

14 Switch the ignition off and connect a BOB between the ECM and the harness multi-plug.

15 Using a switched jumper lead bridge terminals 19 and 50 in the BOB. Operate the switch and the pump motor should run.

> ⚠ **Warning: The test should be made as quickly as possible to avoid damaging the pump.**

16 If the pump does not operate as described, renew the hydraulic control unit.

88-pin ECM (BMW)

Pump resistance test

17 Switch the ignition off and connect a BOB between the ECM and the harness multi-plug.

18 Connect an ohmmeter between terminal 19 in the BOB and a vehicle earth. A reading of approximately 0.1 to 1.0 Ohm should be obtained.

Pump operation test

19 Switch the ignition off and disconnect the ECM multi-plug.

20 Remove the pump motor relay leaving the base exposed **(see illustration 9.29)**.

21 Using a switched jumper lead bridge terminals 30 and 87 in the relay base. Operate the switch and the pump motor should run.

> ⚠ **Warning: The test should be made as quickly as possible to avoid damaging the pump.**

22 If the pump does not operate as described, renew the hydraulic control unit.

9 Brake light switch

Checking the brake light switch (general)

1 Check that the brake light switch is correctly and securely mounted and that the plunger moves smoothly with no trace of binding.

2 Check that the wiring multi-plug is pushed fully home and making good contact.

3 Check that no wires have been disconnected.

4 Faults in any of the above areas are possible reasons for failure or malfunctioning of the switch.

Brake light switch voltage and continuity tests

Voltage test

5 Switch the ignition off and disconnect the ECM multi-plug from the ECM.

6 Connect a voltmeter between a vehicle earth and the following ECM multi-plug pin, according to system **(see illustrations 9.27, 9.30 and 9.31)**:

 26-pin ECM - pin 20
 40-pin ECM - pin 28
 88-pin ECM - pin 48

7 Switch the ignition on and depress the brake pedal. The voltmeter should indicate nbv.

8 If no voltage is found, the fuse, the brake light switch and the wiring are suspect.

9 Release the brake pedal. The voltage should drop to zero as the switch opens.

Continuity test

10 Switch the ignition off and disconnect the brake light switch multi-plug.

11 Connect an ohmmeter between the terminal pins of the brake light switch.

12 Operate the brake light switch and check for continuity. If the test fails, renew the switch.

10 Warning lamp

Checking the warning lamp (general)

1 Inspect the warning lamp bulb holder contacts in the instrument panel.

2 Check that the instrument panel multi-plug terminal pins are pushed fully home and making good contact.

3 Check that no wires have been disconnected.

4 Faults in any of the above areas are possible reasons for failure or malfunctioning of the warning lamp.

Warning lamp operation test

5 With the ignition switched off, the warning lamp should remain off.

6 Switch the ignition on and the warning lamp should illuminate then extinguish after a few seconds. The lamp should then remain off.

7 If the warning lamp comes on and remains on at any time during vehicle operation, carry out the previously described test procedures on the system components.

Pin table - typical 26-pin

Note: *Refer to illustration 9.30*

Pin No.	Connection	Test condition	Voltage
1	-	-	-
2	Supply from battery	Ignition off/on	Nbv
3	Supply from battery	Ignition off/on	Nbv
4*	Supply to ECM (RHD only)	Ignition on	Nbv
5	ABS warning lamp	Engine running: Lamp on Lamp off	 1.0 volts (max) Nbv
6*	Instrument cluster	-	-
7*	Engine management ECM signal	-	-
8*	Traction control warning lamp	-	-
9*	Engine speed signal	-	-
10*	Traction control switch	-	-
11	-	-	-
12	Front left wheel speed sensor signal	Roadwheel rotating	0.1 to 0.5 volts AC (approx)
13	Front left wheel speed sensor earth	Roadwheel rotating	0.25 volts (max)
14	Rear left wheel speed sensor signal	Roadwheel rotating	0.1 to 0.5 volts AC (approx)
15	Rear left wheel speed sensor earth	Roadwheel rotating	0.25 volts (max)
16	Rear right wheel speed sensor signal	Roadwheel rotating	0.1 to 0.5 volts AC (approx)
17	Rear right wheel speed sensor earth	Roadwheel rotating	0.25 volts (max)
18	Front right wheel speed sensor signal	Roadwheel rotating	0.1 to 0.5 volts AC (approx)
19	Front right wheel speed sensor earth	Roadwheel rotating	0.25 volts (max)
20	Brake light switch input	Ignition on: Brake pedal released Brake pedal depressed	 0 volts Nbv
21	SD connector	-	-
22*	Engine management ECM signal	-	-
23	-	-	-
24	-	-	-
25*	Automatic Transmission ECM signal		
26*	Engine management ECM signal	-	-

Traction Control

Pin table - typical 40-pin

Note: *Refer to illustration 9.27*

Pin No.	Connection	Test condition	Voltage
1	ECM earth	Ignition on	0.25 volts (max)
2	Supply from battery	Ignition off/on	Nbv
3	Supply from battery	Ignition off/on	Nbv
4	Supply from ignition switch	Ignition on	Nbv
5	-	-	-
6	-	-	-
7	-	-	-
8	-	-	-
9	-	-	-
10	-	-	-
11	-	-	-
12	-	-	-
13	-	-	-
14	-	-	-
15*	ECM earth	Ignition on	0.25 volts (max)
16	Front left wheel speed sensor signal	Roadwheel rotating	0.1 to 0.5 volts AC (approx)
17	Front left wheel speed sensor earth	Roadwheel rotating	0.25 volts (max)
18	Rear left wheel speed sensor signal	Roadwheel rotating	0.1 to 0.5 volts AC (approx)

9

Pin table - typical 40-pin (continued)

Note: *Refer to illustration 9.27*

Pin No.	Connection	Test condition	Voltage
19	Rear left wheel speed sensor earth	Roadwheel rotating	0.25 volts (max)
20	Rear right wheel speed sensor signal	Roadwheel rotating	0.1 to 0.5 volts AC (approx)
21	Rear right wheel speed sensor earth	Roadwheel rotating	0.25 volts (max)
22	-	-	-
23**	Front right wheel speed sensor signal	Roadwheel rotating	0.1 to 0.5 volts AC (approx)
24	Front right wheel speed sensor earth	Roadwheel rotating	0.25 volts (max)
25	-	-	-
26	-	-	-
27	-	-	-
28	Brake light switch input	Ignition on: Brake pedal released Brake pedal depressed	 0 volts Nbv
29	SD connector	-	-
30	-	-	-
31	-	-	-
32	-	-	-
33	-	-	-
34	-	-	-
35	-	-	-
36	-	-	-
37	-	-	-
38	-	-	-
39	-	-	-
40	ABS warning lamp	Ignition on: Lamp on Lamp off	 1.0 volt (max) Nbv

*Certain versions only
**Pin 22 on certain versions

Pin table - typical 88-pin (Audi)

Note: *Refer to illustration 9.31*

Pin No.	Connection	Test condition	Voltage
1	Supply from ignition switch	Ignition on	Nbv
2	Hydraulic pump relay driver	Ignition on/inactive Actuated	Nbv 1.0 volt (max)
3	-	-	-
4	-	-	-
5	Front left inlet solenoid valve	Ignition on/inactive Actuated	Nbv 1.0 volt (max)
6	Rear right inlet solenoid valve	Ignition on/inactive Actuated	Nbv 1.0 volt (max)
7	Hydraulic pump relay driver	Ignition on/inactive Actuated	Nbv 1.0 volt (max)
8	-	-	-
9	Front left wheel speed sensor earth	Roadwheel rotating	0.25 volts (max)
10	Front left wheel speed sensor signal	Roadwheel rotating	0.2 to 1.0 volts AC (approx)
11	Rear right wheel speed sensor signal	Roadwheel rotating	0.2 to 1.0 volts AC (approx)
12	Rear left wheel speed sensor earth	Roadwheel rotating	0.25 volts (max)
13	Rear left wheel speed sensor signal	Roadwheel rotating	0.2 to 1.0 volts AC (approx)
14	Front right wheel speed sensor earth	Roadwheel rotating	0.25 volts (max)
15	Front right wheel speed sensor signal	Roadwheel rotating	0.2 to 1.0 volts AC (approx)
16	-	-	-
17	-	-	-

18	Engine management control unit signal	-	-
19	Voltage monitoring, hydraulics	Ignition on Pump relay actuated	0 volts Nbv
20	-	-	-
21	-	-	-
22	-	-	-
23	-	-	-
24	-	-	-
25	Rear left outlet solenoid valve	Ignition on/inactive Actuated	Nbv 1.0 volt (max)
26	Front right outlet solenoid valve	Ignition on/inactive Actuated	Nbv 1.0 volt (max)
27	-	-	-
28	ECM earth	Ignition on	0.25 volts (max)
29	ECM earth	Ignition on	0.25 volts (max)
30	-	-	-
31	-	-	-
32	ABS warning lamp	Ignition on: Lamp on Lamp off	 1.0 volt (max) Nbv
33	Front left outlet solenoid valve	Ignition on/inactive Actuated	Nbv 1.0 volt (max)
34	Rear right outlet solenoid valve	Ignition on/inactive Actuated	Nbv 1.0 volt (max)
35	-	-	-
36	-	-	-
37	Solenoid valve relay driver	Ignition on/inactive Actuated	Nbv 1.0 volt (max)
38	Rear right wheel speed sensor earth	Roadwheel rotating	0.25 volts (max)
39	-	-	-
40	-	-	-
41	-	-	-
42	-	-	-
43	-	-	-
44	-	-	-
45	Transmission ECM signal	-	-
46	SD connector	-	-
47	-	-	-
48	Brake light switch input	Ignition on: Brake pedal released Brake pedal depressed	 0 volts Nbv
49	-	-	-
50	Supply from battery	Ignition off/on	Nbv
51	-	-	-
52	-	-	-
53	Rear left inlet solenoid valve	Ignition on/inactive Actuated	Nbv 1.0 volt (max)
54	Front right inlet solenoid valve	Ignition on/inactive Actuated	Nbv 1.0 volt (max)
55 to 88	Unused	-	-

Pin table - typical 88-pin (BMW 5-Series)

Note: *Refer to illustration 9.31*

Pin No.	Connection	Test condition	Voltage
1	Supply from relay	Ignition on	Nbv
2	Supply to relays	Ignition on	Nbv
3	-	-	-
4	-	-	-

9

Pin table - typical 88-pin (BMW 5-Series) (continued)

Note: *Refer to illustration 9.31*

Pin No.	Connection	Test condition	Voltage
5	Front left inlet solenoid valve	Ignition on/inactive Actuated	Nbv 1.0 volt (max)
6	-	-	-
7	Hydraulic pump relay driver	Ignition on/inactive Actuated	Nbv 1.0 volt (max)
8	-	-	-
9	Transmission ECM signal	-	-
10	Front left wheel speed sensor supply	Ignition on	8.0+ volts
11	Rear right wheel speed sensor signal	Ignition on/roadwheel rotating	0.5 or 2.5 volts (switching)
12	Rear left wheel speed sensor supply	Ignition on	8.0+ volts
13	Rear left wheel speed sensor signal	Ignition on/roadwheel rotating	0.5 or 2.5 volts (switching)
14	Front right wheel speed sensor supply	Ignition on	8.0+ volts
15	Front right wheel speed sensor signal	Ignition on/roadwheel rotating	0.5 or 2.5 volts (switching)
16	-	-	-
17	-	-	-
18	-	-	-
19	Voltage monitoring, hydraulics	Ignition on Pump relay actuated	0 volts Nbv
20	-	-	-
21	-	-	-
22	-	-	-
23	-	-	-
24	Alternator voltage input	Engine running	12.0 to 14.0 volts
25	Rear axle outlet solenoid valve	Ignition on/inactive Actuated	Nbv 1.0 volt (max)
26	Front right outlet solenoid valve	Ignition on/inactive Actuated	Nbv 1.0 volt (max)
27	-	-	-
28	ECM earth	Ignition on	0.25 volts (max)
29	ECM earth	Ignition on	0.25 volts (max)
30	ABS warning lamp	Ignition on: Lamp on Lamp off	 1.0 volt (max) Nbv
31	-	-	-
32	-	-	-
33	Front left outlet solenoid valve	Ignition on/inactive Actuated	Nbv 1.0 volt (max)
34	-	-	-
35	Front left wheel speed sensor signal	Ignition on/roadwheel rotating	0.5 or 2.5 volts (switching)
36	Transmission ECM signal	-	-
37	Solenoid valve relay driver	Ignition on/inactive Actuated	Nbv 1.0 volt (max)
38	Rear right wheel speed sensor supply	Ignition on	8.0+ volts
39	-	-	-
40	Transmission ECM signal	-	-
41	-	-	-
42	Transmission ECM signal	-	-
43	-	-	-
44	-	-	-
45	-	-	-
46	SD connector	-	-
47	-	-	-
48	Brake light switch input	Ignition on: Brake pedal released Brake pedal depressed	 0 volts Nbv

49	-	-	-
50	-	-	-
51	-	-	-
52	-	-	-
53	Rear axle inlet solenoid valve	Ignition on/inactive	Nbv
		Actuated	1.0 volt (max)
54	Front right inlet solenoid valve	Ignition on/inactive	Nbv
		Actuated	1.0 volt (max)
55 to 88	Unused	-	-

Pin table - typical 88-pin (BMW 7-Series)
Note: *Refer to illustration 9.31*

Pin No.	Connection	Test condition	Voltage
1	Supply from relay	Ignition on	Nbv
2	Supply to relays	Ignition on	Nbv
3	-	-	-
4	-	-	-
5	Front left inlet solenoid valve	Ignition on/inactive	Nbv
		Actuated	1.0 volt (max)
6	-	-	-
7	Hydraulic pump relay driver	Ignition on/inactive	Nbv
		Actuated	1.0 volt (max)
8	-	-	-
9	Transmission ECM signal	-	-
10	Front left wheel speed sensor earth	Roadwheel rotating	0.25 volts (max)
11	Rear right wheel speed sensor signal	Roadwheel rotating	0.2 to 1.0 volts AC (approx)
12	Rear left wheel speed sensor earth	Roadwheel rotating	0.25 volts (max)
13	Transmission ECM signal	-	-
14	Front right wheel speed sensor earth	Roadwheel rotating	0.25 volts (max)
15	Front right wheel speed sensor signal	Roadwheel rotating	0.2 to 1.0 volts AC (approx)
16	-	-	-
17	-	-	-
18	-	-	-
19	Voltage monitoring, hydraulics	Ignition on	0 volts
		Pump relay actuated	Nbv
20	-	-	-
21	-	-	-
22	-	-	-
23	-	-	-
24	Alternator voltage input	Engine running	12.0 to 14.0 volts
25	Rear axle outlet solenoid valve	Ignition on/inactive	Nbv
		Actuated	1.0 volt (max)
26	Front right outlet solenoid valve	Ignition on/inactive	Nbv
		Actuated	1.0 volt (max)
27	-	-	-
28	ECM earth	Ignition on	0.25 volts (max)
29	ECM earth	Ignition on	0.25 volts (max)
30	ABS warning lamp	Ignition on:	
		Lamp on	1.0 volt (max)
		Lamp off	Nbv
31	-	-	-
32	-	-	-
33	Front left outlet solenoid valve	Ignition on/inactive	Nbv
		Actuated	1.0 volt (max)
34	-	-	-
35	Transmission ECM signal	-	-
36	Front left wheel speed sensor signal	Roadwheel rotating	0.2 to 1.0 volts AC (approx)

9

Pin table - typical 88-pin (BMW 7-Series) (continued)

Note: *Refer to illustration 9.31*

Pin No.	Connection	Test condition	Voltage
37	Solenoid valve relay driver	Ignition on/inactive	Nbv
		Actuated	1.0 volt (max)
38	Rear right wheel speed sensor earth	Roadwheel rotating	0.25 volts (max)
39	-	-	-
40	Rear left wheel speed sensor signal	Roadwheel rotating	0.2 to 1.0 volts AC (approx)
41	-	-	-
42	Transmission ECM signal	-	-
43	-	-	-
44	-	-	-
45	-	-	-
46	SD connector	-	-
47	-	-	-
48	Brake light switch input	Ignition on:	
		Brake pedal released	0 volts
		Brake pedal depressed	Nbv
49	-	-	-
50	-	-	-
51	-	-	-
52	-	-	-
53	Rear axle inlet solenoid valve	Ignition on/inactive	Nbv
		Actuated	1.0 volt (max)
54	Front right inlet solenoid valve	Ignition on/inactive	Nbv
		Actuated	1.0 volt (max)
55 to 88	unused	-	-

Fault codes

11 General fault codes

1 The Bosch 5.0 system requires the use of a FCR for obtaining fault codes. Flash codes are not available for output from this system.

2 If a FCR is available, it should be connected to the SD serial connector (the Vauxhall/Opel term for SD connector is ALDL) and used in accordance with the maker's instructions.

3 The FCR can be used for the following purposes:

a) *Obtaining fault codes.*

b) *Clearing fault codes.*

c) *Obtaining datastream information.*

d) *Testing the system actuators (solenoid valve relay, pump relay and solenoid valves).*

12 BMW, Renault and Rover fault codes

1 On BMW, Renault and Rover models, internal fault codes are used by the ECM to designate faults in the system components and circuits. A proprietary fault code reader (FCR) or system tester (such as the Renault XR25) is required to interrogate the system. No actual fault code numbers are available although the component circuits checked by the ECM are similar to those shown for the other vehicles listed.

Fault code table (26-pin ECM - Vauxhall/Opel)

Code	Item	Fault
16	Front left solenoid valve	Defective (signal interruption, short to earth)
17	Front right solenoid valve	Defective (signal interruption, short to earth)
19	Solenoid valve relay	Defective (signal interruption, short to earth, supply voltage too low)
25	Wheel speed sensor toothed sensor ring	Wrong number of teeth
28	Rear left solenoid valve	Defective (signal interruption, short to earth)
29	Rear right solenoid valve	Defective (signal interruption, short to earth)
31	Engine speed signal	No/poor signal
35	Pump motor	Defective (non-operation of pump after actuation of relay)
37	Brake light switch	Defective (incorrect signal)
39	Front left wheel speed sensor	No/poor signal
41	Front left wheel speed sensor	Signal interrupted
42	Front right wheel speed sensor	No/poor signal
43	Front right wheel speed sensor	Signal interrupted
44	Rear left wheel speed sensor	No/poor signal
45	Rear left wheel speed sensor	Signal interrupted
46	Rear right wheel speed sensor	No/poor signal
47	Rear right wheel speed sensor	Signal interrupted
48	Battery voltage	Voltage too low
49	Battery voltage	Voltage too high
52	ABS warning lamp	Circuit interrupted, short to earth
55	ECM	Defective
65	Traction control version coding	Not programmed
66	Engine ECM throttle valve angle	Malfunction
67	Engine ECM engine torque reduction	Malfunction
68	Resulting throttle valve angle	Malfunction

9

Fault code table (40-pin ECM - Citroën/Peugeot)

Code	Item
015Z	Main (solenoid valve) relay
016Z	Brake light switch
018Z	Wheel speed sensor toothed sensor ring
024Z	Rear left wheel speed sensor
025Z	Front right wheel speed sensor
031Z	Rear right wheel speed sensor
032Z	Front left wheel speed sensor
033Z	Wheel speed sensor signal
042Z	Inlet/outlet solenoid valves
044Z	Inlet/outlet solenoid valves
051Z	Inlet/outlet solenoid valves
052Z	Inlet/outlet solenoid valves
053Z	Hydraulic pump motor
055Z	ECM
057Z	Supply voltage
087Z	SD output fault
091Z	ABS warning lamp
095Z	Hydraulic unions
096Z	Wheel speed sensor wiring transposed
097Z	External signal interference
099Z	No faults found

Fault code table (40-pin ECM - Volkswagen)

Code	Item
00000	No fault detected
00257	Front left inlet solenoid valve
00259	Front right inlet solenoid valve
00265	Front left outlet solenoid valve
00267	Front right outlet solenoid valve
00273	Rear right inlet solenoid valve
00274	Rear left inlet solenoid valve
00275	Rear right outlet solenoid valve
00276	Rear left outlet solenoid valve
00283	Front left wheel speed sensor or circuit
00285	Front right wheel speed sensor or circuit
00287	Rear right wheel speed sensor or circuit
00290	Rear left wheel speed sensor or circuit
00301	Hydraulic pump relay
00302	Solenoid valve relay
00526	Brake light switch
00529	Speed signal information missing
00532	Supply voltage low
00597	Wheel speed sensor signal implausible
00668	Supply voltage circuit fault
00761	Fault code stored in ECM
65535	Internal fault in ECM

Fault code table (40-pin ECM - Volvo)

Code	Item
112	ECM
142	Brake light switch
311	Front left wheel speed sensor or circuit
312	Front right wheel speed sensor or circuit
313	Rear left wheel speed sensor or circuit
314	Rear right wheel speed sensor or circuit
321	Front left wheel speed sensor signal
322	Front right wheel speed sensor signal
323	Rear left wheel speed sensor signal
324	Rear right wheel speed sensor signal
331	Wheel speed sensor implausible signal
411	Front left inlet solenoid valve
412	Front left outlet solenoid valve
413	Front right inlet solenoid valve
414	Front right outlet solenoid valve
421	Rear left inlet solenoid valve
422	Rear left outlet solenoid valve
423	Rear right inlet solenoid valve
424	Rear right outlet solenoid valve
433	Supply voltage - system
441	Supply voltage - solenoid valves
443	Hydraulic pump motor

Fault code table (88-pin ECM - Audi)

Code	Item
00000	No fault detected
00257	Front left inlet solenoid valve
00259	Front right inlet solenoid valve
00265	Front left outlet solenoid valve
00267	Front right outlet solenoid valve
00273	Rear right inlet solenoid valve
00274	Rear left inlet solenoid valve
00275	Rear right outlet solenoid valve
00276	Rear left outlet solenoid valve
00283	Front left wheel speed sensor or circuit
00285	Front right wheel speed sensor or circuit
00287	Rear right wheel speed sensor or circuit
00290	Rear left wheel speed sensor or circuit
00301	Hydraulic pump relay
00302	Solenoid valve relay
00526	Brake light switch
00529	Speed signal information missing
00532	Supply voltage low
00597	Wheel speed sensor signal implausible
00668	Supply voltage circuit fault
00761	Fault code stored in ECM
65535	Internal fault in ECM

Wiring diagrams

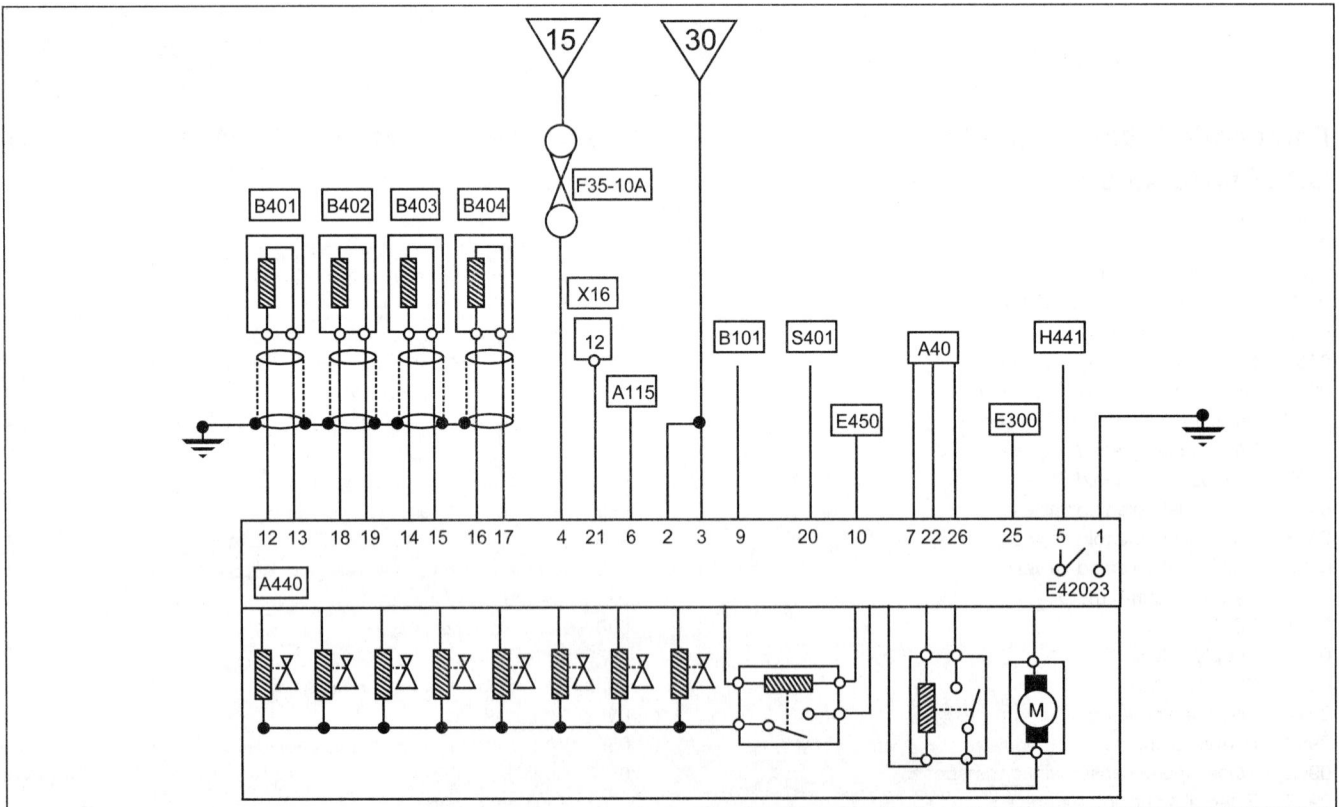

9.33 26-pin wiring diagram, Vauxhall/Opel Vectra-B

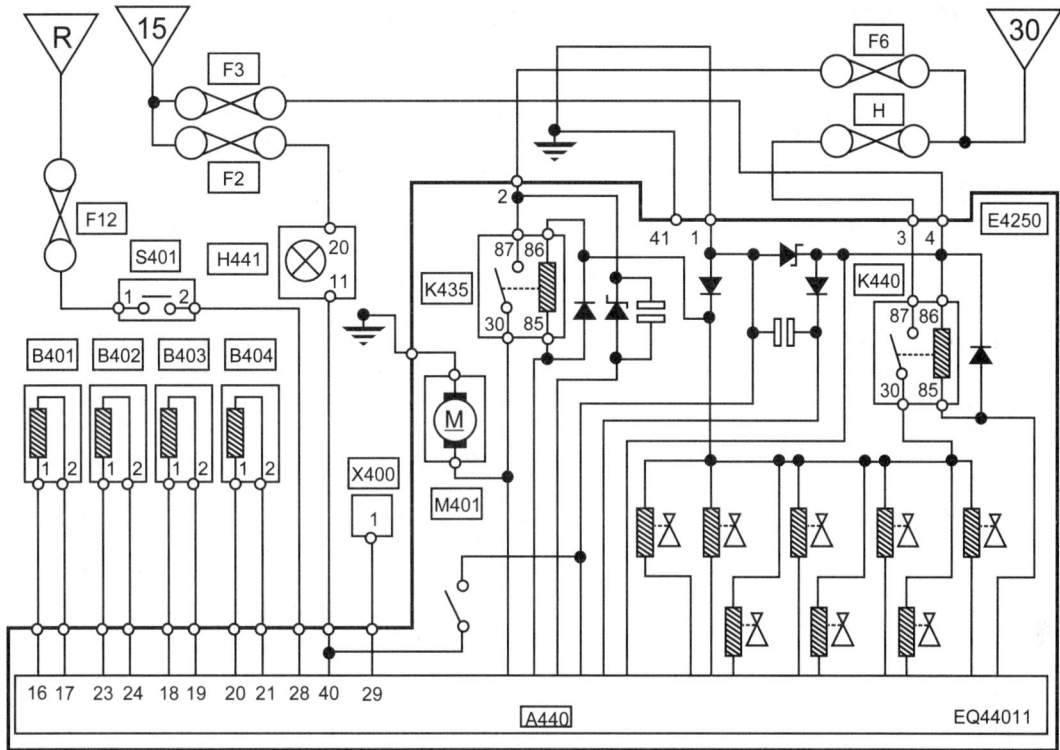

9.34 40-pin wiring diagram, Citroën Berlingo

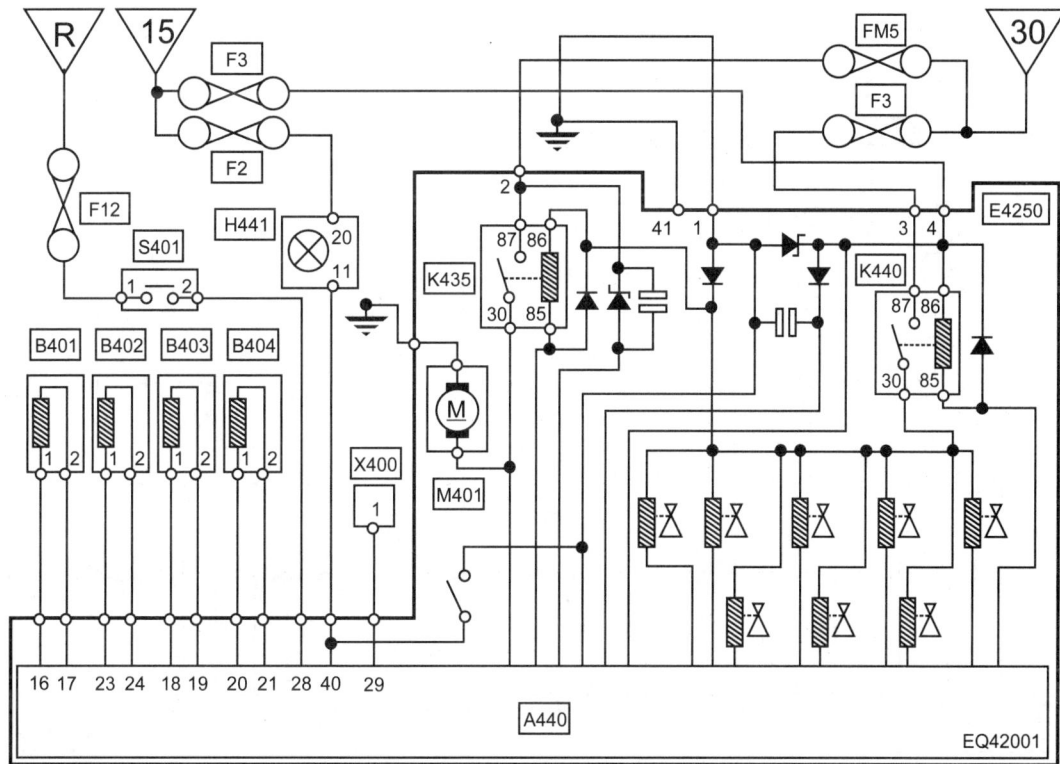

9.35 40-pin wiring diagram, Peugeot 406

9

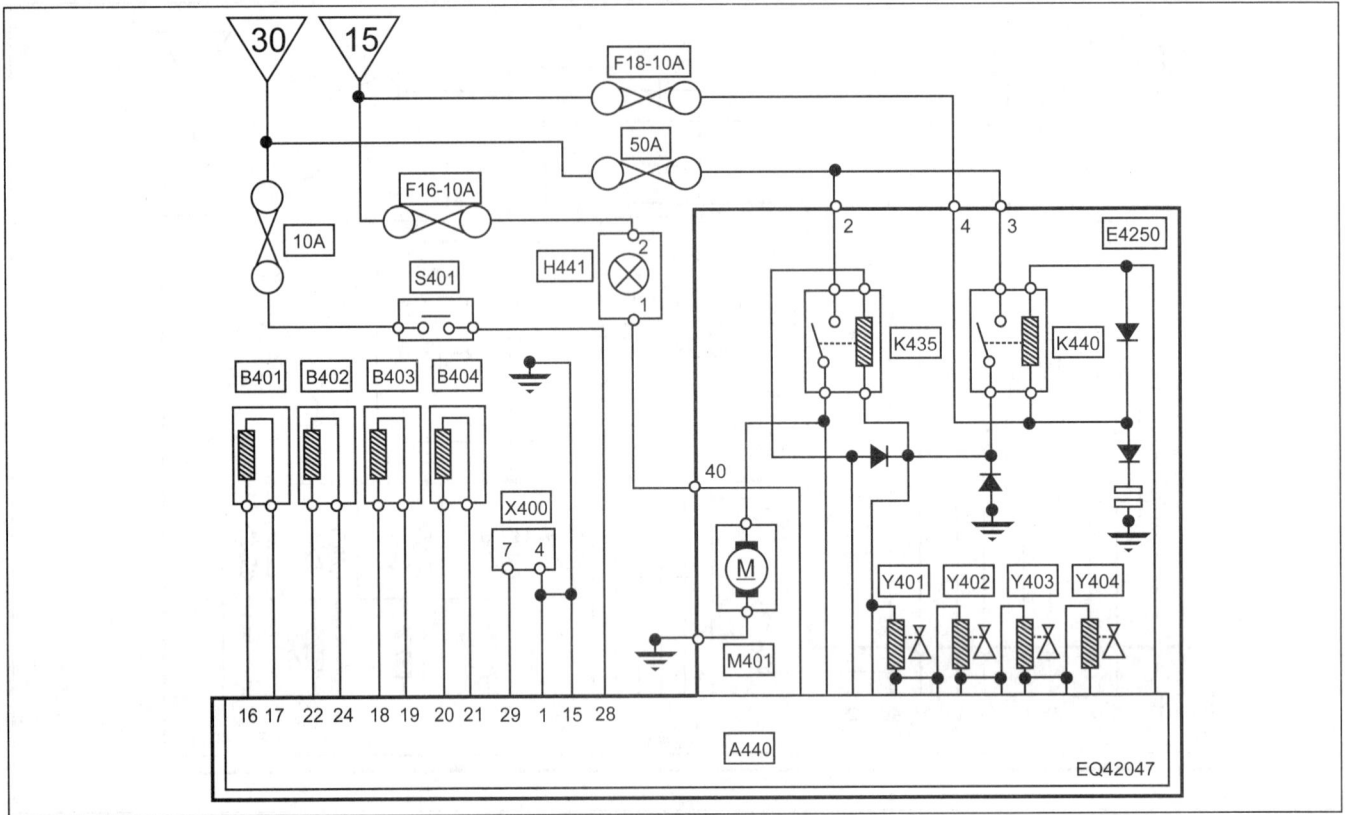

9.36 40-pin wiring diagram, Rover 800

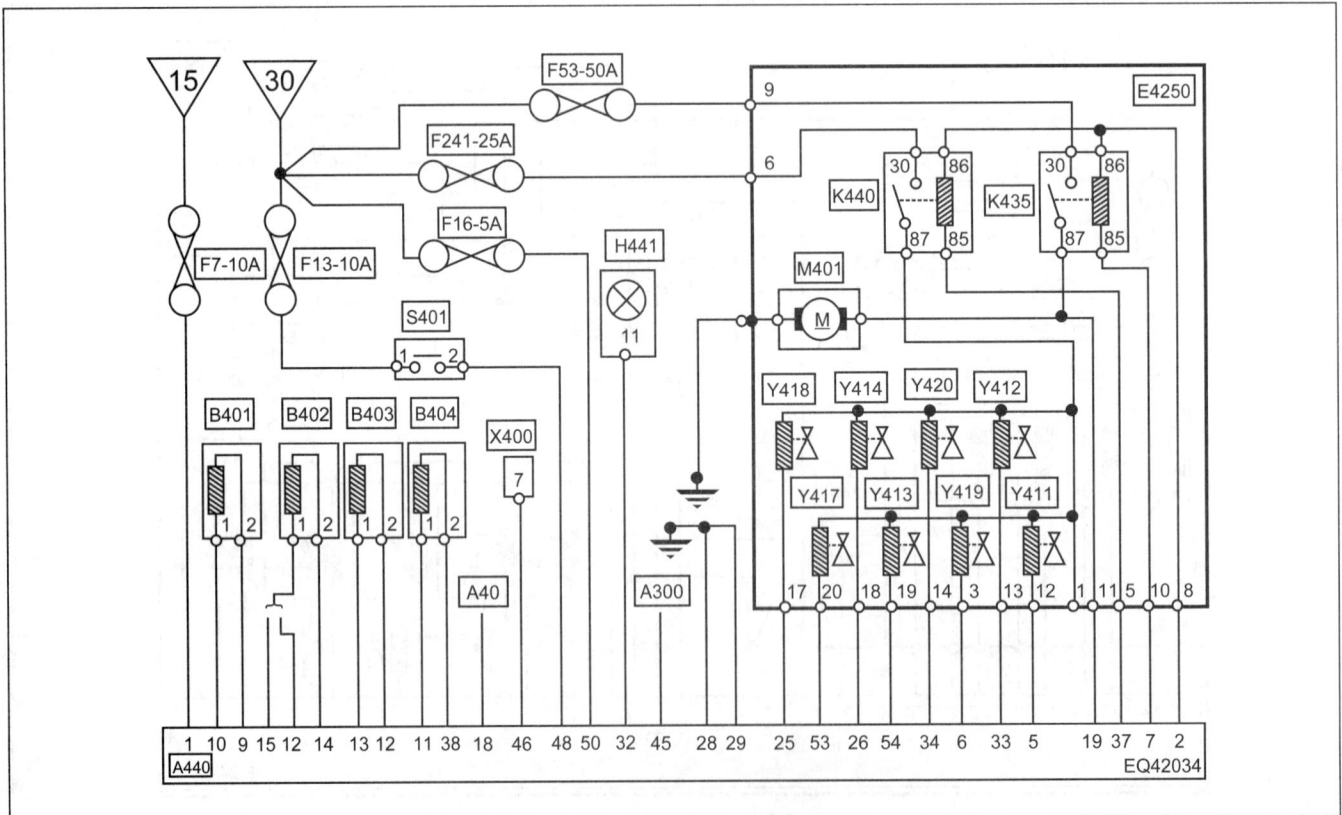

9.37 88-pin wiring diagram, Audi A4

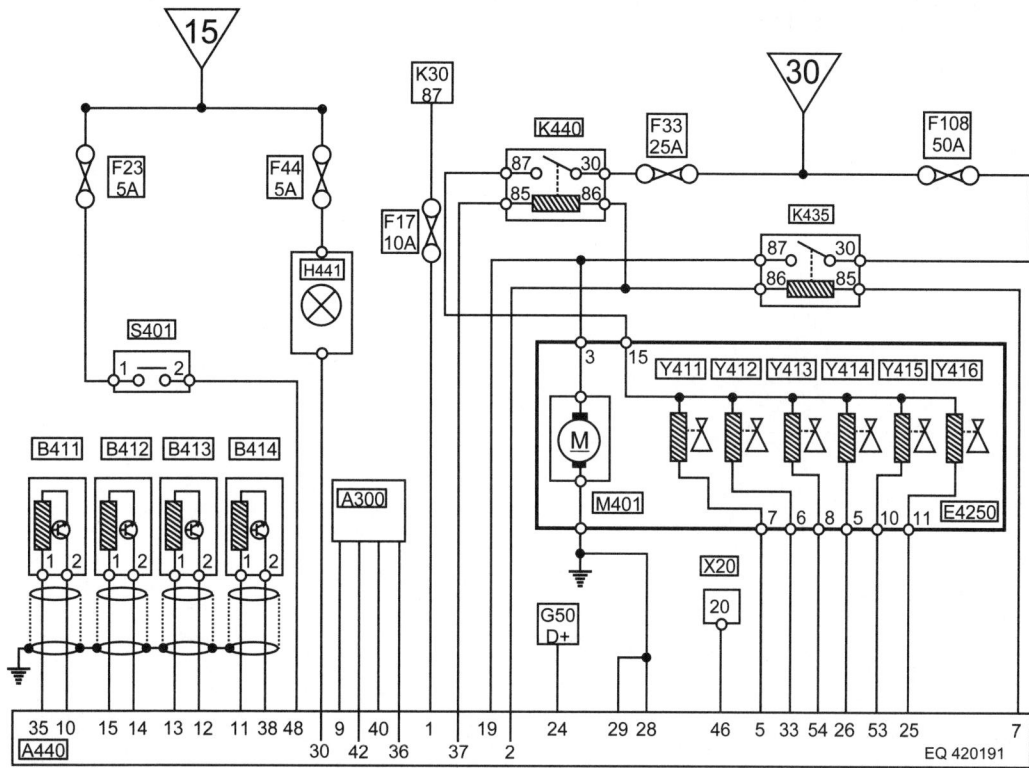

9.38 88-pin wiring diagram, BMW 5-Series (E39)

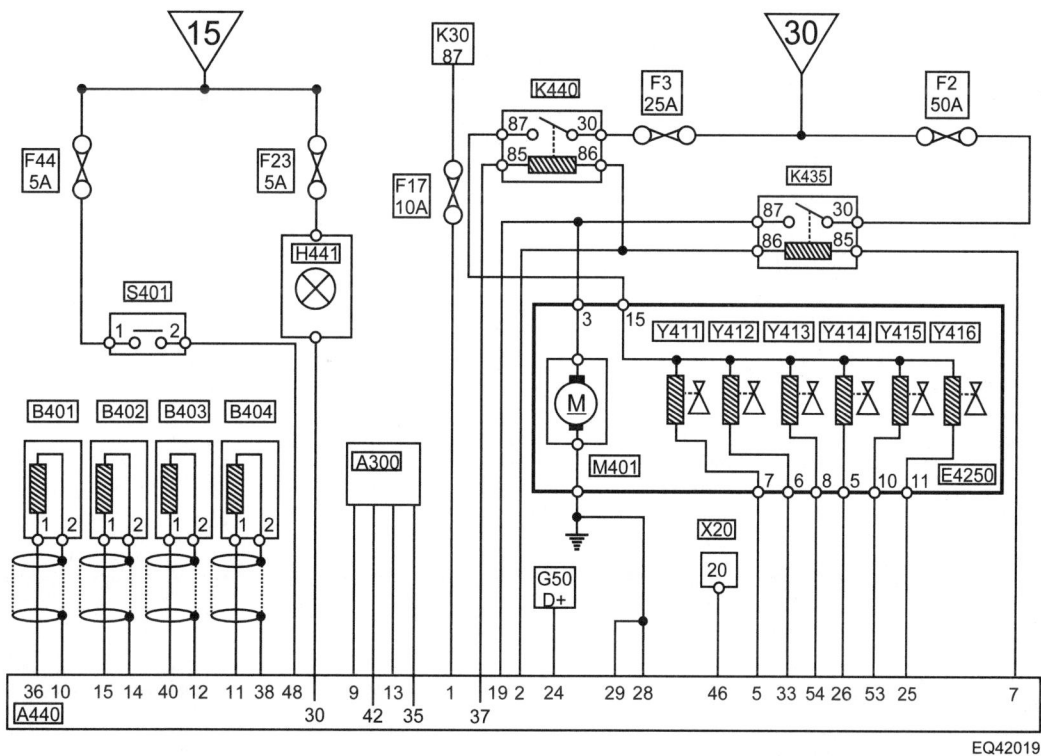

9.39 88-pin wiring diagram, BMW 7-Series (E38)

Chapter 10

Bosch 5.3 ABS

Contents

Vehicle coverage

Model	Year
Alfa Romeo	
145/146	1996-2000
155	1996-1998
156	1996-2000
166	1999-2000
Audi	
A4	1997-2000
A6	1998-2000
A8	1997-2000
Citroën	
Relay	1997-2000
Synergie	1997-1999
Xsara	1997-2000
Fiat	
Barchetta	1997-2000
Fiat Coupe	1998-2000
Fiorino	1997-2000
Punto	1996-2000
Sciecento	1998-2000
Ulysse	1997-2000
Ford	
Cougar	1998-2000
Mondeo	1998-2000
Transit	1997-1999
Peugeot	
306	1997-2000
406	1997-2000
605	1997-1999
806	1997-1999

Vehicle coverage (continued)

Model	Year
Renault	
Clio	1998-2000
Kangoo	1997-2000
Megane	1998-2000
Safrane	1997-2000
Rover	
200	1998-1999
Saab	
900	1996-1998
9-3	1998-2000
9-5	1997-2000
Vauxhall/Opel	
Astra-G	1998-2000
Corsa-B	1996-2000
Omega-B	1998-2000
Tigra	1996-2000
Volkswagen	
LT	1998-2000
Passat	1997-2000
Transporter	1998-2000
Volvo	
S40/V40	1998-2000

Overview of system operation

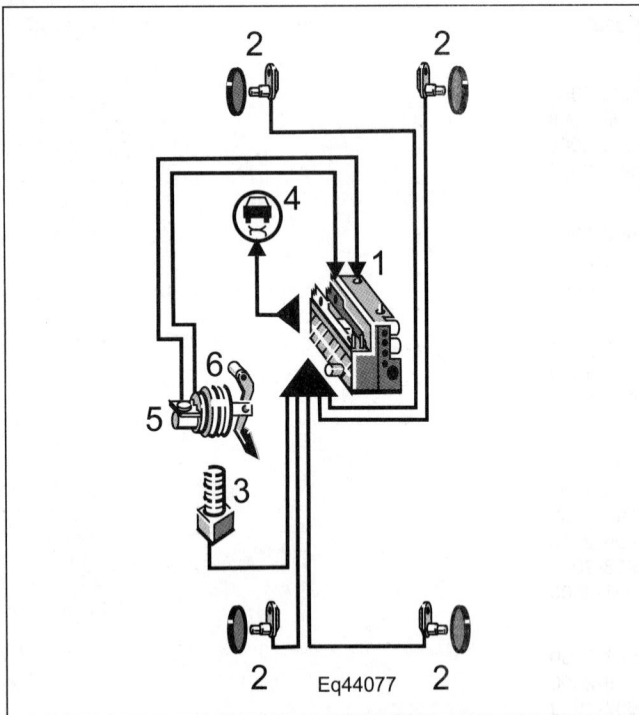

10.1 Typical Bosch 5.3 main components

1 *Hydraulic control unit with integral ECM*
2 *Wheel speed sensors and associated sensor rings*
3 *Brake light switch*
4 *ABS warning lamp*
5 *Tandem master cylinder*
6 *Vacuum servo unit*

1 Basic principles and system identification

Bosch 5.3 is a development of the earlier Bosch 5.0 series ABS, with further improvements to the hydraulic control unit, and the Electronic Control Module (ECM) software. Bosch 5.3 has been fitted to a wide range of passenger vehicles since its introduction in the late 1990s. The system is of the additional or 'add-on' type operating in conjunction with the conventional braking system components.

Depending on application, Bosch 5.3 may be installed as an antilock braking system only, or as an antilock braking system incorporating traction control (Bosch 5.3/TC).

In ABS mode, the purpose of the system is to apply the vehicle brakes at maximum efficiency without wheel lock or loss of directional stability. Inductive, or 'active' sensors (wheel speed sensors) monitor the speed of the wheels by generating an electrical signal as the wheel is rotated. This information is passed to the ABS-ECM which compares the signals received from each wheel and uses the speed of the fastest wheel as a reference value. The ECM continually monitors the speed of each wheel and if the onset of lock at any wheel is detected (a received speed signal being less than the reference value) a signal is sent to the ABS hydraulic control unit which regulates the brake pressure for the relevant wheel(s).

Where the system incorporates traction control, essentially the reverse principle is applied. When the ECM detects that one or more wheels are rotating faster than the reference value, the brake is actually applied on the relevant wheel(s) to reduce the rotational speed. Additionally, when wheel spin is detected, various signals are sent to the engine management ECM to control engine torque and rpm.

Bosch 5.3 ABS is comprised of the following main components **(see illustration 10.1)**:

a) *Hydraulic control unit with integral ABS-ECM.*
b) *Inductive wheel speed sensors and associated sensor rings.*

c) Brake light switch.
d) ABS warning lamp.
e) Diagnostic connector.
f) ABS electrical wiring harness.

In addition, the conventional brake system is comprised of the following components:

a) Tandem brake master cylinder.
b) Vacuum servo unit.
c) Brake calipers/wheel cylinders and hydraulic hoses and pipes.
d) Pressure regulating/load-sensing valve(s) depending on application.

2 Component description and operation

ABS ECM

General

The Bosch 5.3 Electronic Control Module (ECM) continually monitors wheel speed from the signals provided by the wheel speed sensors, and brake application from the brake light switch signal. If the ECM detects the incidence of wheel lock (or wheel spin, if traction control is incorporated) on one or more wheels, a signal is sent to the hydraulic control unit to modulate the hydraulic pressure to the brake of the locking or spinning wheel(s). The ECM contains two microprocessors and uses digital technology to complete this function and other functions such as, fault code memory and power modules for valve and relay activity (see illustration 10.2).

To reduce external electrical connections to a minimum and improve reliability, the ECM is integral with the hydraulic control unit (see illustration 10.3). In most applications the ECM and hydraulic control unit cannot be separated and are only available as a complete sealed unit.

Self-test

The Bosch 5.3 ECM is equipped with a self-test capability that initially examines the ABS system when the ignition is switched on, and then examines the wheel speed sensor signals after a wheel speed of approximately 4 mph is reached from all wheels. The ABS self-test program continues to examine the signals from the various components as long as the ignition is switched on. If self-test determines that faults are not present, the ABS is ready for operation once a specified vehicle speed has been achieved.

If the ECM detects that a fault is present, all ABS functions are switched off and the warning lamp is turned on. The conventional braking system continues to operate as normal without ABS assistance.

Self-diagnostics

If the ECM detects a fault during the self-test routine, an internal

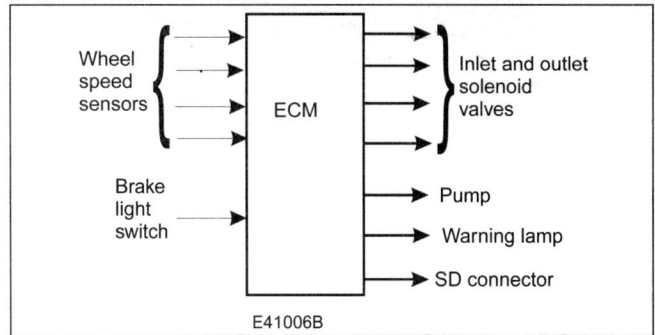

10.2 ECM sensor inputs and control signal outputs

fault code is stored in the ECM memory. Stored fault codes can be retrieved from the SD connector with the aid of a suitable fault code reader. If the fault clears, the code will remain stored until cleared with an FCR.

Hydraulic control unit

Bosch 5.3 is a four-channel system with a separate hydraulic circuit for each brake. The hydraulic control unit consists of an electric motor and radial piston return pump, inlet and outlet solenoid valves pressure accumulators and pulsation dampers (see illustration 10.4). The unit controls the hydraulic pressure applied to the brake for each individual front wheel and each individual rear wheel. The return pump is switched on when the ABS is activated and returns hydraulic fluid, drained off during the pressure reduction phase, back into the brake circuit.

In most installations, the 'select-low' principle is employed for control of the rear brakes during ABS operation. With the 'select-low' principle, the wheel with the lowest adhesion determines the amount of hydraulic pressure to be supplied to both rear brakes during a controlled ABS cycle.

In certain installations, the ABS-ECM contains additional software for rear brake hydraulic fluid pressure regulation when ABS is not in operation. This is generally known as 'Electronic Brake Force Distribution' and in these applications, mechanical pressure

10

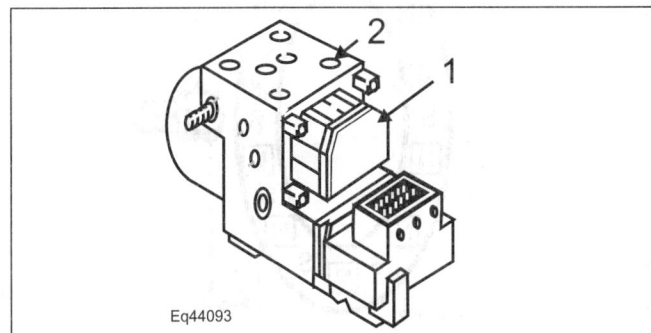

10.4 Bosch 5.3 hydraulic circuit schematic

| 1 | Pump motor | 3 | Pulsation damper |
| 2 | Inlet and outlet solenoid valves | | |

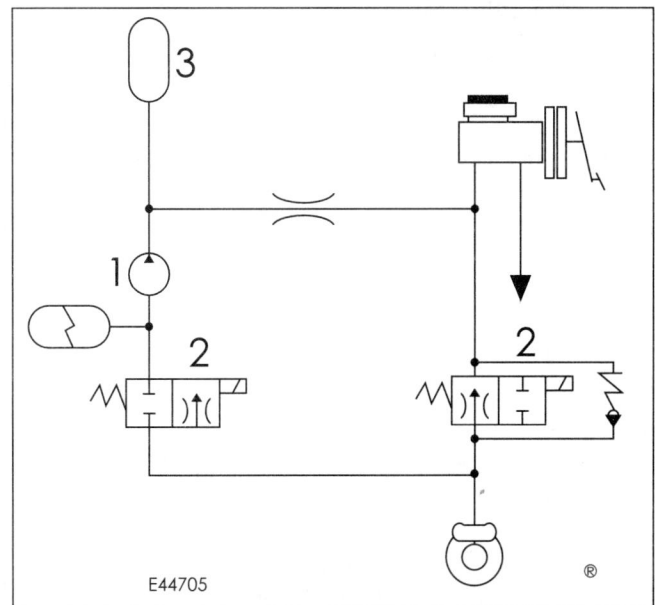

10.3 Bosch 5.3 ECM (1) and integral hydraulic control unit (2)

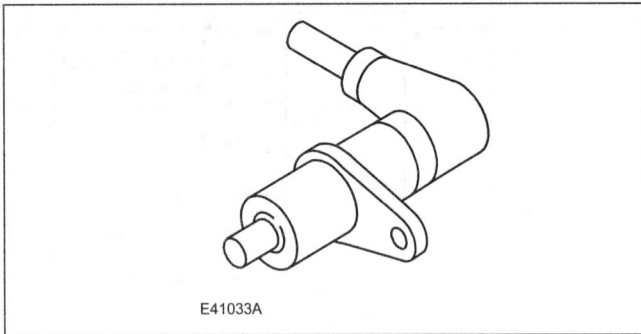

10.5 Typical inductive wheel speed sensor

regulating/load sensing valves are not required as the rear brake hydraulic pressure is controlled by the ABS hydraulic control unit.

Wheel speed sensors

Inductive type wheel speed sensors

The rotational speed of the roadwheels and any changes in the rotational speed are recorded either by inductive, or 'active' wheel speed sensors, one located at each roadwheel **(see illustration 10.5)**.

Where inductive sensors are used, each wheel speed sensor assembly comprises a toothed sensor ring, which rotates at roadwheel speed, and an adjacent sensor mounted a set distance from the sensor ring **(see illustration 10.6)**.

The sensors are permanent magnet pulse generator types producing an AC voltage sine wave as the sensor ring teeth pass through the magnetic field of the sensor **(see illustration 10.7)**.

The frequency of the waveform produced by the wheel speed sensor is proportional to the road speed. This AC voltage signal is continually being delivered to the ABS-ECM for processing.

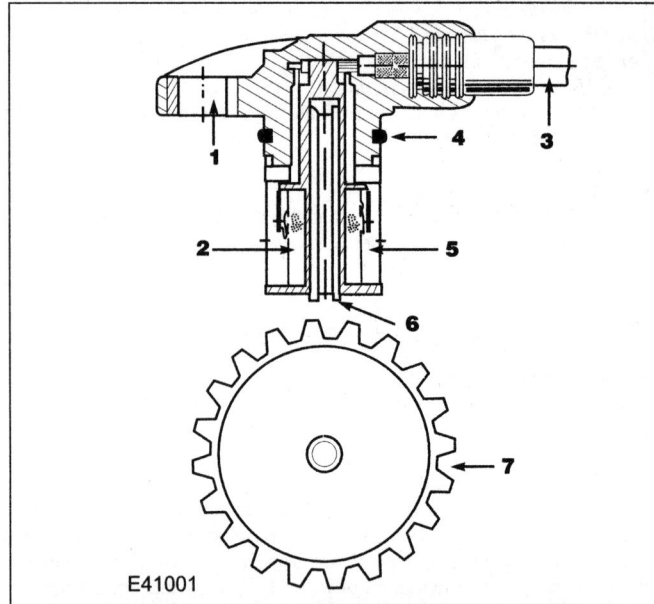

10.7 Inductive wheel speed sensor operation

1	Sensor body
2	Coil
3	Toothed sensor ring
4	AC signal
L	Air gap

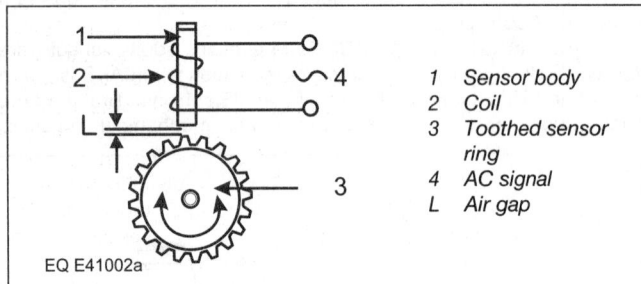

10.8 Inductive wheel speed sensor waveform as viewed on an oscilloscope

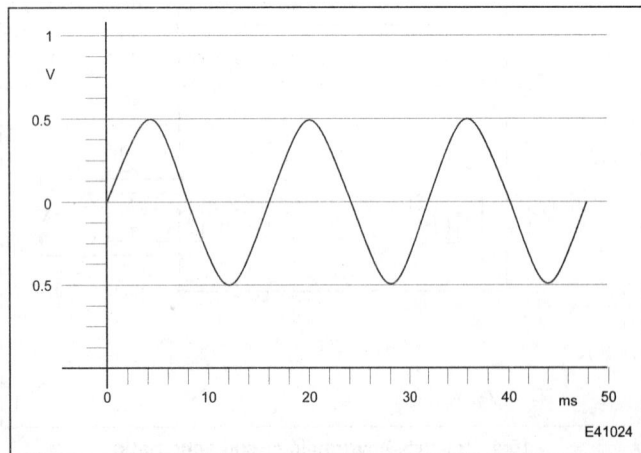

10.6 Sectional view of an inductive wheel speed sensor

1	Mounting bolt location	5	Coil
2	Permanent magnet	6	Sensor tip
3	Wiring harness	7	Toothed sensor ring
4	O-ring		

The peak to peak voltage of the speed signal (when viewed upon an oscilloscope) can vary considerably according to wheel speed. An analogue to digital converter (ADC) in the ECM transforms the AC pulse into a digital signal **(see illustration 10.8)**.

'Active' type wheel speed sensors

On the Bosch 5.3 system fitted to certain Fiat models, the rotational speed of the roadwheels and any changes in the rotational speed are recorded by 'active' wheel speed sensors.

On the majority of antilock brake systems, the wheel speed sensors are of the permanent magnet pulse generator type, as described previously. On the Bosch 5.3 system with 'active' sensors, the internal electrical resistance of the sensor is altered by changes in intensity and direction of the lines of force of an external magnetic field.

This external magnetic field is created by a unique type of sensor ring known as a 'multi-polar ring'. The multi-polar ring consists of a series of magnetic elements with alternating north/south polarities, located around the circumference of the ring **(see illustration 10.9)**.

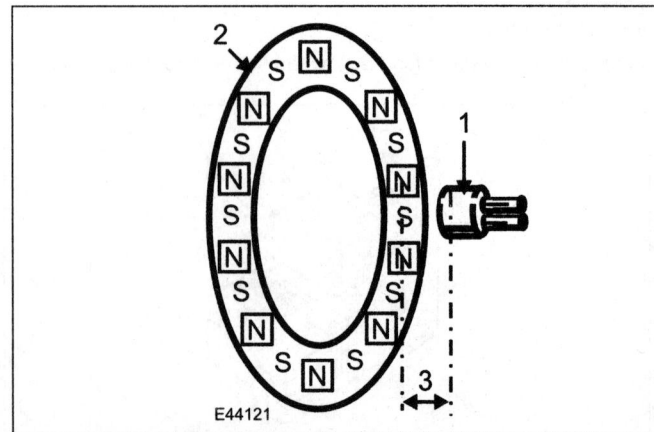

10.9 Active wheel speed sensor multi-polar ring
1 Wheel speed sensor   2 Multi-Polar ring   3 air gap

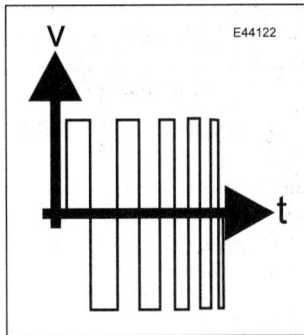

10.10 DC voltage square wave signal produced by the active wheel speed sensors

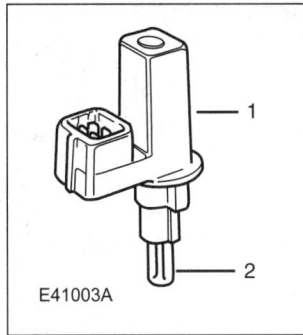

10.11 Typical brake light switch body (1) and contact pin (2)

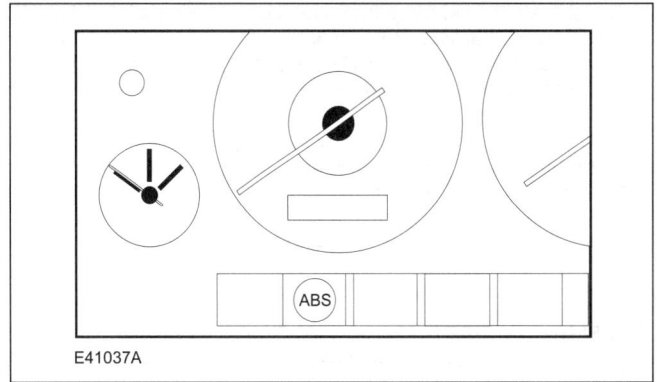

10.12 ABS warning lamp

The multi-polar ring is attached to the driveshaft or wheel hub, with the adjacent sensor mounted a set distances from the ring. As the ring rotates at roadwheel speed, a DC voltage square wave, constant amplitude signal is produced; the frequency of which is proportional to the road speed. This signal is continually being delivered by the sensor to the ABS-ECM for processing **(see illustration 10.10)**.

ABS electrical wiring harness and relays

An integrated 26-pin, or 31-pin main wiring harness is used for ABS-ECM power supply and earth connections, and enables sensor signals to reach the ECM and the ECM, in turn, to send output signals to the ABS warning lamp and diagnostic connector. The main (solenoid valve) relay and return pump relay are an integral part of the ECM and cannot be separately removed. Internal connections between the ECM and hydraulic control unit are used to activate the return pump motor.

Brake light switch

The brake light switch comprises a switch body and contact pin and is located above the brake pedal **(see illustration 10.11)**.

When the brake pedal is depressed, closing the brake light switch, a signal is sent to the ECM indicating that the brakes are being applied. Once this signal is received, the ECM will begin monitoring the wheel speed via the wheel speed sensors and activate the ABS if necessary.

ABS warning lamp

After the ignition is switched on, the ABS warning lamp on the instrument panel is illuminated for approximately 2 to 4 seconds during the system initial self-test cycle, then extinguished. During vehicle operation above a pre-determined wheel speed, the ABS-ECM continues the self-test cycle whereby the system status is continually monitored. If a fault is detected, the ECM illuminates the warning lamp on the instrument panel. The ECM switches off the ABS, however the conventional braking system continues to operate as normal. The warning lamp will remain illuminated until the fault is no longer present **(see illustration 10.12)**.

When the ABS-ECM detects a fault, the fault code is stored and the ABS warning lamp activated. If the fault no longer exists after the next system start (ignition on/off) the ABS warning lamp is extinguished after the self-test cycle, however the fault code remains stored in the ECM memory.

Tandem master cylinder

Typically, the tandem master cylinder comprises a body casting incorporating primary and secondary pressure chambers, primary piston, intermediate piston, floating piston, slotted pin and central valve. The cylinder operates as a conventional master cylinder using vacuum assistance from the vacuum servo unit **(see illustration 10.13)**.

When the brake system is at rest, the central valve in the floating piston rests against the slotted pin. In this condition, the central valve is open and brake fluid can discharge out of the pressure chamber back into the brake fluid reservoir. When the brake pedal is depressed, the build-up of hydraulic pressure in the primary pressure chamber acts on the intermediate piston and floating piston, moving them down the cylinder bore. The floating piston contacts the seal on the central valve, closing the connection between the intermediate and secondary pressure chambers. Brake hydraulic pressure can now also increase in the secondary pressure chamber.

Vacuum servo unit

The vacuum servo unit is located between the brake pedal and tandem master cylinder. When the brake pedal is depressed, the servo unit increases the force applied by the pedal, reducing the effort required to operate the brakes **(see illustration 10.14)**.

10

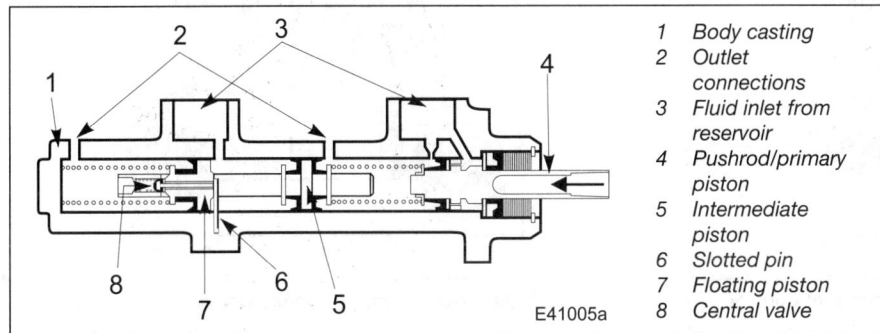

1	Body casting
2	Outlet connections
3	Fluid inlet from reservoir
4	Pushrod/primary piston
5	Intermediate piston
6	Slotted pin
7	Floating piston
8	Central valve

10.13 Sectional view of a typical tandem master cylinder

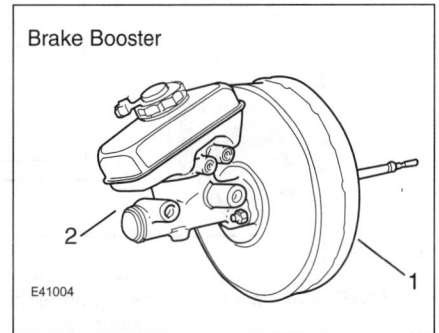

10.14 Vacuum servo unit (1) and tandem master cylinder (2)

The unit is operated by vacuum created in the engine inlet manifold (or from a separate vacuum pump on diesel engines) which is applied to a diaphragm contained within the unit casing. A pushrod connected to the centre of the diaphragm acts directly on the primary piston in the master cylinder.

When the brake pedal is released, vacuum is applied to both sides of the diaphragm. When the pedal is depressed, one side of the diaphragm is opened to atmosphere and the vacuum acting on the other side deflects the diaphragm which in turn operates the master cylinder primary piston. The resulting force applied to the master cylinder piston is therefore significantly greater than the initial force applied to the brake pedal by the driver.

Pressure regulating/load-sensing valve(s)

Depending on vehicle application, pressure regulating valves or load sensing valves may be incorporated to restrict the hydraulic fluid pressure to the rear brakes. The valves may be pressure conscious whereby the hydraulic fluid supply is restricted once a pre-determined pressure is reached, or load conscious whereby the hydraulic pressure is reduced according to vehicle loading.

On certain versions, the ABS-ECM contains additional software for rear brake hydraulic fluid pressure regulation during normal braking, generally known as 'Electronic Brake Force Distribution'. In these applications, mechanical pressure regulating/load sensing valves are not required as the rear brake hydraulic pressure is controlled by the ABS hydraulic control unit.

3 System operation

Brake system at rest

When the system is at rest all the brake components are inoperative. Pressure is non-existent in the hydraulic pipes between the tandem master cylinder and the brake calipers. The inlet solenoid valves in the hydraulic control unit valve block are open and the outlet solenoid valves are closed.

Brake system operating under conventional control without ABS

When the brake pedal is activated, the pedal force is applied to the tandem master cylinder by the vacuum servo unit pushrod. The servo unit pushrod acts directly on the pressure piston in the master cylinder, which pressurises the hydraulic fluid in the brake pipes to the hydraulic control unit. The inlet solenoid valve and outlet solenoid valve both remain in the 'at rest' position (inlet solenoid valve open and outlet solenoid valve closed). Hydraulic pressure is transmitted to

each brake caliper, thus operating the brakes.

When the brake pedal is released, the one-way valve opens allowing the hydraulic pressure in the circuit to rapidly decrease (see illustration 10.15).

Brake system operating in conjunction with ABS control

The ABS-ECM continually monitors wheel speed from the signals provided by the wheel speed sensors. If the ECM detects the incidence of wheel lock on one or more wheels, ABS is automatically initiated in three phases. As Bosch 5.3 ABS operates individually on each wheel, all or any of the wheels could be in any one of the following phases at any particular moment.

First ABS phase, pressure holding

To prevent any further build-up of hydraulic pressure in the circuit being controlled, the ECM closes the inlet solenoid valve and allows the outlet solenoid valve to remain closed. The hydraulic fluid line from the tandem master cylinder to the brake caliper or wheel cylinder is closed, and the hydraulic fluid in the controlled circuit is maintained at a constant pressure. This effectively removes the braking force from the controlled circuit. The pressure cannot now be increased in that circuit by any further application of the brake pedal (see illustration 10.16).

If the wheel speed sensor signals indicate that wheel rotation has now stabilised, the ECM will instigate the pressure build-up phase, allowing braking to continue. If wheel lock is still detected after the pressure holding phase, the ECM instigates the pressure reduction phase.

Second ABS phase, pressure reduction

If the ECM detects wheel instability, a pressure reduction phase is initiated. The inlet solenoid valve remains closed and the outlet solenoid valve is opened by means of a series of short activation pulses. The pressure in the controlled circuit decreases rapidly as the fluid flows from the brake caliper or wheel cylinder into the pressure accumulator. At the same time, the ECM actuates the electric motor to operate the return pump. The hydraulic fluid is then pumped back into the pressure side of the master cylinder. This process creates a pulsation which can be felt in the brake pedal action, but which is softened by the pulsation damper (see illustration 10.17).

Third ABS phase, pressure build-up

The pressure build-up phase is instigated after the wheel rotation has stabilised. The inlet and outlet solenoid valves are returned to the at rest position (inlet solenoid valve open and outlet solenoid valve

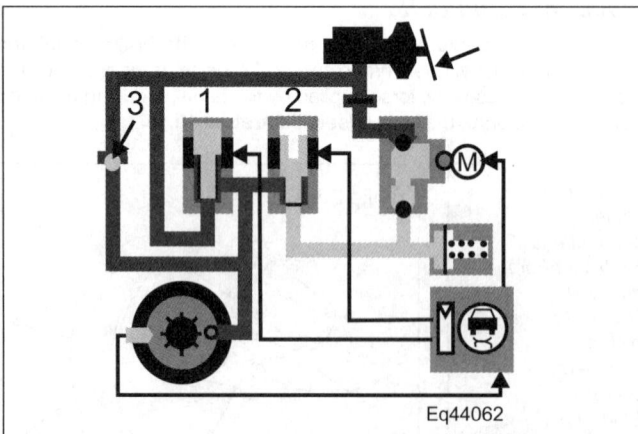

Eq44062

10.15 Brake system operating under conventional control without ABS

1	Inlet solenoid valve	3	One-way valve
2	Outlet solenoid valve		

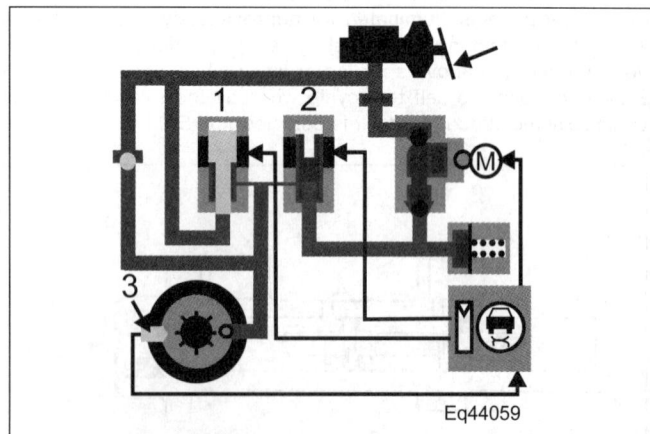

Eq44059

10.16 ABS operation - first phase, pressure holding

1	Inlet solenoid valve	3	Wheel speed sensor
2	Outlet solenoid valve		

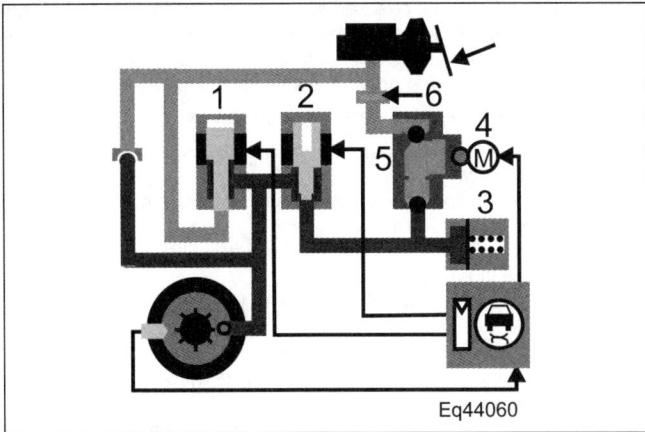

10.17 ABS operation - second phase, pressure reduction

1 *Inlet solenoid valve*	4 *Pump motor*
2 *Outlet solenoid valve*	5 *Return pump*
3 *Pressure accumulator*	6 *Pulsation damper*

closed) which re-opens the hydraulic fluid line from the tandem master cylinder to the brake caliper or wheel cylinder. Hydraulic pressure is reinstated, thus re-introducing operation of the brake. After a brief period, a short pressure holding phase is re-introduced and the ECM continually shifts between pressure build-up and pressure holding until

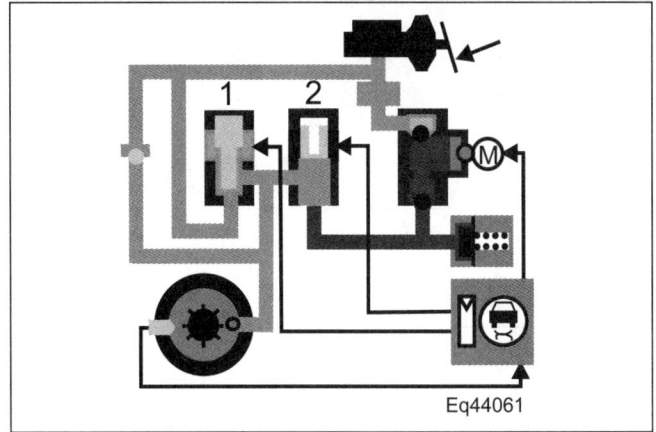

10.18 ABS operation - third phase, pressure build-up

1 *Inlet solenoid valve*	2 *Outlet solenoid valve*

the wheel has decelerated to a sufficient degree where pressure reduction is once more required **(see illustration 10.18)**.

The whole ABS control cycle takes place 4 to 10 times per second for each affected wheel and this ensures maximum braking effect and control during ABS operation.

Test procedures

Important note: *The test procedures, pintables and wiring diagrams contained in this Chapter are necessarily representative of the system depicted. Because of the variations in wiring and other data that often occurs, even between similar vehicles in any particular VM's range, the reader should take great care in identification of ECM pins, and satisfy himself that he has gathered the correct data before failing a particular component.*

4 Wheel speed sensors

Inductive type wheel speed sensors

Checking the wheel speed sensor (general)

1 Inspect the wheel speed sensor for corrosion or damage and check that the sensor is tightly mounted.
2 Check the toothed sensor ring for damage, eccentricity and for broken or missing teeth.
3 Inspect the wheel speed sensor wiring plug for corrosion and damage. One plug for each sensor.
4 Check that the connector terminal pins are pushed fully home and making good contact with the sensor wiring plug.
5 Check the clearance between the sensor and the toothed sensor ring. The clearance is not normally adjustable but is nominally 0.2 to 1.2 mm. If the clearance is excessive, expect a worn sensor tip or problems with the wheel bearings/hub or sensor ring.
6 When carrying out voltage checks with an oscilloscope or voltmeter, the voltage obtained will be proportional to the speed at which the wheel is rotating. In addition to determining that the wheel speed sensors are actually producing a voltage output, it is essential that the output from the sensors on a particular axle is the same for any given wheel speed.

Checking wheel speed sensor output with an oscilloscope

Note: *Refer to the wiring diagrams for specific ECM pin identification according to model.*

7 Switch the ignition off and disconnect the ECM multi-plug or the relevant wheel speed sensor wiring plug.
8 Connect an oscilloscope between the terminal pins for the sensor under test **(see illustration 10.19)**.
9 Select a range to cover 80 Hz on the oscilloscope and a free run time base.
10 Raise the wheel and rotate it by hand at approximately one revolution per second.

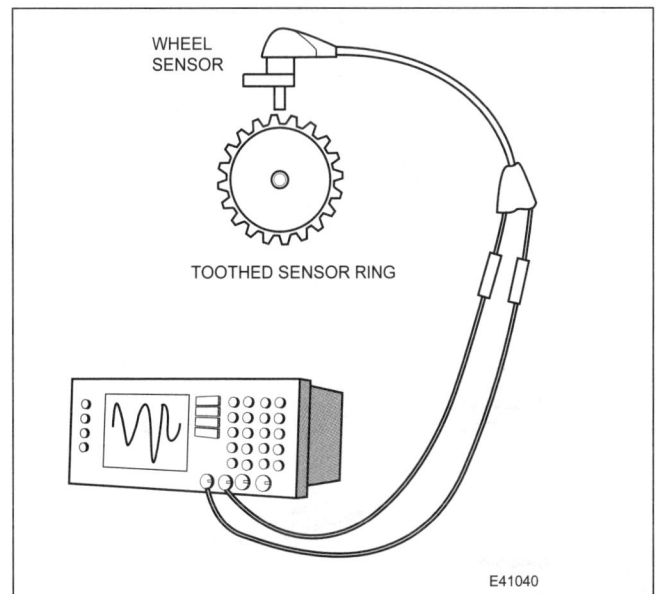

10.19 Checking inductive wheel speed sensor output with an oscilloscope connected to the sensor-wiring plug

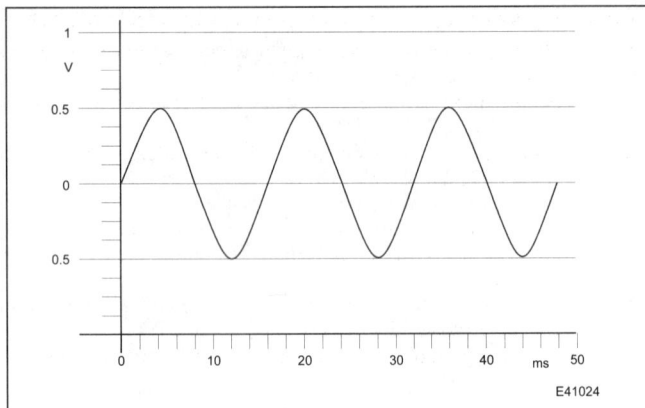

10.20 Typical inductive wheel speed sensor sine wave as displayed on an oscilloscope

11 A sinusoidal wave form should be obtained, with amplitude and duration changing with rotational speed **(see illustration 10.20)**.

12 If there is no signal, or a very weak or intermittent signal at the ECM, repeat the test at the sensor wiring plug. If there is no change in signal status, the sensor is suspect.

13 If the signal is now satisfactory this indicates a fault in the wiring harness which should be checked for continuity.

Checking wheel speed sensor output with an AC voltmeter

Note: *Refer to the wiring diagrams for specific ECM pin identification according to model.*

14 Switch the ignition off and disconnect the ECM multi-plug or the relevant wheel speed sensor wiring plug.

15 Connect an AC voltmeter between the terminal pins for the sensor under test **(see illustration 10.21)**.

16 Raise the wheel and rotate it by hand at approximately one revolution per second.

17 A voltage of approximately 0.5 to 1.5 volts (AC RMS) should be obtained. If there is no signal, or a very weak or intermittent signal at the ECM, repeat the test at the sensor-wiring plug. If there is no change in the signal, the sensor is suspect.

18 If the signal is now satisfactory, this indicates a fault in the wiring

harness which should be checked for continuity. **Note:** *This test at least proves that a signal is being generated by the sensor. However, the voltage produced is an average voltage and does not clearly indicate damage to the sensor ring or that the sinewave is regular in formation.*

Checking wheel speed sensor resistance

Note: *Refer to the wiring diagrams for specific ECM pin identification according to model.*

19 Switch the ignition off and disconnect the ECM multi-plug or the relevant wheel speed sensor-wiring plug.

20 Connect an ohmmeter between the terminal pins for the sensor under test **(see illustration 10.22)**.

21 The readings obtained should be between 0.7 and 2.2 kohms approximately.

22 If the resistance is excessively high, or open circuit at the ECM, repeat the test at the sensor multi-plug. If there is no change in resistance, the sensor is suspect.

23 If the resistance is now satisfactory, this indicates a fault in the wiring harness which should be checked for continuity. **Note:** *Even if the resistance is within the quoted specifications, this does not prove that the speed sensor can generate an acceptable signal.*

'Active' type wheel speed sensors

Checking the wheel speed sensor (general)

24 Inspect the wheel speed sensor for corrosion or damage and check that the sensor is tightly mounted.

25 Check the toothed sensor ring for damage, eccentricity and for broken or missing teeth.

26 Inspect the wheel speed sensor wiring plug for corrosion and damage. One plug for each sensor.

27 Check that the connector terminal pins are pushed fully home and making good contact with the sensor wiring plug.

28 Check the clearance between the sensor and the toothed sensor ring. The clearance is not normally adjustable but is nominally 0.2 to 2.0 mm. If the clearance is excessive, expect a worn sensor tip or problems with the wheel bearings/hub or sensor ring.

Checking wheel speed sensor signal with an oscilloscope

Note: *Refer to a suitable model specific wiring diagram for ECM pin identification.*

29 Switch the ignition off and connect a BOB between the ECM and the harness multi-plug.

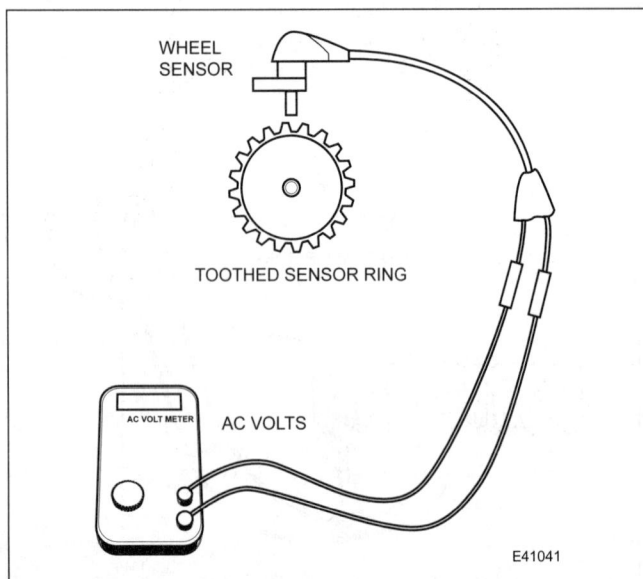

10.21 Checking inductive wheel speed sensor output with a voltmeter connected to the sensor-wiring plug

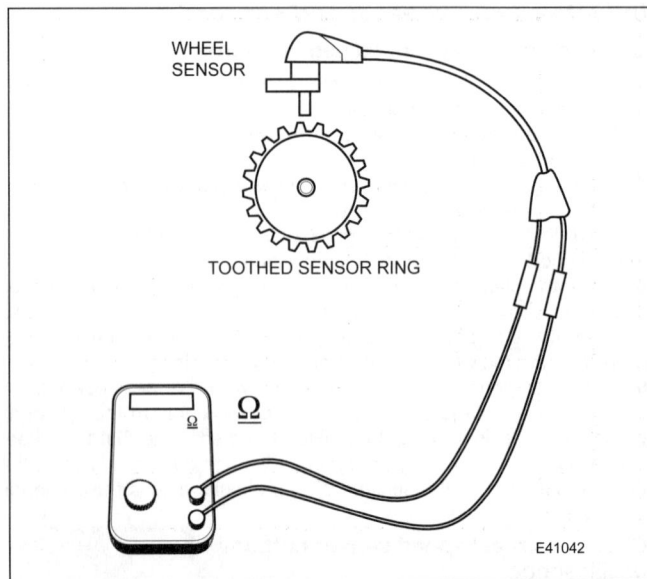

10.22 Checking inductive wheel speed sensor resistance with an ohmmeter connected to the sensor-wiring plug

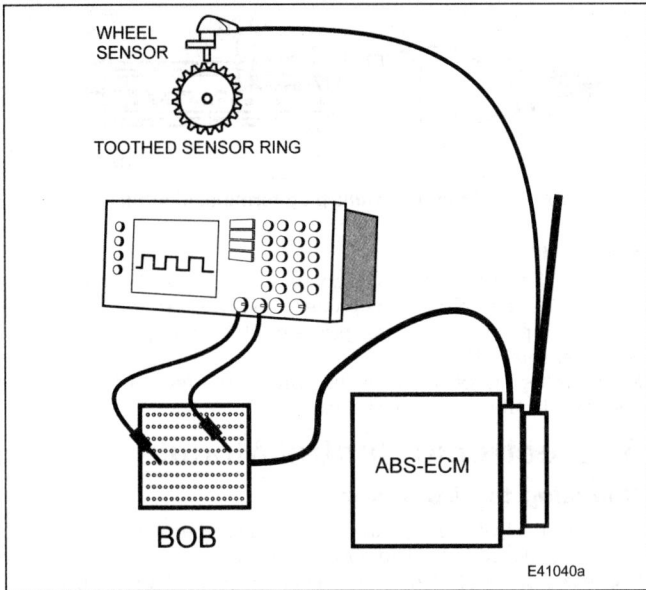

10.23 Checking active wheel speed sensor output with a break-out-box connected between the oscilloscope and ECM multi-plug

30 Connect an oscilloscope to the relevant BOB terminals for the sensor under test, and set the oscilloscope to measure DC voltage **(see illustration 10.23)**.

31 Raise and support the vehicle.

32 Switch the ignition on and rotate the wheel by hand at approximately one revolution per second.

33 A square waveform should be obtained, switching between approximately 0.5 and 2.5 volts as the wheel is rotated **(see illustration 10.24)**.

34 If the output is not as specified, check the supply voltage to the sensor and the wiring harness for continuity.

35 If the supply voltage and wiring are satisfactory, the sensor is suspect.

Checking wheel speed sensor signal with a DC voltmeter

Note: *Refer to a suitable model specific wiring diagram for ECM pin identification.*

36 Switch the ignition off and connect a BOB between the ECM and the harness multi-plug.

37 Connect a DC voltmeter between the relevant BOB terminal for the sensor under test, and earth.

38 Raise and support the vehicle.

39 Switch the ignition on and rotate the wheel very slowly.

40 A voltage switching between approximately 0.5 and 2.5 volts should be obtained as the wheel is rotated.

41 If the voltage is not as specified, check the supply voltage to the sensor and the wiring harness for continuity.

42 If the supply voltage and wiring are satisfactory, the sensor is suspect.

Checking wheel speed sensor supply voltage

43 Switch the ignition off and disconnect the wheel speed sensor wiring harness 2-pin multi-plug, of the sensor under test.

44 Connect a DC voltmeter between a vehicle earth and pin 2 of the multi-plug **(see illustration 10.25)**.

45 Switch the ignition on. A voltage of approximately 8.0 volts should be obtained.

46 If the voltage is not as specified, repeat the test at the ECM multi-plug using the BOB as described previously. If there is no change in the reading, the ABS-ECM is suspect. If the voltage is now satisfactory, this indicates a fault in the wiring harness which should be checked for continuity. **Note:** *This test at least proves that a supply voltage is available for the sensor. However, the integral electronics in the sensor must be active to produce an output.*

Checking wheel speed sensor resistance

Note: *Refer to a suitable model specific wiring diagram for ECM pin identification.*

47 Switch the ignition off and disconnect the ECM multi-plug or the relevant wheel speed sensor 2-pin multi-plug.

48 Connect an ohmmeter between the terminal pins for the sensor under test.

10

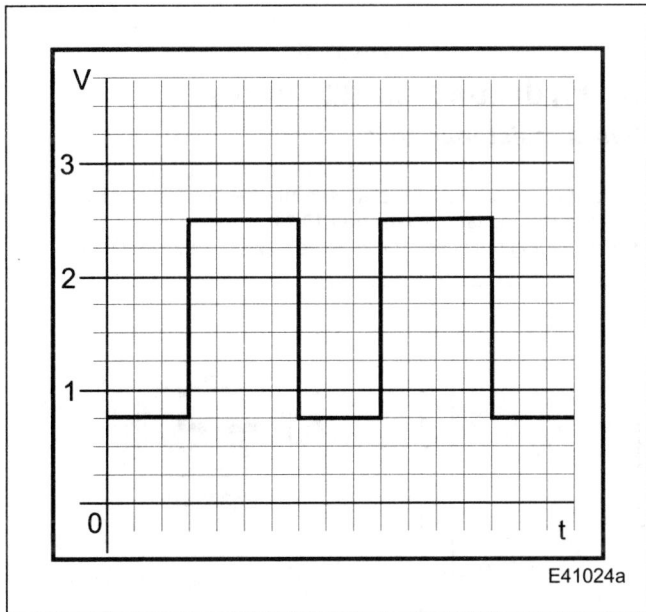

10.24 Active wheel speed sensor square wave output as viewed on an oscilloscope

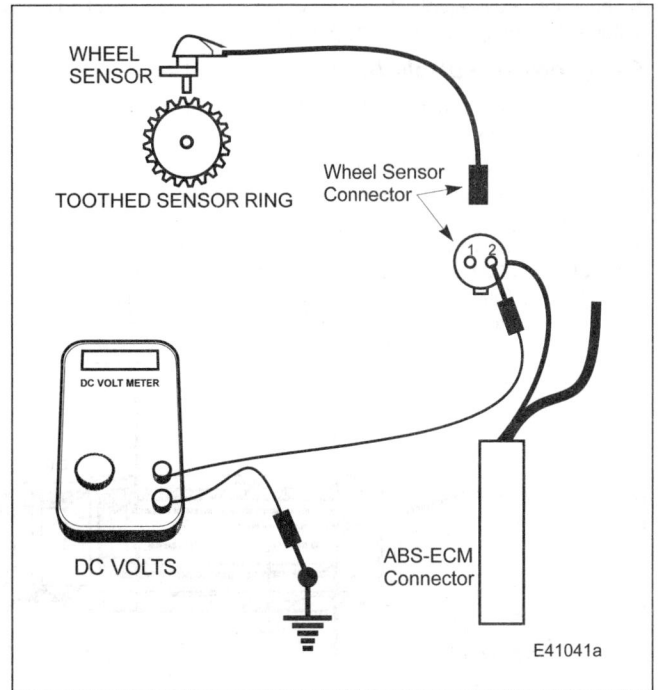

10.25 Checking active wheel speed sensor output using a voltmeter

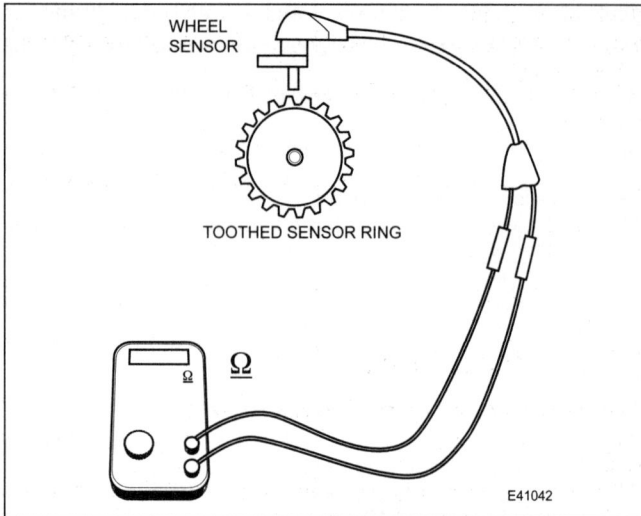

10.26 Checking active wheel speed sensor resistance with an ohmmeter connected to the sensor-wiring plug

49 The readings obtained should be between 2.5 and 4.5 kohms **(see illustration 10.26)**.

50 If the resistance is excessively high, or open circuit at the ECM, repeat the test at the sensor multi-plug. If there is no change in resistance, the sensor is suspect.

51 If the resistance is now satisfactory, this indicates a fault in the wiring harness which should be checked for continuity. **Note:** *Even if the resistance is within the quoted specifications, this does not prove that the speed sensor can generate an acceptable signal.*

5 System relays

Relay operation tests

1 The relays are integral with the ECM and can only be checked using suitable ABS diagnostic test equipment.

Relay power supply tests

Note: *Refer to the wiring diagrams for specific ECM pin identification according to model.*

Main (solenoid valve) relay

2 Switch the ignition off and disconnect the ECM multi-plug.

3 Attach a negative voltmeter probe to a vehicle earth.

4 Attach a positive voltmeter probe to ECM multi-plug pin 18 **(see illustrations 10.27 and 10.28)**. The voltmeter should indicate nbv. If no voltage is found, check the relevant fuse and the supply wiring back to the battery positive terminal.

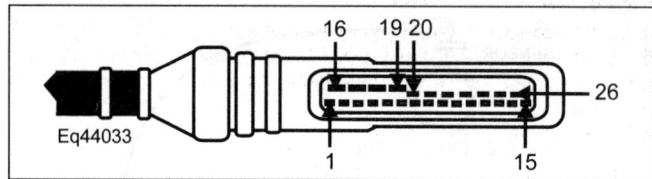

10.27 ECM 26-pin multi-plug terminal identification

Pump relay

5 Switch the ignition off and disconnect the ECM multi-plug.

6 Attach a negative voltmeter probe to a vehicle earth.

7 Attach a positive voltmeter probe to ECM multi-plug pin 17. The voltmeter should indicate nbv.

8 If no voltage is found, check the relevant fuse and the supply wiring back to the battery positive terminal.

6 Electronic Control Module

Checking the ECM (general)

1 Inspect the ECM for corrosion or damage and ensure that the unit is securely attached to the hydraulic control unit.

2 Check that the ECM multi-plug terminals are pushed fully home and making good contact with the ECM pins. Faults in any of the above areas are possible reasons for poor performance in the ABS system.

ECM power supply and earth tests

3 Switch the ignition off and disconnect the ECM multi-plug.

6 Switch the ignition on.

4 Attach a negative voltmeter probe to a vehicle earth.

5 Attach a positive voltmeter probe to ECM multi-plug pin 15 **(see illustrations 10.27 and 10.28)**. The voltmeter should indicate nbv. If no voltage is found, check the relevant fuse and the supply wiring back to the ignition switch.

9 Switch the ignition off.

10 Connect an ohmmeter between a vehicle earth and ECM multi-plug pins 16 and 19 in turn. The ohmmeter should indicate continuity. If not, check the ECM main earth connection and wiring.

7 Solenoid valves

Solenoid valve operation tests

1 The solenoid valves are integral with the ECM and can only be checked using suitable ABS diagnostic test equipment.

8 Hydraulic pump motor

Pump resistance test

1 Switch the ignition off and disconnect the pump motor multi-plug.

2 Connect an ohmmeter between the terminal pins of the pump motor multi-plug **(see illustration 10.29)**.

3 The reading obtained should be between 0.1 and 1.0 Ohms.

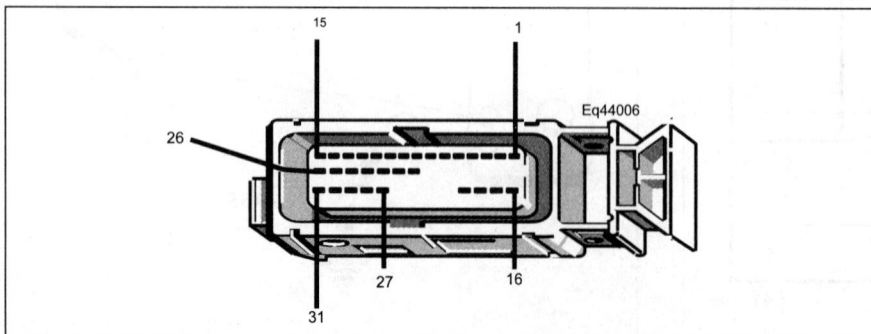

10.28 ECM 31-pin multi-plug terminal identification

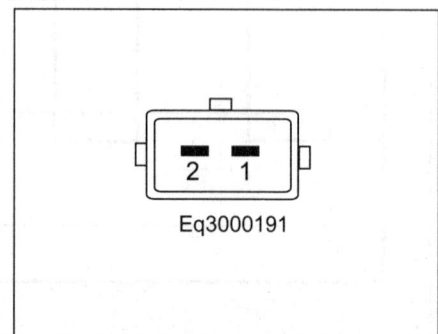

10.29 Hydraulic pump motor multi-plug terminal identification

Pump operation test

4 Switch the ignition off and disconnect the pump motor multi-plug.
5 Connect the positive terminal of a 12 volt supply to terminal 2 of the pump motor multi-plug, and the negative terminal to terminal 1 of the multi-plug.
6 The pump motor should now run.

⚠ **Warning: The test should be made as quickly as possible to avoid damaging the pump.**

7 If the pump does not operate as described, renew the hydraulic control unit.

9 Brake light switch

Checking the brake light switch (general)

1 Check that the brake light switch is correctly and securely mounted and that the plunger moves smoothly with no trace of binding.
2 Check that the wiring multi-plug is pushed fully home and making good contact.
3 Check that no wires have been disconnected.
4 Faults in any of the above areas are possible reasons for failure or malfunctioning of the switch.

Brake light switch voltage and continuity tests

Voltage test

5 Switch the ignition off and disconnect the ECM multi-plug from the ECM.
6 Connect a voltmeter between ECM multi-plug brake light switch input pin 14 and an ECM earth pin **(see illustrations 10.27 and 10.28)**.

7 Switch the ignition on and depress the brake pedal. The voltmeter should indicate nbv.
8 If no voltage is found, the fuse, the brake light switch and the wiring are suspect.
9 Release the brake pedal. The voltage should drop to zero as the switch opens.

Continuity test

10 Switch the ignition off and disconnect the brake light switch multi-plug.
11 Connect an ohmmeter between the terminal pins of the brake light switch.
12 Operate the brake light switch and check for continuity. If the test fails, renew the switch.

10 Warning lamp

Checking the warning lamp (general)

1 Inspect the warning lamp bulb holder contacts in the instrument panel.
2 Check that the instrument panel multi-plug terminal pins are pushed fully home and making good contact.
3 Check that no wires have been disconnected.
4 Faults in any of the above areas are possible reasons for failure or malfunctioning of the warning lamp.

Warning lamp operation test

5 With the ignition switched off, the warning lamp should remain off.
6 Switch the ignition on and the warning lamp should illuminate then extinguish after a few seconds. The lamp should then remain off.
7 If the warning lamp comes on and remains on at any time during vehicle operation, carry out the previously described test procedures on the system components.

10

Pin table - typical 26-pin

Note: *Refer to illustration 10.27*
The following table is applicable to models with inductive wheel speed sensors. At the time of writing, ECM pin connections for models with active wheel speed sensors was not available.

Pin No.	Connection	Test condition	Voltage
1	Rear right wheel speed sensor earth	Roadwheel rotating	0.25 volts (max)
2	Rear right wheel speed sensor signal	Roadwheel rotating	0.2 to 1.0 volts AC (approx)
3	Front right wheel speed sensor earth	Roadwheel rotating	0.25 volts (max)
4*	Front right wheel speed sensor signal	Roadwheel rotating	0.2 to 1.0 volts AC (approx)
5	-	-	-
6	Front left wheel speed sensor earth	Roadwheel rotating	0.25 volts (max)
7	Front left wheel speed sensor signal	Roadwheel rotating	0.2 to 1.0 volts AC (approx)
8	Rear left wheel speed sensor earth	Roadwheel rotating	0.25 volts (max)
9	Rear left wheel speed sensor signal	Roadwheel rotating	0.2 to 1.0 volts AC (approx)
10**	ABS warning lamp	Ignition on: Lamp on Lamp off	 1.0 volt (max) Nbv
11	SD connector	-	-
12***	SD connector	-	-
13	-	-	-
14	Brake light switch input	Ignition on: Brake pedal released Brake pedal depressed	 0 volts Nbv
15	Supply from ignition switch	Ignition on	Nbv
16	Hydraulic pump motor earth	Ignition on	0.25 volts (max)
17	Supply from battery	Ignition off/on	Nbv
18	Supply from battery	Ignition off/on	Nbv

Pin table - typical 26-pin (continued)

Pin No.	Connection	Test condition	Voltage
19	Solenoid valves earth	Ignition on	0.25 volts (max)
20	-	-	-
21	-	-	-
22	-	-	-
23	-	-	-
24	-	-	-
25	-	-	-
26	-	-	-

*Pin 5 on certain versions   **Pin 21 on certain versions   ***Certain versions only*

Pin table - typical 31-pin

Note: *Refer to illustration 10.28*
The following table is applicable to models with inductive wheel speed sensors. At the time of writing, ECM pin connections for models with active wheel speed sensors was not available.

Pin No.	Connection	Test condition	Voltage
1	Rear right wheel speed sensor earth	Roadwheel rotating	0.25 volts (max)
1*	Rear right wheel speed sensor signal	Roadwheel rotating	0.2 to 1.0 volts AC (approx)
2	Rear right wheel speed sensor signal	Roadwheel rotating	0.2 to 1.0 volts AC (approx)
2*	Rear right wheel speed sensor earth	Roadwheel rotating	0.25 volts (max)
3	Front right wheel speed sensor earth	Roadwheel rotating	0.25 volts (max)
4*	Front right wheel speed sensor earth	Roadwheel rotating	0.25 volts (max)
5	Front right wheel speed sensor signal	Roadwheel rotating	0.2 to 1.0 volts AC (approx)
6	Front left wheel speed sensor earth	Roadwheel rotating	0.25 volts (max)
7	Front left wheel speed sensor signal	Roadwheel rotating	0.2 to 1.0 volts AC (approx)
8	Rear left wheel speed sensor earth	Roadwheel rotating	0.25 volts (max)
9	Rear left wheel speed sensor signal	Roadwheel rotating	0.2 to 1.0 volts AC (approx)
10	-	-	-
11	SD connector	-	-
12***	SD connector	-	-
13	-	-	-
14	Brake light switch input	Ignition on: Brake pedal released Brake pedal depressed	 0 volts Nbv
15	Supply from ignition switch	Ignition on	Nbv
16	Hydraulic pump motor earth	Ignition on	0.25 volts (max)
17	Supply from battery	Ignition off/on	Nbv
18	Supply from battery	Ignition off/on	Nbv
19	Solenoid valves earth	Ignition on	0.25 volts (max)
20	ABS warning lamp	Ignition on: Lamp on Lamp off	 1.0 volt (max) Nbv
21*	ABS warning lamp	Ignition on: Lamp on Lamp off	 1.0 volt (max) Nbv
21**	Warning lamp (brake/engine)	-	-
22	-	-	-
23	-	-	-
24	-	-	-
25	-	-	-
26**	Engine management ECM signal	-	-
27	-	-	-
28	-	-	-
29	-	-	-
30	-	-	-
31	-	-	-

*alternative pin on certain versions   **Where fitted*

Pin table - typical 31-pin (Ford Mondeo)

Note: *Refer to illustration 10.28*
The following table is applicable to models with inductive wheel speed sensors. At the time of writing, ECM pin connections for models with active wheel speed sensors was not available.

Pin No.	Connection	Test condition	Voltage
1	-	-	-
2	Rear left wheel speed sensor signal	Roadwheel rotating	0.2 to 1.0 volts AC (approx)
3	Rear left wheel speed sensor earth	Roadwheel rotating	0.25 volts (max)
4	Front right wheel speed sensor signal	Roadwheel rotating	0.2 to 1.0 volts AC (approx)
5	Front right wheel speed sensor earth	Roadwheel rotating	0.25 volts (max)
6	Brake light switch input	Ignition on:	
		Brake pedal released	0 volts
		Brake pedal depressed	Nbv
7	Navigation ECM	-	-
8	Supply from battery	Ignition off/on	Nbv
9	Traction control ECM	-	-
10	-	-	-
11	-	-	-
12	Earth	Ignition on	0.25 volts (max)
13	Supply from battery	Ignition off/on	Nbv
14	Supply from battery	Ignition off/on	Nbv
15	Earth	Ignition on	0.25 volts (max)
16	Warning lamp (low brake fluid/park brake)	-	-
17	Traction control ECM	-	-
18	Traction control ECM	-	-
19	-	-	-
20	Front left wheel speed sensor signal	Roadwheel rotating	0.2 to 1.0 volts AC (approx)
21	Front left wheel speed sensor earth	Roadwheel rotating	0.25 volts (max)
22	Rear right wheel speed sensor signal	Roadwheel rotating	0.2 to 1.0 volts AC (approx)
23	Rear right wheel speed sensor earth	Roadwheel rotating	0.25 volts (max)
24	SD connector	-	-
25	ABS warning lamp	Ignition on:	
		Lamp on	1.0 volt (max)
		Lamp off	Nbv
26	Navigation ECM	-	-
27	-	-	-
28	-	-	-
29	SD connector	-	-
30	SD connector	-	-
31	-	-	-

10

Fault codes

1 The Bosch 5.3 system requires the use of a FCR for obtaining fault codes. Flash codes are not available for output from this system.

2 If a FCR is available, it should be connected to the SD serial connector (the Vauxhall/Opel term for SD connector is ALDL) and used in accordance with the maker's instructions.

3 The FCR can be used for the following purposes:
 a) *Obtaining fault codes.*
 b) *Clearing fault codes.*
 c) *Obtaining datastream information.*
 d) *Testing the system actuators (solenoid valve relay, pump relay and solenoid valves).*

4 On certain models (notably Ford, Renault and Rover), internal fault codes are used by the ECM to designate faults in the system components and circuits. A proprietary fault code reader (FCR) or system tester (such as the Renault XR25) is required to interrogate the system. No actual fault code numbers are available, although the component circuits checked by the ECM are similar to those shown for the other vehicles listed. For the remainder of vehicles covered in this Chapter, where fault code information is unavailable from the manufacturers at this time, the circuits checked by the ECM will also be similar to the other vehicles listed.

Fault code table (Audi, Volkswagen)

Code	Description	Fault
00283	Front left wheel speed sensor	Break, short circuit or poor wiring connection; incorrect air gap; sensor or sensor ring dirty or damaged
00285	Front right wheel speed sensor	Break, short circuit or poor wiring connection; incorrect air gap; sensor or sensor ring dirty or damaged
00287	Rear right wheel speed sensor	Break, short circuit or poor wiring connection; incorrect air gap; sensor or sensor ring dirty or damaged
00290	Rear left wheel speed sensor	Break, short circuit or poor wiring connection; incorrect air gap; sensor or sensor ring dirty or damaged
00301	Hydraulic pump	Pump defective
00526	Brake light switch	Poor wiring connection, switch defective or incorrectly adjusted
00529	Speed information signal absent	Break, short circuit or poor wiring connection between ABS ECM and engine management ECM
00532	Supply voltage	Outside expected values
00597	Wheel speed sensor signal variation	Incorrect wheel/tyre sizes, incorrect sensor air gap, sensor or sensor ring dirty or damaged
01130	ABS operation	Signal external interference, poor wiring connections
01200	Hydraulic unit (solenoid valves/pump)	Supply voltage outside expected values - wiring open circuit
01201	Hydraulic unit (pump)	Supply voltage outside expected values - poor earth connection
01203	Instrument panel connection	Wiring open circuit
65535	ECM	ECM defective

Fault code table (Citroën, Peugeot)

Code	Description	Fault
1	Pump motor	Defective / poor connections
2	Relays	Defective (signal interruption, short to earth, supply voltage too low)
3	Rear left wheel speed sensor	Incorrect resistance
4	Rear left wheel speed sensor	No signal / incorrect signal
5	Front right wheel speed sensor	Incorrect resistance
6	Front right wheel speed sensor	No signal / incorrect signal
7	Rear right wheel speed sensor	Incorrect resistance
8	Rear right wheel speed sensor	No signal / incorrect signal
9	Front left wheel speed sensor	Incorrect resistance
10	Front left wheel speed sensor	No signal / incorrect signal
11	Rear left solenoid valve	Defective (signal interruption, short to earth)
12	Front right solenoid valve	Defective (signal interruption, short to earth)
13	Rear right solenoid valve	Defective (signal interruption, short to earth)
14	Front left solenoid valve	Defective (signal interruption, short to earth)
15	Toothed sensor ring	Wrong number of teeth
16	Brake light switch	Open circuit / poor connections
17	ECM	Defective
18	Battery voltage	Voltage too low

Fault code table (Saab)

Code	Description	Fault
B1371	Front left wheel speed sensor	Faulty signal
B1372	Front left wheel speed sensor	No signal
B1376	Front right wheel speed sensor	Faulty signal
B1377	Front right wheel speed sensor	No signal
B1381	Rear left wheel speed sensor	Faulty signal
B1382	Rear left wheel speed sensor	No signal
B1386	Rear right wheel speed sensor	Faulty signal
B1387	Rear right wheel speed sensor	No signal
B1390	Toothed sensor ring	Wrong number of teeth
B2415	Rear left solenoid valve	Faulty operation
B2450	Front left solenoid valve	Faulty operation
B2455	Front right solenoid valve	Faulty operation
B2485	Rear right solenoid valve	Faulty operation
B1540	Solenoid valves voltage supply	Open circuit
B1605	ECM	Internal fault
B1532	Battery voltage	Voltage too low
B2470	Brake light switch signal	Open circuit
B2465	Pump motor	Faulty

10

Fault code table (Vauxhall, Opel)

Code	Item	Fault
16	Front left solenoid valve	Defective (signal interruption, short to earth)
17	Front right solenoid valve	Defective (signal interruption, short to earth)
19	Solenoid valve relay	Defective (signal interruption, short to earth, supply voltage too low)
25	Wheel speed sensor toothed sensor ring	Wrong number of teeth
28	Rear left solenoid valve	Defective (signal interruption, short to earth)
29	Rear right solenoid valve	Defective (signal interruption, short to earth)
35	Pump motor	Defective (non-operation of pump after actuation of relay)
37	Brake light switch	Defective (incorrect signal)
39	Front left wheel speed sensor	No/poor signal
41	Front left wheel speed sensor	Signal interrupted
42	Front right wheel speed sensor	No/poor signal
43	Front right wheel speed sensor	Signal interrupted
44	Rear left wheel speed sensor	No/poor signal
45	Rear left wheel speed sensor	Signal interrupted
46	Rear right wheel speed sensor	No/poor signal
47	Rear right wheel speed sensor	Signal interrupted
48	Battery voltage	Voltage too low
55	ECM	Defective

Wiring diagrams

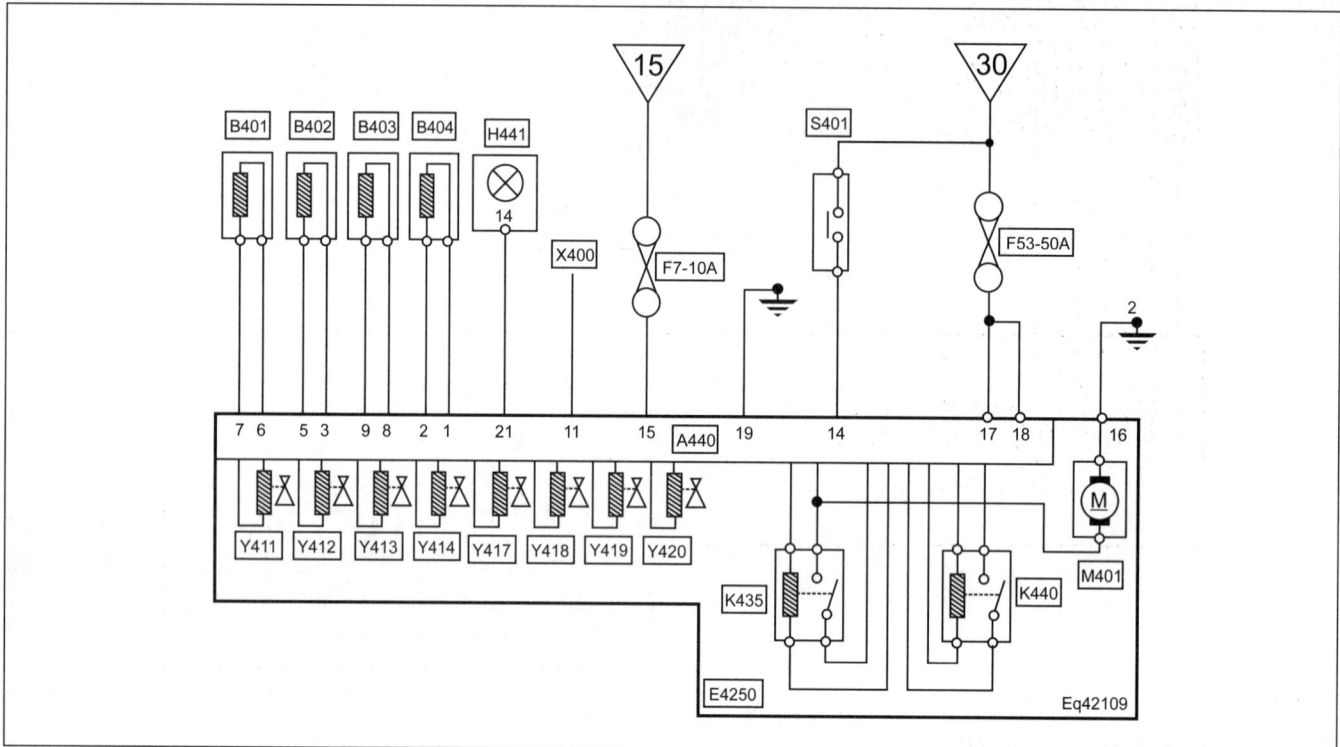

10.30 31-pin Wiring diagram, Audi A4

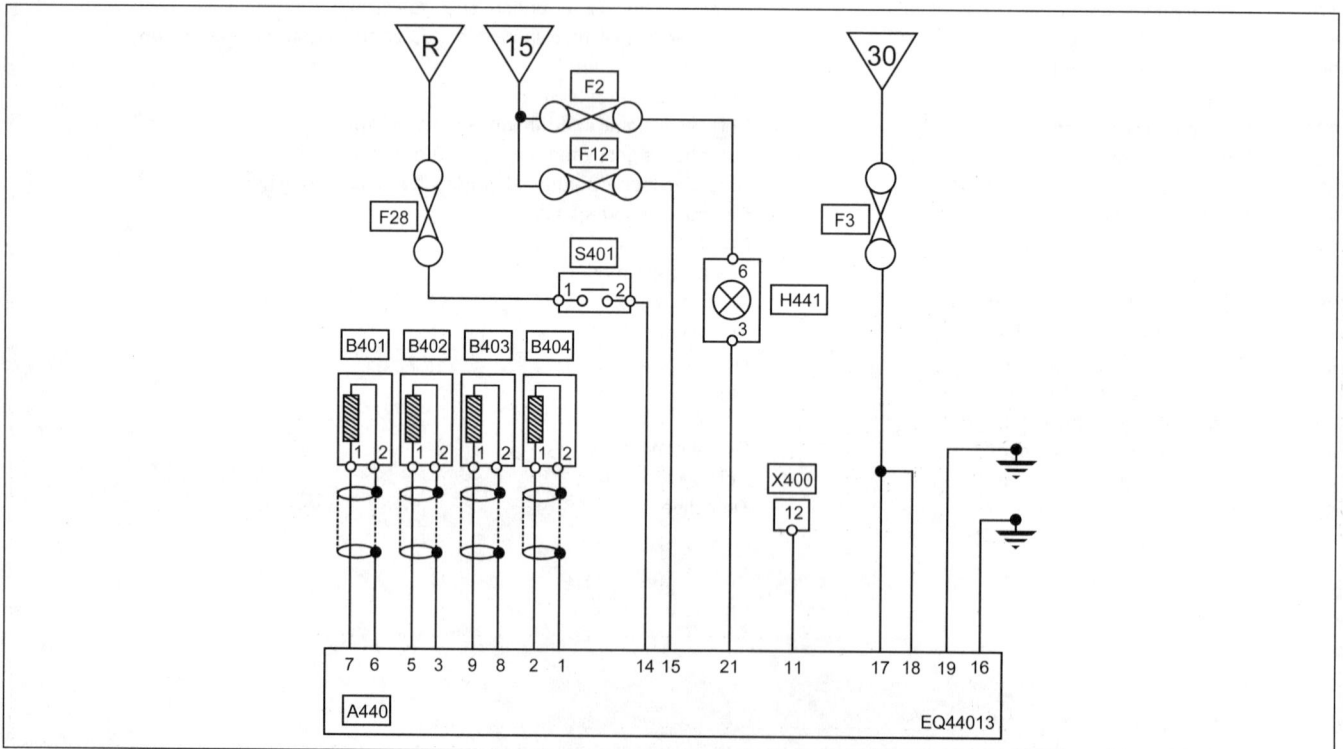

10.31 Wiring diagram, Citroën Xsara

15 / **30**

B401 B402 B403 B404

F24-15A F30-7.5A F22-7.5A F3-60A

S401 X16

H441 H442

E600 E450 S413 A1400

20 21 4 5 2 3 22 23 6 29 30 24 25 9 17 18 16 7 26 15 12 8 13 14 A440

Y401 Y402 Y403 Y404 Y406 Y407

M401 M

K435 K440

E4250 EQ42098

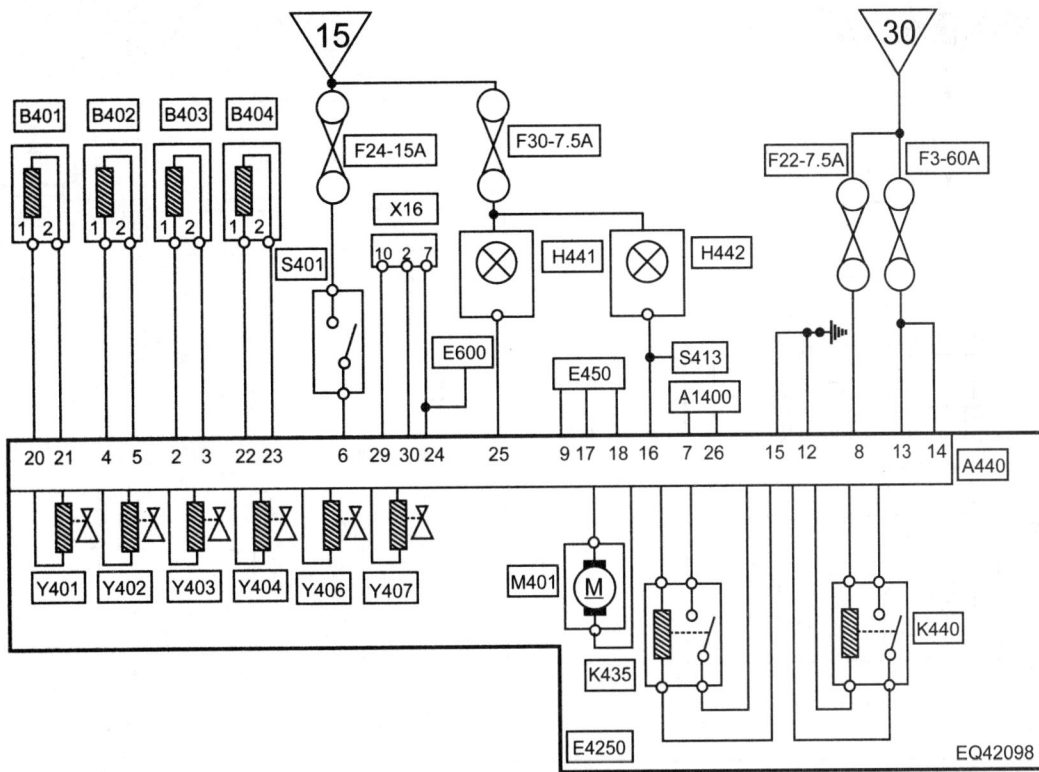

10.32 Wiring diagram, Ford Mondeo (31-pin)

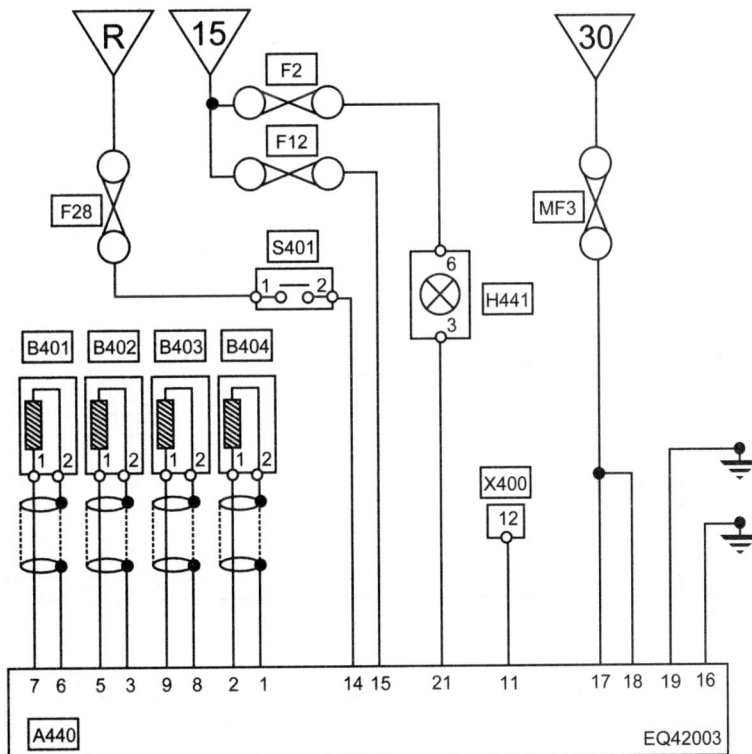

10

R / **15** / **30**

F2

F12

F28 S401

1 2 H441 MF3

6

3

B401 B402 B403 B404

1 2 1 2 1 2 1 2

X400

12

7 6 5 3 9 8 2 1 14 15 21 11 17 18 19 16

A440 EQ42003

10.33 Wiring diagram, Peugeot 306

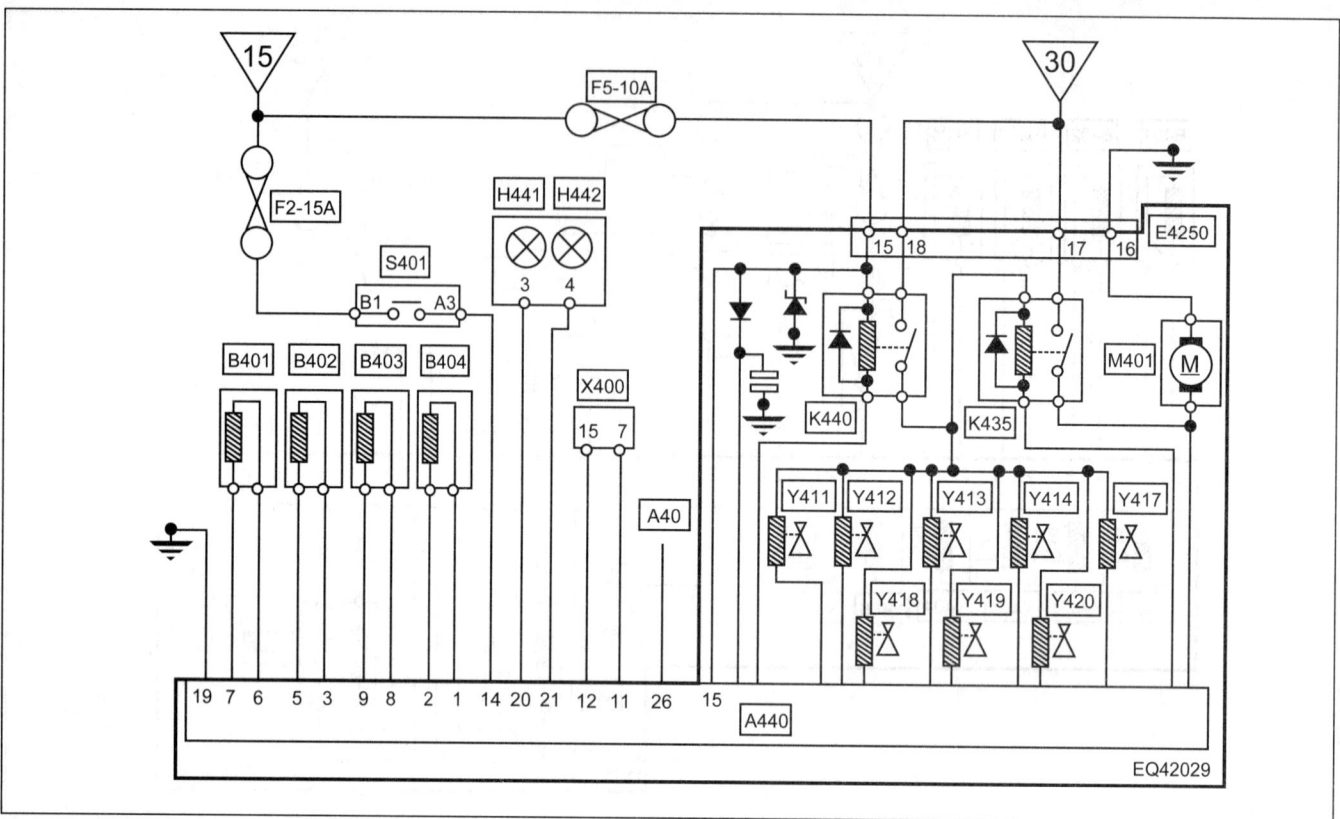

10.34 Wiring diagram, Renault Clio

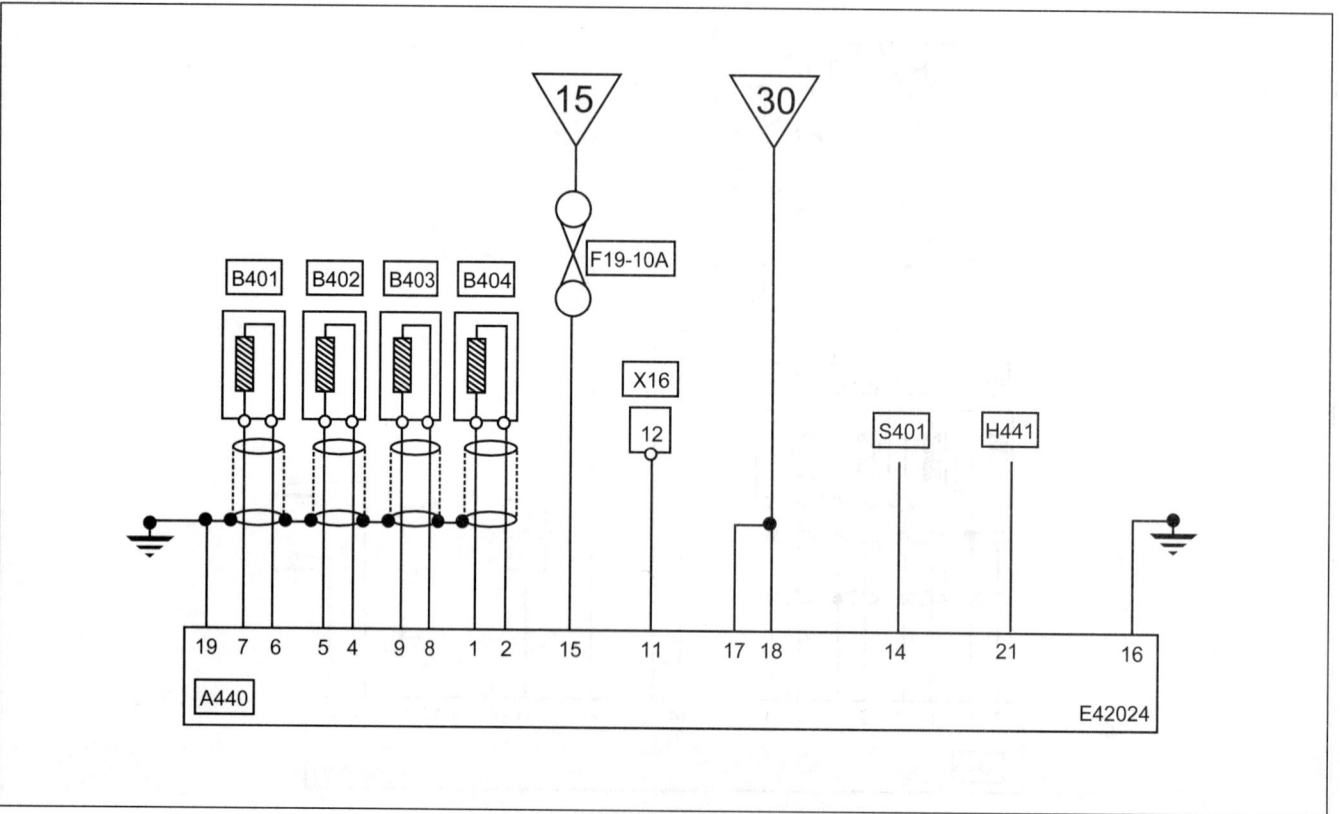

10.35 Wiring diagram, Vauxhall/Opel Omega-B

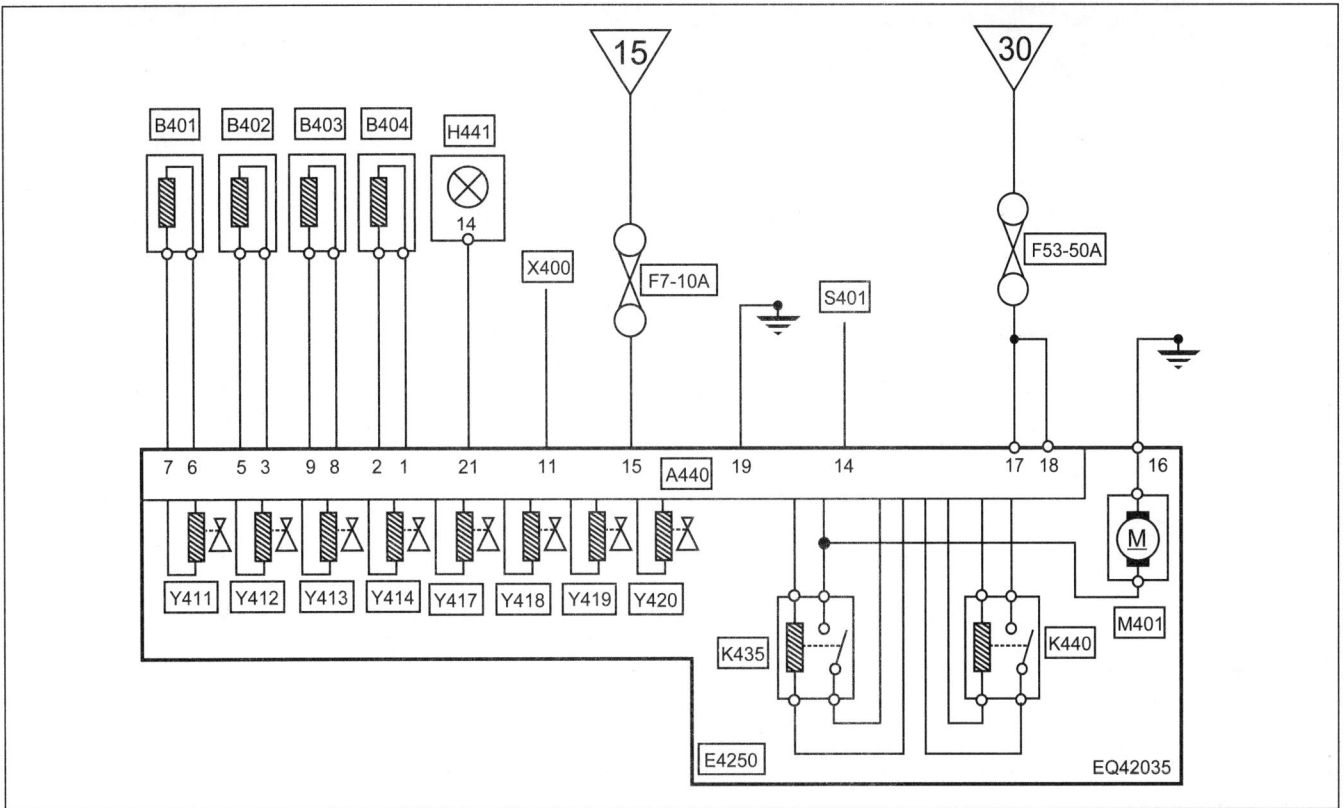

10.36 Wiring diagram, Volkswagen Passat

B401 B402 B403 B404 H441

15 30

X400 F7-10A S401 F53-50A

7 6 5 3 9 8 2 1 21 11 15 A440 19 14 17 18 16

Y411 Y412 Y413 Y414 Y417 Y418 Y419 Y420

K435 K440

M401

E4250 EQ42035

10

Chapter 11 Part A

Clayton Dewandre Wabco Type I ABS

Contents

Vehicle coverage

Overview of system operation

1 Basic principles and system identification

Clayton Dewandre Wabco is an American antilock brake system which has been fitted to Land Rover derived vehicles since its introduction in the late 1980s. There are two distinct versions of the system which, for the purposes of the technical descriptions and test procedures contained in Part A and Part B of this Chapter, will be referred to as Type I and Type II. The Type I system covered in this Part of Chapter 11 is fitted to Range Rover models, and the Type II system covered in Part B, is fitted to the Land Rover Discovery.

The Type I system is often referred to as an integrated type whereby the components of the ABS hydraulic control unit are also required during the operation of the conventional braking system. Type II is of the additional or 'add-on' type operating in conjunction with the conventional braking system components.

Depending on application, Wabco Type I may be installed as an antilock braking system only, or as an antilock braking system incorporating traction control.

The purpose of the system is to apply the vehicle brakes at maximum efficiency without wheel lock or loss of directional stability. Inductive sensors (wheel speed sensors) monitor the speed of the roadwheels by generating an electrical signal as the wheel is rotated. This information is passed to the ABS Electronic Control Module (ECM) which is then able to determine wheel speed, wheel acceleration and wheel deceleration. The ECM compares the signals received from each wheel and if the onset of lock at any wheel is detected, a signal is sent to the ABS hydraulic control unit which regulates the brake pressure for the relevant wheel(s).

Wabco Type I ABS is comprised of the following components:
a) ABS-ECM.
b) Electric pump unit containing pressure switches, non-return valve, pressure relief valve and electric pump.
c) Hydraulic control unit containing master cylinder, power valve, solenoid valves, pressure reducing valve, servo cylinders and fluid reservoir.
d) Pressure accumulator.
e) Four inductive wheel speed sensors and associated sensor rings.
f) ABS electrical wiring harness and relays.
g) ABS warning lamps.
h) Diagnostic connector.
i) Brake calipers and hydraulic hoses and pipes.
j) Pressure conscious reducing valve.

The braking system consists of two entirely separate hydraulic circuits; the main circuit and secondary circuit. The main circuit controls the operation of the rear brake calipers and the upper pistons of the front brake calipers. The secondary circuit controls the operation of the front brake caliper lower pistons. Boosted hydraulic pressure for the main circuit is supplied by the electric pump unit via the master cylinder. Hydraulic pressure for the secondary circuit is supplied by the master cylinder and servo cylinders.

2 Component description and operation

ABS ECM

General

The Clayton Dewandre Wabco ABS Electronic Control Module (ECM) continually monitors wheel speed from the signals provided by the wheel speed sensors. If the ECM detects the incidence of wheel lock on one or more wheels, a signal is sent to the hydraulic control unit to modulate the hydraulic pressure to the brake of the locking wheel.

The ECM contains two microprocessors and uses digital technology to complete this function and other functions such as fault code memory and power modules for valve and relay activity (see illustration 11A.1).

Self-test

The ECM is equipped with a self-test capability that initially examines the ABS system when the ignition is switched on, and then examines the wheel speed sensor signals after a pre-determined wheel speed is reached from all wheels (engine running). The ABS self-test program continues to examine the signals from the various components as long as the ignition is switched on. If self-test determines that faults are not present, the ABS is ready for operation once a specified vehicle speed has been achieved.

If the ECM detects that a fault is present, the ABS functions are restricted or disabled altogether, depending on the nature of the fault. One or both of the warning lamps will also be illuminated indicating the presence of a fault. Conventional braking will still be available, however, if the fault is associated with low system pressure, brake pedal travel will be excessive.

Self-diagnostics

If the ECM detects a fault during the self-test routine, an internal fault code is stored in the ECM memory. Stored fault codes can be retrieved from the SD connector with the aid of a suitable fault code reader, or output manually as flash codes via the ABS warning lamp. If the fault clears, the code will remain stored until cleared with the FCR, or during the flash code retrieval procedure.

Electric pump unit

Instead of using a conventional vacuum servo unit for braking assistance, the Wabco Type I system uses an electric pump, pressure accumulator and pressure switches for operation of the main hydraulic circuit. The pump draws hydraulic fluid from the reservoir and feeds it under high pressure to the accumulator.

The accumulator contains two chambers separated by a flexible diaphragm. The upper chamber is filled with high pressure nitrogen gas and the lower chamber contains the brake hydraulic fluid supplied under pressure from the pump. The hydraulic fluid compresses the nitrogen gas further, thus providing a supply of high pressure hydraulic fluid for the brake system. Pump operation is controlled by a pressure switch located on the pump unit. The switch actuates the pump relay to operate the electric pump if the accumulator pressure drops below a preset value. When the pressure is restored, the switch deactivates the relay to switch off the pump. If the accumulator pressure falls significantly, a signal is sent to the ECM to switch off the ABS operation.

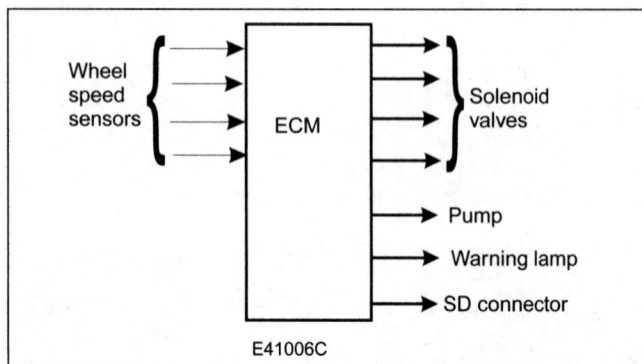

11A.1 ECM sensor inputs and control signal outputs

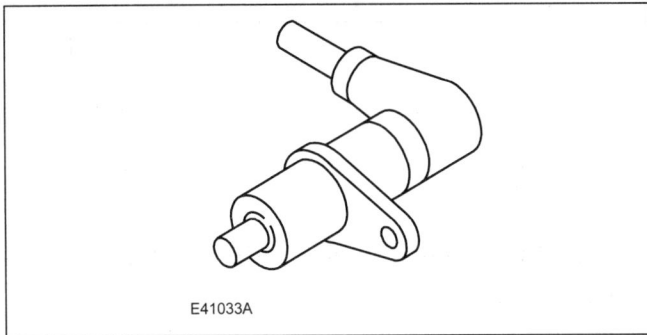

11A.2 Typical inductive wheel speed sensor

The pressurised hydraulic fluid from the accumulator is supplied to the master cylinder and hydraulic control unit for brake system operation either with or without ABS control.

Hydraulic control unit

The hydraulic control unit consists of a master cylinder, a hydraulic valve block containing the ABS solenoid valves, and a fluid reservoir.

The hydraulic valve block contains the inlet and outlet solenoid valves, isolating solenoid valves and rear brake pressure conscious reducing valve. The valve block modulates the pressure applied to the brake caliper for each individual wheel during ABS operation. As Wabco Type I is a four-channel system, there are two solenoid valves (one inlet and one outlet) for each brake, and two isolating solenoid valves. Additional solenoid valves are used where traction control is incorporated in the system.

Wheel speed sensors

The rotational speed of the roadwheels and any changes in the rotational speed are recorded by inductive wheel speed sensors, one located at each roadwheel (see illustration 11A.2).

Each wheel speed sensor assembly comprises a toothed sensor ring which rotates at roadwheel speed, and an adjacent sensor mounted a set distance from the sensor ring (see illustration 11A.3).

The sensors are permanent magnet pulse generator types producing an AC voltage sine wave as the sensor ring teeth pass through the magnetic field of the sensor (see illustration 11A.4).

The frequency of the waveform produced by the wheel speed sensor is proportional to the road speed. This AC voltage signal is continually being delivered to the ABS-ECM for processing.

The peak to peak voltage of the speed signal (when viewed upon an oscilloscope) can vary considerably according to wheel speed. An analogue to digital converter (ADC) in the ECM transforms the AC pulse into a digital signal (see illustration 11A.5).

ABS electrical wiring harness and relays

An integrated 35-pin main wiring harness is used to connect power and earth to various electrical components. This enables sensor

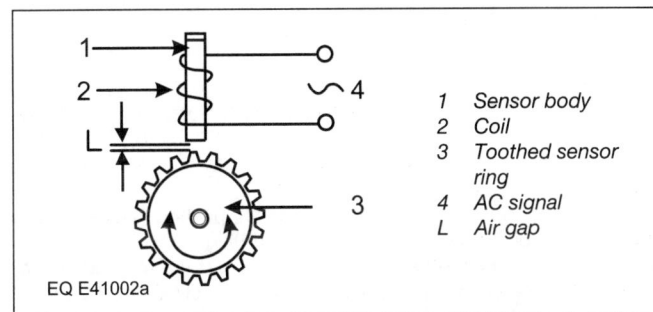

11A.3 Sectional view of an inductive wheel speed sensor

1	Mounting bolt location	5	Coil
2	Permanent magnet	6	Sensor tip
3	Wiring harness	7	Toothed sensor ring
4	O-ring		

signals to reach the ECM and the ECM, in turn, to send command signals to the relays and hydraulic control unit. The ABS main relay, pump relay and the associated fuses, are typically located in the vehicle main fuse/relay box, located behind a trim panel on the front left-hand seat base.

Main relay

Nominal battery voltage is supplied from the battery to the main relay via a fuse. One terminal of the relay solenoid is connected to earth and the relay is activated by the ECM supplying a voltage to the other solenoid terminal. Battery voltage is then supplied to the ECM.

Pump relay

Nominal battery voltage is supplied from the battery to the pump relay via a fuse. The relay is controlled by the pressure switch applying an earth, which activates the relay allowing fused battery voltage to be applied to the pump motor.

11A

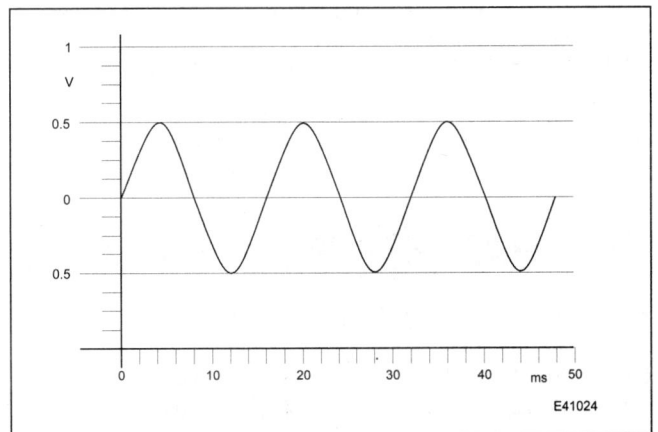

1	Sensor body
2	Coil
3	Toothed sensor ring
4	AC signal
L	Air gap

11A.4 Inductive wheel speed sensor operation

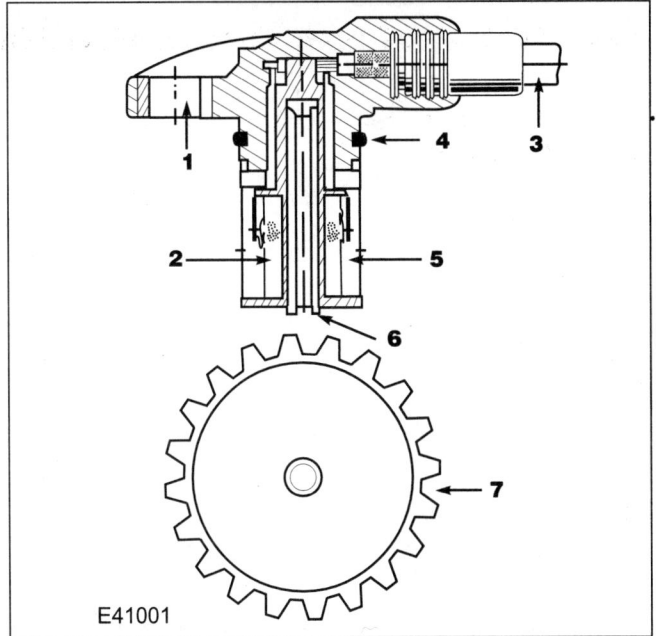

11A.5 Inductive wheel speed sensor waveform as viewed on an oscilloscope

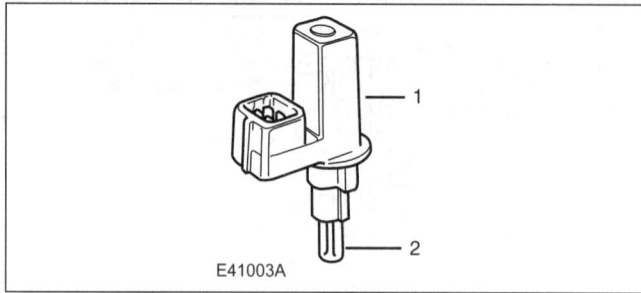

11A.6 Typical brake light switch body (1) and contact pin (2)

Warning lamp relay

The warning lamp relay is used to illuminate the ABS warning lamp on the instrument panel in the event of a system fault being detected by the ECM.

Brake light switch

The brake light switch typically comprises a switch body and contact pin and is located above the brake pedal (see illustration 11A.6).

When the brake pedal is depressed closing the brake light switch, a signal is sent to the ECM indicating that the brakes are being applied. Once this signal is received, the ECM will begin monitoring the wheel speed via the wheel speed sensors and activate the ABS if necessary.

ABS warning lamps

After the ignition is switched on, the ABS warning lamp on the instrument panel is illuminated as the system executes a self-test routine. If satisfactory operation of the system is detected by the ECM, and if the system hydraulic pressure is satisfactory, the lamp is extinguished after the vehicle has been driven above approximately 5 mph. During vehicle operation above a pre-determined wheel speed, the ABS-ECM implements a further self-test cycle whereby ABS operation, hydraulic pressure and wheel speed sensor signals are monitored. If a fault is detected, the warning lamp relay is actuated by the ECM to illuminate the warning lamp on the instrument panel. The ABS function is then restricted, or completely disabled, depending on the nature of the fault. The warning lamp will remain illuminated until the fault is no longer present (see illustration 11A.7).

Depending on application, one or two additional brake system warning lamps are used to indicate that the handbrake is applied or that there is low system pressure or low hydraulic fluid level. These warning lamps are illuminated by the handbrake switch to indicate that the handbrake is on, by the pressure switch to indicate low system pressure, or by the level sensor in the hydraulic fluid reservoir to indicate low fluid level.

When the ABS-ECM detects a fault, the fault code is stored and the ABS warning lamp activated. If the fault no longer exists after the next system start (ignition on/off) the ABS warning lamp is extinguished after the self-test cycle, however the fault code remains stored in the ECM memory.

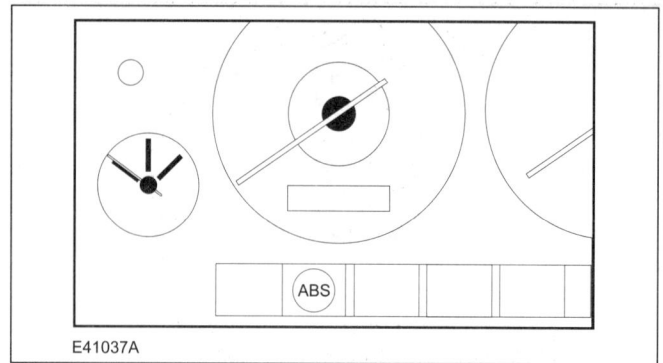

11A.7 ABS warning lamp

Pressure conscious reducing valve

A pressure conscious reducing valve is incorporated into the hydraulic control unit to restrict the hydraulic fluid to the rear brakes once a pre-determined pressure is reached.

3 System operation

There is very little information available from the manufacturer's as to the specific operation of this system, however the following is a brief overview of the basic principles.

When the brake pedal is depressed, the master cylinder allows boosted hydraulic pressure from the electric pump unit and the pressure accumulator to flow through the open inlet solenoid valves in the hydraulic control unit. The pressurised fluid is then supplied to the front brake caliper upper pistons, and to the rear brake calipers via the pressure conscious reducing valve.

At the same time, the master cylinder allows boosted hydraulic pressure to be supplied to the two servo cylinders in the hydraulic control unit. The combination of boosted hydraulic pressure from the servo cylinders and pedal force from the master cylinder is then applied to the front brake caliper lower pistons.

Once a pre-determined vehicle speed has been reached, the ABS-ECM continually monitors wheel speed from the signals provided by the wheel speed sensors. If the ECM detects the incidence of wheel lock on one or more wheels, the ECM closes the relevant inlet solenoid valve and opens the outlet solenoid valve in the hydraulic control unit. Closing the inlet solenoid valve interrupts the supply of hydraulic fluid to the brake, and the open outlet solenoid valve allows the fluid from the brake to be discharged back into the hydraulic fluid reservoir. As the brake hydraulic pressure is now reduced, wheel rotation is allowed to increase.

Once the wheel rotation has stabilised, the outlet solenoid valve is closed and the inlet solenoid valve is opened. Brake hydraulic pressure once again flows through the inlet solenoid valve to the relevant brake caliper.

The whole ABS control cycle takes place 4 to 10 times per second for each affected wheel and this ensures maximum braking effect and control during ABS operation.

Test procedures

4 Wheel speed sensors

Checking the wheel speed sensor (general)

1 Inspect the wheel speed sensor for corrosion or damage and check that the sensor is tightly mounted.
2 Check the toothed sensor ring for damage, eccentricity and for broken or missing teeth.

3 Inspect the wheel speed sensor wiring plug for corrosion and damage. One plug for each sensor.
4 Check that the connector terminal pins are pushed fully home and making good contact with the sensor wiring plug.
5 Check the clearance between the sensor and the toothed sensor ring. The clearance is not adjustable but is nominally 0.6 to 1.3 mm. If the clearance is excessive, expect a worn sensor tip or problems with the wheel bearings/hub or sensor ring.

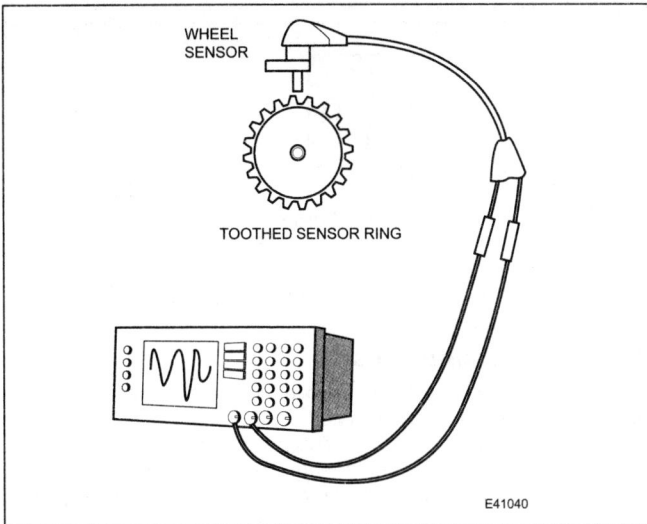

11A.8 Checking wheel speed sensor output with an oscilloscope connected to the sensor-wiring plug

6 When carrying out voltage checks with an oscilloscope or voltmeter, the voltage obtained will be proportional to the speed at which the wheel is rotating. In addition to determining that the wheel speed sensors are actually producing a voltage output, it is essential that the output from the sensors on a particular axle is the same for any given wheel speed.

Checking wheel speed sensor output with an oscilloscope

7 Switch the ignition off and disconnect the ECM multi-plug or the relevant wheel speed sensor wiring plug.
8 Connect an oscilloscope between the terminal pins for the sensor under test (see illustration 11A.8).
9 Select a range to cover 80 Hz on the oscilloscope and a free run time base.
10 Raise the wheel and rotate it by hand at approximately one revolution per second.
11 A sinusoidal wave form should be obtained, with amplitude and duration changing with rotational speed (see illustration 11A.9).
12 If there is no signal, or a very weak or intermittent signal at the ECM, repeat the test at the sensor wiring plug. If there is no change in signal status, the sensor is suspect.

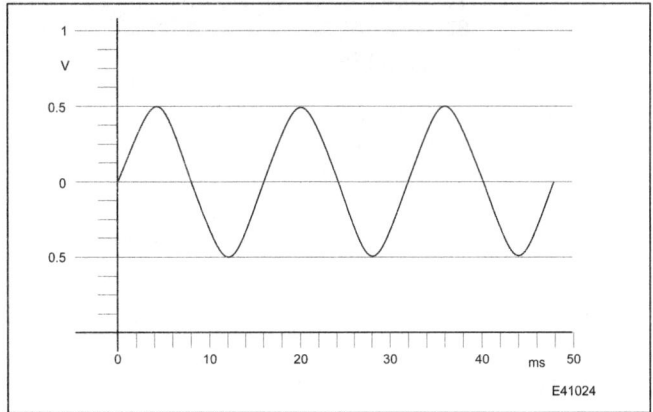

11A.9 Typical wheel speed sensor sine wave as displayed on an oscilloscope

13 If the signal is now satisfactory this indicates a fault in the wiring harness which should be checked for continuity.

Checking wheel speed sensor output with an AC voltmeter

14 Switch the ignition off and disconnect the ECM multi-plug or the relevant wheel speed sensor wiring plug.
15 Connect an AC voltmeter between the terminal pins for the sensor under test (see illustration 11A.10).
16 Raise the wheel and rotate it by hand at approximately one revolution per second.
17 A voltage of approximately 0.4 to 1.5 volts (AC RMS) should be obtained. If there is no signal, or a very weak or intermittent signal at the ECM, repeat the test at the sensor wiring plug. If there is no change in the signal, the sensor is suspect.
18 If the signal is now satisfactory, this indicates a fault in the wiring harness which should be checked for continuity. **Note:** *This test at least proves that a signal is being generated by the sensor. However, the voltage produced is an average voltage and does not clearly indicate damage to the sensor ring or that the sinewave is regular in formation.*

Checking wheel speed sensor resistance

19 Switch the ignition off and disconnect the ECM multi-plug or the relevant wheel speed sensor wiring plug.
20 Connect an ohmmeter between the terminal pins for the sensor under test (see illustration 11A.11).

11A

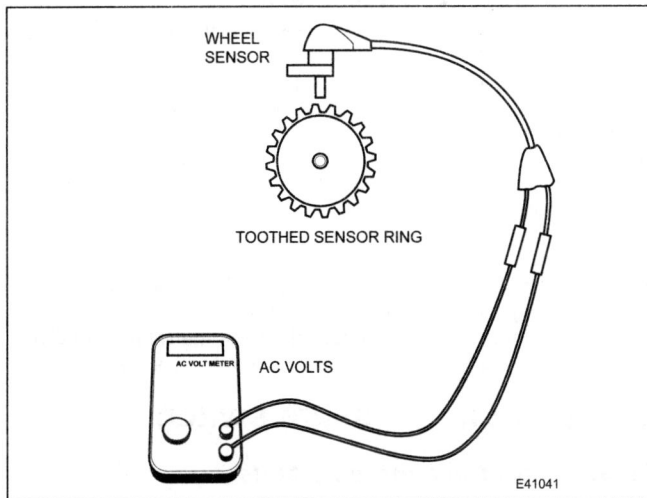

11A.10 Checking wheel speed sensor output with a voltmeter connected to the sensor-wiring plug

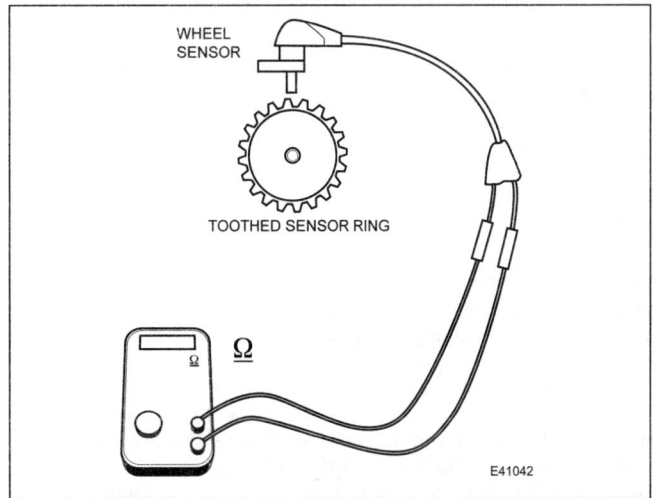

11A.11 Checking wheel speed sensor resistance with an ohmmeter connected to the sensor-wiring plug

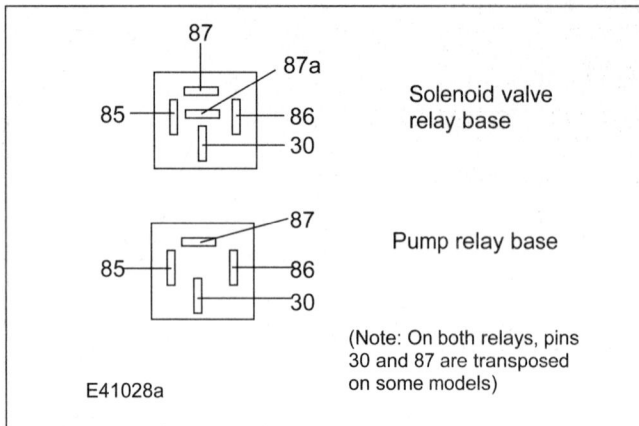

87
87a
85
86
30
Solenoid valve
relay base

87
85
86
30
Pump relay base

(Note: On both relays, pins 30 and 87 are transposed on some models)

E41028a

11A.12 Main (solenoid valve) relay and hydraulic pump relay terminals

21 The readings obtained should be between 0.9 and 1.0 kohms.

22 If the resistance is excessively high, or open circuit at the ECM, repeat the test at the sensor multi-plug. If there is no change in resistance, the sensor is suspect.

23 If the resistance is now satisfactory, this indicates a fault in the wiring harness which should be checked for continuity. **Note:** *Even if the resistance is within the quoted specifications, this does not prove that the speed sensor can generate an acceptable signal.*

5 System relays

Checking relays (general)

1 The following tests are applicable to the ABS main (or solenoid valve) relay, hydraulic pump relay and warning lamp relay.

2 Before carrying out specific tests on the relays or power supplies, inspect the relays and relay base for corrosion and damage.

3 Check that the relays are fully pushed home and making good contact with the base.

4 Faults in any of the above areas are possible reasons for failure or malfunctioning of the relay.

Relay operation tests

Main (solenoid valve) relay and hydraulic pump relay

5 Switch the ignition off, then remove the relevant relay from the relay base.

6 Connect an ohmmeter between the relay output terminal 87 and the main voltage supply terminal 30 **(see illustration 11A.12)**. The ohmmeter should indicate an open circuit.

7 Attach a wire between the relay driver supply terminal 85 (main relay) or 86 (pump relay) and a 12 volt supply. Attach a second wire between the relay driver earth terminal 86 (main relay) or 85 (pump relay) and earth. There should be an unmistakable click from the relay and the ohmmeter should indicate continuity.

8 Renew the relay if it fails these tests. If the relay cover is transparent, inspect the relay for burns. A faulty relay usually has a burnt appearance.

Warning lamp relay

9 Switch the ignition off, then remove the relay from the relay base.

10 Connect an ohmmeter between relay terminals 30 and 87a. The ohmmeter should indicate continuity.

11 Attach a wire between the relay driver supply terminal 85 and a 12 volt supply. Attach a second wire between the relay driver earth terminal 86 and earth. There should be an unmistakable click from the relay and the ohmmeter should indicate an open circuit.

12 Renew the relay if it fails these tests. If the relay cover is transparent, inspect the relay for burns. A faulty relay usually has a burnt appearance.

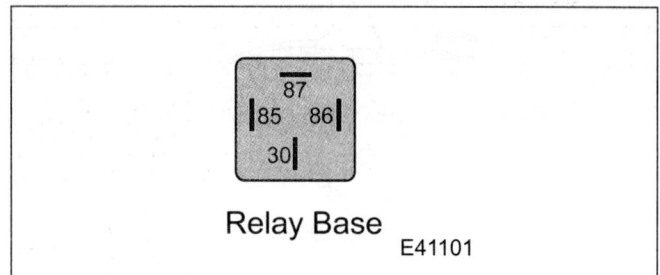

87
85
86
30
Relay Base
E41101

11A.13 Typical relay base terminal identification

Relay power supply tests

13 Switch the ignition off then remove the relevant relay leaving the base exposed **(see illustration 11A.13)**.

14 Using a voltmeter, probe between the main voltage supply terminal 30 in the relay base and earth. The reading should indicate nbv. If no voltage is found, check the wiring back to the voltage supply.

6 Electronic Control Module

Checking the ECM (general)

1 Inspect the ECM for corrosion or damage and ensure that the unit is mounted securely.

2 Check that the ECM multi-plug terminals are pushed fully home and making good contact with the ECM pins. Faults in any of the above areas are possible reasons for poor performance in the ABS system.

ECM power supply and earth tests

3 Switch the ignition off and disconnect the ECM multi-plug from the ECM.

4 Switch the ignition on.

5 Attach a negative voltmeter probe to a vehicle earth and attach a positive voltmeter probe to ECM multi-plug pin 9. The voltmeter should indicate nbv **(see illustration 11A.14)**. If no voltage is found, check the relevant fuse, and the supply wiring back to the ignition switch.

6 Switch the ignition off.

7 Connect an ohmmeter between a vehicle earth and ECM multi-plug earth pin 27. The ohmmeter should indicate continuity. If not, check the ECM main earth connection and wiring.

7 Solenoid valves

Solenoid valve resistance tests

1 Switch the ignition off and disconnect the ECM multi-plug from the ECM.

2 Using an ohmmeter, measure the resistance between pin 27 in the ECM multi-plug, and each of the following multi-plug pins in turn **(see illustration 11A.14)**:

Inlet solenoid valves - 4, 6, 21, 23
Outlet solenoid valves - 5, 7, 22, 24
Inlet isolating solenoid valve - 11.
Outlet isolating solenoid valve - 12.

3 The resistance should be in the order of 6.5 ohms for each inlet solenoid valve, 4.5 ohms for each outlet solenoid valve and 6.5 ohms for the isolating solenoid valves.

4 If the resistance is not as specified, check the wiring and wiring connectors. If satisfactory, the hydraulic control unit is suspect.

8 Hydraulic pressure switch

Pressure switch continuity tests

System depressurised

1 Switch the ignition off.

Front View

19 ▭▭▭▭▭▭▭▭▭▭▭▭▭▭▭▭▭ 35
1 ▭▭▭▭▭▭▭▭▭▭▭▭▭▭▭▭▭ 18

Back View

35 ▭▭▭▭▭▭▭▭▭▭▭▭▭▭▭▭▭ 19
18 ▭▭▭▭▭▭▭▭▭▭▭▭▭▭▭▭▭ 1

EQ 00001a

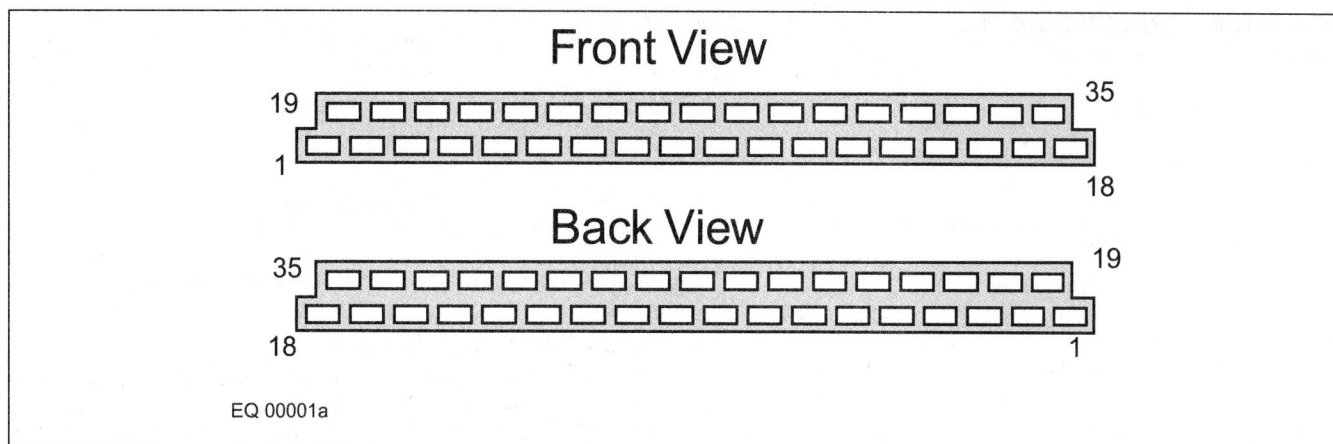

11A.14 ECM 35-pin multi-plug terminal identification

2 Depress the brake pedal at least 30 times to depressurise the system. When the system is fully depressurised, the brake pedal will become firmer and the brake fluid level in the master cylinder reservoir will rise.
3 Disconnect the 5-pin multi-plug from the pressure switch.
4 Connect an ohmmeter between pressure switch terminals 2 and 5. The ohmmeter should indicate continuity.
5 Connect the ohmmeter between pressure switch terminals 4 and 5, and 1 and 3 in turn. The ohmmeter should indicate an open circuit in each case.

System pressurised

6 Reconnect the 5-pin multi-plug to the pressure switch.
7 Switch the ignition on. The pump will run and the ABS warning lamp will be illuminated.
8 Switch the ignition off when the pump switches off. The system is now pressurised to approximately 180 bar.
9 Disconnect the 5-pin multi-plug from the pressure switch.
10 Connect an ohmmeter between pressure switch terminals 2 and 5. The ohmmeter should indicate an open circuit.
11 Connect the ohmmeter between pressure switch terminals 4 and 5, and 1 and 3 in turn. The ohmmeter should indicate continuity.
12 If any of the above readings are not as specified, the pressure switch is suspect.

9 Hydraulic pump motor

Pump operation test

1 Switch the ignition off then remove the pump relay leaving the base exposed.
2 Bridge terminals 30 and 87 in the relay base using a fused jumper lead **(see illustration 11A.13)**. The pump should now operate.

⚠ *Warning: The test should be made as quickly as possible to avoid damaging the pump.*

3 If the pump does not operate as described, carry out the pump relay power supply tests described in Section 5. If satisfactory, suspect a faulty pump.

10 Brake light switch

Checking the brake light switch (general)

1 Check that the brake light switch is correctly and securely mounted and that the plunger moves smoothly with no trace of binding.
2 Check that the wiring multi-plug is pushed fully home and making good contact.
3 Check that no wires have been disconnected.

4 Faults in any of the above areas are possible reasons for failure or malfunctioning of the switch.

Brake light switch voltage tests

5 Switch the ignition off and disconnect the ECM multi-plug from the ECM.
6 Connect a voltmeter between ECM multi-plug pin 10 and a vehicle earth **(see illustration 11A.14)**.
7 Switch the ignition on.
8 Depress the brake pedal. The voltmeter should indicate nbv.
9 If no voltage is found, the fuse, the brake light switch and the wiring are suspect.
10 Release the brake pedal. The voltage should drop to zero as the switch opens.
11 Repeat the test with the voltmeter connected between ECM multi-plug 25 and a vehicle earth. The voltmeter should indicate nbv with the pedal depressed, and zero with the pedal released.

11 Warning lamp

Checking the warning lamps (general)

1 Typically two warning lamps are used in the system, one for "handbrake on/low hydraulic fluid level/low system pressure" and one for "ABS system failure".
2 Inspect the warning lamp bulb holder contacts in the instrument panel.
3 Check that the instrument panel multi-plug terminal pins are pushed fully home and making good contact.
4 Check that no wires have been disconnected.
5 Faults in any of the above areas are possible reasons for failure or malfunctioning of the warning lamp(s).

Warning lamp operation test

6 With the ignition switched off, both warning lamp should remain off.
7 Switch the ignition on. Both warning lamp should illuminate and then extinguish after the vehicle has been driven above approximately 5 mph.
8 If the 'ABS system failure' warning lamp illuminates on its own at any time, this indicates that an electrical fault has been detected by the ECM and the ABS has been turned off.
9 If the 'handbrake on/low fluid level/low system pressure' warning lamp illuminates on its own at any time, with no other noticeable symptoms, this indicates that either the handbrake is applied or the fluid level in the reservoir is very low.
10 If both warning lamps illuminate together at any time, this indicates that the fluid level in the reservoir is very low and that system pressure is reducing to a critical point.

11A

Pin table - typical 35-pin

Note: *Refer to illustration 11A.14*

Pin No.	Connection	Test condition	Voltage
1	Supply from relay	Ignition on	Nbv
2	-	-	-
3	-	-	-
4	Rear right inlet solenoid valve	Ignition on/inactive	0 volts
		Actuated	Nbv
5	Rear right outlet solenoid valve	Ignition on/inactive	0 volts
		Actuated	Nbv
6	Front right inlet solenoid valve	Ignition on/inactive	0 volts
		Actuated	Nbv
7	Front right outlet solenoid valve	Ignition on/inactive	0 volts
		Actuated	Nbv
8	Main (solenoid valve) relay driver	Ignition on, actuated	Nbv
9	Supply from ignition switch	Ignition on	Nbv
10	Brake light switch input terminal 4	Ignition on:	
		Brake pedal not pressed	0 volts
		Brake pedal depressed	Nbv
11	Inlet isolating solenoid valve	Ignition on/inactive	0 volts
		Actuated	Nbv
12	Outlet isolating solenoid valve	Ignition on/inactive	0 volts
		Actuated	Nbv
13	SD connector	-	-
14	SD connector	-	-
15	Front left wheel speed sensor earth	Roadwheel rotating	0.25 volts (max)
16	Rear right wheel speed sensor earth	Roadwheel rotating	0.25 volts (max)
17	Front right wheel speed sensor earth	Roadwheel rotating	0.25 volts (max)
18	Rear left wheel speed sensor earth	Roadwheel rotating	0.25 volts (max)
19	Supply from relay	Ignition on	Nbv
20	-	-	-
21	Rear left inlet solenoid valve	Ignition on/inactive	0 volts
		Actuated	Nbv
22	Rear left outlet solenoid valve	Ignition on/inactive	0 volts
		Actuated	Nbv
23	Front left inlet solenoid valve	Ignition on/inactive	0 volts
		Actuated	Nbv
24	Front left outlet solenoid valve	Ignition on/inactive	0 volts
		Actuated	Nbv
25	Brake light switch input terminal 2	Ignition on:	
		Brake pedal not pressed	0 volts
		Brake pedal depressed	Nbv
26	ABS warning lamp	Ignition on:	
		Lamp on	2.0 volts (max)
		Lamp off	Nbv
27	ECM earth	Ignition on	0.25 volts (max)
28	-	-	-
29	-	-	-
30	Pressure switch terminal 5	-	-
31	Pressure switch terminal 2	-	-
32	Front left wheel speed sensor signal	Roadwheel rotating	0.4 to 1.5 volts AC (approx)
33	Rear right wheel speed sensor signal	Roadwheel rotating	0.4 to 1.5 volts AC (approx)
34	Front right wheel speed sensor signal	Roadwheel rotating	0.4 to 1.5 volts AC (approx)
35	Rear left wheel speed sensor signal	Roadwheel rotating	0.4 to 1.5 volts AC (approx)

Fault codes

12 Early model fault codes

1 If a FCR is available, it should be connected to the SD serial connector and used in accordance with the maker's instructions.

2 The FCR can be used for the following purposes:

a) Obtaining fault codes.

b) Clearing fault codes.

c) Obtaining datastream information.

d) Testing the system actuators (relays and solenoid valves).

3 If a FCR is not available, it is still possible to obtain fault codes which will be displayed as a series of flashes on the dash mounted warning lamp.

4 When the ECM determines that a fault is present, it internally logs a fault code and also illuminates the ABS warning lamp. All of the various fault codes on these vehicles are of the 'slow' variety and can be output as flash codes on the warning lamp.

Obtaining codes without a FCR

5 Switch the ignition off and remove the trim from the front left-hand seat base. The ABS ECM, fuses, relays and SD serial connector are located behind the seat trim.

6 Remove the ABS warning lamp relay from the relay base located to the left of the three fuses on the fuse/relay plate.

7 Switch the ignition on and the ABS warning lamp should illuminate.

8 Use a bridging wire to bridge the black wire and the black/pink wire in the SD connector plug.

9 Five seconds after connecting the bridging wire, the ABS warning lamp will extinguish for 2.5 seconds, then illuminate for 2.5 seconds, then extinguish again for 2.5 seconds then illuminate again for 0.5 seconds, then extinguish again for another 2.5 seconds. At the end of this sequence the flash code output will start.

10 If a fault code is stored in the ECM memory it will now be displayed in two groups of 0.5 second flashes of the warning lamp, with a 2.5 second pause between each group.

11 The ECM will continue to repeat this fault code until the bridging wire is removed from the SD connector plug. When the bridging wire is removed the code will be erased from the ECM memory and the ECM will be ready to display the next fault code, if present.

12 To obtain the next fault code, repeat the procedure described in paragraphs 8 to 11. When all the fault codes have been displayed, the ABS warning lamp will remain extinguished for 7.5 seconds after the start of the output sequence.

13 Later model fault codes

1 On later models (approximately 1994 onwards) internal fault codes are used by the ECM to designate faults in the system components and circuits. A proprietary fault code reader (FCR) or system tester is required to interrogate the system as flash codes are not available for output on later models. No actual fault code numbers are available, although the component circuits checked by the ECM are similar to those shown for early models.

11A

Fault code table

Code	Item	Fault
2 - 6	Brake light switch	No signal, poor wiring connections
2 - 7	ECM supply voltage	Defective main relay, wiring, connections
2 - 8	Solenoid valve supply voltage	Incorrect voltage, defective relay
2 - 12	Front right wheel speed sensor	Low output signal
2 - 13	Rear left wheel speed sensor	Low output signal
2 - 14	Front left wheel speed sensor	Low output signal
2 - 15	Rear right wheel speed sensor	Low output signal
3 - 0	Front right inlet solenoid valve	Open circuit to ECM
3 - 1	Front right outlet solenoid valve	Open circuit to ECM
3 - 2	Front left inlet solenoid valve	Open circuit to ECM
3 - 3	Front left outlet solenoid valve	Open circuit to ECM
3 - 4	Rear right inlet solenoid valve	Open circuit to ECM
3 - 5	Rear right outlet solenoid valve	Open circuit to ECM
3 - 6	Rear left inlet solenoid valve	Open circuit to ECM
3 - 7	Rear left outlet solenoid valve	Open circuit to ECM
3 - 8	Inlet isolating solenoid valve	Open circuit to ECM
3 - 9	Outlet isolating solenoid valve	Open circuit to ECM
4 - 0	Front right inlet solenoid valve	Short circuit to earth
4 - 1	Front right outlet solenoid valve	Short circuit to earth
4 - 2	Front left inlet solenoid valve	Short circuit to earth
4 - 3	Front left outlet solenoid valve	Short circuit to earth
4 - 4	Rear right inlet solenoid valve	Short circuit to earth
4 - 5	Rear right outlet solenoid valve	Short circuit to earth
4 - 6	Rear left inlet solenoid valve	Short circuit to earth
4 - 7	Rear left outlet solenoid valve	Short circuit to earth
4 - 8	Inlet isolating solenoid valve	Short circuit to earth
4 - 9	Outlet isolating solenoid valve	Short circuit to earth
4 - 12	Front right wheel speed sensor	Open circuit
4 - 13	Rear left wheel speed sensor	Open circuit

Fault code table (continued)

Code	Item	Fault
4 - 14	Front left wheel speed sensor	Open circuit
4 - 15	Rear right wheel speed sensor	Open circuit
5 - 0	Front right inlet solenoid valve	Short circuit to battery supply
5 - 1	Front right outlet solenoid valve	Short circuit to battery supply
5 - 2	Front left inlet solenoid valve	Short circuit to battery supply
5 - 3	Front left outlet solenoid valve	Short circuit to battery supply
5 - 4	Rear right inlet solenoid valve	Short circuit to battery supply
5 - 5	Rear right outlet solenoid valve	Short circuit to battery supply
5 - 6	Rear left inlet solenoid valve	Short circuit to battery supply
5 - 7	Rear left outlet solenoid valve	Short circuit to battery supply
5 - 8	Inlet isolating solenoid valve	Short circuit to battery supply
5 - 9	Outlet isolating solenoid valve	Short circuit to battery supply
6 - 0	Front right inlet solenoid valve	Intermittent short circuit between two ECM to solenoid connections
6 - 1	Front right outlet solenoid valve	Intermittent short circuit between two ECM to solenoid connections
6 - 2	Front left inlet solenoid valve	Intermittent short circuit between two ECM to solenoid connections
6 - 3	Front left outlet solenoid valve	Intermittent short circuit between two ECM to solenoid connections
6 - 4	Rear right inlet solenoid valve	Intermittent short circuit between two ECM to solenoid connections
6 - 5	Rear right outlet solenoid valve	Intermittent short circuit between two ECM to solenoid connections
6 - 6	Rear left inlet solenoid valve	Intermittent short circuit between two ECM to solenoid connections
6 - 7	Rear left outlet solenoid valve	Intermittent short circuit between two ECM to solenoid connections
6 - 8	Inlet isolating solenoid valve	Intermittent short circuit between two ECM to solenoid connections
6 - 9	Outlet isolating solenoid valve	Intermittent short circuit between two ECM to solenoid connections
6 - 12	Front right wheel speed sensor	No output signal, incorrect air gap, damaged rotor teeth
6 - 13	Rear left wheel speed sensor	No output signal, incorrect air gap, damaged rotor teeth
6 - 14	Front left wheel speed sensor	No output signal, incorrect air gap, damaged rotor teeth
6 - 15	Rear right wheel speed sensor	No output signal, incorrect air gap, damaged rotor teeth

Wiring diagram

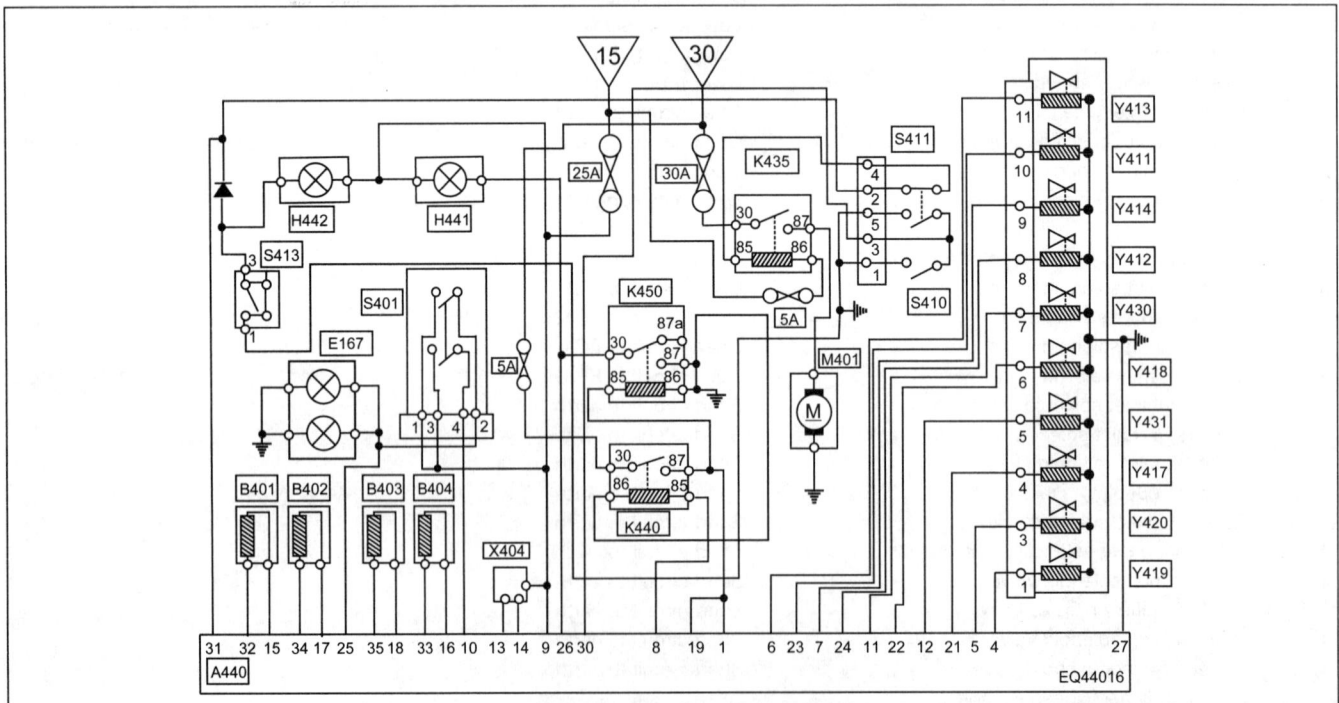

11A.15 Wiring diagram, Land Rover Range Rover

Chapter 11 Part B

Clayton Dewandre Wabco Type II ABS

Contents

Vehicle coverage

Overview of system operation

1 Basic principles and system identification

Clayton Dewandre Wabco is an American antilock brake system which has been fitted to Land Rover derived vehicles since its introduction in the late 1980s. There are two distinct versions of the system which, for the purposes of the technical descriptions and test procedures contained in Part A and Part B of this Chapter, will be referred to as Type I and Type II. The Type II system covered in this Part of Chapter 11 is fitted to the Land Rover Discovery, and the Type I system covered in Part A, is fitted to Range Rover models.

Type II is of the additional or 'add-on' type operating in conjunction with the conventional braking system components. The Type I system is often referred to as an integrated type whereby the components of the ABS hydraulic control unit are also required during the operation of the conventional braking system.

The purpose of the system is to apply the vehicle brakes at maximum efficiency without wheel lock or loss of directional stability. Inductive sensors (wheel speed sensors) monitor the speed of the roadwheels by generating an electrical signal as the wheel is rotated. This information is passed to the ABS Electronic Control Module (ECM) which is then able to determine wheel speed, wheel acceleration and wheel deceleration. The ECM compares the signals received from each wheel and if the onset of lock at any wheel is detected, a signal is sent to the ABS hydraulic control unit which regulates the brake pressure for the relevant wheel(s).

Wabco Type II ABS is comprised of the following components:
a) ABS-ECM.
b) Hydraulic control unit.
c) Four inductive wheel speed sensors and associated sensor rings.
d) Brake light switch.
e) ABS warning lamp.
f) Diagnostic connector.

In addition, the conventional brake system used with the Wabco Type II ABS is comprised of the following components:
a) Tandem brake master cylinder.
b) Vacuum servo unit.
c) Brake calipers/wheel cylinders and hydraulic hoses and pipes.
d) Pressure regulating valves.

2 Component description and operation

ABS ECM

General

The Clayton Dewandre Wabco ABS Electronic Control Module (ECM) continually monitors wheel speed from the signals provided by the wheel speed sensors. If the ECM detects the incidence of wheel lock on one or more wheels, a signal is sent to the hydraulic control unit to modulate the hydraulic pressure to the brake of the locking wheel.

The ECM contains two microprocessors and uses digital technology to complete this function and other functions such as fault code memory and power modules for valve and relay activity (see illustration 11B.1).

Self-test

The ECM is equipped with a self-test capability that initially examines the ABS system when the ignition is switched on, and then examines the wheel speed sensor signals after a pre-determined wheel speed is reached from all wheels (engine running). The ABS self-test program continues to examine the signals from the various

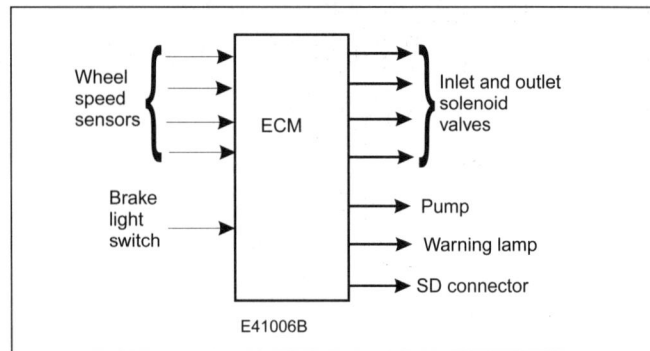

11B.1 ECM sensor inputs and control signal outputs

components as long as the ignition is switched on. If self-test determines that faults are not present, the ABS is ready for operation once a specified vehicle speed has been achieved.

If the ECM detects that a fault is present, all ABS functions are switched off and the warning lamp is turned on. The conventional braking system continues to operate as normal without ABS assistance.

Self-diagnostics

If the ECM detects a fault during the self-test routine, an internal fault code is stored in the ECM memory. Stored fault codes can be retrieved from the SD connector with the aid of a suitable fault code reader, or output manually as flash codes via the ABS warning lamp. If the fault clears, the code will remain stored until cleared with the FCR, or during the flash code retrieval procedure.

Hydraulic control unit

Clayton Dewandre Wabco Type II is a four-channel system with a separate hydraulic circuit for each brake. The hydraulic control unit consists of an electric motor operating a return pump with eccentric drive and twin radial pistons, inlet and outlet solenoid valves, pressure accumulators and pulsation dampers. The unit controls the hydraulic pressure applied to the brake for each individual wheel during ABS operation. The return pump is switched on when the ABS is activated and returns hydraulic fluid, drained off during the pressure reduction phase, back into the brake circuit.

Wheel speed sensors

The rotational speed of the roadwheels and any changes in the rotational speed are recorded by inductive wheel speed sensors, one located at each roadwheel (see illustration 11B.2).

Each wheel speed sensor assembly comprises a toothed sensor

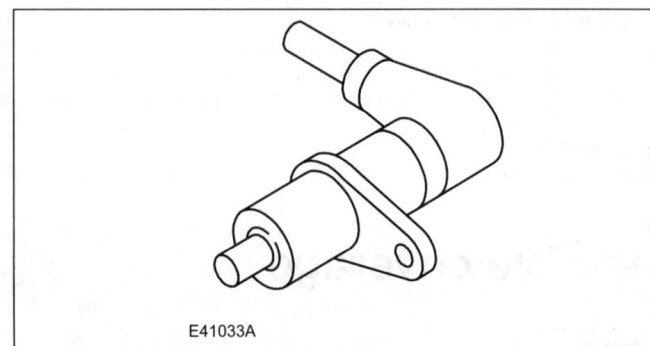

11B.2 Typical inductive wheel speed sensor

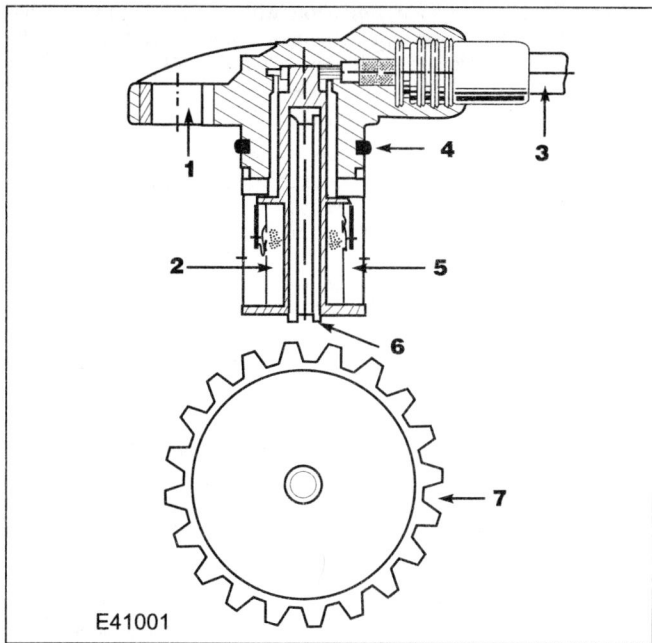

11B.3 Sectional view of an inductive wheel speed sensor

1	Mounting bolt location	5	Coil
2	Permanent magnet	6	Sensor tip
3	Wiring harness	7	Toothed sensor ring
4	O-ring		

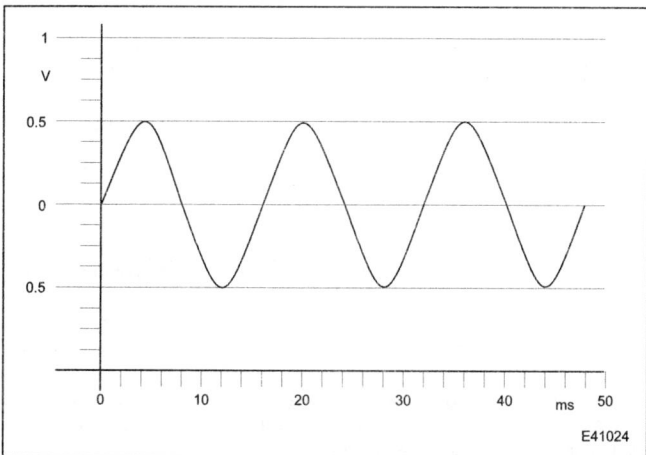

1	Sensor body	
2	Coil	
3	Toothed	
	sensor ring	
4	AC signal	
L	Air gap	

EQ E41002a

11B.4 Inductive wheel speed sensor operation

ring which rotates at roadwheel speed, and an adjacent sensor mounted a set distance from the sensor ring **(see illustration 11B.3)**.

The sensors are permanent magnet pulse generator types producing an AC voltage sine wave as the sensor ring teeth pass through the magnetic field of the sensor **(see illustration 11B.4)**.

The frequency of the waveform produced by the wheel speed sensor is proportional to the road speed. This AC voltage signal is continually being delivered to the ABS-ECM for processing.

The peak to peak voltage of the speed signal (when viewed upon an oscilloscope) can vary considerably according to wheel speed. An analogue to digital converter (ADC) in the ECM transforms the AC pulse into a digital signal **(see illustration 11B.5)**.

ABS electrical wiring harness and relays

An integrated 35-pin main wiring harness is used to connect power

and earth to various electrical components. This enables sensor signals to reach the ECM and the ECM, in turn to send command signals to the relays and hydraulic control unit. The ABS main relay, pump relay and the associated fuses, are typically located in the main fuse/relay box in the vehicle interior.

Main and pump relays

Nominal battery voltage is supplied from the battery to the two relays via a fuse. One terminal of each relay solenoid is connected to earth and the relay is activated by the ECM supplying a voltage to the other solenoid terminal. Battery voltage is then supplied to the ECM or hydraulic pump accordingly.

Warning lamp relay

The warning lamp relay is used to illuminate the ABS warning lamp on the instrument panel in the event of a system fault being detected by the ECM.

ABS warning lamp

After the ignition is switched on, the ABS warning lamp on the instrument panel is illuminated as the system executes a self-test routine **(see illustration 11B.6)**. If satisfactory operation of the system is detected by the ECM, the lamp is extinguished after the vehicle has been driven above approximately 5 mph. During vehicle operation above a pre-determined wheel speed, the ABS-ECM implements a further self-test cycle whereby ABS operation and wheel speed sensor signals are continually monitored. If a fault is detected, the warning lamp relay is actuated by the ECM to illuminate the warning lamp on the instrument panel, and the ABS function is disabled.

When the ABS-ECM detects a fault, the fault code is stored and the ABS warning lamp activated. If the fault no longer exists after the next system start (ignition on/off) the ABS warning lamp is extinguished after the self-test cycle, however the fault code remains stored in the ECM memory.

11B

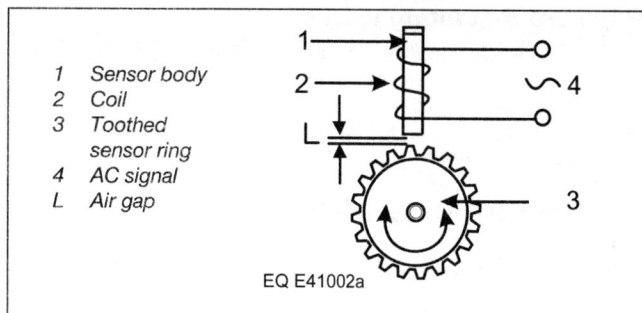

E41024

11B.5 Inductive wheel speed sensor waveform as viewed on an oscilloscope

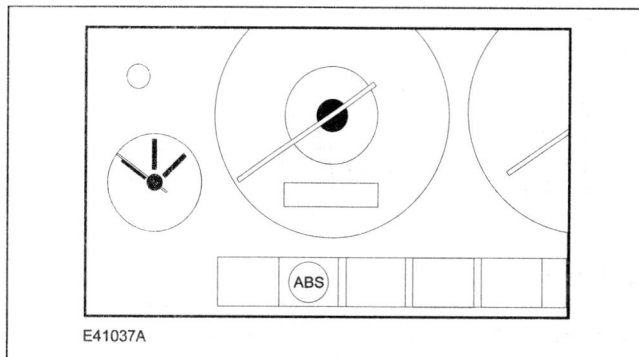

E41037A

11B.6 ABS warning lamp

Pressure regulating valves

A pressure regulating valve is incorporated in the rear brake hydraulic circuit between the master cylinder and hydraulic control unit. The valve is of the pressure conscious type whereby the hydraulic fluid supply to the rear brakes is restricted once a pre-determined pressure is reached.

3 System operation

Brake system at rest

When the ABS system is at rest all the brake components are inoperative. Pressure is non-existent in the hydraulic pipes between the tandem master cylinder and the brake calipers.

Brake system in operation

Phase one, normal braking

When the brake pedal is activated, the pedal force is applied to the tandem master cylinder by the vacuum servo unit pushrod. The servo unit pushrod acts directly on the pressure piston in the master cylinder which pressurises the hydraulic fluid in the brake pipes to the hydraulic control unit. The solenoid valves in the hydraulic control unit are in the 'at rest' position (ie closed) and there is an unrestricted fluid passage between the master cylinder and the relevant brake caliper. Hydraulic pressure can be transmitted directly to each brake caliper, thus operating the brakes.

The ABS-ECM continually monitors wheel speed from the signals provided by the wheel speed sensors. If the ECM detects the incidence of wheel lock on one or more wheels, ABS is automatically initiated.

Phase two, pressure reduction

The ECM opens the solenoid valve on the relevant hydraulic circuit and a return passage from the brake caliper to the hydraulic control unit is opened. Hydraulic fluid now flows from the brake caliper, to the hydraulic control unit resulting in a pressure reduction, thus releasing the brake.

As soon as the solenoid valve is opened, the ECM actuates the electric motor to operate the return pump. The hydraulic fluid is then pumped back into the pressure side of the master cylinder. This process creates a pulsation which can be felt in the brake pedal action, but which is softened by the pulsation damper.

Phase three, pressure build-up

The pressure build-up phase is instigated after the wheel rotation has stabilised. The ECM closes the solenoid valve and switches off the return pump allowing hydraulic pressure to once again increase. Hydraulic fluid now flows to the brake caliper, leading to an increase of pressure at the brake. Normal braking now continues until the wheel has again decelerated to a sufficient degree where pressure reduction is once more required.

The whole ABS control cycle takes place 4 to 10 times per second for each affected wheel and this ensures maximum braking effect and control during ABS operation.

Test procedures

4 Wheel speed sensors

Checking the wheel speed sensor (general)

1 Inspect the wheel speed sensor for corrosion or damage and check that the sensor is tightly mounted.

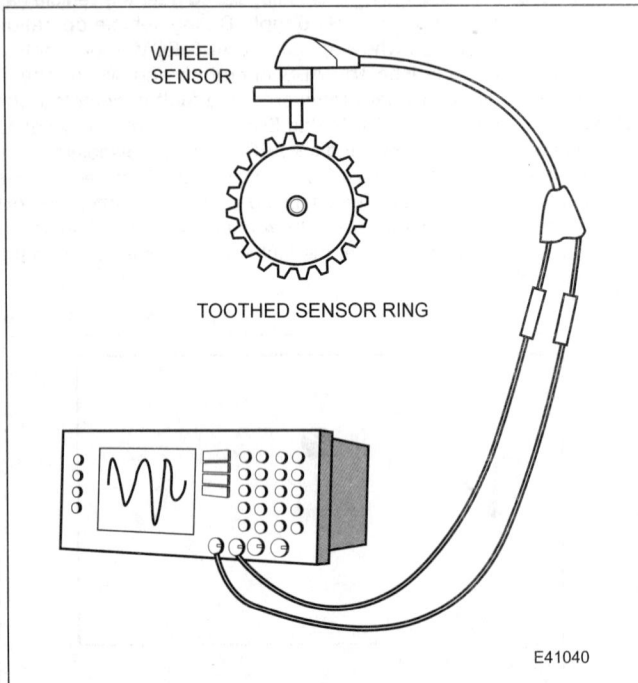

11B.7 Checking wheel speed sensor output with an oscilloscope connected to the sensor wiring plug

2 Check the toothed sensor ring for damage, eccentricity and for broken or missing teeth.
3 Inspect the wheel speed sensor wiring plug for corrosion and damage. One plug for each sensor.
4 Check that the connector terminal pins are pushed fully home and making good contact with the sensor wiring plug.
5 Check the clearance between the sensor and the toothed sensor ring. The clearance is not adjustable but is nominally 0.6 to 1.3 mm. If the clearance is excessive, expect a worn sensor tip or problems with the wheel bearings/hub or sensor ring.
6 When carrying out voltage checks with an oscilloscope or voltmeter, the voltage obtained will be proportional to the speed at which the wheel is rotating. In addition to determining that the wheel speed sensors are actually producing a voltage output, it is essential that the output from the sensors on a particular axle is the same for any given wheel speed.

Checking wheel speed sensor output with an oscilloscope

7 Switch the ignition off and disconnect the ECM multi-plug or the relevant wheel speed sensor wiring plug.
8 Connect an oscilloscope between the terminal pins for the sensor under test (see illustration 11B.7).
9 Select a range to cover 80 Hz on the oscilloscope and a free run time base.
10 Raise the wheel and rotate it by hand at approximately one revolution per second.
11 A sinusoidal wave form should be obtained, with amplitude and duration changing with rotational speed (see illustration 11B.8).
12 If there is no signal, or a very weak or intermittent signal at the ECM, repeat the test at the sensor wiring plug. If there is no change in signal status, the sensor is suspect.
13 If the signal is now satisfactory this indicates a fault in the wiring harness which should be checked for continuity.

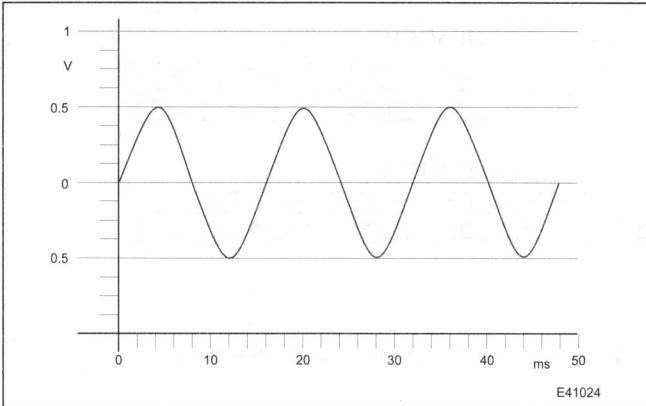

11B.8 Typical wheel speed sensor sine wave as displayed on an oscilloscope

Checking wheel speed sensor output with an AC voltmeter

14 Switch the ignition off and disconnect the ECM multi-plug or the relevant wheel speed sensor wiring plug.

15 Connect an AC voltmeter between the terminal pins for the sensor under test **(see illustration 11B.9)**.

16 Raise the wheel and rotate it by hand at approximately one revolution per second.

17 A voltage of approximately 0.4 to 1.5 volts (AC RMS) should be obtained. If there is no signal, or a very weak or intermittent signal at the ECM, repeat the test at the sensor wiring plug. If there is no change in the signal, the sensor is suspect.

18 If the signal is now satisfactory, this indicates a fault in the wiring harness which should be checked for continuity. **Note:** *This test at least proves that a signal is being generated by the sensor. However, the voltage produced is an average voltage and does not clearly indicate damage to the sensor ring or that the sinewave is regular in formation.*

Checking wheel speed sensor resistance

19 Switch the ignition off and disconnect the ECM multi-plug or the relevant wheel speed sensor wiring plug.

20 Connect an ohmmeter between the terminal pins for the sensor under test **(see illustration 11B.10)**.

21 The readings obtained should be between 0.9 and 1.0 kohms.

22 If the resistance is excessively high, or open circuit at the ECM,

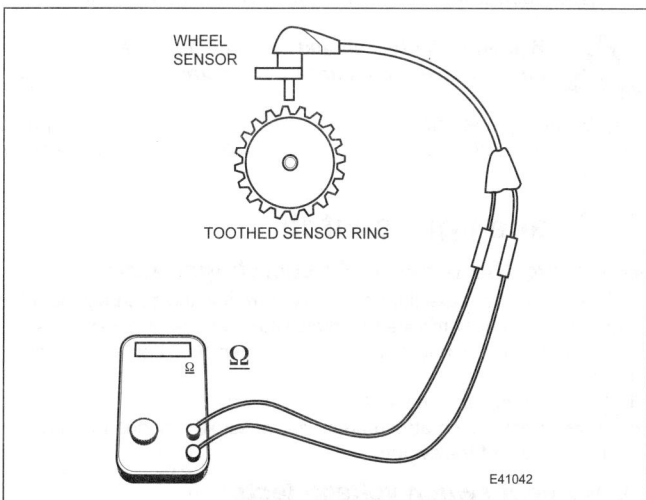

11B.10 Checking wheel speed sensor resistance with an ohmmeter connected to the sensor wiring plug

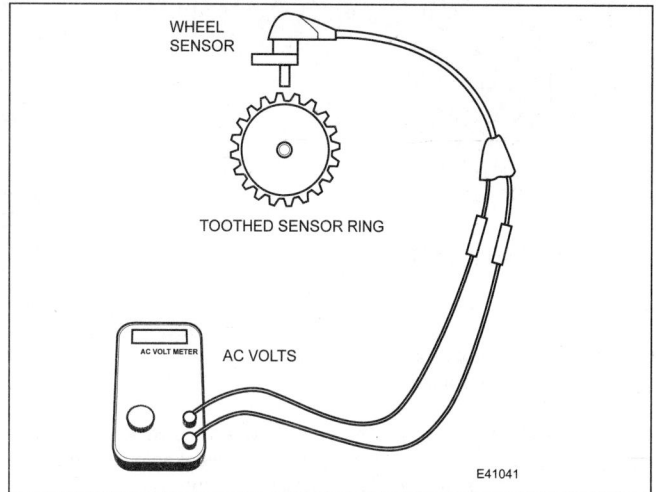

11B.9 Checking wheel speed sensor output with a voltmeter connected to the sensor wiring plug

repeat the test at the sensor multi-plug. If there is no change in resistance, the sensor is suspect.

23 If the resistance is now satisfactory, this indicates a fault in the wiring harness which should be checked for continuity. **Note:** *Even if the resistance is within the quoted specifications, this does not prove that the speed sensor can generate an acceptable signal.*

5 System relays

Checking relays (general)

1 The following tests are applicable to the ABS main (or solenoid valve) relay, hydraulic pump relay and warning lamp relay.

2 Before carrying out specific tests on the relays or power supplies, inspect the relays and relay base for corrosion and damage.

3 Check that the relays are fully pushed home and making good contact with the base.

4 Faults in any of the above areas are possible reasons for failure or malfunctioning of the relay.

Relay operation tests

Main (solenoid valve) relay and hydraulic pump relay

5 Switch the ignition off, then remove the relevant relay from the relay base.

6 Connect an ohmmeter between the relay output terminal 87 and the main voltage supply terminal 30 **(see illustration 11B.11)**. The ohmmeter should indicate an open circuit.

7 Attach a wire between the relay driver supply terminal 86 and a

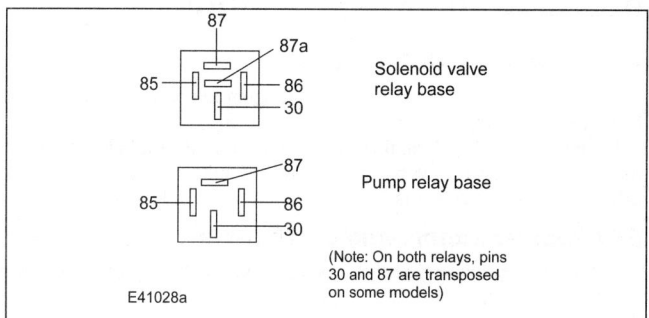

11B.11 Main (solenoid valve) relay and hydraulic pump relay terminals

11B

Relay Base
E41101

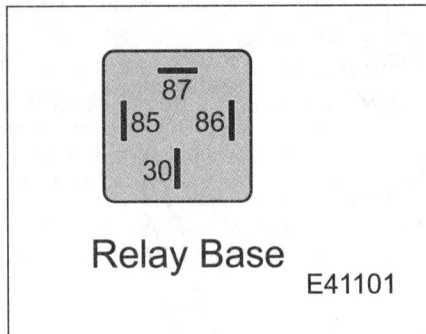

11B.12 Typical relay base terminal identification

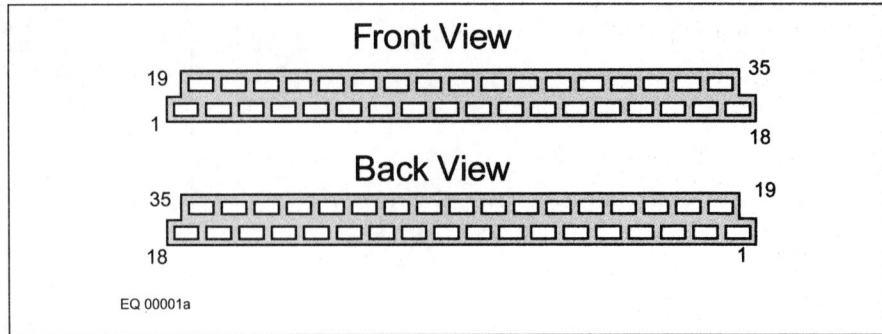

Front View

Back View

EQ 00001a

11B.13 ECM 35-pin multi-plug terminal identification

12 volt supply. Attach a second wire between the relay driver earth terminal 85 and earth. There should be an unmistakable click from the relay and the ohmmeter should indicate continuity.

8 Renew the relay if it fails these tests. If the relay cover is transparent, inspect the relay for burns. A faulty relay usually has a burnt appearance.

Warning lamp relay

9 Switch the ignition off, then remove the relay from the relay base.

10 Connect an ohmmeter between relay terminals 30 and 87a. The ohmmeter should indicate an open circuit.

11 Attach a wire between the relay driver supply terminal 86 and a 12 volt supply. Attach a second wire between the relay driver earth terminal 85 and earth. There should be an unmistakable click from the relay and the ohmmeter should indicate an open circuit.

12 Renew the relay if it fails these tests. If the relay cover is transparent, inspect the relay for burns. A faulty relay usually has a burnt appearance.

Relay power supply tests

Main (solenoid valve) relay and hydraulic pump relay

13 Switch the ignition off then remove the relevant relay leaving the base exposed.

14 Using a voltmeter, probe between the main voltage supply terminal 30 in the relay base and earth **(see illustration 11B.12)**. The reading should indicate nbv. If no voltage is found, check the wiring back to the voltage supply.

Warning lamp relay

13 Switch the ignition off then remove the relevant relay leaving the base exposed.

14 Switch the ignition on.

15 Using a voltmeter, probe between the relay driver supply terminal 86 in the relay base and earth. The reading should indicate nbv. If no voltage is found, check the wiring back to the voltage supply.

6 Electronic Control Module

Checking the ECM (general)

1 Inspect the ECM for corrosion or damage and ensure that the unit is mounted securely.

2 Check that the ECM multi-plug terminals are pushed fully home and making good contact with the ECM pins. Faults in any of the above areas are possible reasons for poor performance in the ABS system.

ECM power supply and earth tests

3 Switch the ignition off and disconnect the ECM multi-plug from the ECM.

4 Switch the ignition on.

5 Attach a negative voltmeter probe to a vehicle earth and attach a positive voltmeter probe to ECM multi-plug pin 9 **(see illus-**

tration **11B.13)**. The voltmeter should indicate nbv. If no voltage is found, check the relevant fuse, and the supply wiring back to the ignition switch.

6 Switch the ignition off.

7 Connect an ohmmeter between a vehicle earth and ECM multi-plug earth pins 27 and 31 in turn. The ohmmeter should indicate continuity in each case. If not, check the ECM main earth connection and wiring.

7 Solenoid valves

Solenoid valve resistance tests

1 Switch the ignition off and disconnect the ECM multi-plug from the ECM.

2 Using an ohmmeter, measure the resistance between pin 27 in the ECM multi-plug, and each of the following multi-plug pins in turn **(see illustration 11B.13)**:

Inlet solenoid valves - 4, 6, 21, 23
Outlet solenoid valves - 5, 7, 22, 24

3 The resistance should be in the order of 6.5 ohms for each inlet solenoid valve and 4.5 ohms for each outlet solenoid valve.

4 If the resistance is not as specified, check the wiring and wiring connectors. If satisfactory, the hydraulic control unit is suspect.

8 Hydraulic pump motor

Pump operation test

1 Switch the ignition off then remove the pump relay leaving the base exposed.

2 Bridge terminals 30 and 87 in the relay base using a fused jumper lead **(see illustration 11B.12)**. The pump should now operate.

> ⚠️ **Warning: The test should be made as quickly as possible to avoid damaging the pump.**

3 If the pump does not operate as described, carry out the pump relay power supply tests described in Section 5. If satisfactory, suspect a faulty pump.

9 Brake light switch

Checking the brake light switch (general)

1 Check that the brake light switch is correctly and securely mounted and that the plunger moves smoothly with no trace of binding.

2 Check that the wiring multi-plug is pushed fully home and making good contact.

3 Check that no wires have been disconnected.

4 Faults in any of the above areas are possible reasons for failure or malfunctioning of the switch.

Brake light switch voltage tests

5 Switch the ignition off and disconnect the ECM multi-plug from the ECM.

6 Connect a voltmeter between ECM multi-plug pin 10 and a vehicle earth **(see illustration 11B.13)**.
7 Switch the ignition on.
8 Depress the brake pedal. The voltmeter should indicate nbv.
9 If no voltage is found, the fuse, the brake light switch and the wiring are suspect.
10 Release the brake pedal. The voltage should drop to zero as the switch opens.
11 Repeat the test with the voltmeter connected between ECM multi-plug 25 and a vehicle earth. The voltmeter should indicate nbv with the pedal depressed, and zero with the pedal released.

10 Warning lamp

Checking the warning lamp (general)

1 Inspect the warning lamp bulb holder contacts in the instrument panel.
2 Check that the instrument panel multi-plug terminal pins are pushed fully home and making good contact.
3 Check that no wires have been disconnected.

4 Faults in any of the above areas are possible reasons for failure or malfunctioning of the warning lamp.

Warning lamp operation test

5 With the ignition switched off, the warning lamp should remain off.
6 Switch the ignition on and the warning lamp should illuminate. Start the engine and drive the vehicle. The lamp should extinguish when a speed of approximately 5 mph has been attained. The lamp should then remain off.
7 If the warning lamp comes on and remains on at any time during vehicle operation, carry out the previously described test procedures on the system components.

Warning lamp circuit test

8 Switch the ignition off then remove the warning lamp relay leaving the base exposed.
9 Bridge terminals 30 and 87a in the relay base using a fused jumper lead.
10 Switch the ignition on and the warning lamp should illuminate.
11 If the lamp fails to illuminate, check the condition of the bulb and wiring.

Pin table - typical 35-pin

Note: *Refer to illustration 11B.13*

Pin No.	Connection	Test condition	Voltage
1	Supply from relay	Ignition on	Nbv
2	-	-	-
3	-	-	-
4	Rear right inlet solenoid valve	Ignition on/inactive	0 volts
		Actuated	Nbv
5	Rear right outlet solenoid valve	Ignition on/inactive	0 volts
		Actuated	Nbv
6	Front right inlet solenoid valve	Ignition on/inactive	0 volts
		Actuated	Nbv
7	Front right outlet solenoid valve	Ignition on/inactive	0 volts
		Actuated	Nbv
8	Main (solenoid valve) relay driver	Ignition on, actuated	Nbv
9	Supply from ignition switch	Ignition on	Nbv
10	Brake light switch input terminal 4	Ignition on:	
		Brake pedal not pressed	0 volts
		Brake pedal depressed	Nbv
11	Hydraulic pump relay driver	Ignition on/inactive	0 volts
		Actuated	Nbv
12	-	-	-
13	SD connector	-	-
14	SD connector	-	-
15	Front left wheel speed sensor signal	Roadwheel rotating	0.4 to 1.5 volts AC (approx)
16	Rear right wheel speed sensor signal	Roadwheel rotating	0.4 to 1.5 volts AC (approx)
17	Front right wheel speed sensor signal	Roadwheel rotating	0.4 to 1.5 volts AC (approx)
18	Rear left wheel speed sensor signal	Roadwheel rotating	0.4 to 1.5 volts AC (approx)
19	Supply from relay	Ignition on	Nbv
20	-	-	-

Pin table - typical 35-pin (continued)

Note: *Refer to illustration 11B.13*

Pin No.	Connection	Test condition	Voltage
21	Rear left inlet solenoid valve	Ignition on/inactive	0 volts
		Actuated	Nbv
22	Rear left outlet solenoid valve	Ignition on/inactive	0 volts
		Actuated	Nbv
23	Front left inlet solenoid valve	Ignition on/inactive	0 volts
		Actuated	Nbv
24	Front left outlet solenoid valve	Ignition on/inactive	0 volts
		Actuated	Nbv
25	Brake light switch input terminal 1	Ignition on:	
		Brake pedal not pressed	0 volts
		Brake pedal depressed	Nbv
26	ABS warning lamp	Ignition on:	
		Lamp on	2.0 volts (max)
		Lamp off	Nbv
27	ECM earth	Ignition on	0.25 volts (max)
28	-	-	-
29	-	-	-
30	Hydraulic pump relay display input	Ignition on/inactive	0 volts
		Actuated	Nbv
31	ECM earth	Ignition on	0.25 volts (max)
32	Front left wheel speed sensor earth	Roadwheel rotating	0.25 volts (max)
33	Rear right wheel speed sensor earth	Roadwheel rotating	0.25 volts (max)
34	Front right wheel speed sensor earth	Roadwheel rotating	0.25 volts (max)
35	Rear left wheel speed sensor earth	Roadwheel rotating	0.25 volts (max)

Fault codes

1 If a FCR is available, it should be connected to the SD serial connector and used in accordance with the maker's instructions.
2 The FCR can be used for the following purposes:
  a) Obtaining fault codes.
  b) Clearing fault codes.
  c) Obtaining datastream information.
  d) Testing the system actuators (relays and solenoid valves).
3 If a FCR is not available, it is still possible to obtain fault codes which will be displayed as a series of flashes on the dash mounted warning lamp.
4 When the ECM determines that a fault is present, it internally logs a fault code and also illuminates the ABS warning lamp. All of the various fault codes on these vehicles are of the 'slow' variety and can be output as flash codes on the warning lamp.

Obtaining codes without a FCR

5 Switch the ignition off and locate the ABS relays and SD serial connector. The relays are typically located below the facia on the left-hand side, with the SD connector located in the vicinity of the centre console at the front.

6 Remove the ABS warning lamp relay from the relay base.
7 Switch the ignition on and the ABS warning lamp should illuminate.
8 Use a bridging wire to bridge terminals 2 and 5 of the 5-pin SD connector plug, or terminals 4 and 7 if the plug is a 16-pin type.
9 Five seconds after connecting the bridging wire, the ABS warning lamp will extinguish for 2.5 seconds, then illuminate for 2.5 seconds, then extinguish again for 2.5 seconds then illuminate again for 0.5 seconds, then extinguish again for another 2.5 seconds. At the end of this sequence the flash code output will start.
10 If a fault code is stored in the ECM memory it will now be displayed in two groups of 0.5 second flashes of the warning lamp, with a 2.5 second pause between each group.
11 The ECM will continue to repeat this fault code until the bridging wire is removed from the SD connector plug. When the bridging wire is removed the code will be erased from the ECM memory and the ECM will be ready to display the next fault code, if present.
12 To obtain the next fault code, repeat the procedure described in paragraphs 8 to 11. When all the fault codes have been displayed, the ABS warning lamp will remain extinguished for 7.5 seconds after the start of the output sequence.

Fault code table

Code	Item	Fault
2 - 6	Brake light switch	No signal, poor wiring connections
2 - 7	ECM supply voltage	Defective main relay, wiring, connections
2 - 8	Solenoid valve supply voltage	Incorrect voltage, defective relay
2 - 12	Front right wheel speed sensor	Low output signal
2 - 13	Rear left wheel speed sensor	Low output signal
2 - 14	Front left wheel speed sensor	Low output signal
2 - 15	Rear right wheel speed sensor	Low output signal
3 - 0	Front right inlet solenoid valve	Open circuit to ECM
3 - 1	Front right outlet solenoid valve	Open circuit to ECM
3 - 2	Front left inlet solenoid valve	Open circuit to ECM
3 - 3	Front left outlet solenoid valve	Open circuit to ECM
3 - 4	Rear right inlet solenoid valve	Open circuit to ECM
3 - 5	Rear right outlet solenoid valve	Open circuit to ECM
3 - 6	Rear left inlet solenoid valve	Open circuit to ECM
3 - 7	Rear left outlet solenoid valve	Open circuit to ECM
4 - 0	Front right inlet solenoid valve	Short circuit to earth
4 - 1	Front right outlet solenoid valve	Short circuit to earth
4 - 2	Front left inlet solenoid valve	Short circuit to earth
4 - 3	Front left outlet solenoid valve	Short circuit to earth
4 - 4	Rear right inlet solenoid valve	Short circuit to earth
4 - 5	Rear right outlet solenoid valve	Short circuit to earth
4 - 6	Rear left inlet solenoid valve	Short circuit to earth
4 - 7	Rear left outlet solenoid valve	Short circuit to earth
4 - 12	Front right wheel speed sensor	Open circuit
4 - 13	Rear left wheel speed sensor	Open circuit
4 - 14	Front left wheel speed sensor	Open circuit
4 - 15	Rear right wheel speed sensor	Open circuit
5 - 0	Front right inlet solenoid valve	Short circuit to battery supply
5 - 1	Front right outlet solenoid valve	Short circuit to battery supply
5 - 2	Front left inlet solenoid valve	Short circuit to battery supply
5 - 3	Front left outlet solenoid valve	Short circuit to battery supply
5 - 4	Rear right inlet solenoid valve	Short circuit to battery supply
5 - 5	Rear right outlet solenoid valve	Short circuit to battery supply
5 - 6	Rear left inlet solenoid valve	Short circuit to battery supply
5 - 7	Rear left outlet solenoid valve	Short circuit to battery supply
6 - 0	Front right inlet solenoid valve	Intermittent short circuit between two ECM to solenoid connections
6 - 1	Front right outlet solenoid valve	Intermittent short circuit between two ECM to solenoid connections
6 - 2	Front left inlet solenoid valve	Intermittent short circuit between two ECM to solenoid connections
6 - 3	Front left outlet solenoid valve	Intermittent short circuit between two ECM to solenoid connections
6 - 4	Rear right inlet solenoid valve	Intermittent short circuit between two ECM to solenoid connections
6 - 5	Rear right outlet solenoid valve	Intermittent short circuit between two ECM to solenoid connections
6 - 6	Rear left inlet solenoid valve	Intermittent short circuit between two ECM to solenoid connections
6 - 7	Rear left outlet solenoid valve	Intermittent short circuit between two ECM to solenoid connections
6 - 12	Front right wheel speed sensor	No output signal, incorrect air gap, damaged rotor teeth
6 - 13	Rear left wheel speed sensor	No output signal, incorrect air gap, damaged rotor teeth
6 - 14	Front left wheel speed sensor	No output signal, incorrect air gap, damaged rotor teeth
6 - 15	Rear right wheel speed sensor	No output signal, incorrect air gap, damaged rotor teeth

11B

Wiring diagram

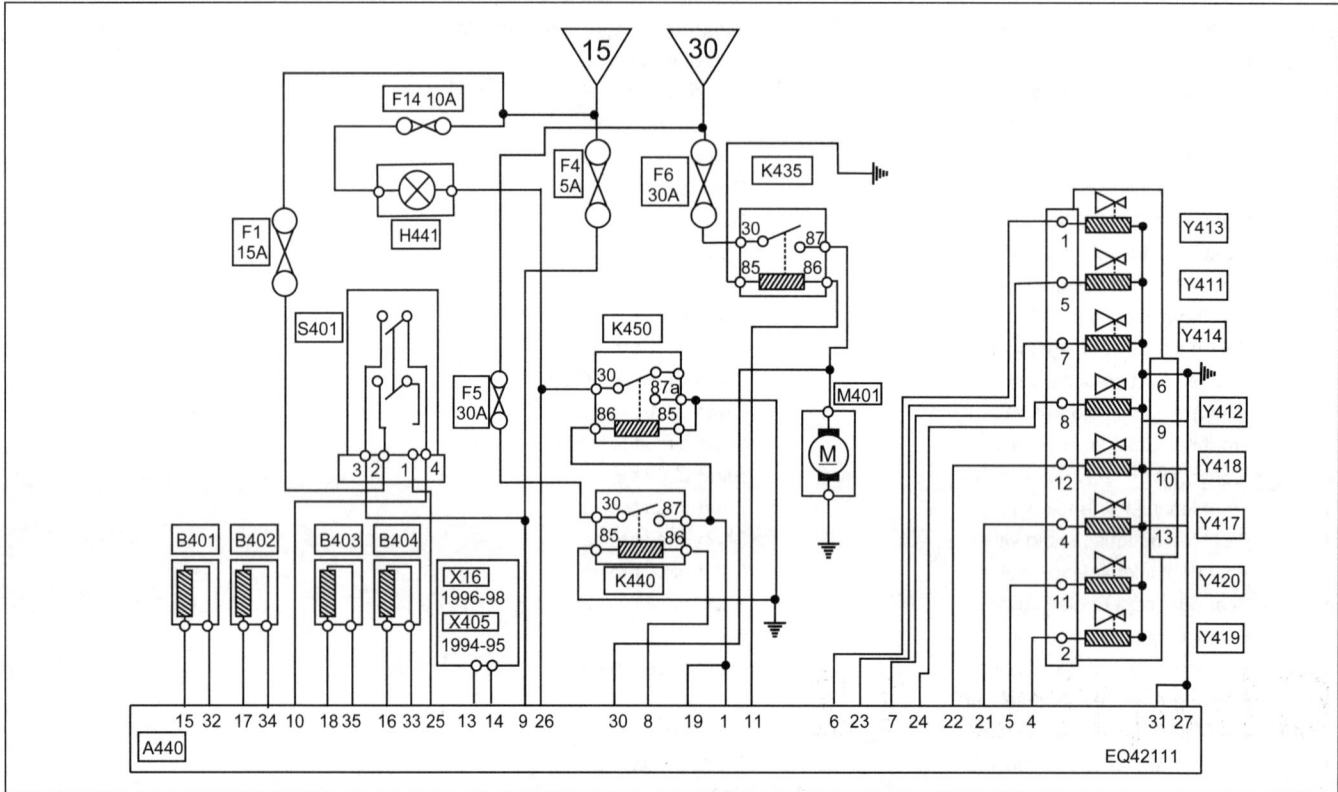

11B.14 Wiring diagram, Land Rover Discovery

Chapter 12
Kelsey Hayes 415 ABS

Contents

Vehicle coverage

Overview of system operation

1 Basic principles and system identification

The Kelsey Hayes 415 Antilock Brake System is a derivative of the Bosch 5.0 series ABS, with minor alterations to the hydraulic control unit, and the Electronic Control Module (ECM) software. Kelsey Hayes 415 has been fitted to a limited number of Vauxhall/Opel passenger vehicles since its introduction in the mid 1990s. The system is of the additional or 'add-on' type operating in conjunction with the conventional braking system components.

The purpose of the system is to apply the vehicle brakes at maximum efficiency without wheel lock or loss of directional stability. Inductive sensors (wheel speed sensors) monitor the speed of the wheels by generating an electrical signal as the wheel is rotated. This information is passed to the ABS-ECM which compares the signals received from each wheel and uses the speed of the fastest wheel as a reference value. The ECM continually monitors the speed of each wheel and if the onset of lock at any wheel is detected (a received speed signal being less than the reference value) a signal is sent to the ABS hydraulic control unit which regulates the brake pressure for the relevant wheel(s).

Kelsey Hayes 415 ABS is comprised of the following main components (see illustration 12.1):
a) Hydraulic control unit with integral ABS-ECM.
b) Inductive wheel speed sensors and associated sensor rings.
c) Brake light switch.
d) ABS warning lamp.
e) Diagnostic connector.

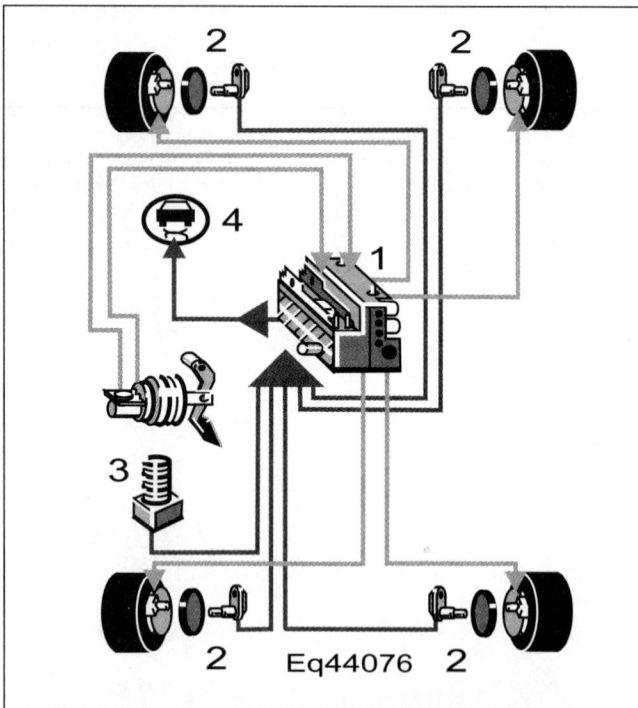

12.1 Kelsey Hayes 415 main components

1   Hydraulic control unit with integral ECM
2   Wheel speed sensors and associated sensor rings
3   Brake light switch
4   ABS warning lamp

In addition, the conventional brake system is comprised of the following components:
a) Tandem brake master cylinder.
b) Vacuum servo unit.
c) Brake calipers/wheel cylinders and hydraulic hoses and pipes.

2 Component description and operation

ABS ECM

General

The Kelsey Hayes 415 ECM continually monitors wheel speed from the signals provided by the wheel speed sensors, and brake application from the brake light switch signal. If the ECM detects the incidence of wheel lock on one or more wheels, a signal is sent to the hydraulic control unit to modulate the hydraulic pressure to the brake of the locking wheel(s). The ECM contains two microprocessors and uses digital technology to complete this function and other functions such as, fault code memory and power modules for valve and relay activity (see illustration 12.2).

To reduce external electrical connections to a minimum and improve reliability, the ECM is integral with the hydraulic control unit. In certain applications the ECM and hydraulic control unit cannot be separated and are only available as a complete sealed unit.

Self-test

Kelsey Hayes 415 ABS is equipped with a self-test capability that begins to examine the ABS signals after a wheel speed of approximately 2 mph is reached from all wheels. ABS self-test continually examines the signals from the various components. If self-test determines that faults are not present, ABS is ready for operation. If the ECM detects that a fault is present, all ABS functions are switched off and the ABS warning lamp is turned on.

Self-diagnostics

If the ECM detects a fault during the self-test routine, an internal fault code is stored in the ECM memory. Stored fault codes can be retrieved from the SD connector with the aid of a suitable fault code reader. If the fault clears, the code will remain stored until cleared with an FCR.

Hydraulic control unit

Kelsey Hayes 415 is a four-channel system with a separate hydraulic circuit for each brake. The hydraulic control unit consists of an electric motor and radial piston return pump, inlet and outlet solenoid valves, pressure accumulators and pulsation dampers. The

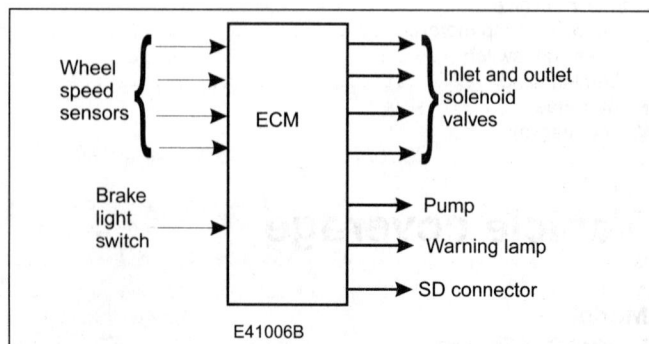

12.2 ECM sensor inputs and control signal outputs

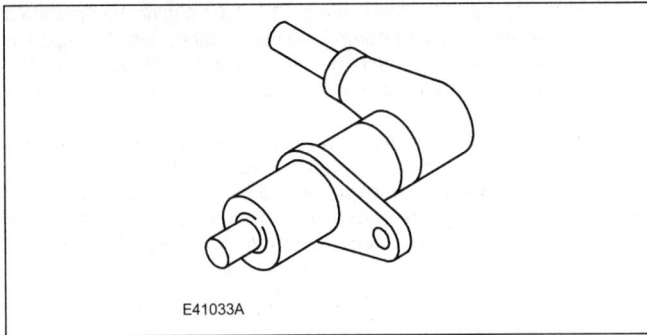

12.3 Typical wheel speed sensor

unit controls the hydraulic pressure applied to the brake for each individual front wheel and each individual rear wheel. The return pump is switched on when the ABS is activated and returns hydraulic fluid, drained off during the pressure reduction phase, back into the brake circuit.

On some applications, the ABS-ECM contains additional software for rear brake hydraulic fluid pressure regulation, generally known as 'Electronic Brake Force Distribution'. In these applications, mechanical pressure regulating/load sensing valves are not required as the rear brake hydraulic pressure is controlled by the ABS hydraulic control unit.

Wheel speed sensors

The rotational speed of the roadwheels and any changes in the rotational speed are recorded by inductive wheel speed sensors, one located at each roadwheel **(see illustration 12.3)**.

Each wheel speed sensor assembly comprises a toothed sensor ring which rotates at roadwheel speed, and an adjacent sensor mounted a set distance from the sensor ring **(see illustration 12.4)**.

The sensors are permanent magnet pulse generator types producing an AC voltage sine wave as the sensor ring teeth pass through the magnetic field of the sensor **(see illustration 12.5)**.

The frequency of the waveform produced by the wheel speed sensor is proportional to the road speed. This AC voltage signal is continually being delivered to the ABS-ECM for processing.

The peak to peak voltage of the speed signal (when viewed upon an oscilloscope) can vary considerably according to wheel speed. An analogue to digital converter (ADC) in the ECM transforms the AC pulse into a digital signal **(see illustration 12.6)**.

ABS electrical wiring harness and relays

An integrated main wiring harness is used for ABS-ECM power supply and earth connections, and enables sensor signals to reach

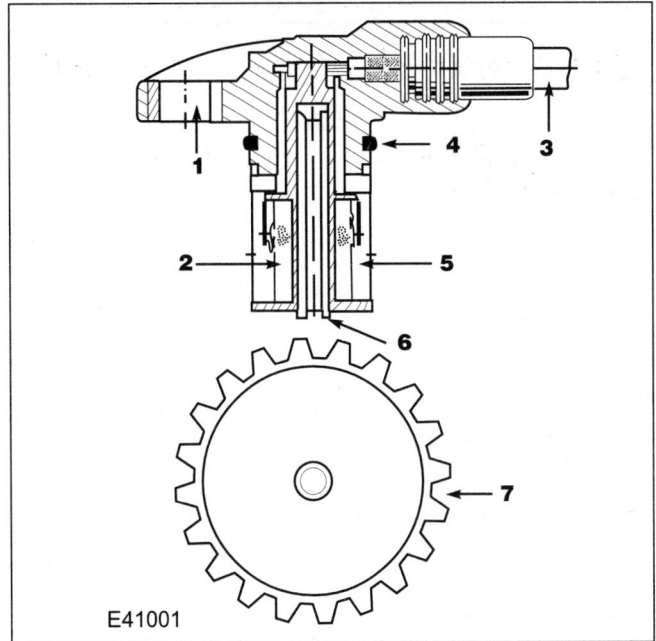

12.4 Sectional view of a wheel speed sensor

1	Mounting bolt location	5	Coil
2	Permanent magnet	6	Sensor tip
3	Wiring harness	7	Toothed sensor ring
4	O-ring		

the ECM and the ECM, in turn, to send output signals to the ABS warning lamp and diagnostic connector. The main (solenoid valve) relay and return pump relay are an integral part of the ECM and cannot be separately removed. Internal connections between the ECM and hydraulic control unit are used to activate the return pump.

Brake light switch

The brake light switch comprises a switch body and contact pin and is located above the brake pedal **(see illustration 12.7)**.

When the brake pedal is depressed, closing the brake light switch, a signal is sent to the ECM indicating that the brakes are being applied. Once this signal is received, the ECM will begin monitoring the wheel speed via the wheel speed sensors and activate the ABS if necessary.

ABS warning lamp

After the ignition is switched on, the ABS warning lamp on the instrument panel is illuminated for approximately 4 seconds, then

12

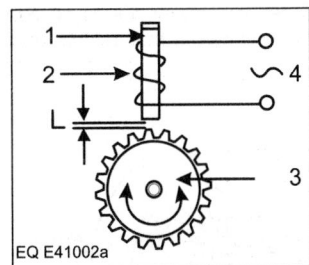

12.5 Wheel speed sensor operation

1 Sensor body
2 Coil
3 Toothed sensor ring
4 AC signal
L Air gap

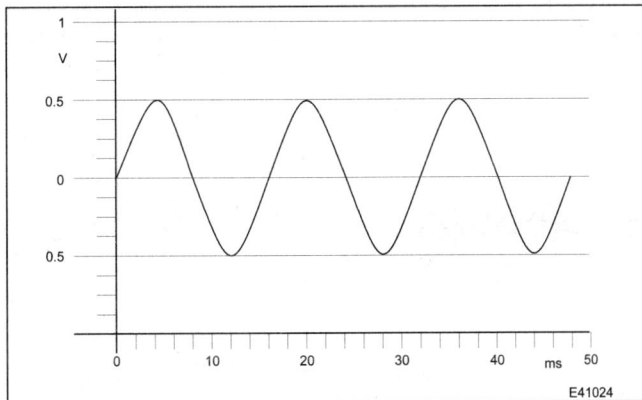

12.6 Wheel speed sensor waveform as viewed on an oscilloscope

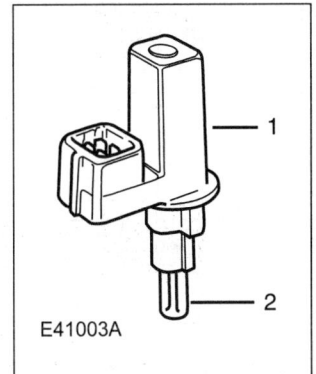

12.7 Typical brake light switch body (1) and contact pin (2)

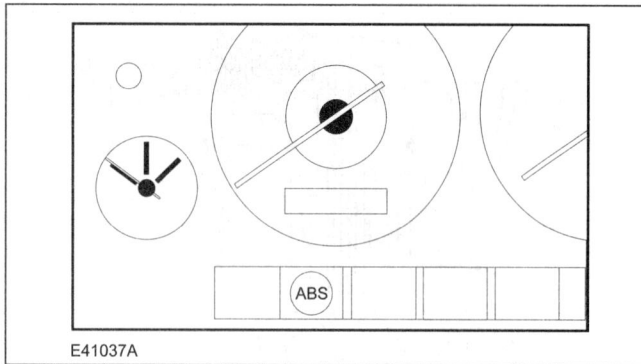

12.8 ABS warning lamp

extinguished. During vehicle operation above a pre-determined wheel speed, the ABS-ECM implements a self-test cycle whereby ABS operation is monitored. If a fault is detected, the ECM illuminates the warning lamp on the instrument panel. The ECM switches off the ABS, however the conventional braking system continues to operate as normal. The warning lamp will remain illuminated until the fault is no longer present **(see illustration 12.8)**.

When the ABS-ECM detects a fault, the fault code is stored and the ABS warning lamp activated. If the fault no longer exists after the next system start (ignition on/off) the ABS warning lamp is extinguished after the self-test cycle, however the fault code remains stored in the ECM memory.

Tandem master cylinder

Typically, the tandem master cylinder comprises a body casting incorporating primary and secondary pressure chambers, primary piston, intermediate piston, floating piston, slotted pin and central valve. The cylinder operates as a conventional master cylinder using vacuum assistance from the vacuum servo unit **(see illustration 12.9)**.

When the brake system is at rest, the central valve in the floating piston rests against the slotted pin. In this condition the central valve is open and brake fluid can discharge out of the pressure chamber back into the brake fluid reservoir. When the brake pedal is depressed, the build-up of hydraulic pressure in the primary pressure chamber acts on the intermediate piston and floating piston, moving them down the cylinder bore. The floating piston contacts the seal on the central valve, closing the connection between the intermediate and secondary pressure chambers. Brake hydraulic pressure can now also increase in the secondary pressure chamber.

Vacuum servo unit

The vacuum servo unit is located between the brake pedal and tandem master cylinder. When the brake pedal is depressed, the servo unit increases the force applied by the pedal, reducing the effort required to operate the brakes **(see illustration 12.10)**.

The unit is operated by vacuum created in the engine inlet manifold (or from a separate vacuum pump on diesel engines) which is applied to a diaphragm contained within the unit casing. A pushrod connected to the centre of the diaphragm acts directly on the primary piston in the master cylinder.

When the brake pedal is released, vacuum is applied to both sides of the diaphragm. When the pedal is depressed, one side of the diaphragm is opened to atmosphere and the vacuum acting on the other side deflects the diaphragm which in turn operates the master cylinder primary piston. The resulting force applied to the master cylinder piston is therefore significantly greater than the initial force applied to the brake pedal by the driver.

3   System operation

Brake system at rest

When the ABS system is at rest all the brake components are inoperative. Pressure is non-existent in the hydraulic pipes between the tandem master cylinder and the brake calipers. The inlet solenoid valves in the hydraulic control unit valve block are open and the outlet solenoid valves are closed.

Brake system operating under conventional control without ABS

When the brake pedal is activated, the pedal force is applied to the tandem master cylinder by the vacuum servo unit pushrod. The servo unit pushrod acts directly on the pressure piston in the master cylinder which pressurises the hydraulic fluid in the brake pipes to the hydraulic control unit. The inlet solenoid valve and outlet solenoid valve both remain in the 'at rest' position (inlet solenoid valve open and outlet solenoid valve closed). Hydraulic pressure is transmitted to each brake caliper, thus operating the brakes.

When the brake pedal is released, the one-way valve opens allowing the hydraulic pressure in the circuit to rapidly decrease **(see illustration 12.11)**.

Brake system operating in conjunction with ABS control

The ABS-ECM continually monitors wheel speed from the signals provided by the wheel speed sensors. If the ECM detects the incidence of wheel lock on one or more wheels, ABS is automatically initiated in three phases. As Kelsey Hayes 415 ABS operates individually on each wheel, all or any of the wheels could be in any one of the following phases at any particular moment.

First ABS phase, pressure holding

To prevent any further build-up of hydraulic pressure in the circuit being controlled, the ECM closes the inlet solenoid valve and allows the outlet solenoid valve to remain closed. The hydraulic fluid line from the tandem master cylinder to the brake caliper or wheel cylinder is closed, and the hydraulic fluid in the controlled circuit is maintained at a

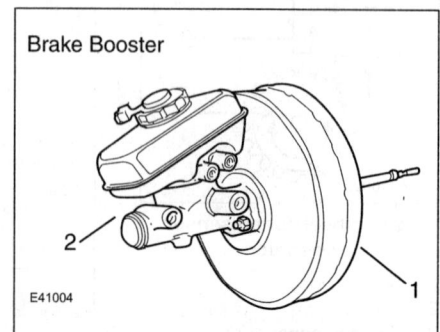

1 *Body casting*
2 *Outlet connections*
3 *Fluid inlet from reservoir*
4 *Pushrod/primary piston*
5 *Intermediate piston*
6 *Slotted pin*
7 *Floating piston*
8 *Central valve*

12.9 Sectional view of a typical tandem master cylinder

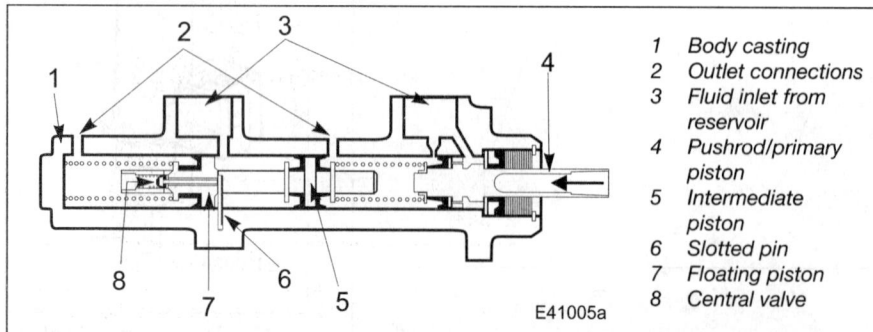

Brake Booster

12.10 Vacuum servo unit (1) and tandem master cylinder (2)

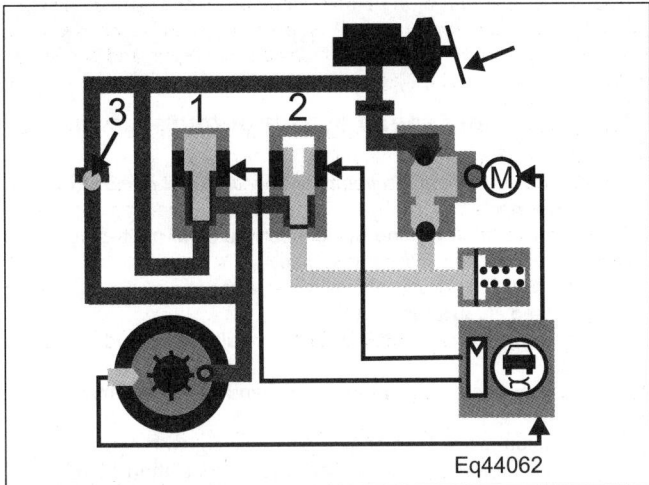

Eq44062

12.11 Brake system operating under conventional control without ABS

1   Inlet solenoid valve   3   One-way valve
2   Outlet solenoid valve

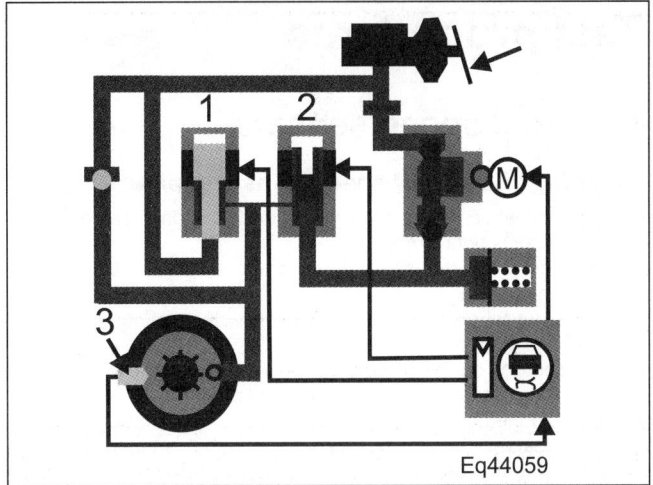

Eq44059

12.12 ABS operation - first phase, pressure holding

1   Inlet solenoid valve   3   Wheel speed sensor
2   Outlet solenoid valve

constant pressure. This effectively removes the braking force from the controlled circuit. The pressure cannot now be increased in that circuit by any further application of the brake pedal (see illustration 12.12).

If the wheel speed sensor signals indicate that wheel rotation has now stabilised, the ECM will instigate the pressure build-up phase, allowing braking to continue. If wheel lock is still detected after the pressure holding phase, the ECM instigates the pressure reduction phase.

Second ABS phase, pressure reduction

If the ECM detects wheel instability, a pressure reduction phase is initiated. The inlet solenoid valve remains closed and the outlet solenoid valve is opened by means of a series of short activation pulses. The pressure in the controlled circuit decreases rapidly as the fluid flows from the brake caliper or wheel cylinder into the pressure accumulator. At the same time, the ECM actuates the electric motor to operate the return pump. The hydraulic fluid is then pumped back into the pressure side of the master cylinder. This process creates a

pulsation which can be felt in the brake pedal action, but which is softened by the pulsation damper (see illustration 12.13).

Third ABS phase, pressure build-up

The pressure build-up phase is instigated after the wheel rotation has stabilised. The inlet and outlet solenoid valves are returned to the at rest position (inlet solenoid valve open and outlet solenoid valve closed) which re-opens the hydraulic fluid line from the tandem master cylinder to the brake caliper or wheel cylinder. Hydraulic pressure is reinstated, thus re-introducing operation of the brake. After a brief period, a short pressure holding phase is re-introduced and the ECM continually shifts between pressure build-up and pressure holding until the wheel has decelerated to a sufficient degree where pressure reduction is once more required (see illustration 12.14).

The whole ABS control cycle takes place 4 to 10 times per second for each affected wheel and this ensures maximum braking effect and control during ABS operation.

12

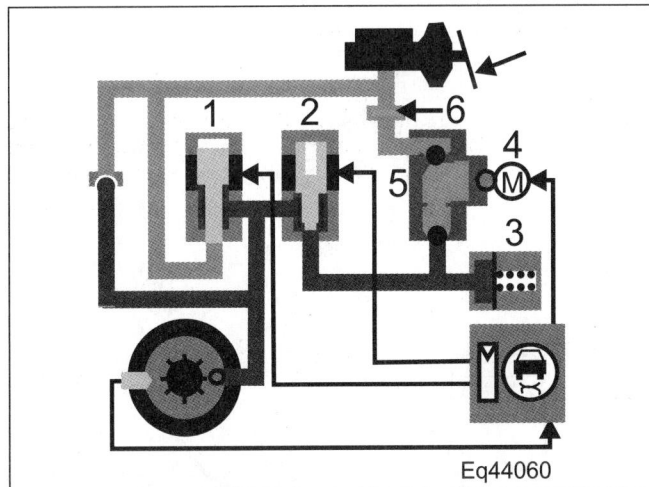

Eq44060

12.13 ABS operation - second phase, pressure reduction

1   Inlet solenoid valve   4   Pump motor
2   Outlet solenoid valve   5   Return pump
3   Pressure accumulator   6   Pulsation damper

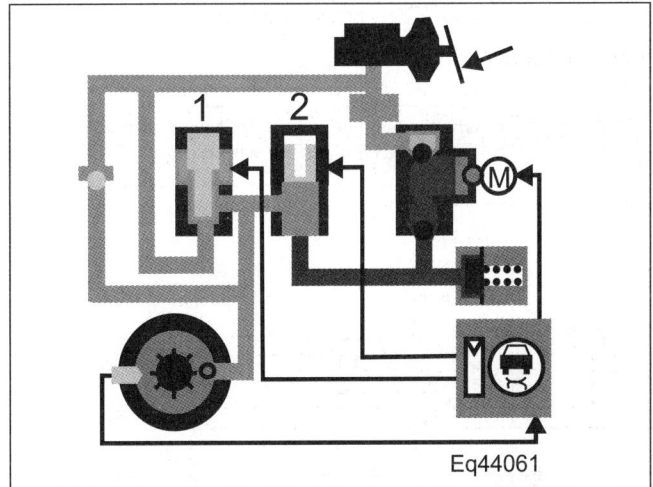

Eq44061

12.14 ABS operation - third phase, pressure build-up

1   Inlet solenoid valve   2   Outlet solenoid valve

Test procedures

4 Wheel speed sensors

Checking the wheel speed sensor (general)

1 Inspect the wheel speed sensor for corrosion or damage and check that the sensor is tightly mounted.

2 Check the toothed sensor ring for damage, eccentricity and for broken or missing teeth.

3 Inspect the wheel speed sensor wiring plug for corrosion and damage. One plug for each sensor.

4 Check that the connector terminal pins are pushed fully home and making good contact with the sensor wiring plug.

5 Check the clearance between the sensor and the toothed sensor ring. The clearance is not normally adjustable but is nominally 0.3 to 0.5 mm. If the clearance is excessive, expect a worn sensor tip or problems with the wheel bearings/hub or sensor ring.

6 When carrying out voltage checks with an oscilloscope or voltmeter, the voltage obtained will be proportional to the speed at which the wheel is rotating. In addition to determining that the wheel speed sensors are actually producing a voltage output, it is essential that the output from the sensors on a particular axle is the same for any given wheel speed.

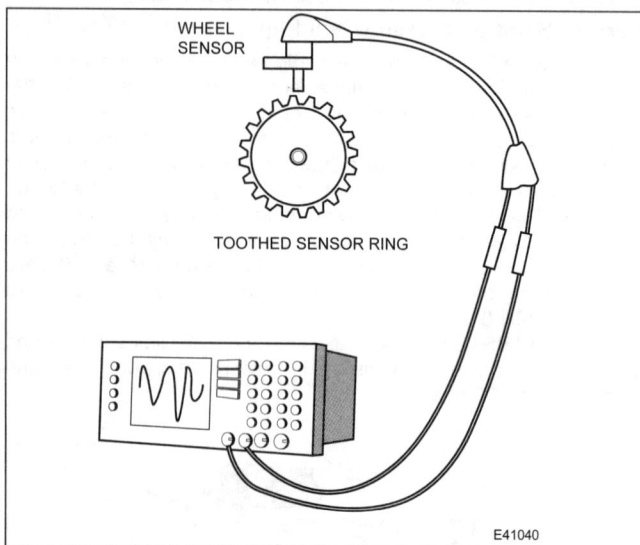

12.15 Checking wheel speed sensor output with an oscilloscope connected to the sensor wiring plug

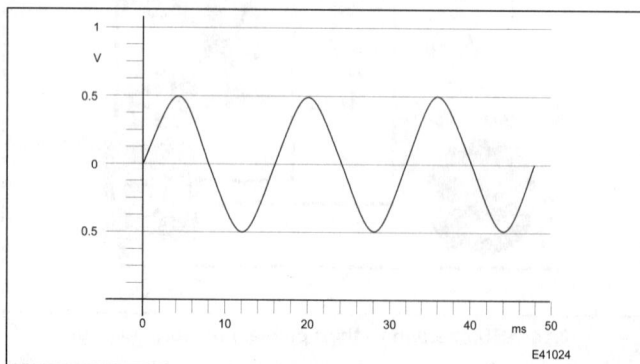

12.16 Typical wheel speed sensor sine wave as displayed on an oscilloscope

Checking wheel speed sensor output with an oscilloscope

Note: Refer to the wiring diagrams for specific ECM pin identification according to model.

7 Switch the ignition off and disconnect the ECM multi-plug or the relevant wheel speed sensor wiring plug.

8 Connect an oscilloscope between the terminal pins for the sensor under test (see illustration 12.15).

9 Select a range to cover 80 Hz on the oscilloscope and a free run time base.

10 Raise the wheel and rotate it by hand at approximately one revolution per second.

11 A sinusoidal wave form should be obtained, with amplitude and duration changing with rotational speed (see illustration 12.16).

12 If there is no signal, or a very weak or intermittent signal at the ECM, repeat the test at the sensor wiring plug. If there is no change in signal status, the sensor is suspect.

13 If the signal is now satisfactory this indicates a fault in the wiring harness which should be checked for continuity.

Checking wheel speed sensor output with an AC voltmeter

Note: Refer to the wiring diagrams for specific ECM pin identification according to model.

14 Switch the ignition off and disconnect the ECM multi-plug or the relevant wheel speed sensor wiring plug.

15 Connect an AC voltmeter between the terminal pins for the sensor under test (see illustration 12.17).

16 Raise the wheel and rotate it by hand at approximately one revolution per second.

17 A voltage of approximately 0.1 to 0.5 volts (AC RMS) should be obtained. If there is no signal, or a very weak or intermittent signal at the ECM, repeat the test at the sensor wiring plug. If there is no change in the signal, the sensor is suspect.

18 If the signal is now satisfactory, this indicates a fault in the wiring harness which should be checked for continuity. Note: This test at least proves that a signal is being generated by the sensor. However,

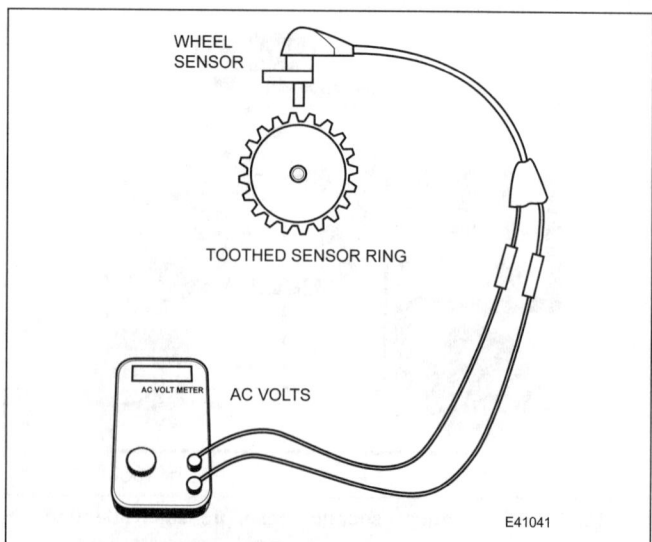

12.17 Checking wheel speed sensor output with a voltmeter connected to the sensor wiring plug

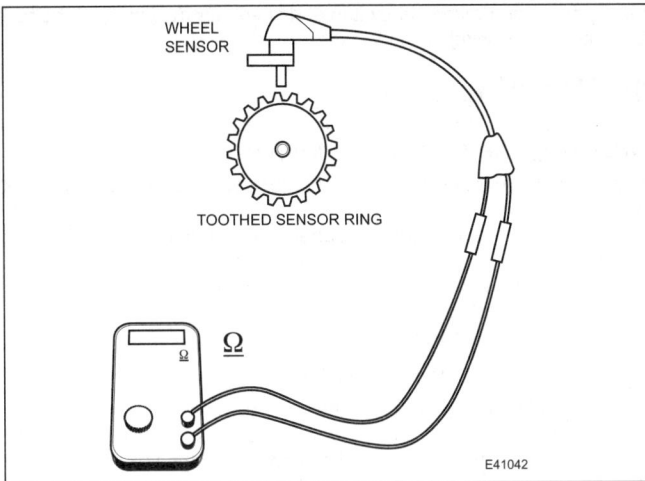

12.18 Checking wheel speed sensor resistance with an ohmmeter connected to the sensor wiring plug

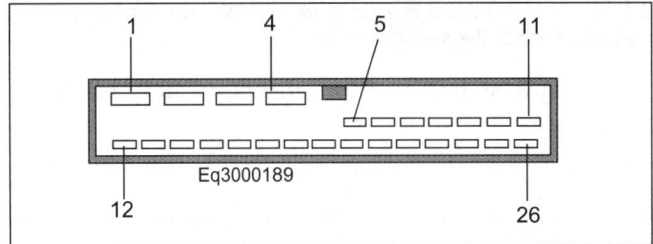

12.19 ECM 26-pin multi-plug terminal identification

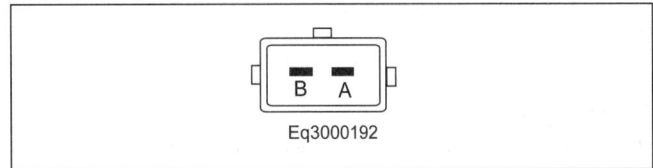

12.20 Hydraulic pump motor multi-plug terminal identification

the voltage produced is an average voltage and does not clearly indicate damage to the sensor ring or that the sinewave is regular in formation.

Checking wheel speed sensor resistance

Note: *Refer to the wiring diagrams for specific ECM pin identification according to model.*

19 Switch the ignition off and disconnect the ECM multi-plug or the relevant wheel speed sensor wiring plug.

20 Connect an ohmmeter between the terminal pins for the sensor under test **(see illustration 12.18)**.

21 The readings obtained should be between 0.9 and 1.7 kohms approximately.

22 If the resistance is excessively high, or open circuit at the ECM, repeat the test at the sensor multi-plug. If there is no change in resistance, the sensor is suspect.

23 If the resistance is now satisfactory, this indicates a fault in the wiring harness which should be checked for continuity. **Note:** *Even if the resistance is within the quoted specifications, this does not prove that the speed sensor can generate an acceptable signal.*

5 System relays

Relay operation and power supply tests

1 The relays are integral with the ECM and can only be checked using suitable ABS diagnostic test equipment.

6 Electronic Control Module

Checking the ECM (general)

1 Inspect the ECM for corrosion or damage and ensure that the unit is securely attached to the hydraulic control unit.

2 Check that the ECM multi-plug terminals are pushed fully home and making good contact with the ECM pins. Faults in any of the above areas are possible reasons for poor performance in the ABS system.

ECM power supply and earth tests

3 Switch the ignition off and disconnect the ECM multi-plug.

4 Attach a negative voltmeter probe to a vehicle earth.

5 Attach a positive voltmeter probe to ECM multi-plug pin 2, and pin 3 in turn **(see illustration 12.19)**. The voltmeter should indicate nbv in each case. If no voltage is found, check the relevant fuse and the supply wiring back to the battery.

6 Switch the ignition on.

7 Attach a negative voltmeter probe to a vehicle earth.

8 Attach a positive voltmeter probe to ECM multi-plug pin 25. The voltmeter should indicate nbv. If no voltage is found, check the relevant fuse and the supply wiring back to the ignition switch.

9 Switch the ignition off.

10 Connect an ohmmeter between a vehicle earth and ECM multi-plug pins 1, 4, and 26 in turn. The ohmmeter should indicate continuity in each case. If not, check the ECM main earth connection and wiring.

7 Solenoid valves

Solenoid valve operation tests

1 The solenoid valves are integral with the ECM and can only be checked using suitable ABS diagnostic test equipment.

8 Hydraulic pump motor

Pump resistance test

1 Switch the ignition off and disconnect the pump motor multi-plug.

2 Connect an ohmmeter between the terminal pins of the pump motor multi-plug **(see illustration 12.20)**.

3 The reading obtained should be between 0.1 and 1.0 Ohms.

Pump operation test

4 Switch the ignition off and disconnect the pump motor multi-plug.

5 Connect the positive terminal of a 12 volt supply to terminal A of the pump motor multi-plug, and the negative terminal to terminal B of the multi-plug.

6 The pump motor should now run.

Warning: The test should be made as quickly as possible to avoid damaging the pump.

7 If the pump does not operate as described, renew the hydraulic control unit.

9 Brake light switch

Checking the brake light switch (general)

1 Check that the brake light switch is correctly and securely mounted and that the plunger moves smoothly with no trace of binding.

2 Check that the wiring multi-plug is pushed fully home and making good contact.

3 Check that no wires have been disconnected.

12

4 Faults in any of the above areas are possible reasons for failure or malfunctioning of the switch.

Brake light switch voltage and continuity tests

Voltage test

5 Switch the ignition off and disconnect the ECM multi-plug from the ECM.
6 Connect a voltmeter between ECM multi-plug brake light switch input pin 20 and an ECM earth pin **(see illustration 12.19)**.
7 Switch the ignition on and depress the brake pedal. The voltmeter should indicate nbv.
8 If no voltage is found, the fuse, the brake light switch and the wiring are suspect.
9 Release the brake pedal. The voltage should drop to zero as the switch opens.

Continuity test

10 Switch the ignition off and disconnect the brake light switch multi-plug.
11 Connect an ohmmeter between the terminal pins of the brake light switch.

12 Operate the brake light switch and check for continuity. If the test fails, renew the switch.

10 Warning lamp

Checking the warning lamp (general)

1 Inspect the warning lamp bulb holder contacts in the instrument panel.
2 Check that the instrument panel multi-plug terminal pins are pushed fully home and making good contact.
3 Check that no wires have been disconnected.
4 Faults in any of the above areas are possible reasons for failure or malfunctioning of the warning lamp.

Warning lamp operation test

5 With the ignition switched off, the warning lamp should remain off.
6 Switch the ignition on and the warning lamp should illuminate then extinguish after a few seconds. The lamp should then remain off.
7 If the warning lamp comes on and remains on at any time during vehicle operation, carry out the previously described test procedures on the system components.

Pin table - typical 26-pin

Note: *Refer to illustration 12.19*

Pin No.	Connection	Test condition	Voltage
1	ECM earth	Ignition on	0.25 volts (max)
2	Supply from battery	Ignition off/on	Nbv
3	Supply from battery	Ignition off/on	Nbv
4	ECM earth	Ignition on	0.25 volts (max)
5	ABS warning lamp	Engine running:	
		Lamp on	1.0 volts (max)
		Lamp off	Nbv
6	-	-	-
7	-	-	-
8	-	-	-
9	-	-	-
10	-	-	-
11	-	-	-
12	Front left wheel speed sensor signal	Roadwheel rotating	0.1 to 0.5 volts AC (approx)
13	Front left wheel speed sensor earth	Roadwheel rotating	0.25 volts (max)
14	Rear left wheel speed sensor signal	Roadwheel rotating	0.1 to 0.5 volts AC (approx)
15	Rear left wheel speed sensor earth	Roadwheel rotating	0.25 volts (max)
16	Rear right wheel speed sensor signal	Roadwheel rotating	0.1 to 0.5 volts AC (approx)
17	Rear right wheel speed sensor earth	Roadwheel rotating	0.25 volts (max)
18	Front right wheel speed sensor signal	Roadwheel rotating	0.1 to 0.5 volts AC (approx)
19	Front right wheel speed sensor earth	Roadwheel rotating	0.25 volts (max)
20	Brake light switch input	Ignition on:	
		Brake pedal released	0 volts
		Brake pedal depressed	Nbv
21	SD connector	-	-
22	-	-	-
23	-	-	-
24	-	-	-
25	Supply from ignition switch	Ignition on	Nbv
26	ECM earth	Ignition on	0.25 volts (max)

Fault codes

1 The Kelsey Hayes 415 system requires the use of a FCR for obtaining fault codes. Flash codes are not available for output from this system.

2 If a FCR is available, it should be connected to the SD serial connector (the Vauxhall/Opel term for SD connector is ALDL) and used in accordance with the maker's instructions.

3 The FCR can be used for the following purposes:
   a) Obtaining fault codes.
   b) Clearing fault codes.
   c) Testing the system actuators (valve relay, pump relay and solenoid valves).

Fault code table

Code	Item	Fault
16	Front left solenoid valve	Defective (signal interruption, short to earth)
17	Front right solenoid valve	Defective (signal interruption, short to earth)
19	Solenoid valve relay	Defective (signal interruption, short to earth, supply voltage too low)
25	Toothed sensor ring	Wrong number of teeth
28	Rear left solenoid valve	Defective (signal interruption, short to earth)
29	Rear right solenoid valve	Defective (signal interruption, short to earth)
32	Wheel speed sensor circuits	Unexpected (implausible) signal
35	Pump motor	Defective (non-operation of pump after actuation of relay)
37	Brake light switch	Defective (incorrect signal)
39	Front left wheel speed sensor	No/poor signal
41	Front left wheel speed sensor	Signal interrupted
42	Front right wheel speed sensor	No/poor signal
43	Front right wheel speed sensor	Signal interrupted
44	Rear left wheel speed sensor	No/poor signal
45	Rear left wheel speed sensor	Signal interrupted
46	Rear right wheel speed sensor	No/poor signal
47	Rear right wheel speed sensor	Signal interrupted
48	Battery voltage	Voltage too low
49	Battery voltage	Voltage too high
52	ABS warning lamp	Circuit interrupted, short to earth
55	ECM	Defective
65	ECM	Not programmed

12

Wiring diagram

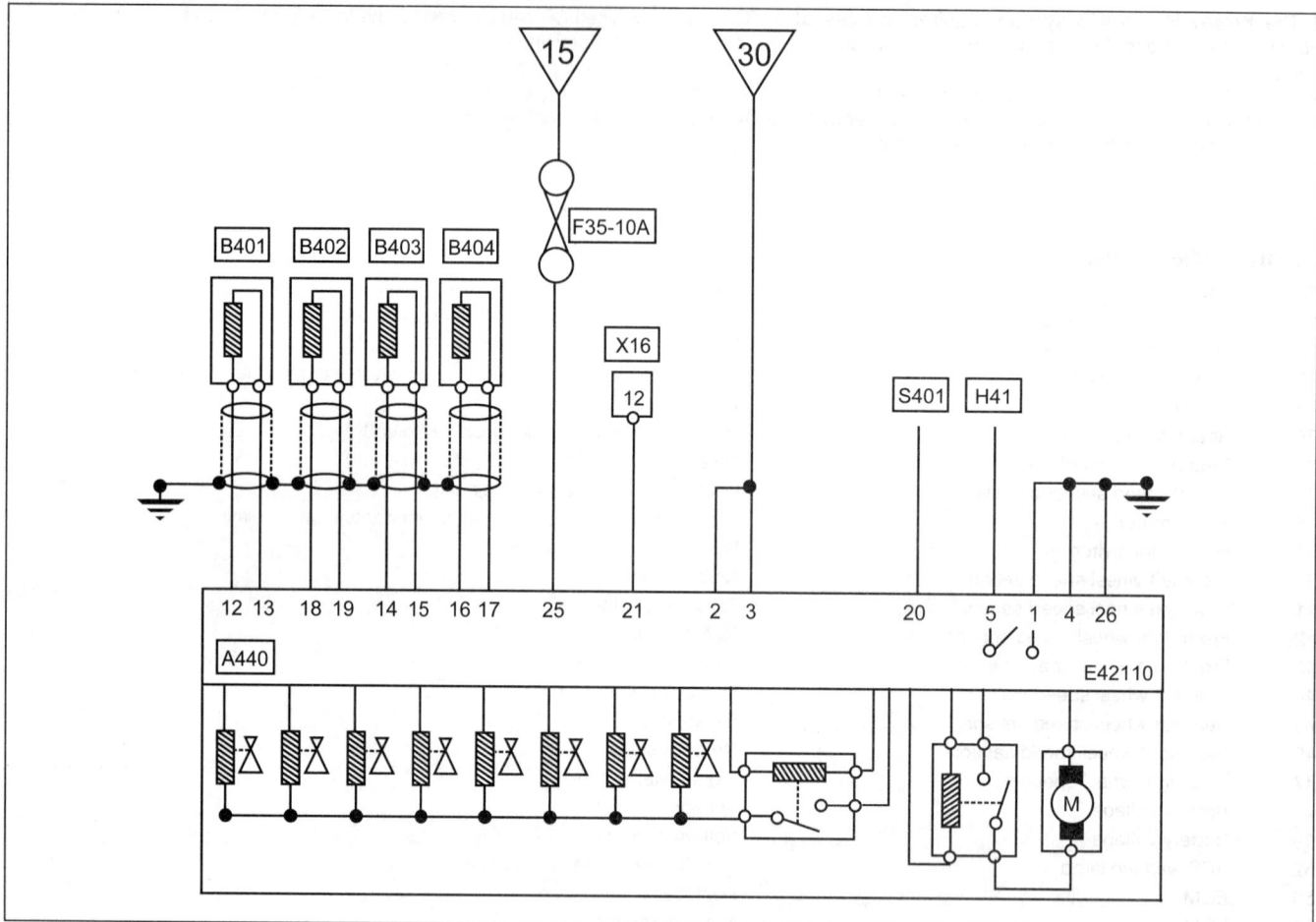

12.21 26-pin wiring diagram, Vauxhall/Opel Vectra-B

Chapter 13
Lucas/Girling 2/2 ABS

Contents

13

Vehicle coverage

Model	Year
Fiat	
Tempra (1.6/1.8)	1988-1993
Tipo (1.4/1.6)	1988-1993
Lancia	
Dedra	1992-1994

Overview of system operation

1 Basic principles and system identification

The Lucas/Girling 2/2 Antilock Brake System has been fitted to a limited number of Fiat/Lancia vehicles since its introduction in the late 1980s. The system is of the additional or 'add-on' type operating in conjunction with the conventional braking system components.

The purpose of the system is to apply the vehicle brakes at maximum efficiency without wheel lock or loss of directional stability. Two inductive sensors (wheel speed sensors) monitor the speed of the front wheels by generating an electrical signal as the wheel is rotated. This information is passed to the ABS Electronic Control Module (ECM) which is then able to determine wheel speed, wheel acceleration and wheel deceleration. The ECM compares the signals received from each wheel and if the onset of lock at any wheel is detected, a signal is sent to the ABS hydraulic control unit which regulates the brake pressure for the relevant wheel(s).

Lucas/Girling 2/2 ABS is comprised of the following main components:
a) ABS-ECM.
b) Hydraulic control unit.
c) Inductive wheel speed sensors and associated sensor rings.
d) Brake light switch.
e) ABS warning lamp.
f) Diagnostic connector.

In addition, the conventional brake system is comprised of the following components:
a) Tandem brake master cylinder.
b) Vacuum servo unit.
c) Brake calipers/wheel cylinders and hydraulic hoses and pipes.
d) Pressure regulating valves.

2 Component description and operation

ABS ECM

General

The Lucas Electronic Control Module (ECM) continually monitors wheel speed from the signals provided by the wheel speed sensors, and brake application from the brake light switch signal. If the ECM detects the incidence of wheel lock on one or more wheels, a signal is sent to the hydraulic control unit to modulate the hydraulic pressure to the brake of the locking wheel(s).

The ECM contains two microprocessors and uses digital technology to complete this function and other functions such as, fault code memory and power modules for valve and relay activity (see illustration 13.1).

The ECM is located remotely within the vehicle, typically in the luggage compartment.

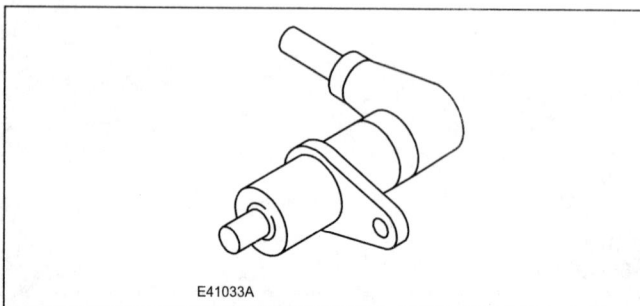

13.2 Typical wheel speed sensor

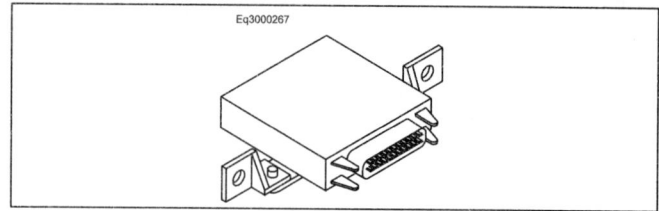

13.1 Lucas Girling 2/2 ECM

Self-test

Lucas/Girling 2/2 ABS is equipped with a self-test capability that begins to examine the ABS signals after a wheel speed of approximately 3 to 4 mph is reached from the front wheels. ABS self-test continually examines the signals from the various components. If self-test determines that faults are not present, ABS is ready for operation. If the ECM detects that a fault is present, all ABS functions are switched off and the ABS warning lamp is turned on.

Self-diagnostics

If the ECM detects a fault during the self-test routine, an internal fault code is stored in the ECM memory. Stored fault codes can be retrieved from the diagnostic connector with the aid of a suitable fault code reader. If the fault clears, the code will remain stored until cleared with an FCR.

Hydraulic control unit

Lucas/Girling 2/2 is a two-channel system with a separate hydraulic circuit for each pair of brakes. The hydraulic control unit consists of an electric motor and radial piston return pump, hydraulic valves actuated by solenoid valves, and two pressure accumulators. The unit controls the hydraulic pressure applied to the two brakes on each hydraulic circuit. The return pump is switched on when the ABS is activated and returns hydraulic fluid, drained off during the pressure reduction phase, back into the brake circuit.

Wheel speed sensors

The rotational speed of the roadwheels and any changes in the rotational speed are recorded by two inductive wheel speed sensors, one located at each front roadwheel (see illustration 13.2).

Each wheel speed sensor assembly comprises a toothed sensor ring which rotates at roadwheel speed, and an adjacent sensor mounted a set distance from the sensor ring (see illustration 13.3).

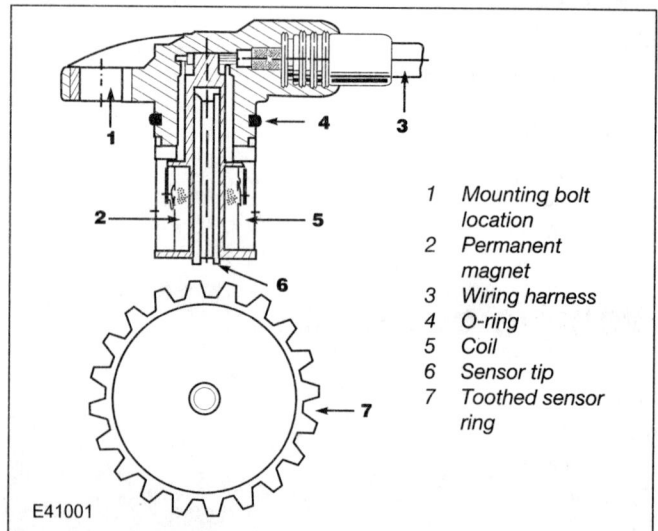

1 Mounting bolt location
2 Permanent magnet
3 Wiring harness
4 O-ring
5 Coil
6 Sensor tip
7 Toothed sensor ring

13.3 Sectional view of a wheel speed sensor

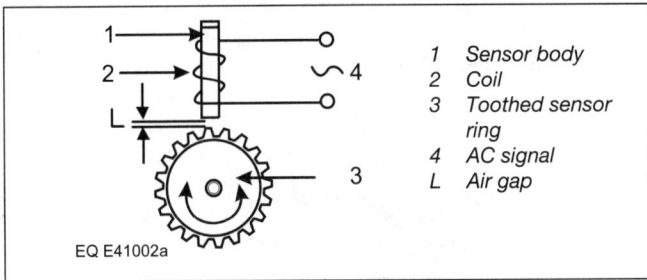

13.4 Wheel speed sensor operation

1 Sensor body
2 Coil
3 Toothed sensor ring
4 AC signal
L Air gap

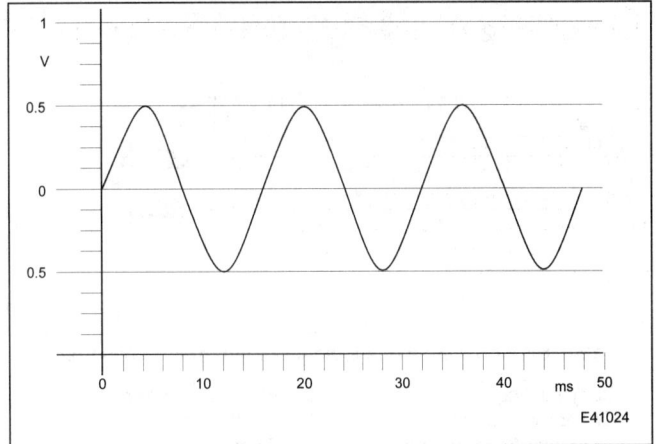

13.5 Wheel speed sensor waveform as viewed on an oscilloscope

The sensors are permanent magnet pulse generator types producing an AC voltage sine wave as the sensor ring teeth pass through the magnetic field of the sensor **(see illustration 13.4)**.

The frequency of the waveform produced by the wheel speed sensor is proportional to the road speed. This AC voltage signal is continually being delivered to the ABS-ECM for processing.

The peak to peak voltage of the speed signal (when viewed upon an oscilloscope) can vary considerably according to wheel speed. An analogue to digital converter (ADC) in the ECM transforms the AC pulse into a digital signal **(see illustration 13.5)**.

ABS electrical wiring harness and relays

An integrated 25-pin main wiring harness is used to connect power and earth to various electrical components. This enables sensor signals to reach the ECM and the ECM, in turn, to send command signals to the relays, return pump and hydraulic control unit. The hydraulic control unit harness connector, valve relay and pump relay are typically located under a cover on the hydraulic control unit.

Pump relay

The pump relay is supplied with a permanent voltage from the battery. The pump relay is controlled by the ECM applying an earth which activates the relay allowing battery voltage to be applied to the pump. The hydraulic control unit provides the earth for the pump motor.

Valve relay

The solenoid valve relay is supplied with a permanent voltage from the battery. The valve relay is controlled by the ECM applying an earth which activates the relay allowing battery voltage to be applied to the solenoid valves, the ECM controlling the earth.

Brake light switch

The brake light switch comprises a switch body and contact pin and is located above the brake pedal **(see illustration 13.6)**.

When the brake pedal is depressed, closing the brake light switch, a signal is sent to the ECM indicating that the brakes are being applied. Once this signal is received, the ECM will begin monitoring the wheel speed via the wheel speed sensors and activate the ABS if necessary.

ABS warning lamp

After the ignition is switched on, the ABS warning lamp on the instrument panel is illuminated, and remains illuminated until a vehicle speed of approximately 3 to 4 mph is reached **(see illustration 13.7)**. During this time, the ABS-ECM implements a self-test cycle whereby ABS operation is monitored. If satisfactory operation of the system is detected by the ECM, the warning light is extinguished.

During vehicle operation, the ABS-ECM implements a further series of self-test cycles whereby ABS operation and wheel speed sensor signals are continually monitored. If a fault is detected, the ECM will illuminate the warning lamp on the instrument panel, and disable ABS operation, however the conventional braking system will continue to operate as normal. The warning lamp will remain illuminated until the fault is no longer present

When the ABS-ECM detects a fault, an internal fault code is logged by the ECM. If the fault no longer exists after the next system start (ignition on/off) the ABS warning lamp is extinguished after the self-test cycle, however the fault code remains stored in the ECM memory.

Tandem master cylinder

Typically, the tandem master cylinder comprises a body casting incorporating primary and secondary pressure chambers, primary

13

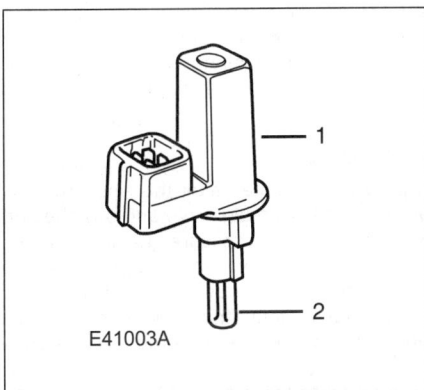

13.6 Typical brake light switch body (1) and contact pin (2)

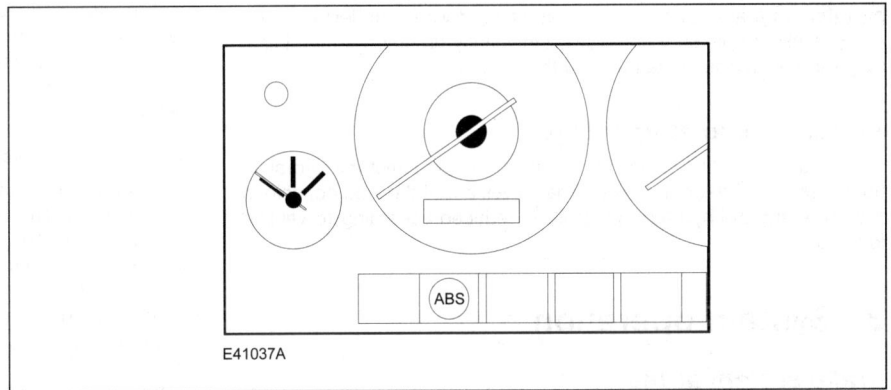

13.7 ABS warning lamp

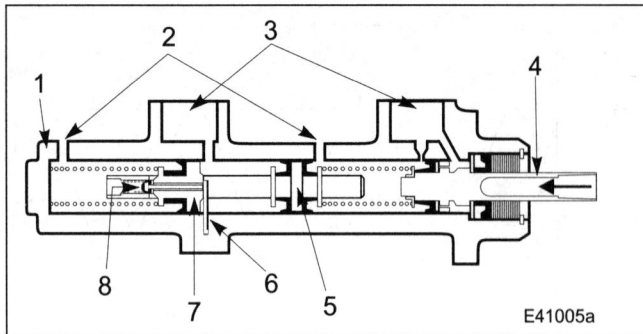

13.8 Sectional view of a typical tandem master cylinder

1 Body casting
2 Outlet connections
3 Fluid inlet from reservoir
4 Pushrod/primary piston
5 Intermediate piston
6 Slotted pin
7 Floating piston
8 Central valve

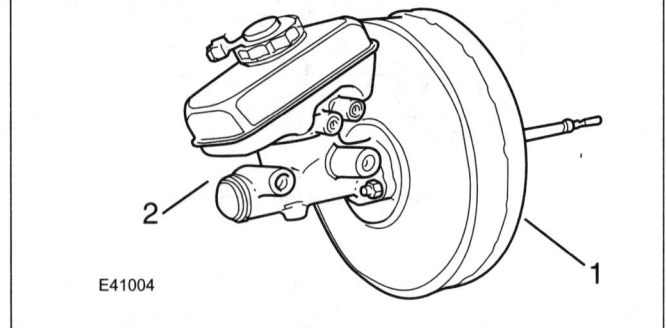

13.9 Vacuum servo unit (1) and tandem master cylinder (2)

piston, intermediate piston, floating piston, slotted pin and central valve. The cylinder operates as a conventional master cylinder using vacuum assistance from the vacuum servo unit **(see illustration 13.8)**.

When the brake system is at rest, the central valve in the floating piston rests against the slotted pin. In this condition the central valve is open and brake fluid can discharge out of the pressure chamber back into the brake fluid reservoir. When the brake pedal is depressed, the build-up of hydraulic pressure in the primary pressure chamber acts on the intermediate piston and floating piston, moving them down the cylinder bore. The floating piston contacts the seal on the central valve, closing the connection between the intermediate and secondary pressure chambers. Brake hydraulic pressure can now also increase in the secondary pressure chamber.

Vacuum servo unit

The vacuum servo unit is located between the brake pedal and tandem master cylinder. When the brake pedal is depressed, the servo unit increases the force applied by the pedal, reducing the effort required to operate the brakes **(see illustration 13.9)**.

The unit operates by means of engine inlet manifold vacuum applied to a diaphragm contained within the unit casing. A pushrod connected to the centre of the diaphragm acts directly on the primary piston in the master cylinder.

When the brake pedal is released, vacuum is applied to both sides of the diaphragm. When the pedal is depressed, one side of the diaphragm is opened to atmosphere and the vacuum acting on the other side deflects the diaphragm which in turn operates the master cylinder primary piston. The resulting force applied to the master cylinder piston is therefore significantly greater than the initial force applied to the brake pedal by the driver.

Pressure regulating valves

Pressure regulating valves are incorporated to restrict the hydraulic fluid pressure to the rear brakes. The valves are of the load conscious type whereby the hydraulic pressure is reduced according to vehicle loading.

3 System operation

Brake system at rest

When the ABS system is at rest all the brake components are inoperative. Pressure is non-existent in the hydraulic pipes

between the tandem master cylinder and the brake calipers or wheel cylinders.

Brake system in operation

Phase one, normal braking

When the brake pedal is activated, the pedal force is applied to the tandem master cylinder by the vacuum servo unit pushrod. The servo unit pushrod acts directly on the pressure piston in the master cylinder which pressurises the hydraulic fluid in the brake pipes to the hydraulic control unit. The solenoid valves in the hydraulic control unit are in the 'at rest' position (ie closed) and there is an unrestricted fluid passage between the master cylinder and the relevant brake caliper or wheel cylinder. Hydraulic pressure can be transmitted directly to each brake caliper or wheel cylinder, thus operating the brakes.

The ABS-ECM continually monitors wheel speed from the signals provided by the wheel speed sensors. If the ECM detects the incidence of wheel lock on one or both front wheels, ABS is automatically initiated.

Phase two, pressure reduction

The ECM opens the solenoid valve on the relevant hydraulic circuit and a return passage from the brake caliper or wheel cylinder to the pressure accumulator is opened. Hydraulic fluid now flows from the brake caliper or wheel cylinder, through a restrictor and into the pressure accumulator. This results in a pressure reduction, thus releasing the brake.

As soon as the solenoid valve is opened, the ECM actuates the electric motor to operate the return pump. The hydraulic fluid is then pumped back into the pressure side of the master cylinder. This process creates a pulsation which can be felt in the brake pedal action, but which is softened by the pulsation damper.

Phase three, pressure build-up

The pressure build-up phase is instigated after the wheel rotation has stabilised. The ECM closes the solenoid valve and switches off the return pump allowing hydraulic pressure to once again increase. Hydraulic fluid now flows to the brake caliper or wheel cylinder, leading to an increase of pressure at the brake. Normal braking now continues until the wheel has again decelerated to a sufficient degree where pressure reduction is once more required.

The whole ABS control cycle takes place 4 to 10 times per second for each affected wheel and this ensures maximum braking effect and control during ABS operation.

Test procedures

4 Wheel speed sensors

Checking the wheel speed sensor (general)

1 Inspect the wheel speed sensor for corrosion or damage and check that the sensor is tightly mounted.
2 Check the toothed sensor ring for damage, eccentricity and for broken or missing teeth.
3 Inspect the wheel speed sensor wiring plug for corrosion and damage. One plug for each sensor.
4 Check that the connector terminal pins are pushed fully home and making good contact with the sensor wiring plug.
5 Check the clearance between the sensor and the toothed sensor ring. The clearance is not normally adjustable but is nominally 0.6 to 1.3 mm. If the clearance is excessive, expect a worn sensor tip or problems with the wheel bearings/hub or sensor ring.

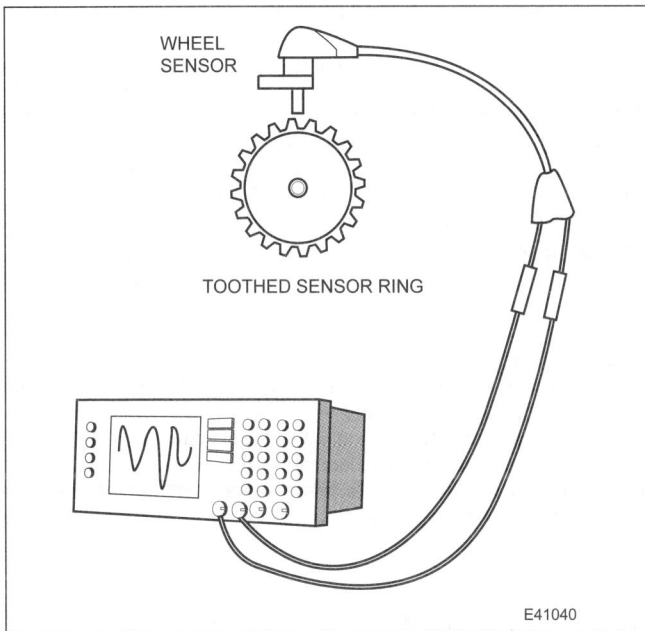

13.10 Checking wheel speed sensor output with an oscilloscope connected to the sensor wiring plug

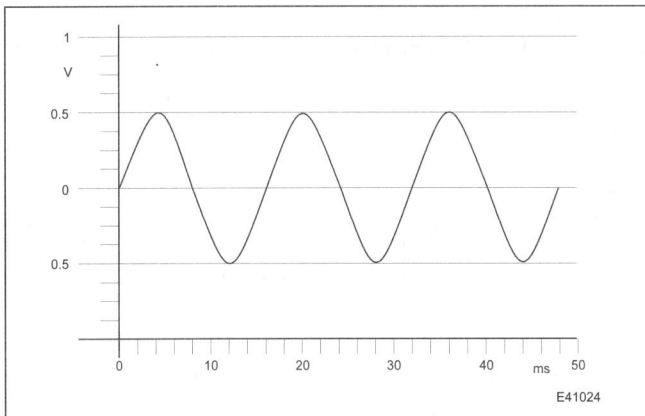

13.11 Typical wheel speed sensor sine wave as displayed on an oscilloscope

6 When carrying out voltage checks with an oscilloscope or voltmeter, the voltage obtained will be proportional to the speed at which the wheel is rotating. In addition to determining that the wheel speed sensors are actually producing a voltage output, it is essential that the output from the sensors on a particular axle is the same for any given wheel speed.

Checking wheel speed sensor output with an oscilloscope

Note: *Refer to the wiring diagrams for specific ECM pin identification according to model.*
7 Switch the ignition off and disconnect the ECM multi-plug or the relevant wheel speed sensor wiring plug.
8 Connect an oscilloscope between the terminal pins for the sensor under test **(see illustration 13.10)**.
9 Select a range to cover 80 Hz on the oscilloscope and a free run time base.
10 Raise the wheel and rotate it by hand at approximately one revolution per second.
11 A sinusoidal wave form should be obtained, with amplitude and duration changing with rotational speed **(see illustration 13.11)**.
12 If there is no signal, or a very weak or intermittent signal at the ECM, repeat the test at the sensor wiring plug. If there is no change in signal status, the sensor is suspect.
13 If the signal is now satisfactory this indicates a fault in the wiring harness which should be checked for continuity.

Checking wheel speed sensor output with an AC voltmeter

Note: *Refer to the wiring diagrams for specific ECM pin identification according to model.*
14 Switch the ignition off and disconnect the ECM multi-plug or the relevant wheel speed sensor wiring plug.
15 Connect an AC voltmeter between the terminal pins for the sensor under test **(see illustration 13.12)**.

13

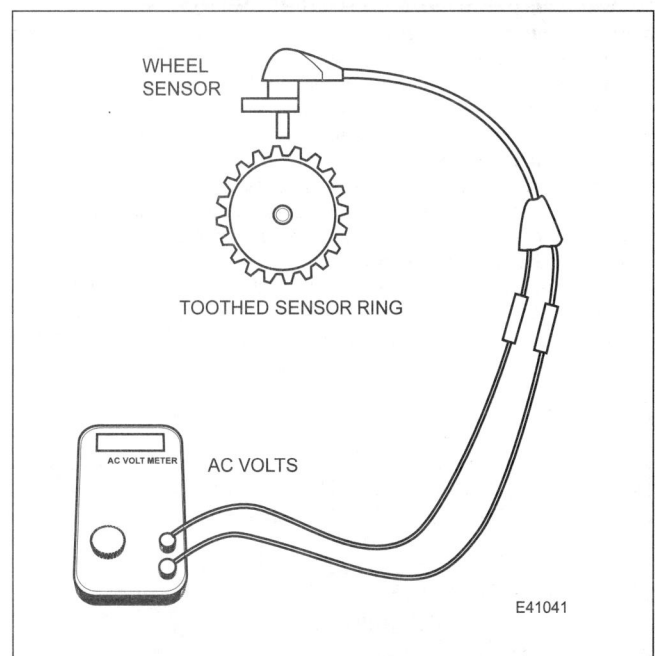

13.12 Checking wheel speed sensor output with a voltmeter connected to the sensor wiring plug

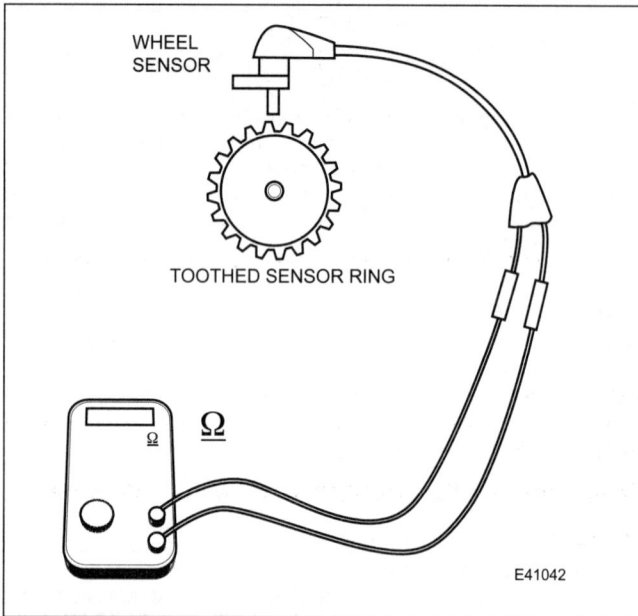

13.13 Checking wheel speed sensor resistance with an ohmmeter connected to the sensor wiring plug

16 Raise the wheel and rotate it by hand at approximately one revolution per second.

17 A voltage of approximately 0.4 to 1.5 volts (AC RMS) should be obtained. If there is no signal, or a very weak or intermittent signal at the ECM, repeat the test at the sensor wiring plug. If there is no change in the signal, the sensor is suspect.

18 If the signal is now satisfactory, this indicates a fault in the wiring harness which should be checked for continuity. **Note:** *This test at least proves that a signal is being generated by the sensor. However, the voltage produced is an average voltage and does not clearly indicate damage to the sensor ring or that the sinewave is regular in formation.*

Checking wheel speed sensor resistance

Note: *Refer to the wiring diagrams for specific ECM pin identification according to model.*

19 Switch the ignition off and disconnect the ECM multi-plug or the relevant wheel speed sensor wiring plug.

20 Connect an ohmmeter between the terminal pins for the sensor under test **(see illustration 13.13)**.

21 The readings obtained should be between 1.0 and 2.0 kohms.

22 If the resistance is excessively high, or open circuit at the ECM, repeat the test at the sensor multi-plug. If there is no change in resistance, the sensor is suspect.

23 If the resistance is now satisfactory, this indicates a fault in the wiring harness which should be checked for continuity. **Note:** *Even if the resistance is within the quoted specifications, this does not prove that the speed sensor can generate an acceptable signal.*

5 System relays

Checking relays (general)

1 The following tests are applicable to the ABS main (or solenoid valve) relay and hydraulic pump relay.

2 Before carrying out specific tests on the relays or power supplies, inspect the relays and relay base for corrosion and damage.

3 Check that the relays are fully pushed home and making good contact with the base.

4 Faults in any of the above areas are possible reasons for failure or malfunctioning of the relay.

Relay operation tests

Main (solenoid valve) relay

5 Switch the ignition off, then remove the relay from the relay base.

6 Connect an ohmmeter between the relay output terminal 30 and earth terminal 87a **(see illustration 13.14)**. The ohmmeter should indicate continuity.

7 Connect an ohmmeter between the relay main voltage supply terminal 87 and the output terminal 30. The ohmmeter should indicate an open circuit.

8 Attach a wire between the relay driver supply terminal 85 and a 12 volt supply. Attach a second wire between the relay driver earth terminal 86 and earth. There should be an unmistakable click from the relay and the ohmmeter should indicate continuity.

9 Renew the relay if it fails these tests. If the relay cover is transparent, inspect the relay for burns. A faulty relay usually has a burnt appearance.

Hydraulic pump relay

10 Switch the ignition off, then remove the relay from the relay base.

11 Connect an ohmmeter between the relay main voltage supply terminal 87 and the output terminal 30 **(see illustration 13.15)**. The ohmmeter should indicate an open circuit.

12 Attach a wire between the relay driver supply terminal 86 and a 12 volt supply. Attach a second wire between the relay driver earth terminal 85 and earth. There should be an unmistakable click from the relay and the ohmmeter should indicate continuity.

13 Renew the relay if it fails these tests. If the relay cover is transparent, inspect the relay for burns. A faulty relay usually has a burnt appearance.

6 Electronic Control Module

Checking the ECM (general)

1 Inspect the ECM for corrosion or damage and ensure that the unit is mounted securely.

2 Check that the ECM multi-plug terminals are pushed fully home and making good contact with the ECM pins. Faults in any of the above areas are possible reasons for poor performance in the ABS system.

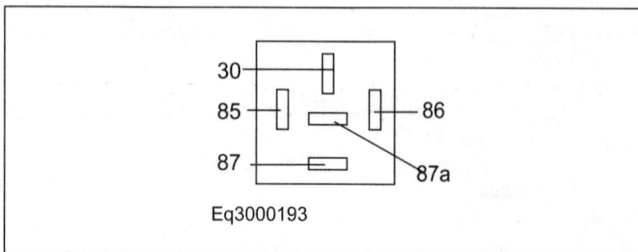

Eq3000193

13.14 Main (solenoid valve) relay terminals

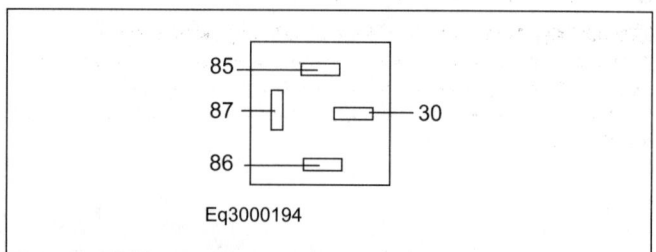

Eq3000194

13.15 Hydraulic pump relay terminals

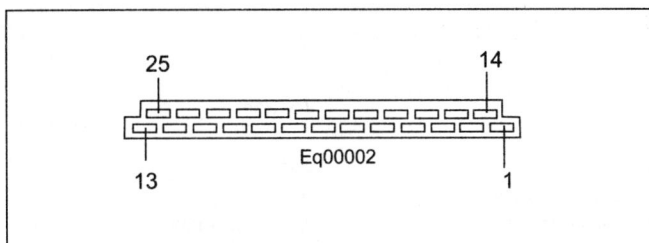

13.16 ECM 25-pin multi-plug terminal identification

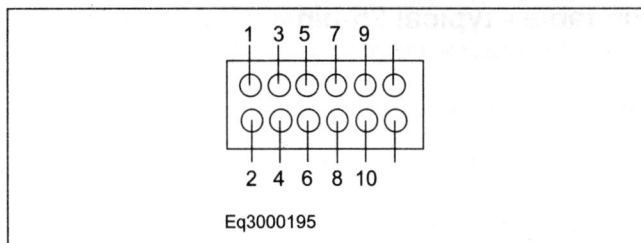

13.17 Hydraulic control unit multi-plug terminal identification

ECM power supply and earth tests

3 Switch the ignition off and disconnect the ECM multi-plug from the ECM.

4 Switch the ignition on.

5 Attach a negative voltmeter probe to an ECM multi-plug earth pin and attach a positive voltmeter probe to ECM multi-plug pin 10 **(see illustration 13.16)**. The voltmeter should indicate nbv. If no voltage is found, check the relevant fuse, and the supply wiring back to the ignition switch.

6 Switch the ignition off.

7 Connect an ohmmeter between a vehicle earth and ECM multi-plug earth, pin 8. The ohmmeter should indicate continuity. If not, check the ECM main earth connection and wiring.

7 Solenoid valves

Solenoid valve resistance tests

1 Switch the ignition off and disconnect the ECM multi-plug from the ECM.

2 Using an ohmmeter, measure the resistance between pin 23 in the multi-plug, and pins 24 and 25 in turn **(see illustration 13.16)**. The resistance should be in the order of 3 to 5 ohms for each solenoid valve. If the resistance is not as specified, the hydraulic control unit is suspect.

8 Hydraulic pump motor

Pump motor operation test

1 Switch the ignition off and disconnect the hydraulic control unit multi-plug.

2 Connect a 12 volt supply to terminals 8 and 4 in the hydraulic control unit **(see illustration 13.17)**. The pump motor should now run.

⚠️ **Warning: The test should be made as quickly as possible to avoid damaging the pump.**

3 If the pump does not operate as described, the pump motor is suspect.

9 Brake light switch

Checking the brake light switch (general)

1 Check that the brake light switch is correctly and securely mounted and that the plunger moves smoothly with no trace of binding.

2 Check that the wiring multi-plug is pushed fully home and making good contact.

3 Check that no wires have been disconnected.

4 Faults in any of the above areas are possible reasons for failure or malfunctioning of the switch.

Brake light switch voltage and continuity tests

Voltage test

5 Switch the ignition off and disconnect the ECM multi-plug from the ECM.

6 Connect a voltmeter between ECM multi-plug pin 2 and a vehicle earth **(see illustration 13.16)**.

7 Depress the brake pedal. The voltmeter should indicate nbv.

8 If no voltage is found, the fuse, the brake light switch and the wiring are suspect.

9 Release the brake pedal. The voltage should drop to zero as the switch opens.

Continuity test

10 Switch the ignition off and disconnect the brake light switch multi-plug.

11 Connect an ohmmeter between the terminal pins of the brake light switch.

12 Operate the brake light switch and check for continuity. If the test fails, renew the switch.

10 Warning lamp

Checking the warning lamp (general)

1 Inspect the warning lamp bulb holder contacts in the instrument panel.

2 Check that the instrument panel multi-plug terminal pins are pushed fully home and making good contact.

3 Check that no wires have been disconnected.

4 Faults in any of the above areas are possible reasons for failure or malfunctioning of the warning lamp.

Warning lamp operation test

5 With the ignition switched off, the warning lamp should remain off.

6 Switch the ignition on and the warning lamp should illuminate. Start the engine and drive the vehicle. The lamp should extinguish when a speed of approximately 3 to 4 mph has been attained. The lamp should then remain off.

7 If the warning lamp comes on and remains on at any time during vehicle operation, carry out the previously described test procedures on the system components.

13

Pin table - typical 25-pin

Note: *Refer to illustration 13.16*

Pin No.	Connection	Test condition	Voltage
1	SD connector	-	-
2	Brake light switch input	Brake pedal released	0 volts
		Brake pedal depressed	Nbv
3	Front left wheel speed sensor earth	Roadwheel rotating	0.25 volts (max)
4	Front right wheel speed sensor earth	Roadwheel rotating	0.25 volts (max)
5	-	-	-
6	-	-	-
7	SD connector	-	-
8	ECM earth	Ignition on	0.25 volts (max)
9	-	-	-
10	Supply from ignition switch	Ignition on	Nbv
11	-	-	-
12	-	-	-
13	-	-	-
14	SD connector	-	-
15	Front left wheel speed sensor signal	Roadwheel rotating	0.4 to 1.5 volts AC (approx)
16	Front right wheel speed sensor signal	Roadwheel rotating	0.4 to 1.5 volts AC (approx)
17	-	-	-
18	-	-	-
19	-	-	-
20	ABS warning lamp	Engine running:	
		Lamp on	1.0 volt (max)
		Lamp off	Nbv
21	ABS pump signal output from relay (in Hydraulic unit)	Ignition on/active	Nbv
22	ABS main relay driver (in Hydraulic unit)	Ignition on	Nbv
		Actuated	1.0 volt (max)
23	ABS pump relay driver (in Hydraulic unit)	Ignition on	Nbv
		Actuated	1.0 volt (max)
24	Right solenoid valve	Ignition on/inactive	Nbv
		Actuated	1.0 volt (max)
25	Left solenoid valve	Ignition on/inactive	Nbv
		Actuated	1.0 volt (max)

Fault codes

1 The Lucas/Girling 2/2 system requires the use of a FCR for obtaining fault codes. Flash codes are not available for output from this system.

2 If a FCR is available, it should be connected to the SD serial connector and used in accordance with the maker's instructions.

3 The FCR can be used for the following purposes:

a) *Obtaining fault codes.*

b) *Clearing fault codes.*

c) *Obtaining datastream information.*

d) *Testing the system actuators (solenoid valve relay, pump relay and solenoid valves).*

4 Internal fault codes are used by the ECM to designate faults in the system components and circuits. A proprietary fault code reader (FCR) or system tester is required to interrogate the system. No actual fault code numbers are available although the components or circuits checked by the ECM typically include the following.

Fault code table

Code	Item	Fault
-	Supply voltage	Too high/low
-	Relays	Defective (signal interruption, short to earth, supply voltage)
-	Pump motor	Defective (poor connections)
-	Front left wheel speed sensor	Incorrect resistance/signal interruption
-	Front right wheel speed sensor	Incorrect resistance/signal interruption
-	ECM	Defective

Wiring diagram

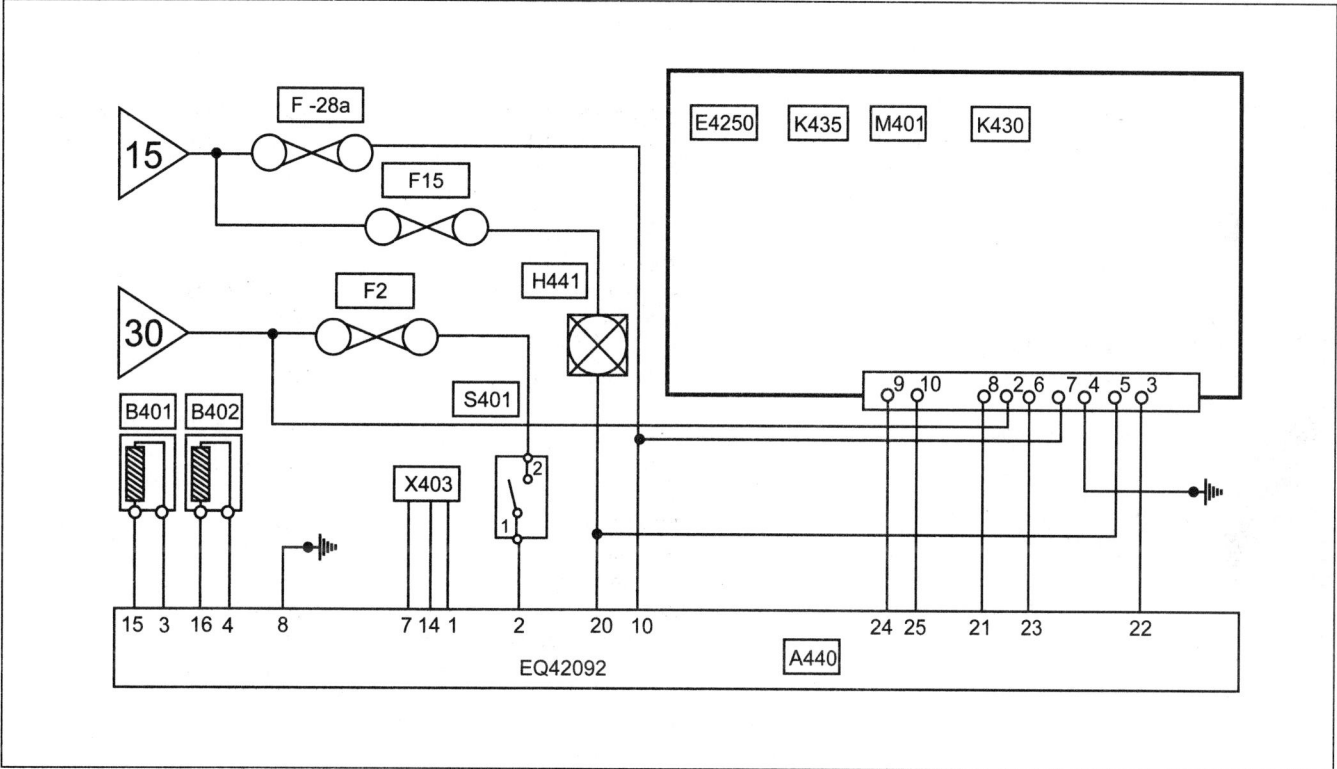

13.18 25-pin Lucas 2/2 wiring diagram, Fiat Tempra

Chapter 14

Lucas 4/4F ABS

Contents

Vehicle coverage

Model	Year
Vauxhall/Opel Corsa-B	1993-1996
Vauxhall/Opel Tigra	1994-1996

Overview of system operation

1 Basic principles and system identification

The Lucas 4/4F Antilock Brake System has been fitted to a limited number of passenger vehicles since its introduction in the mid 1990s. The system is of the additional or 'add-on' type operating in conjunction with the conventional braking system components.

The purpose of the system is to apply the vehicle brakes at maximum efficiency without wheel lock or loss of directional stability. Inductive sensors (wheel speed sensors) monitor the speed of the wheels by generating an electrical signal as the wheel is rotated. This information is passed to the ABS Electronic Control Module (ECM) which is then able to determine wheel speed, wheel acceleration and wheel deceleration. The ECM compares the signals received from each wheel and if the onset of lock at any wheel is detected, a signal is sent to the ABS hydraulic control unit which regulates the brake pressure for the relevant wheel(s).

Lucas 4/4F ABS is comprised of the following main components:
a) ABS-ECM.
b) Hydraulic control unit.
c) Inductive wheel speed sensors and associated sensor rings.
d) Brake light switch.
e) ABS warning lamp.
f) Diagnostic connector.

In addition, the conventional brake system is comprised of the following components:
a) Tandem brake master cylinder.
b) Vacuum servo unit.
c) Brake calipers/wheel cylinders and hydraulic hoses and pipes.
d) Pressure regulating valves.

2 Component description and operation

ABS ECM

General

The Lucas Electronic Control Module (ECM) continually monitors wheel speed from the signals provided by the wheel speed sensors, and brake application from the brake light switch signal. If the ECM detects the incidence of wheel lock on one or more wheels, a signal is sent to the hydraulic control unit to modulate the hydraulic pressure to the brake of the locking wheel(s).

The ECM contains two microprocessors and uses digital technology to complete this function and other functions such as, fault code memory and power modules for valve and relay activity **(see illustration 14.1)**.

The ECM is located remotely within the passenger compartment, typically behind the trim panelling in the lower left-hand side footwell **(see illustration 14.2)**.

Self-test

Lucas 4/4F ABS is equipped with a self-test capability that begins to examine the ABS signals after a wheel speed of approximately 5 mph is reached from all wheels. ABS self-test continually examines the signals from the various components. If self-test determines that faults are not present, ABS is ready for operation. If the ECM detects that a fault is present, all ABS functions are switched off and the ABS warning lamp is turned on.

Self-diagnostics

If the ECM detects a fault during the self-test routine, an internal fault code is stored in the ECM memory. Stored fault codes can be retrieved from the diagnostic connector with the aid of a suitable fault code reader. If the fault clears, the code will remain stored until cleared with an FCR.

Hydraulic control unit

Lucas 4/4F is a four-channel system with a separate hydraulic circuit for each brake. The hydraulic control unit consists of an electric motor and radial piston return pump, 4 hydraulic valves actuated by 4

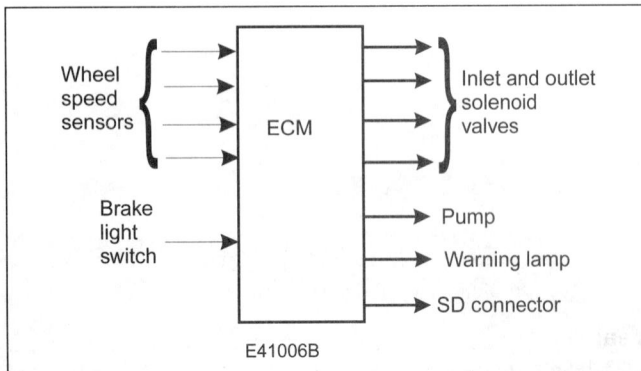

14.1 ECM sensor inputs and control signal outputs

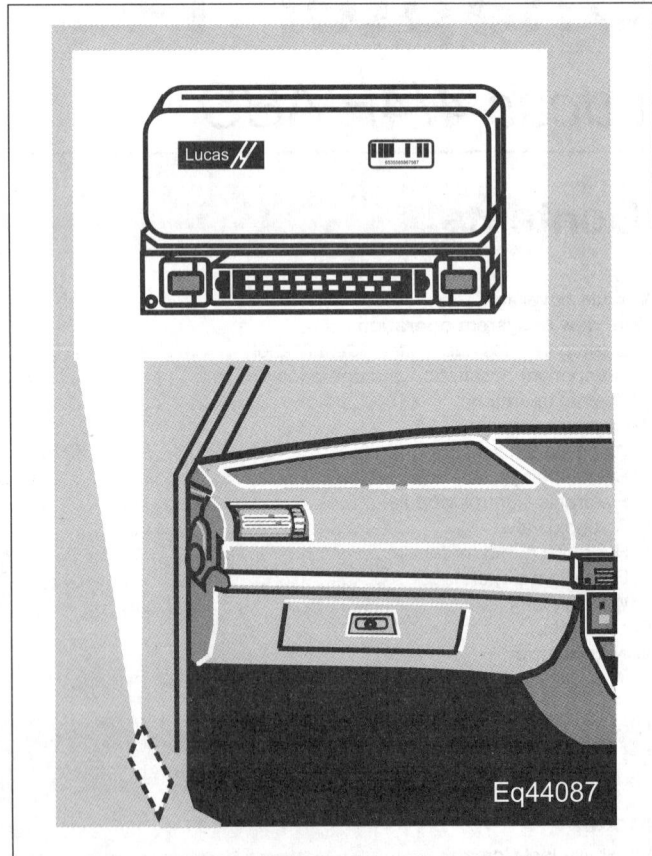

14.2 ECM location

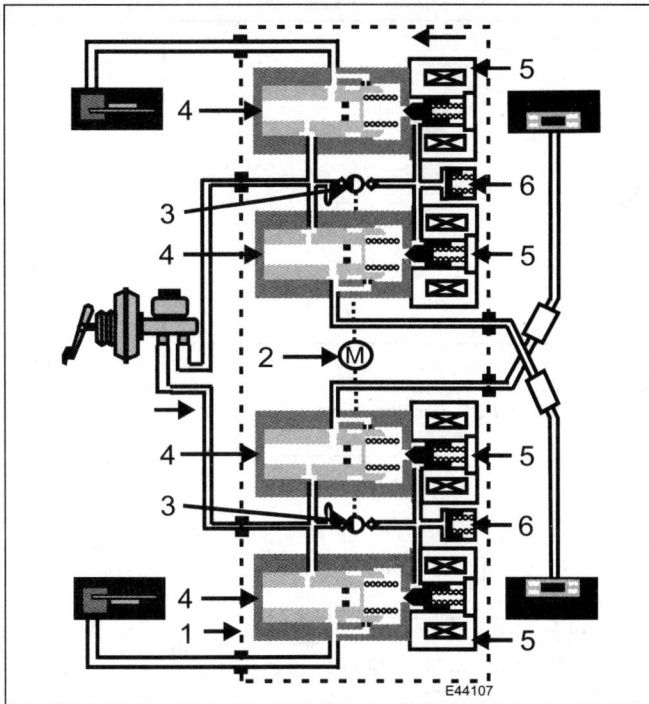

14.3 Hydraulic control circuit layout

1 Hydraulic control unit   4 Hydraulic valves
2 Electric motor   5 Solenoid valves
3 Return pump   6 Pressure accumulators

solenoid valves, and 2 pressure accumulators (see illustration 14.3). The unit controls the hydraulic pressure applied to the brake for each individual front wheel and each individual rear wheel. The return pump is switched on when the ABS is activated and returns hydraulic fluid, drained off during the pressure reduction phase, back into the brake circuit.

Wheel speed sensors

The rotational speed of the roadwheels and any changes in the rotational speed are recorded by inductive wheel speed sensors, one located at each roadwheel (see illustration 14.4).

Each wheel speed sensor assembly comprises a toothed sensor ring which rotates at roadwheel speed, and an adjacent sensor mounted a set distance from the sensor ring (see illustration 14.5).

The sensors are permanent magnet pulse generator types producing an AC voltage sine wave as the sensor ring teeth pass through the magnetic field of the sensor (see illustration 14.6).

The frequency of the waveform produced by the wheel speed sensor is proportional to the road speed. This AC voltage signal is continually being delivered to the ABS-ECM for processing.

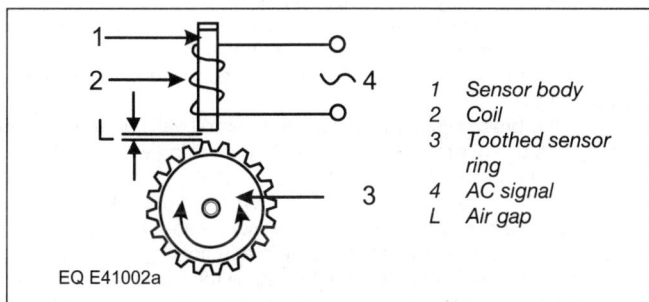

14.6 Wheel speed sensor operation

1 Sensor body
2 Coil
3 Toothed sensor ring
4 AC signal
L Air gap

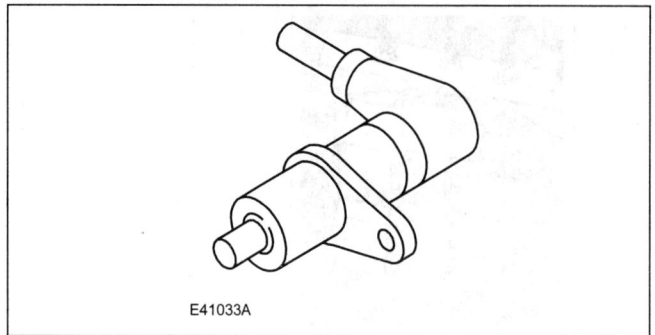

14.4 Typical wheel speed sensor

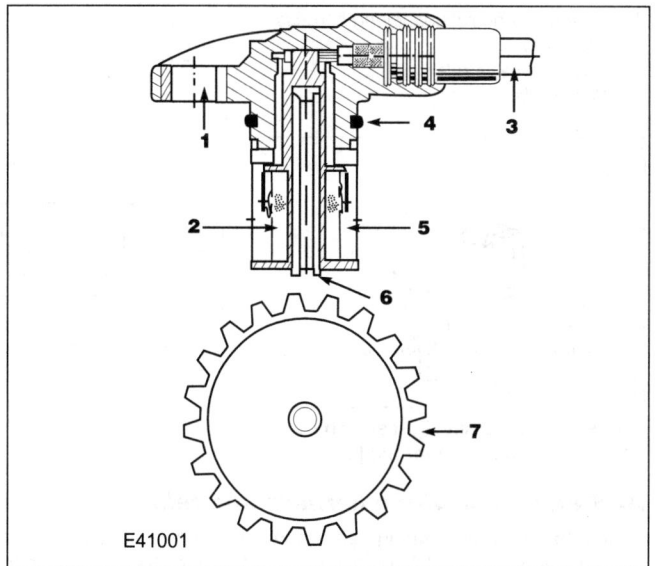

14.5 Sectional view of a wheel speed sensor

1 Mounting bolt location   5 Coil
2 Permanent magnet   6 Sensor tip
3 Wiring harness   7 Toothed sensor ring
4 O-ring

The peak to peak voltage of the speed signal (when viewed upon an oscilloscope) can vary considerably according to wheel speed. An analogue to digital converter (ADC) in the ECM transforms the AC pulse into a digital signal (see illustration 14.7).

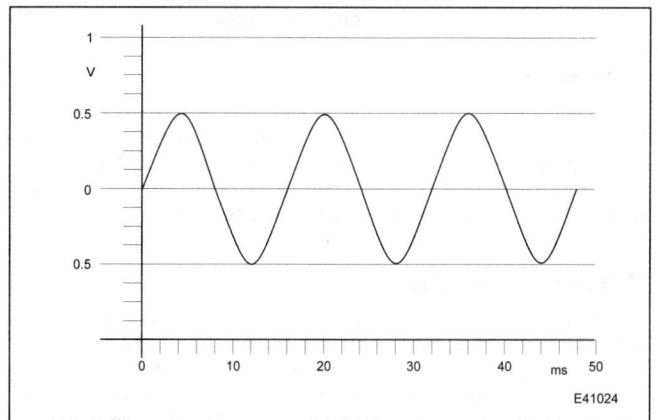

14.7 Wheel speed sensor waveform as viewed on an oscilloscope

14

14.8 Relays and wiring harness connector locations on the hydraulic control unit

1 Hydraulic control unit wiring harness connector
2 Solenoid valve relay
3 Pump relay

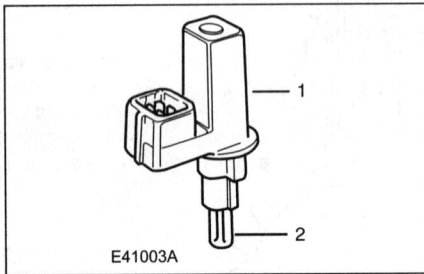

14.9 Typical brake light switch body (1) and contact pin (2)

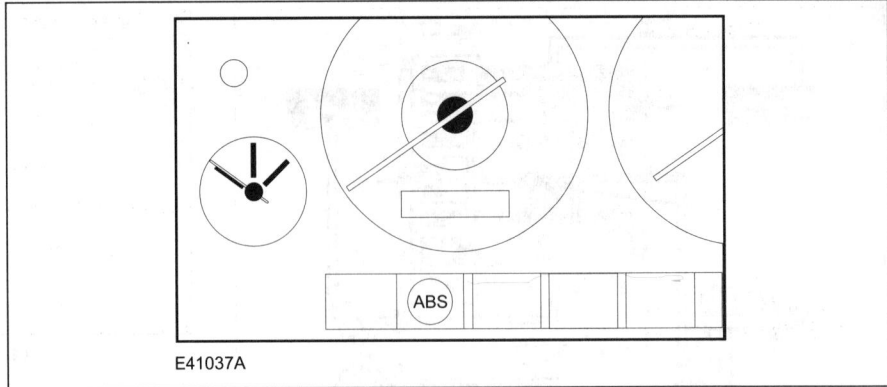

14.10 ABS warning lamp

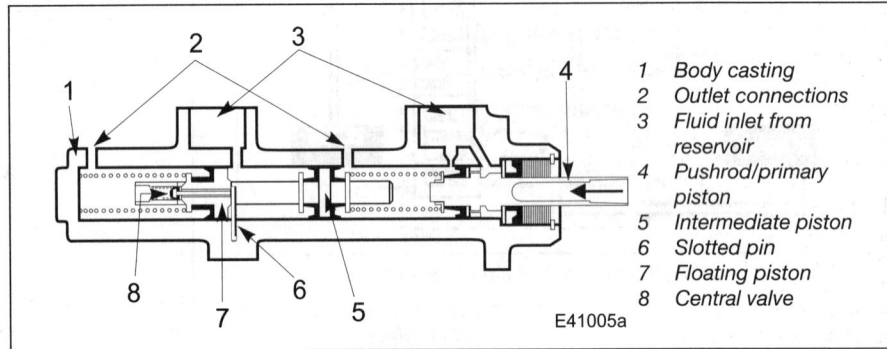

1 Body casting
2 Outlet connections
3 Fluid inlet from reservoir
4 Pushrod/primary piston
5 Intermediate piston
6 Slotted pin
7 Floating piston
8 Central valve

14.11 Sectional view of a typical tandem master cylinder

ABS electrical wiring harness and relays

An integrated main wiring harness is used to connect power and earth to various electrical components. This enables sensor signals to reach the ECM and the ECM, in turn, to send command signals to the relays, return pump and hydraulic control unit. The hydraulic control unit harness connector, valve relay and pump relay are located under a cover on the hydraulic control unit (see illustration 14.8).

Pump relay

The pump relay is the lower of the two relays mounted on the hydraulic control unit and is supplied with a permanent voltage from the battery. The pump relay is controlled by the ECM applying an earth which activates the relay allowing battery voltage to be applied to the pump. The hydraulic control unit provides the earth for the pump motor.

Valve relay

The solenoid valve relay is the upper of the two relays mounted on the hydraulic control unit and is supplied with a permanent voltage from the battery. The valve relay is controlled by the ECM applying an earth which activates the relay allowing battery voltage to be applied to the solenoid valves, the ECM controlling the earth.

Brake light switch

The brake light switch comprises a switch body and contact pin and is located above the brake pedal (see illustration 14.9).

When the brake pedal is depressed, closing the brake light switch, a signal is sent to the ECM indicating that the brakes are being applied. Once this signal is received, the ECM will begin monitoring the wheel speed via the wheel speed sensors and activate the ABS if necessary.

ABS warning lamp

After the ignition is switched on, the ABS warning lamp on the instrument panel is illuminated for approximately 4 seconds, then extinguished. During vehicle operation above a pre-determined wheel speed, the ABS-ECM implements a self-test cycle whereby ABS operation is monitored. If a fault is detected, the ECM illuminates the warning lamp on the instrument panel. The ECM switches off the ABS, however the conventional braking system continues to operate as normal. The warning lamp will remain illuminated until the fault is no longer present (see illustration 14.10).

When the ABS-ECM detects a fault, the fault code is stored and the ABS warning lamp activated. If the fault no longer exists after the next system start (ignition on/off) the ABS warning lamp is extinguished after the self-test cycle, however the fault code remains stored in the ECM memory.

Tandem master cylinder

Typically, the tandem master cylinder comprises a body casting incorporating primary and secondary pressure chambers, primary piston, intermediate piston, floating piston, slotted pin and central valve. The cylinder operates as a conventional master cylinder using vacuum assistance from the vacuum servo unit (see illustration 14.11).

When the brake system is at rest, the central valve in the floating piston rests against the slotted pin. In this condition the central valve is open and brake fluid can discharge out of the pressure chamber back into the brake fluid reservoir. When the brake pedal is depressed, the build-up of hydraulic pressure in the primary pressure chamber acts on the intermediate piston and floating piston, moving them down the cylinder bore. The floating piston contacts the seal on the central valve, closing the connection between the intermediate and secondary pressure chambers. Brake hydraulic pressure can also increase in the secondary pressure chamber.

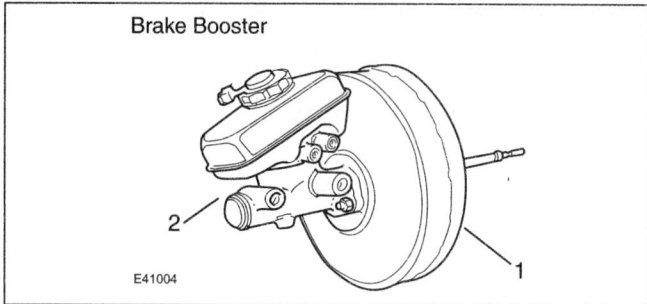

14.12 Vacuum servo unit (1) and tandem master cylinder (2)

Vacuum servo unit

The vacuum servo unit is located between the brake pedal and tandem master cylinder. When the brake pedal is depressed, the servo unit increases the force applied by the pedal, reducing the effort required to operate the brakes (see illustration 14.12).

The unit operates by means of engine inlet manifold vacuum applied to a diaphragm contained within the unit casing. A pushrod connected to the centre of the diaphragm acts directly on the primary piston in the master cylinder.

When the brake pedal is released, vacuum is applied to both sides of the diaphragm. When the pedal is depressed, one side of the diaphragm is opened to atmosphere and the vacuum acting on the other side deflects the diaphragm which in turn operates the master cylinder primary piston. The resulting force applied to the master cylinder piston is therefore significantly greater than the initial force applied to the brake pedal by the driver.

Pressure regulating/load sensing valve(s)

Pressure regulating valves are incorporated to restrict the hydraulic fluid pressure to the rear brakes. The valves are of the pressure conscious type whereby the hydraulic fluid supply is restricted once a pre-determined pressure is reached.

3 System operation

Brake system at rest

When the ABS system is at rest all the brake components are inoperative. Pressure is non-existent in the hydraulic pipes between the tandem master cylinder and the brake calipers or wheel cylinders.

Brake system in operation

Phase one, normal braking

When the brake pedal is activated, the pedal force is applied to the tandem master cylinder by the vacuum servo unit pushrod (see illustration 14.13). The servo unit pushrod acts directly on the pressure piston in the master cylinder which pressurises the hydraulic fluid in the brake pipes to the hydraulic control unit. The solenoid valve is in the 'at rest' position (ie closed) and the volume governor is also 'at rest' (closed) due to the force of the return spring.

Through internal drillings in the hydraulic control unit and volume governor, there is an unrestricted fluid passage between the master cylinder and the relevant brake caliper or wheel cylinder. Hydraulic pressure can be transmitted directly to each brake caliper or wheel cylinder, thus operating the brakes.

The ABS-ECM continually monitors wheel speed from the signals provided by the wheel speed sensors. If the ECM detects the incidence of wheel lock on one or more wheels, ABS is automatically initiated. As Lucas 4/4F ABS operates individually on each wheel, all or any of the wheels could be in either of the following phases at any particular moment.

Phase two, pressure reduction

The ECM opens the solenoid valve on the relevant hydraulic circuit which creates a difference in hydraulic pressure on either side of the main restrictor in the volume governor (see illustration 14.14). As the greater hydraulic pressure is on the inlet side of the main restrictor,

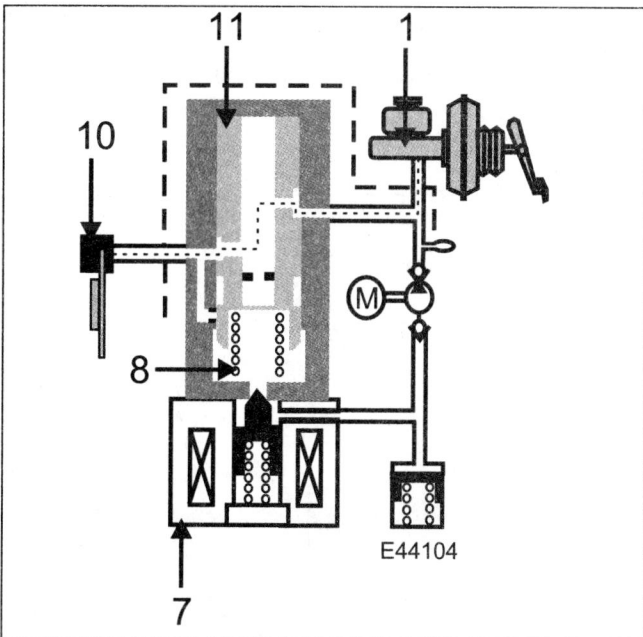

14.14 Brake system operation - phase two, pressure reduction

1	Master cylinder	10	Brake caliper
5	Return pump	11	Volume governor
6	Pressure accumulator	12	Main restrictor
7	Solenoid valve	13	Corner passage
8	Return spring	15	Hydraulic fluid return flow
9	Secondary restrictor	16	Additional hydraulic pressure

14

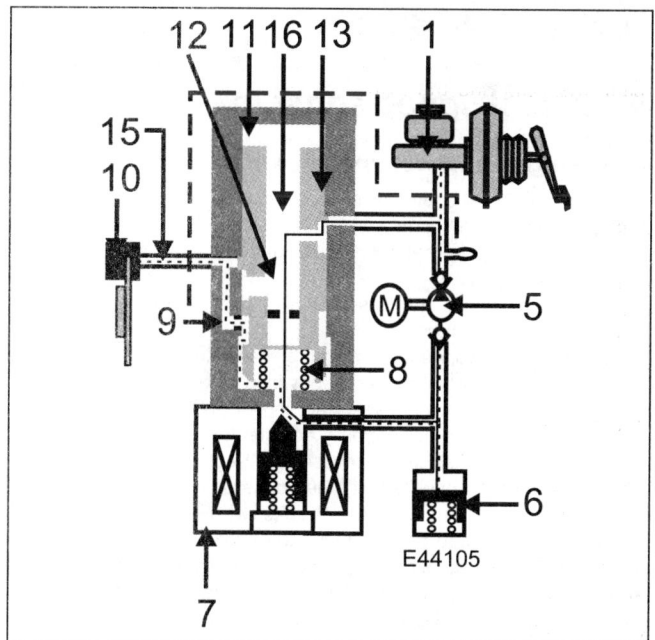

14.13 Brake system operation - phase one, normal braking

1	Master cylinder	10	Brake caliper
7	Solenoid valve	11	Volume governor
8	Return spring		

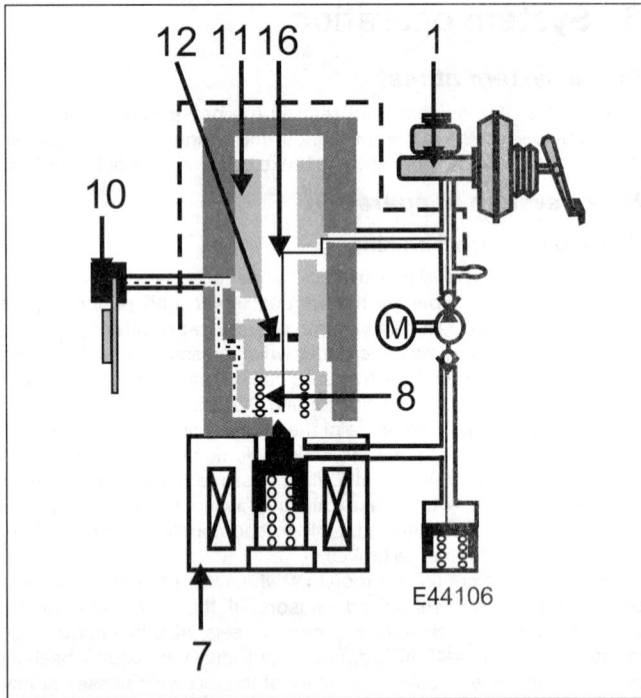

14.15 Brake system operation - phase three, pressure build-up

1 Master cylinder	11 Volume governor
7 Solenoid valve	12 Main restrictor
8 Return spring	16 Hydraulic fluid flow to brake
10 Brake caliper	caliper

the volume governor is forced down against the force of the return spring. As the volume governor moves down, the fluid passage between the master cylinder and brake caliper or wheel cylinder is closed, and a return passage from the brake caliper or wheel cylinder to the pressure accumulator is opened. Hydraulic fluid now flows from the brake caliper or wheel cylinder, through the secondary restrictor and into the pressure accumulator. This results in a pressure reduction, thus releasing the brake.

As soon as the solenoid valve is opened, the ECM actuates the electric motor to operate the return pump. The hydraulic fluid is then pumped back into the pressure side of the master cylinder. This process creates a pulsation which can be felt in the brake pedal action, but which is softened by the pulsation damper.

While a pressure difference remains on either side of the main restrictor, the volume governor will remain open and any additional hydraulic pressure generated by the master cylinder will flow through the corner passage in the volume governor and into the pressure accumulator to be pumped back to the master cylinder.

Phase three, pressure build-up

The pressure build-up phase is instigated after the wheel rotation has stabilised. The ECM closes the solenoid valve and switches off the return pump allowing hydraulic pressure to once again increase below the main restrictor (see illustration 14.15). Hydraulic fluid now flows to the brake caliper or wheel cylinder, leading to a gradual increase of pressure at the brake. Once the pressure on either side of the main restrictor has equalised, the return spring closes the volume governor thus fully re-opening the direct hydraulic passage between the master cylinder and the brake caliper or wheel cylinder. Normal braking now continues until the wheel has again decelerated to a sufficient degree where pressure reduction is once more required.

The whole ABS control cycle takes place 4 to 10 times per second for each affected wheel and this ensures maximum braking effect and control during ABS operation.

Test procedures

4 Wheel speed sensors

Checking the wheel speed sensor (general)

1 Inspect the wheel speed sensor for corrosion or damage and check that the sensor is tightly mounted.
2 Check the toothed sensor ring for damage, eccentricity and for broken or missing teeth.
3 Inspect the wheel speed sensor wiring plug for corrosion and damage. One plug for each sensor.
4 Check that the connector terminal pins are pushed fully home and making good contact with the sensor wiring plug.
5 Check the clearance between the sensor and the toothed sensor ring. The clearance is not normally adjustable but is nominally 0.2 to 1.3 mm. If the clearance is excessive, expect a worn sensor tip or problems with the wheel bearings/hub or sensor ring.
6 When carrying out voltage checks with an oscilloscope or voltmeter, the voltage obtained will be proportional to the speed at which the wheel is rotating. In addition to determining that the wheel speed sensors are actually producing a voltage output, it is essential that the output from the sensors on a particular axle is the same for any given wheel speed.

Checking wheel speed sensor output with an oscilloscope

Note: Refer to the wiring diagrams for specific ECM pin identification according to model.
7 Switch the ignition off and disconnect the ECM multi-plug or the relevant wheel speed sensor wiring plug.

8 Connect an oscilloscope between the terminal pins for the sensor under test (see illustration 14.16).
9 Select a range to cover 80 Hz on the oscilloscope and a free run time base.

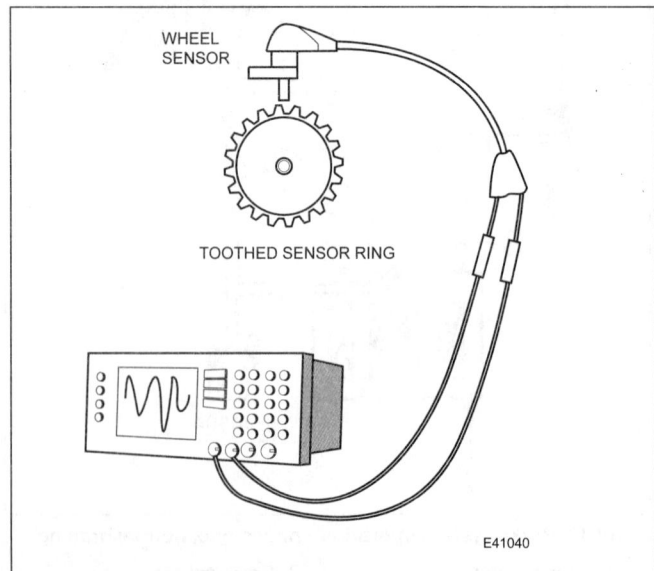

14.16 Checking wheel speed sensor output with an oscilloscope connected to the sensor wiring plug

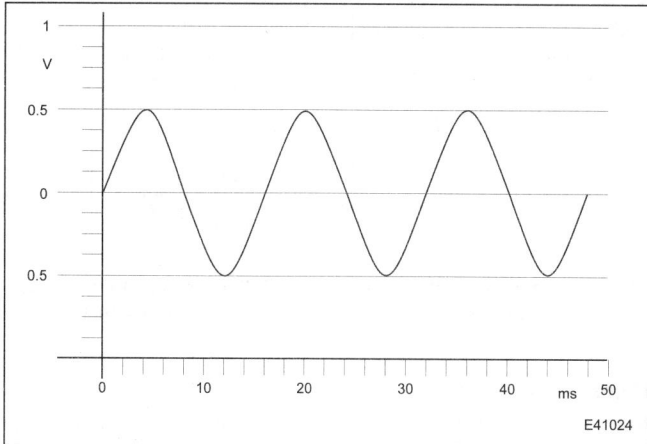

14.17 Typical wheel speed sensor sine wave as displayed on an oscilloscope

10 Raise the wheel and rotate it by hand at approximately one revolution per second.

11 A sinusoidal wave form should be obtained, with amplitude and duration changing with rotational speed (see illustration 14.17).

12 If there is no signal, or a very weak or intermittent signal at the ECM, repeat the test at the sensor wiring plug. If there is no change in signal status, the sensor is suspect.

13 If the signal is now satisfactory this indicates a fault in the wiring harness which should be checked for continuity.

Checking wheel speed sensor output with an AC voltmeter

Note: Refer to the wiring diagrams for specific ECM pin identification according to model.

14 Switch the ignition off and disconnect the ECM multi-plug or the relevant wheel speed sensor wiring plug.

15 Connect an AC voltmeter between the terminal pins for the sensor under test (see illustration 14.18).

16 Raise the wheel and rotate it by hand at approximately one revolution per second.

17 A voltage of approximately 0.4 to 1.5 volts (AC RMS) should be obtained. If there is no signal, or a very weak or intermittent signal at

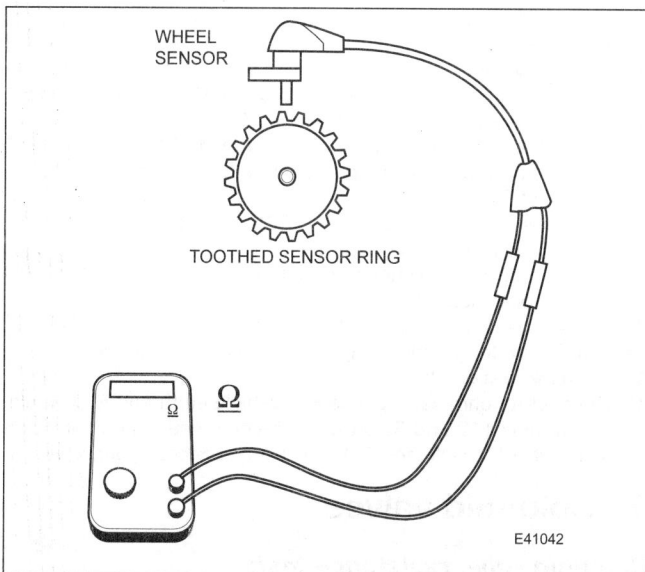

14.19 Checking wheel speed sensor resistance with an ohmmeter connected to the sensor wiring plug

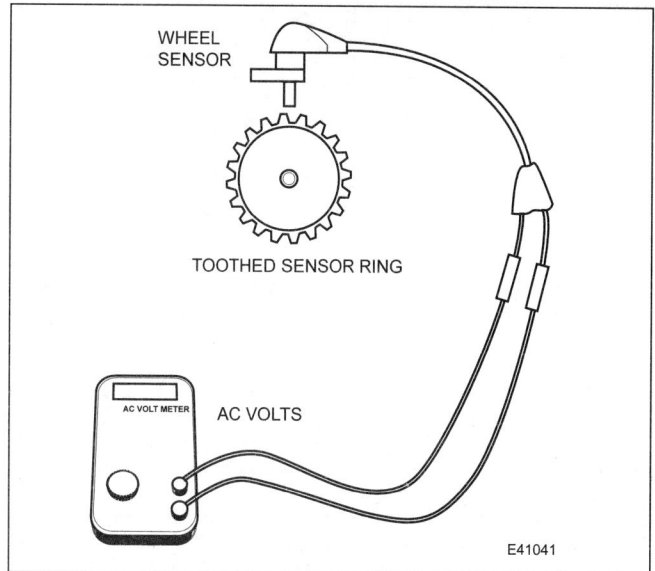

14.18 Checking wheel speed sensor output with a voltmeter connected to the sensor wiring plug

the ECM, repeat the test at the sensor wiring plug. If there is no change in the signal, the sensor is suspect.

18 If the signal is now satisfactory, this indicates a fault in the wiring harness which should be checked for continuity. Note: This test at least proves that a signal is being generated by the sensor. However, the voltage produced is an average voltage and does not clearly indicate damage to the sensor ring or that the sinewave is regular in formation.

Checking wheel speed sensor resistance

Note: Refer to the wiring diagrams for specific ECM pin identification according to model.

19 Switch the ignition off and disconnect the ECM multi-plug or the relevant wheel speed sensor wiring plug.

20 Connect an ohmmeter between the terminal pins for the sensor under test (see illustration 14.19).

21 The readings obtained should be between 1.1 and 1.6 kohms.

22 If the resistance is excessively high, or open circuit at the ECM, repeat the test at the sensor multi-plug. If there is no change in resistance, the sensor is suspect.

23 If the resistance is now satisfactory, this indicates a fault in the wiring harness which should be checked for continuity. Note: Even if the resistance is within the quoted specifications, this does not prove that the speed sensor can generate an acceptable signal.

5 System relays

Checking relays (general)

1 The following tests are applicable to the ABS main (or solenoid valve) relay and hydraulic pump relay.

2 Before carrying out specific tests on the relays or power supplies, inspect the relays and relay base for corrosion and damage.

3 Check that the relays are fully pushed home and making good contact with the base.

4 Faults in any of the above areas are possible reasons for failure or malfunctioning of the relay.

Relay operation tests

Main (solenoid valve) relay

5 Switch the ignition off, then remove the relay from the relay base.

6 Connect an ohmmeter between the relay output terminal 30 and

14

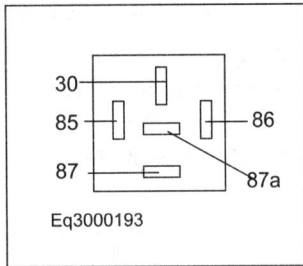

14.20 Main (solenoid valve) relay terminals

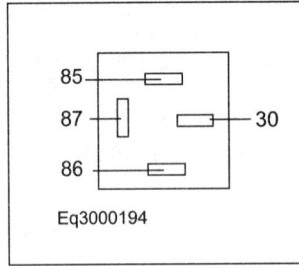

14.21 Hydraulic pump relay terminals

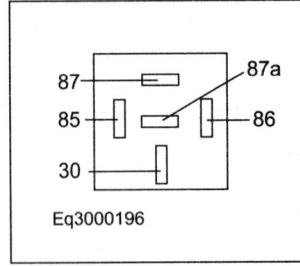

14.22 Main (solenoid valve) relay base terminals

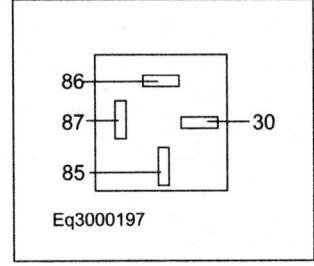

14.23 Hydraulic pump relay base terminals

earth terminal 87a (see illustration 14.20). The ohmmeter should indicate continuity.

7 Connect an ohmmeter between the relay main voltage supply terminal 87 and the output terminal 30. The ohmmeter should indicate an open circuit.

8 Attach a wire between the relay driver supply terminal 85 and a 12 volt supply. Attach a second wire between the relay driver earth terminal 86 and earth. There should be an unmistakable click from the relay and the ohmmeter should indicate continuity.

9 Renew the relay if it fails these tests. If the relay cover is transparent, inspect the relay for burns. A faulty relay usually has a burnt appearance.

Hydraulic pump relay

10 Switch the ignition off, then remove the relay from the relay base.

11 Connect an ohmmeter between the relay main voltage supply terminal 87 and the output terminal 30 (see illustration 14.21). The ohmmeter should indicate an open circuit.

12 Attach a wire between the relay driver supply terminal 86 and a 12 volt supply. Attach a second wire between the relay driver earth terminal 85 and earth. There should be an unmistakable click from the relay and the ohmmeter should indicate continuity.

13 Renew the relay if it fails these tests. If the relay cover is transparent, inspect the relay for burns. A faulty relay usually has a burnt appearance.

Relay power supply tests

Main (solenoid valve) relay

14 Switch the ignition off then remove the main relay leaving the base exposed (see illustration 14.22).

15 Using a voltmeter, probe between the main voltage supply terminal 87 in the relay base and earth. The reading should indicate nbv. If no voltage is found, check the wiring back to the voltage supply.

16 Switch the ignition on and using a voltmeter, probe between the relay driver supply terminal 85 in the relay base and earth. The reading should indicate nbv. If no voltage is found, check the relevant fuse and the supply wiring back to the ignition switch.

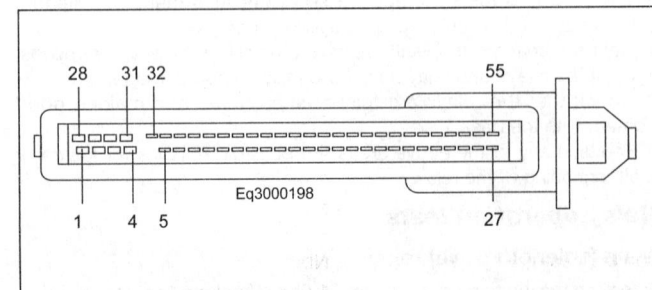

14.24 ECM 55-pin multi-plug terminal identification

Pump relay

Note: *For the following test the ECM and main (solenoid valve) relay must be connected and operating correctly.*

17 Switch the ignition off then remove the pump relay leaving the base exposed (see illustration 14.23).

18 Using a voltmeter, probe between the main voltage supply terminal 87 in the relay base and earth. The reading should indicate nbv. If no voltage is found, check the wiring back to the voltage supply.

19 Switch the ignition on and using a voltmeter, probe between the relay driver supply terminal 86 in the relay base and earth. The reading should indicate nbv. If no voltage is found, check the relevant fuse, the main (solenoid valve) relay and the supply wiring back to the ignition switch.

6 Electronic Control Module

Checking the ECM (general)

1 Inspect the ECM for corrosion or damage and ensure that the unit is mounted securely.

2 Check that the ECM multi-plug terminals are pushed fully home and making good contact with the ECM pins. Faults in any of the above areas are possible reasons for poor performance in the ABS system.

ECM power supply and earth tests

Note: *For the following tests the main (solenoid valve) relay must be connected and operating correctly.*

3 Switch the ignition off and disconnect the ECM multi-plug from the ECM.

4 Switch the ignition on.

5 Attach a negative voltmeter probe to an ECM multi-plug earth pin and attach a positive voltmeter probe to ECM multi-plug pin 33. The voltmeter should indicate nbv (see illustration 14.24). If no voltage is found, check the relevant fuse, and the supply wiring back to the ignition switch.

6 Using a fused jumper lead, connect ECM multi-plug pin 55 to earth.

7 Attach a negative voltmeter probe to an ECM multi-plug earth pin and attach a positive voltmeter probe to ECM multi-plug pin 40. The voltmeter should indicate nbv.

8 If no voltage is found, check the relevant fuse, the main (solenoid valve) relay and the supply wiring back to the ignition switch.

9 Switch the ignition off.

10 Connect an ohmmeter between a vehicle earth and ECM multi-plug earth pins 1, 2 and 32, in turn. The ohmmeter should indicate continuity. If not, check the ECM main earth connection and wiring.

7 Solenoid valves

Solenoid valve resistance tests

1 Switch the ignition off and disconnect the hydraulic control unit 14-pin multi-plug.

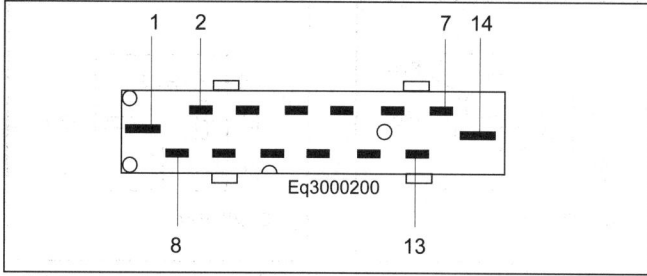

14.25 Hydraulic control unit 14-pin multi-plug terminal identification

2 Using an ohmmeter, measure the resistance between pin 2 in the multi-plug, and each of the following multi-plug pins in turn - 3, 4, 5, and 6. The resistance should be in the order of 3.0 to 4.0 ohms for each solenoid valve **(see illustration 14.25)**. If the resistance is not as specified, the hydraulic control unit is suspect.

8 Hydraulic pump motor

Pump motor resistance test

1 Switch the ignition off and disconnect the hydraulic control unit 14-pin multi-plug.
2 Using an ohmmeter, measure the resistance between pin 9 in the multi-plug and a vehicle earth **(see illustration 14.25)**. The resistance should be in the order of 1.0 ohm. If the resistance is not as specified, the hydraulic control unit is suspect.

Pump motor operation test

3 Switch the ignition off then remove the pump relay leaving the base exposed.
4 Using a fused jumper lead, bridge terminals 30 and 87 in the relay base **(see illustration 14.23)**. The pump motor should now run.

⚠️ **Warning: The test should be made as quickly as possible to avoid damaging the pump.**

5 If the pump does not operate as described, check the hydraulic control unit earth connection. If satisfactory, renew the hydraulic control unit.

9 Brake light switch

Checking the brake light switch (general)

1 Check that the brake light switch is correctly and securely mounted and that the plunger moves smoothly with no trace of binding.

2 Check that the wiring multi-plug is pushed fully home and making good contact.
3 Check that no wires have been disconnected.
4 Faults in any of the above areas are possible reasons for failure or malfunctioning of the switch.

Brake light switch voltage and continuity tests

Voltage test

5 Switch the ignition off and disconnect the ECM multi-plug from the ECM.
6 Connect a voltmeter between ECM multi-plug pin 13 and a vehicle earth **(see illustration 14.24)**.
7 Switch the ignition on and depress the brake pedal. The voltmeter should indicate nbv.
8 If no voltage is found, the fuse, the brake light switch and the wiring are suspect.
9 Release the brake pedal. The voltage should drop to zero as the switch opens.

Continuity test

10 Switch the ignition off and disconnect the brake light switch multi-plug.
11 Connect an ohmmeter between the terminal pins of the brake light switch.
12 Operate the brake light switch and check for continuity. If the test fails, renew the switch.

10 Warning lamp

Checking the warning lamp (general)

1 Inspect the warning lamp bulb holder contacts in the instrument panel.
2 Check that the instrument panel multi-plug terminal pins are pushed fully home and making good contact.
3 Check that no wires have been disconnected.
4 Faults in any of the above areas are possible reasons for failure or malfunctioning of the warning lamp.

Warning lamp operation test

5 With the ignition switched off, the warning lamp should remain off.
6 Switch the ignition on and the warning lamp should illuminate for approximately 4 seconds then extinguish. The lamp should then remain off.
7 If the warning lamp comes on and remains on at any time during vehicle operation, carry out the previously described test procedures on the system components.

14

Pin table - typical 55-pin

Note: *Refer to illustration 14.24*

Pin No.	Connection	Test condition	Voltage
1	ECM earth	Ignition on	0.25 volts (max)
2	ECM earth	Ignition on	0.25 volts (max)
3	Rear right solenoid valve	Ignition on/inactive	Nbv
		Actuated	1.0 volt (max)
4	-	-	-
5	-	-	-
6	-	-	-
7	-	-	-
8	-	-	-
9	Hydraulic pump relay driver	Ignition on/inactive	Nbv
		Actuated	1.0 volt (max)
10	-	-	-

Pin table - typical 55-pin (continued)

Note: *Refer to illustration 14.24*

Pin No.	Connection	Test condition	Voltage
11	ABS warning lamp	Engine running: Lamp on Lamp off	 1.0 volt (max) Nbv
12	-	-	-
13	Brake light switch input	Ignition on: Brake pedal released Brake pedal depressed	 0 volts Nbv
14	-	-	-
15	-	-	-
16	-	-	-
17	-	-	-
18	-	-	-
19	-	-	-
20	-	-	-
21	-	-	-
22	-	-	-
23	-	-	-
24	-	-	-
25	Rear left wheel speed sensor earth	Roadwheel rotating	0.25 volts (max)
26	-	-	-
27	Front right wheel speed sensor earth	Roadwheel rotating	0.25 volts (max)
28	-	-	-
29	Front left solenoid valve	Ignition on/inactive Actuated	Nbv 1.0 volt (max)
30	Rear left solenoid valve	Ignition on/inactive Actuated	Nbv 1.0 volt (max)
31	Front right solenoid valve	Ignition on/inactive Actuated	Nbv 1.0 volt (max)
32	ECM earth	Ignition on	0.25 volts (max)
33	Supply from ignition switch	Ignition on	Nbv
34	-	-	-
35	-	-	-
36	-	-	-
37	-	-	-
38	-	-	-
39	SD connector	-	-
40	Solenoid valve relay display input	Ignition on/inactive Actuated	Nbv 1.0 volt (max)
41	Pump relay display input	Ignition on	0 volts
42	-	-	-
43	-	-	-
44	-	-	-
45	-	-	-
46	Rear right wheel speed sensor earth	Roadwheel rotating	0.25 volts (max)
47	Rear right wheel speed sensor signal	Roadwheel rotating	0.4 volts AC (approx)
48	-	-	-
49	Front left wheel speed sensor earth	Roadwheel rotating	0.25 volts (max)
50	-	-	-
51	Front left wheel speed sensor signal	Roadwheel rotating	0.4 volts AC (approx)
52	Front right wheel speed sensor signal	Roadwheel rotating	0.4 volts AC (approx)
53	-	-	-
54	Rear left wheel speed sensor signal	Roadwheel rotating	0.4 volts AC (approx)
55	Solenoid valve relay driver	Ignition on/inactive Actuated	Nbv 1.0 volt (max)

Fault codes

1 The Lucas 4/4F system requires the use of a FCR for obtaining fault codes. Flash codes are not available for output from this system.

2 If a FCR is available, it should be connected to the SD serial connector (the Vauxhall/Opel term for SD connector is ALDL) and used in accordance with the maker's instructions.

3 The FCR can be used for the following purposes:
a) *Obtaining fault codes.*
b) *Clearing fault codes.*
c) *Obtaining datastream information.*
d) *Testing the system actuators (solenoid valve relay, pump relay and solenoid valves).*

Fault code table

Code	Item	Fault
16	Front left solenoid valve	Defective (signal interruption, short to earth)
17	Front right solenoid valve	Defective (signal interruption, short to earth)
19	Solenoid valve relay	Defective (signal interruption, short to earth, supply voltage too low)
25	Toothed sensor ring	Wrong number of teeth
28	Rear left solenoid valve	Defective (signal interruption, short to earth)
29	Rear right solenoid valve	Defective (signal interruption, short to earth)
35	Pump motor	Defective (non-operation of pump after actuation of relay)
39	Front left wheel speed sensor	No/poor signal
41	Front left wheel speed sensor	Signal interrupted
42	Front right wheel speed sensor	No/poor signal
43	Front right wheel speed sensor	Signal interrupted
44	Rear left wheel speed sensor	No/poor signal
45	Rear left wheel speed sensor	Signal interrupted
46	Rear right wheel speed sensor	No/poor signal
47	Rear right wheel speed sensor	Signal interrupted
48	Battery voltage	Voltage too low
49	Battery voltage	Voltage too high
52	Warning lamp	Defective
55	ECM	Defective
59	Overvoltage relay diode	Defective

Wiring diagram

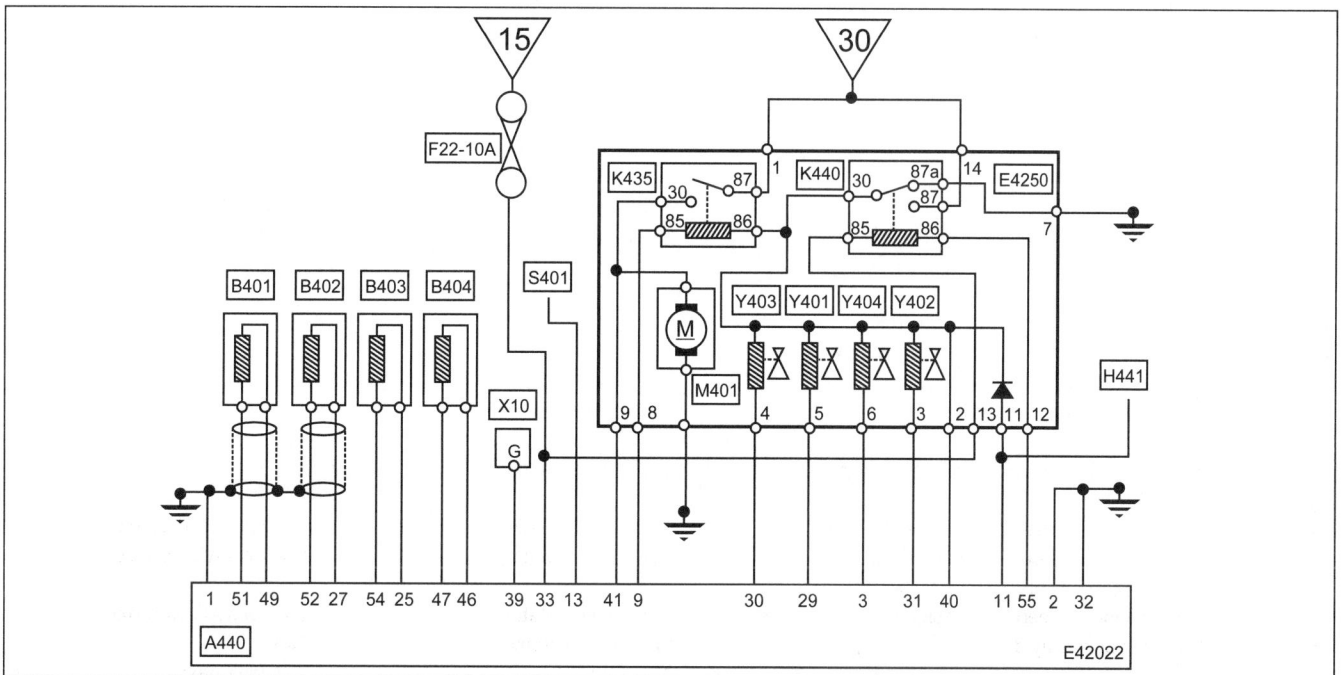

14.26 55-pin wiring diagram, Vauxhall/Opel Corsa-B

Chapter 15

Lucas EBC 430 ABS

Contents

15

Vehicle coverage

Model	Year
Daewoo	
Matiz	1998-2000
Fiat	
Marea/Weekend	1996-2000

Overview of system operation

1 Basic principles and system identification

The Lucas EBC 430 Antilock Brake System has been fitted to a limited number of passenger vehicles since its introduction in the mid 1990s. The system is of the additional or 'add-on' type operating in conjunction with the conventional braking system components.

The purpose of the system is to apply the vehicle brakes at maximum efficiency without wheel lock or loss of directional stability. Inductive, or 'active' sensors (wheel speed sensors) monitor the speed of the wheels by generating an electrical signal as the wheel is rotated. This information is passed to the ABS Electronic Control Module which compares the signals received from each wheel and uses the speed of the fastest wheel as a reference value. The ECM continually monitors the speed of each wheel and if the onset of lock at any wheel is detected (a received speed signal being less than the reference value) a signal is sent to the ABS hydraulic control unit which regulates the brake pressure for the relevant wheel(s).

Lucas EBC 430 ABS is comprised of the following main components (see illustration 15.1):
a) Hydraulic control unit with integral ABS-ECM.
b) Inductive, or 'active' wheel speed sensors and associated sensor rings.
c) Brake light switch.
d) ABS warning lamp.
e) Diagnostic connector.
f) ABS electrical wiring harness.

In addition, the conventional brake system is comprised of the following components:
a) Tandem brake master cylinder.
b) Vacuum servo unit.
c) Brake calipers/wheel cylinders and hydraulic hoses and pipes.

2 Component description and operation

ABS ECM

General

The Lucas EBC 430 ECM continually monitors wheel speed from the signals provided by the wheel speed sensors, and brake application from the brake light switch signal. If the ECM detects the incidence of wheel lock on one or more wheels, a signal is sent to the hydraulic control unit to modulate the hydraulic pressure to the brake of the locking wheel(s). The ECM contains two microprocessors and uses digital technology to complete this function and other functions such as, fault code memory and power modules for valve and relay activity (see illustration 15.2).

To reduce external electrical connections to a minimum and improve reliability, the ECM is integral with the hydraulic control unit (see illustration 15.3). The ECM and hydraulic control unit can be separated for renewal, but the internal components of each unit are sealed.

Self-test

Lucas EBC 430 ABS is equipped with a self-test capability that begins to examine the ABS signals after a wheel speed of approximately 2 mph is reached from all wheels. ABS self-test

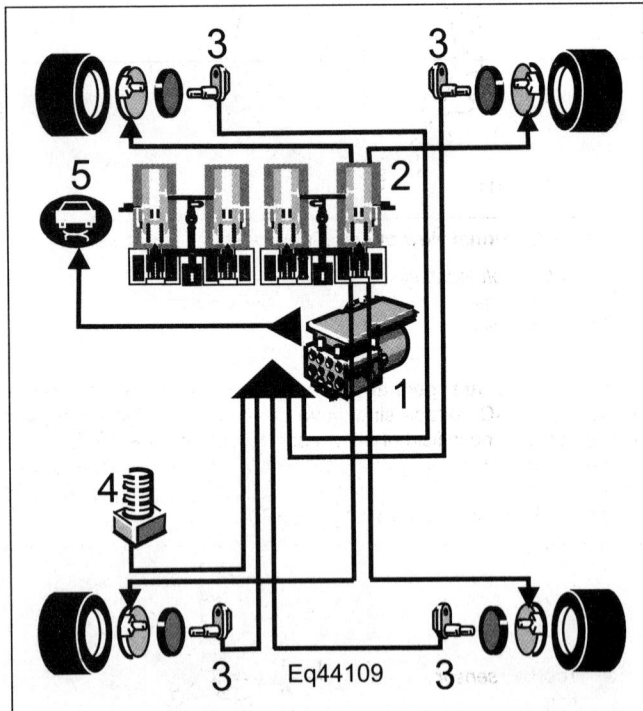

15.1 Typical Lucas EBC 430 main components

1 Hydraulic control unit with integral ECM
2 Hydraulic control unit solenoid valves
3 Wheel speed sensors and associated sensor rings
4 Brake light switch
5 ABS warning lamp

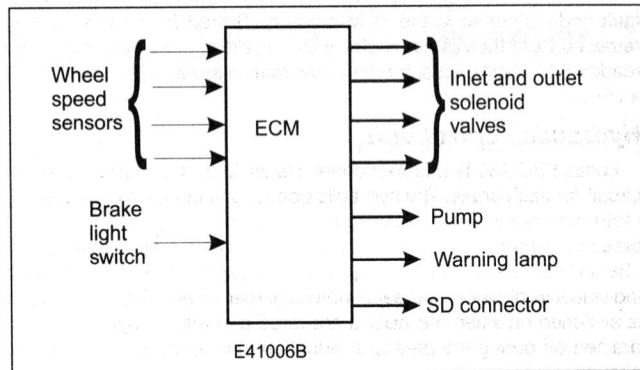

15.2 ECM sensor inputs and control signal outputs

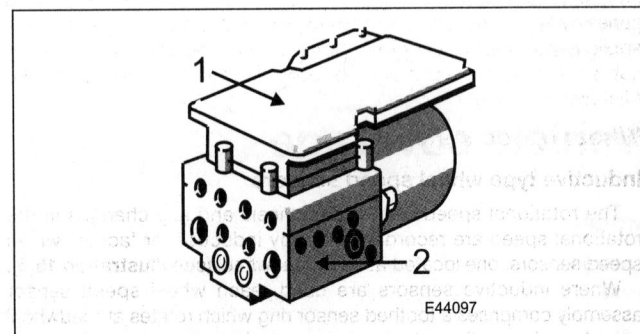

15.3 Lucas EBC 430 ECM (1) and hydraulic control unit (2)

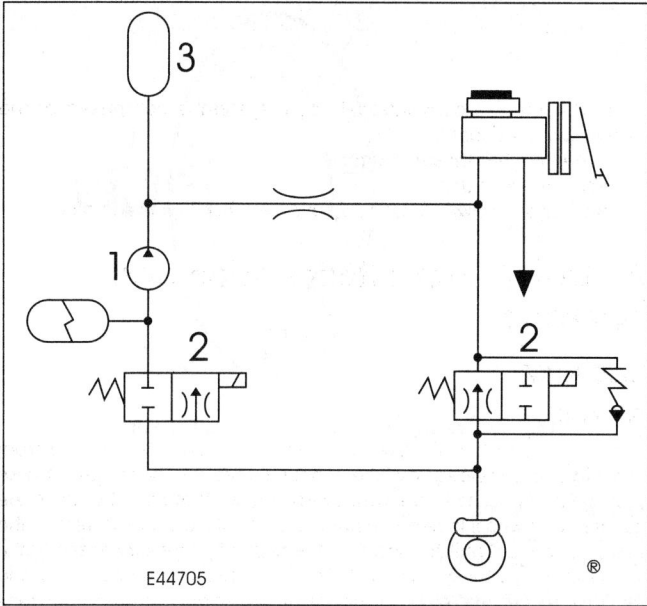

15.4 Lucas EBC 430 hydraulic circuit schematic

1 Pump motor
2 Inlet and outlet solenoid valves
3 Pulsation damper

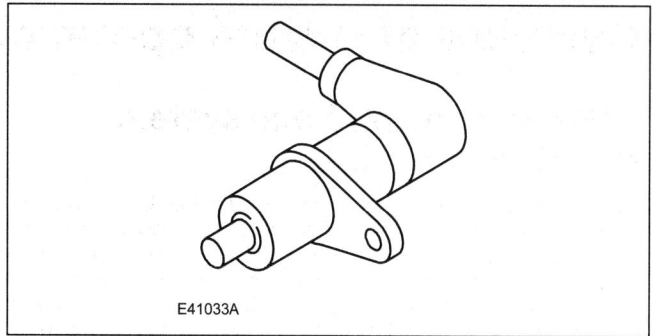

15.5 Typical inductive wheel speed sensor

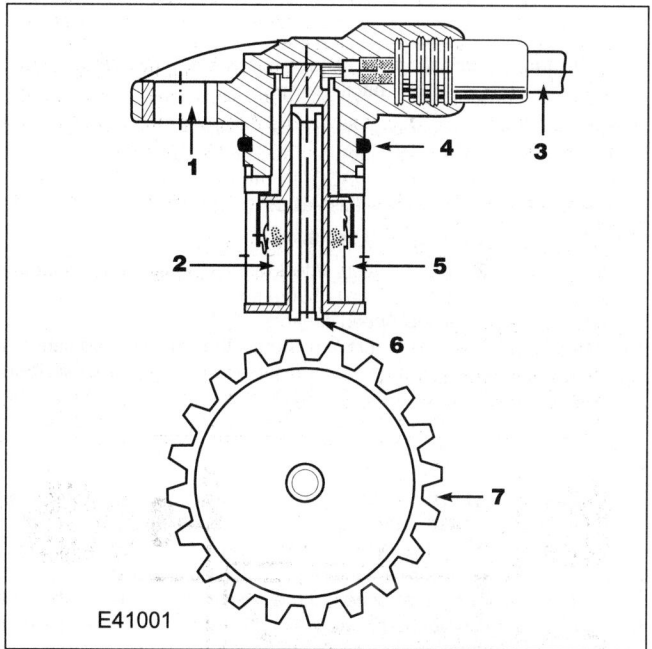

15.6 Sectional view of an inductive wheel speed sensor

1	Mounting bolt location	5	Coil
2	Permanent magnet	6	Sensor tip
3	Wiring harness	7	Toothed sensor ring
4	O-ring		

continually examines the signals from the various components. If self-test determines that faults are not present, ABS is ready for operation. If the ECM detects that a fault is present, all ABS functions are switched off and the ABS warning lamp is turned on.

Self-diagnostics

If the ECM detects a fault during the self-test routine, an internal fault code is stored in the ECM memory. Stored fault codes can be retrieved from the SD connector with the aid of a suitable fault code reader. If the fault clears, the code will remain stored until cleared with an FCR.

Hydraulic control unit

Lucas EBC 430 is a four-channel system with a separate hydraulic circuit for each brake. The hydraulic control unit consists of an electric motor and radial piston return pump, inlet and outlet solenoid valves pressure accumulators and pulsation dampers **(see illustration 15.4)**. The unit controls the hydraulic pressure applied to the brake for each individual front wheel and each individual rear wheel. The return pump is switched on when the ABS is activated and returns hydraulic fluid, drained off during the pressure reduction phase, back into the brake circuit.

The ABS-ECM contains additional software for rear brake hydraulic fluid pressure regulation when ABS is not in operation. This is generally known as 'Electronic Brake Force Distribution' and in these applications, mechanical pressure regulating/load sensing valves are not required as the rear brake hydraulic pressure is controlled by the ABS hydraulic control unit.

Wheel speed sensors

Inductive type wheel speed sensors

The rotational speed of the roadwheels and any changes in the rotational speed are recorded either by inductive, or 'active' wheel speed sensors, one located at each roadwheel **(see illustration 15.5)**.

Where inductive sensors are used, each wheel speed sensor assembly comprises a toothed sensor ring which rotates at roadwheel speed, and an adjacent sensor mounted a set distance from the sensor ring **(see illustration 15.6)**.

The sensors are permanent magnet pulse generator types producing an AC voltage sine wave as the sensor ring teeth pass through the magnetic field of the sensor **(see illustration 15.7)**.

The frequency of the waveform produced by the wheel speed sensor is proportional to the road speed. This AC voltage signal is continually being delivered to the ABS-ECM for processing.

15

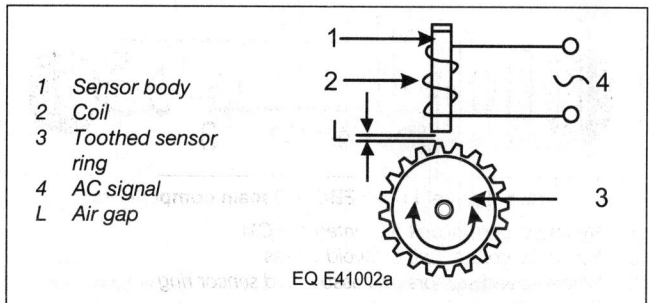

1	Sensor body
2	Coil
3	Toothed sensor ring
4	AC signal
L	Air gap

15.7 Inductive wheel speed sensor operation

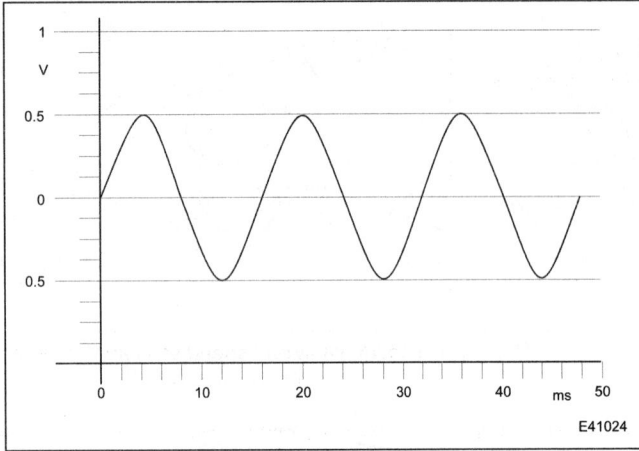

15.8 Inductive wheel speed sensor waveform as viewed on an oscilloscope

The peak to peak voltage of the speed signal (when viewed upon an oscilloscope) can vary considerably according to wheel speed. An analogue to digital converter (ADC) in the ECM transforms the AC pulse into a digital signal **(see illustration 15.8)**.

'Active' type wheel speed sensors

On the Lucas EBC 430 system fitted to Fiat models, the rotational speed of the roadwheels and any changes in the rotational speed are recorded by 'active' wheel speed sensors.

On the majority of antilock brake systems, the wheel speed sensors are of the permanent magnet pulse generator type, as described previously. On the Lucas EBC 430 system with 'active' sensors, the internal electrical resistance of the sensor is altered by changes in intensity and direction of the lines of force of an external magnetic field.

This external magnetic field is created by a unique type of sensor ring known as a 'multi-polar ring'. The multi-polar ring consists of a series of magnetic elements with alternating north/south polarities, located around the circumference of the ring **(see illustration 15.9)**.

The multi-polar ring is attached to the driveshaft or wheel hub, with the adjacent sensor mounted a set distances from the ring. As the ring rotates at roadwheel speed, an AC voltage square wave, constant amplitude signal is produced, the frequency of which is proportional to the road speed. This signal is continually being delivered by the sensor to the ABS-ECM for processing **(see illustration 15.10)**.

ABS electrical wiring harness and relays

An integrated main wiring harness is used for ABS-ECM power supply and earth connections, and enables sensor signals to reach the ECM and the ECM, in turn, to send output signals to the ABS

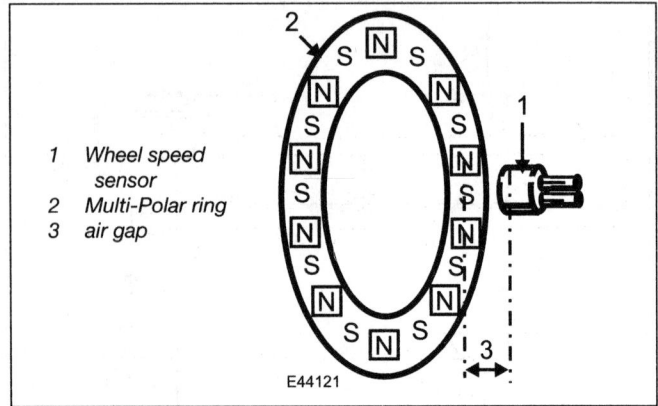

1 Wheel speed
   sensor
2 Multi-Polar ring
3 air gap

15.9 Active wheel speed sensor multi-polar ring

warning lamp and diagnostic connector. The main (solenoid valve) relay and return pump relay are soldered to the ECM circuit board and are an integral part of the ECM. Internal connections between the ECM and hydraulic control unit are used to activate the return pump.

Brake light switch

The brake light switch comprises a switch body and contact pin and is located above the brake pedal **(see illustration 15.11)**.

When the brake pedal is depressed, closing the brake light switch, a signal is sent to the ECM indicating that the brakes are being applied. Once this signal is received, the ECM will begin monitoring the wheel speed via the wheel speed sensors and activate the ABS if necessary.

ABS warning lamp

After the ignition is switched on, the ABS warning lamp on the instrument panel is illuminated for approximately 4 seconds, then extinguished. During vehicle operation above a pre-determined wheel speed, the ABS-ECM implements a self-test cycle whereby ABS operation is monitored. If a fault is detected, the ECM illuminates the warning lamp on the instrument panel. The ECM switches off the ABS, however the conventional braking system continues to operate as normal. The warning lamp will remain illuminated until the fault is no longer present **(see illustration 15.12)**.

When the ABS-ECM detects a fault, the fault code is stored and the ABS warning lamp activated. If the fault no longer exists after the next system start (ignition on/off) the ABS warning lamp is extinguished after the self-test cycle, however the fault code remains stored in the ECM memory.

Tandem master cylinder

Typically, the tandem master cylinder comprises a body casting incorporating primary and secondary pressure chambers, primary

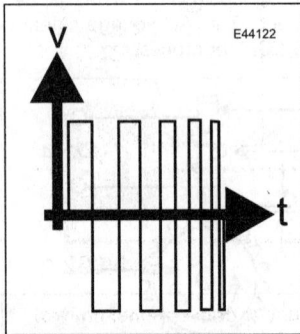

15.10 DC voltage square wave signal produced by the active wheel speed sensors

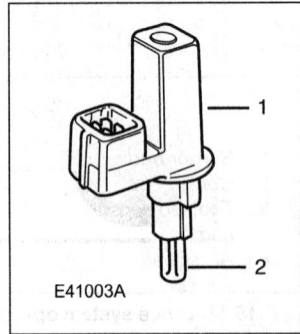

15.11 Typical brake light switch body (1) and contact pin (2)

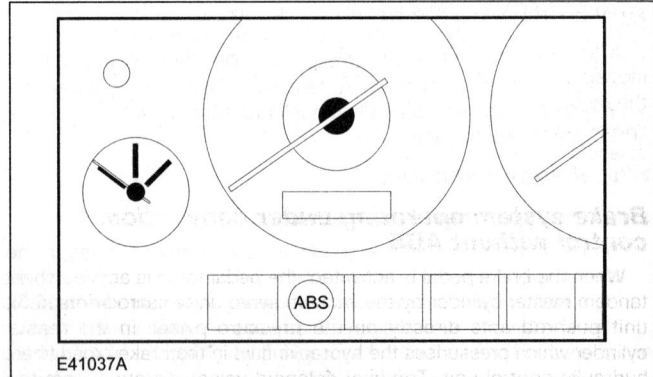

15.12 ABS warning lamp

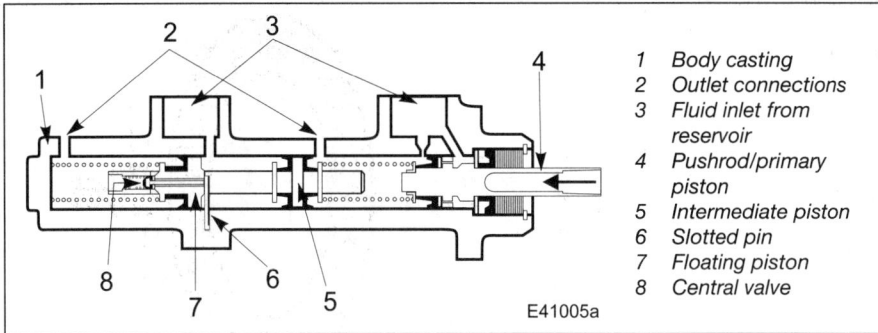

1	Body casting
2	Outlet connections
3	Fluid inlet from reservoir
4	Pushrod/primary piston
5	Intermediate piston
6	Slotted pin
7	Floating piston
8	Central valve

E41005a

15.13 Sectional view of a typical tandem master cylinder

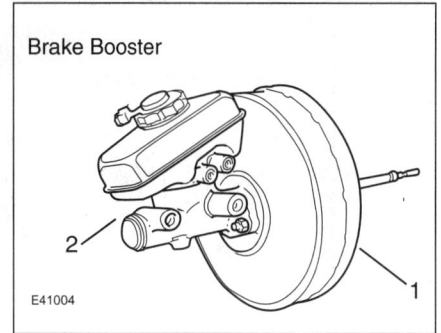

E41004

15.14 Vacuum servo unit (1) and tandem master cylinder (2)

piston, intermediate piston, floating piston, slotted pin and central valve. The cylinder operates as a conventional master cylinder using vacuum assistance from the vacuum servo unit **(see illustration 15.13)**.

When the brake system is at rest, the central valve in the floating piston rests against the slotted pin. In this condition the central valve is open and brake fluid can discharge out of the pressure chamber back into the brake fluid reservoir. When the brake pedal is depressed, the build-up of hydraulic pressure in the primary pressure chamber acts on the intermediate piston and floating piston, moving them down the cylinder bore. The floating piston contacts the seal on the central valve, closing the connection between the intermediate and secondary pressure chambers. Brake hydraulic pressure can now also increase in the secondary pressure chamber.

Vacuum servo unit

The vacuum servo unit is located between the brake pedal and tandem master cylinder. When the brake pedal is depressed, the servo unit increases the force applied by the pedal, reducing the effort required to operate the brakes **(see illustration 15.14)**.

The unit operates by means of engine inlet manifold vacuum applied to a diaphragm contained within the unit casing. A pushrod connected to the centre of the diaphragm acts directly on the primary piston in the master cylinder.

When the brake pedal is released, vacuum is applied to both sides of the diaphragm. When the pedal is depressed, one side of the diaphragm is opened to atmosphere and the vacuum acting on the other side deflects the diaphragm which in turn operates the master cylinder primary piston. The resulting force applied to the master cylinder piston is therefore significantly greater than the initial force applied to the brake pedal by the driver.

3 System operation

Brake system at rest

When the system is at rest all the brake components are inoperative. Pressure is non-existent in the hydraulic pipes between the tandem master cylinder and the brake calipers. The inlet solenoid valves in the hydraulic control unit valve block are open and the outlet solenoid valves are closed.

Brake system operating under conventional control without ABS

When the brake pedal is activated, the pedal force is applied to the tandem master cylinder by the vacuum servo unit pushrod. The servo unit pushrod acts directly on the pressure piston in the master cylinder which pressurises the hydraulic fluid in the brake pipes to the hydraulic control unit. The inlet solenoid valve and outlet solenoid valve both remain in the 'at rest' position (inlet solenoid valve open

and outlet solenoid valve closed). Hydraulic pressure is transmitted to each brake caliper, thus operating the brakes.

When the brake pedal is released, the one-way valve opens allowing the hydraulic pressure in the circuit to rapidly decrease **(see illustration 15.15)**.

Brake system operating in conjunction with ABS control

The ABS-ECM continually monitors wheel speed from the signals provided by the wheel speed sensors. If the ECM detects the incidence of wheel lock on one or more wheels, ABS is automatically initiated in three phases. As Lucas EBC 430 ABS operates individually on each wheel, all or any of the wheels could be in any one of the following phases at any particular moment.

First ABS phase, pressure holding

To prevent any further build-up of hydraulic pressure in the circuit being controlled, the ECM closes the inlet solenoid valve and allows the outlet solenoid valve to remain closed. The hydraulic fluid line from the tandem master cylinder to the brake caliper or wheel cylinder is closed, and the hydraulic fluid in the controlled circuit is maintained at a constant pressure. This effectively removes the braking force from the controlled circuit. The pressure cannot now be increased in that

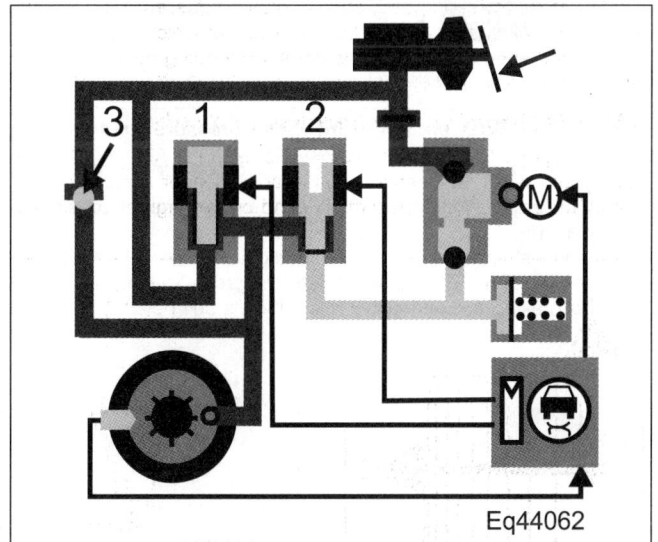

Eq44062

15.15 Brake system operating under conventional control without ABS

1 Inlet solenoid valve	3 One-way valve
2 Outlet solenoid valve	

15

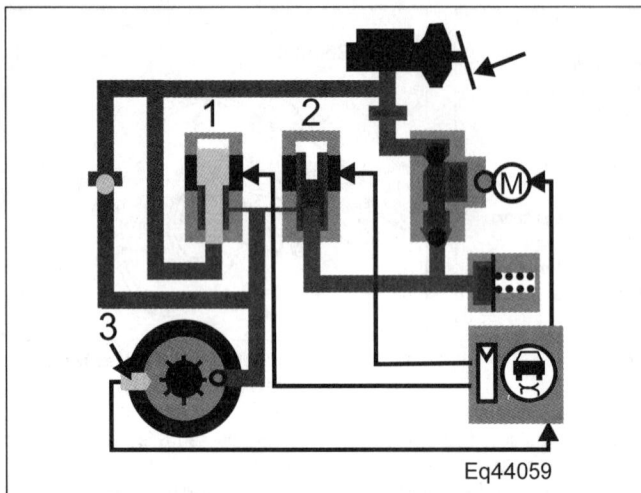

15.16 ABS operation - first phase, pressure holding

1	Inlet solenoid valve	3	Wheel speed sensor
2	Outlet solenoid valve		

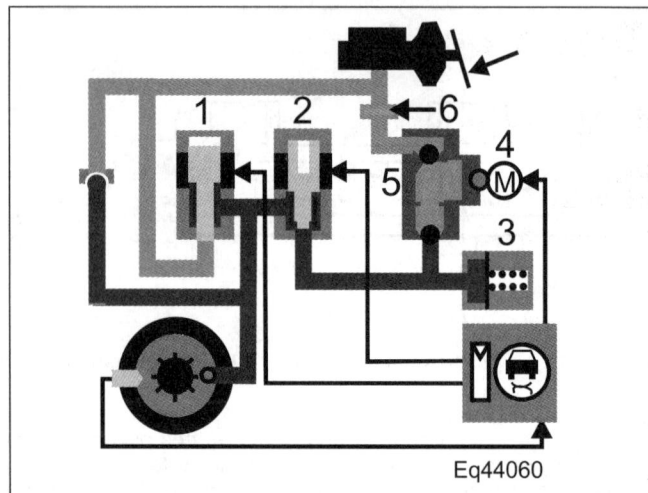

15.17 ABS operation - second phase, pressure reduction

1	Inlet solenoid valve	4	Pump motor
2	Outlet solenoid valve	5	Return pump
3	Pressure accumulator	6	Pulsation damper

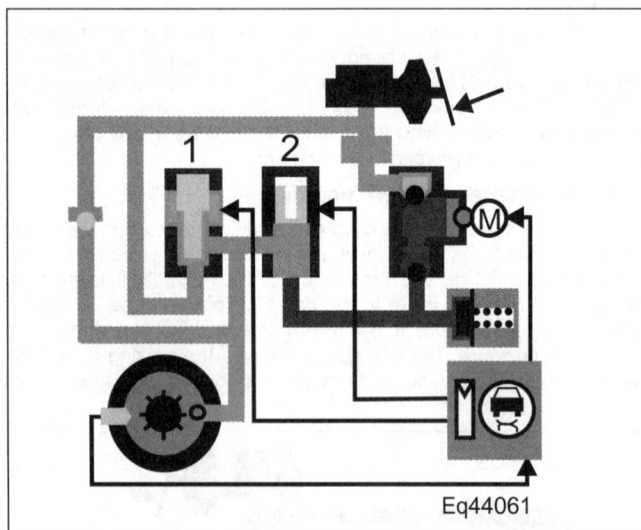

15.18 ABS operation - third phase, pressure build-up

1	Inlet solenoid valve	2	Outlet solenoid valve

circuit by any further application of the brake pedal **(see illustration 15.16)**.

If the wheel speed sensor signals indicate that wheel rotation has now stabilised, the ECM will instigate the pressure build-up phase, allowing braking to continue. If wheel lock is still detected after the pressure holding phase, the ECM instigates the pressure reduction phase.

Second ABS phase, pressure reduction

If the ECM detects wheel instability, a pressure reduction phase is initiated. The inlet solenoid valve remains closed and the outlet solenoid valve is opened by means of a series of short activation pulses. The pressure in the controlled circuit decreases rapidly as the fluid flows from the brake caliper or wheel cylinder into the pressure accumulator. At the same time, the ECM actuates the electric motor to operate the return pump. The hydraulic fluid is then pumped back into the pressure side of the master cylinder. This process creates a pulsation which can be felt in the brake pedal action, but which is softened by the pulsation damper **(see illustration 15.17)**.

Third ABS phase, pressure build-up

The pressure build-up phase is instigated after the wheel rotation has stabilised. The inlet and outlet solenoid valves are returned to the at rest position (inlet solenoid valve open and exhaust solenoid valve closed) which re-opens the hydraulic fluid line from the tandem master cylinder to the brake caliper or wheel cylinder. Hydraulic pressure is reinstated, thus re-introducing operation of the brake. After a brief period, a short pressure holding phase is re-introduced and the ECM continually shifts between pressure build-up and pressure holding until the wheel has decelerated to a sufficient degree where pressure reduction is once more required **(see illustration 15.18)**.

The whole ABS control cycle takes place 4 to 10 times per second for each affected wheel and this ensures maximum braking effect and control during ABS operation.

Test procedures

4 Wheel speed sensors

Inductive type wheel speed sensors (Daewoo models)

Checking the wheel speed sensor (general)

1 Inspect the wheel speed sensor for corrosion or damage and check that the sensor is tightly mounted.

2 Check the toothed sensor ring for damage, eccentricity and for broken or missing teeth.

3 Inspect the wheel speed sensor wiring plug for corrosion and damage. One plug for each sensor.

4 Check that the connector terminal pins are pushed fully home and making good contact with the sensor wiring plug.

5 Check the clearance between the sensor and the toothed sensor ring. The clearance is not normally adjustable but is nominally 0.3 to 0.5 mm. If the clearance is excessive, expect a worn sensor tip or problems with the wheel bearings/hub or sensor ring.

6 When carrying out voltage checks with an oscilloscope or voltmeter, the voltage obtained will be proportional to the speed at which the wheel is rotating. In addition to determining that the wheel speed sensors are actually producing a voltage output, it is essential that the output from the sensors on a particular axle is the same for any given wheel speed.

Checking wheel speed sensor output with an oscilloscope

Note: *Refer to the wiring diagrams for specific ECM pin identification according to model.*

7 Switch the ignition off and disconnect the ECM multi-plug or the relevant wheel speed sensor wiring plug.

8 Connect an oscilloscope between the terminal pins for the sensor under test **(see illustration 15.19)**.

9 Select a range to cover 80 Hz on the oscilloscope and a free run time base.

10 Raise the wheel and rotate it by hand at approximately one revolution per second.

11 A sinusoidal wave form should be obtained, with amplitude and duration changing with rotational speed **(see illustration 15.20)**.

12 If there is no signal, or a very weak or intermittent signal at the ECM, repeat the test at the sensor wiring plug. If there is no change in signal status, the sensor is suspect.

13 If the signal is now satisfactory this indicates a fault in the wiring harness which should be checked for continuity.

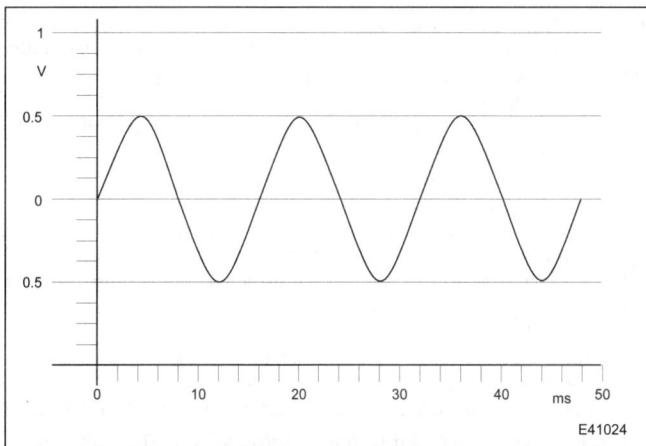

15.19 Checking inductive wheel speed sensor output with an oscilloscope connected to the sensor wiring plug

Checking wheel speed sensor output with an AC voltmeter

Note: *Refer to the wiring diagrams for specific ECM pin identification according to model.*

14 Switch the ignition off and disconnect the ECM multi-plug or the relevant wheel speed sensor wiring plug.

15 Connect an AC voltmeter between the terminal pins for the sensor under test **(see illustration 15.21)**.

16 Raise the wheel and rotate it by hand at approximately one revolution per second.

17 A voltage of approximately 0.1 to 0.5 volts (AC RMS) should be obtained. If there is no signal, or a very weak or intermittent signal at

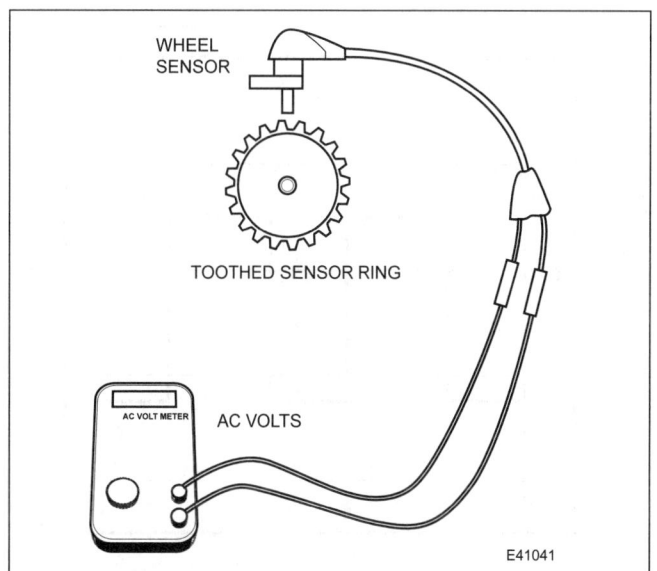

15.20 Typical inductive wheel speed sensor sine wave as displayed on an oscilloscope

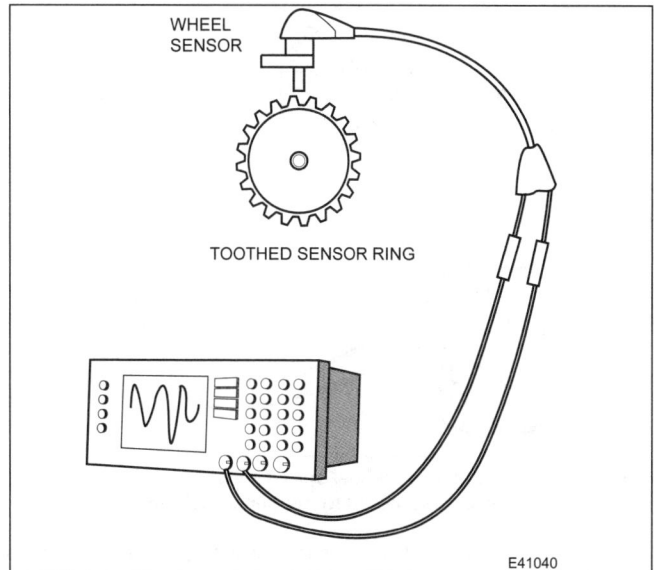

15.21 Checking inductive wheel speed sensor output with a voltmeter connected to the sensor wiring plug

15

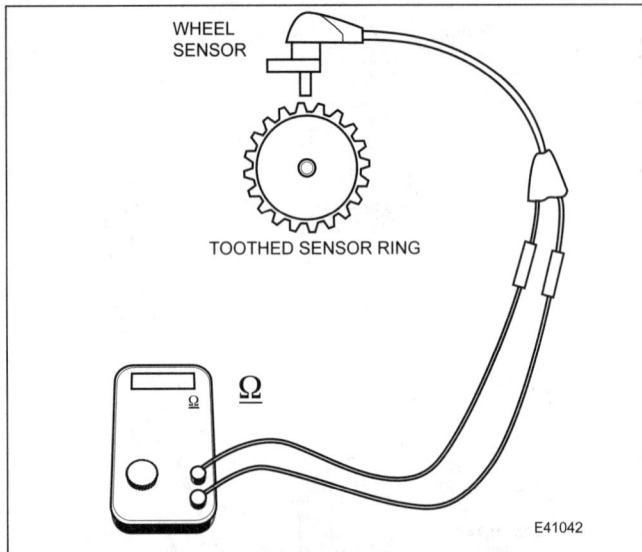

15.22 Checking inductive wheel speed sensor resistance with an ohmmeter connected to the sensor wiring plug

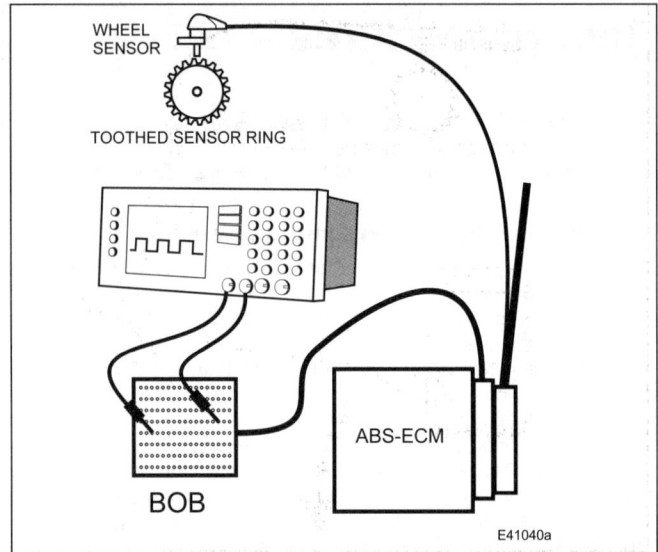

15.23 Checking active wheel speed sensor output with a break-out-box connected between the oscilloscope and ECM multi-plug

the ECM, repeat the test at the sensor wiring plug. If there is no change in the signal, the sensor is suspect.

18 If the signal is now satisfactory, this indicates a fault in the wiring harness which should be checked for continuity. **Note:** *This test at least proves that a signal is being generated by the sensor. However, the voltage produced is an average voltage and does not clearly indicate damage to the sensor ring or that the sinewave is regular in formation.*

Checking wheel speed sensor resistance

Note: *Refer to the wiring diagrams for specific ECM pin identification according to model.*
19 Switch the ignition off and disconnect the ECM multi-plug or the relevant wheel speed sensor wiring plug.
20 Connect an ohmmeter between the terminal pins for the sensor under test **(see illustration 15.22)**.
21 The readings obtained should be between 1.0 and 1.5 kohms.
22 If the resistance is excessively high, or open circuit at the ECM,

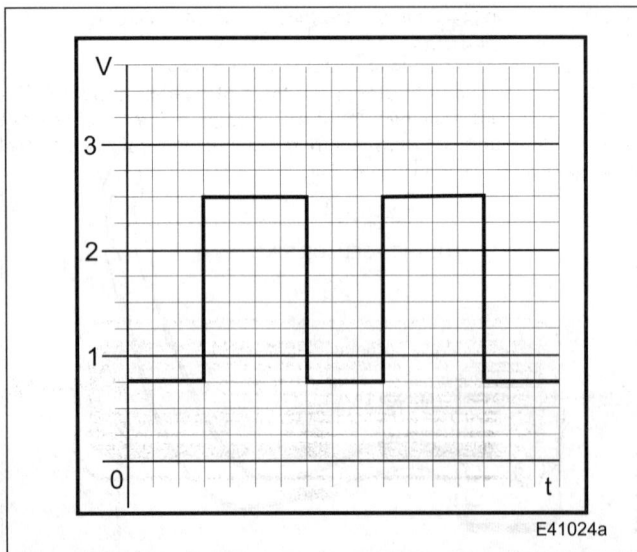

15.24 Active wheel speed sensor square wave output as viewed on an oscilloscope

repeat the test at the sensor multi-plug. If there is no change in resistance, the sensor is suspect.

23 If the resistance is now satisfactory, this indicates a fault in the wiring harness which should be checked for continuity. **Note:** *Even if the resistance is within the quoted specifications, this does not prove that the speed sensor can generate an acceptable signal.*

'Active' type wheel speed sensors (Fiat models)

Checking the wheel speed sensor (general)

24 Inspect the wheel speed sensor for corrosion or damage and check that the sensor is tightly mounted.
25 Check the toothed sensor ring for damage, eccentricity and for broken or missing teeth.
26 Inspect the wheel speed sensor wiring plug for corrosion and damage. One plug for each sensor.
27 Check that the connector terminal pins are pushed fully home and making good contact with the sensor wiring plug.
28 Check the clearance between the sensor and the toothed sensor ring. The clearance is not normally adjustable but is nominally 0.2 to 2.0 mm. If the clearance is excessive, expect a worn sensor tip or problems with the wheel bearings/hub or sensor ring.

Checking wheel speed sensor signal with an oscilloscope

Note: *Refer to the wiring diagrams for specific ECM pin identification according to model.*
29 Switch the ignition off and connect a BOB between the ECM and the harness multi-plug.
30 Connect an oscilloscope to the relevant BOB terminals for the sensor under test, and set the oscilloscope to measure DC voltage **(see illustration 15.23)**.
31 Raise and support the vehicle.
32 Switch the ignition on and rotate the wheel by hand at approximately one revolution per second.
33 A square wave form should be obtained, switching between approximately 0.5 and 2.5 volts as the wheel is rotated **(see illustration 15.24)**.
34 If the output is not as specified, check the supply voltage to the sensor and the wiring harness for continuity.
35 If the supply voltage and wiring are satisfactory, the sensor is suspect.

Checking wheel speed sensor signal with a DC voltmeter

Note: *Refer to the wiring diagrams for specific ECM pin identification according to model.*

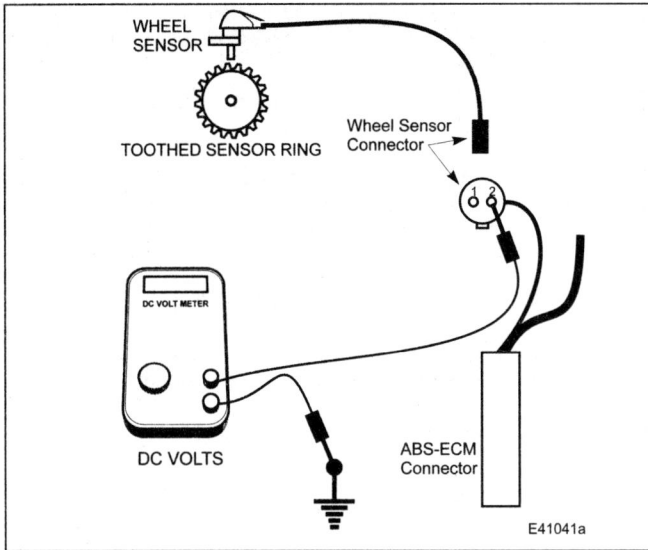

15.25 Checking active wheel speed sensor output using a voltmeter

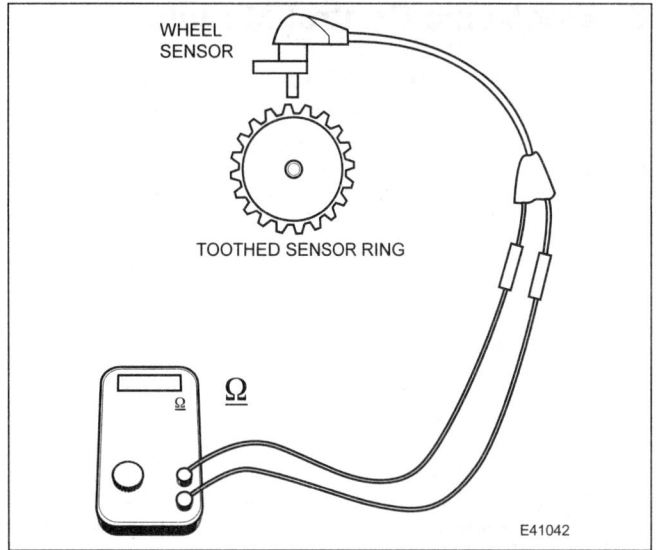

15.26 Checking active wheel speed sensor resistance with an ohmmeter connected to the sensor wiring plug

36 Switch the ignition off and connect a BOB between the ECM and the harness multi-plug.

37 Connect a DC voltmeter between the relevant BOB terminal for the sensor under test, and earth.

38 Raise and support the vehicle.

39 Switch the ignition on and rotate the wheel very slowly.

40 A voltage switching between approximately 0.5 and 2.5 volts should be obtained as the wheel is rotated.

41 If the voltage is not as specified, check the supply voltage to the sensor and the wiring harness for continuity.

42 If the supply voltage and wiring are satisfactory, the sensor is suspect.

Checking wheel speed sensor supply voltage

43 Switch the ignition off and disconnect the wheel speed sensor wiring harness 2-pin multi-plug, of the sensor under test.

44 Connect a DC voltmeter between a vehicle earth and pin 2 of the multi-plug **(see illustration 15.25)**.

45 Switch the ignition on. A voltage of approximately 8.0 volts should be obtained.

46 If the voltage is not as specified, repeat the test at the ECM multi-plug using the BOB as described previously. If there is no change in the reading, the ABS-ECM is suspect. If the voltage is now satisfactory, this indicates a fault in the wiring harness which should be checked for continuity. **Note:** *This test at least proves that a supply voltage is available for the sensor. However, the integral electronics in the sensor must be active to produce an output.*

Checking wheel speed sensor resistance

Note: *Refer to the wiring diagrams for specific ECM pin identification according to model.*

47 Switch the ignition off and disconnect the ECM multi-plug or the relevant wheel speed sensor 2-pin multi-plug.

48 Connect an ohmmeter between the terminal pins for the sensor under test.

49 The readings obtained should be between 2.5 and 4.5 kohms **(see illustration 15.26)**.

50 If the resistance is excessively high, or open circuit at the ECM, repeat the test at the sensor multi-plug. If there is no change in resistance, the sensor is suspect.

51 If the resistance is now satisfactory, this indicates a fault in the wiring harness which should be checked for continuity. **Note:** *Even if the resistance is within the quoted specifications, this does not prove that the speed sensor can generate an acceptable signal.*

5 System relays

Relay operation tests

1 The relays are integral with the ECM and can only be checked using suitable ABS diagnostic test equipment.

Relay power supply tests

Main (solenoid valve) relay

2 Switch the ignition off and disconnect the ECM multi-plug.

3 Switch the ignition on.

4 Attach a negative voltmeter probe to a vehicle earth.

5 Attach a positive voltmeter probe to ECM multi-plug pin 15 **(see illustration 15.27)**. The voltmeter should indicate nbv. If no voltage is found, check the relevant fuse and the supply wiring back to the ignition switch.

Pump relay

6 Switch the ignition off and disconnect the ECM multi-plug.

7 Switch the ignition on.

8 Attach a negative voltmeter probe to a vehicle earth.

9 Attach a positive voltmeter probe to ECM multi-plug pin 17. The voltmeter should indicate nbv. If no voltage is found, check the relevant fuse and the supply wiring back to the battery positive terminal.

15

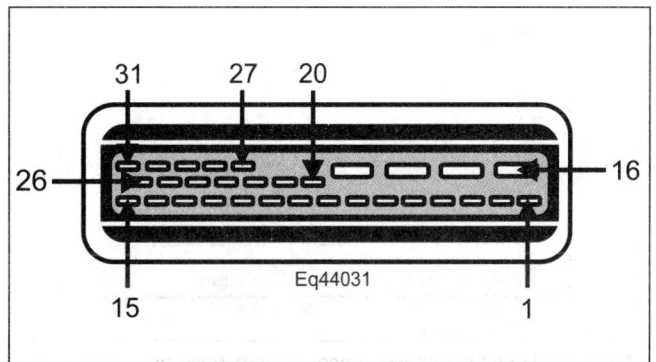

15.27 ECM 31-pin multi-plug terminal identification

6 Electronic Control Module

Checking the ECM (general)

1 Inspect the ECM for corrosion or damage and ensure that the unit is mounted securely.
2 Check that the ECM multi-plug terminals are pushed fully home and making good contact with the ECM pins. Faults in any of the above areas are possible reasons for poor performance in the ABS system.

ECM power supply and earth tests

3 Switch the ignition off and disconnect the ECM multi-plug.
4 Switch the ignition on.
5 Attach a negative voltmeter probe to a vehicle earth.
6 Attach a positive voltmeter probe to ECM multi-plug pin 18 (see illustration 15.27). The voltmeter should indicate nbv. If no voltage is found, check the relevant fuse and the supply wiring back to the ignition switch.
7 Switch the ignition off.
8 Connect an ohmmeter between a vehicle earth and ECM multi-plug pin 19. The ohmmeter should indicate continuity. If not, check the ECM main earth connection and wiring.

7 Solenoid valves

Solenoid valve operation tests

1 The solenoid valves are integral with the ECM and can only be checked using suitable ABS diagnostic test equipment.

8 Hydraulic pump motor

Pump operation test

1 The pump motor is integral with the ECM and hydraulic control unit and can only be checked using suitable ABS diagnostic test equipment.

9 Brake light switch

Checking the brake light switch (general)

1 Check that the brake light switch is correctly and securely mounted and that the plunger moves smoothly with no trace of binding.
2 Check that the wiring multi-plug is pushed fully home and making good contact.

3 Check that no wires have been disconnected.
4 Faults in any of the above areas are possible reasons for failure or malfunctioning of the switch.

Brake light switch voltage and continuity tests

Voltage test

Note: Refer to the wiring diagrams for specific ECM pin identification according to model.
5 Switch the ignition off and disconnect the ECM multi-plug from the ECM.
6 Connect a voltmeter between ECM multi-plug brake light input pin 24 and an ECM earth pin (see illustration 15.27).
7 Switch the ignition on and depress the brake pedal. The voltmeter should indicate nbv.
8 If no voltage is found, the fuse, the brake light switch and the wiring are suspect.
9 Release the brake pedal. The voltage should drop to zero as the switch opens.

Continuity test

10 Switch the ignition off and disconnect the brake light switch multi-plug.
11 Connect an ohmmeter between the terminal pins of the brake light switch.
12 Operate the brake light switch and check for continuity. If the test fails, renew the switch.

10 Warning lamp

Checking the warning lamp (general)

1 Inspect the warning lamp bulb holder contacts in the instrument panel.
2 Check that the instrument panel multi-plug terminal pins are pushed fully home and making good contact.
3 Check that no wires have been disconnected.
4 Faults in any of the above areas are possible reasons for failure or malfunctioning of the warning lamp.

Warning lamp operation test

5 With the ignition switched off, the warning lamp should remain off.
6 Switch the ignition on and the warning lamp should illuminate then extinguish after a few seconds. The lamp should then remain off.
7 If the warning lamp comes on and remains on at any time during vehicle operation, carry out the previously described test procedures on the system components.

Pin table - typical 31-pin (Daewoo)

Note: Refer to illustration 15.27

Pin No.	Connection	Test condition	Voltage
1	-	-	-
2	-	-	-
3	-	-	-
4	-	-	-
5	-	-	-
6	-	-	-
7	-	-	-
8	-	-	-
9	Rear right wheel speed sensor earth	Roadwheel rotating	0.25 volts (max)
10	Rear right wheel speed sensor signal	Roadwheel rotating	0.1 to 0.5 volts AC (approx)
11	SD connector	-	-
12	Front right wheel speed sensor signal	Roadwheel rotating	0.1 to 0.5 volts AC (approx)
13	Front right wheel speed sensor earth	Roadwheel rotating	0.25 volts (max)

Pin No.	Connection	Test condition	Voltage
14	Front left wheel speed sensor signal	Roadwheel rotating	0.1 to 0.5 volts AC (approx)
15	Supply from ignition switch	Ignition on	Nbv
16	ECM earth	Ignition on	0.25 volts (max)
17	Supply from battery	Ignition off/on	Nbv
18	Supply from battery	Ignition off/on	Nbv
19	ECM earth	Ignition on	0.25 volts (max)
20	ABS warning lamp	Ignition on:	
		Lamp on	1.0 volt (max)
		Lamp off	Nbv
21	-	-	-
22	Rear left wheel speed sensor earth	Roadwheel rotating	0.25 volts (max)
23	Rear left wheel speed sensor signal	Roadwheel rotating	0.1 to 0.5 volts AC (approx)
24	Brake light switch input	Ignition on:	
		Brake pedal released	0 volts
		Brake pedal depressed	Nbv
25	Front left wheel speed sensor earth	Roadwheel rotating	0.25 volts (max)
26	-	-	-
27	-	-	-
28	-	-	-
29	-	-	-
30	-	-	-
31	-	-	-

Pin table - typical 31-pin (Fiat)

Note: *Refer to illustration 15.27*

Pin No.	Connection	Test condition	Voltage
1	-	-	-
2	-	-	-
3	-	-	-
4	-	-	-
5	-	-	-
6	-	-	-
7	-	-	-
8	-	-	-
9	Rear right wheel speed sensor supply	Ignition on	8.0+ volts
10	Rear right wheel speed sensor signal	Ignition on/roadwheel rotating	0.5 or 2.5 volts (switching)
11	SD connector	-	-
12	Front right wheel speed sensor supply	Ignition on	8.0+ volts
13	Front right wheel speed sensor signal	Ignition on/roadwheel rotating	0.5 or 2.5 volts (switching)
14	Front left wheel speed sensor supply	Ignition on	8.0+ volts
15	Supply from ignition switch	Ignition on	Nbv
16	ECM earth	Ignition on	0.25 volts (max)
17	Supply from battery	Ignition off/on	Nbv
18	Supply from battery	Ignition off/on	Nbv
19	ECM earth	Ignition on	0.25 volts (max)
20	ABS warning lamp	Ignition on:	
		Lamp on	1.0 volt (max)
		Lamp off	Nbv
21	Brake fluid level warning lamp/switch	-	-
		-	-
22	Rear left wheel speed sensor supply	Ignition on	8.0+ volts
23	Rear left wheel speed sensor signal	Ignition on/roadwheel rotating	0.5 or 2.5 volts (switching)
24	Brake light switch input	Ignition on:	
		Brake pedal released	0 volts
		Brake pedal depressed	Nbv
25	Front left wheel speed sensor signal	Ignition on/roadwheel rotating	0.5 or 2.5 volts (switching)
26	-	-	-
27	-	-	-
28	-	-	-
29	-	-	-
30	-	-	-
31	-	-	-

15

Fault codes

1 The Lucas EBC 430 system requires the use of a FCR for obtaining fault codes. Flash codes are not available for output from this system. For the sake of completeness, we have provided a fault code table for Daewoo vehicles. At this time, fault code tables for Fiat vehicles are not available

2 If a FCR is available, it should be connected to the SD serial connector and used in accordance with the maker's instructions.

3 The FCR can be used for the following purposes:
 a) *Obtaining fault codes.*
 b) *Clearing fault codes.*
 c) *Obtaining datastream information.*
 d) *Testing the system actuators (solenoid valve relay, pump relay and solenoid valves).*

Fault code table

Code	Item	Fault
0354	Front left wheel speed sensor	Open or short circuit
0355	Front left wheel speed sensor	Incorrect air gap, damaged or missing sensor ring
0356	Front left wheel speed sensor	Intermittent short circuit
0404	Front right wheel speed sensor	Open or short circuit
0405	Front right wheel speed sensor	Incorrect air gap, damaged or missing sensor ring
0406	Front right wheel speed sensor	Intermittent short circuit
0454	Rear left wheel speed sensor	Open or short circuit
0455	Rear left wheel speed sensor	Incorrect air gap, damaged or missing sensor ring
0456	Rear left wheel speed sensor	Intermittent short circuit
0504	Rear right wheel speed sensor	Open or short circuit
0505	Rear right wheel speed sensor	Incorrect air gap, damaged or missing sensor ring
0506	Rear right wheel speed sensor	Intermittent short circuit
0601	Front left outlet solenoid valve	Supply voltage short circuit or poor earth connection
0602	Front left outlet solenoid valve	Supply voltage open circuit or poor earth connection
0651	Front left inlet solenoid valve	Supply voltage short circuit or poor earth connection
0652	Front left inlet solenoid valve	Supply voltage open circuit or poor earth connection
0701	Front right outlet solenoid valve	Supply voltage short circuit or poor earth connection
0702	Front right outlet solenoid valve	Supply voltage open circuit or poor earth connection
0751	Front right inlet solenoid valve	Supply voltage short circuit or poor earth connection
0752	Front right inlet solenoid valve	Supply voltage open circuit or poor earth connection
0801	Rear left outlet solenoid valve	Supply voltage short circuit or poor earth connection
0802	Rear left outlet solenoid valve	Supply voltage open circuit or poor earth connection
0851	Rear left inlet solenoid valve	Supply voltage short circuit or poor earth connection
0852	Rear left inlet solenoid valve	Supply voltage open circuit or poor earth connection
0901	Rear right outlet solenoid valve	Supply voltage short circuit or poor earth connection
0902	Rear right outlet solenoid valve	Supply voltage open circuit or poor earth connection
0951	Rear right inlet solenoid valve	Supply voltage short circuit or poor earth connection
0952	Rear right inlet solenoid valve	Supply voltage open circuit or poor earth connection
1102	Pump motor	Open circuit
1103	Pump motor relay	Defective
1104	Pump motor	Short circuit
1211	ECM main relay	Short circuit
1212	ECM main relay	Open circuit
1213	ECM main relay	Defective
1610	Brake light switch	Open circuit
2321	ABS warning lamp	Short to battery
2322	ABS warning lamp	Short to earth
2458	Wheel speed sensor intermittent fault	Poor wiring connection; incorrect sensor air gap
2459	Wheel speed sensor signal variation	Incorrect tyre sizes or pressures
2520	ECM	Internal fault
5503	ECM back-up circuitry	Internal fault
5560	ECM back-up circuitry	Inoperative
5610	ECM memory (8-bit RAM/ROM)	Internal fault
5630	ECM memory (16-bit ROM)	Internal fault
5640	ECM memory (16-bit RAM)	Internal fault
8001	Battery voltage high	Voltage exceeds 16.0 volts
8002	Battery voltage low	Voltage below 9.0 volts
8003	Battery voltage low	Voltage below 9.5 volts

Wiring diagrams

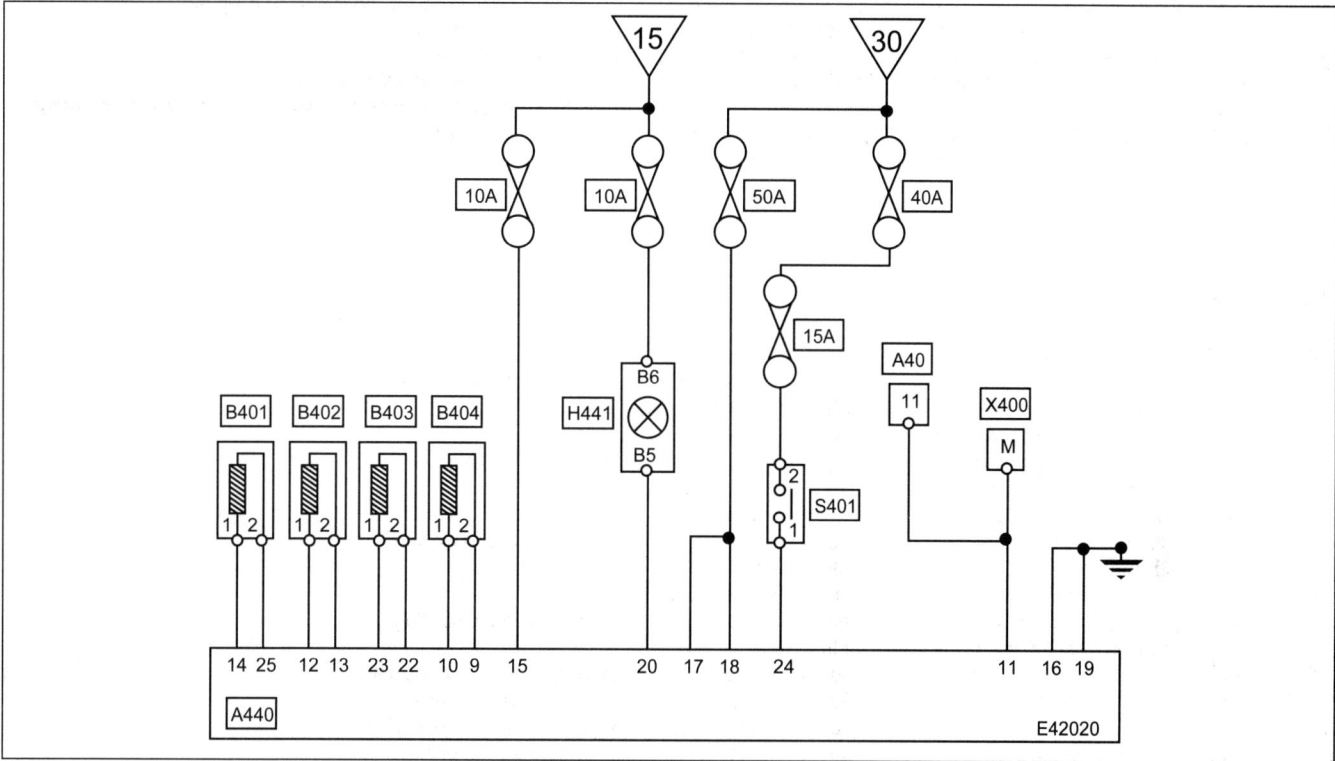

15.28 31-pin wiring diagram, Daewoo Matiz

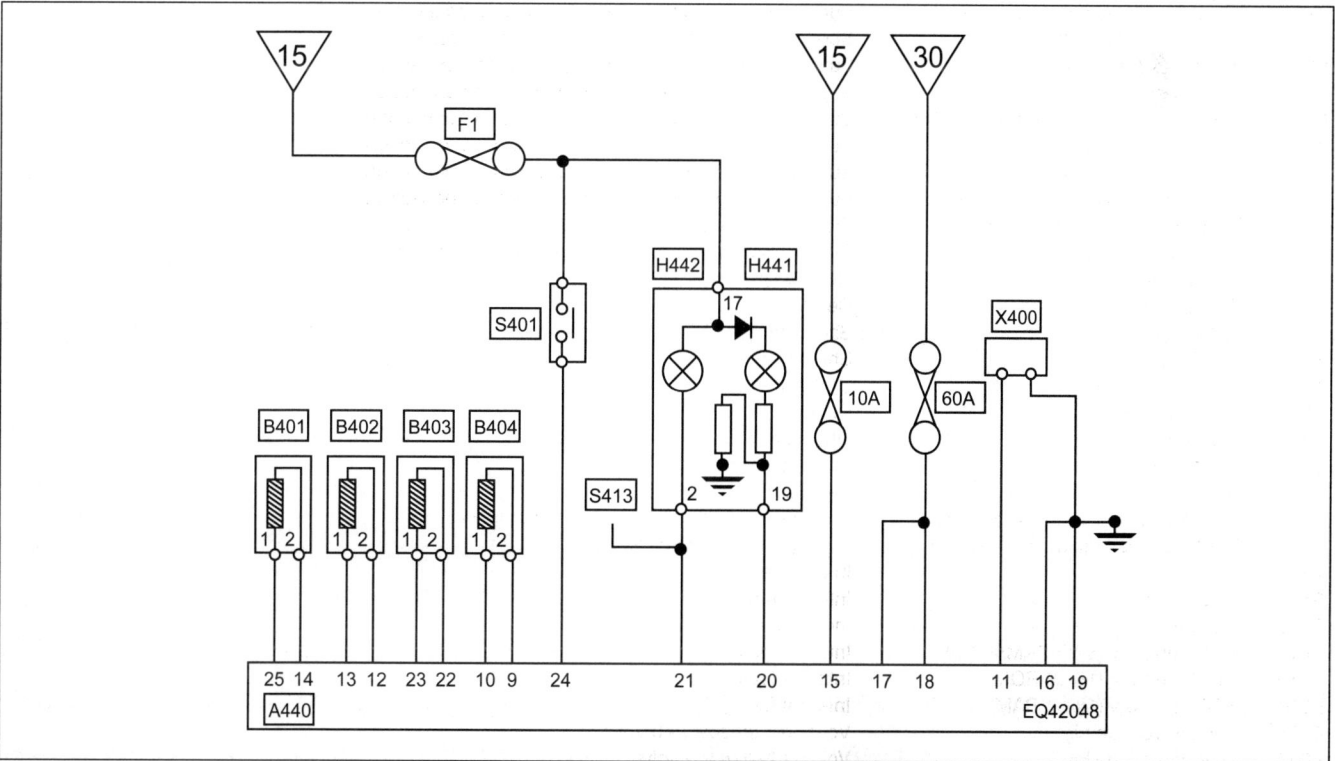

15.29 31-pin wiring diagram, Fiat Marea/Weekend

15

Chapter 16
Lucas SCS ABS

Contents

Vehicle coverage

Model	Year
Ford Escort/Orion	1985-1990
Ford Fiesta/Fiesta RS	1989-1996

Overview of system operation

1 Basic principles and system identification

The SCS (Stop Control System) was developed by Lucas Girling in conjunction with Ford, and fitted to Ford Fiesta and Escort/Orion vehicles in the late 1980s, early 1990s. The system is unique in that it is an entirely mechanical form of antilock brake system. The system is of the additional or 'add-on' type operating in conjunction with the conventional braking system components.

Lucas SCS is a two-channel system with one front brake and one diagonally opposite rear brake on each channel. During ABS operation, hydraulic fluid pressure to the front and rear brakes is regulated by two mechanical/hydraulic modulators, one for each circuit, with additional regulation of rear brake hydraulic pressure by means of load apportioning valves.

The modulators are located above the inner end of each driveshaft and driven from the driveshaft inner constant velocity joint by a toothed belt.

The Lucas Stop Control System is comprised of the following components (see illustration 16.1):

a) Master cylinder and vacuum servo unit.
b) Modulator units.
c) Modulator drivebelts.
d) Load apportioning valves.
e) Hydraulic circuit - front left, rear right.
f) Hydraulic circuit - front right, rear left.

2 Component description and operation

Modulator units

The two modulator units are the prime components of the system and each control the hydraulic fluid pressure to one front brake and one diagonally opposite rear brake. The pressure applied to the rear brakes being additionally regulated by two individual load apportioning valves (see illustration 16.2).

The modulators consist of a shaft, flywheel and friction clutch and a series of internal valves, pistons and control levers. An external sprocket is attached to one end of the shaft and is driven by a toothed belt from the vehicle driveshaft inner constant velocity joint. Under normal driving and braking, the shaft, flywheel and friction clutch rotate together as a unit at roadwheel speed.

When a front wheel is about to lock, it decelerates rapidly and consequently the modulator shaft also decelerates rapidly. The inertia of the flywheel however, is such that it will over-run the friction clutch and operate an internal valve within the modulator. This allows hydraulic fluid pressure in the controlled circuit to be released, reducing the locking tendency of the wheel.

Tandem master cylinder

The tandem master cylinder comprises a body casting incorporating primary and secondary pressure chambers, primary piston, intermediate piston, floating piston, slotted pin and central valve. The cylinder operates as a conventional master cylinder

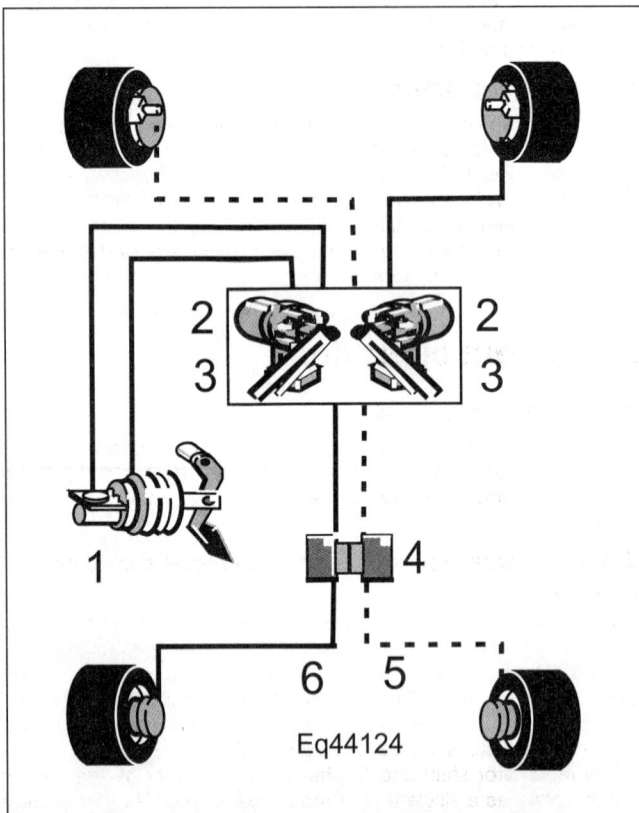

16.1 Lucas SCS main components

1. Master cylinder and vacuum servo unit
2. Modulator units
3. Modulator drivebelts
4. Load apportioning valves
5. Hydraulic circuit - front left, rear right
6. Hydraulic circuit - front right, rear left

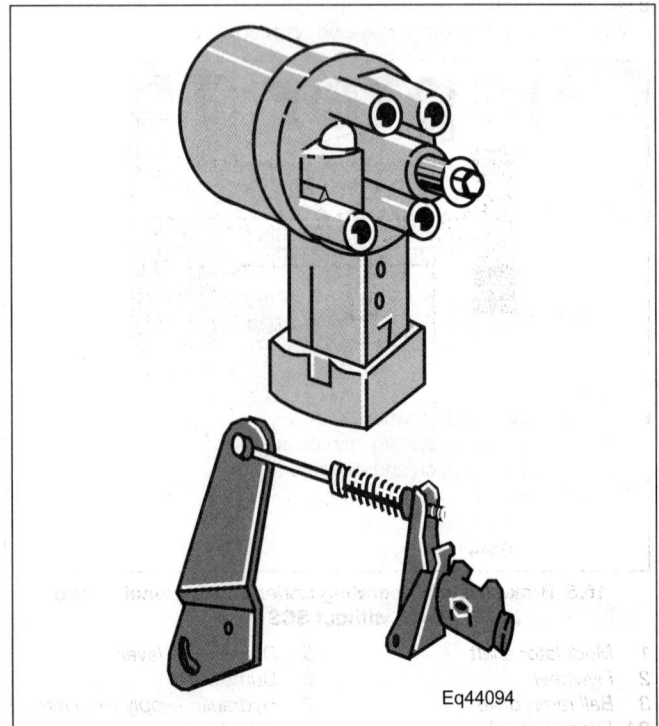

16.2 Lucas SCS modulator unit and load apportioning valve

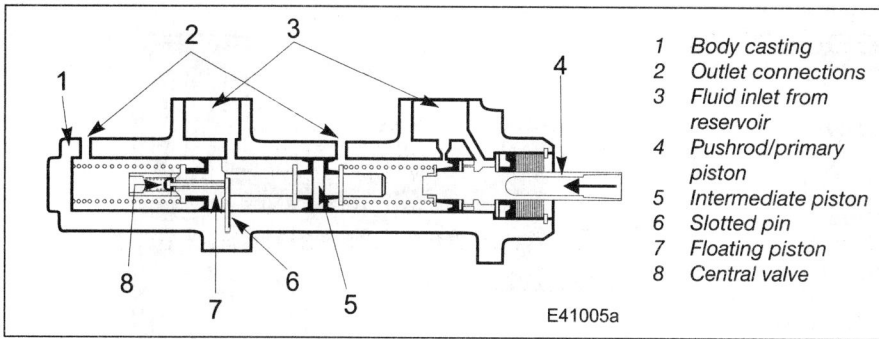

1	Body casting
2	Outlet connections
3	Fluid inlet from reservoir
4	Pushrod/primary piston
5	Intermediate piston
6	Slotted pin
7	Floating piston
8	Central valve

E41005a

16.3 Sectional view of a typical tandem master cylinder

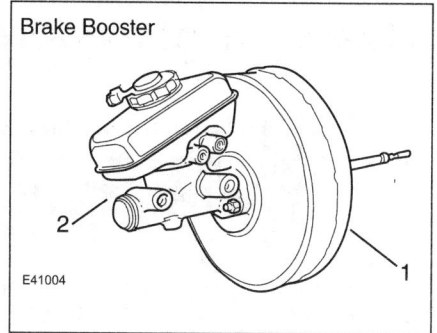

E41004

16.4 Vacuum servo unit (1) and tandem master cylinder (2)

using vacuum assistance from the vacuum servo unit (see illustration 16.3).

When the brake system is at rest, the central valve in the floating piston rests against the slotted pin. In this condition the central valve is open and brake fluid can discharge out of the pressure chamber back into the brake fluid reservoir. When the brake pedal is depressed, the build-up of hydraulic pressure in the primary pressure chamber acts on the intermediate piston and floating piston, moving them down the cylinder bore. The floating piston contacts the seal on the central valve, closing the connection between the intermediate and secondary pressure chambers. Brake hydraulic pressure can now also increase in the secondary pressure chamber.

Vacuum servo unit

The vacuum servo unit is located between the brake pedal and tandem master cylinder. When the brake pedal is depressed, the servo unit increases the force applied by the pedal, reducing the effort required to operate the brakes (see illustration 16.4).

The unit operates by means of engine inlet manifold vacuum applied to a diaphragm contained within the unit casing. A pushrod connected to the centre of the diaphragm acts directly on the primary piston in the master cylinder.

When the brake pedal is released, vacuum is applied to both sides

of the diaphragm. When the pedal is depressed, one side of the diaphragm is opened to atmosphere and the vacuum acting on the other side deflects the diaphragm which in turn operates the master cylinder primary piston. The resulting force applied to the master cylinder piston is therefore significantly greater than the initial force applied to the brake pedal by the driver.

Load apportioning valves

Two load apportioning valves, one for each hydraulic circuit, are incorporated to restrict the hydraulic fluid pressure to the rear brakes. The valves are mounted on the vehicle underbody and connected to brackets on the rear suspension by a lever, rod and spring arrangement.

The valves work on the load conscious principle whereby the hydraulic pressure to the relevant rear brake is regulated according to vehicle loading and attitude.

SCS warning lamps

Warning lamps are used in the Stop Control System to indicate low hydraulic fluid level, application of the handbrake, or a broken modulator drivebelt. The hydraulic fluid level is monitored by a float in the fluid reservoir. If the level falls, the float switch contacts close which illuminates the warning lamp. Switches on the handbrake lever and on the side of the modulator drivebelt covers are used to indicate application of the handbrake or broken modulator drivebelt, respectively.

3 System operation

Brake system at rest

When the system is at rest, all the brake components are inoperative and pressure is non-existent in the hydraulic pipes between the tandem master cylinder and the brake calipers or wheel cylinders.

Brake system operating conventionally without SCS control

When the brake pedal is depressed, the pedal force is applied to the tandem master cylinder by the vacuum servo unit pushrod. The servo unit pushrod acts directly on the pressure piston in the master cylinder. The resulting increase in hydraulic pressure moves the intermediate piston in the cylinder, and the pressure which increases in front of the pistons flows out to the modulators.

The modulator shaft and flywheel, held together by the friction clutch, rotate as a single unit at roadwheel speed. The dump valve lever in the modulator is at rest and consequently the dump valve and all subsequent valves are closed. This allows an unrestricted hydraulic passage from the master cylinder through the modulator to the brake caliper/wheel cylinder, thus operating the brake. The passage of fluid in the rear brake circuit is through the load apportioning valve which regulates the pressure according to vehicle loading and attitude (see illustration 16.5).

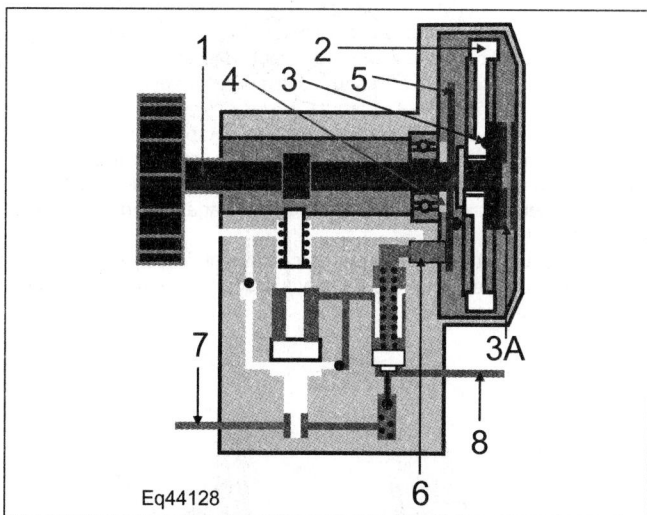

Eq44128

16.5 Brake system operating under conventional control without SCS

1	Modulator shaft	5	Dump valve lever
2	Flywheel	6	Dump valve
3	Ball ramp drive	7	Hydraulic supply from master cylinder
3A	Friction clutch		
4	Dump valve lever spring	8	Hydraulic supply to brakes

16

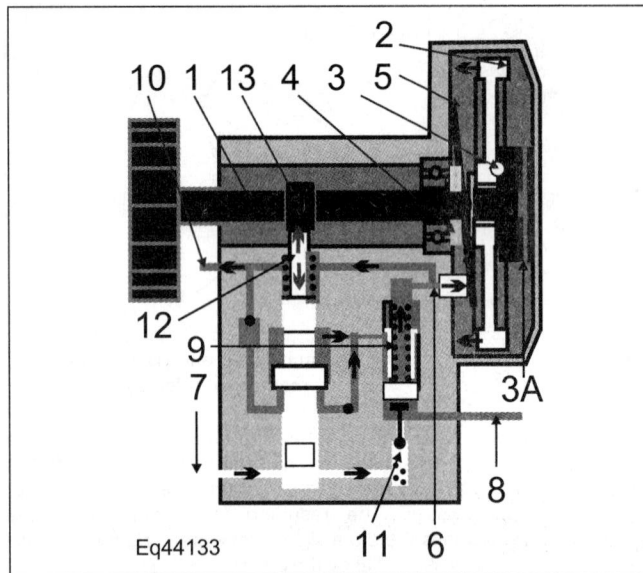

16.6 SCS operation - phase one

1	Modulator shaft	7	Hydraulic supply from master
2	Flywheel		cylinder
3	Ball ramp drive	8	Hydraulic supply to brakes
3A	Friction clutch	9	De-boost piston
4	Dump valve lever	10	Fluid return to reservoir
	pivot	11	Cut-off valve
5	Dump valve lever	12	Pump piston
6	Dump valve	13	Modulator shaft eccentric

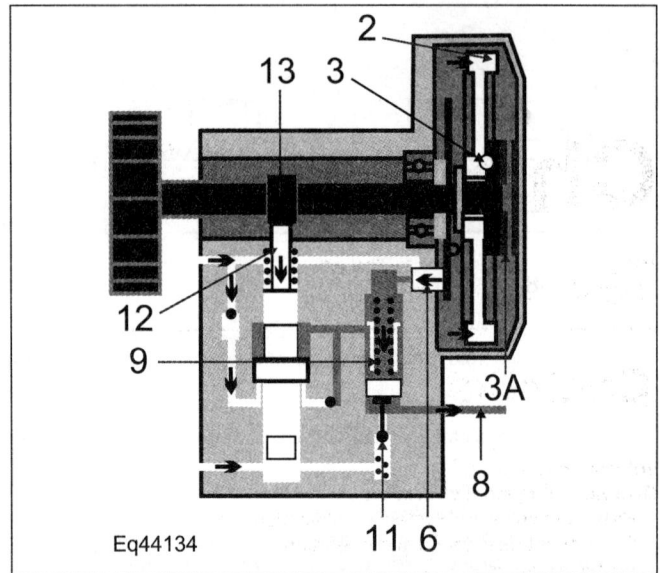

16.7 SCS operation - phase two

2	Flywheel	8	Hydraulic supply to brakes
3	Ball ramp drive	9	De-boost piston
3A	Friction clutch	11	Cut-off valve
6	Dump valve	12	Pump piston

Brake system operating with SCS control

Phase one

When a front wheel is about to lock, it decelerates rapidly and consequently the modulator shaft also decelerates rapidly. The inertia of the flywheel, however, is such that it will over-run the friction clutch and continue rotating. This over-run moves the flywheel against the dump valve lever by means of a ball and ramp assembly. The dump valve lever opens the dump valve allowing a proportion of the hydraulic fluid in the modulator to be returned to the reservoir. Hydraulic pressure in the modulator is reduced, which allows the de-boost piston to lift. Master cylinder input hydraulic pressure is now greater on the underside of the cut-off valve which closes the valve and restricts any further increase in pressure to the controlled brake. Hydraulic pressure moves the pump piston into contact with the eccentric on the modulator shaft, thus creating a hydraulic pump. However, as the dump valve is still open, there is no hydraulic pressure created by the piston at this stage (see illustration 16.6).

Phase two

As the hydraulic pressure to the controlled brake is now reduced, the wheel accelerates and the rotating components in the modulator (modulator shaft, flywheel and friction clutch) return to the normal position, ie, brake system operating conventionally without SCS control.

This allows the dump valve to close and the pump piston to generate hydraulic pressure which is supplied to the controlled circuit. The de-boost piston also starts to close increasing hydraulic pressure still further.

The brake of the controlled wheel is now re-applied and continued closure of the de-boost piston will ultimately restore full hydraulic pressure from the master cylinder to the controlled brake. If however, the wheel begins to lock again, phase one will again be instigated (see illustration 16.7).

The whole control cycle takes place several times per second for each affected wheel until the vehicle is brought to a halt or the driver releases the brakes.

Test procedures

The Lucas Stop Control System is entirely mechanical in operation and therefore there are no electrical test procedures, pin-tables or fault codes.

Chapter 17
Teves II ABS

Contents

Vehicle coverage

Model	Year
Citroën	
BX	1987-1992
Ford	
Escort Cosworth	1992-1996
Granada/Scorpio	1985-1991
Granada/Scorpio 2.5 D	1992-1994
Sierra/Sapphire	1985-1992
Jaguar/Daimler	
XJ6/XJ12	1993-1994
XJS	1992-1996
Peugeot	
505	1983-1990
Renault	
21	1988-1992
Saab	
900	1989-1993
9000	1987-1993
Volkswagen	
Corrado	1988-1992
Golf	1989-1992
Jetta	1989-1992
Passat	1990-1992
Volvo	
440/460/480	1987-1992

17

Overview of system operation

1 Basic principles and system identification

The Teves II Antilock Brake System has been fitted to a wide range of passenger vehicles since its introduction in the mid 1980s. The system is often referred to as an integrated type, whereby the components of the ABS hydraulic control unit are also required during the operation of the conventional braking system.

The purpose of the system is to apply the vehicle brakes at maximum efficiency without wheel lock or loss of directional stability. Inductive sensors (wheel speed sensors) monitor the speed of the wheels by generating an electrical signal as the wheel is rotated. This information is passed to the ABS Electronic Control Module (ECM) which is then able to determine wheel speed, wheel acceleration and wheel deceleration. The ECM compares the signals received from each wheel and if the onset of lock at any wheel is detected, a signal is sent to the ABS hydraulic control unit which regulates the brake pressure for the relevant wheel(s).

Initially, Teves II ABS did not have a self-diagnostic (SD) capability, however versions with self-diagnostics were introduced from approximately 1992 onward (approximately 1989 on Saab models). The two types can be identified by the configuration of the ECM multi-plug connector; models with a 35-pin connector are generally non-SD up to 1992, and SD from 1992 onward. Models fitted with a 55-pin connector are all SD.

Teves II ABS is comprised of the following components **(see illustration 17.1)**:

a) *ABS-ECM.*
b) *Hydraulic control unit.*
c) *Four inductive wheel speed sensors and associated sensor rings.*
d) *ABS electrical wiring harness and relays.*

e) *Brake light switch.*
f) *ABS warning lamps.*
g) *Brake calipers and hydraulic hoses and pipes.*
h) *Pressure regulating/load sensing valve(s) depending on application.*

2 Component description and operation

ABS ECM

General

The Teves Electronic Control Module (ECM) continually monitors wheel speed from the signals provided by the wheel speed sensors. If the ECM detects the incidence of wheel lock on one or more wheels, a signal is sent to the hydraulic control unit to modulate the hydraulic pressure to the brake of the locking wheel **(see illustration 17.2)**.

The ECM contains two microprocessors and uses digital technology to complete this function and other functions such as fault code memory (self-diagnostic versions) and power modules for valve and relay activity.

Self-test

The Teves II ECM is equipped with a self-test capability that initially examines the ABS system when the ignition is switched on, and then examines the wheel speed sensor signals after a wheel speed of approximately 4 mph is reached from all wheels (engine running). The ABS self-test program continues to examine the signals from the various components as long as the ignition is switched on. If self-test determines that faults are not present, the ABS is ready for operation once a specified vehicle speed has been achieved.

If the ECM detects that a fault is present, all ABS functions are switched off and one or both of the warning lamps are turned on. Depending on the nature of the fault detected, conventional braking will still be available on all wheels, or on the front wheels only.

Self-diagnostics

On systems with self-diagnostics, if the ECM detects a fault during the self-test routine, an internal fault code is stored in the ECM memory. Stored fault codes can be retrieved from the SD connector with the aid of a suitable fault code reader, or displayed as flash codes on the ABS warning lamp on certain systems. If the fault clears, the code will remain stored until cleared with the FCR.

Hydraulic control unit

The hydraulic control unit consists of three main assemblies **(see illustration 17.3)**:

Pump assembly
Consisting of a high pressure electric pump, pressure accumulator and pressure switch.

Master cylinder
Incorporating the hydraulic fluid reservoir and the ABS main solenoid valve.

Hydraulic valve block
Containing the ABS inlet and outlet solenoid valves.

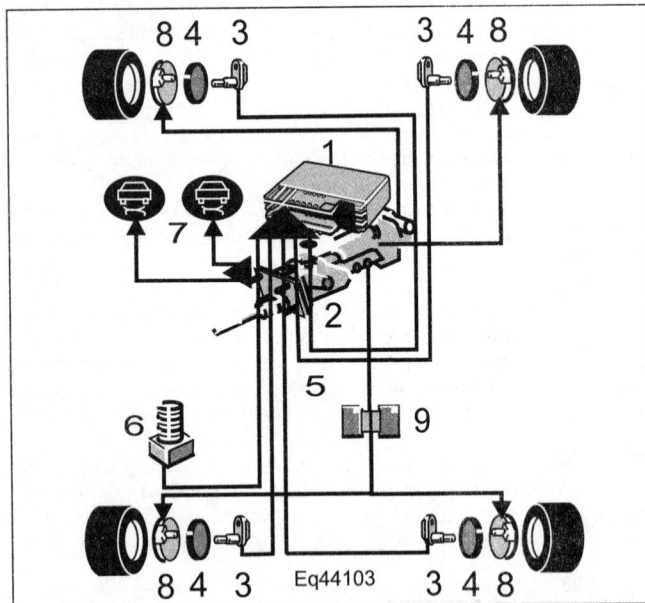

17.1 Teves II main components

1	ABS-ECM
2	Hydraulic control unit
3	Inductive wheel speed sensors
4	Sensor rings
5	ABS electrical wiring harness and relays
6	Brake light switch
7	ABS warning lamps
8	Brake calipers
9	Pressure regulating/load sensing valve(s) depending on application

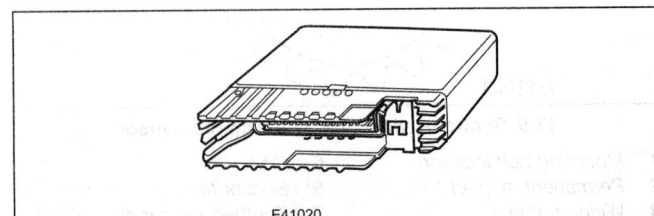

E41020

17.2 Teves II ECM

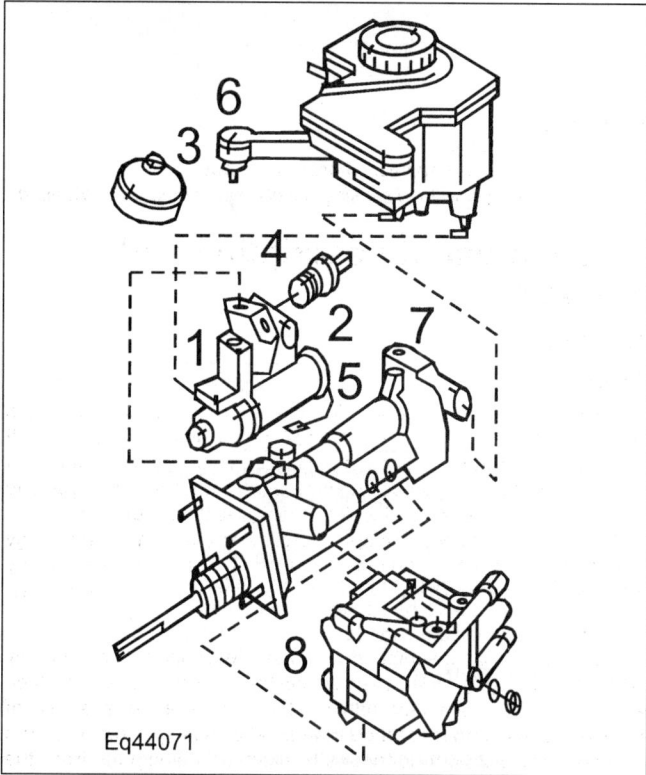

Eq44071

17.3 Teves II hydraulic control unit components

1	Pump assembly	5	Master cylinder assembly
2	Electric pump	6	Fluid reservoir
3	Pressure accumulator	7	Main solenoid valve
4	Pressure switch	8	Hydraulic valve block

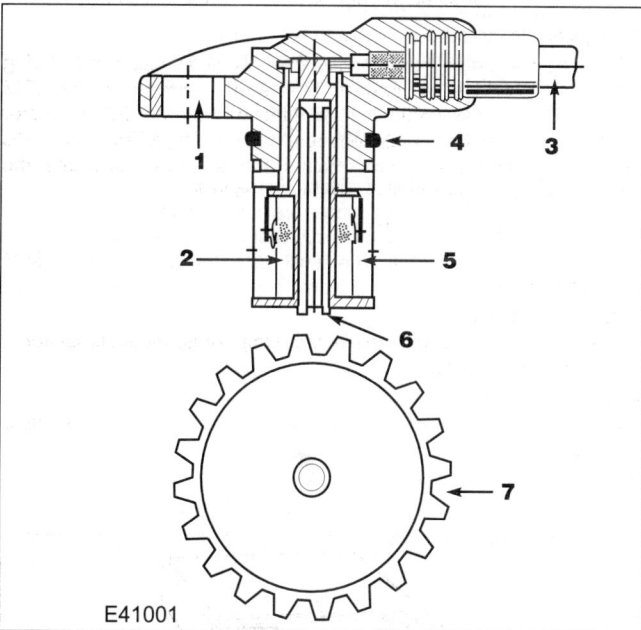

E41001

17.5 Sectional view of a wheel speed sensor

1	Mounting bolt location	5	Coil
2	Permanent magnet	6	Sensor tip
3	Wiring harness	7	Toothed sensor ring
4	O-ring		

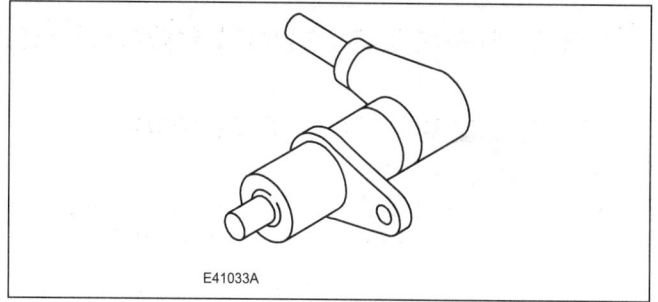

E41033A

17.4 Typical wheel speed sensor

Typically, the three assemblies are housed together as a unit and joined by a series of short connecting hydraulic pipes and pressure hoses. However, in some installations, the pump assembly may be located separately from the master cylinder and hydraulic valve block. On certain Citroën vehicles, an integrated hydraulic system is used whereby hydraulic pressure generated by a mechanical pump is used for both the braking system and the vehicle suspension system. On these vehicles, only the ABS hydraulic valve block is used, the pump assembly and master cylinder being replaced by a compensator/brake control valve which is part of the vehicle main hydraulic system and is unique to Citroën.

Instead of using a conventional vacuum servo unit for braking assistance, the Teves II system uses an electric pump and pressure accumulator for hydraulic operation. The pump draws hydraulic fluid from the reservoir and feeds it under high pressure to the accumulator.

The accumulator contains two chambers separated by a flexible diaphragm. The upper chamber is filled with high pressure nitrogen gas and the lower chamber contains the brake hydraulic fluid, supplied under pressure from the pump. The hydraulic fluid compresses the nitrogen gas further, thus providing a supply of high pressure hydraulic fluid for the brake system. Pump operation is controlled by a pressure switch located below the accumulator. The switch actuates the pump relay to operate the electric pump if the accumulator pressure drops below 140 bar. When the pressure reaches 180 bar, the switch deactivates the relay to switch off the pump. If the accumulator pressure falls below 105 bar, a signal is sent to the ECM to switch off the ABS operation.

The pressurised hydraulic fluid from the accumulator is supplied to the master cylinder for brake system operation either with or without ABS control.

The hydraulic valve block contains the inlet and outlet solenoid valves for ABS operation. The valve block modulates the pressure applied to the brake caliper for each individual front wheel, and jointly to the rear wheels during ABS operation. As Teves II is a three-channel system, there are two solenoid valves for each front brake, and two solenoid valves controlling the rear brakes as a pair (six in total).

Because both the rear brakes are on a single hydraulic channel, the 'select-low' principle is employed. With the 'select-low' principle, the wheel with the lowest adhesion determines the amount of hydraulic pressure to both brakes on that channel during ABS operation.

Wheel speed sensors

The rotational speed of the roadwheels and any changes in the rotational speed are recorded by inductive wheel speed sensors located at the roadwheels. Depending on system type, there may be a separate sensor for each front wheel only, or a separate sensor for all four wheels **(see illustration 17.4)**.

Each wheel speed sensor assembly comprises a toothed sensor ring which rotates at roadwheel speed, and an adjacent sensor mounted a set distance from the sensor ring **(see illustration 17.5)**.

17

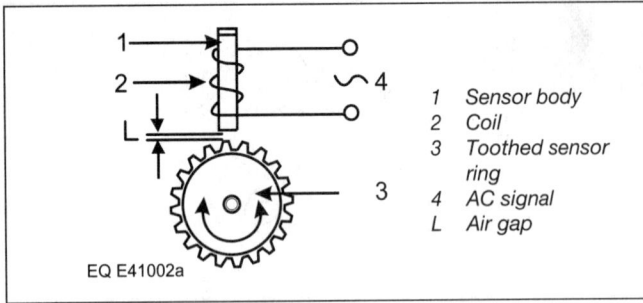

17.6 Wheel speed sensor operation

1 Sensor body
2 Coil
3 Toothed sensor ring
4 AC signal
L Air gap

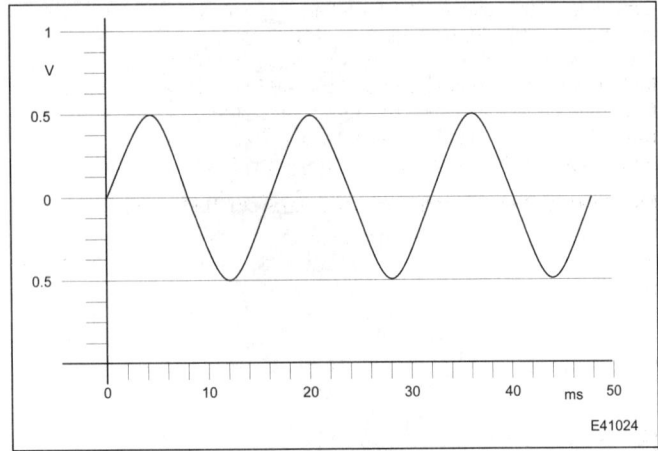

17.7 Wheel speed sensor waveform as viewed on an oscilloscope

The sensors are permanent magnet pulse generator types producing an AC voltage sine wave as the sensor ring teeth pass through the magnetic field of the sensor (see illustration 17.6).

The frequency of the waveform produced by the wheel speed sensor is proportional to the road speed. This AC voltage signal is continually being delivered to the ABS-ECM for processing.

The peak to peak voltage of the speed signal (when viewed upon an oscilloscope) can vary considerably according to wheel speed. An analogue to digital converter (ADC) in the ECM transforms the AC pulse into a digital signal (see illustration 17.7).

ABS electrical wiring harness and relays

An integrated main wiring harness is used to connect power and earth to various electrical components. This enables sensor signals to reach the ECM, and the ECM in turn, to send command signals to the relays and hydraulic control unit. The ABS overvoltage relay, main relay and pump relay, and the associated diodes and fuses, are typically located in the vehicle main fuse/relay box.

Overvoltage relay

The overvoltage relay (commonly orange) is only fitted to certain systems and is used to protect the ECM from voltage surges during engine starting. When the starter is in operation, the relay cuts off the supply from the ignition switch to the ECM.

Main relay

On the 35-pin system, nominal battery voltage is supplied from the battery to the main relay (commonly green) via a fuse. One terminal of the relay solenoid is connected to earth and the relay is activated by the ECM supplying a voltage to the other solenoid terminal. Fused battery voltage is then supplied to the ECM.

On the 55-pin system, nominal battery voltage is supplied from the battery to the main relay (commonly green) via a fuse. The relay is controlled by the ECM applying an earth, which activates the relay, allowing fused battery voltage to be applied to the main valve and solenoid valves. The earth for the solenoid valves is controlled by the ECM.

Pump relay

Nominal battery voltage is supplied from the battery to the pump

relay (commonly purple) via a fuse. The relay is controlled by the pressure switch applying an earth, which activates the relay allowing fused battery voltage to be applied to the pump motor. The hydraulic control unit provides the earth for the pump motor.

Brake light switch

The brake light switch comprises a switch body and contact pin and is located above the brake pedal (see illustration 17.8).

When the brake pedal is depressed closing the brake light switch, a signal is sent to the ECM indicating that the brakes are being applied. Once this signal is received, the ECM will begin monitoring the wheel speed via the wheel speed sensors and activate the ABS if necessary.

ABS warning lamps

After the ignition is switched on, the ABS warning lamp on the instrument panel is illuminated for approximately 2-4 seconds as the system executes a self-test routine. If satisfactory operation of the system is detected by the ECM, and if the system hydraulic pressure is greater than 105 bar, the light is extinguished. During vehicle operation above a pre-determined wheel speed, the ABS-ECM implements a further self-test cycle whereby ABS operation and wheel speed sensor signals are monitored. If a fault is detected, the relevant ECM pin is earthed to illuminate the warning lamp on the instrument panel and the ABS function is disabled. The warning lamp will remain illuminated until the fault is no longer present (see illustration 17.9).

The Teves II system also monitors brake hydraulic fluid level via a level sensor in the reservoir. The sensor is a dual circuit switch which will illuminate the low fluid level warning lamp on the instrument panel

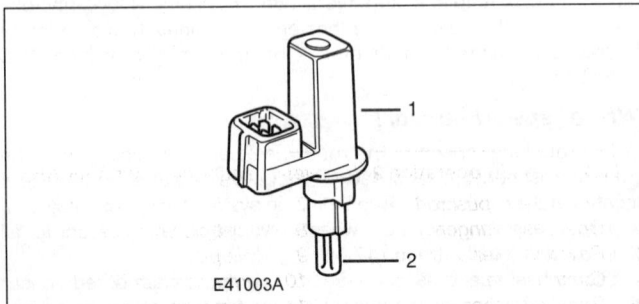

17.8 Typical brake light switch body (1) and contact pin (2)

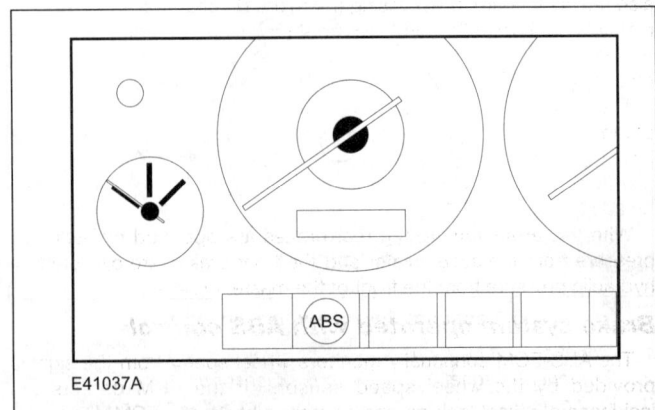

17.9 ABS warning lamp

if the level in the reservoir falls, and also signal the ECM. The ECM will then disable the ABS operation and illuminate the ABS warning lamp.

On systems with self-diagnostics, when the ABS-ECM detects a fault, the fault code is stored and the ABS warning lamp activated. I f the fault no longer exists after the next system start (ignition on/off) the ABS warning lamp is extinguished after the self-test cycle, however the fault code remains stored in the ECM memory.

Pressure regulating/load sensing valve(s)

Depending on vehicle application, pressure regulating valves or load sensing valves may be incorporated to restrict the hydraulic fluid pressure to the rear brakes. The valves may be pressure conscious whereby the hydraulic fluid supply is restricted once a pre-determined pressure is reached, or load conscious whereby the hydraulic pressure is reduced according to vehicle loading.

3 System operation

There are three distinct phases of Teves II ABS operation:
Phase 1 - Brake system at rest
Phase 2 - Brake system operating under conventional control
  without ABS
Phase 3 - Brake system operated with ABS control

Brake system at rest

When the system is at rest prior to operation of the brake pedal, hydraulic fluid under pressure from the accumulator is present at the closed control valve in the master cylinder **(see illustration 17.10)**. The remainder of the master cylinder passages and braking system components contain hydraulic fluid at atmospheric pressure. The inlet and outlet solenoid valves in the hydraulic valve block are in the 'at rest' position (inlet valves open and outlet valves closed). The electric pump maintains a reserve of hydraulic pressure in the accumulator.

Brake system operating under conventional control without ABS

When the brake pedal is depressed, the pedal pushrod pushes the operating plunger in the master cylinder which in turn moves the reaction lever **(see illustration 17.11)**. As the reaction lever pivots, the control valve is opened allowing hydraulic fluid under pressure from the accumulator to enter the boost chamber. The pressurised fluid flows through the outlet port in the master cylinder, through the open inlet solenoid valve in the hydraulic valve block and on to the rear brake calipers to operate the rear brakes.

At the same time, the hydraulic fluid under pressure in the boost chamber acts on the rear of the boost piston assisting the force being applied to the piston via the brake pedal. As the boost piston moves forward under the combined force of brake pedal effort and hydraulic pressure from the boost chamber, the master cylinder main piston also moves forward, pressurising the hydraulic chamber in front of the main piston. Hydraulic fluid flows from the master cylinder outlet port, through the open inlet solenoid valves in the hydraulic valve block and on to the front brake calipers to operate the front brakes.

The interconnection between the master cylinder operating plunger, reaction lever, boost piston and control valve are such hat the amount of hydraulic fluid under pressure in the boost chamber is directly proportional to the effort being applied to the brake pedal.

With this arrangement, the rear brakes are operated by hydraulic pressure from the accumulator and the front brakes are operated by hydraulic pressure from the front of the master cylinder.

Brake system operated with ABS control

The ABS-ECM continually monitors wheel speed from the signals provided by the wheel speed sensors. If the ECM detects the incidence of wheel lock on one or more wheels, the ECM closes the relevant inlet solenoid valve and opens the outlet solenoid valve in the

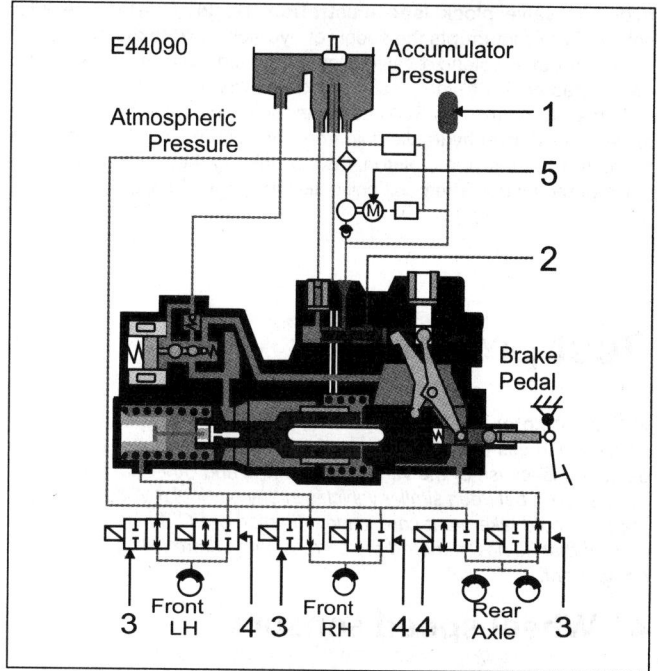

17.10 System operation 1 - brake system at rest

1 Hydraulic fluid pressure from accumulator
2 Control valve
3 and 4 Inlet and outlet solenoid valves
5 Pump

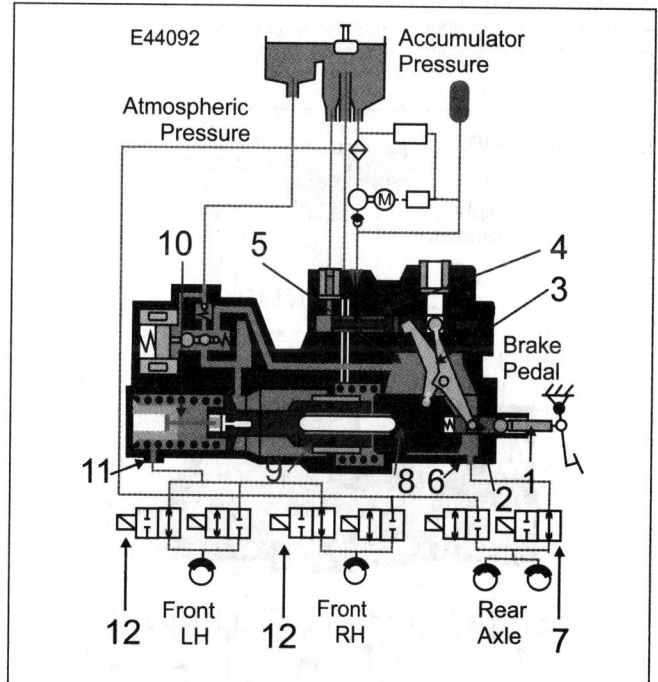

17.11 System operation 2 - conventional braking without ABS

1 Brake pedal pushrod	7 Rear inlet solenoid valve
2 Operating plunger	8 Boost piston
3 Reaction lever	9 Main piston
4 Control valve	10 Hydraulic chamber
5 Boost chamber	11 Outlet port
6 Outlet port	12 Front inlet solenoid valves

17

hydraulic valve block (see illustration 17.12). Closing the inlet solenoid valve interrupts the supply of hydraulic fluid to the brake, and the open outlet solenoid valve allows the fluid from the brake to be discharged back into the master cylinder reservoir.

At the same time, the ECM actuates the main solenoid valve which allows pressurised hydraulic fluid from the accumulator to be applied to the front of the boost piston, thus equalising the pressure on both sides of the piston. The positioning sleeve is pushed back against the boost piston and consequently the brake pedal height is maintained.

Once the wheel rotation has stabilised, the outlet solenoid valve is closed and the inlet solenoid valve is opened. Brake hydraulic pressure from the master cylinder once again flows through the inlet solenoid valve to the relevant brake caliper.

The whole ABS control cycle takes place 4 to 10 times per second for each affected wheel and this ensures maximum braking effect and control during ABS operation.

Test procedures

Important note: *The test procedures, pin-tables and wiring diagrams contained in this Chapter are necessarily representative of the system depicted. Because of the variations in wiring and other data that often occurs, even between similar vehicles in any particular VM's range, the reader should take great care in identification of ECM pins, and satisfy himself that he has gathered the correct data before failing a particular component.*

4 Wheel speed sensors

Checking the wheel speed sensor (general)

1 Inspect the wheel speed sensor for corrosion or damage and check that the sensor is tightly mounted.
2 Check the toothed sensor ring for damage, eccentricity and for broken or missing teeth.
3 Inspect the wheel speed sensor wiring plug for corrosion and damage. One plug for each sensor.
4 Check that the connector terminal pins are pushed fully home and making good contact with the sensor wiring plug.

5 Check the clearance between the sensor and the toothed sensor ring. The clearance is not normally adjustable but is nominally 0.2 to 1.2 mm. If the clearance is excessive, expect a worn sensor tip or problems with the wheel bearings/hub or sensor ring.
6 When carrying out voltage checks with an oscilloscope or voltmeter, the voltage obtained will be proportional to the speed at which the wheel is rotating. In addition to determining that the wheel speed sensors are actually producing a voltage output, it is essential that the output from the sensors on a particular axle is the same for any given wheel speed.

Checking wheel speed sensor output with an oscilloscope

Note: *Refer to the wiring diagrams for specific ECM pin identification according to model.*

7 Switch the ignition off and disconnect the ECM multi-plug or the relevant wheel speed sensor wiring plug.
8 Connect an oscilloscope between the terminal pins for the sensor under test (see illustration 17.13).
9 Select a range to cover 80 Hz on the oscilloscope and a free run time base.
10 Raise the wheel and rotate it by hand at approximately one revolution per second.

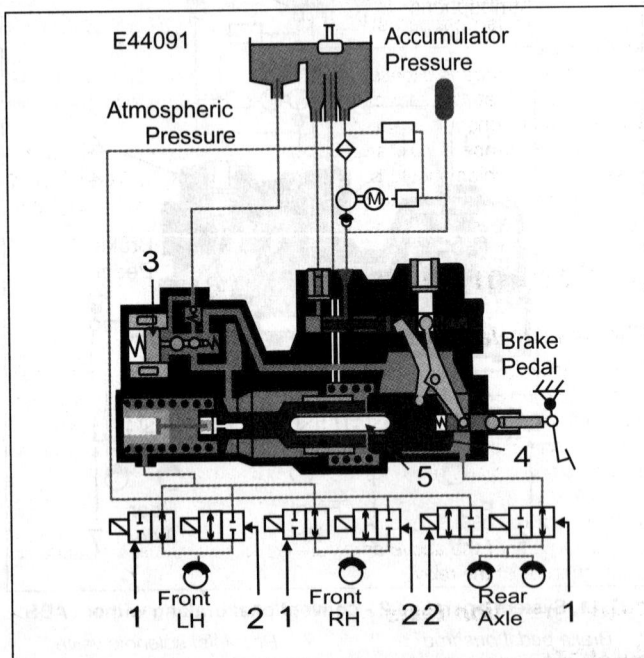

17.12 System operation 3 - braking with ABS

1	Inlet solenoid valve	4	Boost piston
2	Outlet solenoid valve	5	Positioning sleeve
3	Main solenoid valve		

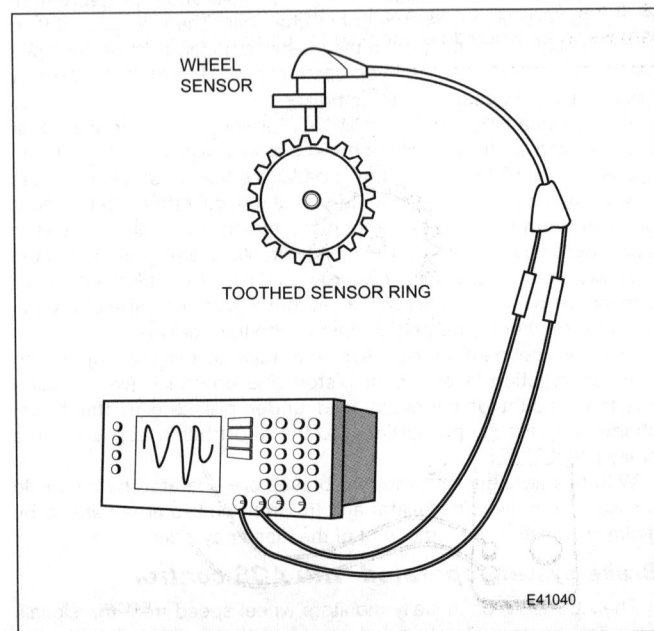

17.13 Checking wheel speed sensor output with an oscilloscope connected to the sensor wiring plug

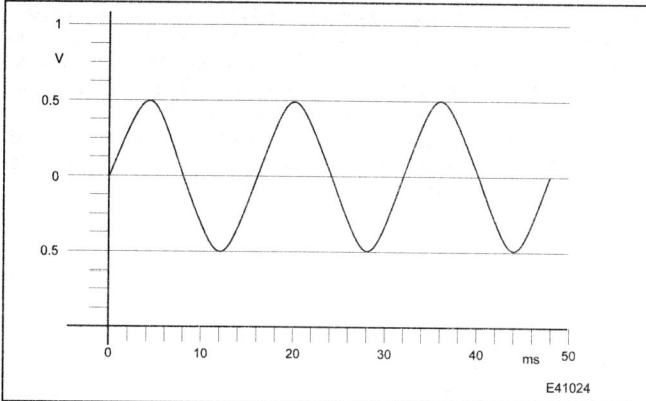

17.14 Typical wheel speed sensor sine wave as displayed on an oscilloscope

11 A sinusoidal wave form should be obtained, with amplitude and duration changing with rotational speed (see illustration 17.14).

12 If there is no signal, or a very weak or intermittent signal at the ECM, repeat the test at the sensor wiring plug. If there is no change in signal status, the sensor is suspect.

13 If the signal is now satisfactory this indicates a fault in the wiring harness which should be checked for continuity.

Checking wheel speed sensor output with an AC voltmeter

Note: *Refer to the wiring diagrams for specific ECM pin identification according to model.*

14 Switch the ignition off and disconnect the ECM multi-plug or the relevant wheel speed sensor wiring plug.

15 Connect an AC voltmeter between the terminal pins for the sensor under test (see illustration 17.15).

16 Raise the wheel and rotate it by hand at approximately one revolution per second.

17 A voltage of approximately 0.5 to 1.5 volts (AC RMS) should be obtained. If there is no signal, or a very weak or intermittent signal at the ECM, repeat the test at the sensor wiring plug. If there is no change in the signal, the sensor is suspect.

18 If the signal is now satisfactory, this indicates a fault in the wiring harness which should be checked for continuity. **Note:** *This test at*

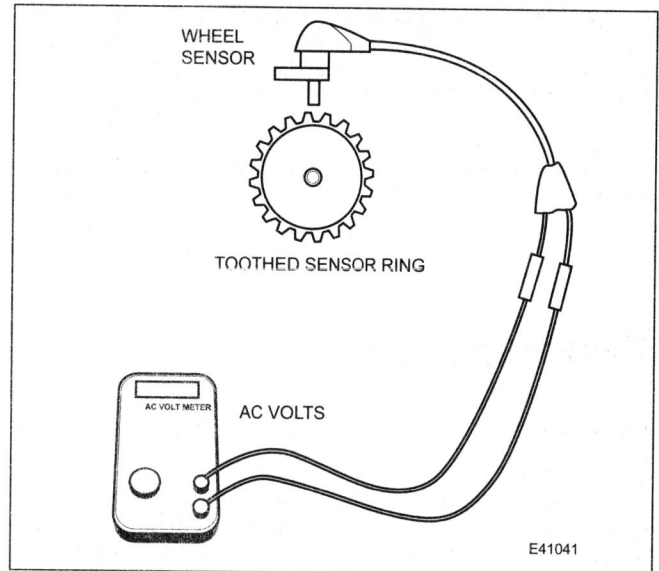

17.16 Checking wheel speed sensor resistance with an ohmmeter connected to the sensor wiring plug

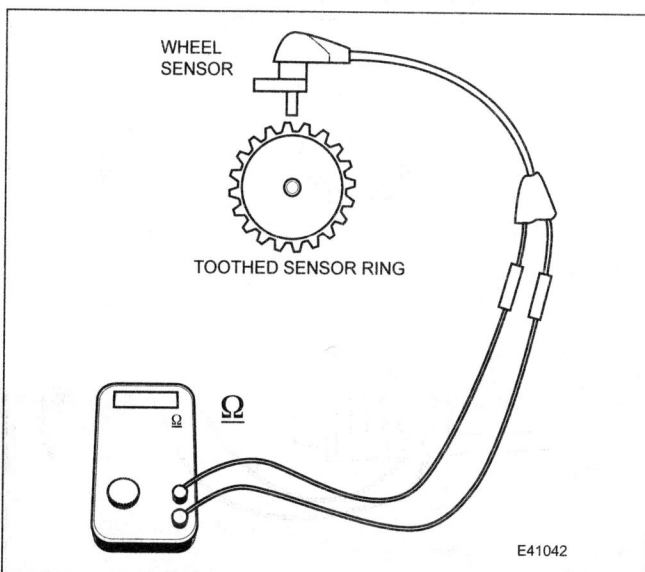

17.15 Checking wheel speed sensor output with a voltmeter connected to the sensor wiring plug

least proves that a signal is being generated by the sensor. However, the voltage produced is an average voltage and does not clearly indicate damage to the sensor ring or that the sinewave is regular in formation.

Checking wheel speed sensor resistance

Note: *Refer to the wiring diagrams for specific ECM pin identification according to model.*

19 Switch the ignition off and disconnect the ECM multi-plug or the relevant wheel speed sensor wiring plug.

20 Connect an ohmmeter between the terminal pins for the sensor under test (see illustration 17.16).

21 The readings obtained should be between 0.7 and 2.2 kohms approximately.

22 If the resistance is excessively high, or open circuit at the ECM, repeat the test at the sensor multi-plug. If there is no change in resistance, the sensor is suspect.

23 If the resistance is now satisfactory, this indicates a fault in the wiring harness which should be checked for continuity. **Note:** *Even if the resistance is within the quoted specifications, this does not prove that the speed sensor can generate an acceptable signal.*

5 System relays

Checking relays (general)

1 The following tests are applicable to the ABS main relay, hydraulic pump relay and, where fitted, overvoltage protection relay. The relays are commonly located in the facia or in the power distribution box.

2 Before carrying out specific tests on the relays or power supplies, inspect the relays and relay base for corrosion and damage.

3 Check that the relays are pushed fully home and making good contact with the base.

4 A fault in any of the above areas are possible reasons for failure or malfunctioning of the relay.

Relay operation tests

Main relay

Note: *In the following test the relay terminal numbers refer to a typical relay as used on Ford vehicles and shown in the illustration. Refer to the wiring diagrams for specific relay terminal numbers according to model.*

17

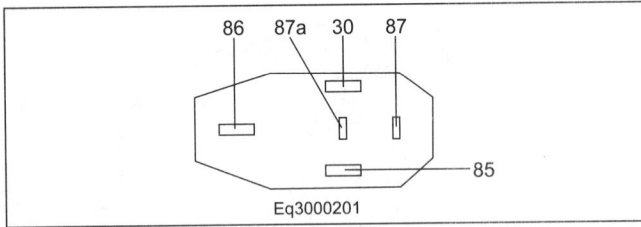

17.17 Typical relay terminal identification

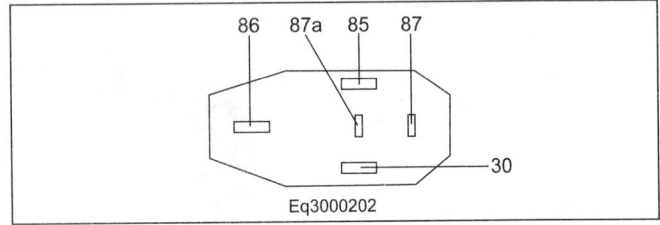

17.18 Typical relay base terminal identification

5 Switch the ignition off, then remove the relay from the relay base.
6 Connect an ohmmeter between the relay output terminal 30, and earth terminal 87a **(see illustration 17.17)**. The ohmmeter should indicate continuity.
7 Connect an ohmmeter between the relay main voltage supply terminal 87, and the output terminal 30. The ohmmeter should indicate an open circuit.
8 Attach a wire between the relay driver supply terminal 86, and a 12 volt supply. Attach a second wire between the relay driver earth terminal 85, and earth. There should be an unmistakable click from the relay and the ohmmeter should indicate continuity.
9 Renew the relay if it fails these tests. If the relay cover is transparent, inspect the relay for burns. A faulty relay usually has a burnt appearance.

Hydraulic pump relay

Note: *In the following test the relay terminal numbers refer to a typical relay as used on Ford vehicles and shown in the illustration. Refer to the wiring diagrams for specific relay terminal numbers according to model. On certain Ford vehicles, two pump relays are used and terminals 30 and 87 are reversed on the second relay.*
10 Switch the ignition off, then remove the relay from the relay base.
11 Connect an ohmmeter between the relay main voltage supply terminal 30, and the output terminal 87 **(see illustration 17.17)**. The ohmmeter should indicate an open circuit.
12 Attach a wire between the relay driver supply terminal 86, and a 12 volt supply. Attach a second wire between the relay driver earth terminal 85, and earth. There should be an unmistakable click from the relay and the ohmmeter should indicate continuity.
13 Renew the relay if it fails these tests. If the relay cover is transparent, inspect the relay for burns. A faulty relay usually has a burnt appearance.

Overvoltage (surge) relay

Note: *In the following test the relay operation and terminal numbers refer to the surge relay as used in certain Ford vehicles.*
14 Switch the ignition off, then remove the relay from the relay base.
15 Connect an ohmmeter between the relay main voltage supply terminal 30, and output terminal 87a **(see illustration 17.17)**. The ohmmeter should indicate continuity.
16 Attach a wire between the relay driver supply terminal 86, and a 12 volt supply. Attach a second wire between the relay driver earth terminal 85, and earth. There should be an unmistakable click from the relay and the ohmmeter should indicate open circuit.
17 Renew the relay if it fails these tests. If the relay cover is transparent, inspect the relay for burns. A faulty relay usually has a burnt appearance.

Relay power supply tests

Main relay

Note: *In the following test the relay terminal numbers refer to a typical relay as used on Ford vehicles and shown in the illustration. Refer to the wiring diagrams for specific relay terminal numbers according to model.*
18 Switch the ignition off then remove the main relay leaving the base exposed.
19 Using a voltmeter, probe between the main voltage supply

terminal 87, in the relay base and earth. The reading should indicate nbv **(see illustration 17.18)**.
20 If no voltage is found, check the relevant fuse and the supply wiring back to the voltage supply.
21 Switch the ignition on.
22 Using a voltmeter, probe between the relay driver supply terminal 86, in the relay base and earth. The reading should indicate nbv.
23 If no voltage is found, check the relevant fuse and the supply wiring back to the ignition switch.

Pump relay

Note: *In the following test the relay terminal numbers refer to a typical relay as used on Ford vehicles and shown in the illustration. Refer to the wiring diagrams for specific relay terminal numbers according to model. On certain Ford vehicles, two pump relays are used and terminals 30 and 87 are reversed on the second relay.*
24 Switch the ignition off then remove the pump relay leaving the base exposed.
25 Using a voltmeter, probe between the main voltage supply terminal 30, in the relay base and earth. The reading should indicate nbv **(see illustration 17.18)**.
26 If no voltage is found, check the relevant fuse and the supply wiring back to the voltage supply.
27 Switch the ignition off.
28 Using a voltmeter, probe between the relay driver supply terminal 86, in the relay base and earth. The reading should indicate nbv.
29 f no voltage is found, check the supply wiring back to the ignition switch.

Overvoltage (surge) relay

Note: *In the following test the relay operation and terminal numbers refer to the surge relay as used in certain Ford vehicles.*
30 Switch the ignition off, then remove the relay from the relay base.
31 Switch the ignition on.
32 Using a voltmeter, probe between the main voltage supply terminal 30, in the relay base and earth. The reading should indicate nbv **(see illustration 17.18)**.
33 If no voltage is found, check the supply wiring back to the ignition switch.

6 Electronic Control Module

Checking the ECM (general)

1 Inspect the ECM for corrosion or damage and ensure that the unit is mounted securely.
2 Check that the ECM multi-plug terminals are fully pushed home and making good contact with the ECM pins.
3 A fault in any of the above areas are possible reasons for poor performance in the ABS system.

ECM power supply and earth tests

Note: *Depending on application, the ECM may have either a 35-pin or a 55-pin multi-plug. Refer to the wiring diagrams for specific ECM pin identification according to model.*
4 Switch the ignition off and disconnect the ECM multi-plug from the ECM.
5 Switch the ignition on.

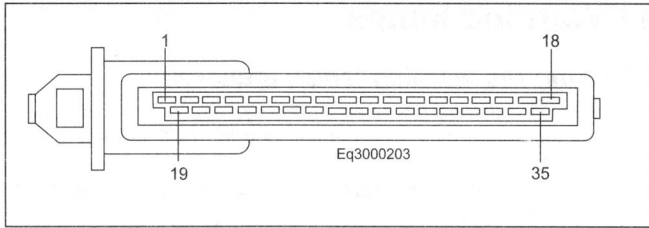

17.19 ECM 35-pin multi-plug

6 Attach a negative voltmeter probe to an ECM earth pin.

7 Attach a positive voltmeter probe to the ECM power supply pin 2 (35-pin ECM multi-plug) or pin 33 (55-pin multi-plug). The voltmeter should indicate nbv **(see illustrations 17.19 and 17.20)**.

8 If no voltage is found, check the ECM main earth connection, and the supply wiring back to the main relay and/or ignition switch.

9 Switch the ignition off.

10 Connect an ohmmeter between a vehicle earth and ECM earth pin 1. The ohmmeter should indicate continuity. If not, check the ECM main earth connection and wiring.

7 Solenoid valves

Solenoid valve resistance tests

35-pin ECM multi-plug systems

Note: *Refer to the wiring diagrams for specific ECM pin identification according to model.*

1 Switch the ignition off and disconnect the ECM multi-plug from the ECM.

2 Using an ohmmeter, measure the resistance between pin 11 in the ECM multi-plug, and each of the following multi-plug pins in turn **(see illustration 17.19)**:

Inlet solenoid valves - 15, 17, 35.
Outlet solenoid valves - 16, 33, 34.
Main solenoid valve - 11.

3 The resistance should be in the order of 6.0 to 10.0 ohms for each inlet solenoid valve, 3.0 to 6.0 ohms for each outlet solenoid valve and 2.0 to 5.0 ohms for the main solenoid valve.

4 If the resistance is not as specified, check the wiring and wiring connectors. If satisfactory, the hydraulic control unit is suspect.

55-pin ECM multi-plug systems

Note: *Refer to the wiring diagrams for specific ECM pin identification according to model.*

5 Switch the ignition off and disconnect the ECM multi-plug from the ECM.

6 Using an ohmmeter, measure the resistance between pin 33 in the ECM multi-plug, and each of the following multi-plug pins in turn **(see illustration 17.20)**:

Inlet solenoid valves - 20, 38, 54.
Outlet solenoid valves - 2, 21, 36.
Main solenoid valve - 39.

7 The resistance should be in the order of 6.0 to 10.0 ohms for each inlet solenoid valve, 3.0 to 6.0 ohms for each outlet solenoid valve and 2.0 to 5.0 ohms for the main solenoid valve.

8 If the resistance is not as specified, check the wiring and wiring connectors. If satisfactory, the hydraulic control unit is suspect.

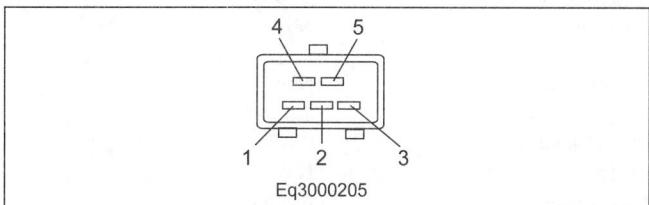

17.21 Pressure/warning switch terminal identification

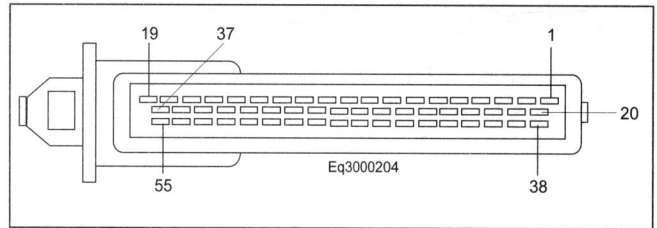

17.20 ECM 55-pin multi-plug

8 Hydraulic pressure/warning switch

Pressure/warning switch continuity tests

System depressurised

1 Switch the ignition off.

2 Depress the brake pedal at least 20 times to depressurise the system. When the system is fully depressurised, the brake pedal will become firmer and the brake fluid level in the master cylinder reservoir will rise.

3 Disconnect the 5-pin multi-plug from the pressure/warning switch, located below the accumulator.

4 Connect an ohmmeter between pressure switch terminals 1 and 2, 1 and 4, and 2 and 4 in turn **(see illustration 17.21)**. The ohmmeter should indicate continuity.

5 Connect the ohmmeter between pressure switch terminals 3 and 5. The ohmmeter should indicate an open circuit.

System pressurised

6 Reconnect the 5-pin multi-plug to the pressure/warning switch.

7 Switch the ignition on. The pump will run and the ABS warning lamp will be illuminated.

8 Switch the ignition off as soon as the warning lamp extinguishes. The system is now pressurised to 105 bar.

9 Disconnect the 5-pin multi-plug from the pressure/warning switch.

10 Connect an ohmmeter between pressure switch terminals 1 and 4, and 3 and 5 in turn **(see illustration 17.21)**. The ohmmeter should indicate continuity.

11 Connect the ohmmeter between pressure switch terminals 1 and 2. The ohmmeter should indicate an open circuit.

12 Reconnect the 5-pin multi-plug to the pressure/warning switch.

13 Switch the ignition on.

14 Switch the ignition off when the pump switches off. The system is now pressurised to 180 bar.

15 Disconnect the 5-pin multi-plug from the pressure/warning switch.

16 Connect an ohmmeter between pressure switch terminals 3 and 5 **(see illustration 17.21)**. The ohmmeter should indicate continuity.

17 Connect the ohmmeter between pressure switch terminals 1 and 2, and 1 and 4 in turn. The ohmmeter should indicate an open circuit.

18 If any of the above readings are not as specified, the pressure/warning switch is suspect.

9 Hydraulic pump motor

Pump operation test

1 Switch the ignition off then remove the pump relay leaving the base exposed.

2 Bridge terminals 30 and 87 in the relay base using a fused jumper lead **(see illustration 17.18)**. The pump should now operate.

⚠️ *Warning: The test should be made as quickly as possible to avoid damaging the pump.*

3 If the pump does not operate as described, carry out the pump relay power supply tests described in Section 5. If satisfactory, suspect a faulty pump.

17

10 Brake light switch

Checking the brake light switch (general)

1 Inspect the brake light switch for corrosion, and damage.
2 Check that the multi-plug terminal pins are pushed fully home and making good contact.
3 Check that no wires have been disconnected.
4 A fault in any of the above areas are possible reasons for failure or malfunctioning of the switch.

Brake light switch voltage and continuity tests

Voltage test

Note: *Refer to the wiring diagrams for specific ECM pin identification according to model.*
5 Switch the ignition off and disconnect the ECM multi-plug from the ECM.
6 Connect a voltmeter between the ECM multi-plug brake light switch input (pin 12 for 35-pin ECM systems, or pin 32 for 55-pin ECM systems) and an ECM earth pin **(see illustrations 17.19 and 17.20)**.
7 Switch the ignition on and depress the brake pedal. The voltmeter should indicate nbv.
8 If no voltage is found, the fuse, the brake light switch and the wiring are suspect.
9 Release the pedal, the voltage should drop to zero as the switch opens.

Continuity test

10 Switch the ignition off and disconnect the brake light switch multi-plug.
11 Connect an ohmmeter between the terminal pins of the brake light switch.
12 Operate the brake light switch and check for continuity. If the test fails, renew the switch.

11 Warning lamps

Checking the warning lamps (general)

1 Two warning lamps are used in the system, one for 'low fluid level/handbrake on' (typically red) and one for 'ABS system failure' (typically amber).
2 Inspect the warning lamp bulb holder contacts in the instrument panel.
3 Check that the instrument panel multi-plug terminal pins are pushed fully home and making good contact.
4 Check that no wires have been disconnected.
5 A fault in any of the above areas are possible reasons for failure or malfunctioning of the warning lamp(s).

Warning lamp operation test

6 With the ignition switched off, both warning lamp should remain off.
7 Switch the ignition on. Both warning lamp should illuminate and then extinguish after approximately 60 seconds. The warning lamps should now remain off.
8 If the 'ABS system failure' warning lamp illuminates on its own at any time, this indicates that the system has been partially or fully turned off by the ECM.
9 If the 'low fluid level/handbrake on' warning lamp illuminates on its own at any time, this indicates that either the handbrake is applied or the fluid level in the reservoir is very low.
10 If both warning lamps illuminate together at any time, this indicates that the fluid level in the reservoir is very low and that system pressure is reducing to the point that ABS operation is about to be, or has been disabled.
11 If the warning lamp comes on and remains on at any time during vehicle operation, carry out the previously described test procedures on the system components.

Pin table - typical 35-pin (without SD)

Note: *Refer to illustration 17.19*

Pin No.	Connection	Test condition	Voltage
1	ECM earth	Ignition on	0.25 volts (max)
2	Supply from ignition switch	Ignition on	Nbv
3	Supply from relay	Ignition on	Nbv
		Ignition off	0 volts
4	Rear right wheel speed sensor earth	Roadwheel rotating	0.25 volts (max)
5	Front left wheel speed sensor earth	Roadwheel rotating	0.25 volts (max)
6	Rear left wheel speed sensor earth	Roadwheel rotating	0.25 volts (max)
7	Front right wheel speed sensor earth	Roadwheel rotating	0.25 volts (max)
8	Relay supply	Ignition on	Nbv
9	Brake fluid level switch	Ignition on, level correct	9.0 to 10.0 volts
10	Accumulator pressure switch	Ignition on	9.0 to 10.0 volts
11	ECM earth	Ignition on	0.25 volts (max)
12	Brake light switch input	Ignition on:	
		Brake pedal released	0 volts
		Brake pedal depressed	Nbv
13	-	-	-
14*	Hydraulic pump relay 2 driver	Ignition on/inactive	Nbv
		Actuated	1.0 volt (max)
15	Front right inlet solenoid valve	Ignition on/inactive	0 volts
		Actuated	Nbv
16	Front left outlet solenoid valve	Ignition on/inactive	0 volts
		Actuated	Nbv

Pin No.	Connection	Test condition	Voltage
17	Rear inlet solenoid valve	Ignition on/inactive	0 volts
		Actuated	Nbv
18	Main solenoid valve	Ignition on/inactive	0 volts
		Actuated	Nbv
19**	Supply from ignition switch	Ignition on, engine cranking	Nbv
20	Supply from relay	Ignition on	Nbv
		Ignition off	0 volts
21	-	-	-
22	Rear right wheel speed sensor signal	Roadwheel rotating	0.5 to 1.5 volts AC (approx)
23	Front left wheel speed sensor signal	Roadwheel rotating	0.5 to 1.5 volts AC (approx)
24	Rear left wheel speed sensor signal	Roadwheel rotating	0.5 to 1.5 volts AC (approx)
25	Front right wheel speed sensor signal	Roadwheel rotating	0.5 to 1.5 volts AC (approx)
26	-	-	-
27	ABS warning lamp	Ignition on:	
		Lamp on	1.0 volts (max)
		Lamp off	Nbv
28	-	-	-
29	-	-	-
30	-	-	-
31	-	-	-
32*	Supply from hydraulic pump relay 1	Ignition on/pump inactive	0 volts
		Ignition on/pump operating	Nbv
33	Rear outlet solenoid valve	Ignition on/inactive	0 volts
		Actuated	Nbv
34	Front right outlet solenoid valve	Ignition on/inactive	0 volts
		Actuated	Nbv
35	Front left inlet solenoid valve	Ignition on/inactive	0 volts
		Actuated	Nbv

*Certain Ford models with two pump relays
**Later Ford Sierra models

Pin table - typical 35-pin (with SD)

Note: *Refer to illustration 17.19*

Pin No.	Connection	Test condition	Voltage
1	ECM earth	Ignition on	0.25 volts (max)
2	Supply from ignition switch	Ignition on	Nbv
3	Supply from relay	Ignition on	Nbv
		Ignition off	0 volts
4	Rear right wheel speed sensor earth	Roadwheel rotating	0.25 volts (max)
5	Front left wheel speed sensor earth	Roadwheel rotating	0.25 volts (max)
6	Rear left wheel speed sensor earth	Roadwheel rotating	0.25 volts (max)
7	Front right wheel speed sensor earth	Roadwheel rotating	0.25 volts (max)
8	Relay supply	Ignition on	Nbv
9	Brake fluid level switch	Ignition on, level correct	9.0 to 10.0 volts
10	Accumulator pressure switch	Ignition on	9.0 to 10.0 volts
11	Solenoid valve earth	Ignition on	0.25 volts (max)
12	Brake light switch input	Ignition on:	
		Brake pedal released	0 volts
		Brake pedal depressed	Nbv
13	-	-	-
14	Hydraulic pump relay driver	Ignition on/inactive	Nbv
		Actuated	1.0 volt (max)
15	Front right inlet solenoid valve	Ignition on/inactive	0 volts
		Actuated	Nbv

Pin table - typical 35-pin (with SD) (continued)

Note: *Refer to illustration 17.19*

Pin No.	Connection	Test condition	Voltage
16	Front left outlet solenoid valve	Ignition on/inactive	0 volts
		Actuated	Nbv
17	Rear inlet solenoid valve	Ignition on/inactive	0 volts
		Actuated	Nbv
18	Main solenoid valve	Ignition on/inactive	0 volts
		Actuated	Nbv
19*	Supply from ignition switch	Ignition on, engine cranking	Nbv
20	Supply from relay	Ignition on	Nbv
		Ignition off	0 volts
21	-	-	-
22	Rear right wheel speed sensor signal	Roadwheel rotating	0.5 to 1.5 volts AC (approx)
23	Front left wheel speed sensor signal	Roadwheel rotating	0.5 to 1.5 volts AC (approx)
24	Rear left wheel speed sensor signal	Roadwheel rotating	0.5 to 1.5 volts AC (approx)
25	Front right wheel speed sensor signal	Roadwheel rotating	0.5 to 1.5 volts AC (approx)
26	SD connector	-	-
27	ABS warning lamp	Ignition on:	
		Lamp on	1.0 volts (max)
		Lamp off	Nbv
28	-	-	-
29	-	-	-
30	-	-	-
31	-	-	-
32	Hydraulic pump relay display input	Ignition on/inactive	0 volts
		Actuated	Nbv
33	Rear outlet solenoid valve	Ignition on/inactive	0 volts
		Actuated	Nbv
34	Front right outlet solenoid valve	Ignition on/inactive	0 volts
		Actuated	Nbv
35	Front left inlet solenoid valve	Ignition on/inactive	0 volts
		Actuated	Nbv

* some models

Pin table - typical 55-pin

Note: *Refer to illustration 17.20*

Pin No.	Connection	Test condition	Voltage
1	ECM earth	Ignition on	0.25 volts (max)
2	Front left outlet valve driver	Ignition on/inactive	Nbv
		Actuated	1.0 volt (max)
3	Voltage monitoring, hydraulics	Ignition on	Nbv
3*	Supply from relay	Ignition on	Nbv
4	-	-	-
5	-	-	-
6	-	-	-
7	-	-	-
8	Brake fluid level sensor	-	-
9	-	-	-
10	-	-	-
11	-	-	-
12	-	-	-
13	ABS pressure control switch	-	-
14	-	-	-
15	-	-	-

Pin No.	Connection	Test condition	Voltage
16	-	-	-
17	-	-	-
18	-	-	-
19	ECM earth	Ignition on	0.25 volts (max)
20	Front left inlet valve driver	Ignition on/inactive	Nbv
		Actuated	1.0 volt (max)
21	Front right outlet valve driver	Ignition on/inactive	Nbv
		Actuated	1.0 volt (max)
22	-	-	-
23	SD connector	-	-
24	-	-	-
25	-	-	-
26	ABS pressure control switch	-	-
27	Rear right wheel speed sensor earth	Roadwheel rotating	0.25 volts (max)
28	Rear left wheel speed sensor earth	Roadwheel rotating	0.25 volts (max)
29	Front right wheel speed sensor earth	Roadwheel rotating	0.25 volts (max)
30	Front left wheel speed sensor earth	Roadwheel rotating	0.25 volts (max)
31	-	-	-
32	Brake light switch input	Ignition on:	
		Brake pedal not pressed	0 V
		Brake pedal depressed	Nbv
33	Supply from relay	Ignition on	Nbv
33*	Voltage monitoring, hydraulics	Ignition on	Nbv
34	ABS relay driver	Ignition on, actuated	1.0 volt (max)
35	Supply from battery	Ignition on/off	Nbv
36	Rear outlet valve driver	Ignition on/inactive	Nbv
		Actuated	1.0 volt (max)
37	-	-	-
38	Front right inlet valve driver	Ignition on/inactive	Nbv
		Actuated	1.0 volt (max)
39	Main solenoid valve driver	Ignition on/inactive	Nbv
		Actuated	1.0 volt (max)
40	-	-	-
41	-	-	-
42	SD connector	-	-
43	-	-	-
44	-	-	-
45	Rear right wheel speed sensor signal	Roadwheel rotating	0.5 to 1.5 volts AC (approx)
46	Rear left wheel speed sensor signal	Roadwheel rotating	0.5 to 1.5 volts AC (approx)
47	Front right wheel speed sensor signal	Roadwheel rotating	0.5 to 1.5 volts AC (approx)
48	Front left wheel speed sensor signal	Roadwheel rotating	0.5 to 1.5 volts AC (approx)
49	-	-	-
50	Hydraulic pump relay display input	Ignition on/inactive	0 volts
		Actuated	Nbv
51	ABS pressure warning switch	-	-
52	ABS warning lamp	Ignition on:	
		Lamp on	2.0 volts (max)
		Lamp off	Nbv
53	Supply from ignition switch	Ignition on	Nbv
54	Rear inlet valve driver	Ignition on/inactive	Nbv
		Actuated	1.0 volt (max)
55	-	-	-

* alternative

17

Fault codes

12 Jaguar/Daimler and Volvo fault codes

1 If a FCR is available, it should be connected to the SD serial connector and used in accordance with the maker's instructions.
2 The FCR can be used for the following purposes:
 a) *Obtaining fault codes.*
 b) *Clearing fault codes.*
 c) *Obtaining datastream information.*
 d) *Testing the system actuators (relays and solenoid valves).*
3 If a FCR is not available, it is still possible to obtain fault codes which will be displayed as a series of flashes on the dash mounted warning lamp.
4 When the ECM determines that a fault is present, it internally logs a fault code and also illuminates the ABS warning lamp(s) if the fault is regarded by the system as major. Faults regarded as minor will not illuminate the warning lamp(s) although a code will still be logged. All of the various two-digit fault codes on these vehicles are of the 'slow' variety and can be output as flash codes on the warning lamp. The first series of flashes indicates the number of tens, the second series of flashes indicates the single units.

Obtaining codes without an FCR (Jaguar/Daimler vehicles)

5 Switch the ignition off and use a bridging wire to bridge the brown/pink and black wires in the SD connector.
6 Switch the ignition on and observe the dash mounted warning lamp.
7 After six seconds the lamp will begin to flash if fault codes are present. The codes are displayed as a series of one second flashes, with a two second pause between the first and second digits. The codes are output in numerical sequence and each code is separated by a six second pause.
8 When all the codes have been recorded, switch the ignition off and remove the bridging wire from the SD connector.

Obtaining codes without an FCR (Volvo vehicles)

9 Switch the ignition off and remove the ABS ECM from its location.

10 Disconnect the wiring harness three-pin multi-plug located next to the ECM.
11 Using a bridging wire, bridge the white/green wire (terminal 1) and the brown wire (terminal 2) in the multi-plug.
12 Switch the ignition on and observe the dash mounted warning lamp.
13 The lamp will illuminate for four seconds and will then will begin to flash if fault codes are present. The codes are displayed as a series of short flashes, with a two second pause between the first and second digits. The codes are output in numerical sequence and each code is separated by a four second pause.
14 When all the codes have been recorded, switch the ignition off and remove the bridging wire from the multi-plug. Reconnect the multi-plug and refit the ECM.

Clearing fault codes from the ECM memory

15 Once all the indicated faults have been repaired, road test the vehicle and carry out the procedure described previously to ensure that all the faults have been rectified.
16 If all faults have been cleared, the warning lamp will not illuminate when the bridging wire is connected (Jaguar/Daimler vehicles), or will illuminate for two seconds then extinguish (Volvo vehicles).
17 On completion, switch the ignition off and disconnect the bridging wire. Reconnect/refit all disturbed components.

13 Ford, Saab and Volkswagen fault codes

1 On these vehicles, the use of a FCR is required for obtaining fault codes. Flash codes are not available for output from the system in these applications.
2 If a FCR is available, it should be connected to the SD serial connector and used in accordance with the maker's instructions.
3 The FCR can be used for the following purposes:
 a) *Obtaining fault codes.*
 b) *Clearing fault codes.*
 c) *Obtaining datastream information.*
 d) *Testing the system actuators (relays and solenoid valves).*

Fault code table (Ford, Jaguar/Daimler, Volvo)

Note: *Not all codes applicable to all models.*

Code	Item	Fault
00	ECM	Power supply failure
10	Ignore if obtained	-
11	Electrical interference	Poor earth or wiring connections
12	ECM	Defective
17	Main relay	Incorrect voltage, defective wiring and connections
21	Main solenoid	Poor connections, defective wiring or solenoid coil
22	Front left inlet solenoid	Poor connections, defective wiring or solenoid coil
23	Front left outlet solenoid	Poor connections, defective wiring or solenoid coil
24	Front right inlet solenoid	Poor connections, defective wiring or solenoid coil
25	Front right outlet solenoid	Poor connections, defective wiring or solenoid coil
26	Rear inlet solenoid	Poor connections, defective wiring or solenoid coil
27	Rear outlet solenoid	Poor connections, defective wiring or solenoid coil
31	Front left wheel speed sensor	Wiring open/short circuit
32	Front right wheel speed sensor	Wiring open/short circuit
33	Rear right wheel speed sensor	Wiring open/short circuit
34	Rear left wheel speed sensor	Wiring open/short circuit

Code	Item	Fault
35	Front left wheel speed sensor signal variation	Incorrect air gap, damaged rotor teeth, defective wiring
36	Front right wheel speed sensor signal variation	Incorrect air gap, damaged rotor teeth, defective wiring
37	Rear right wheel speed sensor signal variation	Incorrect air gap, damaged rotor teeth, defective wiring
38	Rear left wheel speed sensor signal variation	Incorrect air gap, damaged rotor teeth, defective wiring
41	Front left wheel speed sensor intermittent signal loss	Sensor/rotor damaged, incorrect air gap, defective wiring
42	Front right wheel speed sensor intermittent signal loss	Sensor/rotor damaged, incorrect air gap, defective wiring
43	Rear right wheel speed sensor intermittent signal loss	Sensor/rotor damaged, incorrect air gap, defective wiring
44	Rear left wheel speed sensor intermittent signal loss	Sensor/rotor damaged, incorrect air gap, defective wiring
45	All wheel speed sensors intermittent signal loss	Sensor/rotor damaged, incorrect air gap, defective wiring
46	FR, RR, RL wheel speed sensor intermittent signal loss	Sensor/rotor damaged, incorrect air gap, defective wiring
47	Rear wheel speed sensors intermittent signal loss	Sensor/rotor damaged, incorrect air gap, defective wiring
48	Three wheel speed sensors intermittent signal loss	Sensor/rotor damaged, incorrect air gap, defective wiring
51	Front left outlet solenoid	Hydraulic pressure reduction
52	Front right outlet solenoid	Hydraulic pressure reduction
53	Rear outlet solenoid	Hydraulic pressure reduction
54	Rear outlet solenoid	Hydraulic pressure reduction
55	Front left wheel speed sensor permanent signal loss	Sensor/rotor damaged, incorrect air gap, defective wiring
56	Front right wheel speed sensor permanent signal loss	Sensor/rotor damaged, incorrect air gap, defective wiring
57	Rear right wheel speed sensor permanent signal loss	Sensor/rotor damaged, incorrect air gap, defective wiring
58	Rear left wheel speed sensor permanent signal loss	Sensor/rotor damaged, incorrect air gap, defective wiring
61	Pressure switch, fluid level switch	Poor wiring connection, switch faulty, incorrect voltage
71	Front left wheel speed sensor electrical interference	Defective wiring, ECM fault
72	Front right wheel speed sensor electrical interference	Defective wiring, ECM fault
73	Rear right wheel speed sensor electrical interference	Defective wiring, ECM fault
74	Rear left wheel speed sensor electrical interference	Defective wiring, ECM fault
75	Front left wheel speed sensor signal interference	Incorrect installation, wheel bearings, sensor/rotor damaged
76	Front right wheel speed sensor signal interference	Incorrect installation, wheel bearings, sensor/rotor damaged
77	Rear right wheel speed sensor signal interference	Incorrect installation, wheel bearings, sensor/rotor damaged
78	Rear left wheel speed sensor signal interference	Incorrect installation, wheel bearings, sensor/rotor damaged
79	Ignore if obtained	-

Fault code table (Saab)

Code	Item	Fault
E001	ECM earth	Poor main connection
E002	ECM voltage supply	Supply voltage low or absent
E320	Main relay	Incorrect supply voltage, defective wiring and connections
E422	Rear right wheel speed sensor permanent signal loss	Sensor/rotor damaged, incorrect air gap, defective wiring
E523	Front left wheel speed sensor permanent signal loss	Sensor/rotor damaged, incorrect air gap, defective wiring
E624	Rear left wheel speed sensor permanent signal loss	Sensor/rotor damaged, incorrect air gap, defective wiring
E725	Front right wheel speed sensor permanent signal loss	Sensor/rotor damaged, incorrect air gap, defective wiring
E008	Main relay	Signal voltage absent, defective wiring and connections
E009	Fluid level, hydraulic pressure	Fluid level low, hydraulic pressure low
E010	ECM	Incorrect voltage, defective wiring, ECM defective
E011	ECM earth	Poor connection to HCU
E014	Main relay	Faulty relay, faulty pressure switch, defective wiring
E015	Front right inlet solenoid inoperative	Poor connections, defective wiring or solenoid coil
E016	Rear outlet solenoid inoperative	Poor connections, defective wiring or solenoid coil
E017	Rear inlet solenoid inoperative	Poor connections, defective wiring or solenoid coil
E018	Main solenoid inoperative	Poor connections, defective wiring or solenoid coil
EE22	Rear right wheel speed sensor signal variation	Incorrect air gap, damaged rotor teeth, wheel bearings
EE23	Front left wheel speed sensor signal variation	Incorrect air gap, damaged rotor teeth, wheel bearings
EE24	Rear left wheel speed sensor signal variation	Incorrect air gap, damaged rotor teeth, wheel bearings
EE25	Front right wheel speed sensor signal variation	Incorrect air gap, damaged rotor teeth, wheel bearings
E032	Pump relay	Inoperative
E132	Pump motor	Continuous operation
E033	Front left outlet solenoid inoperative	Poor connections, defective wiring or solenoid coil
E034	Front right outlet solenoid inoperative	Poor connections, defective wiring or solenoid coil
E035	Front left inlet solenoid inoperative	Poor connections, defective wiring or solenoid coil

17

Fault code table (Volkswagen)

Code	Item	Fault
0000	End of code output	
1111	ECM	Poor earth or wiring connections, incorrect voltage, ECM defective
1112	Front left inlet solenoid	Poor connections, defective wiring or solenoid coil
1114	Front right inlet solenoid	Poor connections, defective wiring or solenoid coil
1122	Rear inlet solenoid	Poor connections, defective wiring or solenoid coil
1132	Front left outlet solenoid	Poor connections, defective wiring or solenoid coil
1134	Front right outlet solenoid	Poor connections, defective wiring or solenoid coil
1142	Rear outlet solenoid	Poor connections, defective wiring or solenoid coil
1222	Main solenoid	Poor connections, defective wiring or solenoid coil
1233	Front left wheel speed sensor	Break, short circuit or poor wiring connection; incorrect air gap
1241	Front right wheel speed sensor	Break, short circuit or poor wiring connection; incorrect air gap
1243	Rear right wheel speed sensor	Break, short circuit or poor wiring connection; incorrect air gap
1311	Rear left wheel speed sensor	Break, short circuit or poor wiring connection; incorrect air gap
1312	Pressure switch, fluid level switch	Poor wiring connection, switch faulty, incorrect voltage
4444	No faults recognised	

Wiring diagrams

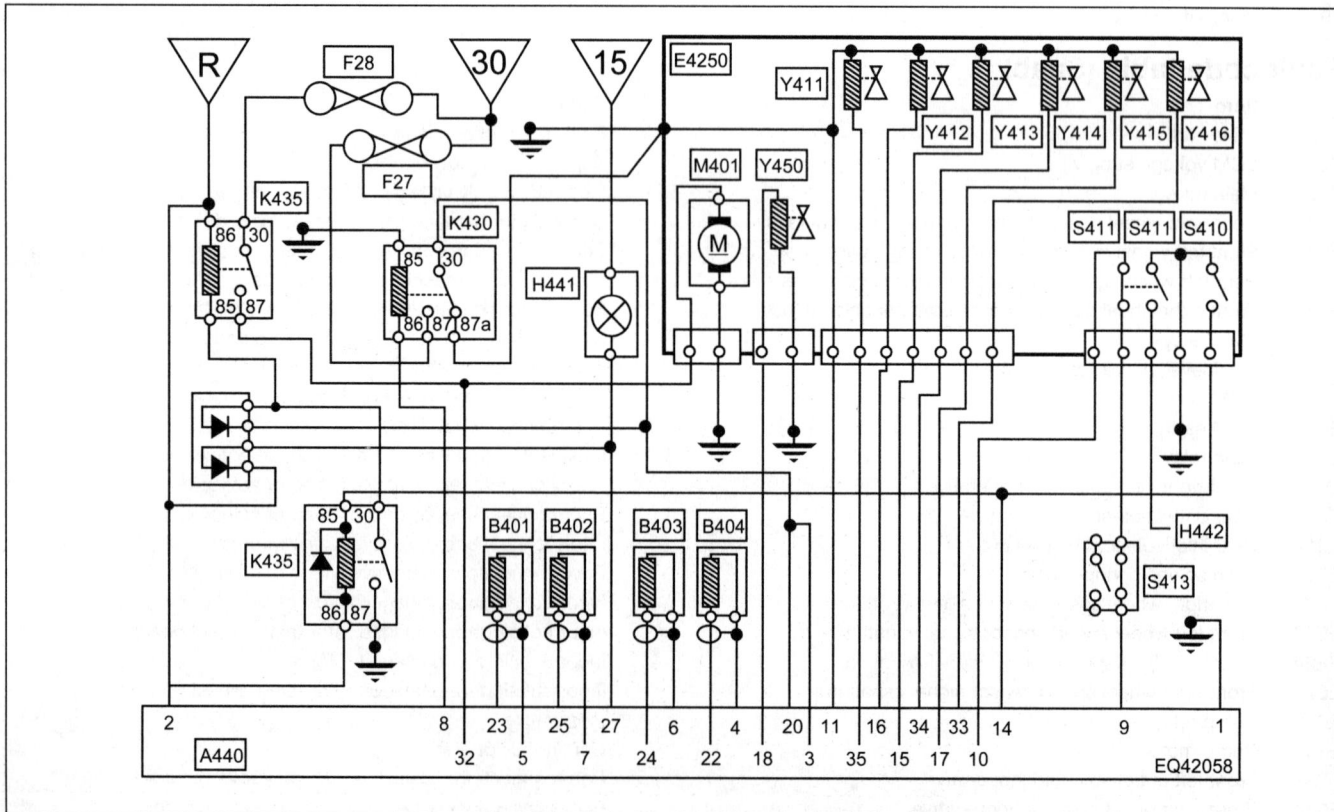

17.22   35-pin wiring diagram, Ford Granada/Scorpio (non-SD)

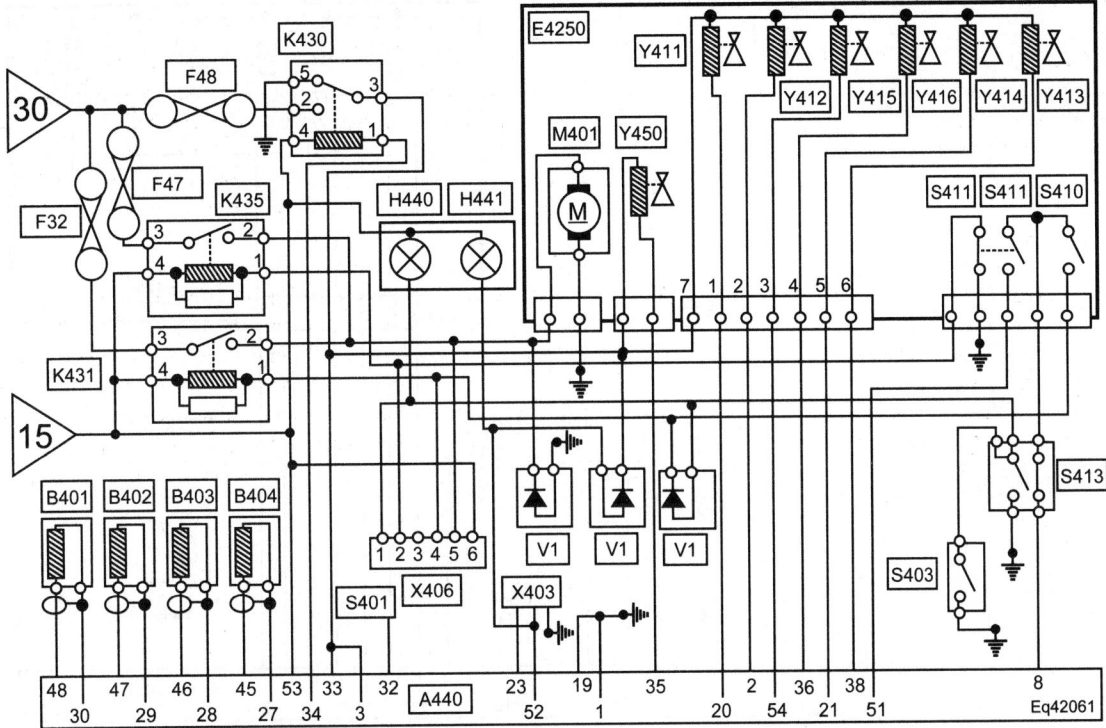

17.23 55-pin wiring diagram, Ford Granada/Scorpio 2.5D

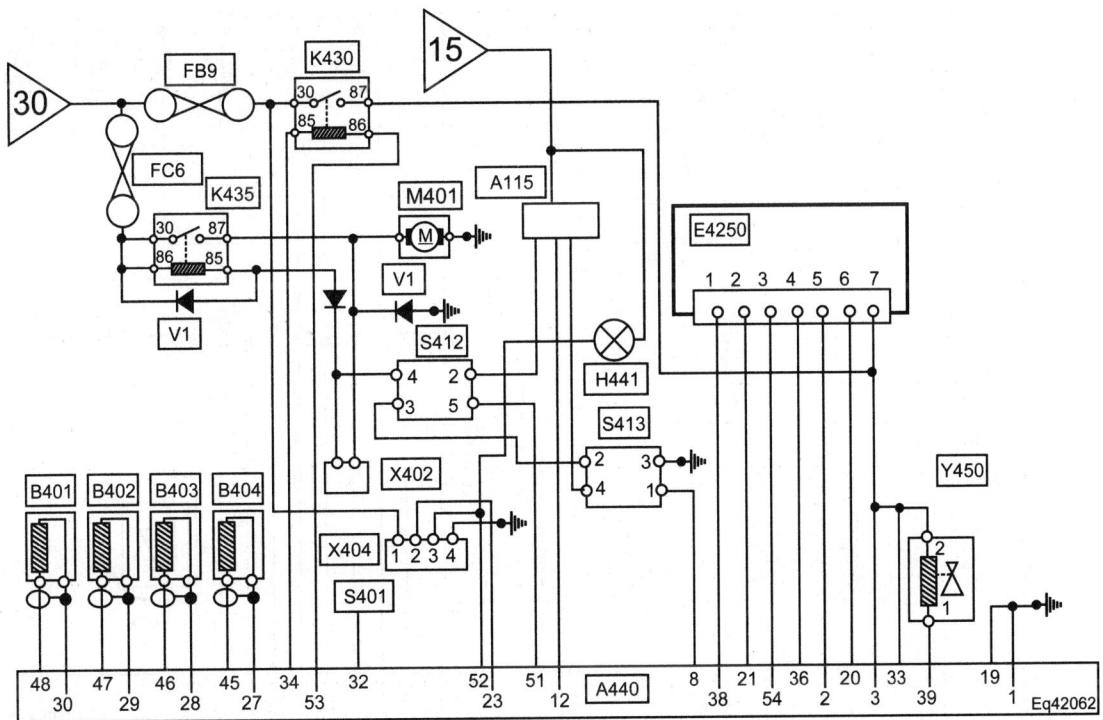

17.24 55-pin wiring diagram, Jaguar/Daimler XJ6/XJ12 (SD)

17

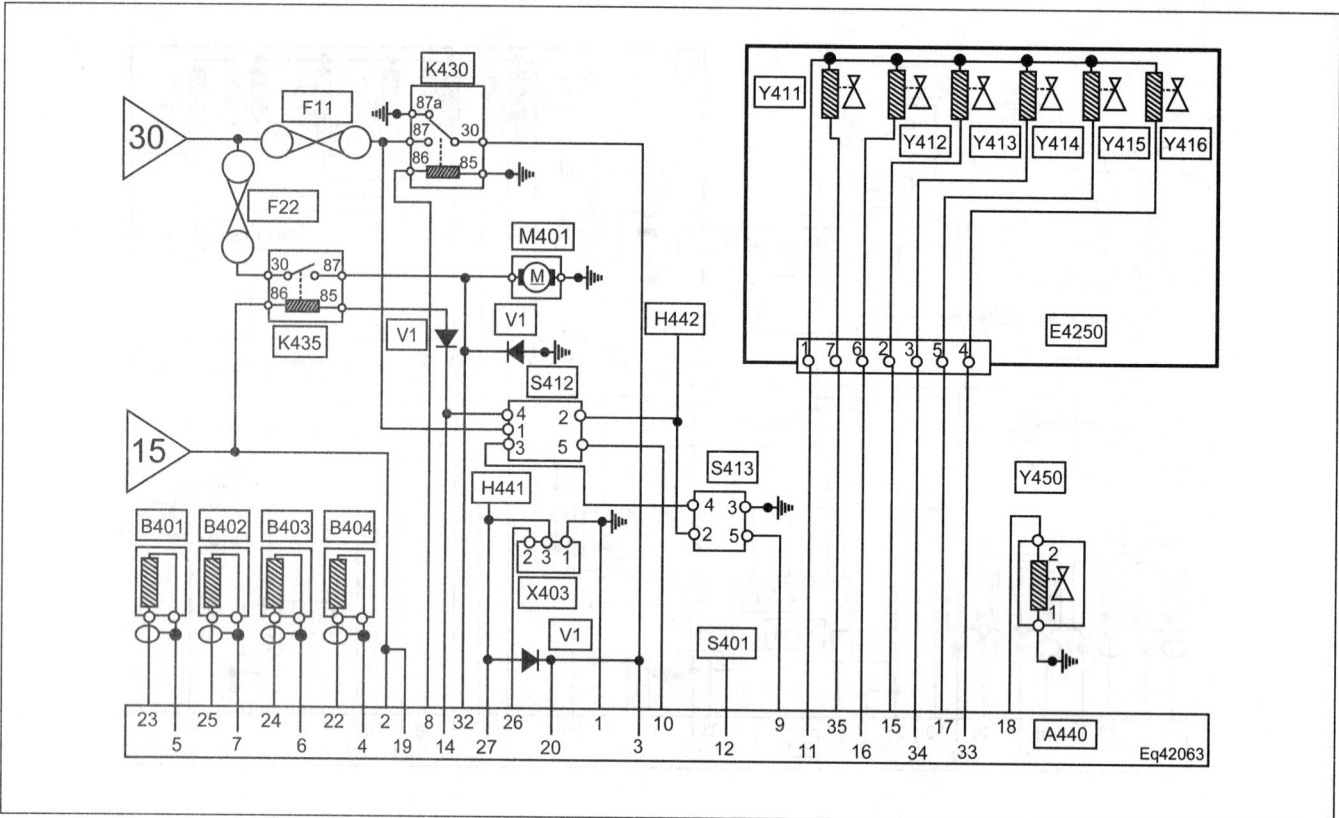

17.25 35-pin wiring diagram, Jaguar/Daimler XJS (SD)

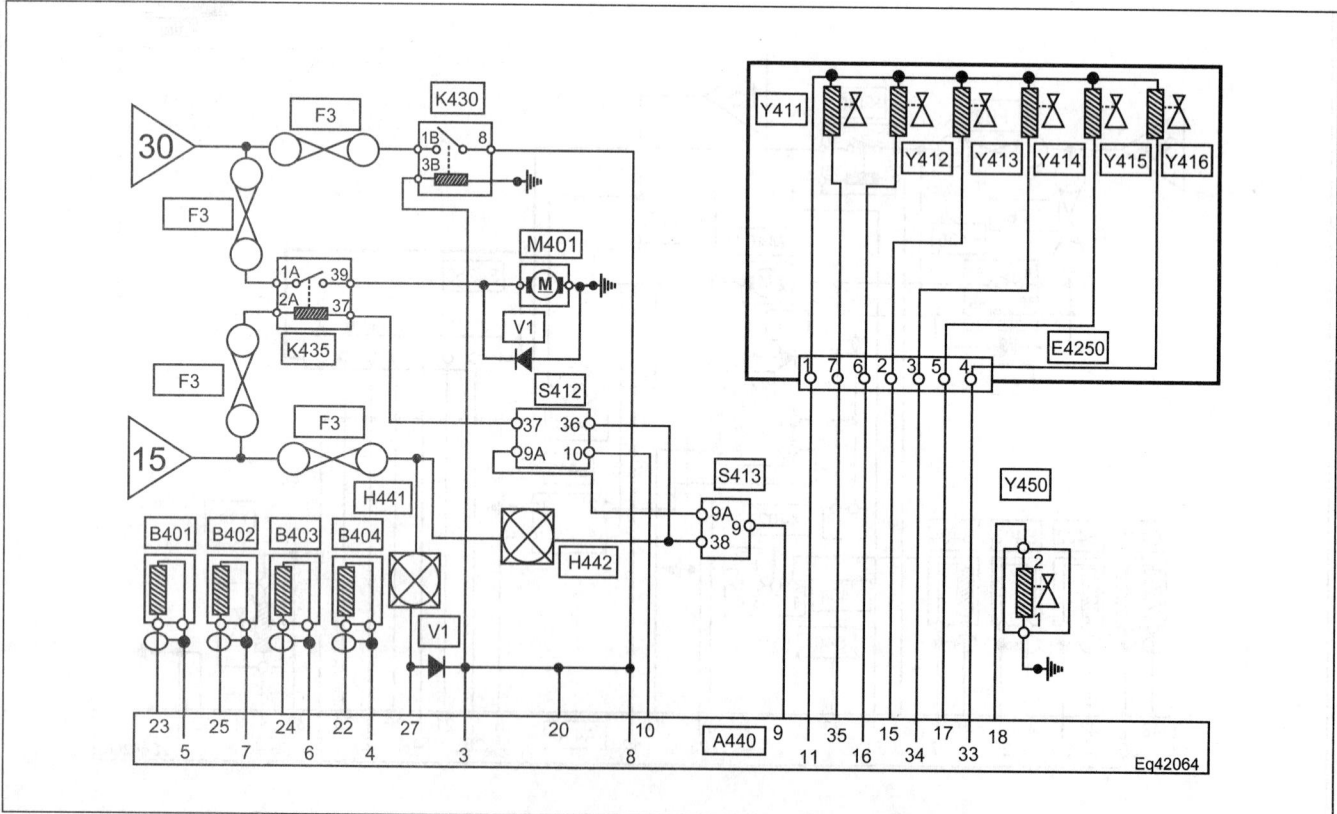

17.26 35-pin wiring diagram, Peugeot 505 (non-SD)

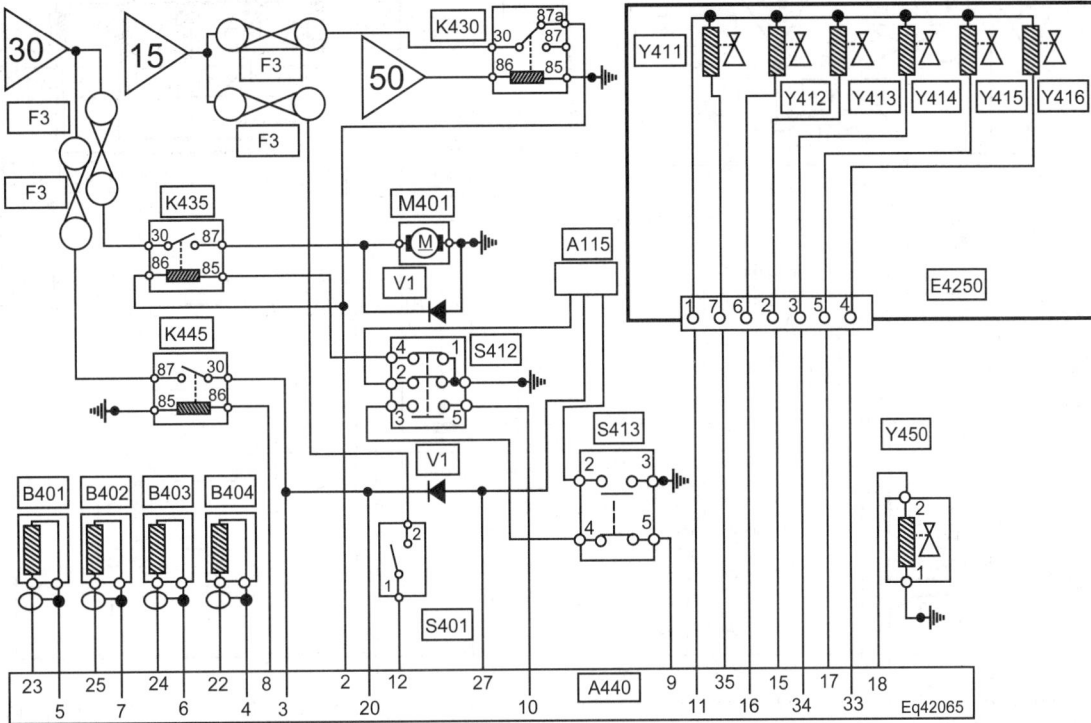

17.27 35-pin wiring diagram, Renault 21 (non-SD)

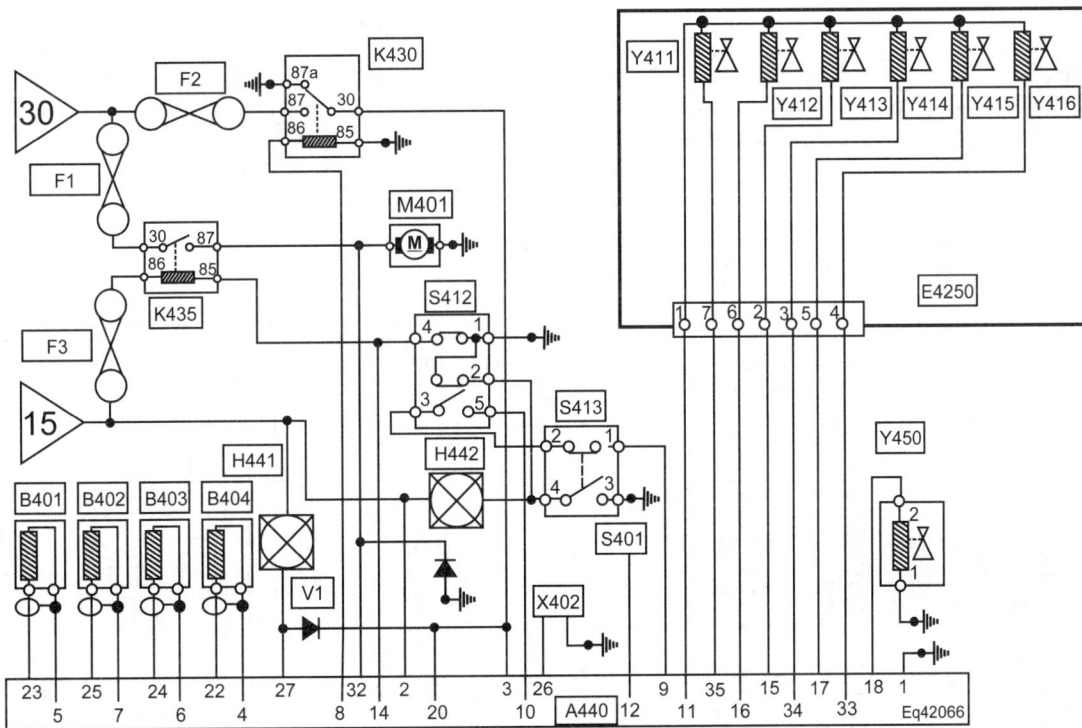

17.28 35-pin wiring diagram, Saab 900/9000 (SD)

17

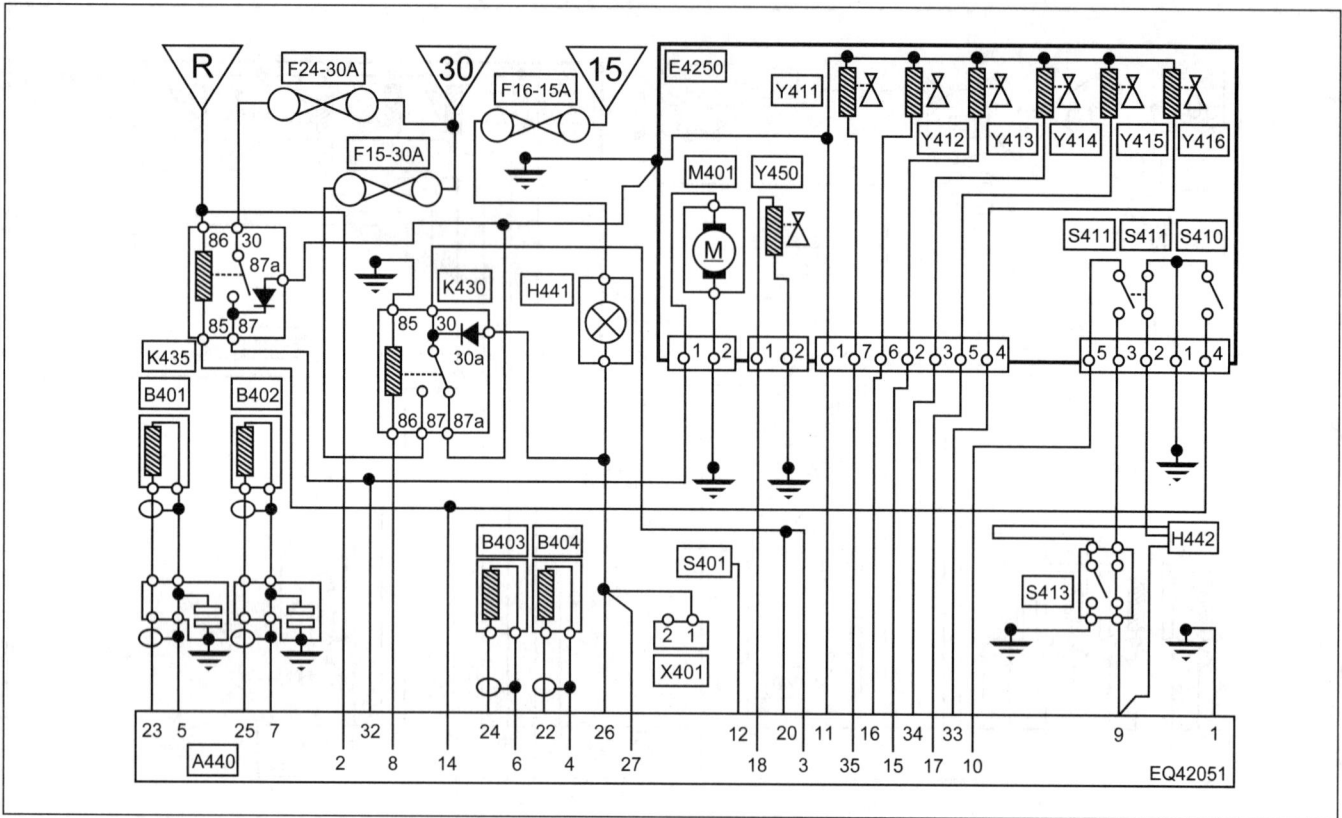

17.29 35-pin wiring diagram, Volkswagen Corrado/Golf/Jetta/Passat

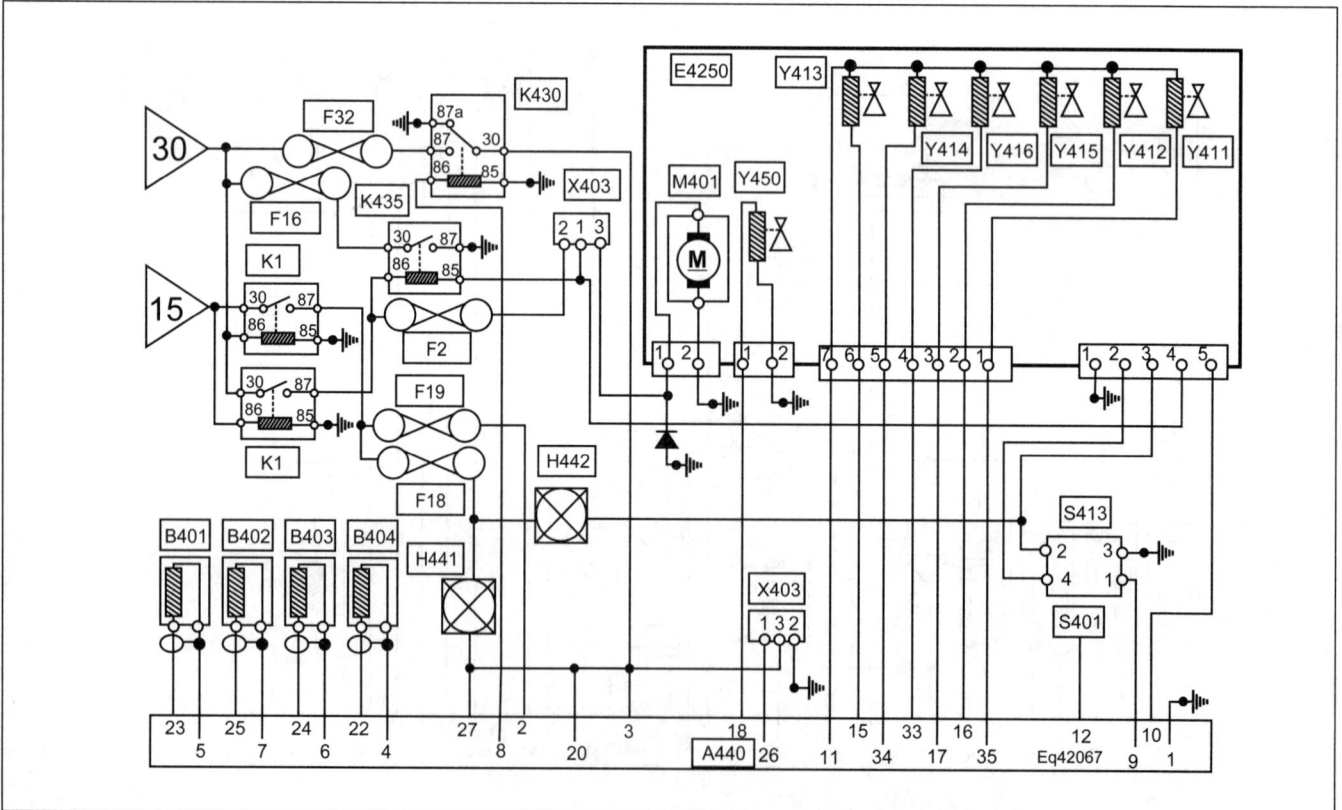

17.30 35-pin wiring diagram, Volvo 440/460/480 (SD)

Chapter 18 Part A

Teves Mk.IV (ITT 04) - 37-pin ABS

Contents

Vehicle coverage

Model	Year
Citroën	
Xantia	1993-1998
XM (2.0 and Diesel)	1996-2000
Renault	
Laguna	1995-1997
Volvo	
440/460/480	1993-1997

Overview of system operation

1 Basic principles and system identification

The Teves Mk.IV Antilock Brake System has been fitted to a wide range of passenger vehicles since its introduction in the early 1990s. The system is of the additional or 'add-on' type operating in conjunction with the conventional braking system components. Later versions of the system may carry the designation ITT 04 although their operation is unchanged from the earlier Teves Mk.IV types.

The Teves Mk.IV system has been refined and uprated over the years in accordance with advances in electrical technology and demands from vehicle manufacturers. Essentially there are three distinct versions of the system, with the main differences being in the type and location of certain ECM input sensors, the ECM software, and the ECM multi-plug which may be a 37-pin, 42-pin or 55-pin type. For clarity, 37-pin ECM multi-plug systems are covered in this Part of Chapter 18, with 42-pin, and 55-pin systems being covered in Parts B and C respectively.

Depending on application, Teves Mk.IV may be installed as an antilock braking system only, or as an antilock braking system incorporating traction control.

In ABS mode, the purpose of the system is to apply the vehicle brakes at maximum efficiency without wheel lock or loss of directional stability. Inductive sensors (wheel speed sensors) monitor the speed of the wheels by generating an electrical signal as the wheel is rotated. This information is passed to the ABS Electronic Control Module (ECM) which is then able to determine wheel speed, wheel acceleration and wheel deceleration. The ECM compares the signals received from each wheel and if the onset of lock at any wheel is detected, a signal is sent to the ABS hydraulic control unit which regulates the brake pressure for the relevant wheel(s).

Where the system incorporates traction control, essentially the reverse principle is applied. When the ECM detects that one or more wheels are rotating faster than the reference value, the brake is actually applied on the relevant wheel(s) to reduce the rotational speed.

Typically, Teves Mk.IV ABS is comprised of the following components (see illustration 18A.1):

a) ABS-ECM.
b) Hydraulic control unit.
c) Four inductive wheel speed sensors and associated sensor rings.
d) ABS electrical wiring harness and relays.
e) Brake light switch.
f) Pedal position sensor.
g) ABS warning lamp.

In addition, the conventional braking system is comprised of the following components:

a) Tandem brake master cylinder.
b) Vacuum servo unit.
c) Brake calipers/wheel cylinders and hydraulic hoses and pipes.
d) Pressure regulating/load sensing valve(s) depending on application.

2 Component description and operation

ABS ECM
General

The Teves Electronic Control Module (ECM) continually monitors wheel speed from the signals provided by the wheel speed sensors. If the ECM detects the incidence of wheel lock on one or more wheels, a signal is sent to the hydraulic control unit to modulate the hydraulic pressure to the brake of the locking wheel (see illustration 18A.2).

The ECM contains two microprocessors and uses digital technology to complete this function and other functions such as fault

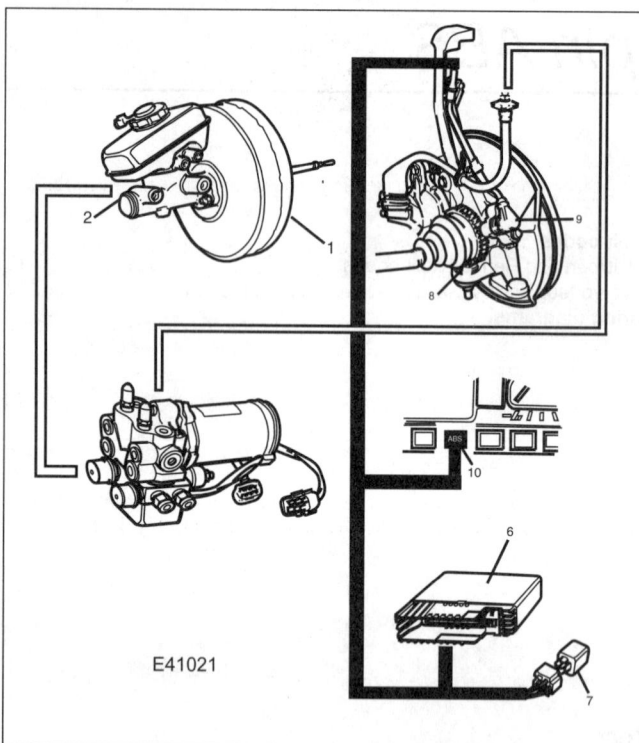

E41021

18A.1 Teves Mk.IV main components

1 ABS-ECM	6 Brake light switch
2 Hydraulic control unit	7 Pedal position sensor
3 Inductive wheel speed	(depending on application)
sensors	8 ABS warning lamp
4 Sensor rings	9 Tandem master cylinder
5 ABS electrical wiring	10 Vacuum servo unit
harness and relays	

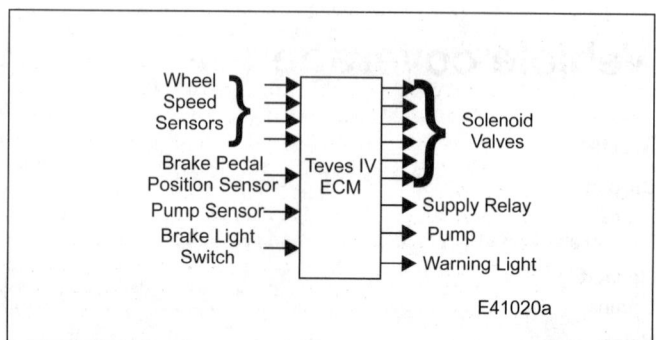

E41020a

18A.2 ECM sensor inputs and control signal outputs

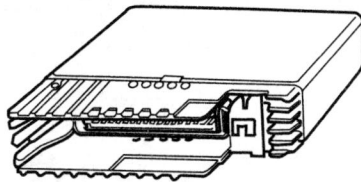

E41020

18A.3 Teves Mk.IV ECM

code memory and power modules for valve and relay activity **(see illustration 18A.3)**.

Self-test

The Teves Mk.IV ECM is equipped with a self-test capability that initially examines the ABS system when the ignition is switched on, and then examines the wheel speed sensor signals after a wheel speed of approximately 4 mph is reached from all wheels (engine running). The ABS self-test program continues to examine the signals from the various components as long as the ignition is switched on. If self-test determines that faults are not present, the ABS is ready for operation once a specified vehicle speed has been achieved.

If the ECM detects that a fault is present, all ABS functions are switched off and the warning lamp is turned on. The conventional braking system continues to operate as normal without ABS assistance.

Self-diagnostics

If the ECM detects a fault during the self-test routine, an internal fault code is stored in the ECM memory. Stored fault codes can be retrieved from the SD connector with the aid of a suitable fault code reader or, on Volvo models, by means of the on-board diagnostic unit. If the fault clears, the code will remain stored until cleared with the FCR, or diagnostic unit.

Hydraulic control unit

The hydraulic control unit consists of an electric motor with flange-mounted two-circuit radial piston return pump, and the flange-mounted valve block containing the inlet and outlet solenoid valves. The hydraulic control unit controls the pressure applied to the brake caliper for each individual wheel or pair of wheels during the three phases of ABS operation. During a controlled ABS cycle, the pump returns brake fluid, drained off during the pressure reduction phase, back into the brake circuit.

The pump only runs intermittently and is controlled by the ECM depending on the position of the brake pedal position sensor. A rotation sensor, integral with the pump, signals the ECM during pump operation.

On Citroën vehicles an integrated hydraulic system is used whereby hydraulic pressure generated by a mechanical pump is used for both the braking system and the vehicle suspension system. On these vehicles the ABS hydraulic control unit consists of the solenoid valves only - the return pump assembly and conventional brake master cylinder being replaced by a compensator/brake control valve which is part of the vehicle main hydraulic system and is unique to Citroën.

According to vehicle application, Teves Mk.IV may be either a three- or four-channel system according to the number of solenoid valves used in the hydraulic control unit. On a four-channel system, there are two solenoid valves for each brake (eight in total). On a three-channel system there are two solenoid valves for each front brake, and two solenoid valves controlling the rear brakes as a pair (six in total). Additional solenoid valves are used on systems incorporating traction control.

The layout of the hydraulic control unit varies according to the number of hydraulic channels used in the system and whether the

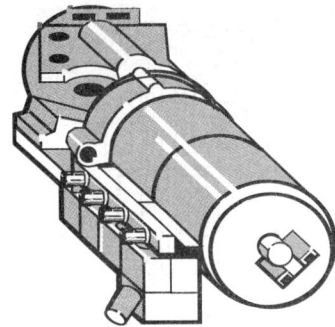

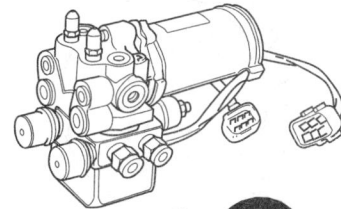

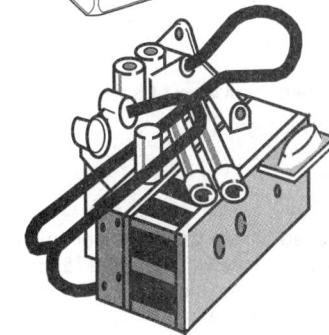

Eq44053

18A.4 Typical Teves Mk.IV hydraulic control unit variations

installation is right-hand drive or left-hand drive format. Although the units may be visually different, their operation is the same **(see illustration 18A.4)**.

Where a pair of brakes are controlled by a single hydraulic channel, the 'select-low' principle is often employed. With the 'select-low' principle, the wheel with the lowest adhesion determines the amount of hydraulic pressure to both brakes on that channel during ABS operation.

Wheel speed sensors

The rotational speed of the roadwheels and any changes in the rotational speed are recorded by inductive wheel speed sensors, one located at each roadwheel **(see illustration 18A.5)**.

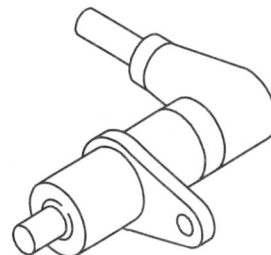

E41033A

18A.5 Typical wheel speed sensor

18A

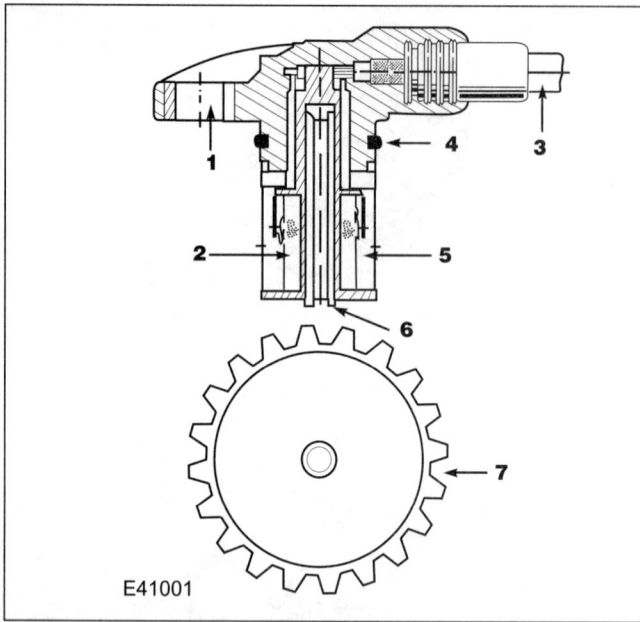

18A.6 Sectional view of a wheel speed sensor

1	Mounting bolt location	5	Coil
2	Permanent magnet	6	Sensor tip
3	Wiring harness	7	Toothed sensor ring
4	O-ring		

Each wheel speed sensor assembly comprises a toothed sensor ring which rotates at roadwheel speed, and an adjacent sensor mounted a set distance from the sensor ring (see illustration 18A.6).

The sensors are permanent magnet pulse generator types producing an AC voltage sine wave as the sensor ring teeth pass through the magnetic field of the sensor (see illustration 18A.7).

The frequency of the waveform produced by the wheel speed sensor is proportional to the road speed. This AC voltage signal is continually being delivered to the ABS-ECM for processing.

The peak to peak voltage of the speed signal (when viewed upon an oscilloscope) can vary considerably according to wheel speed. An analogue to digital converter (ADC) in the ECM transforms the AC pulse into a digital signal (see illustration 18A.8).

ABS electrical wiring harness and relays

An integrated main wiring harness is used to connect power and

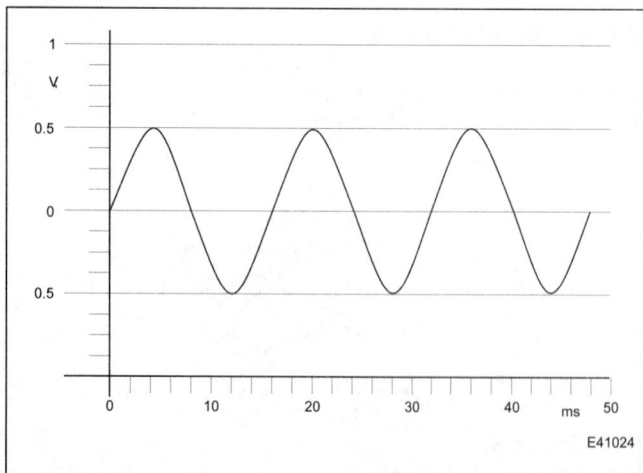

18A.8 Wheel speed sensor waveform as viewed on an oscilloscope

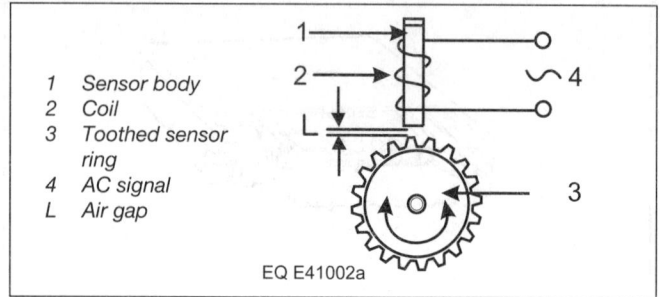

18A.7 Wheel speed sensor operation

1	Sensor body
2	Coil
3	Toothed sensor ring
4	AC signal
L	Air gap

earth to various electrical components. This enables sensor signals to reach the ECM and the ECM, in turn to send command signals to the relays, return pump and hydraulic control unit. The ABS relays are commonly located in the vicinity of the hydraulic control unit, with the associated fuses typically located in the vehicle fuse/relay box.

Main relay

Nominal battery voltage is supplied from the ignition switch to the relay via a fuse. The relay is controlled by the ECM applying an earth, which activates the relay allowing fused battery voltage to be applied to the solenoid valves. The earth for the solenoid valves is controlled by the ECM.

Pump relay

Nominal battery voltage is supplied via a fuse, from the battery, or overvoltage protection relay, to the pump relay. The relay is controlled by the ECM applying an earth, which activates the relay allowing fused battery voltage to be applied to the pump motor. The hydraulic control unit provides the earth for the pump motor.

Overvoltage protection relay

Where an overvoltage protection relay is used, nominal battery voltage is supplied from the battery to the relay via a fuse. The relay is controlled by the ECM applying an earth, which activates the relay allowing fused battery voltage to be applied to the pump relay.

Brake light switch

The brake light switch comprises a switch body and contact pin and is located above the brake pedal (see illustration 18A.9).

When the brake pedal is depressed, closing the brake light switch, a signal is sent to the ECM indicating that the brakes are being applied. Once this signal is received, the ECM will begin monitoring the wheel speed via the wheel speed sensors and activate the ABS if necessary.

Pedal position sensor

The pedal position sensor is used on Renault and Volvo

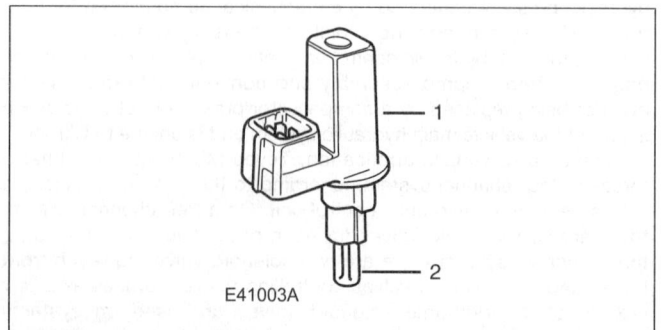

18A.9 Typical brake light switch body (1) and contact pin (2)

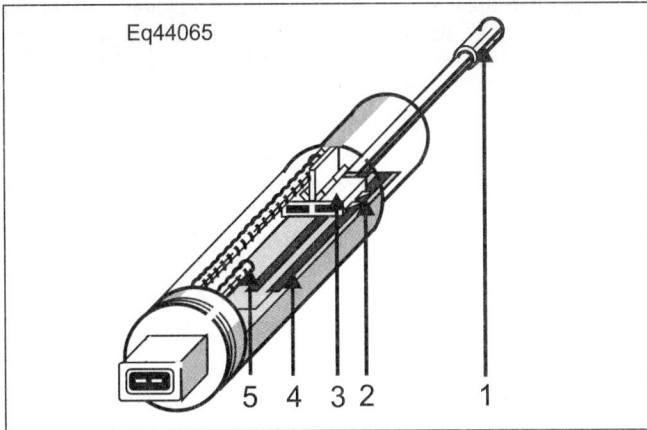

18A.10 Cutaway view of the brake pedal position sensor

1 *Plunger rod*
2 *Contact brush*
3 *Sliding segment*
4 *Track*
5 *Return spring*

vehicles and is located on the vacuum servo unit **(see illustration 18A.10)**.

The sensor contains a plunger rod, contact brush, sliding segment, track and return spring located within the switch body. The plunger rod is held in contact with the vacuum servo unit diaphragm by the pressure of the return spring. As the position of the servo diaphragm changes with brake application, the plunger rod moves accordingly, thus altering the position of the contact brush on the track. The track is divided into a series of stepped resistances corresponding to various pedal positions. The ECM supplies the track with a continuous current and determines pedal position by the drop in voltage according to the position of the contact brush on the track.

When the brake pedal is activated with the ABS in operation, the ECM receives a variable voltage signal from the sensor according to pedal position. If the brake pedal begins to creep down during a controlled pressure reduction phase, the changing voltage signal from the sensor is detected by the ECM and the return pump is activated. Hydraulic fluid is pumped from the reservoir back into the brake circuit, forcing back the master cylinder piston (and the brake pedal). When the ECM detects that the pedal has returned to its initial position, by means of the pedal position sensor signal, the return pump is switched off.

ABS warning lamp

After the ignition is switched on, the ABS warning lamp on the instrument panel is illuminated for approximately 2 to 4 seconds as the system executes a self-test routine. If satisfactory operation of the system is detected by the ECM, the light is extinguished. During vehicle operation above a pre-determined wheel speed, the ABS-

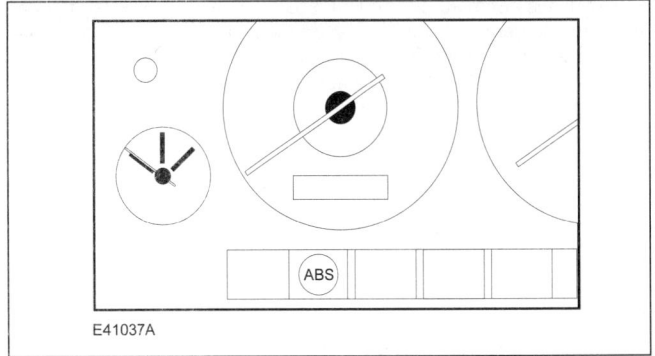

18A.11 ABS warning lamp

ECM implements a further self-test cycle whereby ABS operation and wheel speed sensor signals are continually monitored. If a fault is detected, the relevant ECM pin is earthed to illuminate the warning lamp on the instrument panel, and the ABS function is disabled. The warning lamp will remain illuminated until the fault is no longer present **(see illustration 18A.11)**.

When the ABS-ECM detects a fault, the fault code is stored and the ABS warning lamp activated. If the fault no longer exists after the next system start (ignition on/off) the ABS warning lamp is extinguished after the self-test cycle, however the fault code remains stored in the ECM memory.

Tandem master cylinder

Typically, the tandem master cylinder comprises a body casting incorporating primary and secondary pressure chambers, primary piston, intermediate piston, floating piston, slotted pin and central valve. The cylinder operates as a conventional master cylinder using vacuum assistance from the vacuum servo unit **(see illustration 18A.12)**.

When the brake system is at rest, the central valve in the floating piston rests against the slotted pin. In this condition the central valve is open and brake fluid can discharge out of the pressure chamber back into the brake fluid reservoir. When the brake pedal is depressed, the build-up of hydraulic pressure in the primary pressure chamber acts on the intermediate piston and floating piston, moving them down the cylinder bore. The floating piston contacts the seal on the central valve, closing the connection between the intermediate and secondary pressure chambers. Brake hydraulic pressure can now also increase in the secondary pressure chamber.

Vacuum servo unit

The vacuum servo unit is located between the brake pedal and tandem master cylinder. When the brake pedal is depressed, the servo unit increases the force applied by the pedal, reducing the effort required to operate the brakes **(see illustration 18A.13)**.

18A

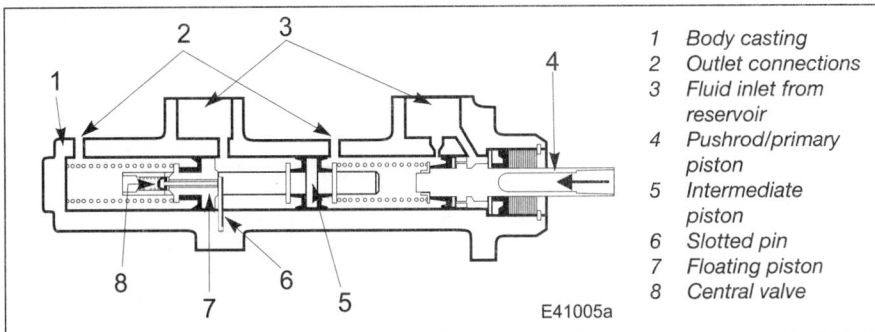

1 *Body casting*
2 *Outlet connections*
3 *Fluid inlet from reservoir*
4 *Pushrod/primary piston*
5 *Intermediate piston*
6 *Slotted pin*
7 *Floating piston*
8 *Central valve*

18A.12 Sectional view of a typical tandem master cylinder

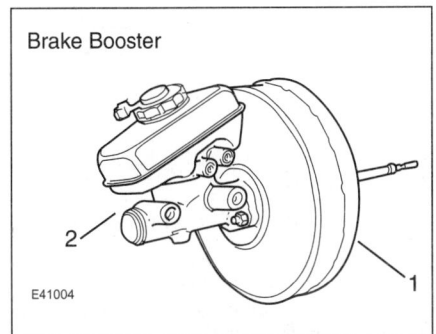

18A.13 Vacuum servo unit (1) and tandem master cylinder (2)

The unit is operated by vacuum created in the engine inlet manifold (or from a separate vacuum pump on diesel engines) which is applied to a diaphragm contained within the unit casing. A pushrod connected to the centre of the diaphragm acts directly on the primary piston in the master cylinder.

When the brake pedal is released, vacuum is applied to both sides of the diaphragm. When the pedal is depressed, one side of the diaphragm is opened to atmosphere and the vacuum acting on the other side deflects the diaphragm which in turn operates the master cylinder primary piston. The resulting force applied to the master cylinder piston is therefore significantly greater than the initial force applied to the brake pedal by the driver.

Pressure regulating/load sensing valve(s)

Depending on vehicle application, pressure regulating valves or load sensing valves may be incorporated to restrict the hydraulic fluid pressure to the rear brakes. The valves may be pressure conscious whereby the hydraulic fluid supply is restricted once a pre-determined pressure is reached, or load conscious whereby the hydraulic pressure is reduced according to vehicle loading.

3 System operation

Brake system at rest

The illustrations in this Section depict the operation of a typical three-channel hydraulic system with two solenoid valves for each front brake, and two solenoid valves controlling the rear brakes as a pair. Four-channel systems operate in the same way apart from the number of solenoid valves used.

When the ABS system is at rest all the brake components are inoperative. Pressure is non-existent in the hydraulic pipes between the tandem master cylinder and the brake calipers. The inlet solenoid valves in the valve block are open and the outlet solenoid valves are shut (see illustration 18A.14).

Brake system operating under conventional control without ABS

When the brake pedal is activated, the pedal force is applied to the tandem master cylinder by the vacuum servo unit pushrod. The servo unit pushrod acts directly on the pressure piston in the master cylinder. The resulting increase in hydraulic pressure moves the intermediate piston in the cylinder, and the pressure which increases in front of the intermediate piston flows through the open inlet solenoid valves in the valve block, to the brake calipers. The outlet solenoid valves in the valve block remain shut (see illustration 18A.15).

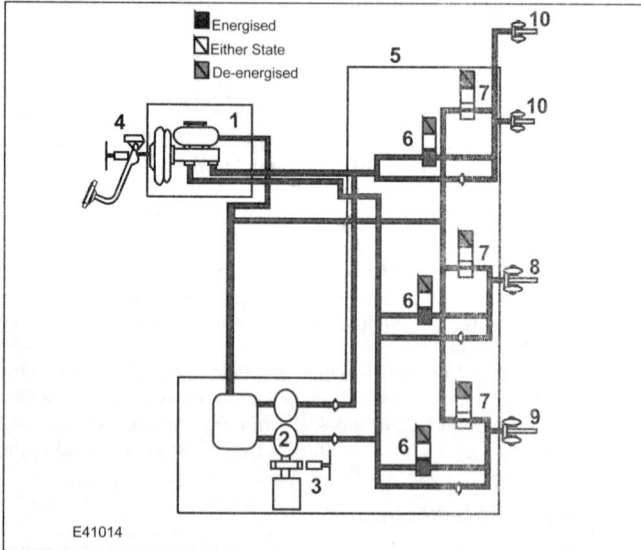

18A.14 Brake system at rest

1	Vacuum servo unit and tandem master cylinder	5	Hydraulic control unit
2	ABS return pump	6	Inlet solenoid valve
3	Pump motor sensor	7	Outlet solenoid valve
4	Brake pedal position sensor	8	Front left brake caliper
		9	Front right brake caliper
		10	Rear brake calipers

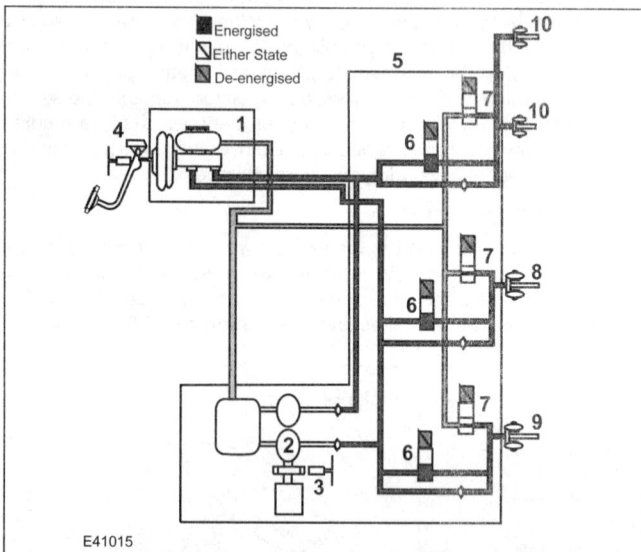

18A.15 Conventional braking without ABS

1	Vacuum servo unit and tandem master cylinder	5	Hydraulic control unit
2	ABS return pump	6	Inlet solenoid valve
3	Pump motor sensor	7	Outlet solenoid valve
4	Brake pedal position sensor	8	Front left brake caliper
		9	Front right brake caliper
		10	Rear brake calipers

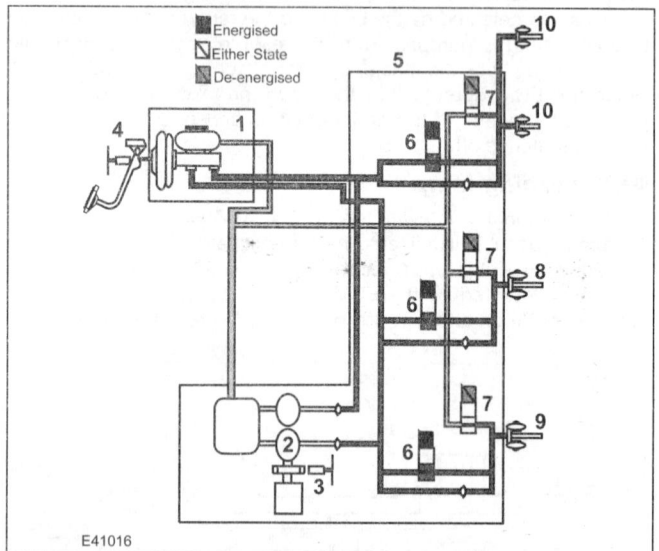

18A.16 ABS operation - first phase, pressure holding

1	Vacuum servo unit and tandem master cylinder	5	Hydraulic control unit
2	ABS return pump	6	Inlet solenoid valve
3	Pump motor sensor	7	Outlet solenoid valve
4	Brake pedal position sensor	8	Front left brake caliper
		9	Front right brake caliper
		10	Rear brake calipers

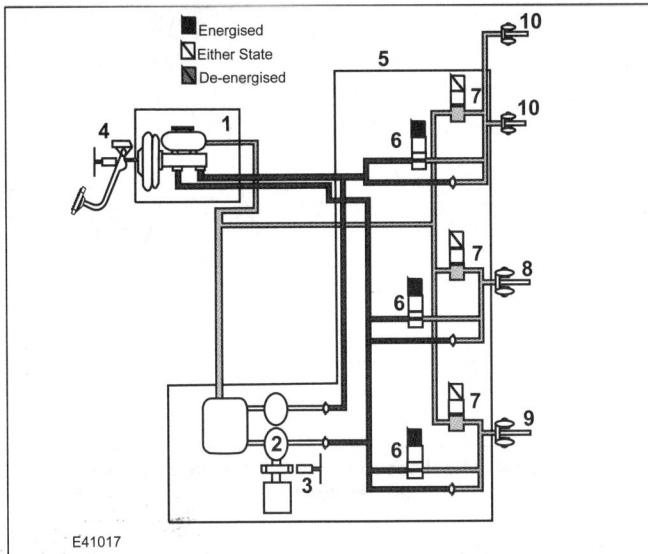

■ Energised
□ Either State
▨ De-energised

E41017

18A.17 ABS operation - second phase, pressure reduction

1 Vacuum servo unit and
  tandem master cylinder
2 ABS return pump
3 Pump motor sensor
4 Brake pedal position
  sensor

5 Hydraulic control unit
6 Inlet solenoid valve
7 Outlet solenoid valve
8 Front left brake caliper
9 Front right brake caliper
10 Rear brake calipers

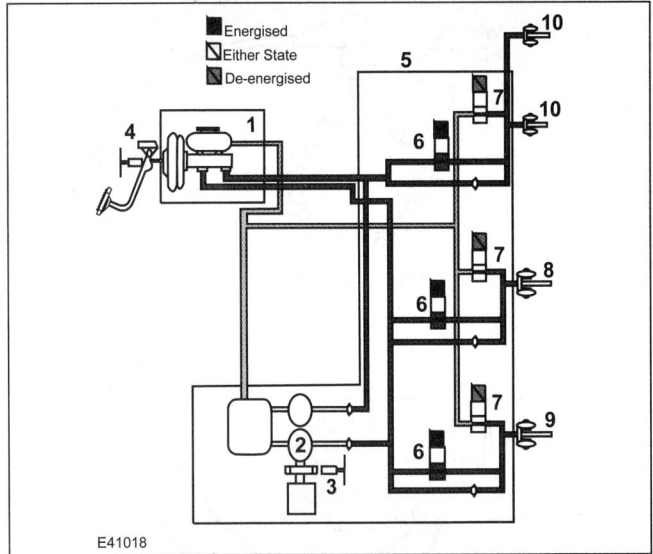

■ Energised
□ Either State
▨ De-energised

E41018

18A.18 ABS operation - third phase, pressure build-up

1 Vacuum servo unit and
  tandem master cylinder
2 ABS return pump
3 Pump motor sensor
4 Brake pedal position
  sensor

5 Hydraulic control unit
6 Inlet solenoid valve
7 Outlet solenoid valve
8 Front left brake caliper
9 Front right brake caliper
10 Rear brake calipers

Brake system operated with ABS control

The ABS-ECM continually monitors wheel speed from the signals provided by the wheel speed sensors. If the ECM detects the incidence of wheel lock on one or more wheels, ABS is automatically initiated in three phases. Typically, ABS operates individually on each front wheel and either individually or jointly on the rear wheels. In which case, all or any of the wheels could be in any one of the following phases at any particular moment.

First ABS phase, pressure holding

To prevent any further build-up of hydraulic pressure in the circuit being controlled, the inlet solenoid valve is shut and the outlet solenoid valve remains shut. The pressure cannot now be increased in the controlled circuit by any further application of the brake pedal **(see illustration 18A.16)**.

If the wheel speed sensor signals indicate that wheel rotation has now stabilised, the ECM will instigate the pressure build-up phase, allowing braking to continue. If wheel lock is still detected after the pressure holding phase, the ECM instigates the pressure reduction phase.

Second ABS phase, pressure reduction

A controlled braking operation can be instigated with a pressure reduction phase or it can follow a pressure holding phase. The inlet solenoid valve is, or remains shut and the outlet solenoid valve is opened. As long as the outlet solenoid valve is open, the brake fluid discharges back into the reservoir of the tandem master cylinder. The return pump then pumps the brake fluid out of the reservoir back into the control circuit, forcing back the master cylinder piston (and the brake pedal) **(see illustration 18A.17)**.

Third ABS phase, pressure-build up

The pressure build up phase is instigated after the wheel rotation has stabilised. The outlet solenoid valve is shut and the inlet solenoid valve is opened. The brake pressure from the tandem master cylinder flows through the inlet solenoid valve to the relevant brake caliper allowing braking to continue **(see illustration 18A.18)**.

The whole ABS control cycle takes place 4 to 10 times per second for each affected wheel and this ensures maximum braking effect and control during ABS operation.

Test procedures

18A

Important note: The test procedures, pin-tables and wiring diagrams contained in this Chapter are necessarily representative of the system depicted. Because of the variations in wiring and other data that often occurs, even between similar vehicles in any particular VM's range, the reader should take great care in identification of ECM pins, and satisfy himself that he has gathered the correct data before failing a particular component.

4 Wheel speed sensors

Checking the wheel speed sensor (general)

1 Inspect the wheel speed sensor for corrosion or damage and check that the sensor is tightly mounted.

2 Check the toothed sensor ring for damage, eccentricity and for broken or missing teeth.
3 Inspect the wheel speed sensor wiring plug for corrosion and damage. One plug for each sensor.
4 Check that the connector terminal pins are pushed fully home and making good contact with the sensor wiring plug.
5 Check the clearance between the sensor and the toothed sensor ring. The clearance is not normally adjustable but is nominally 0.2 to 1.2 mm. If the clearance is excessive, expect a worn sensor tip or problems with the wheel bearings/hub or sensor ring.
6 When carrying out voltage checks with an oscilloscope or voltmeter, the voltage obtained will be proportional to the speed at which the wheel is rotating. In addition to determining that the wheel

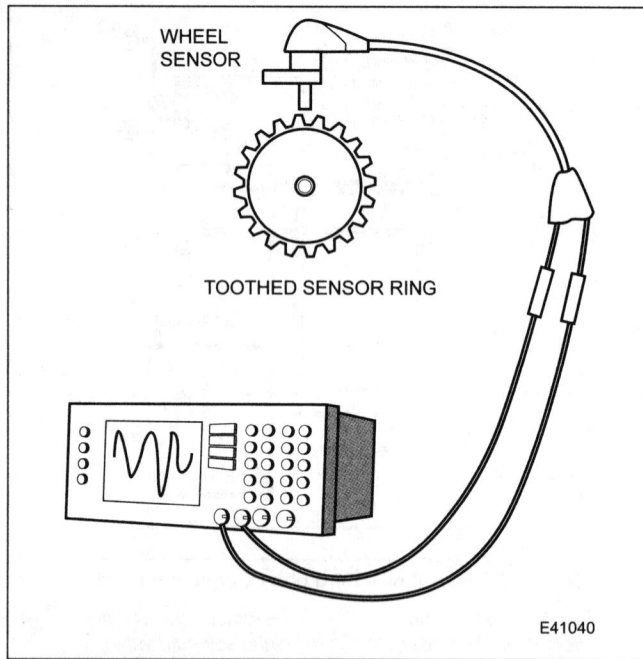

18A.19 Checking wheel speed sensor output with an oscilloscope connected to the sensor wiring plug

speed sensors are actually producing a voltage output, it is essential that the output from the sensors on a particular axle is the same for any given wheel speed.

Checking wheel speed sensor output with an oscilloscope

Note: *Refer to the wiring diagrams for specific ECM pin identification according to model.*
7 Switch the ignition off and disconnect the ECM multi-plug or the relevant wheel speed sensor wiring plug.
8 Connect an oscilloscope between the terminal pins for the sensor under test (see illustration 18A.19).
9 Select a range to cover 80 Hz on the oscilloscope and a free run time base.

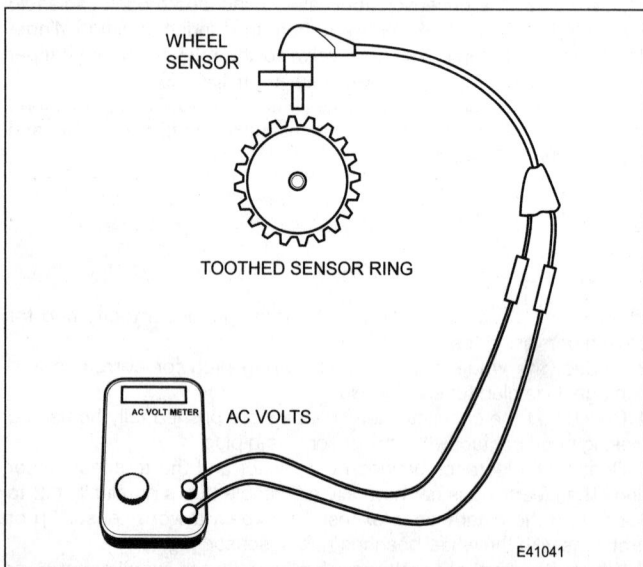

18A.21 Checking wheel speed sensor output with a voltmeter connected to the sensor wiring plug

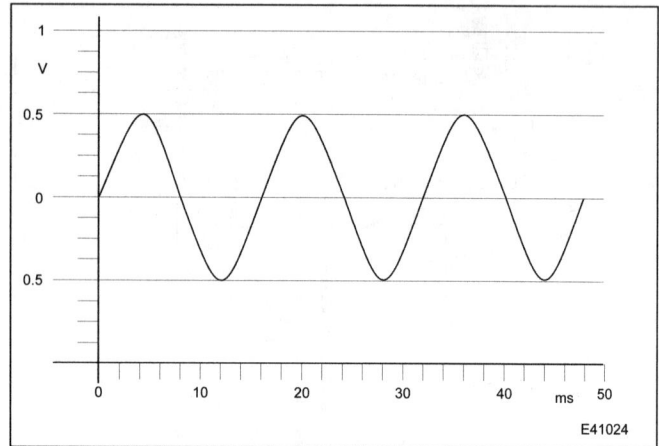

18A.20 Typical wheel speed sensor sine wave as displayed on an oscilloscope

10 Raise the wheel and rotate it by hand at approximately one revolution per second.
11 A sinusoidal wave form should be obtained, with amplitude and duration changing with rotational speed (see illustration 18A.20).
12 If there is no signal, or a very weak or intermittent signal at the ECM, repeat the test at the sensor wiring plug. If there is no change in signal status, the sensor is suspect.
13 If the signal is now satisfactory this indicates a fault in the wiring harness which should be checked for continuity.

Checking wheel speed sensor output with an AC voltmeter

Note: *Refer to the wiring diagrams for specific ECM pin identification according to model.*
14 Switch the ignition off and disconnect the ECM multi-plug or the relevant wheel speed sensor wiring plug.
15 Connect an AC voltmeter between the terminal pins for the sensor under test (see illustration 18A.21).
16 Raise the wheel and rotate it by hand at approximately one revolution per second.
17 A voltage of approximately 0.3 to 1.5 volts (AC RMS) should be obtained. If there is no signal, or a very weak or intermittent signal at the ECM, repeat the test at the sensor wiring plug. If there is no change in the signal, the sensor is suspect.
18 If the signal is now satisfactory, this indicates a fault in the wiring harness which should be checked for continuity. **Note:** *This test at least proves that a signal is being generated by the sensor. However, the voltage produced is an average voltage and does not clearly indicate damage to the sensor ring or that the sinewave is regular in formation.*

Checking wheel speed sensor resistance

Note: *Refer to the wiring diagrams for specific ECM pin identification according to model.*
19 Switch the ignition off and disconnect the ECM multi-plug or the relevant wheel speed sensor wiring plug.
20 Connect an ohmmeter between the terminal pins for the sensor under test (see illustration 18A.22).
21 The readings obtained should be between 0.7 and 1.5 kohms approximately.
22 If the resistance is excessively high, or open circuit at the ECM, repeat the test at the sensor multi-plug. If there is no change in resistance, the sensor is suspect.
23 If the resistance is now satisfactory, this indicates a fault in the wiring harness which should be checked for continuity. **Note:** *Even if the resistance is within the quoted specifications, this does not prove that the speed sensor can generate an acceptable signal.*

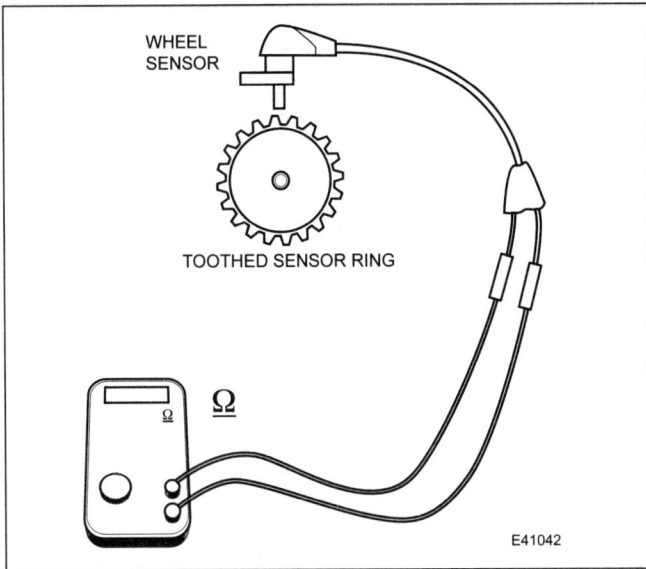

18A.22 Checking wheel speed sensor resistance with an ohmmeter connected to the sensor wiring plug

5 System relays

Relay operation tests (Renault and Volvo)

Main relay

1 Switch the ignition off and disconnect the 15-pin multi-plug from the relay board adjacent to the hydraulic control unit.
2 Connect an ohmmeter between terminals 2 and 10 in the relay board **(see illustration 18A.23)**. The ohmmeter should indicate continuity.
3 Connect an ohmmeter between terminals 4 and 10 in the relay board. The ohmmeter should indicate an open circuit.
4 Attach the negative lead of a 12 volt supply to relay board terminal 3 and the positive lead relay board terminal 12.
5 The ohmmeter should now indicate continuity.
6 If the results are not as specified, the relay board is suspect.

Hydraulic pump relay

7 Switch the ignition off and disconnect the 15-pin multi-plug and 4-pin multi-plug from the relay board adjacent to the hydraulic control unit.
8 Connect an ohmmeter between the 15-pin multi-plug terminal 15 in the relay board, and the 4-pin multi-plug terminal 1 in the relay board **(see illustration 18A.23)**. The ohmmeter should indicate an open circuit.
9 Attach the negative lead of a 12 volt supply to the 15-pin multi-plug terminal 13 in the relay board and attach the positive lead to the 15-pin multi-plug terminal 10 in the relay board.
10 The ohmmeter should now indicate continuity.
11 If the results are not as specified, the relay board is suspect.

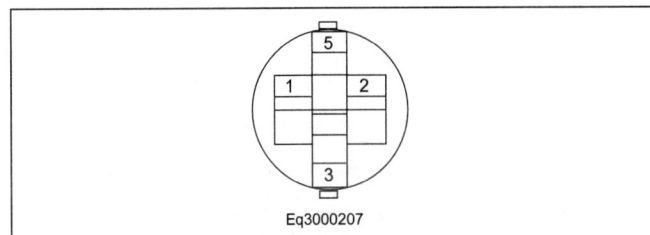

18A.24 Overvoltage protection relay terminal identification (Renault)

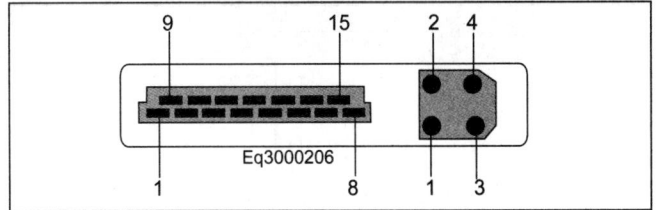

18A.23 Relay board 15-pin and 4-pin terminal identification

Overvoltage protection relay (Renault only)

12 Switch the ignition off, then remove the overvoltage protection relay from the wiring harness multi-plug connector.
13 Connect an ohmmeter between the relay terminals 3 and 5 **(see illustration 18A.24)**. The ohmmeter should indicate an open circuit.
14 Attach the negative lead of a 12 volt supply to the relay terminal 2, and the positive lead to the relay terminal 1.
15 There should be an unmistakable click from the relay and the ohmmeter should indicate continuity.
16 If the results are not as specified, the relay is suspect.

Relay operation tests (Citroën)

Main relay

17 Switch the ignition off and remove the main relay from the relay base.
18 Connect an ohmmeter between the relay output terminal 30 and earth terminal 87a **(see illustration 18A.25)**. The ohmmeter should indicate continuity.
19 Connect an ohmmeter between the relay main voltage supply terminal 87 and the output terminal 30. The ohmmeter should indicate an open circuit.
20 Attach the positive lead of a 12 volt supply to the relay driver supply terminal 86, and the negative lead to the relay driver earth terminal 85.
21 There should be an unmistakable click from the relay and the ohmmeter should indicate continuity.
22 Connect an ohmmeter or a diode tester between the relay output terminal 30 and ABS warning lamp terminal 30a.
23 Measure the circuit and then reverse the connections. The instrument should indicate continuity in one direction (diode conducting), but not the other (diode blocking).
24 Renew the relay if it fails these tests. If the relay cover is transparent, inspect the relay for burns. A faulty relay usually has a burnt appearance.

Relay power supply tests (Renault)

25 Switch the ignition off and remove the overvoltage protection relay from the wiring harness multi-plug connector.
26 Using a voltmeter, probe between the main voltage supply terminal 3 in the multi-plug and earth **(see illustration 18A.24)**. The reading should indicate nbv. If no voltage is found, check the relevant fuse and the supply wiring back to the voltage supply.
27 Disconnect the 15-pin multi-plug from the relay board adjacent to the hydraulic control unit.

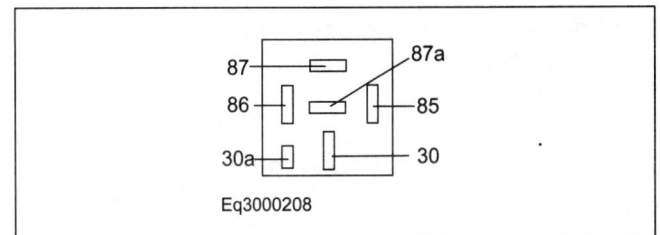

18A.25 Main relay terminal identification (Citroen)

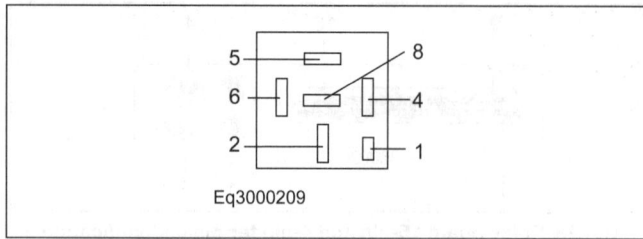

18A.26 Main relay base terminal identification (Citroën)

28 Connect a voltmeter between the multi-plug terminal 4 and earth, and terminal 5 and earth in turn **(see illustration 18A.23)**. The reading should indicate nbv in each case.

29 Connect a voltmeter between the multi-plug terminal 3 and earth.

30 Switch the ignition on. The voltmeter should indicate nbv.

31 If no voltage is found, check the relevant fuse and the supply wiring back to the ignition switch.

Relay power supply tests (Volvo)

32 Switch the ignition off and disconnect the 15-pin multi-plug from the relay board adjacent to the hydraulic control unit.

33 Connect a voltmeter between the multi-plug terminal 4 and earth, and terminal 15 and earth in turn **(see illustration 18A.23)**. The reading should indicate nbv in each case.

34 Connect a voltmeter between the multi-plug terminal 3 and earth, and terminal 6 and earth in turn.

35 Switch the ignition on. The voltmeter should indicate nbv in each case.

36 If no voltage is found, check the relevant fuse and the supply wiring back to the ignition switch.

Relay power supply tests (Citroën)

37 Switch the ignition off and remove the main relay leaving the base exposed.

38 Using a voltmeter, probe between the main voltage supply terminal 8 in the relay base and earth. The reading should indicate nbv **(see illustration 18A.26)**.

39 If no voltage is found, check the relevant fuse and the supply wiring back to the voltage supply.

40 Switch the ignition on.

41 Using a voltmeter, probe between the relay driver supply terminal 4 in the relay base and earth. The reading should indicate nbv.

42 If no voltage is found, check the relevant fuse and the supply wiring back to the ignition switch.

6 Electronic Control Module

Checking the ECM (general)

1 Inspect the ECM for corrosion or damage and ensure that the unit is securely attached to the hydraulic control unit.

2 Check that the ECM multi-plug terminals are pushed fully home and making good contact with the ECM pins. A fault in any of the above areas are possible reasons for poor performance in the ABS system.

ECM power supply and earth tests

3 Switch the ignition off and disconnect the ECM multi-plug.

4 Switch the ignition on.

5 Attach a negative voltmeter probe to a vehicle earth.

6 Attach a positive voltmeter probe to ECM multi-plug pin 33. The voltmeter should indicate nbv **(see illustration 18A.27)**.

7 If no voltage is found, check the relevant fuse and the supply wiring back to the ignition switch.

8 Switch the ignition off.

9 Connect an ohmmeter between a vehicle earth and ECM multi-plug earth pins 11and 24 in turn.

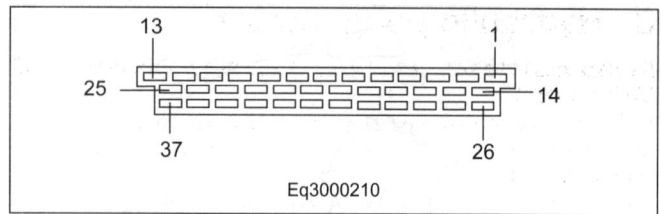

18A.27 ECM 37-pin multi-plug

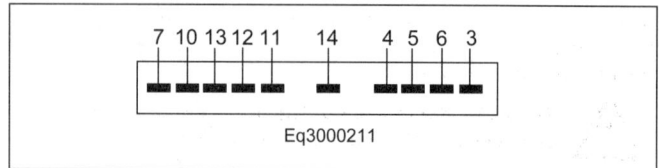

18A.28 Hydraulic control unit terminal identification (Renault, Volvo)

10 The ohmmeter should indicate continuity in each case. If not, check the ECM main earth connection and wiring.

7 Solenoid valves

Solenoid valve resistance tests (Renault and Volvo)

1 Switch the ignition off and disconnect the ECM multi-plug from the ECM.

2 Remove the ECM from the hydraulic control unit.

3 Using an ohmmeter, measure the resistance between pin 7 in the hydraulic control unit, and each of the following pins in turn **(see illustration 18A.28)**:

Inlet solenoid valves - 3, 4, 5, 6.
Outlet solenoid valves - 10, 11, 12, 13.

4 The resistance should be in the order of 6.0 to 10.0 ohms for each inlet solenoid valve, and 3.0 to 6.0 ohms for each outlet solenoid valve.

5 If the resistance is not as specified, check the wiring and wiring connectors. If satisfactory, the hydraulic control unit is suspect.

Solenoid valve resistance tests (Citroën)

6 Switch the ignition off and disconnect the ECM multi-plug from the ECM.

7 Remove the ECM from the hydraulic control unit.

8 Using an ohmmeter, measure the resistance between pin 7 in the hydraulic control unit, and each of the following pins in turn **(see illustration 18A.29)**:

Inlet solenoid valves - 3, 5, 6.
Outlet solenoid valves - 10, 12, 13.

9 The resistance should be in the order of 6.0 to 10.0 ohms for each inlet solenoid valve, and 3.0 to 6.0 ohms for each outlet solenoid valve.

10 If the resistance is not as specified, check the wiring and wiring connectors. If satisfactory, the hydraulic control unit is suspect.

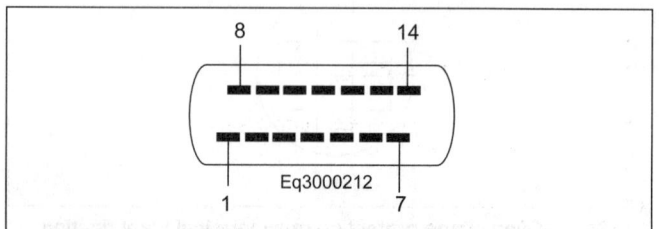

18A.29 Hydraulic control unit terminal identification (Citroën)

8 Hydraulic pump motor

Pump resistance and operation test (Renault and Volvo only)

1 Switch the ignition off and disconnect the 4-pin multi-plug from the relay board adjacent to the hydraulic control unit **(see illustration 18A.23)**.

2 Using an ohmmeter, measure the resistance between pins 1 and 3 in the multi-plug. The resistance should be in the order of 0.4 to 1.6 ohms.

3 Connect the ohmmeter between pins 2 and 4 in the multi-plug. The resistance should be in the order of 30 ohms.

4 Attach the negative lead of a 12 volt supply to terminal 3 in the multi-plug, and the positive lead to terminal 1 in the multi-plug. The pump should now operate.

⚠️ **Warning: The test should be made as quickly as possible to avoid damaging the pump.**

5 If the pump does not operate, or if the resistance readings were unsatisfactory, the hydraulic control unit is suspect.

9 Brake light switch

Checking the brake light switch (general)

1 Check that the brake light switch is correctly and securely mounted and that the plunger moves smoothly with no trace of binding.

2 Check that the wiring multi-plug is pushed fully home and making good contact.

3 Check that no wires have been disconnected.

4 A fault in any of the above areas are possible reasons for failure or malfunctioning of the switch.

Brake light switch voltage and continuity tests

Voltage test

5 Switch the ignition off and disconnect the ECM multi-plug from the ECM.

6 Connect a voltmeter between a vehicle earth and the ECM multi-plug brake light switch input pin 22 **(see illustration 18A.27)**.

7 Switch the ignition on and depress the brake pedal. The voltmeter should indicate nbv.

8 If no voltage is found, the fuse, the brake light switch and the wiring are suspect.

9 Release the brake pedal. The voltage should drop to zero as the switch opens.

Continuity test

10 Switch the ignition off and disconnect the brake light switch multi-plug.

11 Connect an ohmmeter between the terminal pins of the brake light switch.

12 Operate the brake light switch and check for continuity. If the test fails, renew the switch.

10 Brake pedal position sensor

Pedal position sensor resistance test

Note: *The brake pedal position sensor is not fitted to all systems.*

1 Switch the ignition off and disconnect the ECM multi-plug from the ECM.

2 Connect an ohmmeter between ECM multi-plug pins 10 and 36 **(see illustration 18A.27)**.

3 Measure the resistance of the position sensor against brake pedal travel.

4 The resistance should change in stages according to the distance travelled by the brake pedal.

5 If the test fails, renew the sensor.

11 Warning lamp

Checking the warning lamp (general)

1 Inspect the warning lamp bulb holder contacts in the instrument panel.

2 Check that the instrument panel multi-plug terminal pins are pushed fully home and making good contact.

3 Check that no wires have been disconnected.

4 A fault in any of the above areas are possible reasons for failure or malfunctioning of the warning lamp.

Warning lamp operation test

5 With the ignition switched off, the warning lamp should remain off.

6 Switch the ignition on and the warning lamp should illuminate then extinguish after a few seconds. The lamp should then remain off.

7 If the warning lamp comes on and remains on at any time during vehicle operation, carry out the previously described test procedures on the system components.

Pin table - typical 37-pin (Citroën)

Note: *Refer to illustration 18A.27*

Pin No.	Connection	Test condition	Voltage
1	Front right wheel speed sensor signal	Roadwheel rotating	0.3 to 1.5 volts AC (approx)
2	-	-	-
3	-	-	-
4	-	-	-
5	-	-	-
6	Front right wheel speed sensor earth	Roadwheel rotating	0.25 volts (max)
7	-	-	-
8	-	-	-
9	-	-	-
10	-	-	-
11	ECM earth	Ignition on	0.25 volts (max)
12	-	-	-
13	Supply from relay	Ignition on	Nbv
14	-	-	-
15	Front left wheel speed sensor signal	Roadwheel rotating	0.3 to 1.5 volts AC (approx)
16	ABS warning lamp	-	-
17	-	-	-
18	-	-	-
19	Rear left wheel speed sensor earth	Roadwheel rotating	0.25 volts (max)
20	-	-	-

18A

Pin table - typical 37-pin (Citroën) (continued)

Note: *Refer to illustration 18A.27*

Pin No.	Connection	Test condition	Voltage
21	-	-	-
22	Brake light switch input	Ignition on: Brake pedal released Brake pedal depressed	 0 volts Nbv
23	-	-	-
24	ECM earth	Ignition on	0.25 volts (max)
25	Supply from relay	Ignition on	Nbv
26	-	-	-
27	SD connector	-	-
28	Rear left wheel speed sensor signal	Roadwheel rotating	0.3 to 1.5 volts AC (approx)
29	Rear right wheel speed sensor signal	Roadwheel rotating	0.3 to 1.5 volts AC (approx)
30	Front left wheel speed sensor earth	Roadwheel rotating	0.25 volts (max)
31	Rear right wheel speed sensor earth	Roadwheel rotating	0.25 volts (max)
32	Main relay driver	Ignition on, actuated	1.0 volt (max)
33	Supply from ignition switch	Ignition on	Nbv
34	-	-	-
35	-	-	-
36	-	-	-
37	Supply from relay	Ignition on	Nbv

Pin table - typical 37-pin (Renault, Volvo)

Note: *Refer to illustration 18A.27*

Pin No.	Connection	Test condition	Voltage
1	Front right wheel speed sensor signal	Roadwheel rotating	0.3 to 1.5 volts AC (approx)
2	-	-	-
3	Hydraulic pump relay driver	Ignition on/inactive Actuated	Nbv 1.0 volt (max)
4	-	-	-
5	SD connector	-	-
6	Front right wheel speed sensor earth	Roadwheel rotating	0.25 volts (max)
7	-	-	-
8	Pump motor sensor earth	Pump motor actuated	0.25 volts (max)
9	-	-	-
10	Pedal position sensor	Pedal at rest Pedal fully depressed	1.0 volt 3.0 volts
11	ECM earth	Ignition on	0.25 volts (max)
12	-	-	-
13	Supply from relay	Ignition on	Nbv
14	-	-	-
15	Front left wheel speed sensor earth	Roadwheel rotating	0.25 volts (max)
16	ABS warning lamp	-	-
17	-	-	-
18	-	-	-
19	Rear left wheel speed sensor earth	Roadwheel rotating	0.25 volts (max)
20	-	-	-
21	Pump motor sensor signal	Pump motor actuated	0.5 volts AC (approx)
22	Brake light switch input	Ignition on: Brake pedal released Brake pedal depressed	 0 volts Nbv
23	-	-	-
24	ECM earth	Ignition on	0.25 volts (max)
25	-	-	-
26	-	-	-
27	SD connector	-	-
28	Rear left wheel speed sensor signal	Roadwheel rotating	0.3 to 1.5 volts AC (approx)
29	Rear right wheel speed sensor signal	Roadwheel rotating	0.3 to 1.5 volts AC (approx)
30	Front left wheel speed sensor signal	Roadwheel rotating	0.3 to 1.5 volts AC (approx)
31	Rear right wheel speed sensor earth	Roadwheel rotating	0.25 volts (max)
32	Main relay driver	Ignition on, actuated	1.0 volt (max)
33	Supply from ignition switch	Ignition on	Nbv
34	-	-	-
35	-	-	-
36	Pedal position sensor earth	Ignition on	0.25 volts (max)
37	Supply from relay	Ignition on	Nbv

Fault codes

12 Citroën and Renault fault codes

1 The Teves Mk.IV system as fitted to Citroën and Renault vehicles requires the use of a proprietary fault code reader (FCR) or system tester (such as the Renault XR25) to interrogate the system. Flash codes are not available for output from this system in these applications.

2 If a FCR is available, it should be connected to the SD serial connector and used in accordance with the maker's instructions.

3 The FCR can be used for the following purposes:

  a) *Obtaining fault codes.*
  b) *Clearing fault codes.*
  c) *Obtaining datastream information.*
  d) *Testing the system actuators (relays and solenoid valves).*

4 Internal fault codes are used by the ECM to designate faults in the system components and circuits. No actual fault code numbers are available although for the sake of completeness a list of components that will provide errors for readout upon a FCR are included.

13 Volvo fault codes

1 On Volvo vehicles, it is possible to obtain fault codes by means of the on-board diagnostic unit. The codes are displayed as a series of flashes of the red LED on the diagnostic unit.

Obtaining codes using the on-board diagnostic unit

2 Remove the cover from the diagnostic unit located on the left-hand side of the engine compartment.

3 Unclip the flylead from the holder on the side of the unit and insert it into socket 3 of the unit **(see illustration 18A.30)**. The three-digit codes will be displayed as a series of blinks of the red LED (located on the top face of the unit, next to the test button) with a slight pause between each digit.

4 With the flylead inserted, switch on the ignition. Press the test button on the top of the unit once, for about one second, then release it and wait for the LED to flash. As the LED flashes, copy down the fault code. Press the button again and copy down the next fault code, if there is one. Continue until the first fault code is displayed again

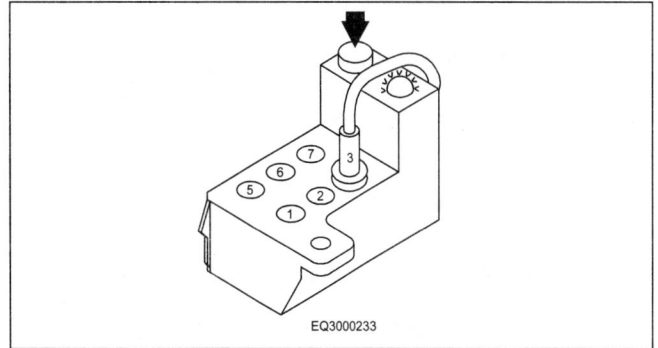

18A.30 Volvo on-board diagnostic unit showing test button (arrowed) and flylead sockets

indicating that all the stored codes have been accessed, then switch off the ignition.

5 If code 1-1-1 is obtained, this indicates that there are no fault codes stored in the diagnostic unit and the system is operating correctly.

Clearing fault codes from the ECM memory

6 Once all the fault codes have been recorded they should be deleted from the ECM memory, using the diagnostic unit. Note that the codes cannot be deleted until all of them have been displayed at least once, and the first one is displayed again.

7 With the flylead still inserted in position 3 of the diagnostic unit, switch on the ignition, press the test button and hold it down for approximately five seconds. Release the test button and after three seconds the LED will light. When the LED lights, press and hold the test button down for a further five seconds then release it - the LED will go out.

8 Switch off the ignition and check that all the fault codes have been deleted by switching the ignition on again and pressing the test button for one second - code 1-1-1 should appear. If a code other than 1-1-1 appears, record the code then repeat the deleting procedure. When all the codes have been deleted, switch off the ignition, locate the flylead in its holder and refit the unit cover.

Fault conditions (Citroën and Renault)

Code	Item	Fault
-	Supply voltage	Too high/low
-	Relays	Defective (signal interruption, short to earth, supply voltage)
-	Pump motor	Defective (poor connections)
-	Front left wheel speed sensor	Incorrect resistance/signal interruption
-	Front right wheel speed sensor	Incorrect resistance/signal interruption
-	Rear left wheel speed sensor	Incorrect resistance/signal interruption
-	Rear right wheel speed sensor	Incorrect resistance/signal interruption
-	ECM	Defective

18A

Fault code table (Volvo)

Code	Item	Fault
1-1-1	No fault detected	
1-2-1	Front left wheel speed sensor	Faulty signal (below 25 mph)
1-2-2	Front right wheel speed sensor	Faulty signal (below 25 mph)
1-2-3	Rear left wheel speed sensor	Faulty signal (below 25 mph)
1-2-4	Rear right wheel speed sensor	Faulty signal (below 25 mph)
1-4-1	Brake pedal position sensor	Shorted to earth or supply
1-4-2	Brake light switch	Open or short circuit
1-4-3	ABS ECM	Memory fault
2-1-1	Front left wheel speed sensor	No signal when moving off
2-1-2	Front right wheel speed sensor	No signal when moving off
2-1-3	Rear left wheel speed sensor	No signal when moving off
2-1-4	Rear right wheel speed sensor	No signal when moving off
2-2-1	Front left wheel speed sensor	No signal
2-2-2	Front right wheel speed sensor	No signal
2-2-3	Rear left wheel speed sensor	No signal
2-2-4	Rear right wheel speed sensor	No signal
3-1-1	Front left wheel speed sensor	Open or short circuit
3-1-2	Front right wheel speed sensor	Open or short circuit
3-1-3	Rear left wheel speed sensor	Open or short circuit
3-1-4	Rear right wheel speed sensor	Open or short circuit
3-2-1	Front left wheel speed sensor	Intermittent signal (over 25 mph)
3-2-2	Front right wheel speed sensor	Intermittent signal (over 25 mph)
3-2-3	Rear left wheel speed sensor	Intermittent signal (over 25 mph)
3-2-4	Rear right wheel speed sensor	Intermittent signal (over 25 mph)
4-1-1	Front left inlet solenoid valve	Open or short circuit
4-1-2	Front left outlet solenoid valve	Open or short circuit
4-1-3	Front right inlet solenoid valve	Open or short circuit
4-1-4	Front right outlet solenoid valve	Open or short circuit
4-2-1	Rear left inlet solenoid valve	Open or short circuit
4-2-2	Rear left outlet solenoid valve	Open or short circuit
4-2-3	Traction control valve (where fitted)	Open or short circuit
4-2-4	Traction control pressure sensor (where fitted)	Faulty or short circuit
4-3-1	Rear right inlet solenoid valve	Open or short circuit
4-3-2	Rear right outlet solenoid valve	Open or short circuit
4-4-1	ABS ECM	Processing fault
4-4-2	Hydraulic control unit	Pump pressure low
4-4-3	Hydraulic pump motor	Electrical or mechanical fault
4-4-4	Solenoid valves	Supply voltage absent

Wiring diagrams

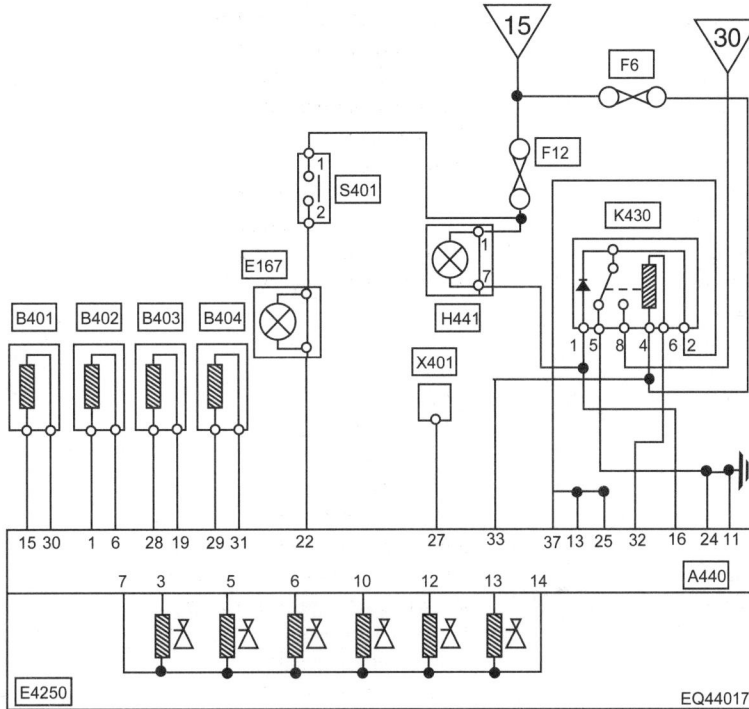

18A.31 37-pin wiring diagram, Citroën Xantia/XM (2.0 and Diesel)

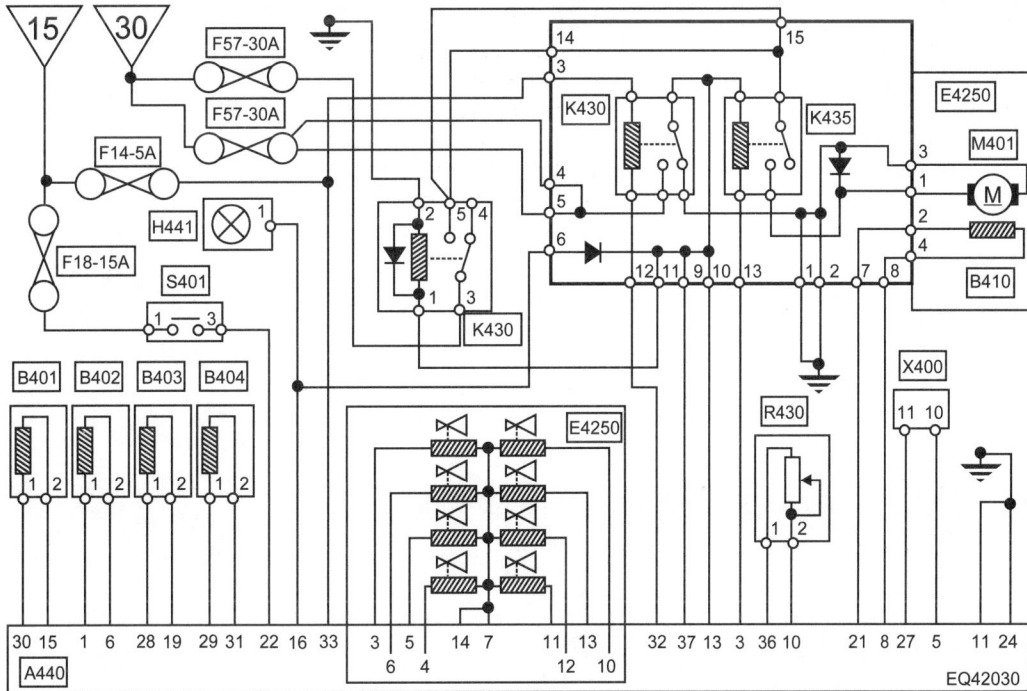

18A.32 37-pin wiring diagram, Renault Laguna

18A

Chapter 18 Part B

Teves Mk.IV (ITT 04) - 42-pin ABS

Contents

18B

Vehicle coverage

Overview of system operation

1 Basic principles and system identification

The Teves Mk.IV Antilock Brake System has been fitted to a wide range of passenger vehicles since its introduction in the early 1990s. The system is of the additional or 'add-on' type operating in conjunction with the conventional braking system components. Later versions of the system may carry the designation ITT 04 although their operation is unchanged from the earlier Teves Mk.IV types.

The Teves Mk.IV system has been refined and uprated over the years in accordance with advances in electrical technology and demands from vehicle manufacturers. Essentially there are three distinct versions of the system, with the main differences being in the type and location of certain ECM input sensors, the ECM software, and the ECM multi-plug which may be a 37-pin, 42-pin or 55-pin type. For clarity, 42-pin ECM multi-plug systems are covered in this Part of Chapter 18, with 37-pin, and 55-pin systems being covered in Parts A and Part C respectively.

Depending on application, Teves Mk.IV may be installed as an antilock braking system only, or as an antilock braking system incorporating traction control.

In ABS mode, the purpose of the system is to apply the vehicle brakes at maximum efficiency without wheel lock or loss of directional stability. Inductive sensors (wheel speed sensors) monitor the speed of the wheels by generating an electrical signal as the wheel is rotated. This information is passed to the ABS Electronic Control Module (ECM) which is then able to determine wheel speed, wheel acceleration and wheel deceleration. The ECM compares the signals received from each wheel and if the onset of lock at any wheel is detected, a signal is sent to the ABS hydraulic control unit which regulates the brake pressure for the relevant wheel(s).

Where the system incorporates traction control, essentially the reverse principle is applied. When the ECM detects that one or more wheels are rotating faster than the reference value, the brake is actually applied on the relevant wheel(s) to reduce the rotational speed.

Typically, Teves Mk.IV ABS is comprised of the following components (see illustration 18B.1). Note that the illustration depicts the 37-pin and 55-pin systems, however, apart from the pedal position sensor, the system components are virtually the same.

a) ABS-ECM.
b) Hydraulic control unit.
c) Four inductive wheel speed sensors and associated sensor rings.
d) ABS electrical wiring harness and relays.
e) Brake light switch.
f) ABS warning lamp.

In addition, the conventional braking system is comprised of the following components:

a) Tandem brake master cylinder.
b) Vacuum servo unit.
c) Brake calipers and hydraulic hoses and pipes.

2 Component description and operation

ABS ECM

General

The Teves Electronic Control Module (ECM) continually monitors wheel speed from the signals provided by the wheel speed sensors. If the ECM detects the incidence of wheel lock on one or more wheels, a signal is sent to the hydraulic control unit to modulate the hydraulic pressure to the brake of the locking wheel.

The ECM contains two microprocessors and uses digital technology to complete this function and other functions such as fault code memory and power modules for valve and relay activity (see illustration 18B.2).

Self-test

The Teves Mk.IV ECM is equipped with a self-test capability that initially examines the ABS system when the ignition is switched on, and then examines the wheel speed sensor signals after a wheel speed of approximately 4 mph is reached from all wheels (engine running). The ABS self-test program continues to examine the signals from the various components as long as the ignition is switched on. If

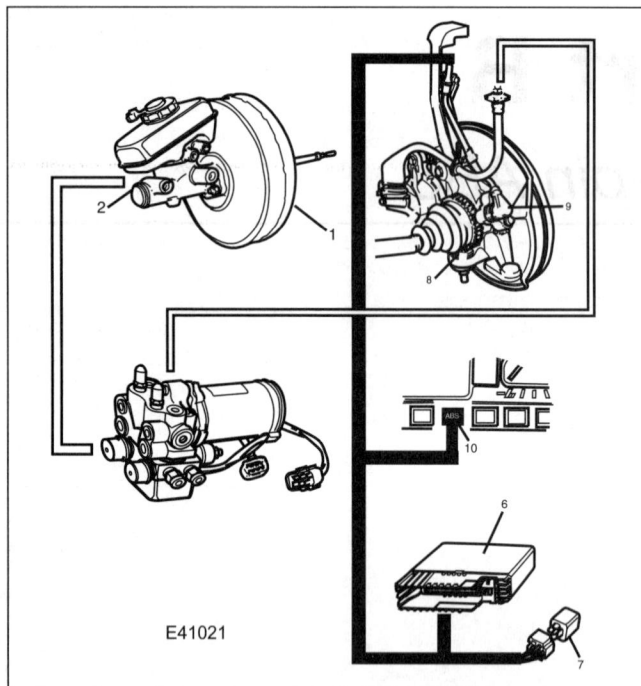

E41021

18B.1 Teves Mk.IV main components

1	ABS-ECM	6	Brake light switch
2	Hydraulic control unit	7	Pedal position sensor
3	Inductive wheel speed		(depending on application)
	sensors	8	ABS warning lamp
4	Sensor rings	9	Tandem master cylinder
5	ABS electrical wiring harness	10	Vacuum servo unit
	and relays		

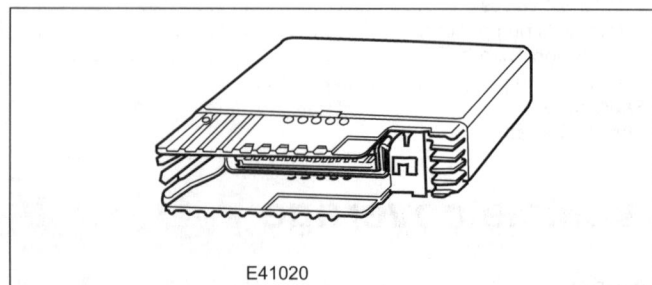

E41020

18B.2 Teves Mk.IV ECM

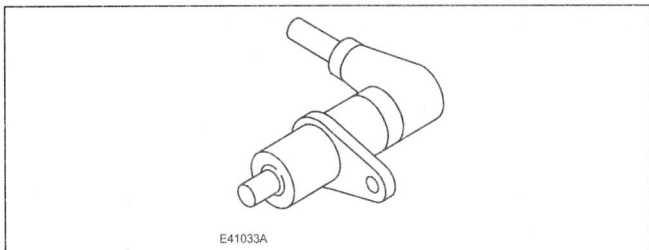

18B.3 Typical wheel speed sensor

self-test determines that faults are not present, the ABS is ready for operation once a specified vehicle speed has been achieved.

If the ECM detects that a fault is present, all ABS functions are switched off and the warning lamp is turned on. The conventional braking system continues to operate as normal without ABS assistance.

Self-diagnostics

If the ECM detects a fault during the self-test routine, an internal fault code is stored in the ECM memory. Stored fault codes can be retrieved from the SD connector with the aid of a suitable fault code reader. If the fault clears, the code will remain stored until cleared with the FCR.

Hydraulic control unit

The hydraulic control unit consists of an electric motor with flange-mounted two-circuit radial piston return pump, and the flange-mounted valve block containing the inlet and outlet solenoid valves. The hydraulic control unit controls the pressure applied to the brake caliper for each individual wheel or pair of wheels during the three phases of ABS operation. During a controlled ABS cycle, the pump returns brake fluid, drained off during the pressure reduction phase, back into the brake circuit.

Teves Mk.IV in 42-pin configuration is a three-channel system incorporating six solenoid valves in the hydraulic control unit (two solenoid valves for each front brake, and two solenoid valves controlling the rear brakes as a pair).

Where a pair of brakes are controlled by a single hydraulic channel, the 'select-low' principle is employed. With the 'select-low' principle, the wheel with the lowest adhesion determines the amount of hydraulic pressure to both brakes on that channel during ABS operation.

Wheel speed sensors

The rotational speed of the roadwheels and any changes in the rotational speed are recorded by inductive wheel speed sensors, one located at each roadwheel (see illustration 18B.3).

Each wheel speed sensor assembly comprises a toothed sensor ring which rotates at roadwheel speed, and an adjacent sensor mounted a set distance from the sensor ring (see illustration 18B.4).

The sensors are permanent magnet pulse generator types producing an AC voltage sine wave as the sensor ring teeth pass through the magnetic field of the sensor (see illustration 18B.5).

The frequency of the waveform produced by the wheel speed sensor is proportional to the road speed. This AC voltage signal is continually being delivered to the ABS-ECM for processing.

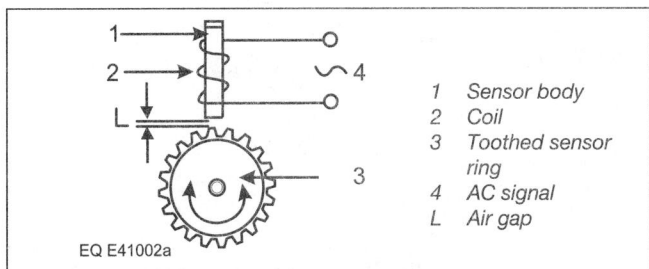

18B.5 Wheel speed sensor operation

1	Sensor body
2	Coil
3	Toothed sensor ring
4	AC signal
L	Air gap

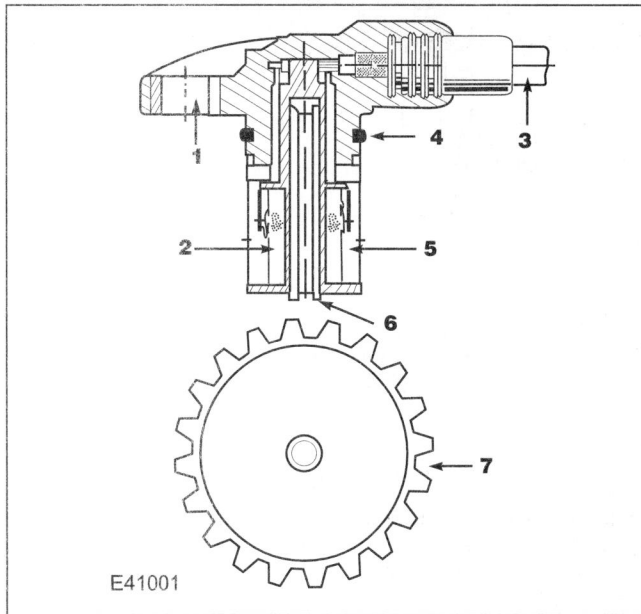

E41001

18B.4 Sectional view of a wheel speed sensor

1	Mounting bolt location	5	Coil
2	Permanent magnet	6	Sensor tip
3	Wiring harness	7	Toothed sensor ring
4	O-ring		

The peak to peak voltage of the speed signal (when viewed upon an oscilloscope) can vary considerably according to wheel speed. An analogue to digital converter (ADC) in the ECM transforms the AC pulse into a digital signal (see illustration 18B.6).

ABS electrical wiring harness and relays

An integrated main wiring harness is used to connect power and earth to various electrical components. This enables sensor signals to reach the ECM and the ECM, in turn to send command signals to the relays, return pump and hydraulic control unit. The ABS overvoltage relay, pump relay and associated fuses are located in the engine compartment fuse/relay box.

Overvoltage protection relay

Nominal battery voltage is supplied from the ignition switch to the relay via a fuse. The relay is controlled by the ECM applying an earth, which activates the relay allowing fused battery voltage to be applied to the solenoid valves. The earth for the solenoid valves is controlled by the ECM.

18B

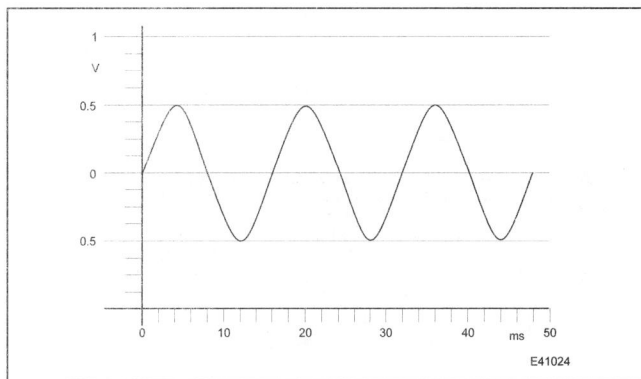

18B.6 Wheel speed sensor waveform as viewed on an oscilloscope

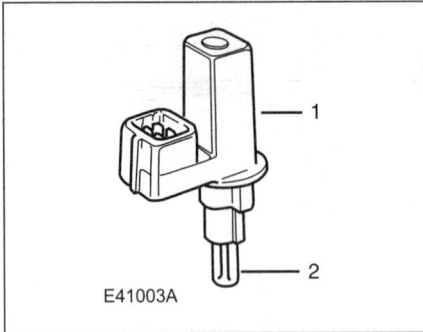

18B.7 Typical brake light switch body (1) and contact pin (2)

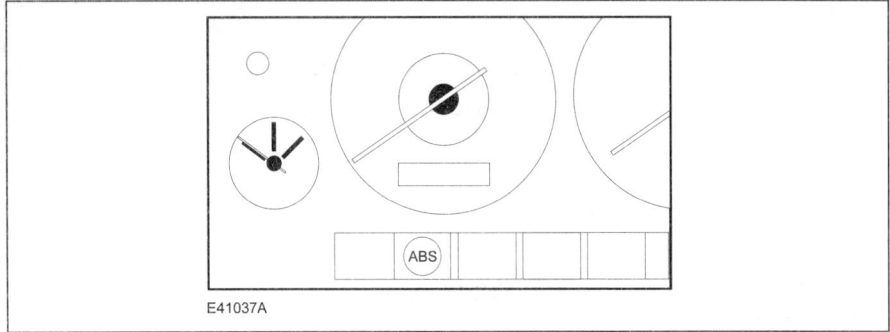

18B.8 ABS warning lamp

Pump relay

Nominal battery voltage is supplied via from the overvoltage protection relay to the pump relay. The relay is controlled by the ECM applying an earth, which activates the relay allowing fused battery voltage to be applied to the pump motor. The hydraulic control unit provides the earth for the pump motor.

Brake light switch

The brake light switch comprises a switch body and contact pin and is located above the brake pedal (see illustration 18B.7).

When the brake pedal is depressed, closing the brake light switch, a signal is sent to the ECM indicating that the brakes are being applied. Once this signal is received, the ECM will begin monitoring the wheel speed via the wheel speed sensors and activate the ABS if necessary.

ABS warning lamp

After the ignition is switched on, the ABS warning lamp on the instrument panel is illuminated for approximately 2 to 4 seconds as the system executes a self-test routine. If satisfactory operation of the system is detected by the ECM, the light is extinguished. During vehicle operation above a pre-determined wheel speed, the ABS-ECM implements a further self-test cycle whereby ABS operation and wheel speed sensor signals are continually monitored. If a fault is detected, the relevant ECM pin is earthed to illuminate the warning lamp on the instrument panel, and the ABS function is disabled. The warning lamp will remain illuminated until the fault is no longer present (see illustration 18B.8).

When the ABS-ECM detects a fault, the fault code is stored and the ABS warning lamp activated. If the fault no longer exists after the next system start (ignition on/off) the ABS warning lamp is extinguished after the self-test cycle, however the fault code remains stored in the ECM memory.

Tandem master cylinder

Typically, the tandem master cylinder comprises a body casting incorporating primary and secondary pressure chambers, primary piston, intermediate piston, floating piston, slotted pin and central valve. The cylinder operates as a conventional master cylinder using vacuum assistance from the vacuum servo unit (see illustration 18B.9).

When the brake system is at rest, the central valve in the floating piston rests against the slotted pin. In this condition the central valve is open and brake fluid can discharge out of the pressure chamber back into the brake fluid reservoir. When the brake pedal is depressed, the build-up of hydraulic pressure in the primary pressure chamber acts on the intermediate piston and floating piston, moving them down the cylinder bore. The floating piston contacts the seal on the central valve, closing the connection between the intermediate and secondary pressure chambers. Brake hydraulic pressure can now also increase in the secondary pressure chamber.

Vacuum servo unit

The vacuum servo unit is located between the brake pedal and tandem master cylinder. When the brake pedal is depressed, the servo unit increases the force applied by the pedal, reducing the effort required to operate the brakes (see illustration 18B.10).

The unit is operated by vacuum created in the engine inlet manifold (or from a separate vacuum pump on diesel engines) which is applied to a diaphragm contained within the unit casing. A pushrod connected to the centre of the diaphragm acts directly on the primary piston in the master cylinder.

When the brake pedal is released, vacuum is applied to both sides of the diaphragm. When the pedal is depressed, one side of the diaphragm is opened to atmosphere and the vacuum acting on the other side deflects the diaphragm which in turn operates the master cylinder primary piston. The resulting force applied to the master cylinder piston is therefore significantly greater than the initial force applied to the brake pedal by the driver.

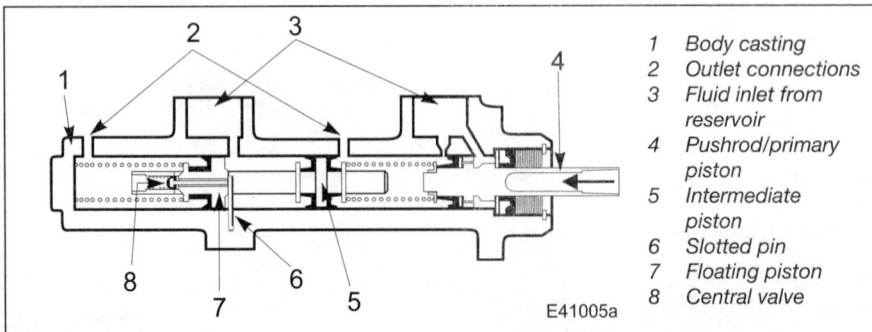

1 Body casting
2 Outlet connections
3 Fluid inlet from reservoir
4 Pushrod/primary piston
5 Intermediate piston
6 Slotted pin
7 Floating piston
8 Central valve

18B.9 Sectional view of a typical tandem master cylinder

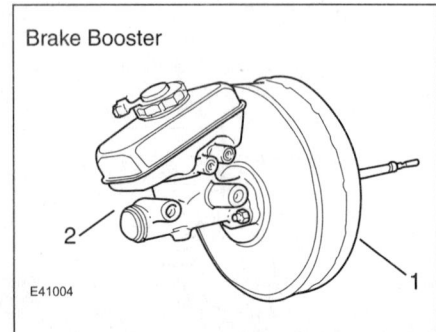

Brake Booster

18B.10 Vacuum servo unit (1) and tandem master cylinder (2)

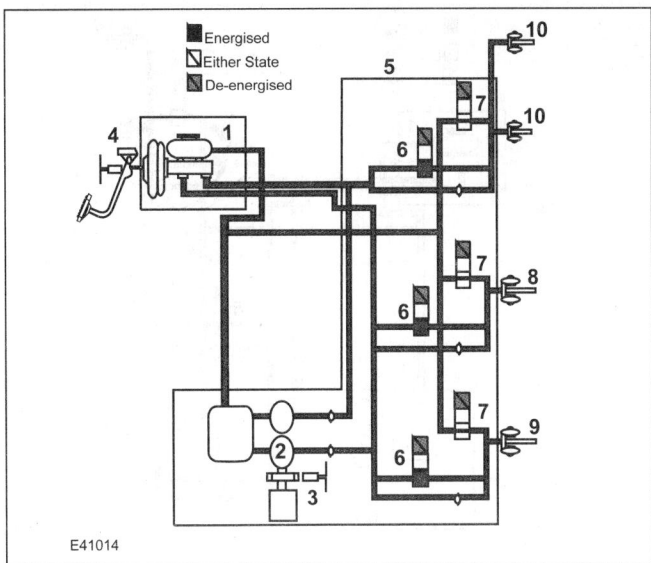

18B.11 Brake system at rest

1	Vacuum servo unit and tandem master cylinder	5	Hydraulic control unit
2	ABS return pump	6	Inlet solenoid valve
3	Pump motor sensor	7	Outlet solenoid valve
4	Brake pedal position sensor	8	Front left brake caliper
		9	Front right brake caliper
		10	Rear brake calipers

3 System operation

Brake system at rest

The illustrations in this Section depict the operation of a typical three-channel hydraulic system with two solenoid valves for each front brake, and two solenoid valves controlling the rear brakes as a pair. Four-channel systems operate in the same way apart from the number of solenoid valves used.

When the ABS system is at rest all the brake components are inoperative. Pressure is non-existent in the hydraulic pipes between the tandem master cylinder and the brake calipers. The inlet solenoid valves in the valve block are open and the outlet solenoid valves are shut (see illustration18B.11).

Brake system operating under conventional control without ABS

When the brake pedal is activated, the pedal force is applied to the tandem master cylinder by the vacuum servo unit pushrod. The servo unit pushrod acts directly on the pressure piston in the master cylinder. The resulting increase in hydraulic pressure moves the intermediate piston in the cylinder, and the pressure which increases in front of the intermediate piston flows through the open inlet solenoid valves in the valve block, to the brake calipers. The outlet solenoid valves in the valve block remain shut (see illustration 18B.12).

Brake system operated with ABS control

The ABS-ECM continually monitors wheel speed from the signals provided by the wheel speed sensors. If the ECM detects the incidence of wheel lock on one or more wheels, ABS is automatically initiated in three phases. Typically, ABS operates individually on each front wheel and either individually or jointly on the rear wheels. In which case, all or any of the wheels could be in any one of the following phases at any particular moment.

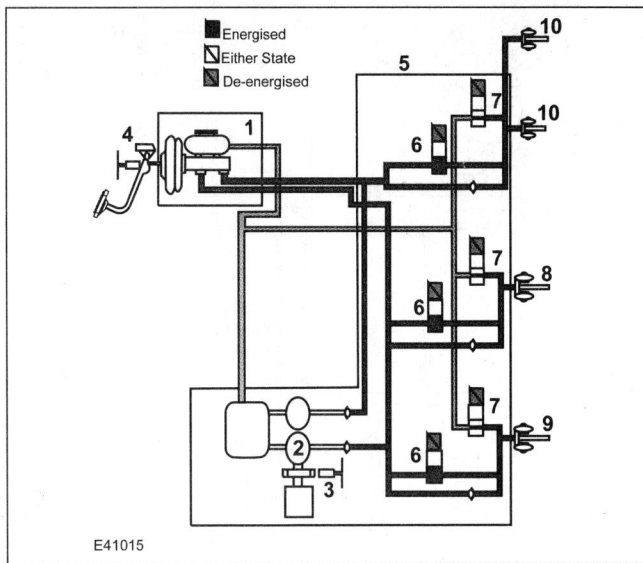

18B.12 Conventional braking without ABS

1	Vacuum servo unit and tandem master cylinder	5	Hydraulic control unit
2	ABS return pump	6	Inlet solenoid valve
3	Pump motor sensor	7	Outlet solenoid valve
4	Brake pedal position sensor	8	Front left brake caliper
		9	Front right brake caliper
		10	Rear brake calipers

First ABS phase, pressure holding

To prevent any further build-up of hydraulic pressure in the circuit being controlled, the inlet solenoid valve is shut and the outlet solenoid valve remains shut. The pressure cannot now be increased in the controlled circuit by any further application of the brake pedal (see illustration 18B.13).

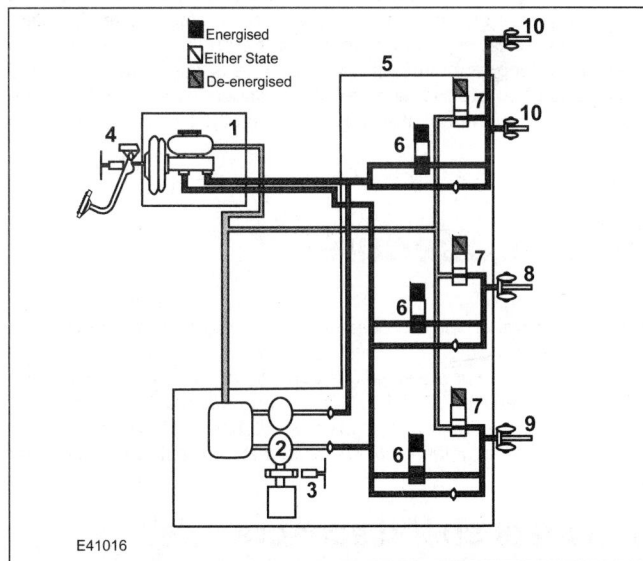

18B.13 ABS operation - first phase, pressure holding

1	Vacuum servo unit and tandem master cylinder	5	Hydraulic control unit
2	ABS return pump	6	Inlet solenoid valve
3	Pump motor sensor	7	Outlet solenoid valve
4	Brake pedal position sensor	8	Front left brake caliper
		9	Front right brake caliper
		10	Rear brake calipers

18B

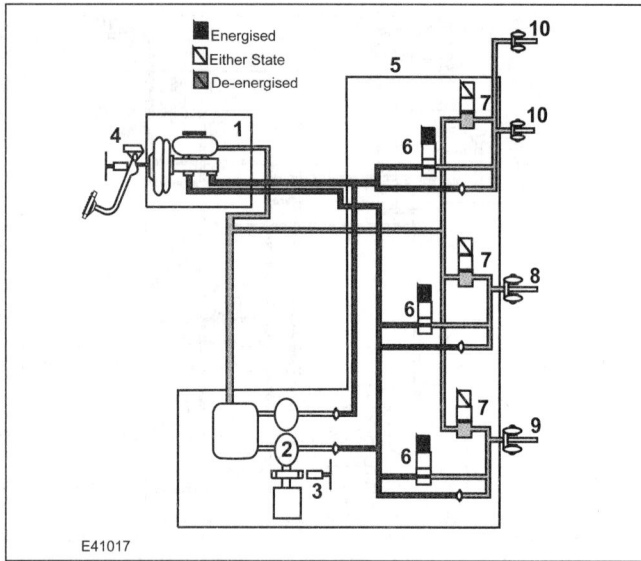

18B.14 ABS operation - second phase, pressure reduction

1 Vacuum servo unit and tandem master cylinder	5 Hydraulic control unit
	6 Inlet solenoid valve
2 ABS return pump	7 Outlet solenoid valve
3 Pump motor sensor	8 Front left brake caliper
4 Brake pedal position sensor	9 Front right brake caliper
	10 Rear brake calipers

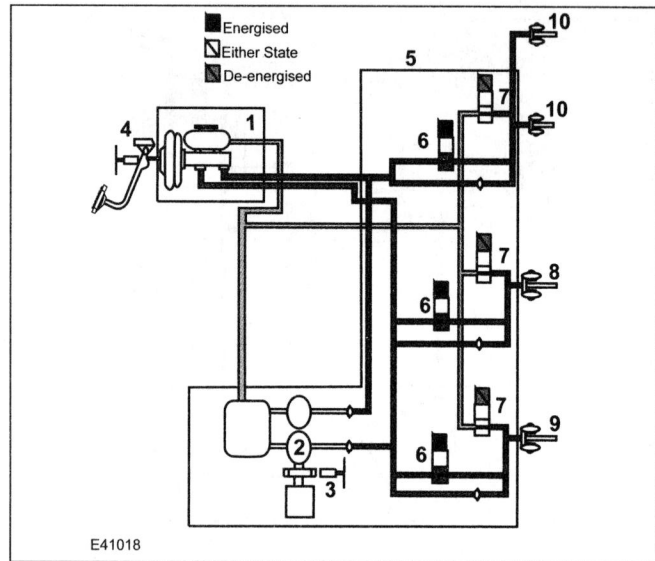

18B.15 ABS operation - third phase, pressure build-up

1 Vacuum servo unit and tandem master cylinder	5 Hydraulic control unit
	6 Inlet solenoid valve
2 ABS return pump	7 Outlet solenoid valve
3 Pump motor sensor	8 Front left brake caliper
4 Brake pedal position sensor	9 Front right brake caliper
	10 Rear brake calipers

If the wheel speed sensor signals indicate that wheel rotation has now stabilised, the ECM will instigate the pressure build-up phase, allowing braking to continue. If wheel lock is still detected after the pressure holding phase, the ECM instigates the pressure reduction phase.

Second ABS phase, pressure reduction

A controlled braking operation can be instigated with a pressure reduction phase or it can follow a pressure holding phase. The inlet solenoid valve is, or remains shut and the outlet solenoid valve is opened. As long as the outlet solenoid valve is open, the brake fluid discharges back into the reservoir of the tandem master cylinder. The return pump then pumps the brake fluid out of the reservoir back into

the control circuit, forcing back the master cylinder piston (and the brake pedal) **(see illustration 18B.14**.

Third ABS phase, pressure-build up

The pressure build up phase is instigated after the wheel rotation has stabilised. The outlet solenoid valve is shut and the inlet solenoid valve is opened. The brake pressure from the tandem master cylinder flows through the inlet solenoid valve to the relevant brake caliper allowing braking to continue **(see illustration 18B.15)**.

The whole ABS control cycle takes place 4 to 10 times per second for each affected wheel and this ensures maximum braking effect and control during ABS operation.

Test procedures

Important note: *The test procedures, pin-tables and wiring diagrams contained in this Chapter are necessarily representative of the system depicted. Because of the variations in wiring and other data that often occurs, even between similar vehicles in any particular VM's range, the reader should take great care in identification of ECM pins, and satisfy himself that he has gathered the correct data before failing a particular component.*

4 Wheel speed sensors

Checking the wheel speed sensor (general)

1 Inspect the wheel speed sensor for corrosion or damage and check that the sensor is tightly mounted.
2 Check the toothed sensor ring for damage, eccentricity and for broken or missing teeth.
3 Inspect the wheel speed sensor wiring plug for corrosion and damage. One plug for each sensor.

4 Check that the connector terminal pins are pushed fully home and making good contact with the sensor wiring plug.
5 Check the clearance between the sensor and the toothed sensor ring. The clearance is not adjustable but is nominally 0.2 to 1.2 mm. If the clearance is excessive, expect a worn sensor tip or problems with the wheel bearings/hub or sensor ring.
6 When carrying out voltage checks with an oscilloscope or voltmeter, the voltage obtained will be proportional to the speed at which the wheel is rotating. In addition to determining that the wheel speed sensors are actually producing a voltage output, it is essential that the output from the sensors on a particular axle is the same for any given wheel speed.

Checking wheel speed sensor output with an oscilloscope

Note: *Refer to the wiring diagrams for specific ECM pin identification.*
7 Switch the ignition off and disconnect the ECM multi-plug or the relevant wheel speed sensor wiring plug.

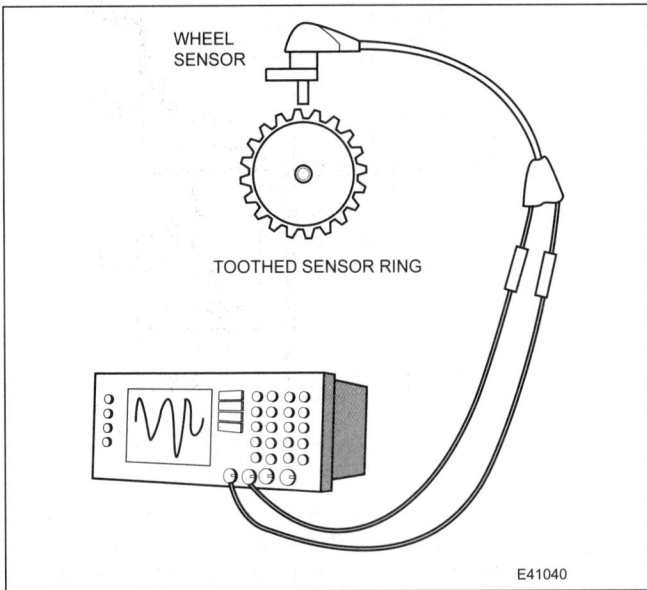

18B.16 Checking wheel speed sensor output with an oscilloscope connected to the sensor wiring plug

8 Connect an oscilloscope between the terminal pins for the sensor under test **(see illustration 18B.16)**.

9 Select a range to cover 80 Hz on the oscilloscope and a free run time base.

10 Raise the wheel and rotate it by hand at approximately one revolution per second.

11 A sinusoidal wave form should be obtained, with amplitude and duration changing with rotational speed **(see illustration 18B.17)**.

12 If there is no signal, or a very weak or intermittent signal at the ECM, repeat the test at the sensor wiring plug. If there is no change in signal status, the sensor is suspect.

13 If the signal is now satisfactory this indicates a fault in the wiring harness which should be checked for continuity.

Checking wheel speed sensor output with an AC voltmeter

Note: *Refer to the wiring diagrams for specific ECM pin identification.*

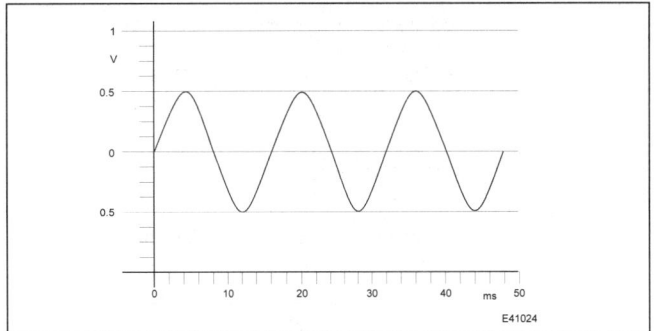

18B.18 Checking wheel speed sensor output with a voltmeter connected to the sensor wiring plug

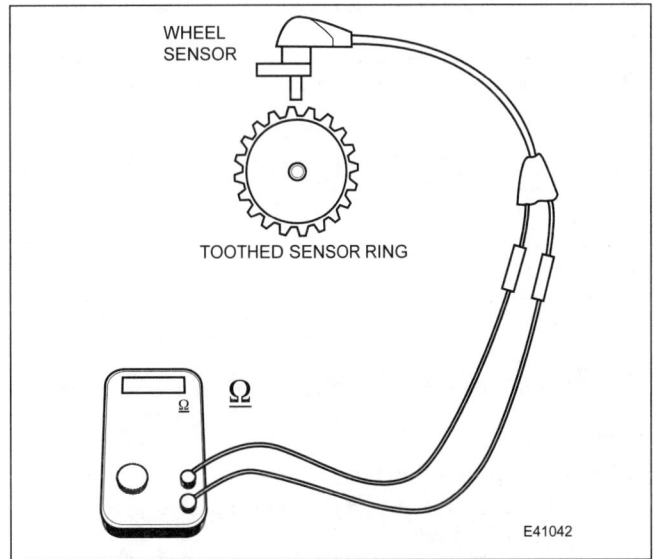

18B.17 Typical wheel speed sensor sine wave as displayed on an oscilloscope

14 Switch the ignition off and disconnect the ECM multi-plug or the relevant wheel speed sensor wiring plug.

15 Connect an AC voltmeter between the terminal pins for the sensor under test **(see illustration 18B.18)**.

16 Raise the wheel and rotate it by hand at approximately one revolution per second.

17 A voltage of approximately 0.3 to 1.0 volts (AC RMS) should be obtained. If there is no signal, or a very weak or intermittent signal at the ECM, repeat the test at the sensor wiring plug. If there is no change in the signal, the sensor is suspect.

18 If the signal is now satisfactory, this indicates a fault in the wiring harness which should be checked for continuity. **Note:** *This test at least proves that a signal is being generated by the sensor. However, the voltage produced is an average voltage and does not clearly indicate damage to the sensor ring or that the sinewave is regular in formation.*

Checking wheel speed sensor resistance

Note: *Refer to the wiring diagrams for specific ECM pin identification.*
19 Switch the ignition off and disconnect the ECM multi-plug or the relevant wheel speed sensor wiring plug.

20 Connect an ohmmeter between the terminal pins for the sensor under test **(see illustration 18B.19)**.

21 The readings obtained should be approximately 1.1 kohms.

22 If the resistance is excessively high, or open circuit at the ECM, repeat the test at the sensor multi-plug. If there is no change in resistance, the sensor is suspect.

23 If the resistance is now satisfactory, this indicates a fault in the

18B

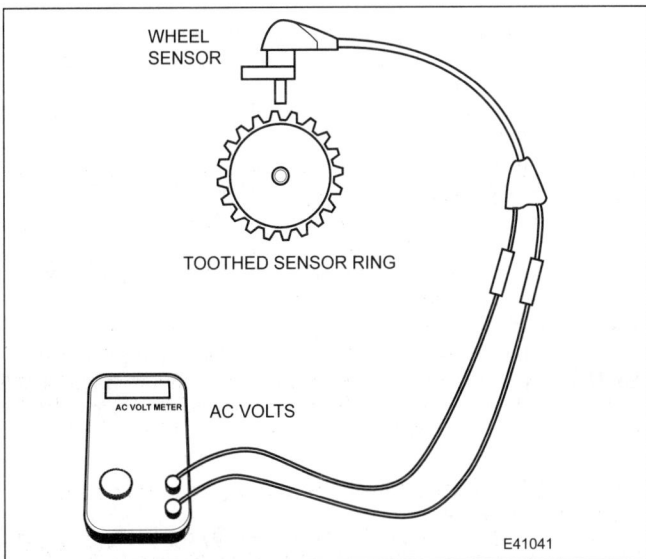

18B.19 Checking wheel speed sensor resistance with an ohmmeter connected to the sensor wiring plug

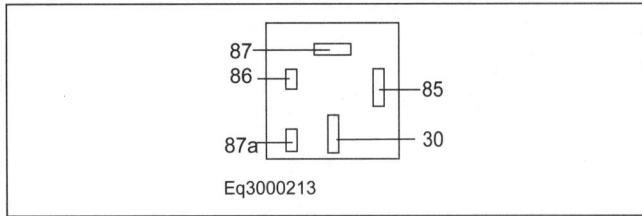

18B.20 Pump relay terminal identification

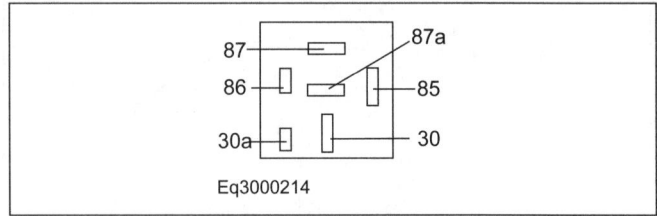

18B.21 Overvoltage protection relay terminal identification

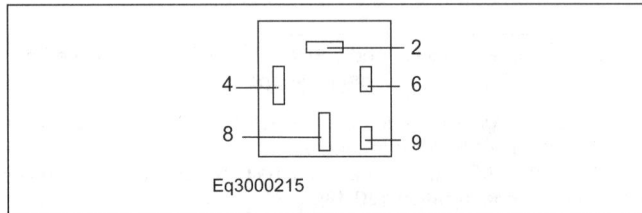

18B.22 Pump relay base terminal identification

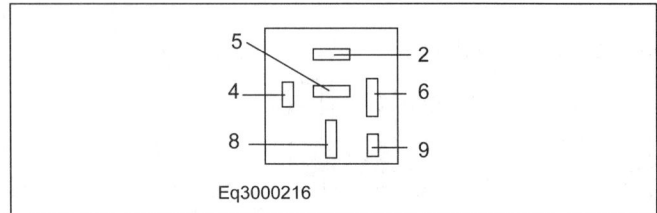

18B.23 Overvoltage protection relay base terminal identification

wiring harness which should be checked for continuity. **Note:** *Even if the resistance is within the quoted specifications, this does not prove that the speed sensor can generate an acceptable signal.*

5 System relays

Checking relays (general)

1 The following tests are applicable to the ABS hydraulic pump relay and overvoltage protection relay. The relays are located in the engine compartment fuse/relay box.

2 Before carrying out specific tests on the relays or power supplies, inspect the relays and relay base for corrosion and damage.

3 Check that the relays are pushed fully home and making good contact with the base.

4 Faults in any of the above areas are possible reasons for failure or malfunctioning of the relay.

Relay operation tests

Pump relay

5 Switch the ignition off and remove the pump relay from the relay base.

6 Connect an ohmmeter between the relay main voltage supply terminal 87 and the output terminal 30 **(see illustration 18B.20)**. The ohmmeter should indicate an open circuit.

7 Attach the positive lead of a 12 volt supply to the relay driver supply terminal 85, and the negative lead to the relay driver earth terminal 86.

8 There should be an unmistakable click from the relay and the ohmmeter should indicate continuity.

9 With the 12 volt supply still connected to relay terminals 85 and 86, connect an ohmmeter or a diode tester between the relay terminals 30 and 87a.

10 Measure the circuit and then reverse the connections. The instrument should indicate continuity in one direction (diode conducting), but not the other (diode blocking).

11 Renew the relay if it fails these tests. If the relay cover is transparent, inspect the relay for burns. A faulty relay usually has a burnt appearance.

Overvoltage protection relay

12 Switch the ignition off and remove the overvoltage protection relay from the relay base.

13 Connect an ohmmeter between the relay main voltage supply terminal 87 and the output terminal 30 **(see illustration 18B.21)**. The ohmmeter should indicate an open circuit.

14 Attach the positive lead of a 12 volt supply to the relay driver supply terminal 85, and the negative lead to the relay driver earth terminal 86.

15 There should be an unmistakable click from the relay and the ohmmeter should indicate continuity.

16 Disconnect the 12 volt supply and connect an ohmmeter or a diode tester between the relay terminals 30 and 30a.

17 Measure the circuit and then reverse the connections. The instrument should indicate continuity in one direction (diode conducting), but not the other (diode blocking).

18 Renew the relay if it fails these tests. If the relay cover is transparent, inspect the relay for burns. A faulty relay usually has a burnt appearance.

Relay power supply tests

Pump relay

Note: *For the following test the ECM and overvoltage protection relay must be connected and operating correctly.*

19 Switch the ignition off and remove the pump relay leaving the base exposed.

20 Using a voltmeter, probe between the main voltage supply terminal 8 in the relay base and earth. The reading should indicate nbv **(see illustration 18B.22)**.

21 If no voltage is found, check the relevant fuse and the supply wiring back to the voltage supply.

22 Switch the ignition on.

23 Using a voltmeter, probe between the relay driver supply terminal 4 in the relay base and earth. The reading should indicate nbv.

24 If no voltage is found, check the relevant fuse, the overvoltage protection relay and the supply wiring back to the ignition switch.

Overvoltage protection relay

25 Switch the ignition off and remove the relay from the relay base.

26 Using a voltmeter, probe between the main voltage supply terminal 2 in the relay base and earth. The reading should indicate nbv **(see illustration 18B.23)**.

27 If no voltage is found, check the relevant fuse and the supply wiring back to the voltage supply.

28 Switch the ignition on.

29 Using a voltmeter, probe between the relay driver supply terminal 4 in the relay base and earth. The reading should indicate nbv.

30 If no voltage is found, check the relevant fuse and the supply wiring back to the ignition switch.

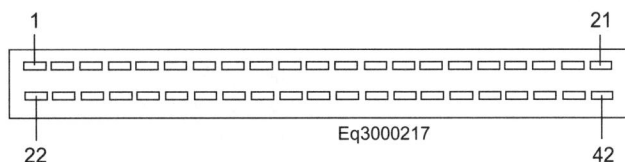

18B.24 ECM 42-pin multi-plug

6 Electronic Control Module

Checking the ECM (general)

1 Inspect the ECM for corrosion or damage and ensure that the unit is securely attached to the hydraulic control unit.
2 Check that the ECM multi-plug terminals are pushed fully home and making good contact with the ECM pins. Faults in any of the above areas are possible reasons for poor performance in the ABS system.

ECM power supply and earth tests

3 Switch the ignition off and disconnect the ECM multi-plug.
4 Switch the ignition on.
5 Attach a negative voltmeter probe to a vehicle earth.
6 Attach a positive voltmeter probe to ECM multi-plug pins 8 and 19 in turn **(see illustration 18B.24)**. The voltmeter should indicate nbv in each case.
7 If no voltage is found, check the relevant fuse and the supply wiring back to the ignition switch.
8 Switch the ignition off.
9 Connect an ohmmeter between a vehicle earth and ECM multi-plug earth pins 17and 26 and 41 in turn.
10 The ohmmeter should indicate continuity in each case. If not, check the ECM main earth connection and wiring.

7 Solenoid valves

Solenoid valve resistance tests

1 Switch the ignition off and disconnect the ECM multi-plug from the ECM.
2 Using an ohmmeter, measure the resistance between pin 27 in the ECM multi-plug, and each of the following multi-plug pins in turn **(see illustration 18B.24)**:
   Inlet solenoid valves - 2, 22, 24.
   Outlet solenoid valves - 25, 39, 40.
3 Resistance should be in the order of 6.0 to 10.0 ohms for each inlet solenoid valve, and 3.0 to 6.0 ohms for each outlet solenoid valve.
4 If the resistance is not as specified, check the wiring and wiring connectors. If satisfactory, the hydraulic control unit is suspect.

8 Hydraulic pump motor

Pump operation test

1 Switch the ignition off and disconnect the ECM multi-plug from the ECM.
2 Using a fused jumper lead, connect ECM pins 1 and 23 to earth **(see illustration 18B.24)**.
3 Switch the ignition on.
4 The overvoltage protection relay, pump relay and the pump should activate (listen for an unmistakable click from the relays and a hum from the pump).

⚠️ **Warning: The test should be made as quickly as possible to avoid damaging the pump.**

5 If the relay(s) fails to activate, carry out the relay test procedures described in Section 5.
6 If the pump does not operate as described, renew the pump.

9 Brake light switch

Checking the brake light switch (general)

1 Check that the brake light switch is correctly and securely mounted and that the plunger moves smoothly with no trace of binding.
2 Check that the wiring multi-plug is pushed fully home and making good contact.
3 Check that no wires have been disconnected.
4 Faults in any of the above areas are possible reasons for failure or malfunctioning of the switch.

Brake light switch voltage and continuity tests

Voltage test

5 Switch the ignition off and disconnect the ECM multi-plug from the ECM.
6 Connect a voltmeter between a vehicle earth and the ECM multi-plug brake light switch input pin 29 **(see illustration 18B.24)**.
7 Switch the ignition on and depress the brake pedal. The voltmeter should indicate nbv.
8 If no voltage is found, the fuse, the brake light switch and the wiring are suspect.
9 Release the brake pedal. The voltage should drop to zero as the switch opens.

Continuity test

10 Switch the ignition off and disconnect the brake light switch multi-plug.
11 Connect an ohmmeter between the terminal pins of the brake light switch.
12 Operate the brake light switch and check for continuity. If the test fails, renew the switch.

10 Warning lamp

Checking the warning lamp (general)

1 Inspect the warning lamp bulb holder contacts in the instrument panel.
2 Check that the instrument panel multi-plug terminal pins are pushed fully home and making good contact.
3 Check that no wires have been disconnected.
4 Faults in any of the above areas are possible reasons for failure or malfunctioning of the warning lamp.

Warning lamp operation test

5 With the ignition switched off, the warning lamp should remain off.
6 Switch the ignition on and the warning lamp should illuminate then extinguish after a few seconds. The lamp should then remain off.
7 If the warning lamp comes on and remains on at any time during vehicle operation, carry out the previously described test procedures on the system components.

18B

Pin table - typical 42-pin

Note: *Refer to illustration 18B.24*

Pin No.	Connection	Test condition	Voltage
1	Overvoltage protection relay driver	Ignition on, actuated	1.0 volt (max)
2	Front left inlet solenoid valve	Ignition on/inactive	Nbv
		Actuated	0 volts
3	-	-	-
4	-	-	-
5	-	-	-
6	-	-	-
7	-	-	-
8	Supply from ignition switch	Ignition on	Nbv
9	Hydraulic pump relay supply	Ignition on:	
		Pump inactive	0 volts
		Pump running	Nbv
10	-	-	-
11	-	-	-
12	-	-	-
13	Rear left wheel speed sensor earth	Roadwheel rotating	0.25 volts (max)
14	Rear right wheel speed sensor earth	Roadwheel rotating	0.25 volts (max)
15	Front right wheel speed sensor signal	Roadwheel rotating	0.3 to 1.0 volts AC (approx)
16	Front left wheel speed sensor earth	Roadwheel rotating	0.25 volts (max)
17	ECM earth	Ignition on	0.25 volts (max)
18	-	-	-
19	Supply from ignition switch	Ignition on	Nbv
20	-	-	-
21	-	-	-
22	Front right inlet solenoid valve	Ignition on/inactive	Nbv
		Actuated	0 volts
23	Hydraulic pump relay driver	Ignition on/inactive	Nbv
		Actuated	0 volts
24	Rear inlet solenoid valve	Ignition on/inactive	Nbv
		Actuated	0 volts
25	Front left outlet solenoid valve	Ignition on/inactive	Nbv
		Actuated	0 volts
26	ECM earth	Ignition on	0.25 volts (max)
27	Voltage monitoring: hydraulics	Ignition on:	
		Pump inactive	0 volts
		Pump running	Nbv
28	SD connector	-	-
29	Brake light switch input	Ignition on	
		brake pedal released	0 volts
		brake pedal depressed	Nbv
30	-	-	-
31	-	-	-
32	Input from engine management ECM	-	-
33	-	-	-
34	Rear left wheel speed sensor signal	Roadwheel rotating	0.3 to 1.0 volts AC (approx)
35	Rear right wheel speed sensor signal	Roadwheel rotating	0.3 to 1.0 volts AC (approx)
36	Front right wheel speed sensor earth	Roadwheel rotating	0.25 volts (max)
37	Front left wheel speed sensor signal	Roadwheel rotating	0.3 to 1.0 volts AC (approx)
38	-	-	-
39	Rear outlet solenoid valve	Ignition on/inactive	Nbv
		Actuated	0 volts
40	Front right outlet solenoid valve	Ignition on/inactive	Nbv
		Actuated	0 volts
41	ECM earth	Ignition on	0.25 volts
42	ABS warning lamp	Ignition on:	
		Lamp on	2.0 volts (max)
		Lamp off	Nbv

Fault codes

1 The Teves Mk.IV, 42-pin system requires the use of a FCR for obtaining fault codes. Flash codes are not available for output from this system.

2 If a FCR is available, it should be connected to the SD serial connector and used in accordance with the maker's instructions.

3 The FCR can be used for the following purposes:
  a) *Obtaining fault codes.*
  b) *Clearing fault codes.*
  c) *Obtaining datastream information.*
  d) *Testing the system actuators (relays and solenoid valves)*

4 Internal fault codes are used by the ECM to designate faults in the system components and circuits. A proprietary fault code reader (FCR) or system tester is required to interrogate the system. No actual fault code numbers are available, although the component circuits checked by the ECM are similar to those for the earlier generation of BMW 3-series vehicles that are shown in the Fault Table below.

Fault table

Item	Fault
Front left inlet solenoid valve	Open or short circuit, relay inoperative
Front left outlet solenoid valve	Open or short circuit, relay inoperative
Front right inlet solenoid valve	Open or short circuit, relay inoperative
Front right outlet solenoid valve	Open or short circuit, relay inoperative
Rear inlet solenoid valve	Open or short circuit, relay inoperative
Rear outlet solenoid valve	Open or short circuit, relay inoperative
ECM	Processor fault
ECM	Internal fault
Front left wheel speed sensor	Open or short circuit
Front right wheel speed sensor	Open or short circuit
Rear left wheel speed sensor	Open or short circuit
Rear right wheel speed sensor	Open or short circuit
Front left wheel speed sensor	Implausible signal
Front right wheel speed sensor	Implausible signal
Rear left wheel speed sensor	Implausible signal
Rear right wheel speed sensor	Implausible signal
Front left wheel speed sensor	No signal, intermittent signal
Front right wheel speed sensor	No signal, intermittent signal
Rear left wheel speed sensor	No signal, intermittent signal
Rear right wheel speed sensor	No signal, intermittent signal
Brake hydraulic system	Hydraulic fault
Hydraulic pump	Inoperative
Front left outlet solenoid valve	Inoperative or leaking
Front right outlet solenoid valve	Inoperative or leaking
Rear outlet solenoid valve	Inoperative or leaking
ECM	Memory access fault

18B

Wiring diagram

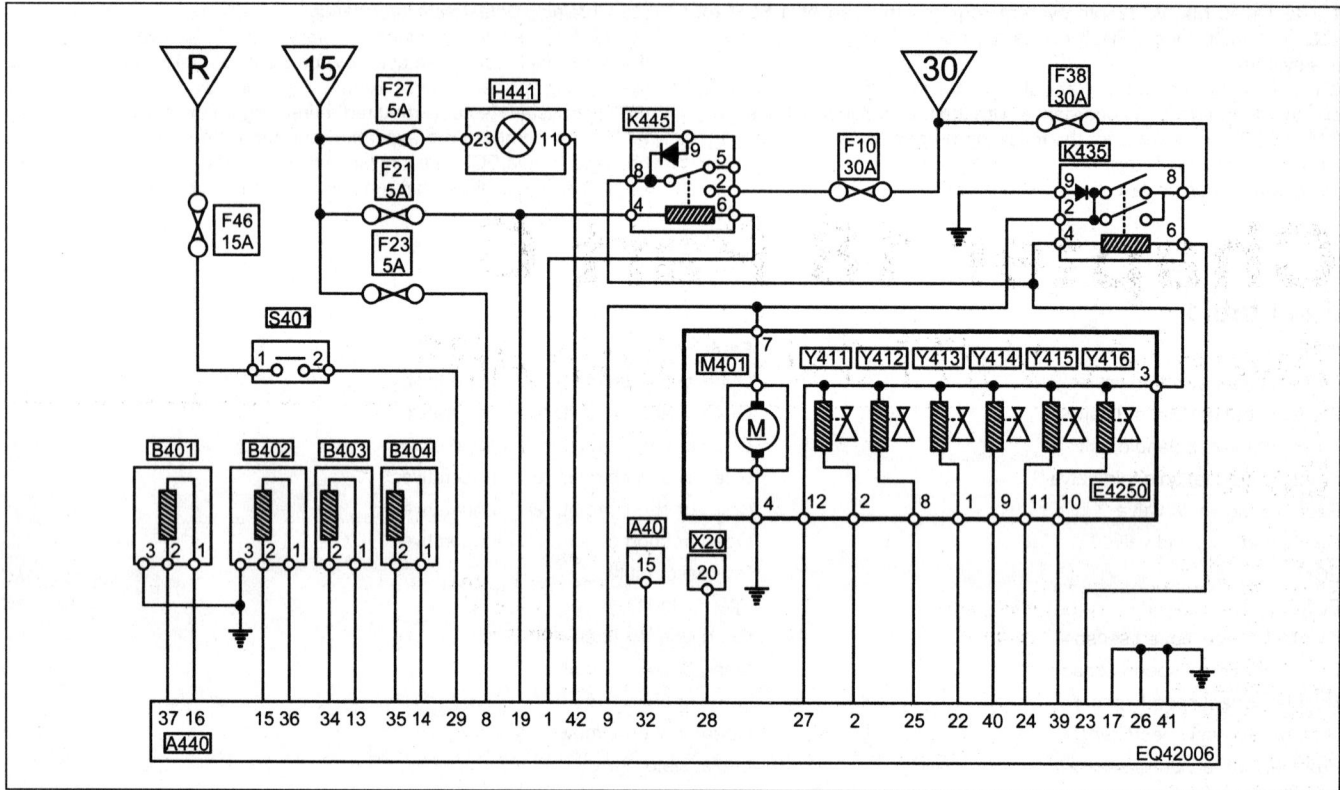

18B.25 42-pin wiring diagram, BMW 3-Series (E36)

Chapter 18 Part C

Teves Mk.IV (ITT 04) - 55-pin ABS

Contents

Vehicle coverage

Model	Year
BMW	
3-Series (E36)	1990-1996
Ford	
Escort/Orion	1990-1996
Granada/Scorpio	1992-1994
Transit	1991-1997
Saab	
9000	1993-1999
Seat	
Cordoba	1993-1995
Ibiza	1993-1995
Toledo	1991-1995
Volkswagen	
Corrado	1992-1996
Golf	1992-1994
Passat	1992-1995
Vento	1992-1994
Volvo	
850	1992-1995

Overview of system operation

1 Basic principles and system identification

The Teves Mk.IV Antilock Brake System has been fitted to a wide range of passenger vehicles since its introduction in the early 1990s. The system is of the additional or 'add-on' type operating in conjunction with the conventional braking system components. Later versions of the system may carry the designation ITT 04 although their operation is unchanged from the earlier Teves Mk.IV types.

The Teves Mk.IV system has been refined and uprated over the years in accordance with advances in electrical technology and demands from vehicle manufacturers. Essentially there are three distinct versions of the system, with the main differences being in the type and location of certain ECM input sensors, the ECM software, and the ECM multi-plug which may be a 37-pin, 42-pin or 55-pin type. For clarity, 55-pin ECM multi-plug systems are covered in this Part of Chapter 18, with 37-pin, and 42-pin systems being covered in Parts A and B respectively.

Depending on application, Teves Mk.IV may be installed as an antilock braking system only, or as an antilock braking system incorporating traction control.

In ABS mode, the purpose of the system is to apply the vehicle brakes at maximum efficiency without wheel lock or loss of directional stability. Inductive sensors (wheel speed sensors) monitor the speed of the wheels by generating an electrical signal as the wheel is rotated. This information is passed to the ABS Electronic Control Module (ECM) which is then able to determine wheel speed, wheel acceleration and wheel deceleration. The ECM compares the signals received from each wheel and if the onset of lock at any wheel is detected, a signal is sent to the ABS hydraulic control unit which regulates the brake pressure for the relevant wheel(s).

Where the system incorporates traction control, essentially the reverse principle is applied. When the ECM detects that one or more wheels are rotating faster than the reference value, the brake is actually applied on the relevant wheel(s) to reduce the rotational speed.

Typically, Teves Mk.IV ABS is comprised of the following components (see illustration 18C.1):

a) ABS-ECM.
b) Hydraulic control unit.
c) Four inductive wheel speed sensors and associated sensor rings.
d) ABS electrical wiring harness and relays.
e) Brake light switch.
f) Pedal position sensor.
g) ABS warning lamp.

In addition, the conventional braking system is comprised of the following components:

a) Tandem brake master cylinder.
b) Vacuum servo unit.
c) Brake calipers/wheel cylinders and hydraulic hoses and pipes.
d) Pressure regulating/load sensing valve(s) depending on application.

2 Component description and operation

ABS ECM

General

The Teves Electronic Control Module (ECM) continually monitors wheel speed from the signals provided by the wheel speed sensors. If the ECM detects the incidence of wheel lock on one or more wheels, a signal is sent to the hydraulic control unit to modulate the hydraulic pressure to the brake of the locking wheel (see illustration 18C.2).

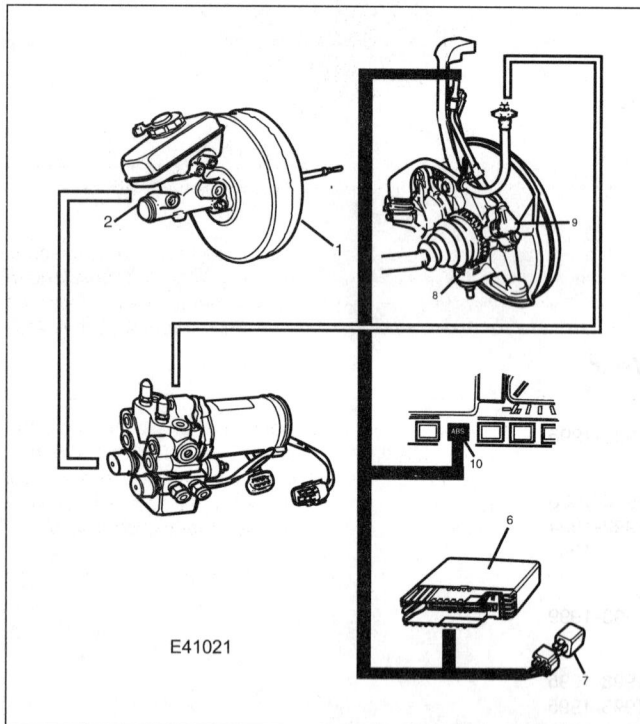

E41021

18C.1 Teves Mk.IV main components

1	ABS-ECM	6	Brake light switch
2	Hydraulic control unit	7	Pedal position sensor
3	Inductive wheel speed		(depending on application)
	sensors	8	ABS warning lamp
4	Sensor rings	9	Tandem master cylinder
5	ABS electrical wiring harness	10	Vacuum servo unit
	and relays		

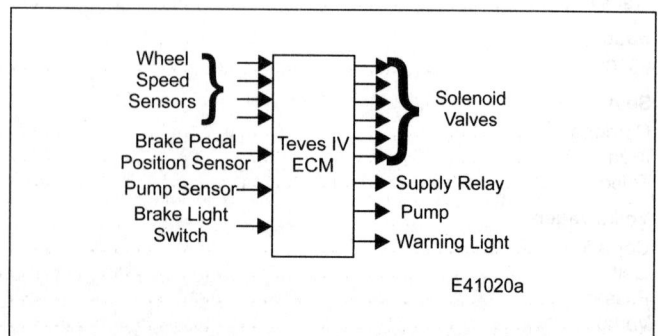

18C.2 ECM sensor inputs and control signal outputs

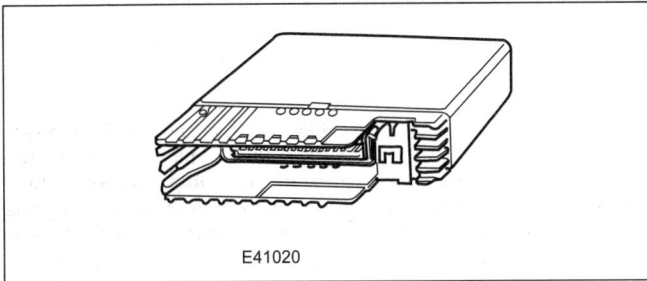

E41020

18C.3 Teves Mk.IV ECM

The ECM contains two microprocessors and uses digital technology to complete this function and other functions such as fault code memory and power modules for valve and relay activity **(see illustration 18C.3)**.

Self-test

The Teves Mk.IV ECM is equipped with a self-test capability that initially examines the ABS system when the ignition is switched on, and then examines the wheel speed sensor signals after a wheel speed of approximately 4 mph is reached from all wheels (engine running). The ABS self-test program continues to examine the signals from the various components as long as the ignition is switched on. If self-test determines that faults are not present, the ABS is ready for operation once a specified vehicle speed has been achieved.

If the ECM detects that a fault is present, all ABS functions are switched off and the warning lamp is turned on. The conventional braking system continues to operate as normal without ABS assistance.

Self-diagnostics

If the ECM detects a fault during the self-test routine, an internal fault code is stored in the ECM memory. Stored fault codes can be retrieved from the SD connector with the aid of a suitable fault code reader or, on Volvo models, by means of the on-board diagnostic unit. If the fault clears, the code will remain stored until cleared with the FCR, or diagnostic unit.

Hydraulic control unit

The hydraulic control unit consists of an electric motor with flange-mounted two-circuit radial piston return pump, and the flange-mounted valve block containing the inlet and outlet solenoid valves. The hydraulic control unit controls the pressure applied to the brake caliper for each individual wheel or pair of wheels during the three phases of ABS operation. During a controlled ABS cycle, the pump returns brake fluid, drained off during the pressure reduction phase, back into the brake circuit.

The pump only runs intermittently and is controlled by the ECM depending on the position of the brake pedal position sensor. A rotation sensor, integral with the pump, signals the ECM during pump operation.

According to vehicle application, Teves Mk.IV may be either a two-, three- or four-channel system according to the number of solenoid valves used in the hydraulic control unit. On a four-channel system, there are two solenoid valves for each brake (eight in total). On a three-channel system, there are two solenoid valves for each front brake, and two solenoid valves controlling the rear brakes as a pair (six in total). On a two-channel system, there are four solenoid valves, two for each controlled brake. Additional solenoid valves are used on systems incorporating traction control.

The layout of the hydraulic control unit varies according to the number of hydraulic channels used in the system and whether the installation is right-hand drive or left-hand drive format. Although the units may be visually different, their operation is the same **(see illustration 18C.4)**.

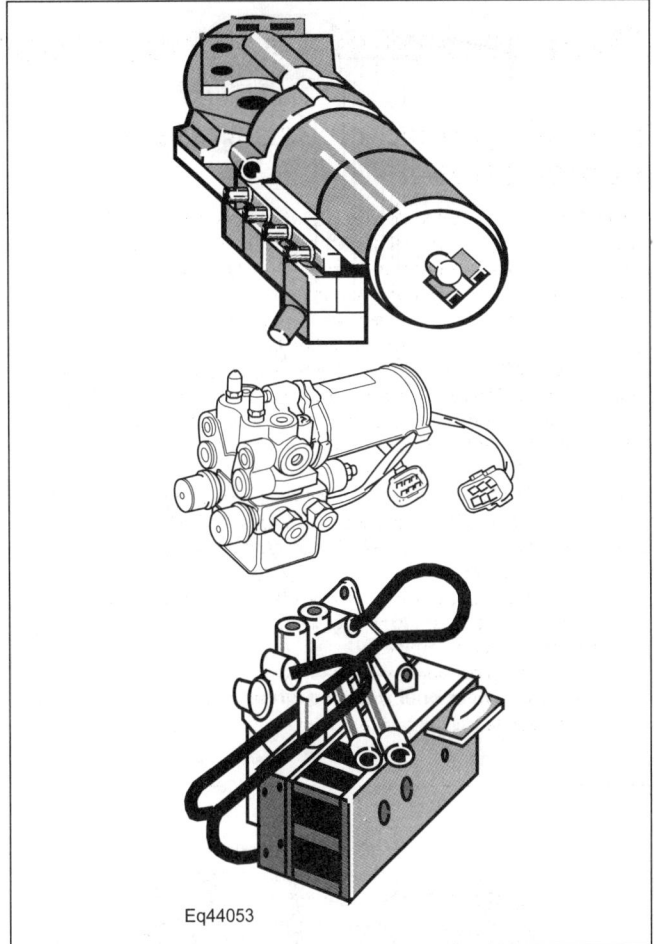

Eq44053

18C.4 Teves Mk.IV hydraulic control unit variations

Where a pair of brakes are controlled by a single hydraulic channel, the 'select-low' principle is often employed. With the 'select-low' principle, the wheel with the lowest adhesion determines the amount of hydraulic pressure to both brakes on that channel during ABS operation.

Wheel speed sensors

The rotational speed of the roadwheels and any changes in the rotational speed are recorded by inductive wheel speed sensors, located at the roadwheels. Depending on system type and vehicle application, there may be a separate sensor for each wheel, or a separate sensor for each front wheel only **(see illustration 18C.5)**.

18C

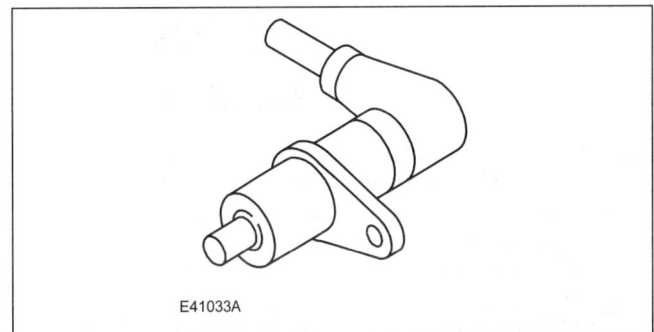

E41033A

18C.5 Typical wheel speed sensor

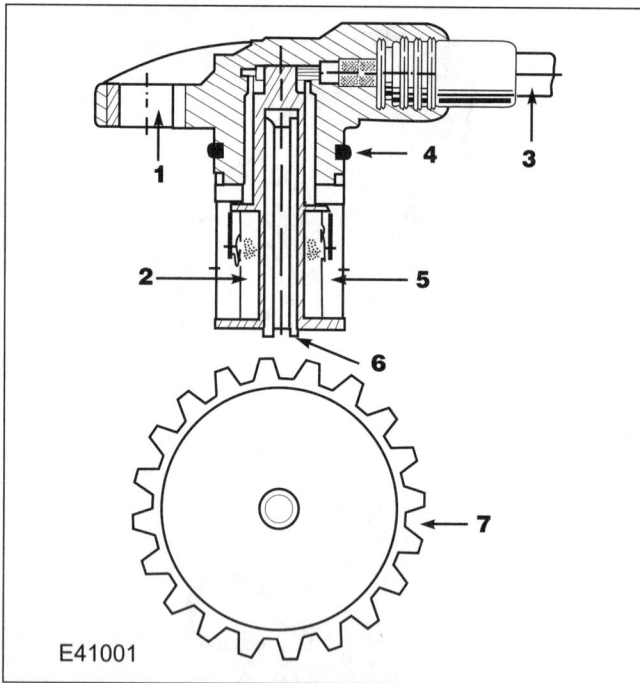

18C.6 Sectional view of a wheel speed sensor

1	Mounting bolt location	5	Coil
2	Permanent magnet	6	Sensor tip
3	Wiring harness	7	Toothed sensor ring
4	O-ring		

Each wheel speed sensor assembly comprises a toothed sensor ring, which rotates at roadwheel speed, and an adjacent sensor mounted a set distance from the sensor ring **(see illustration 18C.6)**.

The sensors are permanent magnet pulse generator types producing an AC voltage sine wave as the sensor ring teeth pass through the magnetic field of the sensor **(see illustration 18C.7)**.

The frequency of the waveform produced by the wheel speed sensor is proportional to the road speed. This AC voltage signal is continually being delivered to the ABS-ECM for processing.

The peak to peak voltage of the speed signal (when viewed upon an oscilloscope) can vary considerably according to wheel speed. An analogue to digital converter (ADC) in the ECM transforms the AC pulse into a digital signal **(see illustration 18C.8)**.

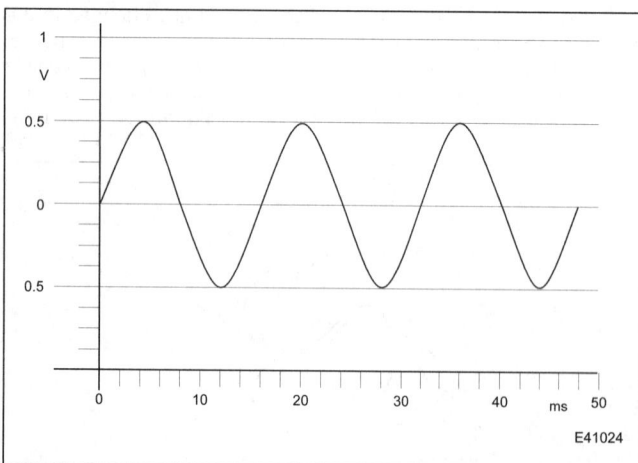

18C.8 Wheel speed sensor waveform as viewed on an oscilloscope

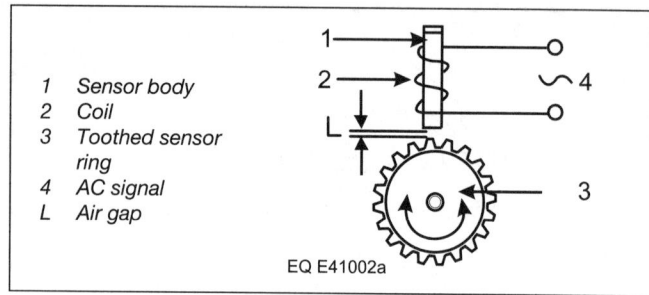

1	Sensor body
2	Coil
3	Toothed sensor ring
4	AC signal
L	Air gap

EQ E41002a

18C.7 Wheel speed sensor operation

ABS electrical wiring harness and relays

An integrated main wiring harness is used to connect power and earth to various electrical components. This enables sensor signals to reach the ECM and the ECM, in turn to send command signals to the relays, return pump and hydraulic control unit. The ABS relays and associated fuses are typically located in the vehicle fuse/relay box.

Main relay

Nominal battery voltage is supplied from the ignition switch to the relay via a fuse. The relay is controlled by the ECM applying an earth, which activates the relay allowing fused battery voltage to be applied to the solenoid valves. The earth for the solenoid valves is controlled by the ECM.

Pump relay

Nominal battery voltage is supplied via a fuse, from the battery, or overvoltage protection relay, to the pump relay. The relay is controlled by the ECM applying an earth, which activates the relay allowing fused battery voltage to be applied to the pump motor. The hydraulic control unit provides the earth for the pump motor.

Overvoltage protection relay

Where an overvoltage protection relay is used, nominal battery voltage is supplied from the battery to the relay via a fuse. The relay is controlled by the ECM applying an earth, which activates the relay allowing fused battery voltage to be applied to the pump relay.

Brake light switch

The brake light switch comprises a switch body and contact pin and is located above the brake pedal **(see illustration 18C.9)**. On some applications, the brake light switch and pedal position sensor are combined to form a single unit.

When the brake pedal is depressed closing the brake light switch, a signal is sent to the ECM indicating that the brakes are being applied. Once this signal is received, the ECM will begin monitoring the wheel speed via the wheel speed sensors and activate the ABS if necessary.

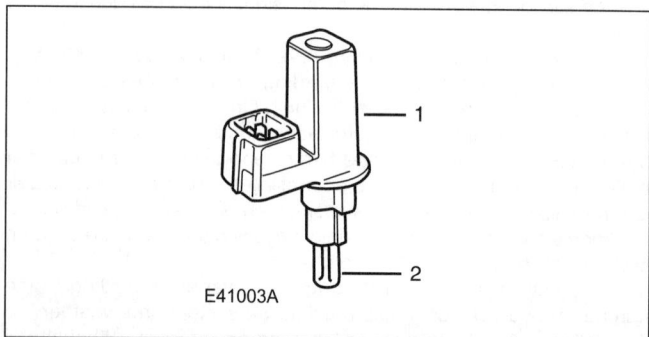

E41003A

18C.9 Typical brake light switch body (1) and contact pin (2)

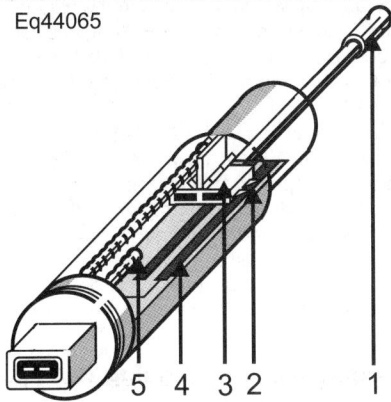

18C.10 Cutaway view of the brake pedal position sensor

1 Plunger rod   3 Sliding segment   5 Return spring
2 Contact brush   4 Track

Pedal position sensor

The pedal position sensor is typically located in the vacuum servo unit, although in some installations it may combined with the brake light switch mounted above the brake pedal **(see illustration 18C.10)**.

The sensor contains a plunger rod, contact brush, sliding segment, track and return spring located within the switch body. The plunger rod is held in contact with the vacuum servo unit diaphragm or brake pedal by the pressure of the return spring. As the position of the servo diaphragm or brake pedal changes with brake application, the plunger rod moves accordingly, thus altering the position of the contact brush on the track. The track is divided into a series of stepped resistances corresponding to various pedal positions. The ECM supplies the track with a continuous current and determines pedal position by the drop in voltage according to the position of the contact brush on the track.

When the brake pedal is activated with the ABS in operation, the ECM receives a variable voltage signal from the sensor according to pedal position. If the brake pedal begins to creep down during a controlled pressure reduction phase, the changing voltage signal from the sensor is detected by the ECM and the return pump is activated. Hydraulic fluid is pumped from the reservoir back into the brake circuit, forcing back the master cylinder piston (and the brake pedal). When the ECM detects that the pedal has returned to its initial position, by means of the pedal position sensor signal, the return pump is switched off.

ABS warning lamp

After the ignition is switched on, the ABS warning lamp on the instrument panel is illuminated for approximately 2 to 4 seconds as the system executes a self-test routine. If satisfactory operation of the system is detected by the ECM, the light is extinguished. During

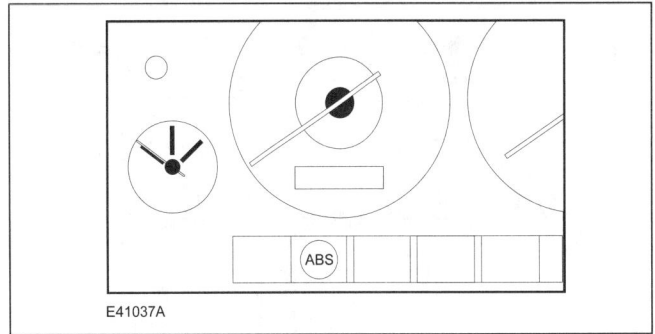

18C.11 ABS warning lamp

vehicle operation above a pre-determined wheel speed, the ABS-ECM implements a further self-test cycle whereby ABS operation and wheel speed sensor signals are continually monitored. If a fault is detected, the relevant ECM pin is earthed to illuminate the warning lamp on the instrument panel, and the ABS function is disabled. The warning lamp will remain illuminated until the fault is no longer present **(see illustration 18C.11)**.

When the ABS-ECM detects a fault, the fault code is stored and the ABS warning lamp activated. If the fault no longer exists after the next system start (ignition on/off) the ABS warning lamp is extinguished after the self-test cycle, however the fault code remains stored in the ECM memory.

Tandem master cylinder

Typically, the tandem master cylinder comprises a body casting incorporating primary and secondary pressure chambers, primary piston, intermediate piston, floating piston, slotted pin and central valve. The cylinder operates as a conventional master cylinder using vacuum assistance from the vacuum servo unit **(see illustration 18C.12)**.

When the brake system is at rest, the central valve in the floating piston rests against the slotted pin. In this condition the central valve is open and brake fluid can discharge out of the pressure chamber back into the brake fluid reservoir. When the brake pedal is depressed, the build-up of hydraulic pressure in the primary pressure chamber acts on the intermediate piston and floating piston, moving them down the cylinder bore. The floating piston contacts the seal on the central valve, closing the connection between the intermediate and secondary pressure chambers. Brake hydraulic pressure can now also increase in the secondary pressure chamber.

Vacuum servo unit

The vacuum servo unit is located between the brake pedal and tandem master cylinder. When the brake pedal is depressed, the servo unit increases the force applied by the pedal, reducing the effort required to operate the brakes **(see illustration 18C.13)**.

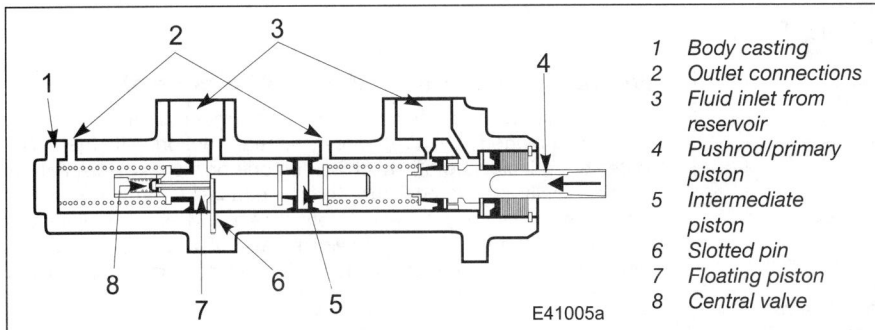

1 Body casting
2 Outlet connections
3 Fluid inlet from reservoir
4 Pushrod/primary piston
5 Intermediate piston
6 Slotted pin
7 Floating piston
8 Central valve

18C.12 Sectional view of a typical tandem master cylinder

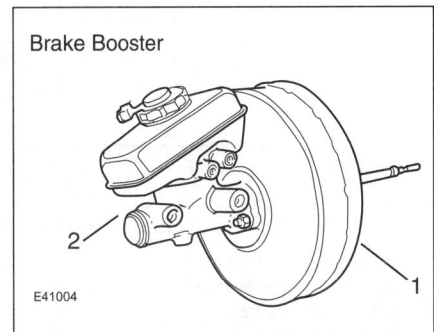

18C.13 Vacuum servo unit (1) and tandem master cylinder (2)

18C

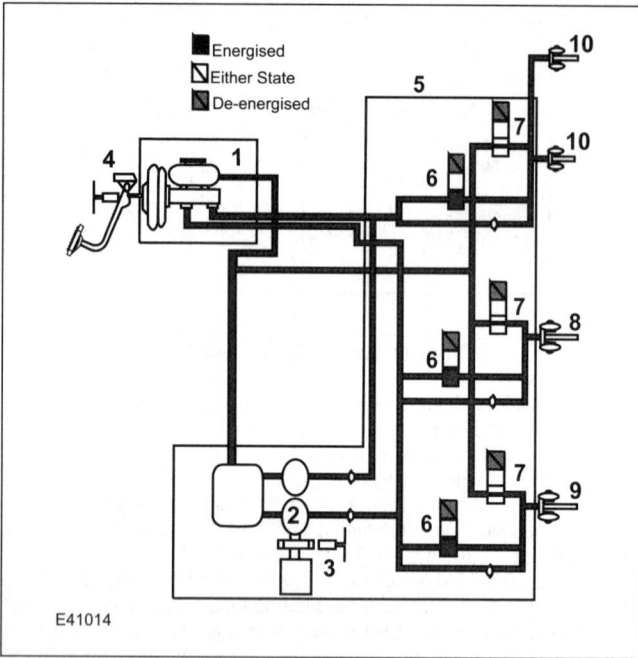

Energised
Either State
De-energised

E41014

18C.14 Brake system at rest

1 Vacuum servo unit and tandem master cylinder	5 Hydraulic control unit
2 ABS return pump	6 Inlet solenoid valve
3 Pump motor sensor	7 Outlet solenoid valve
4 Brake pedal position sensor	8 Front left brake caliper
	9 Front right brake caliper
	10 Rear brake calipers

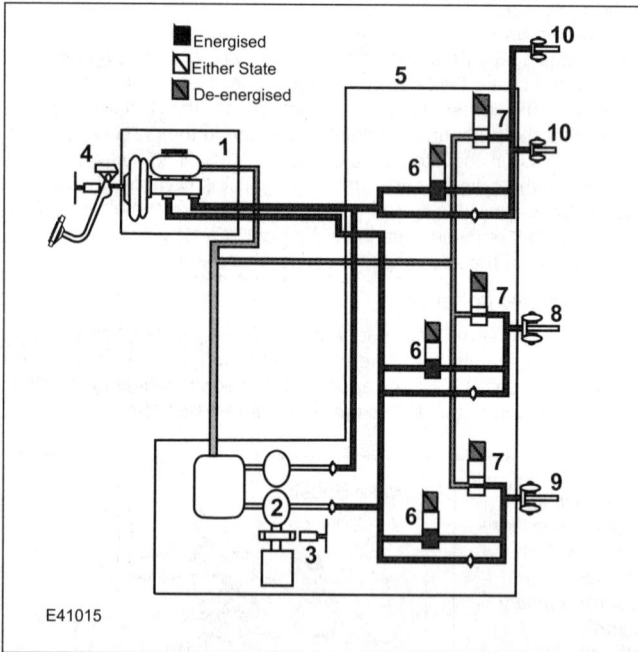

Energised
Either State
De-energised

E41015

18C.15 Conventional braking without ABS

1 Vacuum servo unit and tandem master cylinder	5 Hydraulic control unit
2 ABS return pump	6 Inlet solenoid valve
3 Pump motor sensor	7 Outlet solenoid valve
4 Brake pedal position sensor	8 Front left brake caliper
	9 Front right brake caliper
	10 Rear brake calipers

The unit is operated by vacuum created in the engine inlet manifold (or from a separate vacuum pump on diesel engines) which is applied to a diaphragm contained within the unit casing. A pushrod connected to the centre of the diaphragm acts directly on the primary piston in the master cylinder.

When the brake pedal is released, vacuum is applied to both sides of the diaphragm. When the pedal is depressed, one side of the diaphragm is opened to atmosphere and the vacuum acting on the other side deflects the diaphragm which in turn operates the master cylinder primary piston. The resulting force applied to the master cylinder piston is therefore significantly greater than the initial force applied to the brake pedal by the driver.

Pressure regulating/load sensing valve(s)

Depending on vehicle application, pressure regulating valves or load sensing valves may be incorporated to restrict the hydraulic fluid pressure to the rear brakes. The valves may be pressure conscious whereby the hydraulic fluid supply is restricted once a pre-determined pressure is reached, or load conscious whereby the hydraulic pressure is reduced according to vehicle loading.

3 System operation

Brake system at rest

The illustrations in this Section depict the operation of a typical three-channel hydraulic system with two solenoid valves for each front brake, and two solenoid valves controlling the rear brakes as a pair. Two- and four-channel systems operate in the same way apart from the number of solenoid valves used.

When the ABS system is at rest all the brake components are inoperative. Pressure is non-existent in the hydraulic pipes between the tandem master cylinder and the brake calipers. The inlet solenoid valves in the valve block are open and the outlet solenoid valves are shut (see illustration 18C.14).

Brake system operating under conventional control without ABS

When the brake pedal is activated, the pedal force is applied to the tandem master cylinder by the vacuum servo unit pushrod. The servo unit pushrod acts directly on the pressure piston in the master cylinder. The resulting increase in hydraulic pressure moves the intermediate piston in the cylinder, and the pressure which increases in front of the intermediate piston flows through the open inlet solenoid valves in the valve block, to the brake calipers. The outlet solenoid valves in the valve block remain shut (see illustration 18C.15).

Brake system operated with ABS control

The ABS-ECM continually monitors wheel speed from the signals provided by the wheel speed sensors. If the ECM detects the incidence of wheel lock on one or more wheels, ABS is automatically initiated in three phases. Typically, ABS operates individually on each front wheel and either individually or jointly on the rear wheels. In which case, all or any of the wheels could be in any one of the following phases at any particular moment.

First ABS phase, pressure holding

To prevent any further build-up of hydraulic pressure in the circuit

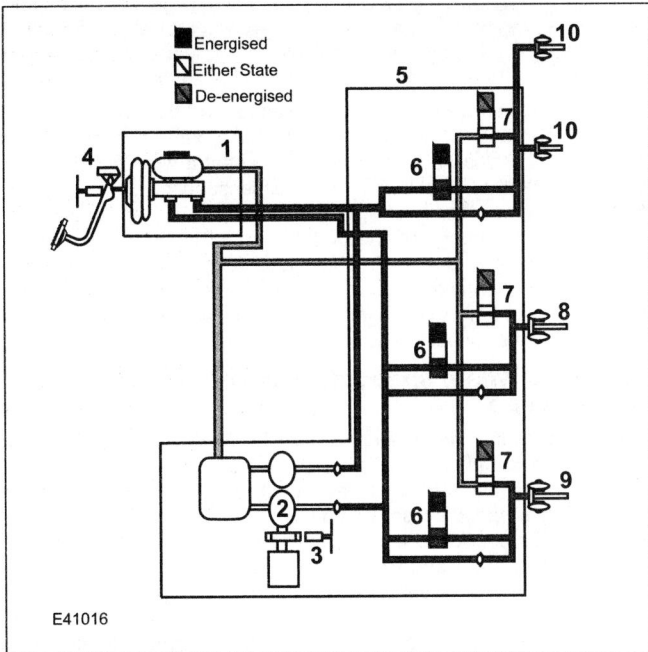

E41016

18C.16 ABS operation - first phase, pressure holding

1	Vacuum servo unit and tandem master cylinder	5	Hydraulic control unit
2	ABS return pump	6	Inlet solenoid valve
3	Pump motor sensor	7	Outlet solenoid valve
4	Brake pedal position sensor	8	Front left brake caliper
		9	Front right brake caliper
		10	Rear brake calipers

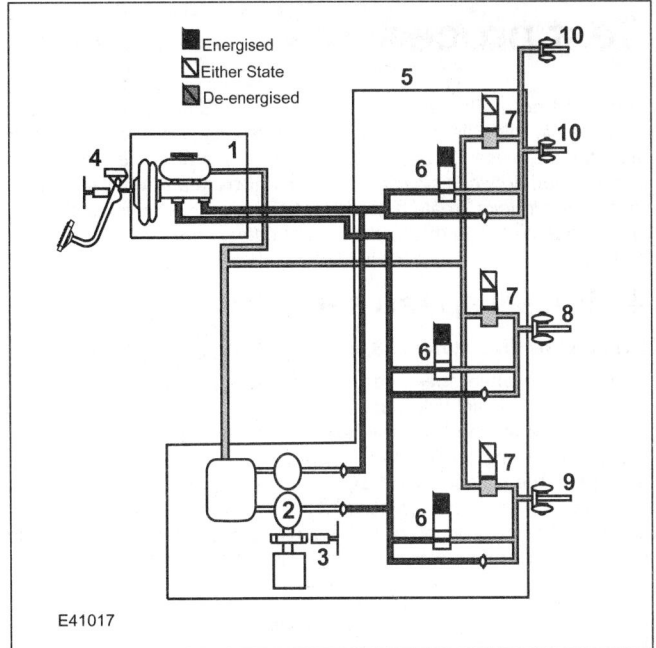

E41017

18C.17 ABS operation - second phase, pressure reduction

1	Vacuum servo unit and tandem master cylinder	5	Hydraulic control unit
2	ABS return pump	6	Inlet solenoid valve
3	Pump motor sensor	7	Outlet solenoid valve
4	Brake pedal position sensor	8	Front left brake caliper
		9	Front right brake caliper
		10	Rear brake calipers

being controlled, the inlet solenoid valve is shut and the outlet solenoid valve remains shut. The pressure cannot now be increased in the controlled circuit by any further application of the brake pedal **(see illustration 18C.16)**.

If the wheel speed sensor signals indicate that wheel rotation has now stabilised, the ECM will instigate the pressure build-up phase, allowing braking to continue. If wheel lock is still detected after the pressure holding phase, the ECM instigates the pressure reduction phase.

Second ABS phase, pressure reduction

A controlled braking operation can be instigated with a pressure reduction phase or it can follow a pressure holding phase. The inlet solenoid valve is, or remains shut and the outlet solenoid valve is opened. As long as the outlet solenoid valve is open, the brake fluid discharges back into the reservoir of the tandem master cylinder. The return pump then pumps the brake fluid out of the reservoir back into the control circuit, forcing back the master cylinder piston (and the brake pedal) **(see illustration 18C.17)**.

Third ABS phase, pressure-build up

The pressure build up phase is instigated after the wheel rotation has stabilised. The outlet solenoid valve is shut and the inlet solenoid valve is opened. The brake pressure from the tandem master cylinder flows through the inlet solenoid valve to the relevant brake caliper allowing braking to continue **(see illustration 18C.18)**.

The whole ABS control cycle takes place 4 to 10 times per second for each affected wheel and this ensures maximum braking effect and control during ABS operation.

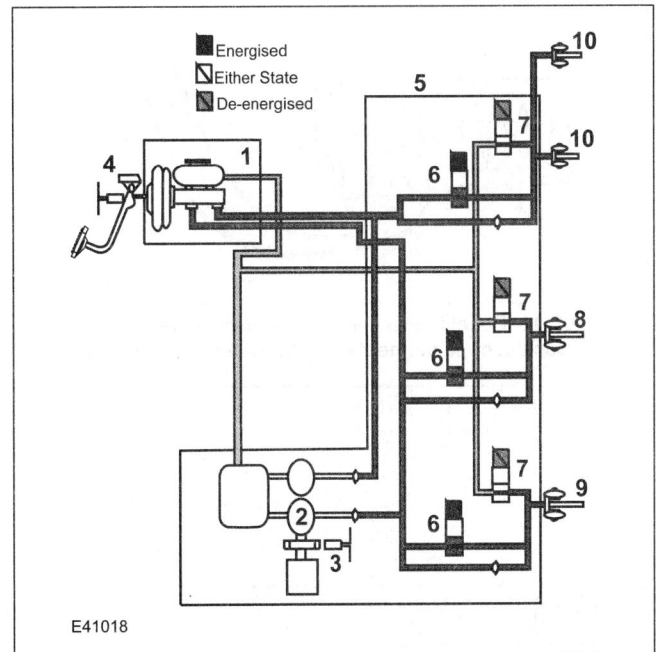

E41018

18C.18 ABS operation - third phase, pressure build-up

1	Vacuum servo unit and tandem master cylinder	5	Hydraulic control unit
2	ABS return pump	6	Inlet solenoid valve
3	Pump motor sensor	7	Outlet solenoid valve
4	Brake pedal position sensor	8	Front left brake caliper
		9	Front right brake caliper
		10	Rear brake calipers

18C

Test procedures

Important note: *The test procedures, pin-tables and wiring diagrams contained in this Chapter are necessarily representative of the system depicted. Because of the variations in wiring and other data that often occurs, even between similar vehicles in any particular VM's range, the reader should take great care in identification of ECM pins, and satisfy himself that he has gathered the correct data before failing a particular component.*

4 Wheel speed sensors

Checking the wheel speed sensor (general)

1 Inspect the wheel speed sensor for corrosion or damage and check that the sensor is tightly mounted.
2 Check the toothed sensor ring for damage, eccentricity and for broken or missing teeth.
3 Inspect the wheel speed sensor wiring plug for corrosion and damage. One plug for each sensor.
4 Check that the connector terminal pins are pushed fully home and making good contact with the sensor wiring plug.

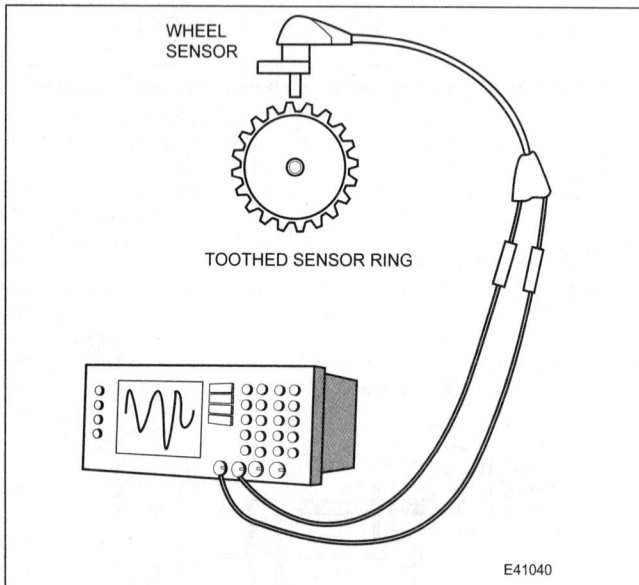

18C.19 Checking wheel speed sensor output with an oscilloscope connected to the sensor wiring plug

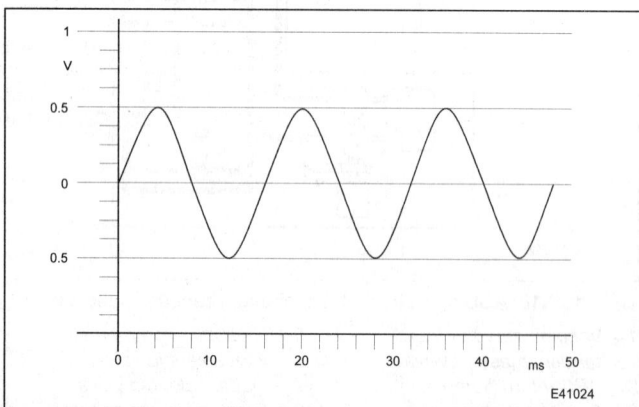

18C.20 Typical wheel speed sensor sine wave as displayed on an oscilloscope

5 Check the clearance between the sensor and the toothed sensor ring. The clearance is not normally adjustable but is nominally 0.2 to 1.2 mm. If the clearance is excessive, expect a worn sensor tip or problems with the wheel bearings/hub or sensor ring.
6 When carrying out voltage checks with an oscilloscope or voltmeter, the voltage obtained will be proportional to the speed at which the wheel is rotating. In addition to determining that the wheel speed sensors are actually producing a voltage output, it is essential that the output from the sensors on a particular axle is the same for any given wheel speed.

Checking wheel speed sensor output with an oscilloscope

Note: *Refer to the wiring diagrams for specific ECM pin identification according to model.*
7 Switch the ignition off and disconnect the ECM multi-plug or the relevant wheel speed sensor wiring plug.
8 Connect an oscilloscope between the terminal pins for the sensor under test **(see illustration 18C.19)**.
9 Select a range to cover 80 Hz on the oscilloscope and a free run time base.
10 Raise the wheel and rotate it by hand at approximately one revolution per second.
11 A sinusoidal wave form should be obtained, with amplitude and duration changing with rotational speed **(see illustration 18C.20)**.
12 If there is no signal, or a very weak or intermittent signal at the ECM, repeat the test at the sensor wiring plug. If there is no change in signal status, the sensor is suspect.
13 If the signal is now satisfactory this indicates a fault in the wiring harness which should be checked for continuity.

Checking wheel speed sensor output with an AC voltmeter

Note: *Refer to the wiring diagrams for specific ECM pin identification according to model.*
14 Switch the ignition off and disconnect the ECM multi-plug or the relevant wheel speed sensor wiring plug.
15 Connect an AC voltmeter between the terminal pins for the sensor under test **(see illustration 18C.21)**.

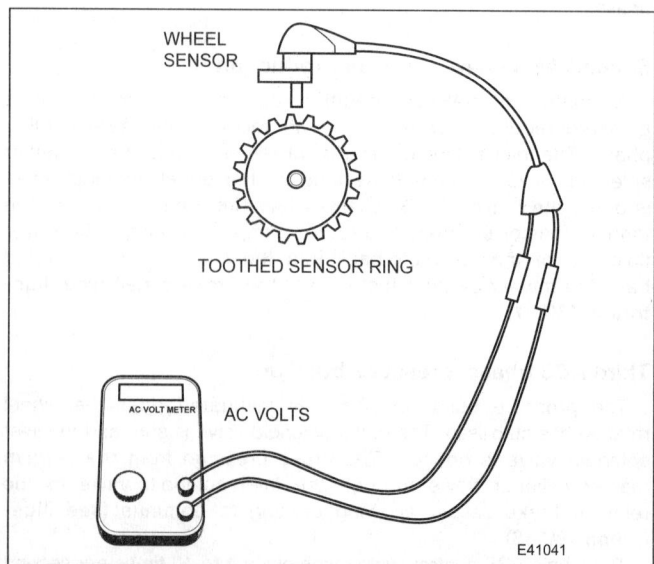

18C.21 Checking wheel speed sensor output with a voltmeter connected to the sensor wiring plug

16 Raise the wheel and rotate it by hand at approximately one revolution per second.

17 A voltage of approximately 0.5 to 1.5 volts (AC RMS) should be obtained. If there is no signal, or a very weak or intermittent signal at the ECM, repeat the test at the sensor wiring plug. If there is no change in the signal, the sensor is suspect.

18 If the signal is now satisfactory, this indicates a fault in the wiring harness which should be checked for continuity. **Note:** *This test at least proves that a signal is being generated by the sensor. However, the voltage produced is an average voltage and does not clearly indicate damage to the sensor ring or that the sinewave is regular in formation.*

Checking wheel speed sensor resistance

Note: *Refer to the wiring diagrams for specific ECM pin identification according to model.*

19 Switch the ignition off and disconnect the ECM multi-plug or the relevant wheel speed sensor wiring plug.

20 Connect an ohmmeter between the terminal pins for the sensor under test **(see illustration 18C.22)**.

21 The readings obtained should be between 0.7 and 2.2 kohms approximately.

22 If the resistance is excessively high, or open circuit at the ECM, repeat the test at the sensor multi-plug. If there is no change in resistance, the sensor is suspect.

23 If the resistance is now satisfactory, this indicates a fault in the wiring harness which should be checked for continuity. **Note:** *Even if the resistance is within the quoted specifications, this does not prove that the speed sensor can generate an acceptable signal.*

5 System relays

Relay operation tests

Main relay

Note: *In the following test the relay terminal numbers refer to a typical relay as shown in the illustration. Refer to the wiring diagrams for specific relay terminal numbers according to model.*

1 Switch the ignition off and remove the relay from the relay base.

2 Connect an ohmmeter between the relay output terminal 30 and earth terminal 87a **(see illustration 18C.23)**. The ohmmeter should indicate continuity.

3 Connect an ohmmeter between the relay main voltage supply terminal 87 and the output terminal 30. The ohmmeter should indicate an open circuit.

4 Attach the positive lead of a 12 volt supply to the relay driver supply terminal 86, and the negative lead to the relay driver earth terminal 85.

5 There should be an unmistakable click from the relay and the ohmmeter should indicate continuity.

6 Where applicable, connect an ohmmeter or a diode tester between the relay output terminal 30 and ABS warning lamp terminal 30a.

7 Measure the circuit and then reverse the connections. The instrument should indicate continuity in one direction (diode conducting), but not the other (diode blocking).

8 Renew the relay if it fails these tests. If the relay cover is transparent, inspect the relay for burns. A faulty relay usually has a burnt appearance.

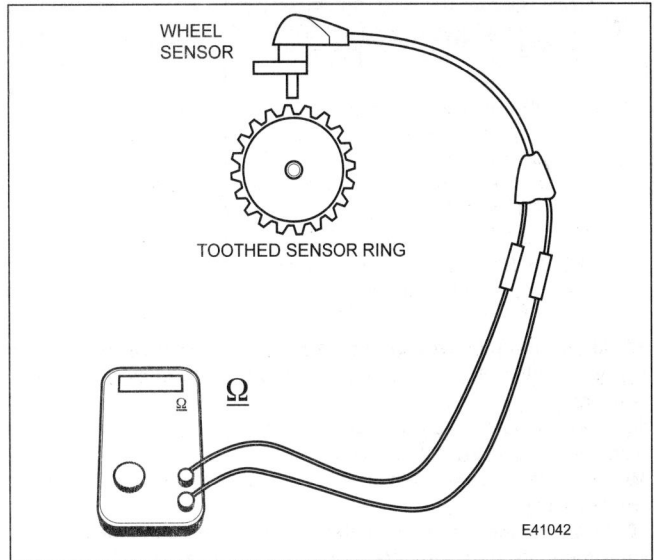

18C.22 Checking wheel speed sensor resistance with an ohmmeter connected to the sensor wiring plug

Hydraulic pump relay

Note: *In the following test the relay terminal numbers refer to a typical relay as shown in the illustration. Refer to the wiring diagrams for specific relay terminal numbers according to model.*

9 Switch the ignition off and remove the relay from the relay base.

10 Connect an ohmmeter between the relay main voltage supply terminal 30 and the output terminal 87 **(see illustration 8C.24)**. The ohmmeter should indicate an open circuit.

11 Attach the positive lead of a 12 volt supply to the relay driver supply terminal 86, and the negative lead to the relay driver earth terminal 85.

12 There should be an unmistakable click from the relay and the ohmmeter should indicate continuity.

13 Where applicable, connect an ohmmeter or a diode tester between the relay output terminal 87 and relay earth terminal 87a.

14 Measure the circuit and then reverse the connections. The instrument should indicate continuity in one direction (diode conducting), but not the other (diode blocking).

15 Renew the relay if it fails these tests. If the relay cover is transparent, inspect the relay for burns. A faulty relay usually has a burnt appearance.

Overvoltage protection relay

Note: *In the following test the relay terminal numbers refer to a typical overvoltage protection relay as shown in the illustration. Refer to the wiring diagrams for specific relay terminal numbers according to model.*

16 Switch the ignition off and remove the relay from the relay base. Connect an ohmmeter across the relay main voltage supply terminal 2 and output terminal 8 **(see illustration 18C.25)**. The ohmmeter should indicate an open circuit.

18C

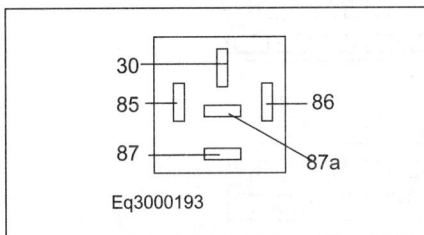

18C.23 Main relay terminal identification

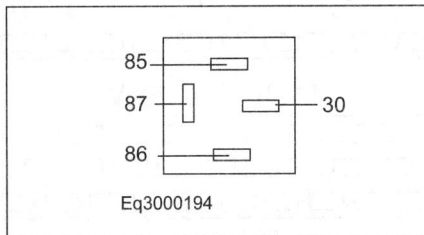

18C.24 Pump relay base terminal identification

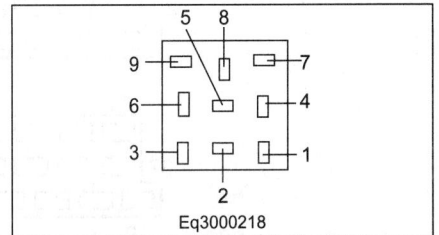

18C.25 Overvoltage protection relay terminal identification

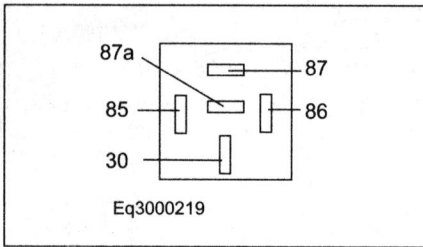

18C.26 Main relay base terminal identification

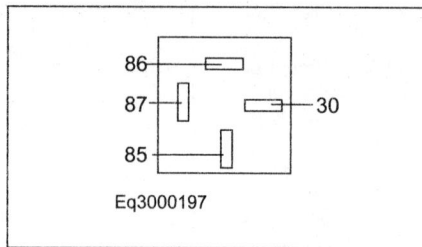

18C.27 Pump relay base terminal identification

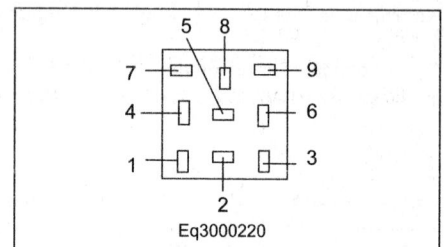

18C.28 Overvoltage relay base terminal identification

17 Attach the positive lead of a 12 volt supply to the relay driver supply terminal 4, and the negative lead to the relay driver earth terminal 6.

18 There should be an unmistakable click from the relay and the ohmmeter should indicate continuity.

19 Connect an ohmmeter or a diode tester across the two diode terminals 8 and 9.

20 Measure the circuit and then reverse the connections. The instrument should indicate continuity in one direction (diode conducting), but not the other (diode blocking).

21 Renew the relay if it fails these tests. If the relay cover is transparent, inspect the relay for burns. A faulty relay usually has a burnt appearance.

Relay power supply tests

Main relay

Note: *In the following test the relay terminal numbers refer to a typical relay as shown in the illustration. Refer to the wiring diagrams for specific relay terminal numbers according to model.*

22 Switch the ignition off and remove the main relay leaving the base exposed.

23 Using a voltmeter, probe between the main voltage supply terminal 87 in the relay base and earth **(see illustration 18C.26)**. The reading should indicate nbv.

24 If no voltage is found, check the relevant fuse and the supply wiring back to the voltage supply.

25 Switch the ignition on.

26 Using a voltmeter, probe between the relay driver supply terminal 86 in the relay base and earth. The reading should indicate nbv.

27 If no voltage is found, check the relevant fuse and the supply wiring back to the ignition switch.

Pump relay

Note: *In the following test the relay terminal numbers refer to a typical relay as shown in the illustration. Refer to the wiring diagrams for specific relay terminal numbers according to model.*

Note: *For the following test the ECM and main relay must be connected and operating correctly.*

28 Switch the ignition off and remove the pump relay leaving the base exposed.

29 Using a voltmeter, probe between the main voltage supply terminal 30 in the relay base and earth. The reading should indicate nbv **(see illustration 18C.27)**.

30 If no voltage is found, check the relevant fuse and the supply wiring back to the voltage supply.

31 Switch the ignition on.

32 Using a voltmeter, probe between the relay driver supply terminal 86 in the relay base and earth. The reading should indicate nbv.

33 If no voltage is found, check the relevant fuse, the main relay or overvoltage protection relay (as applicable) and the supply wiring back to the ignition switch.

Overvoltage protection relay

Note: *In the following test the relay base terminal numbers refer to a typical overvoltage protection relay as shown in the illustration. Refer to the wiring diagrams for specific terminal numbers according to model.*

34 Switch the ignition off and remove the relay from the relay base.

35 Using a voltmeter, probe between the main voltage supply terminal 2 in the relay base and earth. The reading should indicate nbv **(see illustration 18C.28)**.

36 If no voltage is found, check the relevant fuse and the supply wiring back to the voltage supply.

37 Switch the ignition on.

38 Using a voltmeter, probe between the relay driver supply terminal 4 in the relay base and earth. The reading should indicate nbv.

39 If no voltage is found, check the relevant fuse and the supply wiring back to the ignition switch.

6 Electronic Control Module

Checking the ECM (general)

1 Inspect the ECM for corrosion or damage and ensure that the unit is securely attached to the hydraulic control unit.

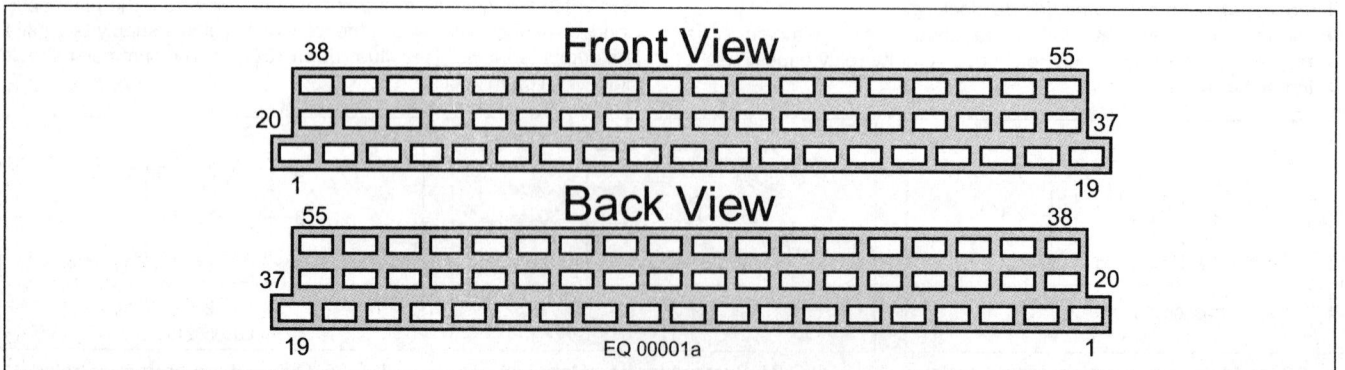

18C.29 ECM 55-pin multi-plug

2 Check that the ECM multi-plug terminals are pushed fully home and making good contact with the ECM pins. Faults in any of the above areas are possible reasons for poor performance in the ABS system.

ECM power supply and earth tests

Note: *Refer to the wiring diagrams for specific ECM pin identification according to model.*

3 Switch the ignition off and disconnect the ECM multi-plug.

4 Switch the ignition on.

5 Attach a negative voltmeter probe to a vehicle earth.

6 Attach a positive voltmeter probe to ECM multi-plug power supply pin 53 **(see illustration 18C.29)**. The voltmeter should indicate nbv.

7 If no voltage is found, check the relevant fuse and the supply wiring back to the ignition switch.

8 Switch the ignition off.

9 Connect an ohmmeter between a vehicle earth and ECM multi-plug earth pins 1 and 19 in turn.

10 The ohmmeter should indicate continuity in each case. If not, check the ECM main earth connection and wiring.

7 Solenoid valves

Solenoid valve resistance tests

Note: *Refer to the wiring diagrams for specific ECM pin identification according to model.*

1 Switch the ignition off and disconnect the ECM multi-plug from the ECM.

3 Using an ohmmeter, measure the resistance between pin 3 in the ECM multi-plug, and each of the following multi-plug pins in turn, noting that pins 55 and 18 are only used on four-channel hydraulic systems **(see illustration 18C.29)**:

Inlet solenoid valves - 20, 38, 54, 55.
Outlet solenoid valves - 2, 18, 21, 36.

4 The resistance should be in the order of 6.0 to 10.0 ohms for each inlet solenoid valve, and 3.0 to 6.0 ohms for each outlet solenoid valve.

5 If the resistance is not as specified, check the wiring and wiring connectors. If satisfactory, the hydraulic control unit is suspect.

8 Hydraulic pump motor

Pump resistance and operation test

1 Switch the ignition off and disconnect the ECM multi-plug from the ECM.

2 Where a pump sensor is fitted, connect an AC voltmeter across ECM multi-plug pins 49 and 31 **(see illustration 18C.29)**.

3 Using a fused jumper lead, connect ECM multi-plug pins 34 and 15 to earth.

4 Switch the ignition on.

5 The main (or overvoltage protection) relay, pump relay and the pump should activate (listen for an unmistakable click from the relays and a hum from the pump).

⚠️ **Warning: The test should be made as quickly as possible to avoid damaging the pump.**

6 If the relay(s) fails to activate, carry out the relay test procedures described in Section 5.

7 If the pump does not operate as described, or if the pump operates but there is no reading on the voltmeter, renew the pump. The sensor is integral with the pump and cannot be renewed separately.

9 Brake light switch

Checking the brake light switch (general)

1 Check that the brake light switch is correctly and securely mounted and that the plunger moves smoothly with no trace of binding.

2 Check that the wiring multi-plug is pushed fully home and making good contact.

3 Check that no wires have been disconnected.

4 Faults in any of the above areas are possible reasons for failure or malfunctioning of the switch.

Brake light switch voltage and continuity tests

Voltage test

Note: *Refer to the wiring diagrams for specific ECM pin identification according to model.*

5 Switch the ignition off and disconnect the ECM multi-plug from the ECM.

6 Connect a voltmeter between a vehicle earth and the ECM multi-plug brake light switch input pin 32 **(see illustration 18C.29)**.

7 Switch the ignition on and depress the brake pedal. The voltmeter should indicate nbv.

8 If no voltage is found, the fuse, the brake light switch and the wiring are suspect.

9 Release the brake pedal. The voltage should drop to zero as the switch opens.

Continuity test

10 Switch the ignition off and disconnect the brake light switch multi-plug.

11 Connect an ohmmeter between the terminal pins of the brake light switch.

12 Operate the brake light switch and check for continuity. If the test fails, renew the switch.

10 Brake pedal position sensor

Pedal position sensor resistance test

1 Switch the ignition off and disconnect the ECM multi-plug from the ECM.

2 Connect an ohmmeter between ECM multi-plug pins 16 and 41 **(see illustration 18C.29)**.

3 Measure the resistance of the position sensor against brake pedal travel.

4 The resistance should change in stages according to the distance travelled by the brake pedal.

5 If the test fails, renew the sensor.

11 Warning lamp

Checking the warning lamp (general)

1 Inspect the warning lamp bulb holder contacts in the instrument panel.

2 Check that the instrument panel multi-plug terminal pins are pushed fully home and making good contact.

3 Check that no wires have been disconnected.

4 Faults in any of the above areas are possible reasons for failure or malfunctioning of the warning lamp.

Warning lamp operation test

5 With the ignition switched off, the warning lamp should remain off.

6 Switch the ignition on and the warning lamp should illuminate then extinguish after a few seconds. The lamp should then remain off.

7 If the warning lamp comes on and remains on at any time during vehicle operation, carry out the previously described test procedures on the system components.

18C

Pin table - typical 55-pin (BMW, Ford Granada/Scorpio, Volvo)

Note: *Refer to illustration 18C.29*

Pin No.	Connection	Test condition	Voltage
1	ECM earth	Ignition on	0.25 V (max)
2	Front left outlet solenoid valve	Ignition on/inactive	Nbv
		Actuated	1.0 volt (max)
3	Voltage monitoring, hydraulics	Ignition on	Nbv
4	-	-	-
5	-	-	-
6	-	-	-
7	-	-	-
8	-	-	-
9	-	-	-
10	-	-	-
11	-	-	-
12	-	-	-
13	-	-	-
14	-	-	-
15	Pump relay driver	Ignition on/inactive	Nbv
		Actuated	1.0 volt (max)
16	Pedal position sensor	Pedal at rest	1.0 volt
		Pedal fully depressed	3.0 volts
17	-	-	-
18	-	-	-
19*	ECM earth	Ignition on	0.25 volts (max)
20	Front left inlet solenoid valve	Ignition on/inactive	Nbv
		Actuated	1.0 volt (max)
21	Front right outlet solenoid valve	Ignition on/inactive	Nbv
		Actuated	1.0 volt (max)
22	-	-	-
23	SD connector	-	-
24	-	-	-
25	-	-	-
26	-	-	-
27	Rear right wheel speed sensor earth	Roadwheel rotating	0.25 volts (max)
28	Rear left wheel speed sensor earth	Roadwheel rotating	0.25 volts (max)
29	Front right wheel speed sensor earth	Roadwheel rotating	0.25 volts (max)
30	Front left wheel speed sensor earth	Roadwheel rotating	0.25 volts (max)
31	Pump motor sensor earth	Ignition on	0.25 volts (approx)
32	Brake light switch input	Ignition on:	
		Brake pedal released	0 volts
		Brake pedal depressed	Nbv
33	Supply from relay	Ignition on	Nbv
34	Main/overvoltage protection relay driver	Ignition on, actuated	1.0 volt (max)
35	-	-	-
36	Rear outlet solenoid valve	Ignition on/inactive	Nbv
		Actuated	1.0 volt (max)
37	-	-	-
38	Front right inlet solenoid valve	Ignition on/inactive	Nbv
		Actuated	1.0 volt (max)
39	-	-	-
40	-	-	-
41	Pedal position sensor earth	Ignition on	0.25 volts (max)
42**	SD connector	-	-
43	-	-	-
44	-	-	-

Pin No.	Connection	Test condition	Voltage
45	Rear right wheel speed sensor signal	Roadwheel rotating	0.5 to 1.5 volts AC (approx)
46	Rear left wheel speed sensor signal	Roadwheel rotating	0.5 to 1.5 volts AC (approx)
47	Front right wheel speed sensor signal	Roadwheel rotating	0.5 to 1.5 volts AC (approx)
48	Front left wheel speed sensor signal	Roadwheel rotating	0.5 to 1.5 volts AC (approx)
49	Pump motor sensor signal	Pump motor actuated	0.5 volts AC
50	-	-	-
51	-	-	-
52	ABS warning lamp	Engine running:	
		Lamp on	2.0 volts (max)
		Lamp off	Nbv
53	Supply from ignition switch	Ignition on	Nbv
54	Rear inlet solenoid valve	Ignition on/inactive	Nbv
		Actuated	1.0 volt (max)
55	-	-	-

*Ford Granada/Scorpio and BMW only
**BMW only

Pin table - typical 55-pin (Ford Escort/Orion)

Note: *Refer to illustration 18C.29*

Pin No.	Connection	Test condition	Voltage
1	ECM earth	Ignition on	0.25 V (max)
2	-	-	-
3	Voltage monitoring, hydraulics	Ignition on	Nbv
4	-	-	-
5	-	-	-
6	-	-	-
7	-	-	-
8	-	-	-
9	-	-	-
10	-	-	-
11	-	-	-
12	-	-	-
13	-	-	-
14	-	-	-
15	Pump relay driver	Ignition on/inactive	Nbv
		Actuated	1.0 volt (max)
16	Pedal position sensor	Pedal at rest	1.0 volt
		Pedal fully depressed	3.0 volts
17	-	-	-
18	Front right outlet solenoid valve	Ignition on/inactive	Nbv
		Actuated	1.0 volt (max)
19	ECM earth	Ignition on	0.25 volts (max)
20	-	-	-
21	-	-	-
22	-	-	-
23	SD connector	-	-
24	-	-	-
25	-	-	-
26	-	-	-
27	-	-	-
28	-	-	-
29	Front left wheel speed sensor signal	Roadwheel rotating	0.5 to 1.5 volts AC (approx)
30	Front right wheel speed sensor signal	Roadwheel rotating	0.5 to 1.5 volts AC (approx)
31	Pump motor sensor earth	Ignition on	0.25 volts (approx)
32	Brake light switch input	Ignition on:	
		Brake pedal released	0 volts
		Brake pedal depressed	Nbv

Pin table - typical 55-pin (Ford Escort/Orion) (continued)

Note: *Refer to illustration 18C.29*

Pin No.	Connection	Test condition	Voltage
33	Supply from relay	Ignition on	Nbv
34	Overvoltage protection relay driver	Ignition on, actuated	1.0 volt (max)
35	-	-	-
36	Front left outlet solenoid valve	Ignition on/inactive	Nbv
		Actuated	1.0 volt (max)
37	-	-	-
38	-	-	-
39	-	-	-
40	-	-	-
41	Pedal position sensor earth	Ignition on	0.25 volts (max)
42	-	-	-
43	-	-	-
44	-	-	-
45	-	-	-
46	-	-	-
47	Front left wheel speed sensor earth	Roadwheel rotating	0.25 volts (max)
48	Front right wheel speed sensor earth	Roadwheel rotating	0.25 volts (max)
49	Pump motor sensor signal	Pump motor actuated	0.5 volts AC
50	-	-	-
51	-	-	-
52	ABS warning lamp	Engine running:	
		Lamp on	2.0 volts (max)
		Lamp off	Nbv
53	Supply from ignition switch	Ignition on	Nbv
54	Front left inlet solenoid valve	Ignition on/inactive	Nbv
		Actuated	1.0 volt (max)
55	Front right inlet solenoid valve	Ignition on/inactive	Nbv
		Actuated	1.0 volt (max)

Pin table - typical 55-pin (Ford Transit, Saab, Seat, Volkswagen)

Note: *Refer to illustration 18C.29*

Pin No.	Connection	Test condition	Voltage
1	ECM earth	Ignition on	0.25 V (max)
2	Front left outlet solenoid valve	Ignition on/inactive	Nbv
		Actuated	1.0 volt (max)
3	Voltage monitoring, hydraulics	Ignition on	Nbv
4	-	-	-
5	-	-	-
6	-	-	-
7	-	-	-
8	-	-	-
9	-	-	-
10	-	-	-
11	-	-	-
12	-	-	-
13	-	-	-
14	-	-	-
15	Pump relay driver	Ignition on/inactive	Nbv
		Actuated	1.0 volt (max)
16	Pedal position sensor	Pedal at rest	1.0 volt
		Pedal fully depressed	3.0 volts
17	-	-	-

Pin No.	Connection	Test condition	Voltage
18	Rear right outlet solenoid valve	Ignition on/inactive	Nbv
		Actuated	1.0 volt (max)
19	ECM earth	Ignition on	0.25 volts (max)
20	Front left inlet solenoid valve	Ignition on/inactive	Nbv
		Actuated	1.0 volt (max)
21	Front right outlet solenoid valve	Ignition on/inactive	Nbv
		Actuated	1.0 volt (max)
22	-	-	-
23*	SD connector	-	-
24	-	-	-
25	-	-	-
26	-	-	-
27**	Rear right wheel speed sensor earth	Roadwheel rotating	0.25 volts (max)
28**	Rear left wheel speed sensor earth	Roadwheel rotating	0.25 volts (max)
29**	Front right wheel speed sensor earth	Roadwheel rotating	0.25 volts (max)
30**	Front left wheel speed sensor earth	Roadwheel rotating	0.25 volts (max)
31	Pump motor sensor earth	Ignition on	0.25 volts (approx)
32*	Brake light switch input	Ignition on:	
		Brake pedal released	0 volts
		Brake pedal depressed	Nbv
33	Supply from relay	Ignition on	Nbv
34	Main relay driver	Ignition on, actuated	1.0 volt (max)
35	-	-	-
36	Rear left outlet solenoid valve	Ignition on/inactive	Nbv
		Actuated	1.0 volt (max)
37	-	-	-
38	Front right inlet solenoid valve	Ignition on/inactive	Nbv
		Actuated	1.0 volt (max)
39	-	-	-
40	-	-	-
41	Pedal position sensor earth	Ignition on	0.25 volts (max)
42	SD connector	-	-
43	-	-	-
44	-	-	-
45**	Rear right wheel speed sensor signal	Roadwheel rotating	0.5 to 1.5 volts AC (approx)
46**	Rear left wheel speed sensor signal	Roadwheel rotating	0.5 to 1.5 volts AC (approx)
47**	Front right wheel speed sensor signal	Roadwheel rotating	0.5 to 1.5 volts AC (approx)
48**	Front left wheel speed sensor signal	Roadwheel rotating	0.5 to 1.5 volts AC (approx)
49	Pump motor sensor signal	Pump motor actuated	0.5 volts AC
50	-	-	-
51	-	-	-
52	ABS warning lamp	Engine running:	
		Lamp on	2.0 volts (max)
		Lamp off	Nbv
53	Supply from ignition switch	Ignition on	Nbv
54	Rear left inlet solenoid valve	Ignition on/inactive	Nbv
		Actuated	1.0 volt (max)
55	Rear right inlet solenoid valve	Ignition on/inactive	Nbv
		Actuated	1.0 volt (max)

*Certain models only
**Refer to wiring diagrams for exact connections according to model

18C

Fault codes

12 General fault codes

1 The Teves Mk.IV, 55-pin system requires the use of a FCR for obtaining fault codes. With the exception of Volvo models, flash codes are not available for output from this system.

2 If a FCR is available, it should be connected to the SD serial connector and used in accordance with the maker's instructions.

3 The FCR can be used for the following purposes:
  a) Obtaining fault codes.
  b) Clearing fault codes.
  c) Obtaining datastream information.
  d) Testing the system actuators (relays and solenoid valves).

13 Volvo fault codes

1 On Volvo vehicles, it is also possible to obtain fault codes by means of the on-board diagnostic unit. The codes are displayed as a series of flashes of the red LED on the on-board diagnostic unit.

Obtaining codes using the on-board diagnostic unit

2 The diagnostic unit, situated in the front right-hand side of the engine compartment, consists of two modules mounted side by side, with a plastic cover over each. Remove the covers and note that the two modules are marked A and B, each having six numbered sockets on their top face.

3 Unclip the flylead from the holder on the side of the unit and insert it into socket 3 of module A (see illustration 18C.30). The three-digit codes will be displayed as a series of blinks of the red LED (located on the top face of the unit, next to the test button) with a slight pause between each digit.

4 With the flylead inserted, switch on the ignition. Press the test button on the top of the unit once, for about one second, then release it and wait for the LED to flash. As the LED flashes, copy down the fault code. Press the button again and copy down the next fault code, if there is one. Continue until the first fault code is displayed again indicating that all the stored codes have been accessed, then switch off the ignition.

5 If code 1-1-1 is obtained, this indicates that there are no fault codes stored in the diagnostic unit and the system is operating correctly.

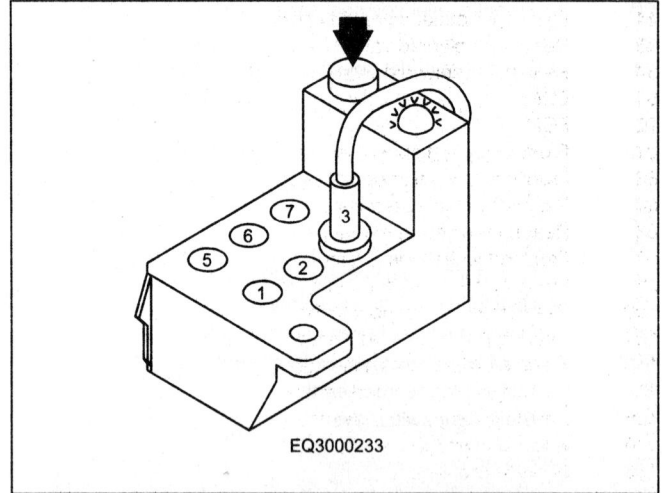

EQ3000233

18C.30 Volvo on-board diagnostic unit module A, showing test button (arrowed) and flylead sockets

Clearing fault codes from the ECM memory

6 Once all the fault codes have been recorded they should be deleted from the ECM memory using the diagnostic unit. Note that the codes cannot be deleted until all of them have been displayed at least once, and the first one is displayed again.

7 With the flylead still inserted in position 3 of module A, switch on the ignition, press the test button and hold it down for approximately five seconds. Release the test button and after three seconds the LED will light. When the LED lights, press and hold the test button down for a further five seconds then release it - the LED will go out.

8 Switch off the ignition and check that all the fault codes have been deleted by switching the ignition on again and pressing the test button for one second - code 1-1-1 should appear. If a code other than 1-1-1 appears, record the code then repeat the deleting procedure. When all the codes have been deleted, switch off the ignition, locate the flylead in its holder and refit the unit covers.

Fault code table (BMW)

Code	Item	Fault
17	Front left inlet solenoid valve	Open or short circuit, relay inoperative
18	Front left outlet solenoid valve	Open or short circuit, relay inoperative
20	Front right inlet solenoid valve	Open or short circuit, relay inoperative
24	Front right outlet solenoid valve	Open or short circuit, relay inoperative
33	Rear inlet solenoid valve	Open or short circuit, relay inoperative
34	Rear outlet solenoid valve	Open or short circuit, relay inoperative
68	ECM	Processor fault
72	ECM	Internal fault
81	Front left wheel speed sensor	Open or short circuit
82	Front right wheel speed sensor	Open or short circuit
84	Rear left wheel speed sensor	Open or short circuit
88	Rear right wheel speed sensor	Open or short circuit
97	Front left wheel speed sensor	Implausible signal
98	Front right wheel speed sensor	Implausible signal
100	Rear left wheel speed sensor	Implausible signal
104	Rear right wheel speed sensor	Implausible signal
113	Front left wheel speed sensor	No signal, intermittent signal
114	Front right wheel speed sensor	No signal, intermittent signal
116	Rear left wheel speed sensor	No signal, intermittent signal
120	Rear right wheel speed sensor	No signal, intermittent signal
136	Brake hydraulic system	Hydraulic fault
145	Hydraulic pump	Inoperative
161	Front left outlet solenoid valve	Inoperative or leaking
162	Front right outlet solenoid valve	Inoperative or leaking
164	Rear outlet solenoid valve	Inoperative or leaking
255	ECM	Memory access fault

Fault code table (Ford)

Note: *Not all codes applicable to all models.*

Code	Item	Fault
00	ECM	Power supply failure
10	Ignore if obtained	-
11	Electrical interference	Poor earth or wiring connections
12	ECM	Defective
17	Main relay	Incorrect voltage, defective wiring and connections
22	Front left inlet solenoid	Poor connections, defective wiring or solenoid coil
23	Front left outlet solenoid	Poor connections, defective wiring or solenoid coil
24	Front right inlet solenoid	Poor connections, defective wiring or solenoid coil
25	Front right outlet solenoid	Poor connections, defective wiring or solenoid coil
26	Rear/rear right inlet solenoid	Poor connections, defective wiring or solenoid coil
27	Rear/rear right outlet solenoid	Poor connections, defective wiring or solenoid coil
31	Front left wheel speed sensor	Wiring open/short circuit
32	Front right wheel speed sensor	Wiring open/short circuit
33	Rear right wheel speed sensor	Wiring open/short circuit
34	Rear left wheel speed sensor	Wiring open/short circuit
35	Front left wheel speed sensor signal variation	Incorrect air gap, damaged rotor teeth, defective wiring
36	Front right wheel speed sensor signal variation	Incorrect air gap, damaged rotor teeth, defective wiring
37	Rear right wheel speed sensor signal	variation Incorrect air gap, damaged rotor teeth, defective wiring
38	Rear left wheel speed sensor signal	variation Incorrect air gap, damaged rotor teeth, defective wiring
41	Front left wheel speed sensor intermittent signal loss	Sensor/rotor damaged, incorrect air gap, defective wiring
42	Front right wheel speed sensor intermittent signal loss	Sensor/rotor damaged, incorrect air gap, defective wiring
43	Rear right wheel speed sensor intermittent signal loss	Sensor/rotor damaged, incorrect air gap, defective wiring
44	Rear left wheel speed sensor intermittent signal loss	Sensor/rotor damaged, incorrect air gap, defective wiring
55	Front left wheel speed sensor permanent signal loss	Sensor/rotor damaged, incorrect air gap, defective wiring
56	Front right wheel speed sensor permanent signal loss	Sensor/rotor damaged, incorrect air gap, defective wiring
57	Rear right wheel speed sensor permanent signal loss	Sensor/rotor damaged, incorrect air gap, defective wiring
58	Rear left wheel speed sensor permanent signal loss	Sensor/rotor damaged, incorrect air gap, defective wiring
75	Front left wheel speed sensor signal interference	Incorrect installation, wheel bearings, sensor/rotor damaged
76	Front right wheel speed sensor signal interference	Incorrect installation, wheel bearings, sensor/rotor damaged
77	Rear right wheel speed sensor signal interference	Incorrect installation, wheel bearings, sensor/rotor damaged
78	Rear left wheel speed sensor signal interference	Incorrect installation, wheel bearings, sensor/rotor damaged
79	Ignore if obtained	-

18C

Fault code table (Saab)

Code	Item	Fault
775B1	ECM earth	Poor main connection
775B2	ECM	Memory fault
42251	Main relay	Incorrect supply voltage, defective wiring and connections
44221	Front left wheel speed sensor signal loss	Sensor/rotor damaged, incorrect air gap, defective wiring
44222	Front right wheel speed sensor signal loss	Sensor/rotor damaged, incorrect air gap, defective wiring
44223	Rear left wheel speed sensor signal loss	Sensor/rotor damaged, incorrect air gap, defective wiring
44224	Rear right wheel speed sensor signal loss	Sensor/rotor damaged, incorrect air gap, defective wiring
2422A	Front left wheel speed sensor	Intermittent signal (over 25 mph)
2422B	Front right wheel speed sensor	Intermittent signal (over 25 mph)
2422C	Rear left wheel speed sensor	Intermittent signal (over 25 mph)
2422D	Rear right wheel speed sensor	Intermittent signal (over 25 mph)
24291	Front left wheel speed sensor	Intermittent signal (under 25 mph)
24292	Front right wheel speed sensor	Intermittent signal (under 25 mph)
24293	Rear left wheel speed sensor	Intermittent signal (under 25 mph)
24294	Rear right wheel speed sensor	Intermittent signal (under 25 mph)
24251	Front left wheel speed sensor	No signal (under 25 mph)
24252	Front right wheel speed sensor	No signal (under 25 mph)
24253	Rear left wheel speed sensor	No signal (under 25 mph)
24254	Rear right wheel speed sensor	No signal (under 25 mph)
53421	Front left inlet solenoid valve	Open or short circuit
53422	Front left outlet solenoid valve	Open or short circuit
53423	Front right inlet solenoid valve	Open or short circuit
53424	Front right outlet solenoid valve	Open or short circuit
53425	Rear left inlet solenoid valve	Open or short circuit
53426	Rear left outlet solenoid valve	Open or short circuit
53427	Rear right inlet solenoid valve	Open or short circuit
53428	Rear right outlet solenoid valve	Open or short circuit
334B1	Front left outlet solenoid valve	Hydraulic fault
334B2	Front right outlet solenoid valve	Hydraulic fault
334B3	Rear left outlet solenoid valve	Hydraulic fault
334B4	Rear right outlet solenoid valve	Hydraulic fault
45721	Brake pedal position sensor	Sensor faulty, wiring open or short circuit
24971	Pump motor inoperative	Faulty relay, wiring or connections
44792	Pump runs continuously	Faulty relay, pedal position sensor, pump motor or wiring
E75B1	Hydraulic system fault	Faulty master cylinder, brake pads/hoses, low fluid level

Fault code table (Seat, Volkswagen)

Code	Item	Fault
0000	End of fault code output	
1111	ECM	Poor earth or wiring connections, incorrect voltage, ECM defective
1112	Front left inlet solenoid valve	Poor connections, defective wiring or solenoid coil
1114	Front right inlet solenoid valve	Poor connections, defective wiring or solenoid coil
1132	Front left outlet solenoid valve	Poor connections, defective wiring or solenoid coil
1134	Front right outlet solenoid valve	Poor connections, defective wiring or solenoid coil
1211	Rear right inlet solenoid valve	Poor connections, defective wiring or solenoid coil
1212	Rear left inlet solenoid valve	Poor connections, defective wiring or solenoid coil
1213	Rear right outlet solenoid valve	Poor connections, defective wiring or solenoid coil
1214	Rear left outlet solenoid valve	Poor connections, defective wiring or solenoid coil
1223	Differential lock solenoid valve 1	Poor connections, defective wiring or solenoid coil
1224	Differential lock solenoid valve 2	Poor connections, defective wiring or solenoid coil
1233	Front left wheel speed sensor	Poor wiring connection; incorrect air gap; sensor or sensor ring dirty or damaged
1241	Front right wheel speed sensor	Poor wiring connection; incorrect air gap; sensor or sensor ring dirty or damaged
1243	Rear right wheel speed sensor	Poor wiring connection; incorrect air gap; sensor or sensor ring dirty or damaged
1311	Rear left wheel speed sensor	Poor wiring connection; incorrect air gap; sensor or sensor ring dirty or damaged
1313	Hydraulic supply pressure	Hydraulic leakage, defective pump
2234	Supply voltage	Incorrect voltage; poor connections, defective wiring, blown fuses
3231	Pedal position sensor	Poor wiring connection, sensor faulty
4133	Hydraulic pump	Poor wiring/relay connections, pump sensor faulty, pump defective
4444	No faults recognised	

Fault code table (Volvo)

Code	Item	Fault
1-1-1	No fault detected	
1-2-1	Front left wheel speed sensor	Faulty signal (below 25 mph)
1-2-2	Front right wheel speed sensor	Faulty signal (below 25 mph)
1-2-3	Rear left wheel speed sensor	Faulty signal (below 25 mph)
1-2-4	Rear right wheel speed sensor	Faulty signal (below 25 mph)
1-4-1	Brake pedal position sensor	Shorted to earth or supply
1-4-2	Brake light switch	Open or short circuit
1-4-3	ABS ECM	Memory fault
2-1-1	Front left wheel speed sensor	No signal when moving off
2-1-2	Front right wheel speed sensor	No signal when moving off
2-1-3	Rear left wheel speed sensor	No signal when moving off
2-1-4	Rear right wheel speed sensor	No signal when moving off
2-2-1	Front left wheel speed sensor	No signal
2-2-2	Front right wheel speed sensor	No signal
2-2-3	Rear left wheel speed sensor	No signal
2-2-4	Rear right wheel speed sensor	No signal
3-1-1	Front left wheel speed sensor	Open or short circuit
3-1-2	Front right wheel speed sensor	Open or short circuit
3-1-3	Rear left wheel speed sensor	Open or short circuit
3-1-4	Rear right wheel speed sensor	Open or short circuit
3-2-1	Front left wheel speed sensor	Intermittent signal (over 25 mph)
3-2-2	Front right wheel speed sensor	Intermittent signal (over 25 mph)
3-2-3	Rear left wheel speed sensor	Intermittent signal (over 25 mph)
3-2-4	Rear right wheel speed sensor	Intermittent signal (over 25 mph)
4-1-1	Front left inlet solenoid valve	Open or short circuit
4-1-2	Front left outlet solenoid valve	Open or short circuit
4-1-3	Front right inlet solenoid valve	Open or short circuit
4-1-4	Front right outlet solenoid valve	Open or short circuit
4-2-1	Rear left inlet solenoid valve	Open or short circuit
4-2-2	Rear left outlet solenoid valve	Open or short circuit
4-2-3	Traction control valve (where fitted)	Open or short circuit
4-2-4	Traction control pressure sensor (where fitted)	Faulty or short circuit
4-3-1	Rear right inlet solenoid valve	Open or short circuit
4-3-2	Rear right outlet solenoid valve	Open or short circuit
4-4-1	ABS ECM	Processing fault
4-4-2	Hydraulic control unit	Pump pressure low
4-4-3	Hydraulic pump motor	Electrical or mechanical fault
4-4-4	Solenoid valves	Supply voltage absent

Wiring diagrams

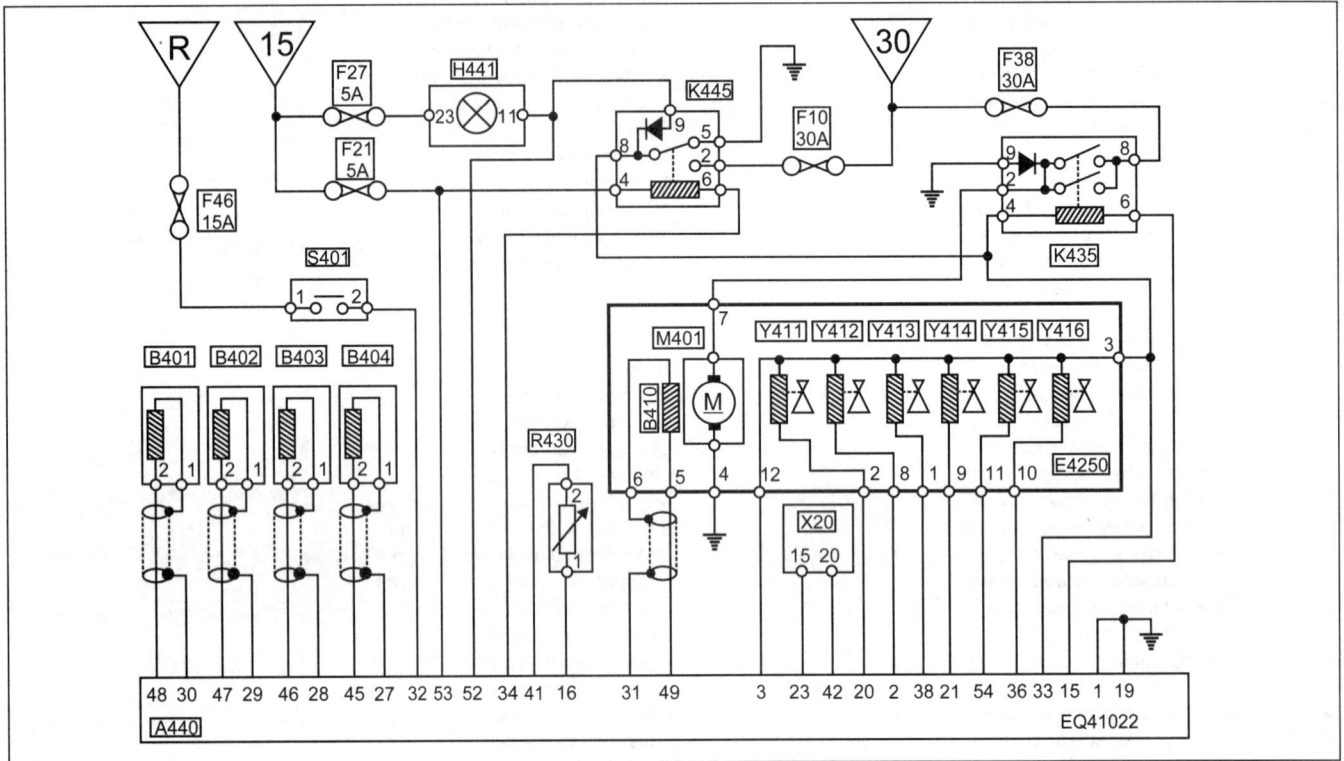

18C.31 55-pin wiring diagram, BMW 3-Series (E36)

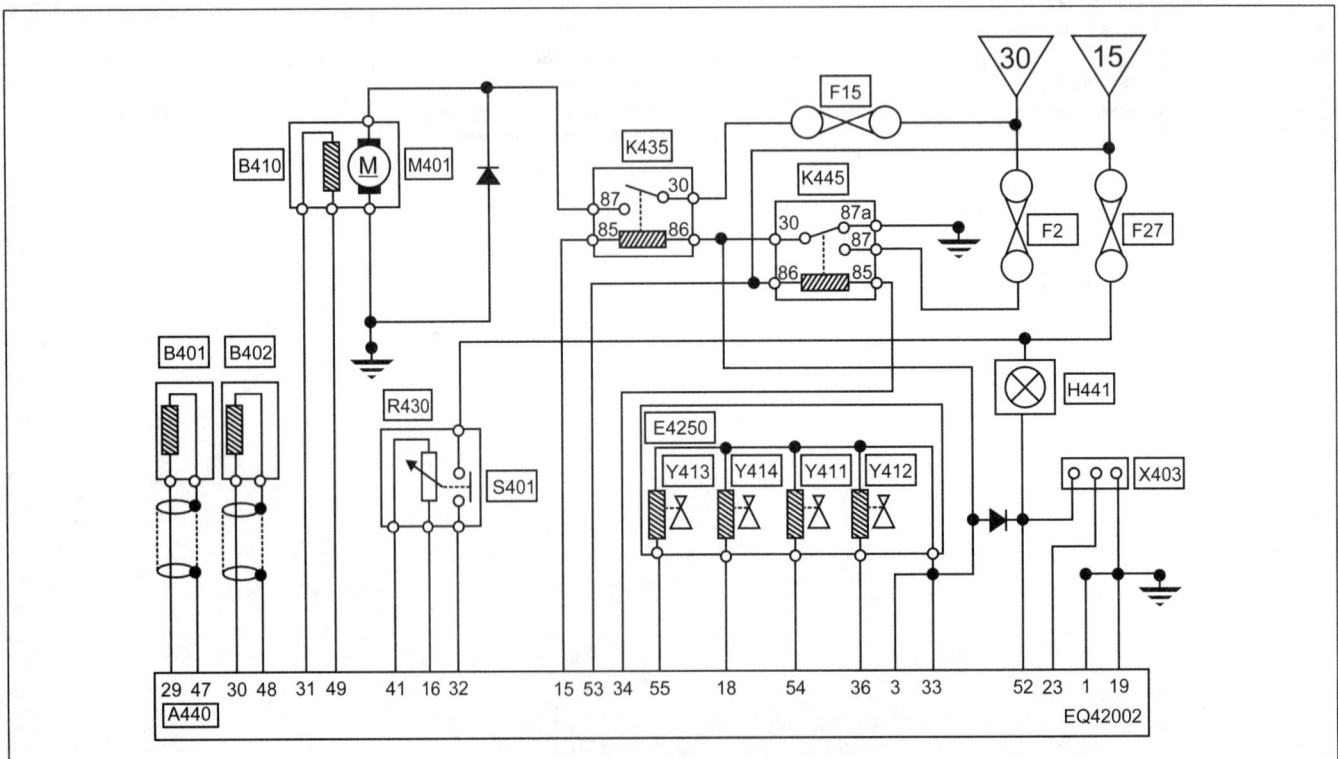

18C.32 55-pin wiring diagram, Ford Escort/Orion

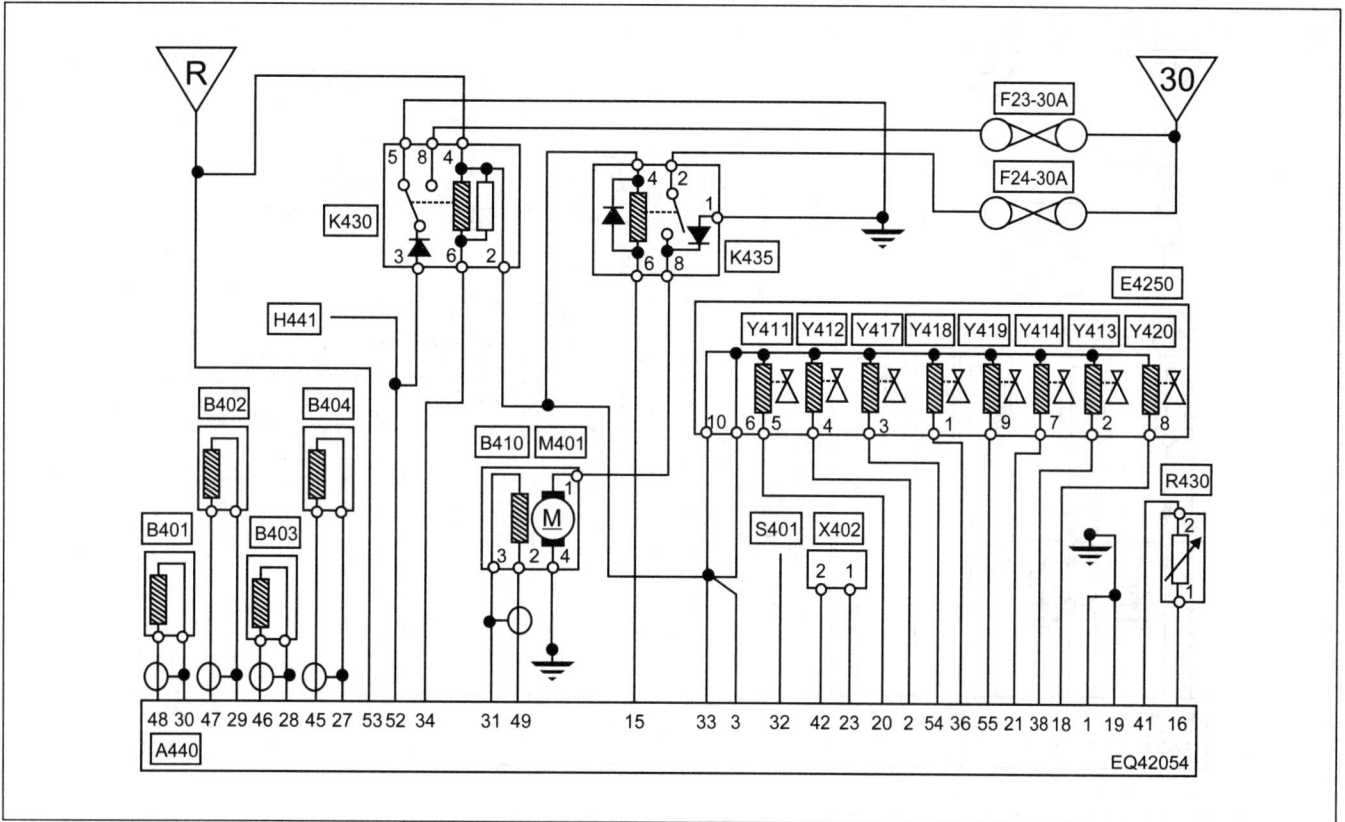

18C.33 55-pin wiring diagram, Saab 9000

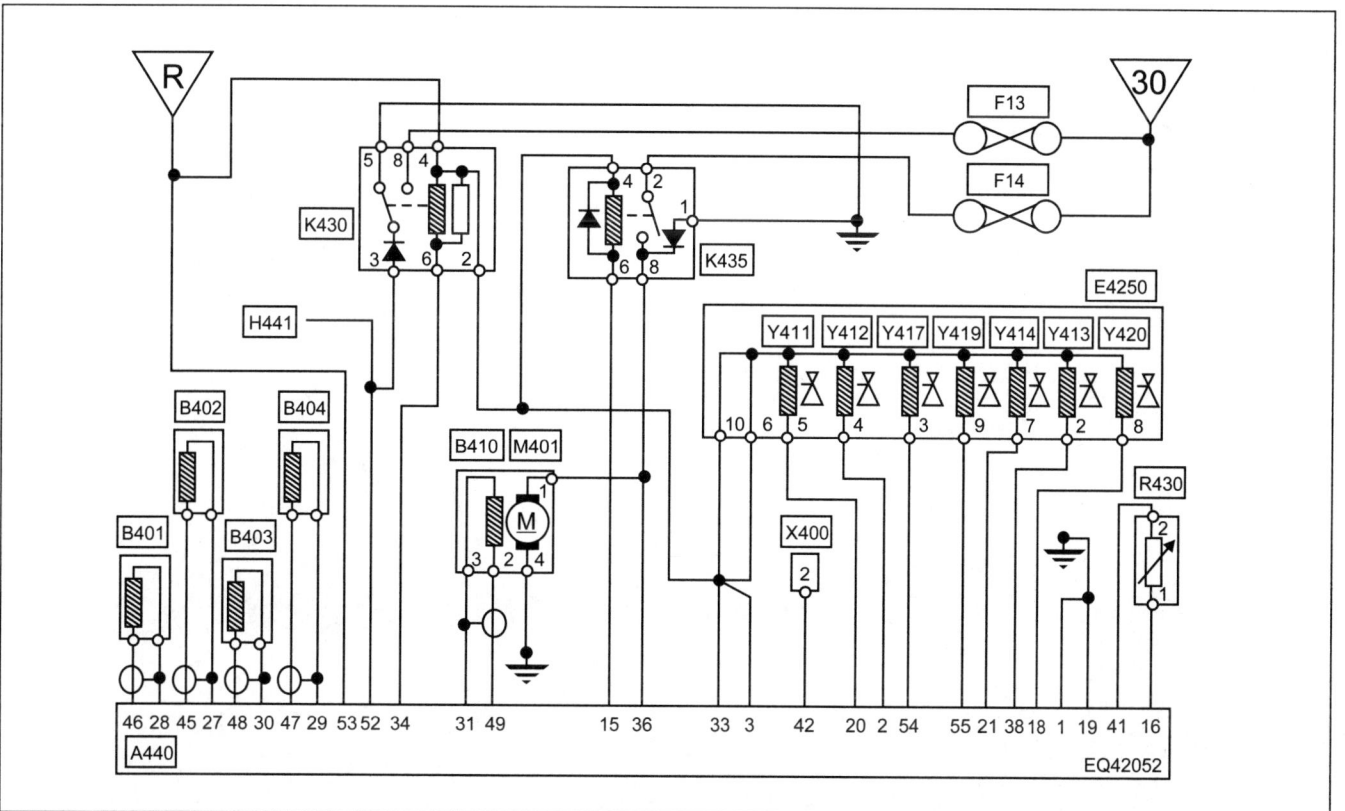

18C.34 55-pin wiring diagram, Seat Cordoba/Ibiza

18C

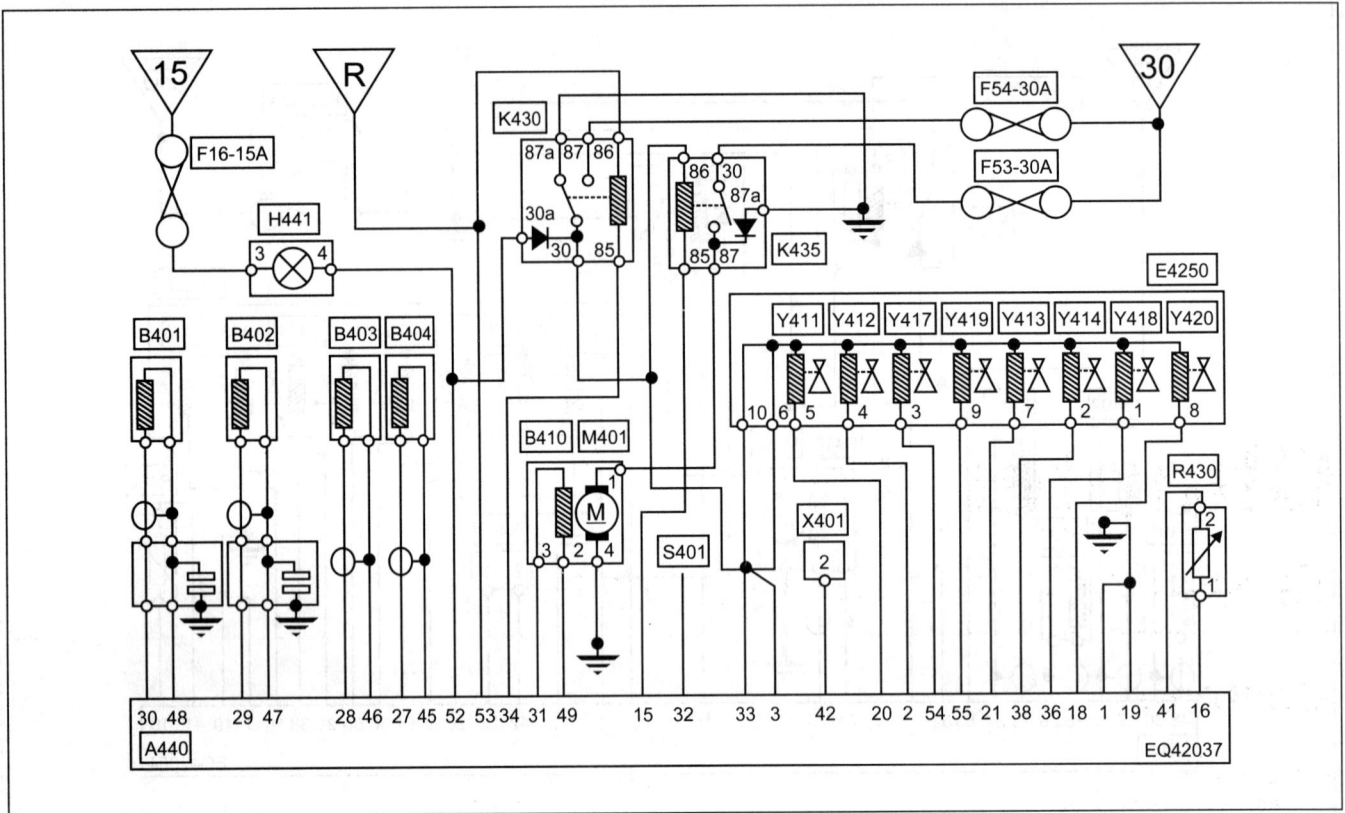

18C.35 55-pin wiring diagram, Volkswagen Corrado/Golf/Passat/Vento

Chapter 19

Teves IV-GI (ITT 04-GI) ABS

Contents

Vehicle coverage

Model	Year
Citroën	
Saxo	1996-1998
Peugeot	
106	1996-2000
Ford	
Galaxy	1995-1997
Scorpio	1995-1997
Renault	
Megane	1995-1998
Seat	
Alhambra	1996-2000
Volkswagen	
Polo	1994-1997
Sharan	1995-1997

19

Overview of system operation

1 Basic principles and system identification

The Teves IV-GI Antilock Brake System is a development of the earlier Teves Mk.IV series ABS, with significant improvements to the hydraulic control unit, and the Electronic Control Module (ECM) software. Teves IV-GI has been fitted to a wide range of passenger vehicles since its introduction in the mid 1990s. The system is of the additional or 'add-on' type operating in conjunction with the conventional braking system components. Later versions of the system may carry the designation ITT 04-GI although their operation is unchanged from the Teves IV-GI types.

Depending on application, Teves IV-GI may be installed as an antilock braking system only, or as an antilock braking system incorporating traction control.

In ABS mode, the purpose of the system is to apply the vehicle brakes at maximum efficiency without wheel lock or loss of directional stability. Inductive sensors (wheel speed sensors) monitor the speed of the wheels by generating an electrical signal as the wheel is rotated. This information is passed to the ABS Electronic Control Module (ECM) which is then able to determine wheel speed, wheel acceleration and wheel deceleration. The ECM compares the signals received from each wheel and if the onset of lock at any wheel is detected, a signal is sent to the ABS hydraulic control unit which regulates the brake pressure for the relevant wheel(s).

Where the system incorporates traction control, essentially the reverse principle is applied. When the ECM detects that one or more wheels are rotating faster than the reference value, the brake is actually applied on the relevant wheel(s) to reduce the rotational speed. Additionally, when wheel spin is detected, various signals are sent to the engine management ECM to control engine torque and rpm.

Typically, Teves IV-GI ABS is comprised of the following components (see illustration 19.1):

a) Hydraulic control unit with integral ABS-ECM.
b) Four inductive wheel speed sensors and associated sensor rings.
c) Brake light switch.
d) ABS warning lamp.
e) Diagnostic connector.

In addition, the conventional brake system is comprised of the following components:

a) Tandem brake master cylinder.
b) Vacuum servo unit.
c) Brake calipers/wheel cylinders and hydraulic hoses and pipes.
d) Pressure regulating/load sensing valve(s) depending on application.

Although the hydraulic operation of the Teves IV-GI system is essentially the same for all models, there are considerable differences in the ECM software and pin connections, and in the construction of the hydraulic control unit. For the purposes of the test procedures described later in this Chapter, Teves IV-GI can be divided into two distinct groups; 25-pin ECM types (all except Ford Scorpio), and 28-pin ECM types (Ford Scorpio).

2 Component description and operation

ABS ECM

General

The Teves IV-GI Electronic Control Module (ECM) continually monitors wheel speed from the signals provided by the wheel speed sensors, and brake application from the brake light switch signal. If the ECM detects the incidence of wheel lock on one or more wheels, a signal is sent to the hydraulic control unit to modulate the hydraulic pressure to the brake of the locking wheel(s). The ECM contains two microprocessors and uses digital technology to complete this function and other functions such as, fault code memory and power modules for valve and pump activation (see illustration 19.2).

To reduce external electrical connections to a minimum and

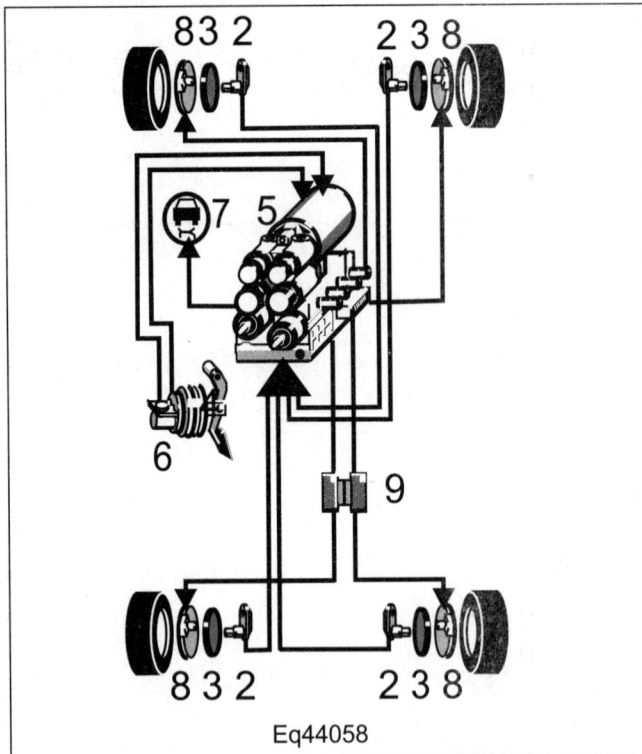

Eq44058

19.1 Typical Teves IV-GI main components

2	Wheel speed sensors
3	Sensor rings
5	Hydraulic control unit with integral ECM
6	Tandem master cylinder and vacuum servo unit
7	ABS warning lamp
8	Brake calipers/wheel cylinders and hydraulic hoses and pipes
9	Pressure regulating/load sensing valve(s)

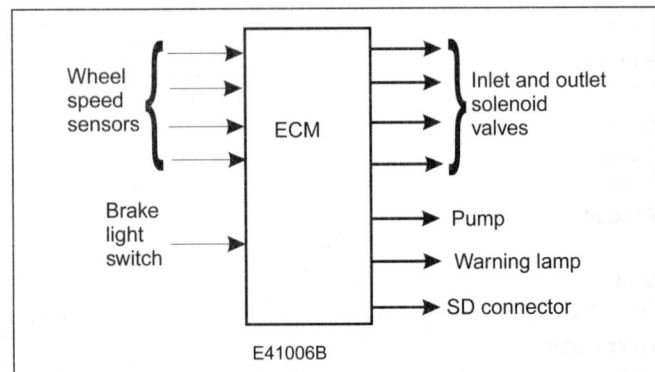

E41006B

19.2 ECM sensor inputs and control signal outputs

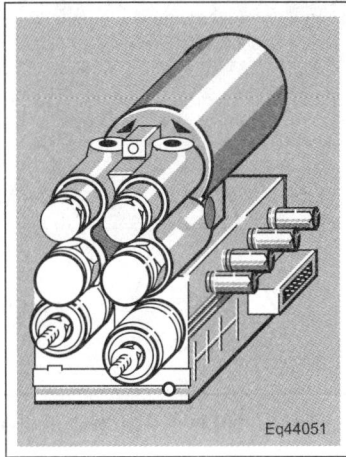

19.3 Teves IV-GI hydraulic control unit and integral ECM

improve reliability, the ECM is integral with the hydraulic control unit (see illustration 19.3).

Self-test

The Teves IV-GI ECM is equipped with a self-test capability that initially examines the ABS system when the ignition is switched on, and then examines the wheel speed sensor signals after a wheel speed of approximately 4 mph is reached from all wheels (engine running). The ABS self-test program continues to examine the signals from the various components as long as the ignition is switched on. If self-test determines that faults are not present, the ABS is ready for operation once a specified vehicle speed has been achieved.

If the ECM detects that a fault is present, all ABS functions are switched off and the warning lamp is turned on. The conventional braking system continues to operate as normal without ABS assistance.

Self-diagnostics

If the ECM detects a fault during the self-test routine, an internal fault code is stored in the ECM memory. Stored fault codes can be retrieved from the SD connector with the aid of a suitable fault code reader. If the fault clears, the code will remain stored until cleared with the FCR.

Hydraulic control unit

Teves IV-GI is typically a four-channel system with a separate hydraulic circuit for each brake. On certain models, however, the system is installed in three-channel configuration with a separate hydraulic circuit for each front brake, but with the rear brakes controlled as a pair.

The hydraulic control unit consists of an electric motor operating a return pump with eccentric drive and twin radial pistons, inlet and outlet solenoid valves, pressure accumulators and pulsation dampers (see illustration 19.4). The unit controls the hydraulic pressure applied to the brake for each individual front wheel and each individual rear wheel, or pair of wheels. The return pump is switched on when the ABS is activated and returns hydraulic fluid, drained off during the pressure reduction phase, back into the brake circuit.

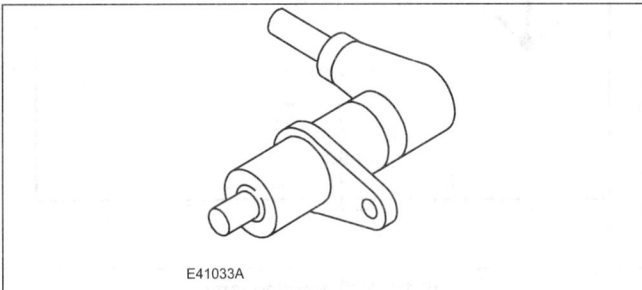

19.5 Typical wheel speed sensor

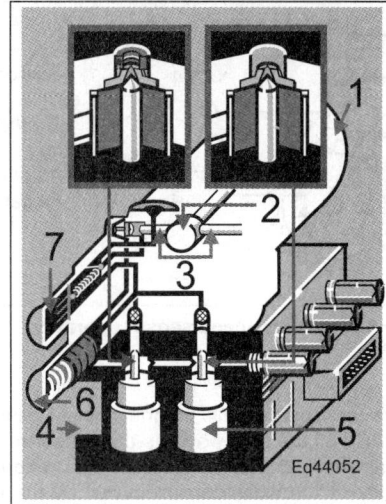

19.4 Teves IV-GI hydraulic control unit components

1	Electric motor	5	Outlet solenoid valve
2	Return pump	6	Pressure accumulator
3	Return pump radial pistons	7	Pulsation damper
4	Inlet solenoid valve		

On some applications, the 'select-low' principle is employed for control of the rear brakes. With the 'select-low' principle, the wheel with the lowest adhesion determines the amount of hydraulic pressure to be supplied to both rear brakes during ABS operation.

On certain versions, the ABS-ECM contains additional software for rear brake hydraulic fluid pressure regulation during normal braking, generally known as 'Electronic Brake Force Distribution'. In these applications, mechanical pressure regulating/load sensing valves are not required as the rear brake hydraulic pressure is controlled by the ABS hydraulic control unit.

Wheel speed sensors

The rotational speed of the roadwheels and any changes in the rotational speed are recorded by inductive wheel speed sensors located at the roadwheels. Depending on system type, there may be a separate sensor for each front wheel only, or a separate sensor for all four wheels (see illustration 19.5).

Each wheel speed sensor assembly comprises a toothed sensor ring which rotates at roadwheel speed, and an adjacent sensor mounted a set distance from the sensor ring (see illustration 19.6).

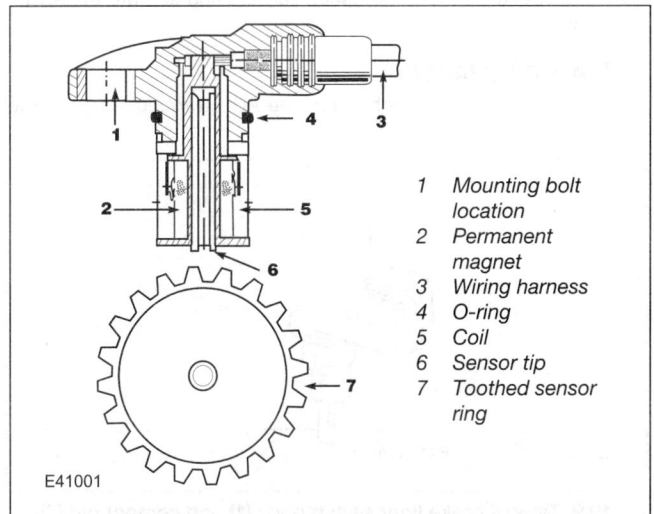

1 Mounting bolt location
2 Permanent magnet
3 Wiring harness
4 O-ring
5 Coil
6 Sensor tip
7 Toothed sensor ring

19.6 Sectional view of a wheel speed sensor

19

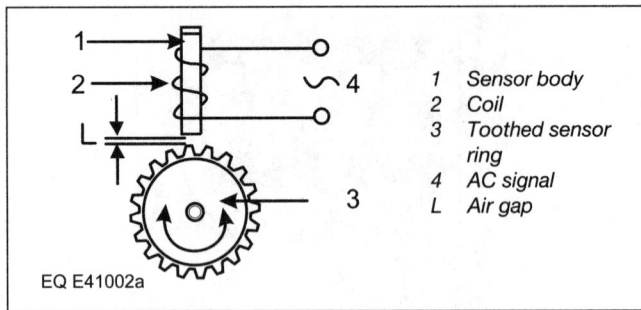

1 Sensor body
2 Coil
3 Toothed sensor ring
4 AC signal
L Air gap

EQ E41002a

19.7 Wheel speed sensor operation

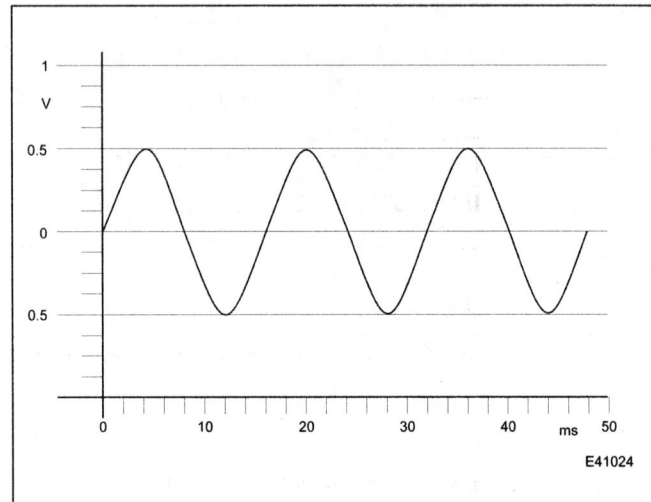

E41024

19.8 Wheel speed sensor waveform as viewed on an oscilloscope

The sensors are permanent magnet pulse generator types producing an AC voltage sine wave as the sensor ring teeth pass through the magnetic field of the sensor (see illustration 19.7).

The frequency of the waveform produced by the wheel speed sensor is proportional to the road speed. This AC voltage signal is continually being delivered to the ABS-ECM for processing.

The peak to peak voltage of the speed signal (when viewed upon an oscilloscope) can vary considerably according to wheel speed. An analogue to digital converter (ADC) in the ECM transforms the AC pulse into a digital signal (see illustration 19.8).

ABS electrical wiring harness and relays

An integrated main wiring harness is used to connect power and earth to various electrical components and to enable sensor signals to reach the ECM. The ECM, in turn, sends command signals to the hydraulic control unit via internal connections between the two units. The vehicle relays and ABS fuses are typically located in the engine compartment fuse/relay box or adjacent to the battery.

Nominal battery voltage is supplied to the ECM via the ABS main fuses. The ECM then supplies nominal battery voltage to the solenoid valves and return pump and controls the earth for activation of the solenoid valves.

Brake light switch

The brake light switch comprises a switch body and contact pin and is located above the brake pedal (see illustration 19.9).

When the brake pedal is depressed, closing the brake light switch, a signal is sent to the ECM indicating that the brakes are being applied. Once this signal is received, the ECM will begin monitoring the wheel speed via the wheel speed sensors and activate the ABS if necessary.

ABS warning lamp

After the ignition is switched on, the ABS warning lamp on the instrument panel is illuminated for approximately 2 to 4 seconds as the system executes a self-test routine. If satisfactory operation of the system is detected by the ECM, the light is extinguished. During vehicle operation above a pre-determined wheel speed, the ABS-ECM implements a further self-test cycle whereby ABS operation and wheel speed sensor signals are continually monitored. If a fault is detected, the relevant ECM pin is earthed to illuminate the warning lamp on the instrument panel, and the ABS function is disabled. The warning lamp will remain illuminated until the fault is no longer present (see illustration 19.10).

When the ABS-ECM detects a fault, the fault code is stored and the ABS warning lamp activated. If the fault no longer exists after the next system start (ignition on/off) the ABS warning lamp is extinguished after the self-test cycle, however the fault code remains stored in the ECM memory.

Tandem master cylinder

Typically, the tandem master cylinder comprises a body casting incorporating primary and secondary pressure chambers, primary piston, intermediate piston, floating piston, slotted pin and central valve. The cylinder operates as a conventional master cylinder using

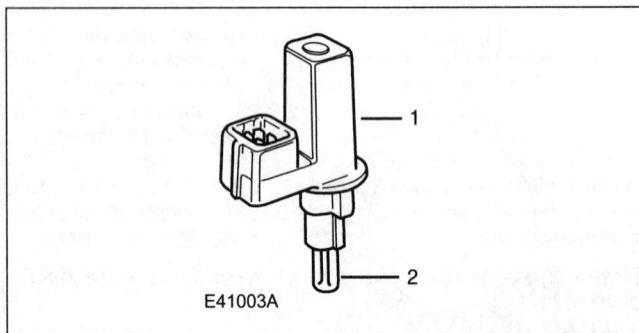

E41003A

19.9 Typical brake light switch body (1) and contact pin (2)

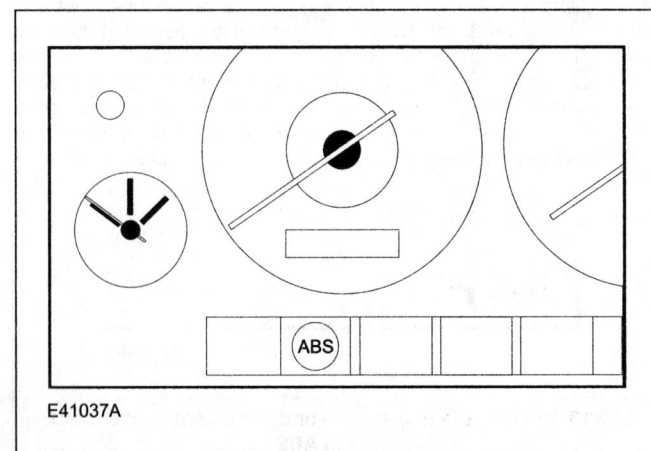

E41037A

19.10 ABS warning lamp

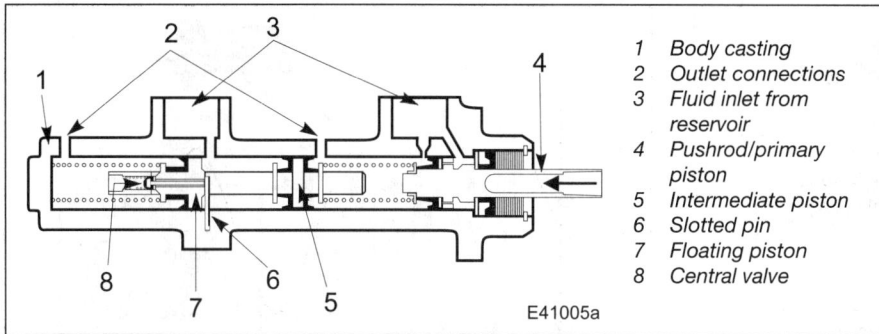

1	Body casting
2	Outlet connections
3	Fluid inlet from reservoir
4	Pushrod/primary piston
5	Intermediate piston
6	Slotted pin
7	Floating piston
8	Central valve

E41005a

19.11 Sectional view of a typical tandem master cylinder

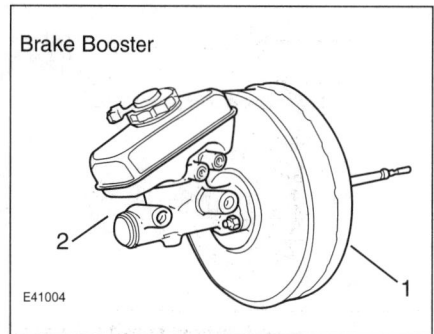

Brake Booster

E41004

19.12 Vacuum servo unit (1) and tandem master cylinder (2)

vacuum assistance from the vacuum servo unit (see illustration 19.11).

When the brake system is at rest, the central valve in the floating piston rests against the slotted pin. In this condition the central valve is open and brake fluid can discharge out of the pressure chamber back into the brake fluid reservoir. When the brake pedal is depressed, the build-up of hydraulic pressure in the primary pressure chamber acts on the intermediate piston and floating piston, moving them down the cylinder bore. The floating piston contacts the seal on the central valve, closing the connection between the intermediate and secondary pressure chambers. Brake hydraulic pressure can now also increase in the secondary pressure chamber.

Vacuum servo unit

The vacuum servo unit is located between the brake pedal and tandem master cylinder. When the brake pedal is depressed, the servo unit increases the force applied by the pedal, reducing the effort required to operate the brakes (see illustration 19.12).

The unit is operated by vacuum created in the engine inlet manifold (or from a separate vacuum pump on diesel engines) which is applied to a diaphragm contained within the unit casing. A pushrod connected to the centre of the diaphragm acts directly on the primary piston in the master cylinder.

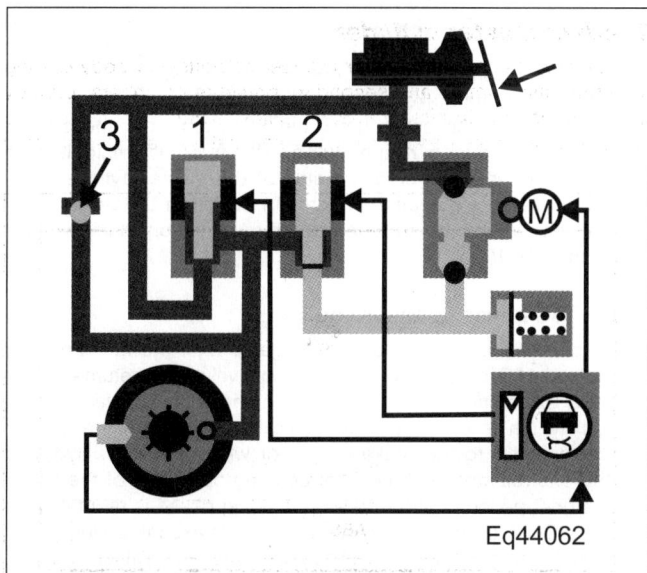

Eq44062

19.13 Brake system operating under conventional control without ABS

1 Inlet solenoid valve
2 Outlet solenoid valve
3 One-way valve

When the brake pedal is released, vacuum is applied to both sides of the diaphragm. When the pedal is depressed, one side of the diaphragm is opened to atmosphere and the vacuum acting on the other side deflects the diaphragm which in turn operates the master cylinder primary piston. The resulting force applied to the master cylinder piston is therefore significantly greater than the initial force applied to the brake pedal by the driver.

Pressure regulating/load sensing valve(s)

Depending on vehicle application, pressure regulating valves or load sensing valves may be incorporated to restrict the hydraulic fluid pressure to the rear brakes. The valves may be pressure conscious whereby the hydraulic fluid supply is restricted once a pre-determined pressure is reached, or load conscious whereby the hydraulic pressure is reduced according to vehicle loading.

On certain versions, the ABS-ECM contains additional software for rear brake hydraulic fluid pressure regulation during normal braking, generally known as 'Electronic Brake Force Distribution'. In these applications, mechanical pressure regulating/load sensing valves are not required as the rear brake hydraulic pressure is controlled by the ABS hydraulic control unit.

3 System operation

Brake system at rest

When the system is at rest all the brake components are inoperative. Pressure is non-existent in the hydraulic pipes between the tandem master cylinder and the brake calipers. The inlet solenoid valves in the hydraulic control unit are open and the outlet solenoid valves are closed.

Brake system operating under conventional control without ABS

When the brake pedal is activated, the pedal force is applied to the tandem master cylinder by the vacuum servo unit pushrod. The servo unit pushrod acts directly on the pressure piston in the master cylinder which pressurises the hydraulic fluid in the brake pipes to the hydraulic control unit. The inlet solenoid valve and outlet solenoid valve both remain in the 'at rest' position (inlet solenoid valve open and outlet solenoid valve closed). Hydraulic pressure is transmitted to each brake caliper, thus operating the brakes.

When the brake pedal is released, the one-way valve opens allowing the hydraulic pressure in the circuit to rapidly decrease (see illustration 19.13).

Brake system operating in conjunction with ABS control

The ABS-ECM continually monitors wheel speed from the signals provided by the wheel speed sensors. If the ECM detects the incidence of wheel lock on one or more wheels, ABS is automatically

19

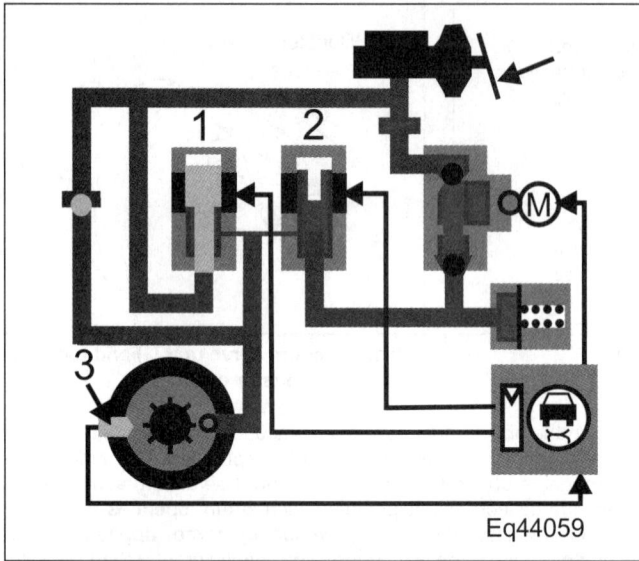

Eq44059

19.14 ABS operation - first phase, pressure holding

1	Inlet solenoid valve	3	Wheel speed sensor
2	Outlet solenoid valve		

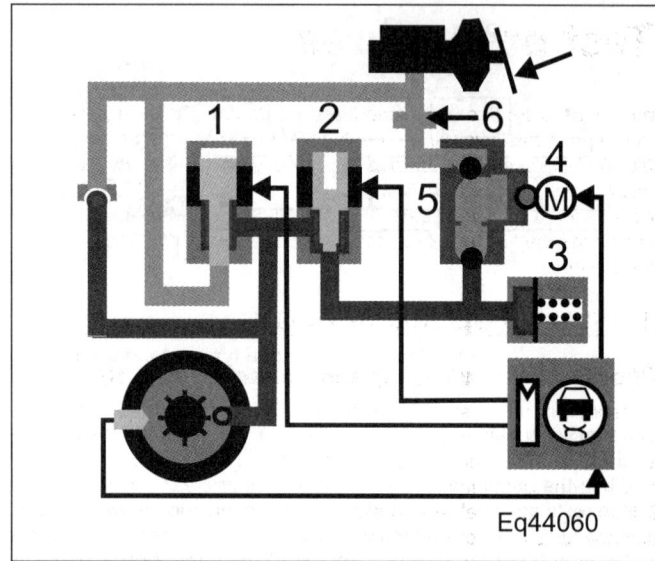

Eq44060

19.15 ABS operation - second phase, pressure reduction

1	Inlet solenoid valve	4	Pump motor
2	Outlet solenoid valve	5	Return pump
3	Pressure accumulator	6	Pulsation damper

initiated in three phases. As Teves IV-GI ABS typically operates individually on each wheel, all or any of the wheels could be in any one of the following phases at any particular moment.

First ABS phase, pressure holding

To prevent any further build-up of hydraulic pressure in the circuit being controlled, the ECM closes the inlet solenoid valve and allows the outlet solenoid valve to remain closed. The hydraulic fluid line from the tandem master cylinder to the brake caliper or wheel cylinder is closed, and the hydraulic fluid in the controlled circuit is maintained at a constant pressure. This effectively removes the braking force from

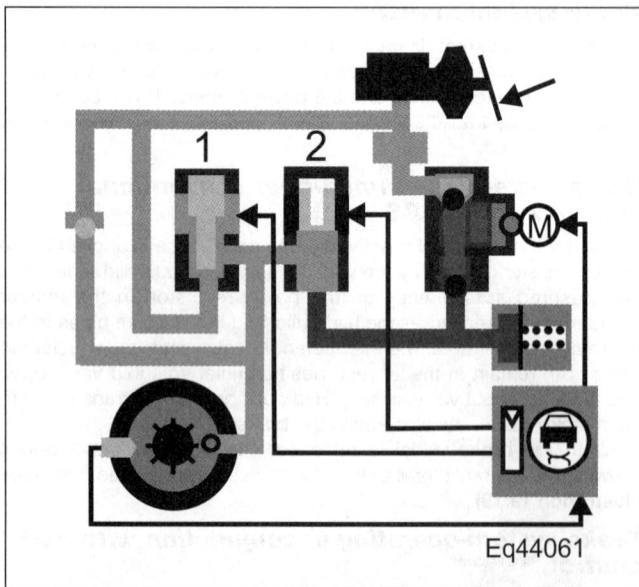

Eq44061

19.16 ABS operation - third phase, pressure build-up

1	Inlet solenoid valve	2	Outlet solenoid valve

the controlled circuit. The pressure cannot now be increased in that circuit by any further application of the brake pedal **(see illustration 19.14)**.

If the wheel speed sensor signals indicate that wheel rotation has now stabilised, the ECM will instigate the pressure build-up phase, allowing braking to continue. If wheel lock is still detected after the pressure holding phase, the ECM instigates the pressure reduction phase.

Second ABS phase, pressure reduction

If the ECM detects wheel instability, a pressure reduction phase is initiated. The inlet solenoid valve remains closed and the outlet solenoid valve is opened by means of a series of short activation pulses. The pressure in the controlled circuit decreases rapidly as the fluid flows from the brake caliper or wheel cylinder into the pressure accumulator. At the same time, the ECM actuates the electric motor to operate the return pump. The hydraulic fluid is then pumped back into the pressure side of the master cylinder. This process creates a pulsation which can be felt in the brake pedal action, but which is softened by the pulsation damper **(see illustration 19.15)**.

Third ABS phase, pressure build-up

The pressure build-up phase is instigated after the wheel rotation has stabilised. The inlet and outlet solenoid valves are returned to the at rest position (inlet solenoid valve open and exhaust solenoid valve closed) which re-opens the hydraulic fluid line from the tandem master cylinder to the brake caliper or wheel cylinder. Hydraulic pressure is reinstated, thus re-introducing operation of the brake. After a brief period, a short pressure holding phase is re-introduced and the ECM continually shifts between pressure build-up and pressure holding until the wheel has decelerated to a sufficient degree where pressure reduction is once more required **(see illustration 19.16)**.

The whole ABS control cycle takes place 4 to 10 times per second for each affected wheel and this ensures maximum braking effect and control during ABS operation.

Test procedures

Important note: *The test procedures, pin-tables and wiring diagrams contained in this Chapter are necessarily representative of the system depicted. Because of the variations in wiring and other data that often occurs, even between similar vehicles in any particular VM's range, the reader should take great care in identification of ECM pins, and satisfy himself that he has gathered the correct data before failing a particular component.*

4 Wheel speed sensors

Checking the wheel speed sensor (general)

1 Inspect the wheel speed sensor for corrosion or damage and check that the sensor is tightly mounted.
2 Check the toothed sensor ring for damage, eccentricity and for broken or missing teeth.
3 Inspect the wheel speed sensor wiring plug for corrosion and damage. One plug for each sensor.
4 Check that the connector terminal pins are pushed fully home and making good contact with the sensor wiring plug.
5 Check the clearance between the sensor and the toothed sensor ring. The clearance is not normally adjustable but is nominally 0.2 to 1.2 mm. If the clearance is excessive, expect a worn sensor tip or problems with the wheel bearings/hub or sensor ring.
6 When carrying out voltage checks with an oscilloscope or voltmeter, the voltage obtained will be proportional to the speed at which the wheel is rotating. In addition to determining that the wheel speed sensors are actually producing a voltage output, it is essential that the output from the sensors on a particular axle is the same for any given wheel speed.

Checking wheel speed sensor output with an oscilloscope

Note: *Refer to the wiring diagrams for specific ECM pin identification according to model.*
7 Switch the ignition off and disconnect the ECM multi-plug or the relevant wheel speed sensor wiring plug.
8 Connect an oscilloscope between the terminal pins for the sensor under test **(see illustration 19.17)**.
9 Select a range to cover 80 Hz on the oscilloscope and a free run time base.
10 Raise the wheel and rotate it by hand at approximately one revolution per second.
11 A sinusoidal wave form should be obtained, with amplitude and duration changing with rotational speed **(see illustration 19.18)**.

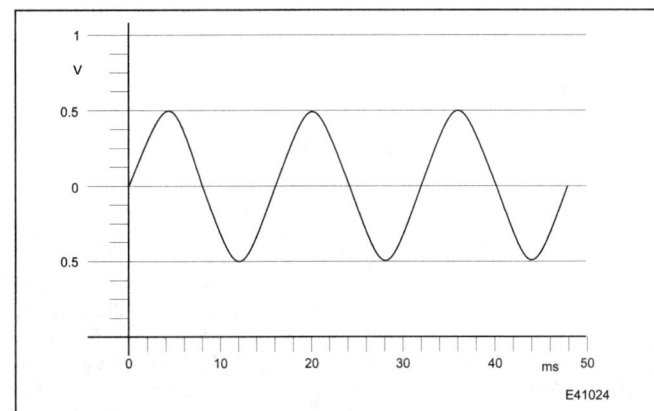

19.17 Checking wheel speed sensor output with an oscilloscope connected to the sensor wiring plug

12 If there is no signal, or a very weak or intermittent signal at the ECM, repeat the test at the sensor wiring plug. If there is no change in signal status, the sensor is suspect.
13 If the signal is now satisfactory this indicates a fault in the wiring harness which should be checked for continuity.

Checking wheel speed sensor output with an AC voltmeter

Note: *Refer to the wiring diagrams for specific ECM pin identification according to model.*
14 Switch the ignition off and disconnect the ECM multi-plug or the relevant wheel speed sensor wiring plug.
15 Connect an AC voltmeter between the terminal pins for the sensor under test **(see illustration 19.19)**.

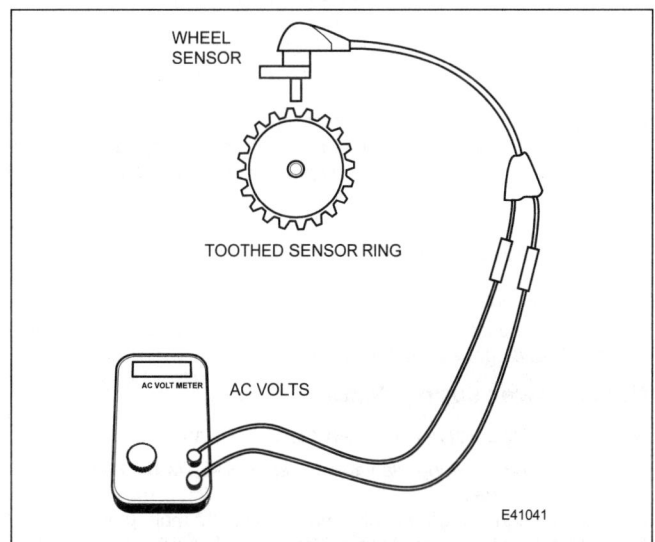

19.18 Typical wheel speed sensor sine wave as displayed on an oscilloscope

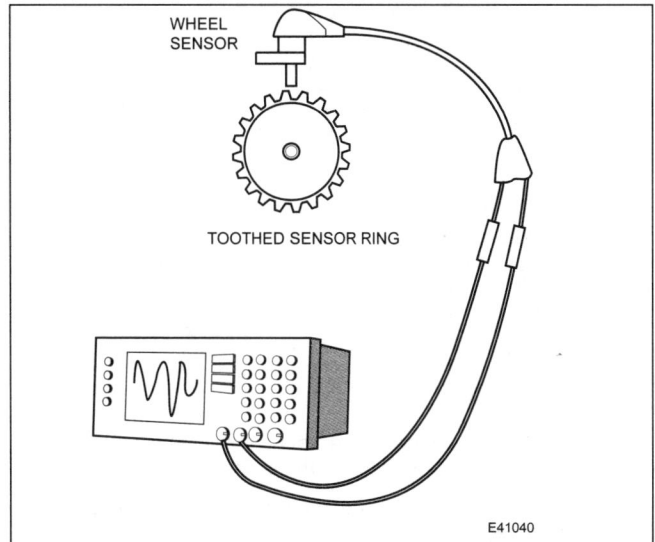

19.19 Checking wheel speed sensor output with a voltmeter connected to the sensor wiring plug

19

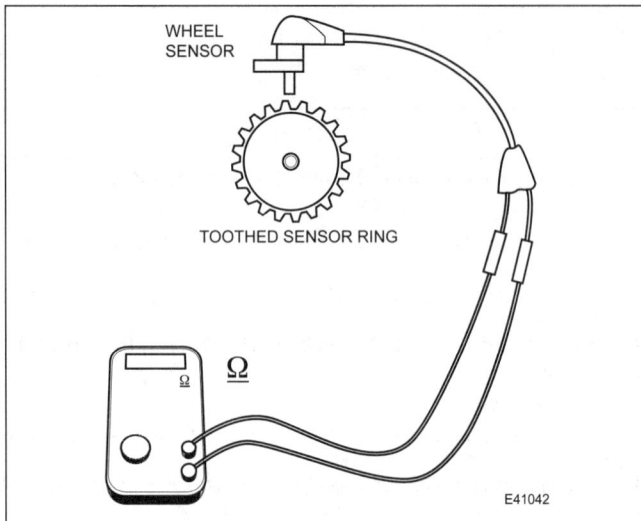

19.20 Checking wheel speed sensor resistance with an ohmmeter connected to the sensor wiring plug

16 Raise the wheel and rotate it by hand at approximately one revolution per second.

17 A voltage of approximately 0.1 to 1.5 volts (AC RMS) should be obtained. If there is no signal, or a very weak or intermittent signal at the ECM, repeat the test at the sensor wiring plug. If there is no change in the signal, the sensor is suspect.

18 If the signal is now satisfactory, this indicates a fault in the wiring harness which should be checked for continuity. **Note:** *This test at least proves that a signal is being generated by the sensor. However, the voltage produced is an average voltage and does not clearly indicate damage to the sensor ring or that the sinewave is regular in formation.*

Checking wheel speed sensor resistance

Note: *Refer to the wiring diagrams for specific ECM pin identification according to model.*

19 Switch the ignition off and disconnect the ECM multi-plug or the relevant wheel speed sensor wiring plug.

20 Connect an ohmmeter between the terminal pins for the sensor under test **(see illustration 19.20)**.

21 The readings obtained should be between 0.7 and 2.2 kohms approximately.

22 If the resistance is excessively high, or open circuit at the ECM, repeat the test at the sensor multi-plug. If there is no change in resistance, the sensor is suspect.

23 If the resistance is now satisfactory, this indicates a fault in the wiring harness which should be checked for continuity. **Note:** *Even if the resistance is within the quoted specifications, this does not prove that the speed sensor can generate an acceptable signal.*

5 System relays

Relay operation tests

1 The relays are integral with the ECM and can only be checked using suitable ABS diagnostic test equipment.

Relay power supply tests

Main (solenoid valve) relay and pump relay

Note: *Refer to the wiring diagrams for specific ECM pin identification according to model.*

2 Switch the ignition off and disconnect the ECM multi-plug.

3 Attach a negative voltmeter probe to a vehicle earth.

4 Attach a positive voltmeter probe to the following ECM multi-plug pins according to system:

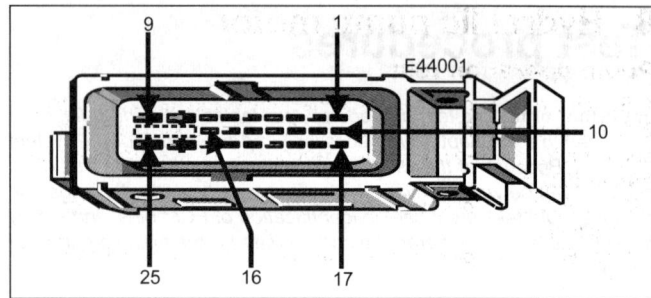

19.21 ECM 25-pin multi-plug

25-pin multi-plug systems - pin 8 and pin 9 in turn **(see illustration 19.21)**

28-pin multi-plug systems - pin 1 and pin 2 in turn **(see illustration 19.22)**

5 The voltmeter should indicate nbv in each case. If no voltage is found, check the relevant fuse and the supply wiring back to the battery positive terminal.

6 Electronic Control Module

Checking the ECM (general)

1 Inspect the ECM for corrosion or damage and ensure that the unit is securely attached to the hydraulic control unit.

2 Check that the ECM multi-plug terminals are pushed fully home and making good contact with the ECM pins. A fault in any of the above areas are possible reasons for poor performance in the ABS system.

ECM power supply and earth tests

Note: *Refer to the wiring diagrams for specific ECM pin identification according to model.*

3 Switch the ignition off and disconnect the ECM multi-plug.

4 Switch the ignition on.

5 Attach a negative voltmeter probe to a vehicle earth.

6 Attach a positive voltmeter probe to the following ECM multi-plug pin according to system:

25-pin multi-plug systems - pin 22 **(see illustration 19.21)**

28-pin multi-plug systems - pin 15 **(see illustration 19.22)**

7 If no voltage is found, check the relevant fuse and the supply wiring back to the ignition switch.

8 Switch the ignition off.

9 Connect an ohmmeter between a vehicle earth and the following ECM multi-plug pins according to system:

25-pin multi-plug systems - pin 24 and pin 25 in turn

28-pin multi-plug systems - pin 13, pin 14 and pin 22 in turn

10 The ohmmeter should indicate continuity in each case. If not, check the ECM main earth connection and wiring.

7 Solenoid valves

Solenoid valve operation tests

1 The solenoid valves are integral with the ECM and can only be checked using suitable ABS diagnostic test equipment.

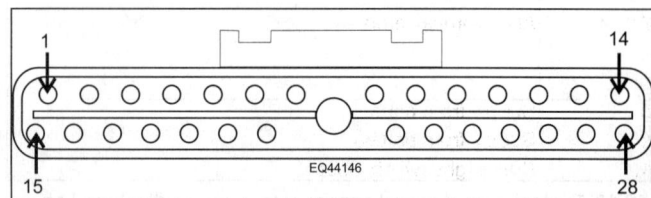

19.22 ECM 28-pin multi-plug

8 Hydraulic pump motor

Pump operation test

1 Switch the ignition off and disconnect the pump motor multi-plug.
2 Connect the positive terminal of a 12 volt supply to the following pump motor multi-plug terminal according to system **(see illustration 19.23)**:
   25-pin ECM multi-plug systems - terminal 2 of the pump motor multi-plug
   28-pin ECM multi-plug systems - terminal 1 of the pump motor multi-plug
3 Connect the negative terminal of the 12 volt supply to the other pump motor multi-plug terminal. The pump motor should now run.

⚠️ *Warning: The test should be made as quickly as possible to avoid damaging the pump.*

4 If the pump does not operate as described, renew the hydraulic control unit.

9 Brake light switch

Checking the brake light switch (general)

1 Check that the brake light switch is correctly and securely mounted and that the plunger moves smoothly with no trace of binding.
2 Check that the wiring multi-plug is pushed fully home and making good contact.
3 Check that no wires have been disconnected.
4 A fault in any of the above areas are possible reasons for failure or malfunctioning of the switch.

Brake light switch voltage and continuity tests

Voltage test

Note: *Refer to the wiring diagrams for specific ECM pin identification according to model.*
5 Switch the ignition off and disconnect the ECM multi-plug from the ECM.
6 Connect a voltmeter between a vehicle earth and the ECM multi-plug brake light switch input pin according to system, as follows:
   25-pin ECM multi-plug systems - typically pin 10 (see illustration 19.21)
   28-pin ECM multi-plug systems - typically pin 20 (see illustration 19.22)

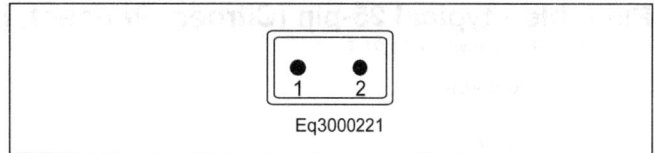

19.23 Pump motor multi-plug terminal identification

7 Switch the ignition on and depress the brake pedal. The voltmeter should indicate nbv.
8 If no voltage is found, the fuse, the brake light switch and the wiring are suspect.
9 Release the brake pedal. The voltage should drop to zero as the switch opens.

Continuity test

10 Switch the ignition off and disconnect the brake light switch multi-plug.
11 Connect an ohmmeter between the terminal pins of the brake light switch.
12 Operate the brake light switch and check for continuity. If the test fails, renew the switch.

10 Warning lamp

Checking the warning lamp (general)

1 Inspect the warning lamp bulb holder contacts in the instrument panel.
2 Check that the instrument panel multi-plug terminal pins are pushed fully home and making good contact.
3 Check that no wires have been disconnected.
4 A fault in any of the above areas are possible reasons for failure or malfunctioning of the warning lamp.

Warning lamp operation test

5 With the ignition switched off, the warning lamp should remain off.
6 Switch the ignition on and the warning lamp should illuminate then extinguish after a few seconds. The lamp should then remain off.
7 If the warning lamp comes on and remains on at any time during vehicle operation, carry out the previously described test procedures on the system components.

Pin table - typical 25-pin (Citroën, Peugeot, Ford Galaxy, Seat, Volkswagen)

Note: *Refer to illustration 19.21*

Pin No.	Connection	Test condition	Voltage
1	Front right wheel speed sensor earth	Roadwheel rotating	0.25 volts (max)
2	-	-	-
2**	Link to pin 12	-	-
3	Front left wheel speed sensor earth	Roadwheel rotating	0.25 volts (max)
4	-	-	-
5	Rear right wheel speed sensor earth	Roadwheel rotating	0.25 volts (max)
6	SD connector		
7	-	-	-
7**	ABS warning lamp	Ignition on: Lamp on Lamp off	 2.0 volts (max) Nbv
8	Supply from battery	Ignition on/off	Nbv
9	Supply from battery	Ignition on/off	Nbv
10	Brake light switch input	Ignition on: Brake pedal released Brake pedal depressed	 0 volts Nbv

19

Pin table - typical 25-pin (Citroën, Peugeot, Ford Galaxy, Seat, Volkswagen) (continued)

Note: *Refer to illustration 19.21*

Pin No.	Connection	Test condition	Voltage
11	-	-	-
11*	Link to pin 12		
12	-	-	-
12*	Link to pin 11		
12**	Link to pin 2	-	-
13	Rear right wheel speed sensor signal	Roadwheel rotating	0.1 to 1.5 volts AC (approx)
14	Rear left wheel speed sensor signal	Roadwheel rotating	0.1 to 1.5 volts AC (approx)
15	-	-	-
16	ABS warning lamp	Ignition on:	
		Lamp on	2.0 volts (max)
		Lamp off	Nbv
17	Front right wheel speed sensor signal	Roadwheel rotating	0.1 to 1.5 volts AC (approx)
18	Front left wheel speed sensor signal	Roadwheel rotating	0.1 to 1.5 volts AC (approx)
19	-	-	-
20	-	-	-
21	Rear left wheel speed sensor earth	Roadwheel rotating	0.25 volts (max)
22	Supply from ignition switch	Ignition on	Nbv
23***	Supply from ignition switch	Ignition on	Nbv
24	ECM earth	Ignition on	0.25 volts (max)
25	ECM earth	Ignition on	0.25 volts (max)

* Seat Alhambra/Ford Galaxy
** VW Polo
***Peugeot 106

Pin table - typical 25-pin (Renault)

Note: *Refer to illustration 19.21*

Pin No.	Connection	Test condition	Voltage
1	Front right wheel speed sensor signal	Roadwheel rotating	0.1 to 1.5 volts AC (approx)
2	-	-	-
3	Front left wheel speed sensor signal	Roadwheel rotating	0.1 to 1.5 volts AC (approx)
4	-	-	-
5	Rear right wheel speed sensor signal	Roadwheel rotating	0.1 to 1.5 volts AC (approx)
6	SD connector		
7	-	-	-
8	Supply from battery	Ignition on/off	Nbv
9	Supply from battery	Ignition on/off	Nbv
10	Brake light switch input	Ignition on:	
		Brake pedal released	0 volts
		Brake pedal depressed	Nbv
11	-	-	-
12	-	-	-
13	Rear right wheel speed sensor earth	Roadwheel rotating	0.25 volts (max)
14	Rear left wheel speed sensor earth	Roadwheel rotating	0.25 volts (max)
15	-	-	-
16	ABS warning lamp	Ignition on:	
		Lamp on	2.0 volts (max)
		Lamp off	Nbv
17	Front right wheel speed sensor earth	Roadwheel rotating	0.25 volts (max)
18	Front left wheel speed sensor earth	Roadwheel rotating	0.25 volts (max)
19	-	-	-
20	SD connector		
21	Rear left wheel speed sensor signal	Roadwheel rotating	0.1 to 1.5 volts AC (approx)
22	Supply from ignition switch	Ignition on	Nbv
23	Supply from ignition switch	Ignition on	Nbv
24	ECM earth	Ignition on	0.25 volts (max)
25	ECM earth	Ignition on	0.25 volts (max)

Pin table - typical 28-pin (Ford Scorpio)

Note: *Refer to illustration 19.22*

Pin No.	Connection	Test condition	Voltage
1	Supply from battery	Ignition on/off	Nbv
2	Supply from battery	Ignition on/off	Nbv
3	-	-	-
4	-	-	-
5	Front left wheel speed sensor signal	Roadwheel rotating	0.1 to 1.5 volts AC (approx)
6	Front left wheel speed sensor earth	Roadwheel rotating	0.25 volts (max)
7	Front right wheel speed sensor signal	Roadwheel rotating	0.1 to 1.5 volts AC (approx)
8	Front right wheel speed sensor earth	Roadwheel rotating	0.25 volts (max)
9	Rear left wheel speed sensor signal	Roadwheel rotating	0.1 to 1.5 volts AC (approx)
10	Rear left wheel speed sensor earth	Roadwheel rotating	0.25 volts (max)
11	Rear right wheel speed sensor signal	Roadwheel rotating	0.1 to 1.5 volts AC (approx)
12	Rear right wheel speed sensor earth	Roadwheel rotating	0.25 volts (max)
13	ECM earth	Ignition on	0.25 volts (max)
14	ECM earth	Ignition on	0.25 volts (max)
15	Supply from ignition switch	Ignition on	Nbv
16	-	-	-
17	-	-	-
18	-	-	-
19	-	-	-
20	Brake light switch input	Ignition on: Brake pedal released Brake pedal depressed	 0 volts Nbv
21	ABS warning lamp	Ignition on: Lamp on Lamp off	 2.0 volts (max) Nbv
22	ECM earth	Ignition on	0.25 volts (max)
23	-	-	-
24	-	-	-
25	-	-	-
26	-	-	-
27	-	-	-
28	SD connector		

Fault codes

11 General fault codes

1 The Teves IV-GI system requires the use of a FCR for obtaining fault codes. Flash codes are not available for output from this system.
2 If a FCR is available, it should be connected to the SD serial connector and used in accordance with the maker's instructions.
3 The FCR can be used for the following purposes:
  a) *Obtaining fault codes.*
  b) *Clearing fault codes.*
  c) *Obtaining datastream information.*
  d) *Testing the system actuators (solenoid valve relay, pump relay and solenoid valves).*

12 Ford and Renault fault codes

1 On Ford and Renault models, internal fault codes are used by the ECM to designate faults in the system components and circuits. A proprietary fault code reader (FCR) or system tester (such as the Ford FDS 2000 or Renault XR25) is required to interrogate the system. No actual fault code numbers are available although the component circuits checked by the ECM are similar to those shown for the other vehicles listed.

19

Fault code table (25-pin ECM - Citroën)

Code	Item	Fault
1	Supply voltage	Voltage too low
2	Supply voltage	Voltage too high
3	Relays	Defective (signal interruption, short to earth, supply voltage too low)
4	Pump motor	Defective / poor connections
5	Front left wheel speed sensor	Incorrect resistance
6	Front left wheel speed sensor	No signal / incorrect signal
7	Front left wheel speed sensor	Signal interruption
8	Front right wheel speed sensor	Incorrect resistance
9	Front right wheel speed sensor	No signal / incorrect signal
10	Front right wheel speed sensor	Signal interruption
11	Rear left wheel speed sensor	Incorrect resistance
12	Rear left wheel speed sensor	No signal / incorrect signal
13	Rear left wheel speed sensor	Signal interruption
14	Rear right wheel speed sensor	Incorrect resistance
15	Rear right wheel speed sensor	No signal / incorrect signal
16	Rear right wheel speed sensor	Signal interruption
17	ECM	Defective

Fault code table (25-pin ECM - Peugeot)

Code	Item
015Z	Main (solenoid valve) relay
018Z	Wheel speed sensor toothed sensor ring
024Z	Rear left wheel speed sensor
025Z	Front right wheel speed sensor
031Z	Rear right wheel speed sensor
032Z	Front left wheel speed sensor
033Z	Wheel speed sensor signal
053Z	Hydraulic pump motor
055Z	ECM
057Z	Supply voltage too low
058Z	Supply voltage too high
066Z	ABS operation
067Z	ABS operation
068Z	ABS operation
069Z	ABS operation
087Z	SD output fault
091Z	ABS warning lamp
095Z	Hydraulic unions
096Z	Wheel speed sensor wiring transposed
097Z	External signal interference
099Z	No faults found

Fault code table (25-pin ECM - Seat, Volkswagen)

Code	Item	Fault
00283	Front left wheel speed sensor	Poor wiring/connections; incorrect air gap; sensor or sensor ring dirty or damaged
00285	Front right wheel speed sensor	Poor wiring/connections; incorrect air gap; sensor or sensor ring dirty or damaged
00287	Rear right wheel speed sensor	Poor wiring/connections; incorrect air gap; sensor or sensor ring dirty or damaged
00290	Rear left wheel speed sensor	Poor wiring/connections; incorrect air gap; sensor or sensor ring dirty or damaged
00668	Supply voltage	Outside expected values
01044	ECM coding	ECM incorrectly coded, poor wiring connection
01130	ABS operation	Signal external interference, poor wiring connections
01276	Hydraulic pump	Faulty operation
65535	ECM	ECM defective

Wiring diagrams

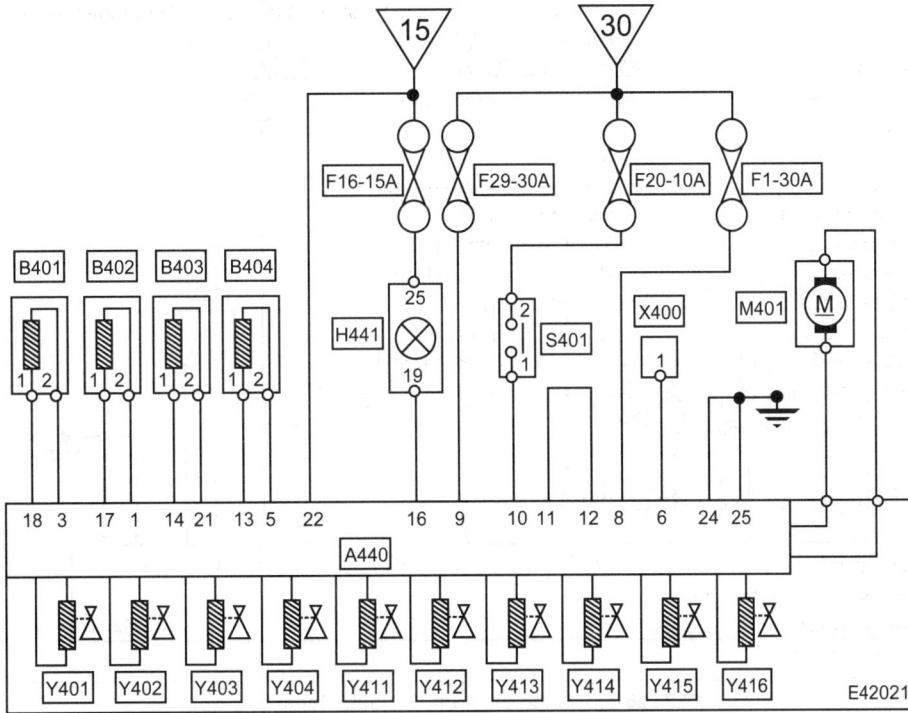

19.24 25-pin wiring diagram, Ford Galaxy

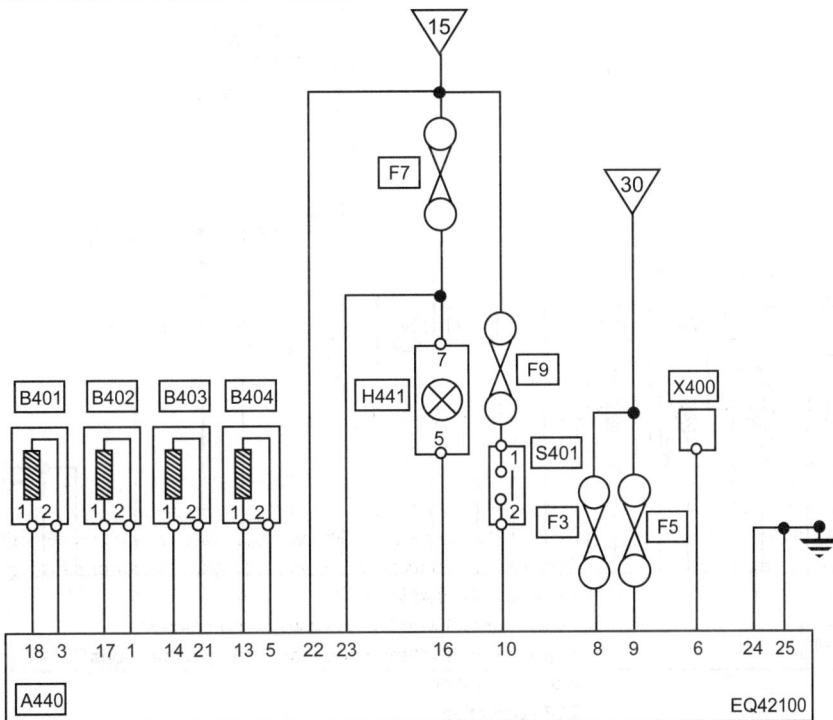

19.25 25-pin wiring diagram, Peugeot 106

19

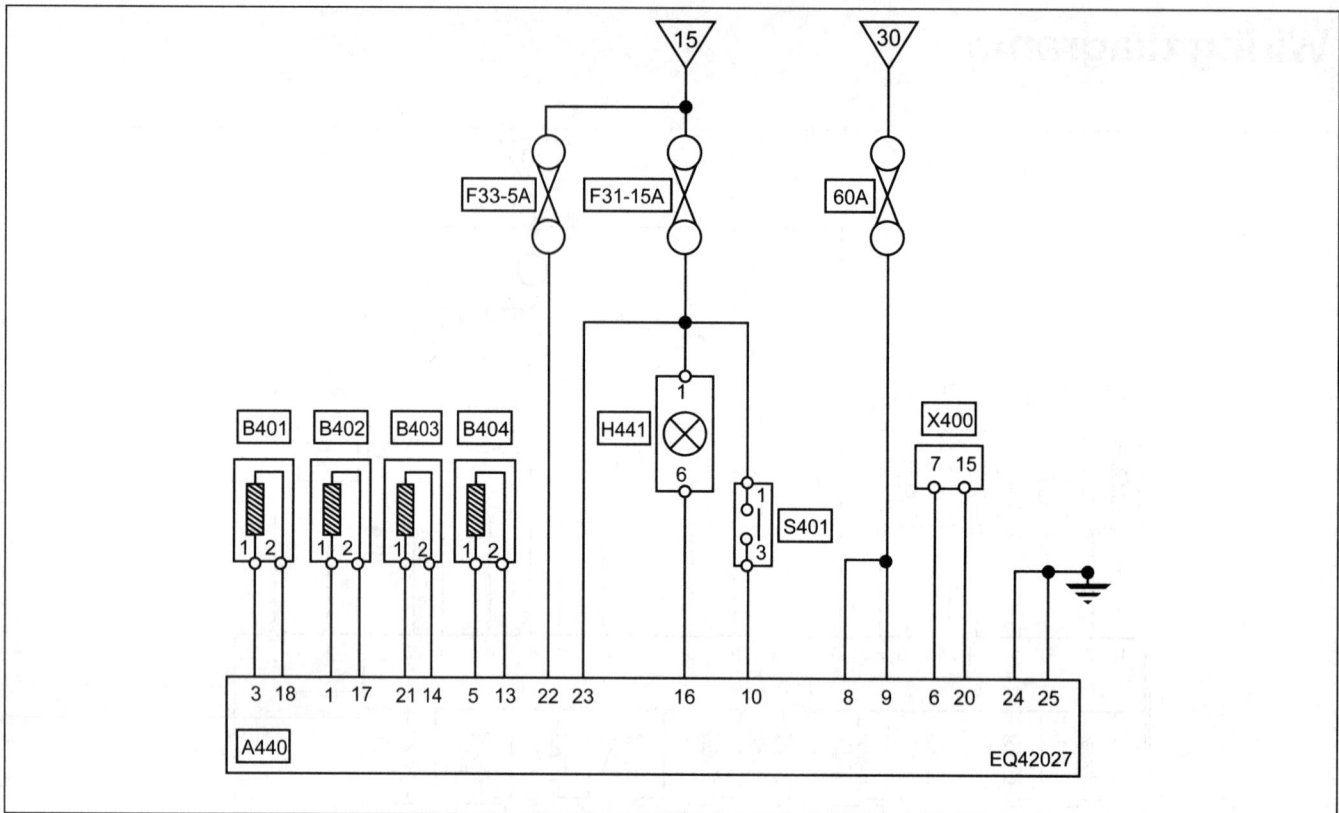

19.26 25-pin wiring diagram, Renault Megane

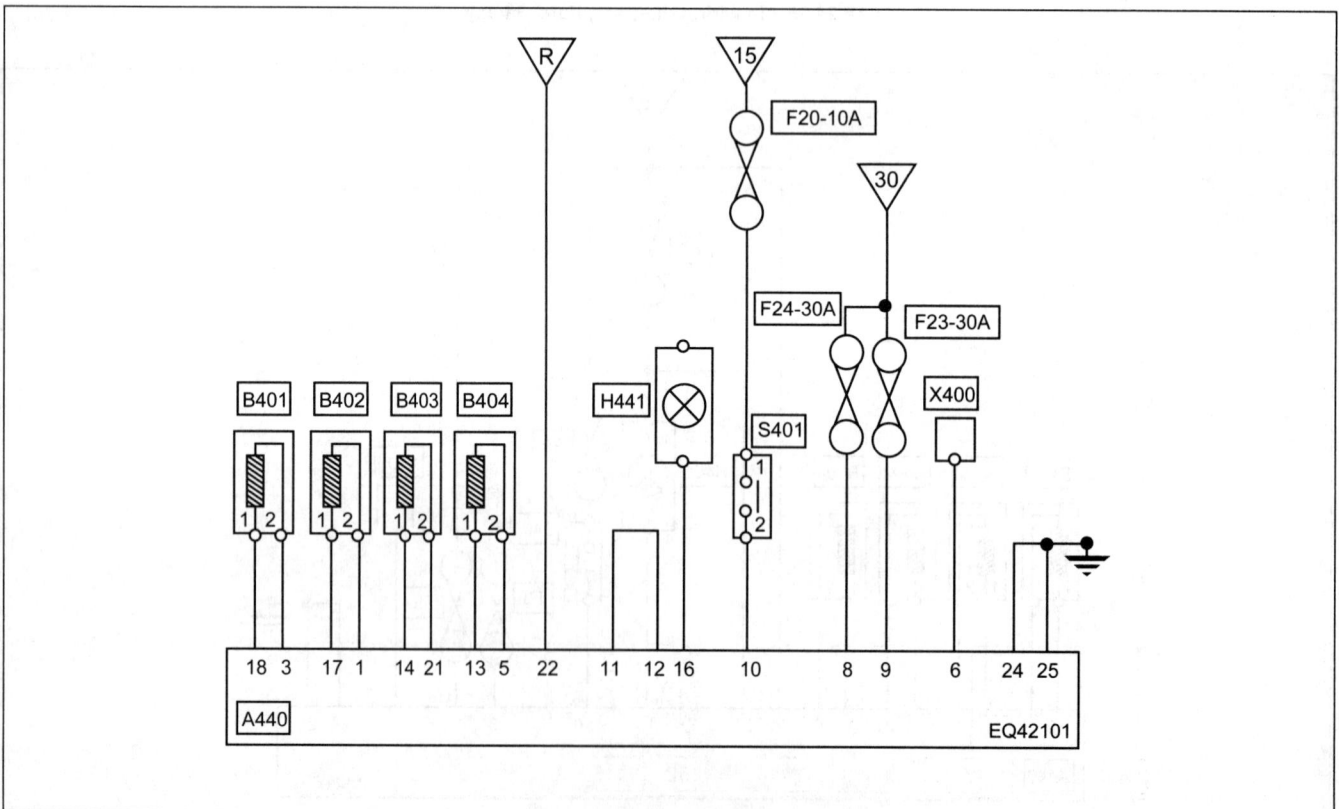

19.27 25-pin wiring diagram, Seat Alhambra

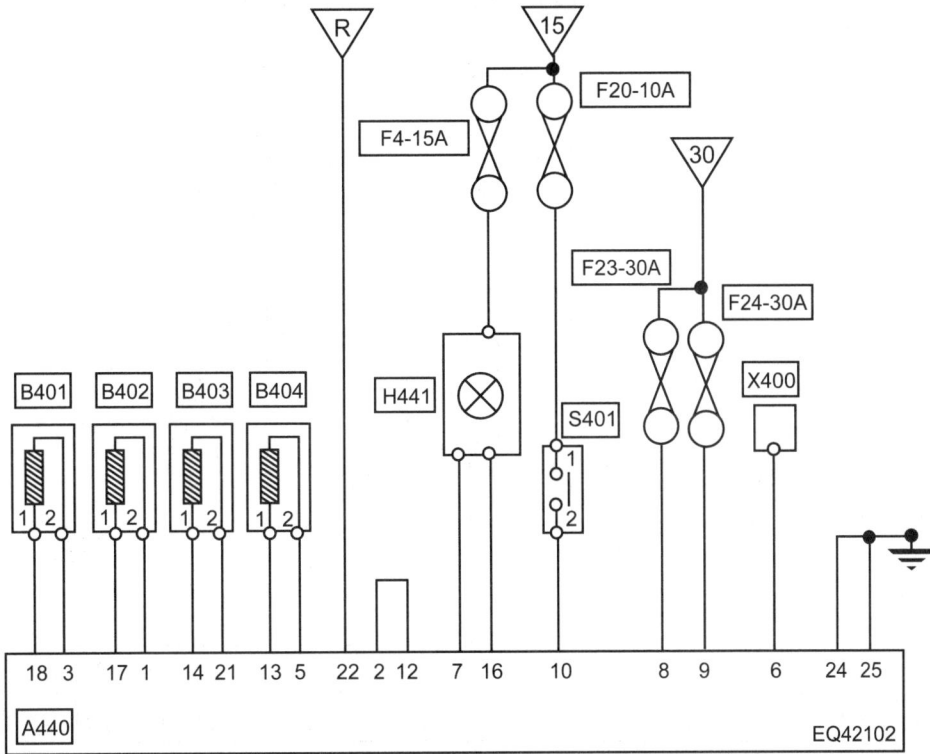

19.28 25-pin wiring diagram, VW Polo

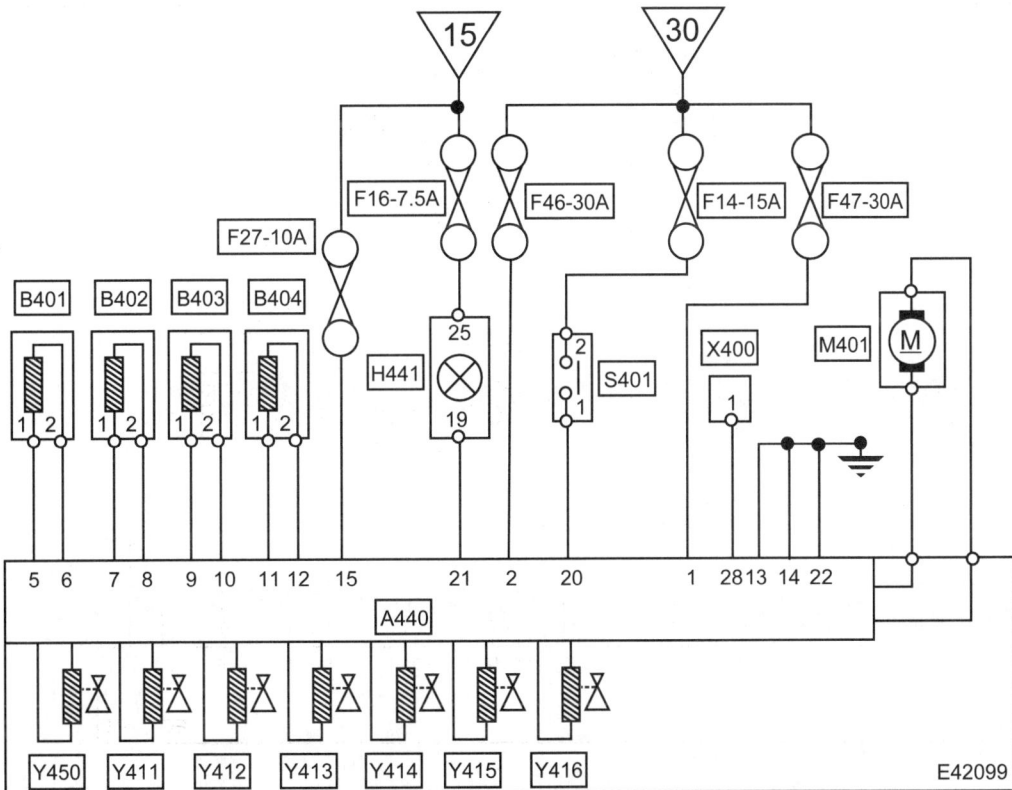

19.29 28-pin wiring diagram, Ford Scorpio

19

Chapter 20
Teves 20-I ABS

Contents

Vehicle coverage

Model	Year
Ford Escort	1996-2000
Ford Fiesta	1996-2000
Ford Focus	1998-2000
Ford Ka	1996-1999
Ford Puma	1997-2000
Ford Scorpio	1997-1999

20

Overview of system operation

1 Basic principles and system identification

The Teves 20-I Antilock Brake System is a development of the Teves (ITT) 20GI series ABS, with revisions to the hydraulic control unit, and enhancement to the Electronic Control Module (ECM) software. Teves 20-I has been fitted to various, Ford derived, passenger vehicles since its introduction in the mid 1990s. The system is of the additional or 'add-on' type operating in conjunction with the conventional braking system components.

Depending on application, Teves 20-I may be installed as an antilock braking system only, or as an antilock braking system incorporating traction control.

In ABS mode, the purpose of the system is to apply the vehicle brakes at maximum efficiency without wheel lock or loss of directional stability. Inductive sensors (wheel speed sensors) monitor the speed of the wheels by generating an electrical signal as the wheel is rotated. This information is passed to the ABS Electronic Control Module (ECM) which is then able to determine wheel speed, wheel

acceleration and wheel deceleration. The ECM compares the signals received from each wheel and if the onset of lock at any wheel is detected, a signal is sent to the ABS hydraulic control unit which regulates the brake pressure for the relevant wheel(s).

Where the system incorporates traction control, essentially the reverse principle is applied. When the ECM detects that one or more wheels are rotating faster than the reference value, the brake is actually applied on the relevant wheel(s) to reduce the rotational speed. Additionally, on some installations, when wheel spin is detected, various signals are sent to the engine management ECM to control engine torque and rpm.

Typically, Teves 20-I ABS is comprised of the following components (see illustration 20.1):

a) Hydraulic control unit with integral ABS-ECM.
b) Four inductive wheel speed sensors and associated sensor rings.
c) Brake light switch.
d) ABS warning lamp.
e) Diagnostic connector.

In addition, the conventional brake system is comprised of the following components:

a) Tandem brake master cylinder.
b) Vacuum servo unit.
c) Brake calipers/wheel cylinders and hydraulic hoses and pipes.
d) Pressure regulating/load sensing valve(s) depending on application.

2 Component description and operation

ABS ECM

General

The Teves 20-I Electronic Control Module (ECM) continually monitors wheel speed from the signals provided by the wheel speed sensors, and brake application from the brake light switch signal. If the ECM detects the incidence of wheel lock on one or more wheels, a signal is sent to the hydraulic control unit to modulate the hydraulic pressure to the brake of the locking wheel(s). The ECM contains two microprocessors and uses digital technology to complete this function and other functions such as, fault code memory and power modules for valve and pump activation (see illustration 20.2).

To reduce external electrical connections to a minimum and

20.1 Typical Teves 20-I main components

1	Hydraulic control unit with integral ABS-ECM	7	Tandem brake master cylinder
2	Inductive wheel speed sensors	8	Vacuum servo unit
3	Sensor rings	9	Brake calipers/wheel cylinders and hydraulic hoses and pipes
4	Brake light switch		
5	ABS warning lamp	10	Pressure regulating/load sensing valve(s) depending on application
6	Diagnostic connector		

20.2 ECM sensor inputs and control signal outputs

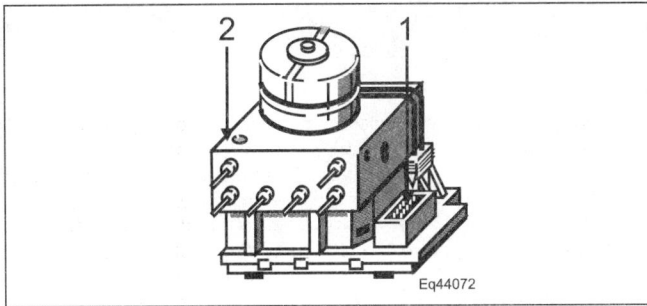

20.3 Teves 20-I ECM (1) and hydraulic control unit (2)

improve reliability, the ECM is integral with the hydraulic control unit **(see illustration 20.3)**.

Self-test

The Teves 20-I ECM is equipped with a self-test capability that initially examines the ABS system when the ignition is switched on, and then examines the wheel speed sensor signals after a wheel speed of approximately 4 mph is reached from all wheels (engine running). The ABS self-test program continues to examine the signals from the various components as long as the ignition is switched on. If self-test determines that faults are not present, the ABS is ready for operation once a specified vehicle speed has been achieved.

If the ECM detects that a fault is present, all ABS functions are switched off and the warning lamp is turned on. The conventional braking system continues to operate as normal without ABS assistance.

Self-diagnostics

If the ECM detects a fault during the self-test routine, an internal fault code is stored in the ECM memory. Stored fault codes can be retrieved from the SD connector with the aid of a suitable fault code reader. If the fault clears, the code will remain stored until cleared with the FCR.

Hydraulic control unit

The hydraulic control unit consists of an electric motor operating a return pump with eccentric drive and twin radial pistons, inlet and outlet solenoid valves, pressure accumulators and pulsation dampers **(see illustration 20.4)**. The unit controls the hydraulic pressure applied to the brake for each individual front wheel and each individual rear wheel, or pair of wheels, according to application. The return pump is switched on when the ABS is activated and returns hydraulic fluid, drained off during the pressure reduction phase, back into the brake circuit.

According to vehicle application, Teves 20-1 ABS may be either a three- or four-channel system according to the number of solenoid valves used in the hydraulic control unit. On a four-channel system, there are two solenoid valves for each brake (eight in total). On a three-channel system there are two solenoid valves for each front

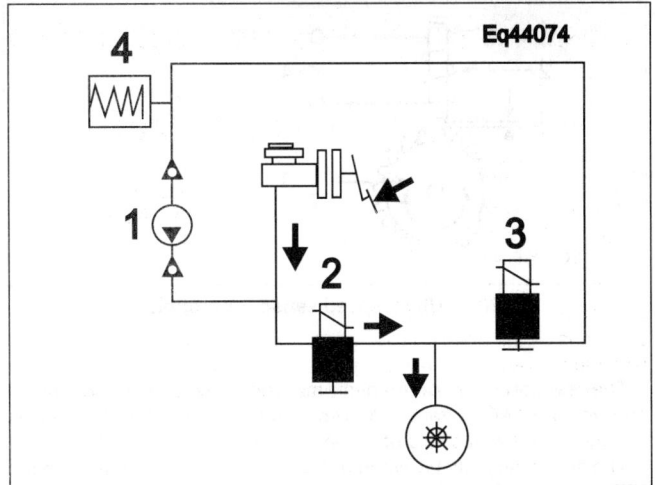

20.4 Hydraulic control unit components

| 1 | Electric motor and return pump | 3 | Outlet solenoid valve |
| 2 | Inlet solenoid valve | 4 | Pressure accumulator |

brake, and two solenoid valves controlling the rear brakes as a pair (six in total). Additional solenoid valves are used on systems incorporating traction control.

On some applications, the 'select-low' principle is employed for control of the rear brakes. With the 'select-low' principle, the wheel with the lowest adhesion determines the amount of hydraulic pressure to be supplied to both rear brakes during ABS operation.

On certain versions, the ABS-ECM contains additional software for rear brake hydraulic fluid pressure regulation during normal braking, generally known as 'Electronic Brake Force Distribution'. In these applications, mechanical pressure regulating/load sensing valves are not required as the rear brake hydraulic pressure is controlled by the ABS hydraulic control unit.

Wheel speed sensors

The rotational speed of the roadwheels and any changes in the rotational speed are recorded by inductive wheel speed sensors, one located at each roadwheel **(see illustration 20.5)**.

Each wheel speed sensor assembly comprises a toothed sensor ring which rotates at roadwheel speed, and an adjacent sensor mounted a set distance from the sensor ring **(see illustration 20.6)**.

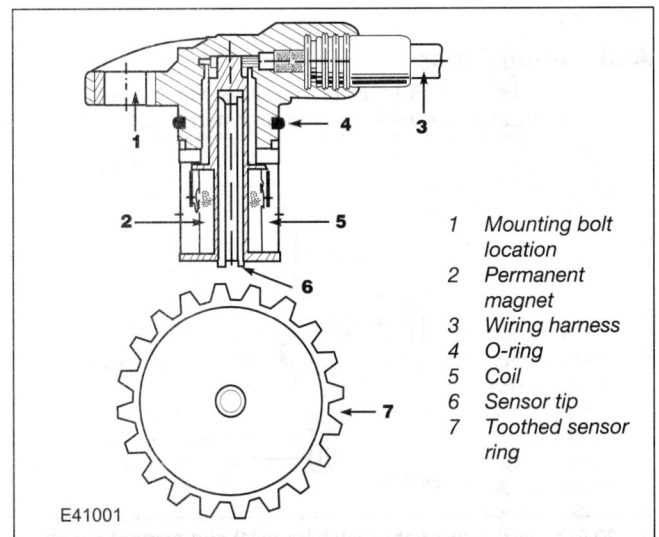

1	Mounting bolt location
2	Permanent magnet
3	Wiring harness
4	O-ring
5	Coil
6	Sensor tip
7	Toothed sensor ring

20

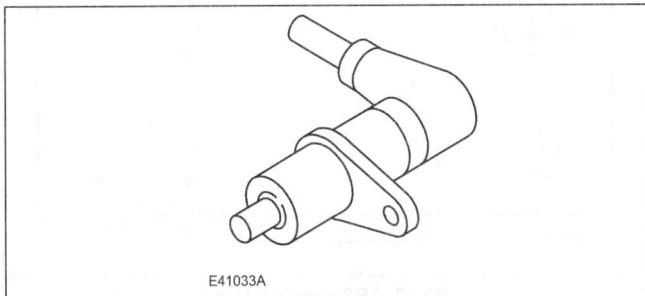

20.5 Typical wheel speed sensor

20.6 Sectional view of a wheel speed sensor

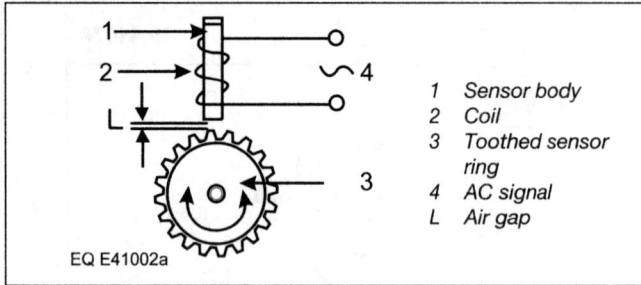

20.7 Wheel speed sensor operation

1 Sensor body
2 Coil
3 Toothed sensor ring
4 AC signal
L Air gap

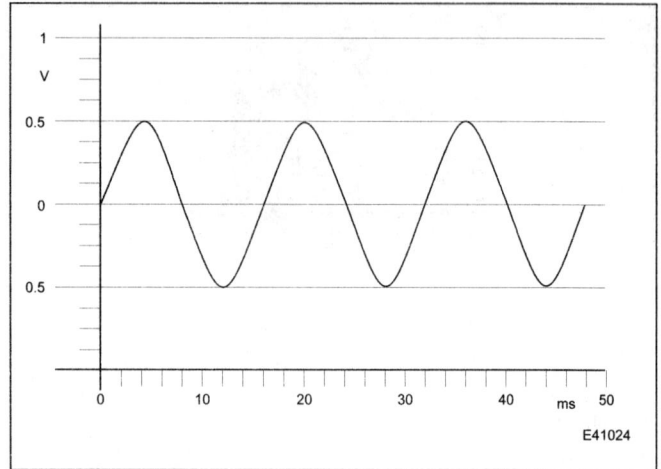

20.8 Wheel speed sensor waveform as viewed on an oscilloscope

The sensors are permanent magnet pulse generator types producing an AC voltage sine wave as the sensor ring teeth pass through the magnetic field of the sensor **(see illustration 20.7)**.

The frequency of the waveform produced by the wheel speed sensor is proportional to the road speed. This AC voltage signal is continually being delivered to the ABS-ECM for processing.

The peak to peak voltage of the speed signal (when viewed upon an oscilloscope) can vary considerably according to wheel speed. An analogue to digital converter (ADC) in the ECM transforms the AC pulse into a digital signal **(see illustration 20.8)**.

ABS electrical wiring harness and relays

An integrated main wiring harness is used to connect power and earth to various electrical components and to enable sensor signals to reach the ECM. The ECM, in turn, sends command signals to the hydraulic control unit via internal connections between the two units. The vehicle relays and ABS fuses are typically located in the engine compartment fuse/relay box .

Nominal battery voltage is supplied to the ECM via the ABS main fuses. The ECM then supplies nominal battery voltage to the solenoid valves and return pump and controls the earth for activation of the solenoid valves.

Brake light switch

The brake light switch comprises a switch body and contact pin and is located above the brake pedal **(see illustration 20.9)**.

When the brake pedal is depressed, closing the brake light switch, a signal is sent to the ECM indicating that the brakes are being applied. Once this signal is received, the ECM will begin monitoring the wheel speed via the wheel speed sensors and activate the ABS if necessary.

ABS warning lamp

After the ignition is switched on, the ABS warning lamp on the instrument panel is illuminated for approximately 2 to 4 seconds as the system executes a self-test routine. If satisfactory operation of the system is detected by the ECM, the light is extinguished. During vehicle operation above a pre-determined wheel speed, the ABS-ECM implements a further self-test cycle whereby ABS operation and wheel speed sensor signals are continually monitored. If a fault is detected, the relevant ECM pin is earthed to illuminate the warning lamp on the instrument panel, and the ABS function is disabled. The warning lamp will remain illuminated until the fault is no longer present **(see illustration 20.10)**.

When the ABS-ECM detects a fault, the fault code is stored and the ABS warning lamp activated. If the fault no longer exists after the next system start (ignition on/off) the ABS warning lamp is extinguished after the self-test cycle, however the fault code remains stored in the ECM memory.

Tandem master cylinder

Typically, the tandem master cylinder comprises a body casting incorporating primary and secondary pressure chambers, primary piston, intermediate piston, floating piston, slotted pin and central valve. The cylinder operates as a conventional master cylinder using vacuum assistance from the vacuum servo unit **(see illustration 20.11)**.

When the brake system is at rest, the central valve in the floating piston rests against the slotted pin. In this condition the central valve

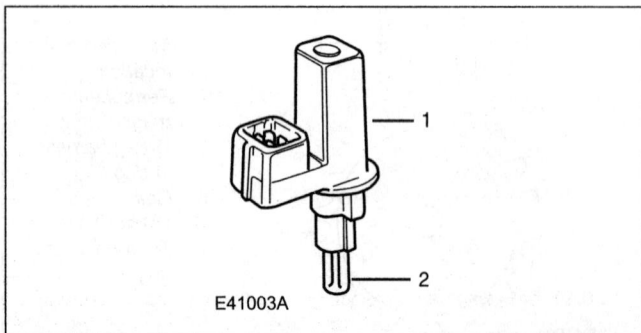

20.9 Typical brake light switch body (1) and contact pin (2)

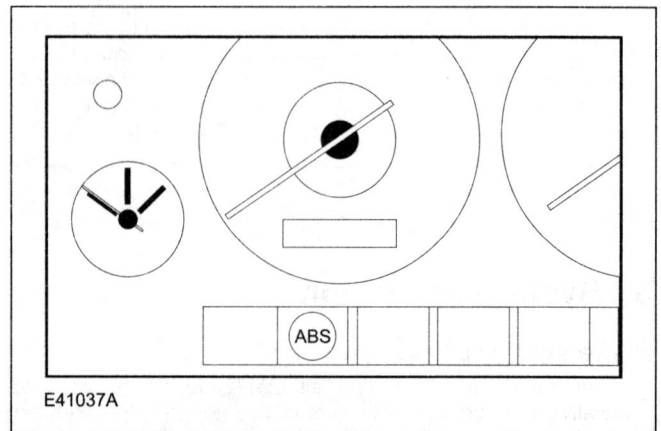

20.10 ABS warning lamp

1	Body casting
2	Outlet connections
3	Fluid inlet from reservoir
4	Pushrod/primary piston
5	Intermediate piston
6	Slotted pin
7	Floating piston
8	Central valve

E41005a

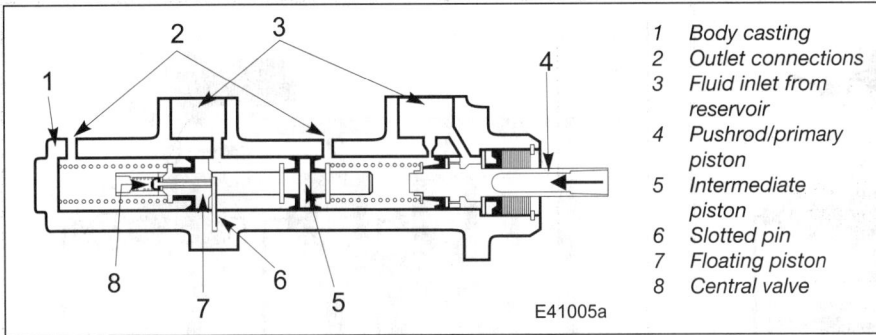

20.11 Sectional view of a typical tandem master cylinder

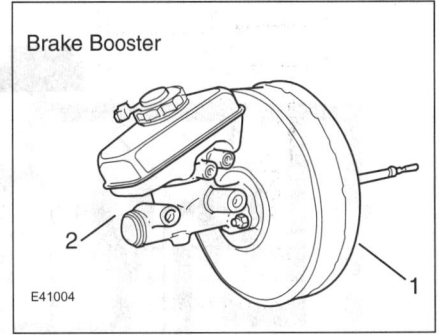

Brake Booster

E41004

20.12 Vacuum servo unit (1) and tandem master cylinder (2)

is open and brake fluid can discharge out of the pressure chamber back into the brake fluid reservoir. When the brake pedal is depressed, the build-up of hydraulic pressure in the primary pressure chamber acts on the intermediate piston and floating piston, moving them down the cylinder bore. The floating piston contacts the seal on the central valve, closing the connection between the intermediate and secondary pressure chambers. Brake hydraulic pressure can now also increase in the secondary pressure chamber.

Vacuum servo unit

The vacuum servo unit is located between the brake pedal and tandem master cylinder. When the brake pedal is depressed, the servo unit increases the force applied by the pedal, reducing the effort required to operate the brakes (see illustration 20.12).

The unit is operated by vacuum created in the engine inlet manifold (or from a separate vacuum pump on diesel engines) which is applied to a diaphragm contained within the unit casing. A pushrod connected to the centre of the diaphragm acts directly on the primary piston in the master cylinder.

When the brake pedal is released, vacuum is applied to both sides of the diaphragm. When the pedal is depressed, one side of the diaphragm is opened to atmosphere and the vacuum acting on the other side deflects the diaphragm which in turn operates the master cylinder primary piston. The resulting force applied to the master cylinder piston is therefore significantly greater than the initial force applied to the brake pedal by the driver.

Pressure regulating/load sensing valve(s)

Depending on vehicle application, pressure regulating valves or load sensing valves may be incorporated to restrict the hydraulic fluid pressure to the rear brakes. The valves may be pressure conscious whereby the hydraulic fluid supply is restricted once a pre-determined pressure is reached, or load conscious whereby the hydraulic pressure is reduced according to vehicle loading.

On certain versions, the ABS-ECM contains additional software for rear brake hydraulic fluid pressure regulation during normal braking, generally known as 'Electronic Brake Force Distribution'. In these applications, mechanical pressure regulating/load sensing valves are not required as the rear brake hydraulic pressure is controlled by the ABS hydraulic control unit.

3 System operation

Brake system at rest

When the system is at rest all the brake components are inoperative. Pressure is non-existent in the hydraulic pipes between the tandem master cylinder and the brake calipers. The inlet solenoid valves in the hydraulic control unit are open and the outlet solenoid valves are closed.

Brake system operating under conventional control without ABS

When the brake pedal is activated, the pedal force is applied to the tandem master cylinder by the vacuum servo unit pushrod. The servo unit pushrod acts directly on the pressure piston in the master cylinder which pressurises the hydraulic fluid in the brake pipes to the hydraulic control unit. The inlet solenoid valve and outlet solenoid valve both remain in the 'at rest' position (inlet solenoid valve open and outlet solenoid valve closed). Hydraulic pressure is transmitted to each brake caliper, thus operating the brakes.

When the brake pedal is released, the one-way valve opens allowing the hydraulic pressure in the circuit to rapidly decrease (see illustration 20.13).

Brake system operating in conjunction with ABS control

The ABS-ECM continually monitors wheel speed from the signals provided by the wheel speed sensors. If the ECM detects the incidence of wheel lock on one or more wheels, ABS is automatically initiated in three phases. As Teves 20-I ABS typically operates

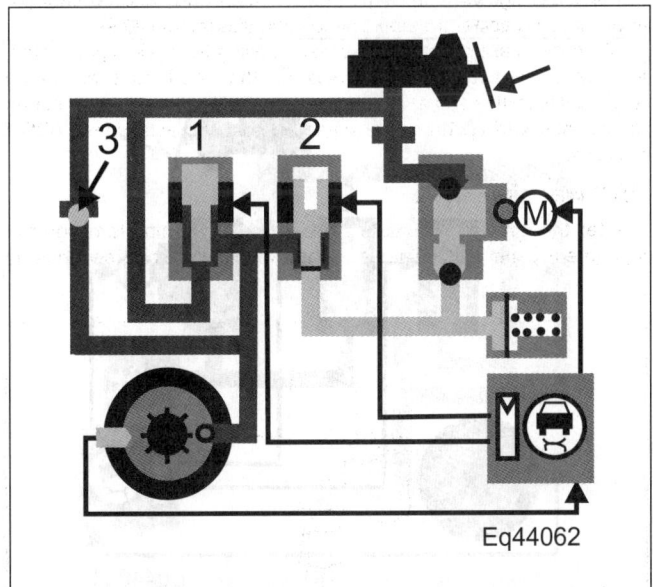

Eq44062

20.13 Brake system operating under conventional control without ABS

1	Inlet solenoid valve	3	One-way valve
2	Outlet solenoid valve		

20

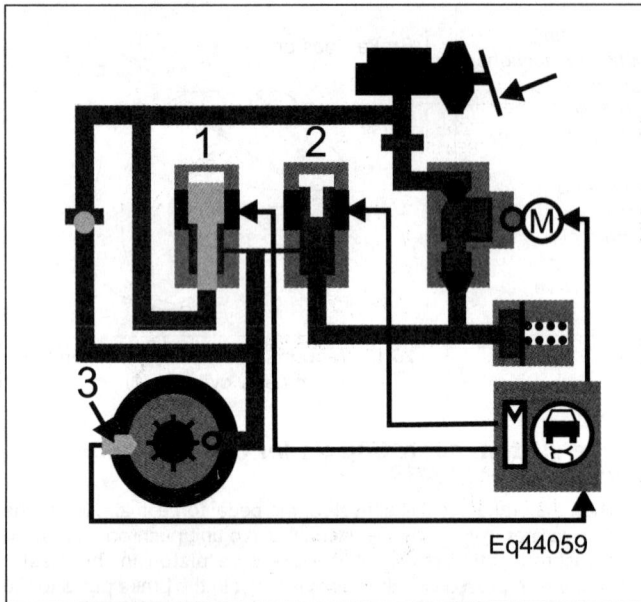

Eq44059

20.14 ABS operation - first phase, pressure holding

1	Inlet solenoid valve	3	Wheel speed sensor
2	Outlet solenoid valve		

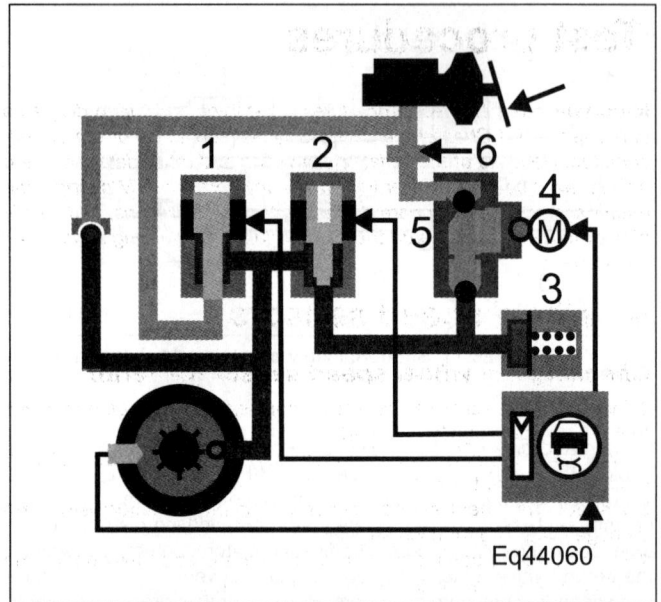

Eq44060

20.15 ABS operation - second phase, pressure reduction

1	Inlet solenoid valve	4	Pump motor
2	Outlet solenoid valve	5	Return pump
3	Pressure accumulator	6	Pulsation damper

individually on each wheel, all or any of the wheels could be in any one of the following phases at any particular moment.

First ABS phase, pressure holding

To prevent any further build-up of hydraulic pressure in the circuit being controlled, the ECM closes the inlet solenoid valve and allows the outlet solenoid valve to remain closed. The hydraulic fluid line from the tandem master cylinder to the brake caliper or wheel cylinder is closed, and the hydraulic fluid in the controlled circuit is maintained at a constant pressure. This effectively removes the braking force

from the controlled circuit. The pressure cannot now be increased in that circuit by any further application of the brake pedal (see illustration 20.14).

If the wheel speed sensor signals indicate that wheel rotation has now stabilised, the ECM will instigate the pressure build-up phase, allowing braking to continue. If wheel lock is still detected after the pressure holding phase, the ECM instigates the pressure reduction phase.

Second ABS phase, pressure reduction

If the ECM detects wheel instability, a pressure reduction phase is initiated. The inlet solenoid valve remains closed and the outlet solenoid valve is opened by means of a series of short activation pulses. The pressure in the controlled circuit decreases rapidly as the fluid flows from the brake caliper or wheel cylinder into the pressure accumulator. At the same time, the ECM actuates the electric motor to operate the return pump. The hydraulic fluid is then pumped back into the pressure side of the master cylinder. This process creates a pulsation which can be felt in the brake pedal action, but which is softened by the pulsation damper (see illustration 20.15).

Third ABS phase, pressure build-up

The pressure build-up phase is instigated after the wheel rotation has stabilised. The inlet and outlet solenoid valves are returned to the at rest position (inlet solenoid valve open and exhaust solenoid valve closed) which re-opens the hydraulic fluid line from the tandem master cylinder to the brake caliper or wheel cylinder. Hydraulic pressure is reinstated, thus re-introducing operation of the brake. After a brief period, a short pressure holding phase is re-introduced and the ECM continually shifts between pressure build-up and pressure holding until the wheel has decelerated to a sufficient degree where pressure reduction is once more required (see illustration 20.16).

The whole ABS control cycle takes place 4 to 10 times per second for each affected wheel and this ensures maximum braking effect and control during ABS operation.

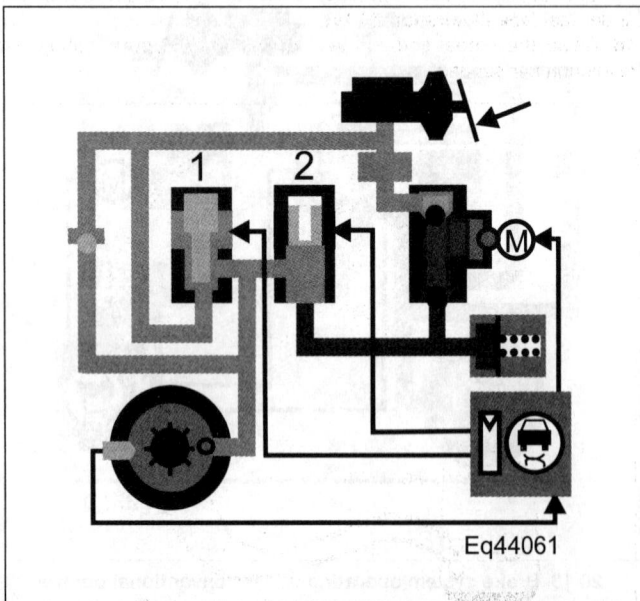

Eq44061

20.16 ABS operation - third phase, pressure build-up

1	Inlet solenoid valve	2	Outlet solenoid valve

Test procedures

Important note: *The test procedures, pin-tables and wiring diagrams contained in this Chapter are necessarily representative of the system depicted. Because of the variations in wiring and other data that often occurs, even between similar vehicles in any particular VM's range, the reader should take great care in identification of ECM pins, and satisfy himself that he has gathered the correct data before failing a particular component.*

4 Wheel speed sensors

Checking the wheel speed sensor (general)

1 Inspect the wheel speed sensor for corrosion or damage and check that the sensor is tightly mounted.
2 Check the toothed sensor ring for damage, eccentricity and for broken or missing teeth.
3 Inspect the wheel speed sensor wiring plug for corrosion and damage. One plug for each sensor.
4 Check that the connector terminal pins are pushed fully home and making good contact with the sensor wiring plug.
5 Check the clearance between the sensor and the toothed sensor ring. The clearance is not normally adjustable but is nominally 0.2 to 1.2 mm. If the clearance is excessive, expect a worn sensor tip or problems with the wheel bearings/hub or sensor ring.
6 When carrying out voltage checks with an oscilloscope or voltmeter, the voltage obtained will be proportional to the speed at which the wheel is rotating. In addition to determining that the wheel speed sensors are actually producing a voltage output, it is essential that the output from the sensors on a particular axle is the same for any given wheel speed.

Checking wheel speed sensor output with an oscilloscope

Note: *Refer to the wiring diagram and pin tables for specific ECM pin identification according to model.*
7 Switch the ignition off and disconnect the ECM multi-plug or the relevant wheel speed sensor wiring plug.
8 Connect an oscilloscope between the terminal pins for the sensor under test **(see illustration 20.17)**.
9 Select a range to cover 80 Hz on the oscilloscope and a free run time base.
10 Raise the wheel and rotate it by hand at approximately one revolution per second.
11 A sinusoidal wave form should be obtained, with amplitude and duration changing with rotational speed **(see illustration 20.18)**.

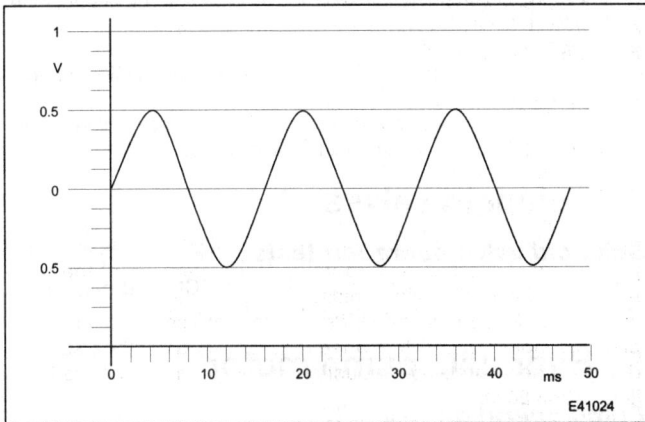

20.17 Checking wheel speed sensor output with an oscilloscope connected to the sensor wiring plug

12 If there is no signal, or a very weak or intermittent signal at the ECM, repeat the test at the sensor wiring plug. If there is no change in signal status, the sensor is suspect.
13 If the signal is now satisfactory this indicates a fault in the wiring harness which should be checked for continuity.

Checking wheel speed sensor output with an AC voltmeter

Note: *Refer to the wiring diagram and pin tables for specific ECM pin identification according to model.*
14 Switch the ignition off and disconnect the ECM multi-plug or the relevant wheel speed sensor wiring plug.
15 Connect an AC voltmeter between the terminal pins for the sensor under test **(see illustration 20.19)**.
16 Raise the wheel and rotate it by hand at approximately one revolution per second.

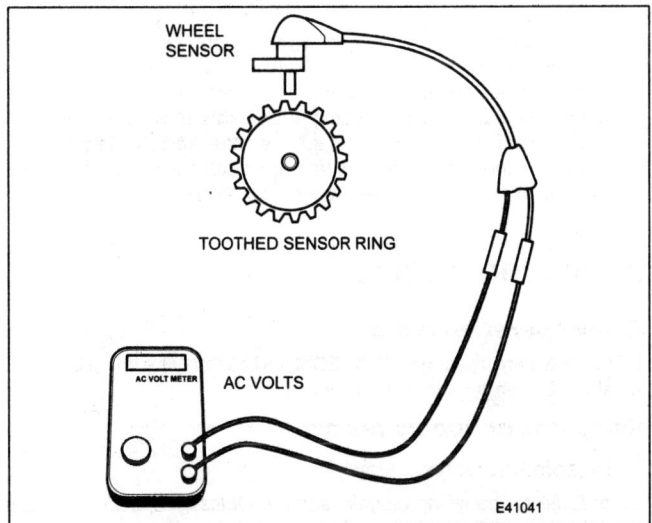

20.18 Typical wheel speed sensor sine wave as displayed on an oscilloscope

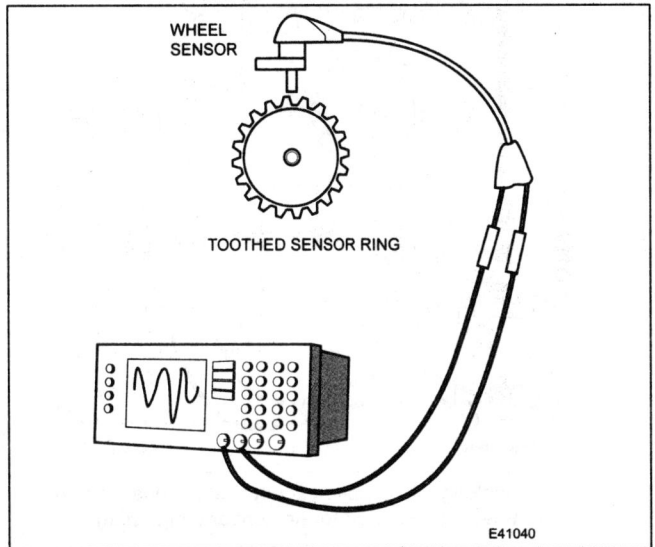

20.19 Checking wheel speed sensor output with a voltmeter connected to the sensor wiring plug

20

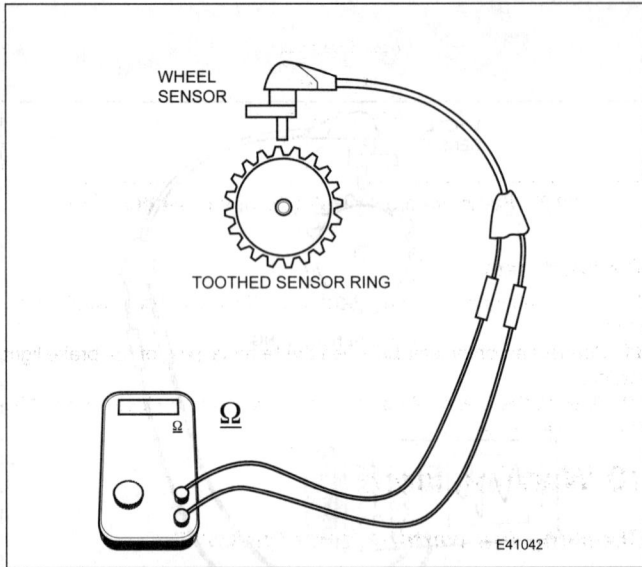

20.20 Checking wheel speed sensor resistance with an ohmmeter connected to the sensor wiring plug

17 A voltage of approximately 0.1 to 1.0 volts (AC RMS) should be obtained. If there is no signal, or a very weak or intermittent signal at the ECM, repeat the test at the sensor wiring plug. If there is no change in the signal, the sensor is suspect.

18 If the signal is now satisfactory, this indicates a fault in the wiring harness which should be checked for continuity. **Note:** *This test at least proves that a signal is being generated by the sensor. However, the voltage produced is an average voltage and does not clearly indicate damage to the sensor ring or that the sinewave is regular in formation.*

Checking wheel speed sensor resistance

Note: *Refer to the wiring diagram and pin tables for specific ECM pin identification according to model.*

19 Switch the ignition off and disconnect the ECM multi-plug or the relevant wheel speed sensor wiring plug.

20 Connect an ohmmeter between the terminal pins for the sensor under test **(see illustration 20.20)**.

21 The readings obtained should be between 0.7 and 1.5 kohms approximately.

22 If the resistance is excessively high, or open circuit at the ECM, repeat the test at the sensor multi-plug. If there is no change in resistance, the sensor is suspect.

23 If the resistance is now satisfactory, this indicates a fault in the wiring harness which should be checked for continuity. **Note:** *Even if the resistance is within the quoted specifications, this does not prove that the speed sensor can generate an acceptable signal.*

5 System relays

Relay operation tests

1 The relays are integral with the ECM and can only be checked using suitable ABS diagnostic test equipment.

Relay power supply tests

Main (solenoid valve) relay

Note: *Refer to the wiring diagram and pin tables for specific ECM pin identification according to model.*

2 Switch the ignition off and disconnect the ECM multi-plug.

3 Attach a negative voltmeter probe to a vehicle earth.

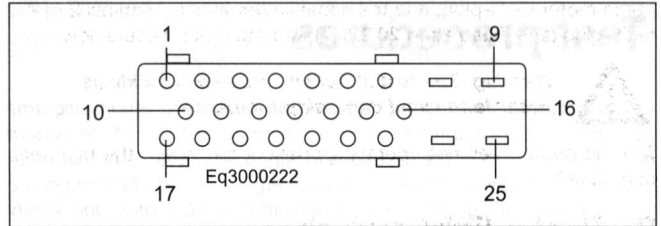

20.21 ECM 25-pin multi-plug

4 Attach a positive voltmeter probe to ECM multi-plug pin 9 **(see illustration 20.21)**. The voltmeter should indicate nbv.

5 If no voltage is found, check the relevant fuse and the supply wiring back to the battery positive terminal.

Pump relay

Note: *Refer to the wiring diagram and pin tables for specific ECM pin identification according to model.*

6 Switch the ignition off and disconnect the ECM multi-plug.

7 Attach a negative voltmeter probe to a vehicle earth.

8 Attach a positive voltmeter probe to ECM multi-plug pin 25 **(see illustration 20.21)**.

9 Where applicable, switch the ignition on.

10 The voltmeter should indicate nbv. If no voltage is found, check the relevant fuse and the supply wiring back to the ignition switch or battery positive terminal.

6 Electronic Control Module

Checking the ECM (general)

1 Inspect the ECM for corrosion or damage and ensure that the unit is securely attached to the hydraulic control unit.

2 Check that the ECM multi-plug terminals are pushed fully home and making good contact with the ECM pins. A fault in any of the above areas are possible reasons for poor performance in the ABS system.

ECM power supply and earth tests

Note: *Refer to the wiring diagram and pin tables for specific ECM pin identification according to model.*

3 Switch the ignition off and disconnect the ECM multi-plug.

4 Switch the ignition on.

5 Attach a negative voltmeter probe to a vehicle earth.

6 Attach a positive voltmeter probe to the following ECM multi-plug pin according to model **(see illustration 20.21)**:

  Focus and Scorpio - pin 20
  All other models - pin 7

7 If no voltage is found, check the relevant fuse and the supply wiring back to the ignition switch.

8 Switch the ignition off.

9 Connect an ohmmeter between a vehicle earth and ECM multi-plug earth pins, typically 8 and 24, in turn.

10 The ohmmeter should indicate continuity in each case. If not, check the ECM main earth connection and wiring.

7 Solenoid valves

Solenoid valve operation tests

1 The solenoid valves are integral with the ECM and can only be checked using suitable ABS diagnostic test equipment.

8 Hydraulic pump motor

Pump operation test

1 Switch the ignition off and disconnect the pump motor multi-plug.

2 Connect the positive terminal of a 12 volt supply to terminal 1 of the pump motor multi-plug, and the negative terminal to terminal 2 of the multi-plug (see illustration 20.22). The pump motor should now run.

⚠ **Warning: The test should be made as quickly as possible to avoid damaging the pump.**

3 If the pump does not operate as described, renew the hydraulic control unit.

9 Brake light switch

Checking the brake light switch (general)

1 Check that the brake light switch is correctly and securely mounted and that the plunger moves smoothly with no trace of binding.
2 Check that the wiring multi-plug is pushed fully home and making good contact.
3 Check that no wires have been disconnected.
4 A fault in any of the above areas are possible reasons for failure or malfunctioning of the switch.

Brake light switch voltage and continuity tests

Voltage test

Note: Refer to the wiring diagram and pin tables for specific ECM pin identification according to model.

5 Switch the ignition off and disconnect the ECM multi-plug from the ECM.
6 Connect a voltmeter between a vehicle earth and the ECM multi-plug brake light switch input pin according to model, as follows (see illustration 20.21):
  Focus and Scorpio - pin 2
  All other models - pin 12
7 Switch the ignition on and depress the brake pedal. The voltmeter should indicate nbv.
8 If no voltage is found, the fuse, the brake light switch and the wiring are suspect.
9 Release the brake pedal. The voltage should drop to zero as the switch opens.

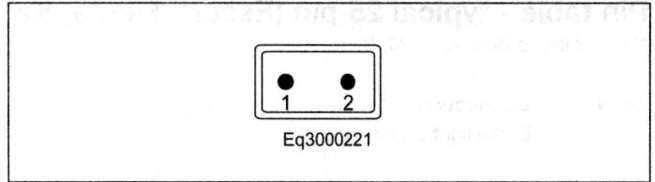

20.22 Pump motor multi-plug terminal identification

Continuity test

10 Switch the ignition off and disconnect the brake light switch multi-plug.
11 Connect an ohmmeter between the terminal pins of the brake light switch.
12 Operate the brake light switch and check for continuity. If the test fails, renew the switch.

10 Warning lamp

Checking the warning lamp (general)

1 Inspect the warning lamp bulb holder contacts in the instrument panel.
2 Check that the instrument panel multi-plug terminal pins are pushed fully home and making good contact.
3 Check that no wires have been disconnected.
4 A fault in any of the above areas are possible reasons for failure or malfunctioning of the warning lamp.

Warning lamp operation test

5 With the ignition switched off, the warning lamp should remain off.
6 Switch the ignition on and the warning lamp should illuminate then extinguish after a few seconds. The lamp should then remain off.
7 If the warning lamp comes on and remains on at any time during vehicle operation, carry out the previously described test procedures on the system components.

Pin table - typical 25-pin (Escort, Fiesta, Ka, Puma)

Note: Refer to illustration 20.21

Pin No.	Connection	Test condition	Voltage
1	Rear right wheel speed sensor earth	Roadwheel rotating	0.25 volts (max)
2	Front right wheel speed sensor signal	Roadwheel rotating	0.1 to 1.0 volts AC (approx)
3	-	-	-
4	-	-	-
5	-	-	-
6	-	-	-
7	Supply from ignition switch	Ignition on	Nbv
8	ECM earth	Ignition on	0.25 volts (max)
9	Supply from battery	Ignition on/off	Nbv
10	Rear left wheel speed sensor earth	Roadwheel rotating	0.25 volts (max)
11	Front left wheel speed sensor signal	Roadwheel rotating	0.1 to 1.0 volts AC (approx)

Pin table - typical 25-pin (Escort, Fiesta, Ka, Puma) (continued)

Note: *Refer to illustration 20.21*

Pin No.	Connection	Test condition	Voltage
12	Brake light switch input	Ignition on: Brake pedal released Brake pedal depressed	 0 volts Nbv
13	SD connector	-	-
14	-	-	-
15	-	-	-
16	ABS warning lamp	Ignition on: Lamp on Lamp off	 2.0 volts (max) Nbv
17	Rear right wheel speed sensor signal	Roadwheel rotating	0.1 to 1.0 volts AC (approx)
18	Rear left wheel speed sensor signal	Roadwheel rotating	0.1 to 1.0 volts AC (approx)
19	Front right wheel speed sensor earth	Roadwheel rotating	0.25 volts (max)
20	Front left wheel speed sensor earth	Roadwheel rotating	0.25 volts (max)
21	-	-	-
22*	Supply from ignition switch	Ignition on	Nbv
23	-	-	-
24	ECM earth	Ignition on	0.25 volts (max)
25	Supply from ignition switch	Ignition on	Nbv

*Early models only

Pin table - typical 25-pin (Focus)

Note: *Refer to illustration 20.21*

Pin No.	Connection	Test condition	Voltage
1	-	-	-
2	Brake light switch input	Ignition on: Brake pedal released Brake pedal depressed	 0 volts Nbv
3	Front right wheel speed sensor signal	Roadwheel rotating	0.1 to 1.0 volts AC (approx)
4	Front right wheel speed sensor earth	Roadwheel rotating	0.25 volts (max)
5	-	-	-
6	Rear right wheel speed sensor earth	Roadwheel rotating	0.25 volts (max)
7	Rear right wheel speed sensor signal	Roadwheel rotating	0.1 to 1.0 volts AC (approx)
8	ECM earth	Ignition on	0.25 volts (max)
9	Supply from battery	Ignition on/off	Nbv
10	-	-	-
11	-	-	-
12	-	-	-
13	Instrument cluster	-	-
14	-	-	-
15	-	-	-
16	ABS warning lamp	Ignition on: Lamp on Lamp off	 2.0 volts (max) Nbv
17	Front left wheel speed sensor signal	Roadwheel rotating	0.1 to 1.0 volts AC (approx)
18	Front left wheel speed sensor earth	Roadwheel rotating	0.25 volts (max)
19	-	-	-
20	Supply from ignition switch	Ignition on	Nbv
21	Rear left wheel speed sensor signal	Roadwheel rotating	0.1 to 1.0 volts AC (approx)
22	Rear left wheel speed sensor earth	Roadwheel rotating	0.25 volts (max)
23	SD connector	-	-
24	-	-	-
25	Supply from battery	Ignition on/off	Nbv

Pin table - typical 25-pin (Scorpio)

Note: *Refer to illustration 20.21*

Pin No.	Connection	Test condition	Voltage
1	-	-	-
2	Brake light switch input	Ignition on: Brake pedal released Brake pedal depressed	 0 volts Nbv
3	Front right wheel speed sensor earth	Roadwheel rotating	0.25 volts (max)
4	Front right wheel speed sensor signal	Roadwheel rotating	0.1 to 1.0 volts AC (approx)
5	-	-	-
6	Rear right wheel speed sensor signal	Roadwheel rotating	0.1 to 1.0 volts AC (approx)
7	Rear right wheel speed sensor earth	Roadwheel rotating	0.25 volts (max)
8	ECM earth	Ignition on	0.25 volts (max)
9	Supply from battery	Ignition on/off	Nbv
10	-	-	-
11	-	-	-
12	-	-	-
13	-	-	-
14	-	-	-
15	-	-	-
16	ABS warning lamp	Ignition on: Lamp on Lamp off	 2.0 volts (max) Nbv
17	Front left wheel speed sensor earth	Roadwheel rotating	0.25 volts (max)
18	Front left wheel speed sensor signal	Roadwheel rotating	0.1 to 1.0 volts AC (approx)
19	-	-	-
20	Supply from ignition switch	Ignition on	Nbv
21	Rear left wheel speed sensor earth	Roadwheel rotating	0.25 volts (max)
22	Rear left wheel speed sensor signal	Roadwheel rotating	0.1 to 1.0 volts AC (approx)
23	SD connector	-	-
24	ECM earth	Ignition on	0.25 volts (max)
25	Supply from battery	Ignition on/off	Nbv

Fault codes

1 The Teves 20-I system requires the use of a FCR for obtaining fault codes. Flash codes are not available for output from this system.
2 If a FCR is available, it should be connected to the SD serial connector and used in accordance with the maker's instructions.
3 The FCR can be used for the following purposes:
a) Obtaining fault codes.
b) Clearing fault codes.
c) Obtaining datastream information.

d) Testing the system actuators (solenoid valve relay, pump relay and solenoid valves).
4 Internal fault codes are used by the ECM to designate faults in the system components and circuits. A proprietary fault code reader (FCR) or system tester (such as the Ford FDS 2000) is required to interrogate the system. No actual fault code numbers are available although the components or circuits checked by the ECM include the following.

Fault code table

Code	Item	Fault
-	Supply voltage	Too high/low
-	Relays	Defective (signal interruption, short to earth, supply voltage)
-	Pump motor	Defective (poor connections)
-	Front left wheel speed sensor	Incorrect resistance/signal interruption
-	Front right wheel speed sensor	Incorrect resistance/signal interruption
-	Rear left wheel speed sensor	Incorrect resistance/signal interruption
-	Rear right wheel speed sensor	Incorrect resistance/signal interruption
-	ECM	Defective

20

Wiring diagrams

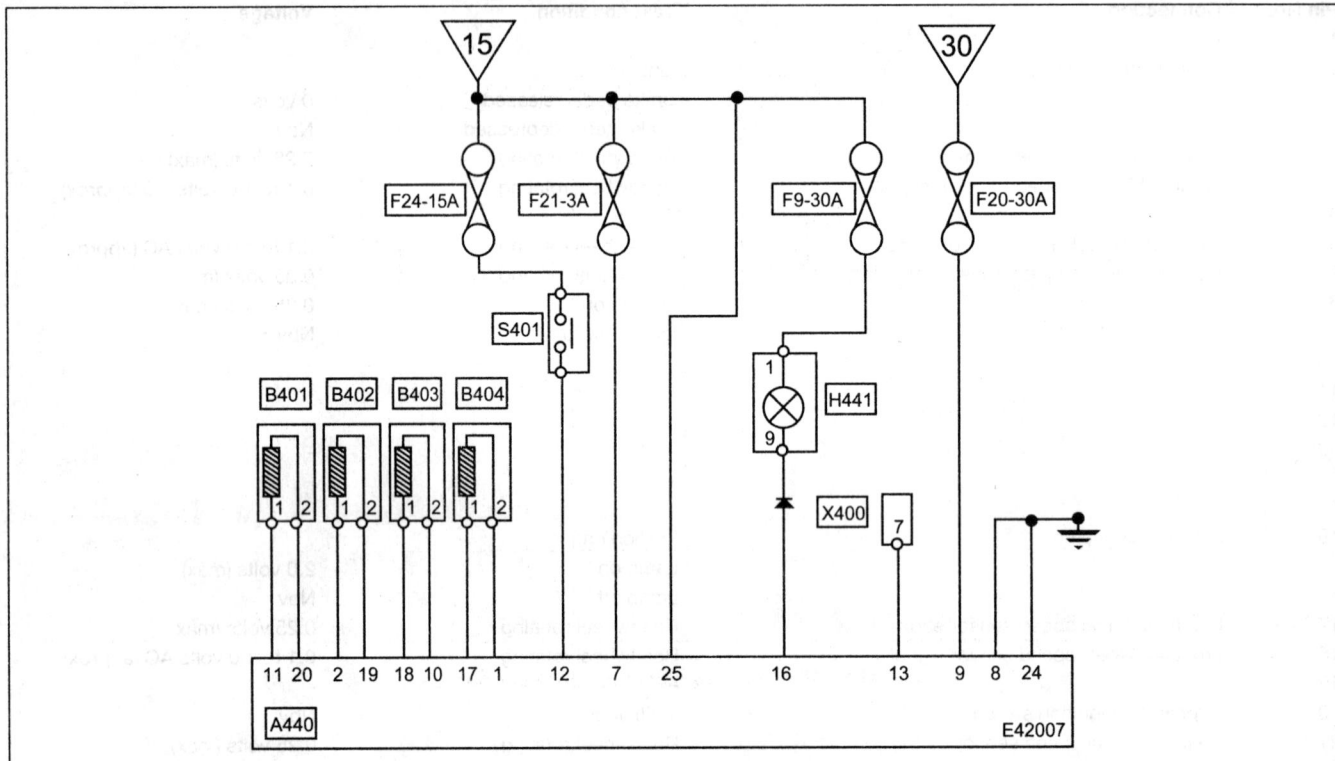

20.23 25-pin wiring diagram, Ford Escort

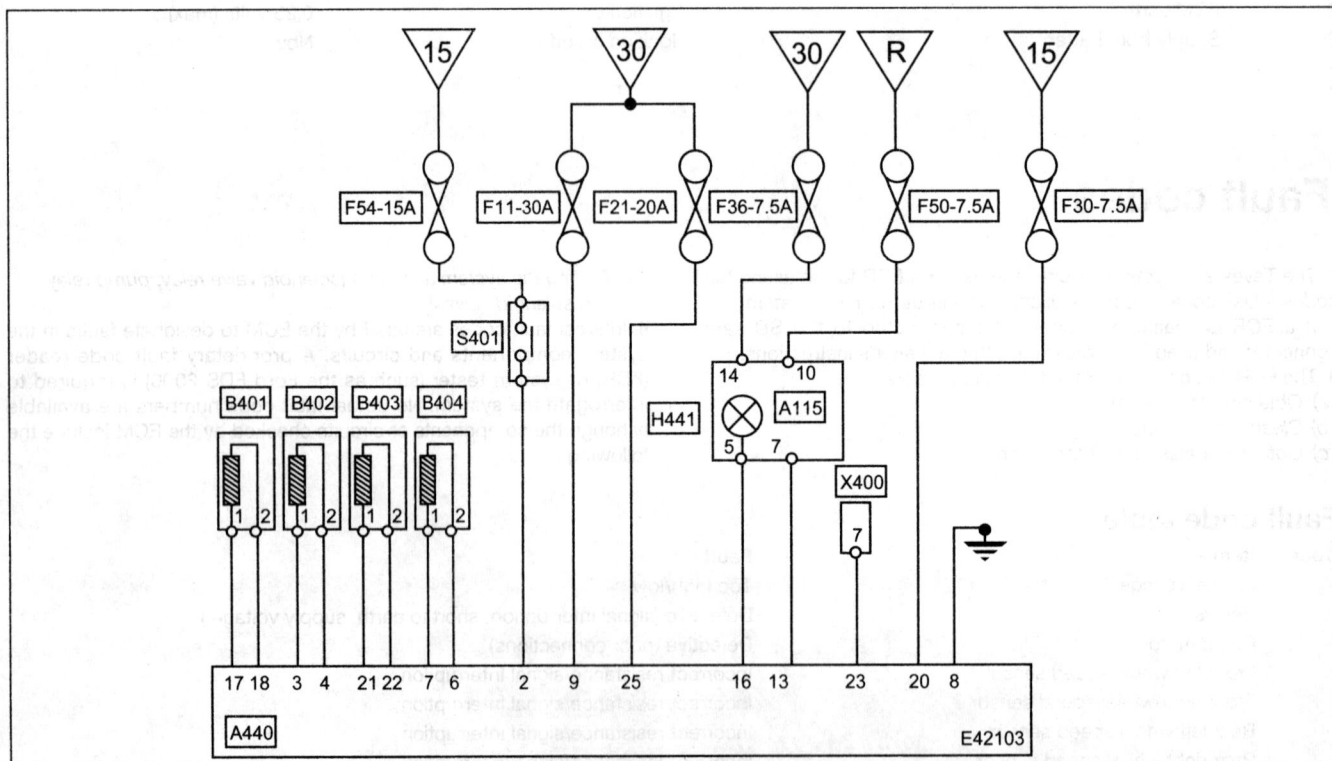

20.24 25-pin wiring diagram, Ford Focus

| F14-15A | F47-30A | F46-30A | F16-7.5A | F27-10A |

S401

B401 B402 B403 B404

33

H441

5

X400

| 18 17 | 4 | 3 | 22 21 | 6 | 7 | 2 | 9 | 25 | 16 | 23 | 20 | 8 | 24 |

A440

E42104

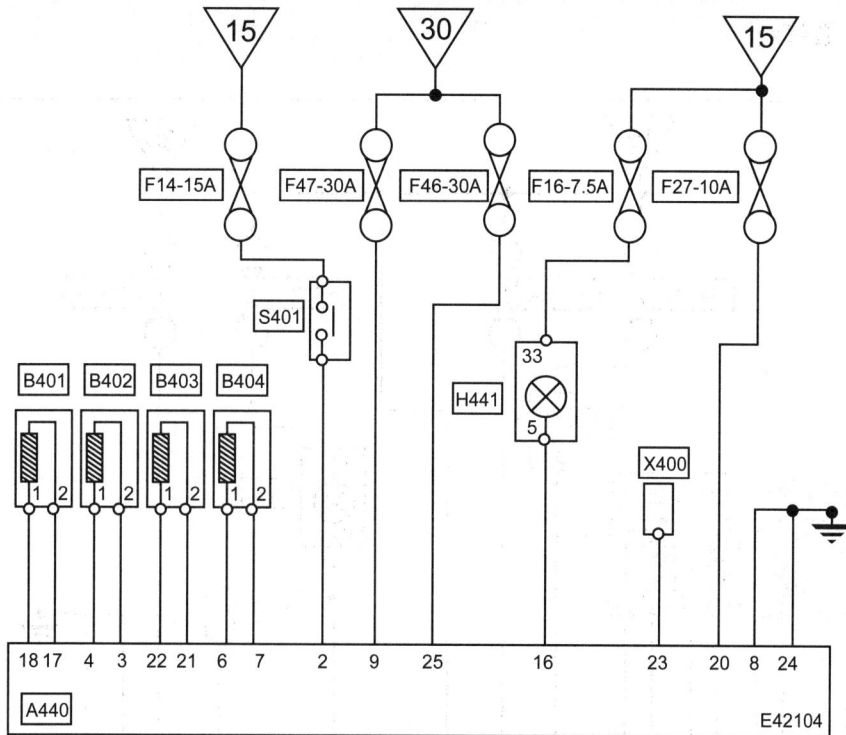

20.25 25-pin wiring diagram, Ford Scorpio

Chapter 21

Teves 20GI (ITT 20GI) ABS

Contents

Vehicle coverage

Model	Year
Fiat	
Bravo/Brava	1995-2000
Seat	
Cordoba	1995-2000
Ibiza	1995-2000
Inca	1995-1997
Toledo	1995-1999
Skoda	
Felicia	1995-2000
Volkswagen	
Caddy	1995-1997
Golf	1995-1998
Passat	1995-1997
Polo	1994-1997
Vento	1995-1998
Volvo	
850	1996
S70/V70/C70	1996-1998

21

Overview of system operation

1 Basic principles and system identification

The Teves 20GI Antilock Brake System is a development of the earlier Teves IV-GI series ABS, with further improvements to the hydraulic control unit, and the Electronic Control Module (ECM) software. Teves 20GI has been fitted to various passenger vehicles since its introduction in the early 1990s. The system is of the additional or 'add-on' type operating in conjunction with the conventional braking system components. Later versions of the system may carry the designation ITT 20GI although their operation is unchanged from the Teves 20GI types.

There are certain discrepancies in vehicle manufacturers technical documentation relating to the identification of Teves (ITT) 20GI and the later ITT 20IE system which replaced it. The two systems are visually and operationally identical apart from different ECM pin connections. In most instances it is only possible to differentiate between the two systems by interrogation of the ECM using a suitable system tester.

Depending on application, Teves 20GI may be installed as an antilock brake system only, or as an antilock brake system incorporating traction control.

In ABS mode, the purpose of the system is to apply the vehicle brakes at maximum efficiency without wheel lock or loss of directional stability. Inductive sensors (wheel speed sensors) monitor the speed of the wheels by generating an electrical signal as the wheel is rotated. This information is passed to the ABS Electronic Control Module (ECM) which is then able to determine wheel speed, wheel acceleration and wheel deceleration. The ECM compares the signals received from each wheel and if the onset of lock at any wheel is detected, a signal is sent to the ABS hydraulic control unit which regulates the brake pressure for the relevant wheel(s).

Where the system incorporates traction control, essentially the reverse principle is applied. When the ECM detects that one or more wheels are rotating faster than the reference value, the brake is actually applied on the relevant wheel(s) to reduce the rotational speed. Additionally, on some installations, when wheel spin is detected, various signals are sent to the engine management ECM to control engine torque and rpm.

Typically, Teves 20GI ABS is comprised of the following components (see illustration 21.1):
a) Hydraulic control unit with integral ABS-ECM.
b) Four inductive wheel speed sensors and associated sensor rings.
c) Brake light switch.
d) ABS warning lamp.
e) Diagnostic connector.

In addition, the conventional brake system is comprised of the following components:
a) Tandem brake master cylinder.
b) Vacuum servo unit.
c) Brake calipers/wheel cylinders and hydraulic hoses and pipes.
d) Pressure regulating/load sensing valve(s) depending on application.

2 Component description and operation

ABS ECM

General

The Teves 20GI Electronic Control Module (ECM) continually monitors wheel speed from the signals provided by the wheel speed sensors, and brake application from the brake light switch signal. If the ECM detects the incidence of wheel lock on one or more wheels, a signal is sent to the hydraulic control unit to modulate the hydraulic pressure to the brake of the locking wheel(s). The ECM contains two microprocessors and uses digital technology to complete this function and other functions such as, fault code memory and power modules for valve and pump activation (see illustration 21.2).

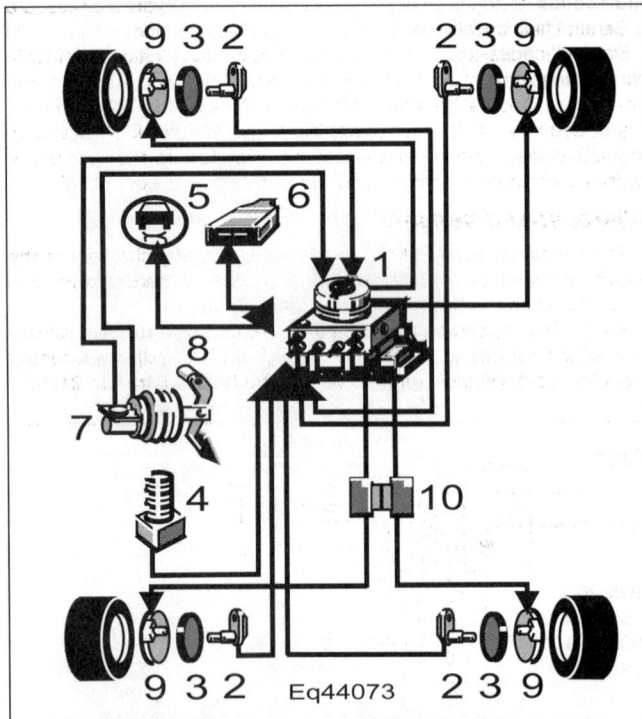

21.1 Typical Teves 20GI main components

1	Hydraulic control unit with integral ABS-ECM	7	Tandem brake master cylinder
2	Inductive wheel speed sensors	8	Vacuum servo unit
3	Sensor rings	9	Brake calipers/wheel cylinders and hydraulic hoses and pipes
4	Brake light switch		
5	ABS warning lamp	10	Pressure regulating/load sensing valve(s) depending on application
6	Diagnostic connector		

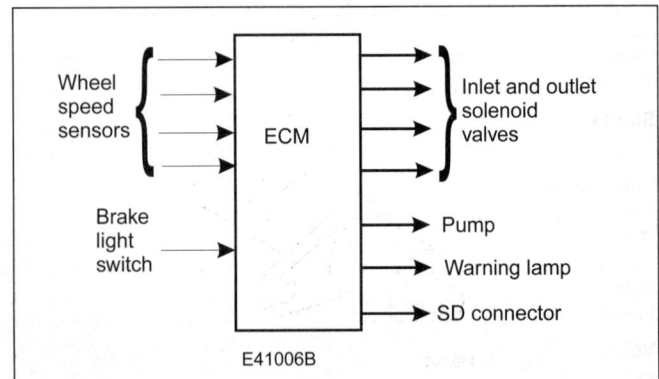

21.2 ECM sensor inputs and control signal outputs

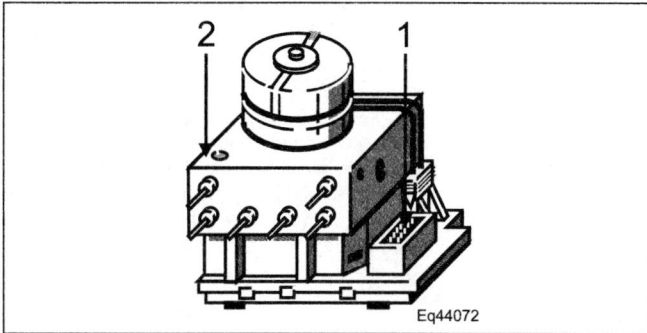

21.3 Teves 20GI ECM (1) and hydraulic control unit (2)

To reduce external electrical connections to a minimum and improve reliability, the ECM is integral with the hydraulic control unit **(see illustration 21.3)**.

Self-test

The Teves 20GI ECM is equipped with a self-test capability that initially examines the ABS system when the ignition is switched on, and then examines the wheel speed sensor signals after a wheel speed of approximately 4 mph is reached from all wheels (engine running). The ABS self-test program continues to examine the signals from the various components as long as the ignition is switched on. If self-test determines that faults are not present, the ABS is ready for operation once a specified vehicle speed has been achieved.

If the ECM detects that a fault is present, all ABS functions are switched off and the warning lamp is turned on. The conventional braking system continues to operate as normal without ABS assistance.

Self-diagnostics

If the ECM detects a fault during the self-test routine, an internal fault code is stored in the ECM memory. Stored fault codes can be retrieved from the SD connector with the aid of a suitable fault code reader. If the fault clears, the code will remain stored until cleared with the FCR.

Hydraulic control unit

Teves 20GI is typically a four-channel system with a separate hydraulic circuit for each brake. On certain models, however, the system is installed in three-channel configuration with a separate hydraulic circuit for each front brake, but with the rear brakes controlled as a pair. The hydraulic control unit consists of an electric motor operating a return pump with eccentric drive and twin radial pistons, inlet and outlet solenoid valves, pressure accumulators and pulsation dampers **(see illustration 21.4)**. The unit controls the hydraulic pressure applied to the brake for each individual front wheel

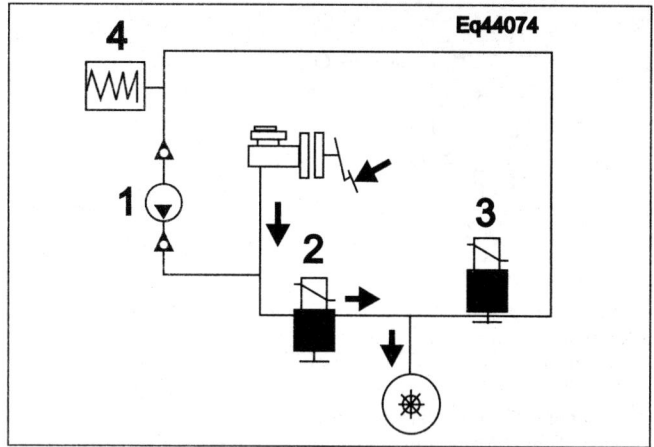

21.4 Hydraulic control unit components

1	Electric motor and return pump	3	Outlet solenoid valve
2	Inlet solenoid valve	4	Pressure accumulator

and each individual rear wheel, or pair of wheels. The return pump is switched on when the ABS is activated and returns hydraulic fluid, drained off during the pressure reduction phase, back into the brake circuit.

On some applications, the 'select-low' principle is employed for control of the rear brakes. With the 'select-low' principle, the wheel with the lowest adhesion determines the amount of hydraulic pressure to be supplied to both rear brakes during ABS operation.

In certain installations, the ABS-ECM contains additional software for rear brake hydraulic fluid pressure regulation when ABS is not in operation. This is generally known as 'Electronic Brake Force Distribution' and in these applications, mechanical pressure regulating/load sensing valves are not required as the rear brake hydraulic pressure is controlled by the ABS hydraulic control unit.

Wheel speed sensors

The rotational speed of the roadwheels and any changes in the rotational speed are recorded by inductive wheel speed sensors, one located at each roadwheel **(see illustration 21.5)**.

Each wheel speed sensor assembly comprises a toothed sensor ring which rotates at roadwheel speed, and an adjacent sensor mounted a set distance from the sensor ring **(see illustration 21.6)**.

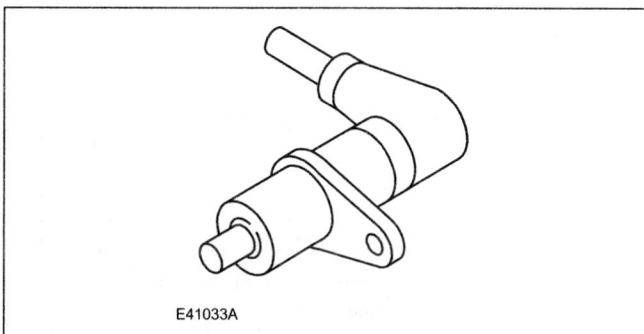

21.5 Typical wheel speed sensor

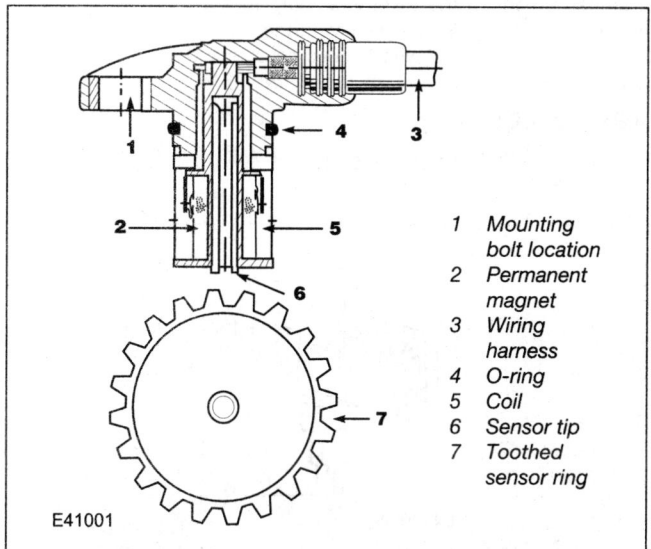

1	Mounting bolt location
2	Permanent magnet
3	Wiring harness
4	O-ring
5	Coil
6	Sensor tip
7	Toothed sensor ring

21.6 Sectional view of a wheel speed sensor

21

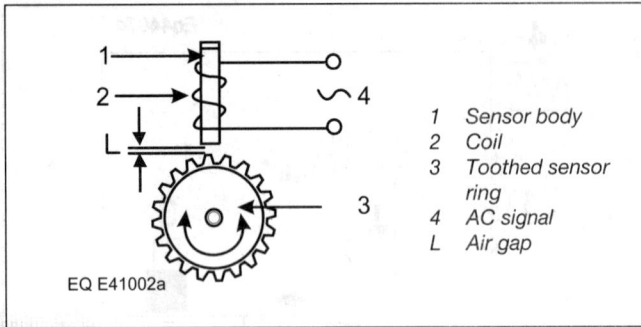

1. Sensor body
2. Coil
3. Toothed sensor ring
4. AC signal
L. Air gap

EQ E41002a

21.7 Wheel speed sensor operation

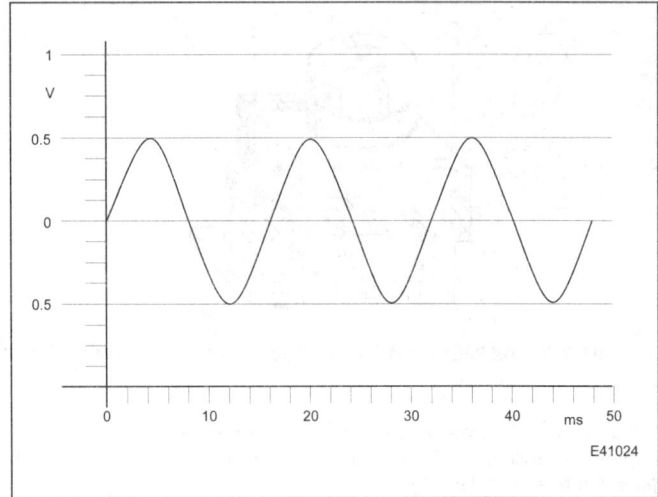

21.8 Wheel speed sensor waveform as viewed on an oscilloscope

The sensors are permanent magnet pulse generator types producing an AC voltage sine wave as the sensor ring teeth pass through the magnetic field of the sensor **(see illustration 21.7)**.

The frequency of the waveform produced by the wheel speed sensor is proportional to the road speed. This AC voltage signal is continually being delivered to the ABS-ECM for processing.

The peak to peak voltage of the speed signal (when viewed upon an oscilloscope) can vary considerably according to wheel speed. An analogue to digital converter (ADC) in the ECM transforms the AC pulse into a digital signal **(see illustration 21.8)**.

ABS electrical wiring harness and relays

An integrated main wiring harness is used to connect power and earth to various electrical components and to enable sensor signals to reach the ECM. The ECM, in turn, sends command signals to the hydraulic control unit via internal connections between the two units. The vehicle relays and ABS fuses are typically located in the engine compartment fuse/relay box .

Nominal battery voltage is supplied to the ECM via the ABS main fuses. The ECM then supplies nominal battery voltage to the solenoid valves and return pump and controls the earth for activation of the solenoid valves.

Brake light switch

The brake light switch comprises a switch body and contact pin and is located above the brake pedal **(see illustration 21.9)**.

When the brake pedal is depressed, closing the brake light switch, a signal is sent to the ECM indicating that the brakes are being applied. Once this signal is received, the ECM will begin monitoring the wheel speed via the wheel speed sensors and activate the ABS if necessary.

ABS warning lamp

After the ignition is switched on, the ABS warning lamp on the instrument panel is illuminated for approximately 2 to 4 seconds as the system executes a self-test routine. If satisfactory operation of the system is detected by the ECM, the light is extinguished. During vehicle operation above a pre-determined wheel speed, the ABS-ECM implements a further self-test cycle whereby ABS operation and wheel speed sensor signals are continually monitored. If a fault is detected, the relevant ECM pin is earthed to illuminate the warning lamp on the instrument panel, and the ABS function is disabled. The warning lamp will remain illuminated until the fault is no longer present **(see illustration 21.10)**.

When the ABS-ECM detects a fault, the fault code is stored and the ABS warning lamp activated. If the fault no longer exists after the next system start (ignition on/off) the ABS warning lamp is extinguished after the self-test cycle, however the fault code remains stored in the ECM memory.

Tandem master cylinder

Typically, the tandem master cylinder comprises a body casting incorporating primary and secondary pressure chambers, primary piston, intermediate piston, floating piston, slotted pin and central valve. The cylinder operates as a conventional master cylinder using

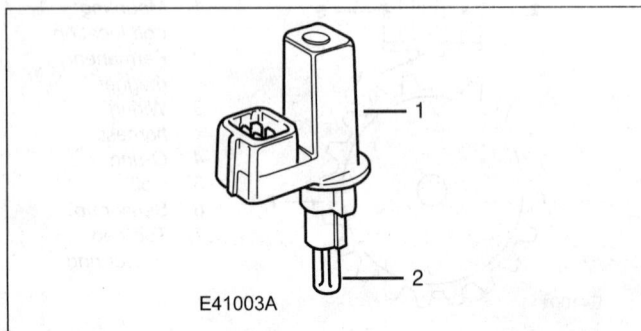

E41003A

21.9 Typical brake light switch body (1) and contact pin (2)

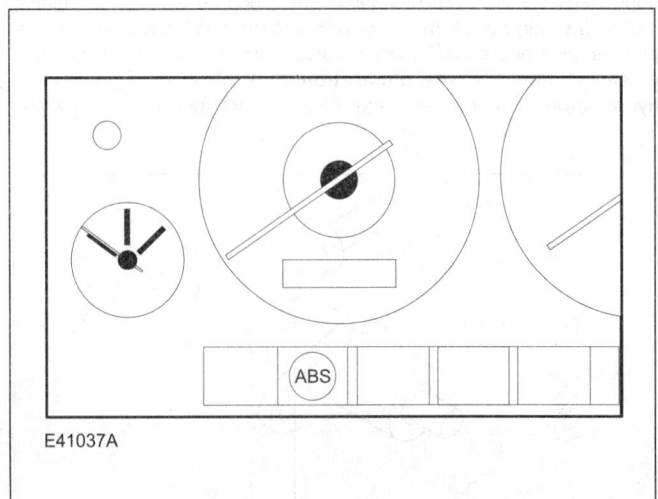

E41037A

21.10 ABS warning lamp

1 Body casting
2 Outlet connections
3 Fluid inlet from reservoir
4 Pushrod/primary piston
5 Intermediate piston
6 Slotted pin
7 Floating piston
8 Central valve

E41005a

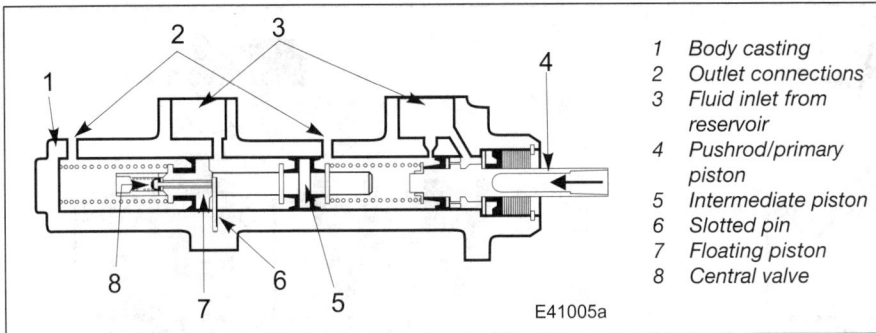

21.11 Sectional view of a typical tandem master cylinder

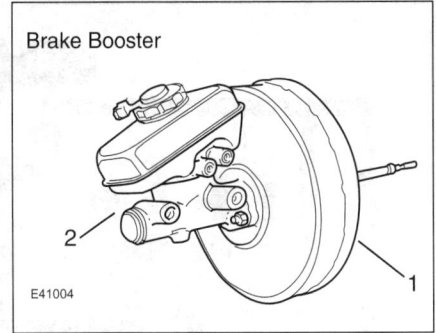

Brake Booster

E41004

21.12 Vacuum servo unit (1) and tandem master cylinder (2)

vacuum assistance from the vacuum servo unit **(see illustration 21.11)**.

When the brake system is at rest, the central valve in the floating piston rests against the slotted pin. In this condition the central valve is open and brake fluid can discharge out of the pressure chamber back into the brake fluid reservoir. When the brake pedal is depressed, the build-up of hydraulic pressure in the primary pressure chamber acts on the intermediate piston and floating piston, moving them down the cylinder bore. The floating piston contacts the seal on the central valve, closing the connection between the intermediate and secondary pressure chambers. Brake hydraulic pressure can now also increase in the secondary pressure chamber.

Vacuum servo unit

The vacuum servo unit is located between the brake pedal and tandem master cylinder. When the brake pedal is depressed, the servo unit increases the force applied by the pedal, reducing the effort required to operate the brakes **(see illustration 21.12)**.

The unit is operated by vacuum created in the engine inlet manifold (or from a separate vacuum pump on diesel engines) which is applied to a diaphragm contained within the unit casing. A pushrod connected to the centre of the diaphragm acts directly on the primary piston in the master cylinder.

When the brake pedal is released, vacuum is applied to both sides of the diaphragm. When the pedal is depressed, one side of the diaphragm is opened to atmosphere and the vacuum acting on the other side deflects the diaphragm which in turn operates the master cylinder primary piston. The resulting force applied to the master cylinder piston is therefore significantly greater than the initial force applied to the brake pedal by the driver.

Pressure regulating/load sensing valve(s)

Depending on vehicle application, pressure regulating valves or load sensing valves may be incorporated to restrict the hydraulic fluid pressure to the rear brakes. The valves may be pressure conscious whereby the hydraulic fluid supply is restricted once a pre-determined pressure is reached, or load conscious whereby the hydraulic pressure is reduced according to vehicle loading.

On certain versions, the ABS-ECM contains additional software for rear brake hydraulic fluid pressure regulation during normal braking, generally known as 'Electronic Brake Force Distribution'. In these applications, mechanical pressure regulating/load sensing valves are not required as the rear brake hydraulic pressure is controlled by the ABS hydraulic control unit.

3 System operation

Brake system at rest

When the system is at rest all the brake components are inoperative. Pressure is non-existent in the hydraulic pipes between the tandem master cylinder and the brake calipers. The inlet solenoid valves in the hydraulic control unit are open and the outlet solenoid valves are closed.

Brake system operating under conventional control without ABS

When the brake pedal is activated, the pedal force is applied to the tandem master cylinder by the vacuum servo unit pushrod. The servo unit pushrod acts directly on the pressure piston in the master cylinder which pressurises the hydraulic fluid in the brake pipes to the hydraulic control unit. The inlet solenoid valve and outlet solenoid valve both remain in the 'at rest' position (inlet solenoid valve open and outlet solenoid valve closed). Hydraulic pressure is transmitted to each brake caliper, thus operating the brakes.

When the brake pedal is released, the one-way valve opens allowing the hydraulic pressure in the circuit to rapidly decrease **(see illustration 21.13)**.

Brake system operating in conjunction with ABS control

The ABS-ECM continually monitors wheel speed from the signals provided by the wheel speed sensors. If the ECM detects the incidence of wheel lock on one or more wheels, ABS is automatically

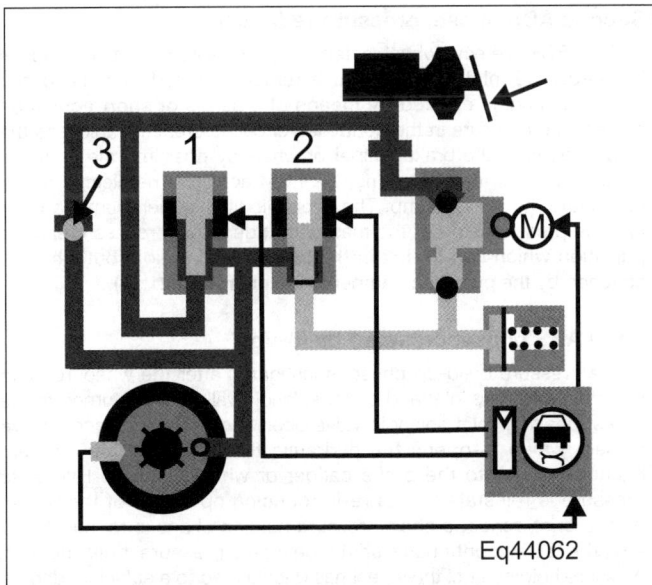

Eq44062

21.13 Brake system operating under conventional control without ABS

1 Inlet solenoid valve
2 Outlet solenoid valve
3 One-way valve

21

Eq44059

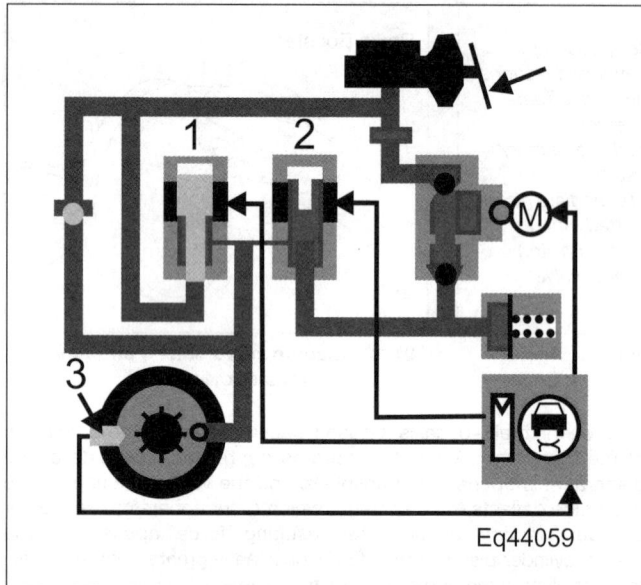

21.14 ABS operation - first phase, pressure holding

1	Inlet solenoid valve	3	Wheel speed sensor
2	Outlet solenoid valve		

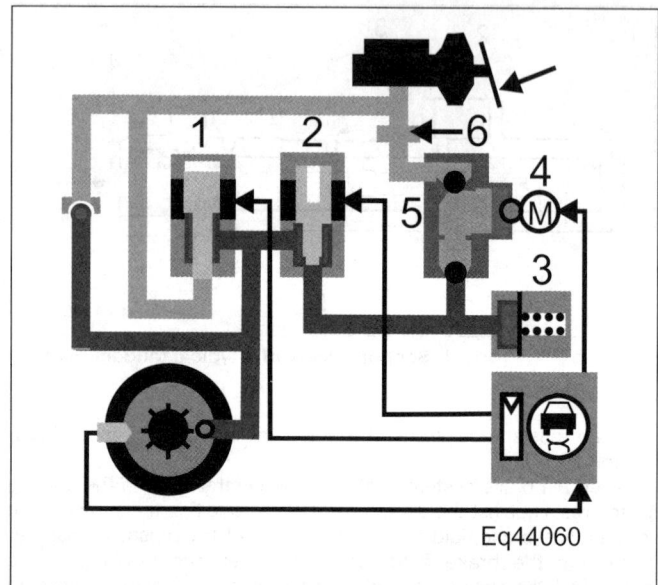

Eq44060

21.15 ABS operation - second phase, pressure reduction

1	Inlet solenoid valve	4	Pump motor
2	Outlet solenoid valve	5	Return pump
3	Pressure accumulator	6	Pulsation damper

initiated in three phases. As Teves 20GI ABS typically operates individually on each wheel, all or any of the wheels could be in any one of the following phases at any particular moment.

First ABS phase, pressure holding

To prevent any further build-up of hydraulic pressure in the circuit being controlled, the ECM closes the inlet solenoid valve and allows the outlet solenoid valve to remain closed. The hydraulic fluid line from the tandem master cylinder to the brake caliper or wheel cylinder is closed, and the hydraulic fluid in the controlled circuit is maintained at

Eq44061

21.16 ABS operation - third phase, pressure build-up

1	Inlet solenoid valve	2	Outlet solenoid valve

a constant pressure. This effectively removes the braking force from the controlled circuit. The pressure cannot now be increased in that circuit by any further application of the brake pedal (see illustration 21.14).

If the wheel speed sensor signals indicate that wheel rotation has now stabilised, the ECM will instigate the pressure build-up phase, allowing braking to continue. If wheel lock is still detected after the pressure holding phase, the ECM instigates the pressure reduction phase.

Second ABS phase, pressure reduction

If the ECM detects wheel instability, a pressure reduction phase is initiated. The inlet solenoid valve remains closed and the outlet solenoid valve is opened by means of a series of short activation pulses. The pressure in the controlled circuit decreases rapidly as the fluid flows from the brake caliper or wheel cylinder into the pressure accumulator. At the same time, the ECM actuates the electric motor to operate the return pump. The hydraulic fluid is then pumped back into the pressure side of the master cylinder. This process creates a pulsation which can be felt in the brake pedal action, but which is softened by the pulsation damper (see illustration 21.15).

Third ABS phase, pressure build-up

The pressure build-up phase is instigated after the wheel rotation has stabilised. The inlet and outlet solenoid valves are returned to the at rest position (inlet solenoid valve open and exhaust solenoid valve closed) which re-opens the hydraulic fluid line from the tandem master cylinder to the brake caliper or wheel cylinder. Hydraulic pressure is reinstated, thus re-introducing operation of the brake. After a brief period, a short pressure holding phase is re-introduced and the ECM continually shifts between pressure build-up and pressure holding until the wheel has decelerated to a sufficient degree where pressure reduction is once more required (see illustration 21.16).

The whole ABS control cycle takes place 4 to 10 times per second for each affected wheel and this ensures maximum braking effect and control during ABS operation.

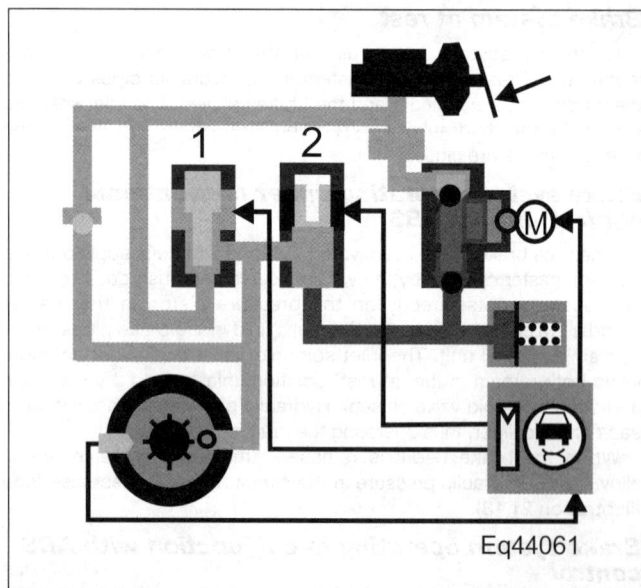

Test procedures

Important note: *The test procedures, pin-tables and wiring diagrams contained in this Chapter are necessarily representative of the system depicted. Because of the variations in wiring and other data that often occurs, even between similar vehicles in any particular VM's range, the reader should take great care in identification of ECM pins, and satisfy himself that he has gathered the correct data before failing a particular component.*

4 Wheel speed sensors

Checking the wheel speed sensor (general)

1 Inspect the wheel speed sensor for corrosion or damage and check that the sensor is tightly mounted.
2 Check the toothed sensor ring for damage, eccentricity and for broken or missing teeth.
3 Inspect the wheel speed sensor wiring plug for corrosion and damage. One plug for each sensor.
4 Check that the connector terminal pins are pushed fully home and making good contact with the sensor wiring plug.
5 Check the clearance between the sensor and the toothed sensor ring. The clearance is not normally adjustable but is nominally 0.2 to 1.2 mm. If the clearance is excessive, expect a worn sensor tip or problems with the wheel bearings/hub or sensor ring.
6 When carrying out voltage checks with an oscilloscope or voltmeter, the voltage obtained will be proportional to the speed at which the wheel is rotating. In addition to determining that the wheel speed sensors are actually producing a voltage output, it is essential that the output from the sensors on a particular axle is the same for any given wheel speed.

Checking wheel speed sensor output with an oscilloscope

Note: *Refer to the wiring diagrams for specific ECM pin identification according to model.*
7 Switch the ignition off and disconnect the ECM multi-plug or the relevant wheel speed sensor wiring plug.
8 Connect an oscilloscope between the terminal pins for the sensor under test **(see illustration 21.17)**.
9 Select a range to cover 80 Hz on the oscilloscope and a free run time base.
10 Raise the wheel and rotate it by hand at approximately one revolution per second.
11 A sinusoidal wave form should be obtained, with amplitude and duration changing with rotational speed **(see illustration 21.18)**.

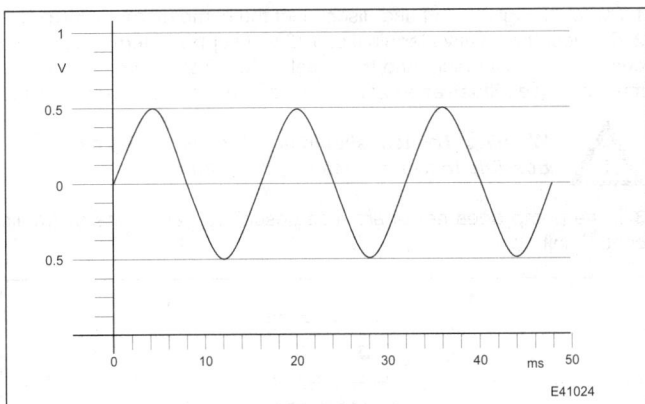

21.17 Checking wheel speed sensor output with an oscilloscope connected to the sensor wiring plug

12 If there is no signal, or a very weak or intermittent signal at the ECM, repeat the test at the sensor wiring plug. If there is no change in signal status, the sensor is suspect.
13 If the signal is now satisfactory this indicates a fault in the wiring harness which should be checked for continuity.

Checking wheel speed sensor output with an AC voltmeter

Note: *Refer to the wiring diagrams for specific ECM pin identification according to model.*
14 Switch the ignition off and disconnect the ECM multi-plug or the relevant wheel speed sensor wiring plug.
15 Connect an AC voltmeter between the terminal pins for the sensor under test **(see illustration 21.19)**.

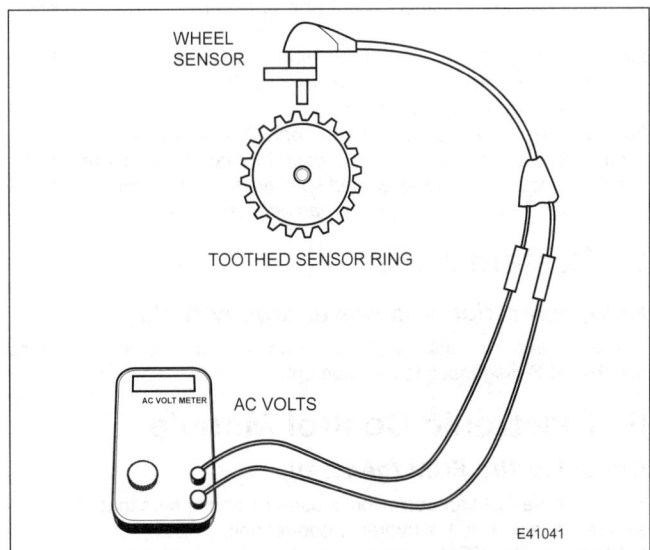

21.18 Typical wheel speed sensor sine wave as displayed on an oscilloscope

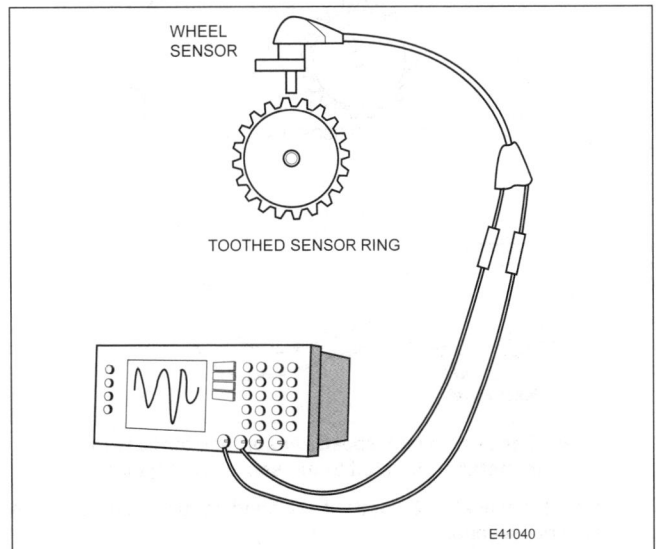

21.19 Checking wheel speed sensor output with a voltmeter connected to the sensor wiring plug

21

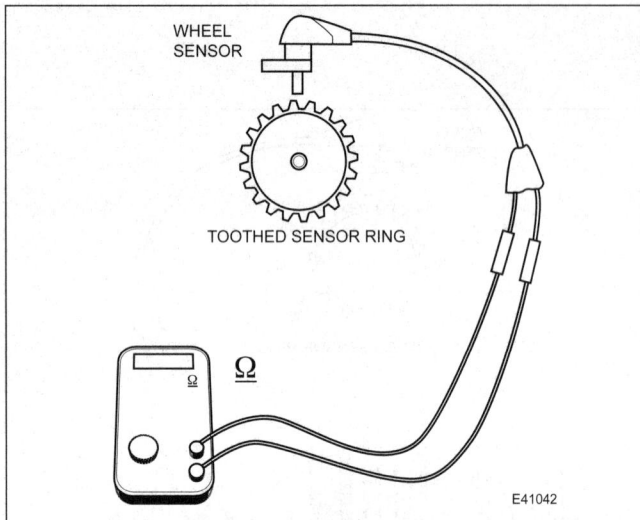

21.20 Checking wheel speed sensor resistance with an ohmmeter connected to the sensor wiring plug

16 Raise the wheel and rotate it by hand at approximately one revolution per second.

17 A voltage of approximately 0.5 to 1.5 volts (AC RMS) should be obtained. If there is no signal, or a very weak or intermittent signal at the ECM, repeat the test at the sensor wiring plug. If there is no change in the signal, the sensor is suspect.

18 If the signal is now satisfactory, this indicates a fault in the wiring harness which should be checked for continuity. **Note:** *This test at least proves that a signal is being generated by the sensor. However, the voltage produced is an average voltage and does not clearly indicate damage to the sensor ring or that the sinewave is regular in formation.*

Checking wheel speed sensor resistance

Note: *Refer to the wiring diagrams for specific ECM pin identification according to model.*

19 Switch the ignition off and disconnect the ECM multi-plug or the relevant wheel speed sensor wiring plug.

20 Connect an ohmmeter between the terminal pins for the sensor under test **(see illustration 21.20)**.

21 The readings obtained should be between 0.7 and 2.2 kohms approximately.

22 If the resistance is excessively high, or open circuit at the ECM, repeat the test at the sensor multi-plug. If there is no change in resistance, the sensor is suspect.

23 If the resistance is now satisfactory, this indicates a fault in the wiring harness which should be checked for continuity. **Note:** *Even if the resistance is within the quoted specifications, this does not prove that the speed sensor can generate an acceptable signal.*

5 System relays

Relay operation and power supply tests

1 The relays are integral with the ECM and can only be checked using suitable ABS diagnostic test equipment.

6 Electronic Control Module

Checking the ECM (general)

1 Inspect the ECM for corrosion or damage and ensure that the unit is securely attached to the hydraulic control unit.

2 Check that the ECM multi-plug terminals are pushed fully home and making good contact with the ECM pins. A fault in any of the above areas are possible reasons for poor performance in the ABS system.

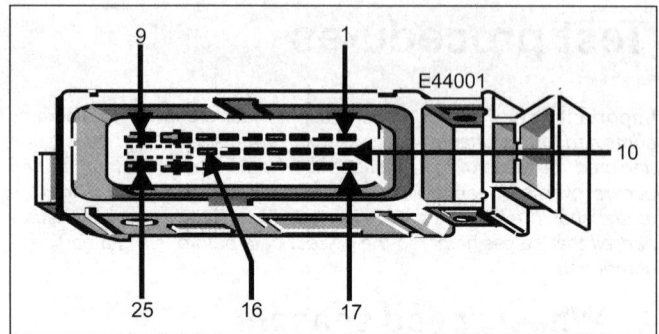

21.21 ECM 25-pin multi-plug

ECM power supply and earth tests

Note: *Refer to the wiring diagrams for specific ECM pin identification according to model.*

3 Switch the ignition off and disconnect the ECM multi-plug.

4 Attach a negative voltmeter probe to a vehicle earth.

5 Attach a positive voltmeter probe to ECM multi-plug pins 9 and 25 in turn **(see illustration 21.21)**. The voltmeter should indicate nbv in each case.

6 If no voltage is found, check the relevant fuse and the supply wiring back to the battery positive terminal.

7 Switch the ignition on.

8 Attach a negative voltmeter probe to a vehicle earth.

9 Attach a positive voltmeter probe to ECM multi-plug pin 23 **(see illustration 21.21)**. The voltmeter should indicate nbv.

10 If no voltage is found, check the supply wiring back to the ignition switch.

11 Switch the ignition off

12 Connect an ohmmeter between a vehicle earth and ECM multi-plug earth pins 8 and 24 in turn.

13 The ohmmeter should indicate continuity in each case. If not, check the ECM main earth connection and wiring.

7 Solenoid valves

Solenoid valve operation tests

1 The solenoid valves are integral with the ECM and can only be checked using suitable ABS diagnostic test equipment.

8 Hydraulic pump motor

Pump operation test

1 Switch the ignition off and disconnect the pump motor multi-plug.

2 Connect the positive terminal of a 12 volt supply to terminal 2 of the pump motor multi-plug, and the negative terminal to terminal 1 of the multi-plug **(see illustration 21.22)**. The pump motor should now run.

⚠️ **Warning: The test should be made as quickly as possible to avoid damaging the pump.**

3 If the pump does not operate as described, renew the hydraulic control unit.

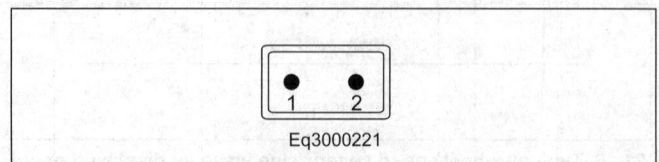

21.22 Pump motor multi-plug terminal identification

9 Brake light switch

Checking the brake light switch (general)

1 Check that the brake light switch is correctly and securely mounted and that the plunger moves smoothly with no trace of binding.
2 Check that the wiring multi-plug is pushed fully home and making good contact.
3 Check that no wires have been disconnected.
4 A fault in any of the above areas are possible reasons for failure or malfunctioning of the switch.

Brake light switch voltage and continuity tests

Voltage test

Note: *Refer to the wiring diagrams for specific ECM pin identification according to model.*
5 Switch the ignition off and disconnect the ECM multi-plug from the ECM.
6 Connect a voltmeter between a vehicle earth and the ECM multi-plug brake light switch input pin (typically pin 12) **(see illustration 21.21)**.
7 Switch the ignition on and depress the brake pedal. The voltmeter should indicate nbv.
8 If no voltage is found, the fuse, the brake light switch and the wiring are suspect.
9 Release the brake pedal. The voltage should drop to zero as the switch opens.

Continuity test

10 Switch the ignition off and disconnect the brake light switch multi-plug.
11 Connect an ohmmeter between the terminal pins of the brake light switch.
12 Operate the brake light switch and check for continuity. If the test fails, renew the switch.

10 Warning lamp

Checking the warning lamp (general)

1 Inspect the warning lamp bulb holder contacts in the instrument panel.
2 Check that the instrument panel multi-plug terminal pins are pushed fully home and making good contact.
3 Check that no wires have been disconnected.
4 A fault in any of the above areas are possible reasons for failure or malfunctioning of the warning lamp.

Warning lamp operation test

5 With the ignition switched off, the warning lamp should remain off.
6 Switch the ignition on and the warning lamp should illuminate then extinguish after a few seconds. The lamp should then remain off.
7 If the warning lamp comes on and remains on at any time during vehicle operation, carry out the previously described test procedures on the system components.

Pin table - typical 25-pin (Seat, Skoda, Volvo)

Note: *Refer to illustration 21.21*

Pin No.	Connection	Test condition	Voltage
1	Rear right wheel speed sensor signal	Roadwheel rotating	0.5 to 1.5 volts AC (approx)
2	Rear left wheel speed sensor signal	Roadwheel rotating	0.5 to 1.5 volts AC (approx)
3	Front right wheel speed sensor earth	Roadwheel rotating	0.25 volts (max)
4	Front left wheel speed sensor earth	Roadwheel rotating	0.25 volts (max)
5*	Brake fluid pressure switch input	Ignition on: Brake pedal released	5.0 to 6.0 volts (approx)
		Ignition off: Brake pedal depressed	10.0 volts (approx)
6*	Vehicle speed signal	-	-
7*	Brake fluid pressure switch earth	Ignition on	0.25 volts (max)
7****	Coding link to pin 15	-	-
8	ECM earth	Ignition on	0.25 volts (max)
9	Supply from battery	Ignition on/off	Nbv
10	Rear left wheel speed sensor earth	Roadwheel rotating	0.25 volts (max)
11	Front left wheel speed sensor signal	Roadwheel rotating	0.5 to 1.5 volts AC (approx)
12	Brake light switch input	Ignition on: Brake pedal released	0 volts
		Brake pedal depressed	Nbv
13	SD connector		
14**	Coding link to pin 22	-	-
15****	Coding link to pin 7	-	-
15***	Coding link to pin21	-	-
16	ABS warning lamp	Ignition on: Lamp on	2.0 volts (max)
		Lamp off	Nbv
17	Rear right wheel speed sensor earth	Roadwheel rotating	0.25 volts (max)
18	Front right wheel speed sensor signal	Roadwheel rotating	0.5 to 1.5 volts AC (approx)
19	-	-	-
20	-	-	-
21***	Coding link to pin 15	-	-
22**	Coding link to pin 14		

21

Pin table - typical 25-pin (Seat, Skoda, Volvo) (continued)

Note: *Refer to illustration 21.21*

Pin No.	Connection	Test condition	Voltage
23	Supply from ignition switch	Ignition on	Nbv
24	ECM earth	Ignition on	0.25 volts (max)
25	Supply from battery	Ignition on/off	Nbv

*Volvo models only
**Seat models only
***VW models only
****Skoda models only

Pin table - typical 25-pin (Fiat, Volkswagen)

Note: *Refer to illustration 21.21*

Pin No.	Connection	Test condition	Voltage
1	Rear right wheel speed sensor earth	Roadwheel rotating	0.25 volts (max)
2	Rear left wheel speed sensor earth	Roadwheel rotating	0.25 volts (max)
3	Front right wheel speed sensor signal	Roadwheel rotating	0.5 to 1.5 volts AC (approx)
4	Front left wheel speed sensor signal	Roadwheel rotating	0.5 to 1.5 volts AC (approx)
5	-	-	-
6	-	-	-
7	-	-	-
8	ECM earth	Ignition on	0.25 volts (max)
9	Supply from battery	Ignition on/off	Nbv
10	Rear left wheel speed sensor signal	Roadwheel rotating	0.5 to 1.5 volts AC (approx)
11	Front left wheel speed sensor earth	Roadwheel rotating	0.25 volts (max)
12	Brake light switch input	Ignition on:	
		Brake pedal released	0 volts
		Brake pedal depressed	Nbv
13	SD connector		
14	-	-	-
15	-	-	-
16	ABS warning lamp	Ignition on:	
		Lamp on	2.0 volts (max)
		Lamp off	Nbv
17	Rear right wheel speed sensor signal	Roadwheel rotating	0.5 to 1.5 volts AC (approx)
18	Front right wheel speed sensor earth	Roadwheel rotating	0.25 volts (max)
19	-	-	-
20	-	-	-
21	-	-	-
22	-	-	-
23	Supply from ignition switch	Ignition on	Nbv
24	ECM earth	Ignition on	0.25 volts (max)
25	Supply from battery	Ignition on/off	Nbv

Fault codes

11 General fault codes

1 The Teves 20GI system requires the use of a FCR for obtaining fault codes. Flash codes are not available for output from this system.
2 If a FCR is available, it should be connected to the SD serial connector and used in accordance with the maker's instructions.
3 The FCR can be used for the following purposes:
  a) Obtaining fault codes.
  b) Clearing fault codes.
  c) Obtaining datastream information.
  d) Testing the system actuators (solenoid valve relay, pump relay and solenoid valves).

11 Fiat fault codes

1 On Fiat models, internal fault codes are used by the ECM to designate faults in the system components and circuits. A proprietary fault code reader (FCR) or system tester is required to interrogate the system. No actual fault code numbers are available although the component circuits checked by the ECM are similar to those shown for the other vehicles listed.

Fault code table (25-pin ECM - Seat, Skoda, Volkswagen)

Code	Item	Fault
00283	Front left wheel speed sensor	Poor wiring/connections; incorrect air gap; sensor or sensor ring dirty or damaged
00285	Front right wheel speed sensor	Poor wiring/connections; incorrect air gap; sensor or sensor ring dirty or damaged
00287	Rear right wheel speed sensor	Poor wiring/connections; incorrect air gap; sensor or sensor ring dirty or damaged
00290	Rear left wheel speed sensor	Poor wiring/connections; incorrect air gap; sensor or sensor ring dirty or damaged
00668	Supply voltage	Outside expected values
01044	ECM coding	ECM incorrectly coded, poor wiring connection
01130	ABS operation	Signal external interference, poor wiring connections
01276	Hydraulic pump	Faulty operation
65535	ECM	ECM defective

Fault code table (25-pin ECM - Volvo)

Code	Item	Fault
141	Brake fluid pressure switch	Circuit fault
142	Brake light switch	No signal, poor wiring connections
143	Vehicle speed signal	Circuit fault
144	Front brake discs	Overheating (models with TRACS only)
211	Front left wheel speed sensor	No signal or signal outside expected values
212	Front right wheel speed sensor	No signal or signal outside expected values
213	Rear left wheel speed sensor	No signal or signal outside expected values
214	Rear right wheel speed sensor	No signal or signal outside expected values
221	Front left wheel speed sensor	ABS control phase too long - sensor/wiring fault, hydraulic control unit fault
222	Front right wheel speed sensor	ABS control phase too long - sensor/wiring fault, hydraulic control unit fault
223	Rear left wheel speed sensor	ABS control phase too long - sensor/wiring fault, hydraulic control unit fault
224	Rear right wheel speed sensor	ABS control phase too long - sensor/wiring fault, hydraulic control unit fault
311	Front left wheel speed sensor	No signal, wiring short circuit
312	Front right wheel speed sensor	No signal, wiring short circuit
313	Rear left wheel speed sensor	No signal, wiring short circuit
314	Rear right wheel speed sensor	No signal, wiring short circuit
321	Front left wheel speed sensor	Intermittent signal
322	Front right wheel speed sensor	Intermittent signal
323	Rear left wheel speed sensor	Intermittent signal
324	Rear right wheel speed sensor	Intermittent signal
411	Front left inlet solenoid valve	Circuit fault
412	Front left outlet solenoid valve	Circuit fault
413	Front right inlet solenoid valve	Circuit fault
414	Front right outlet solenoid valve	Circuit fault
421	Rear inlet solenoid valve	Circuit fault
422	Rear outlet solenoid valve	Circuit fault
423	Traction control solenoid valve	Circuit fault (models with TRACS only)
431	ECM	Internal circuit fault/ECM defective
432	ECM	Signal interference
433	Supply voltage	Voltage too high
441	Main relay	No voltage supply to solenoid valves
442	ECM	Internal circuit fault/ECM defective
443	Hydraulic pump motor	Circuit fault/pump defective
444	ECM	Internal circuit fault/ECM defective
445	ECM	Internal circuit fault/ECM defective

Wiring diagrams

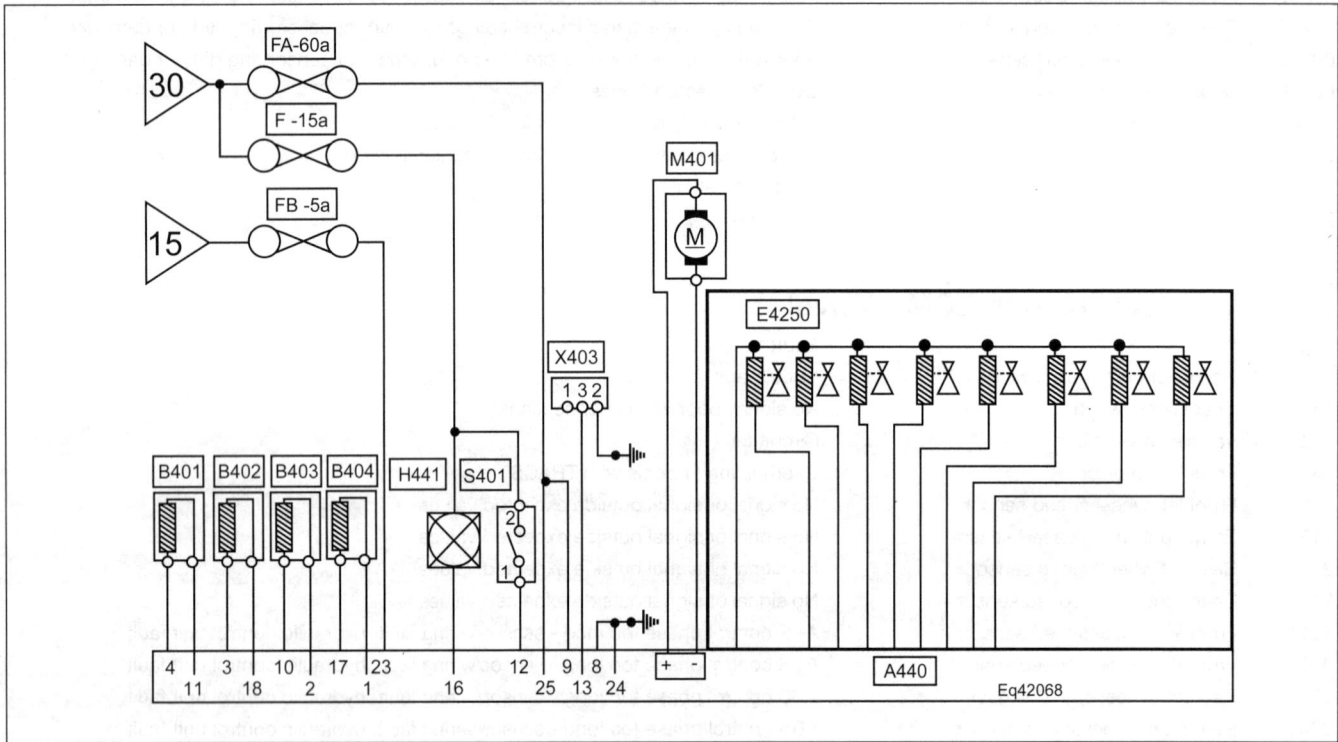

21.23 25-pin wiring diagram, Fiat Bravo/Brava

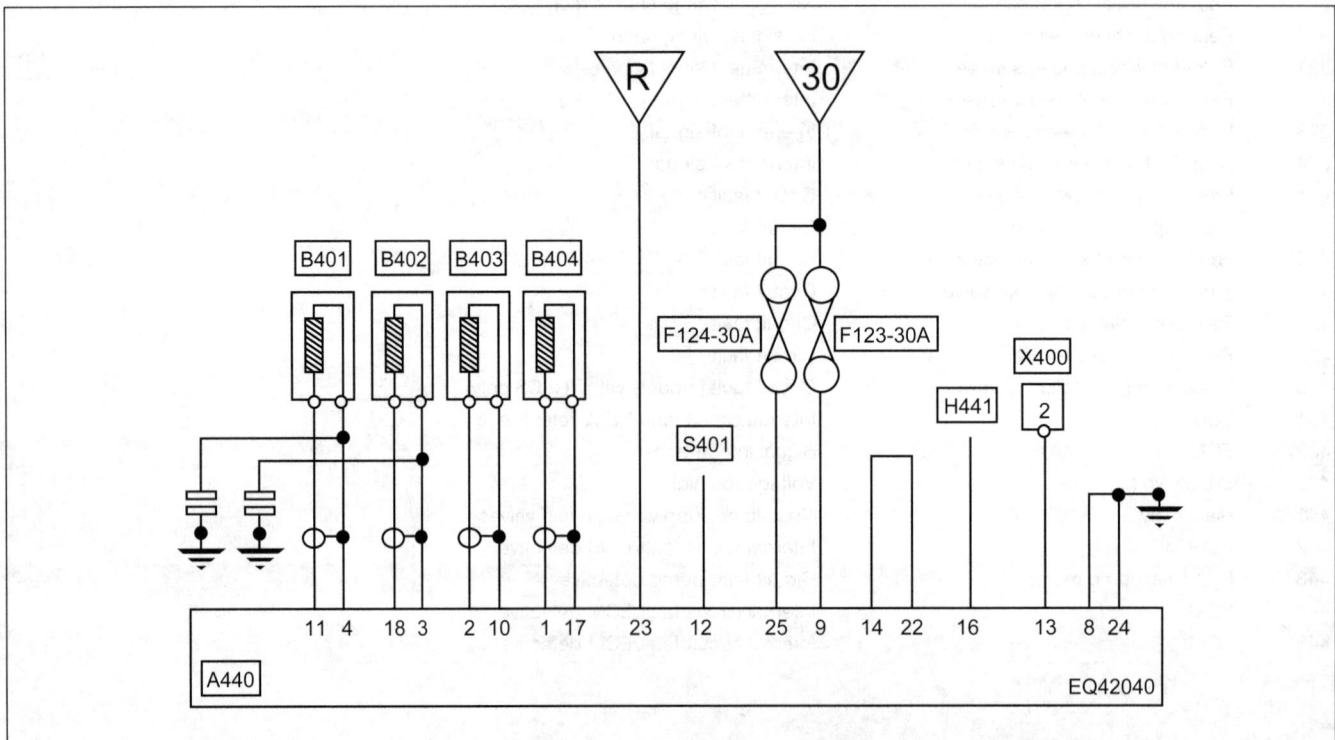

21.24 25-pin wiring diagram, Seat Cordoba/Ibiza

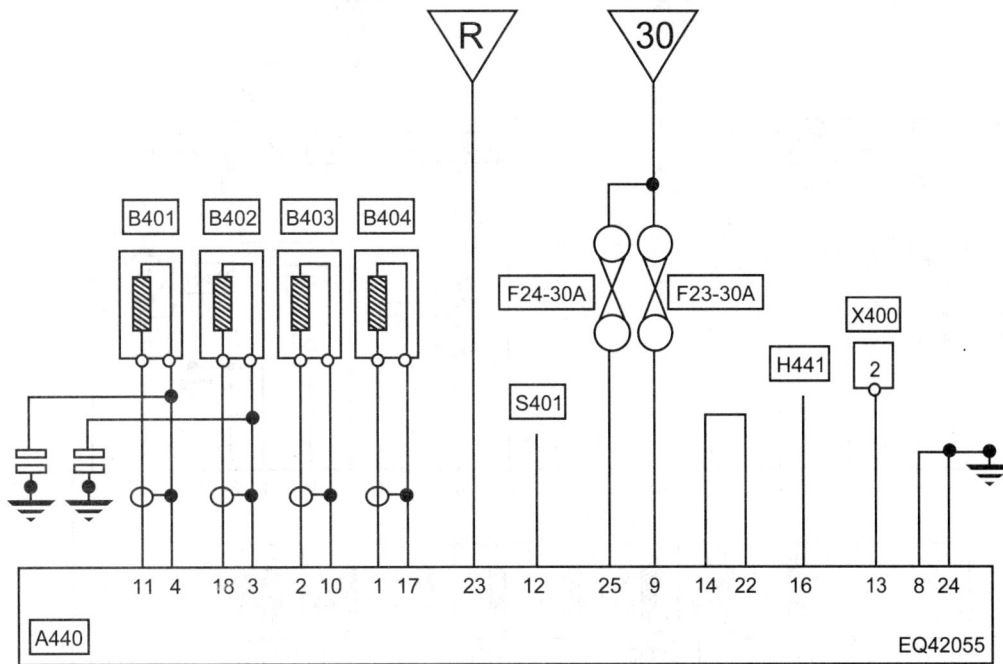

21.25 25-pin wiring diagram, Seat Inca

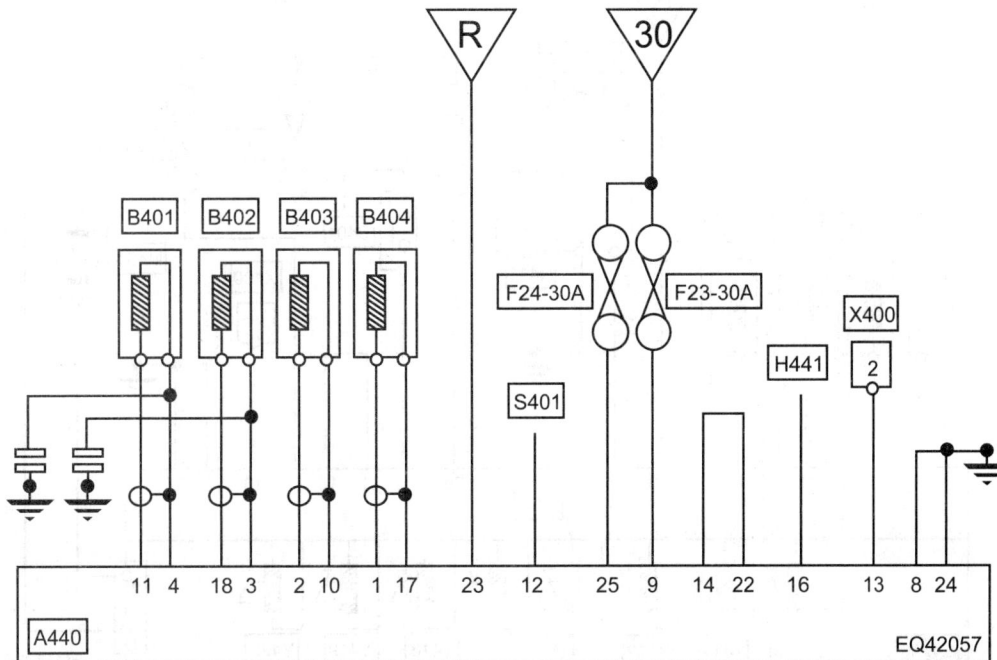

21.26 25-pin wiring diagram, Seat Toledo

21

21.27 25-pin wiring diagram, Skoda Felicia

21.28 25-pin wiring diagram, Volkswagen Golf/Passat/Vento

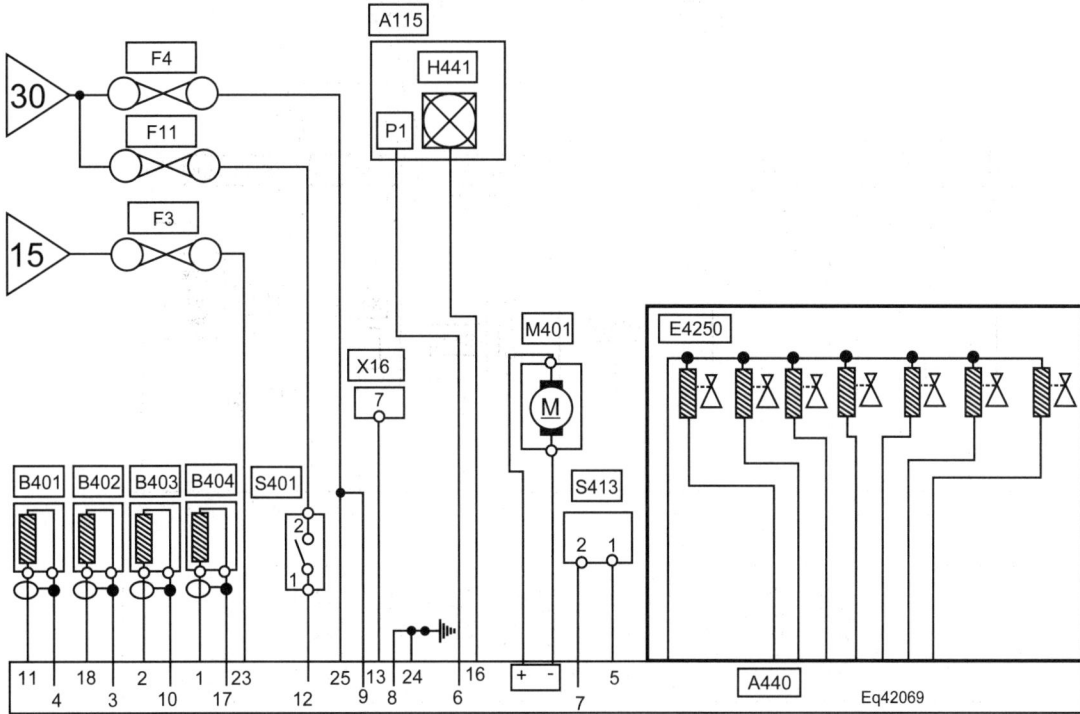

21.29 25-pin wiring diagram, Volvo S70/V70/C70

Chapter 22
Teves 20IE (ITT 20IE) ABS

Contents

Vehicle coverage

Model	Year
Audi	
A3	1996-2000
TT	1999-2000
Citroën	
Saxo	1998-1999
Xantia	1998-1999
Ford	
Galaxy	1997-2000
Peugeot	
206	1998-2000
Renault	
Laguna	1998-2000
Master	1997-2000
Seat	
Arosa	1996-2000
Inca	1997-2000
Skoda	
Octavia	1998-2000
Volkswagen	
Bora	1999-2000
Caddy	1997-2000
Golf	1998-2000
Lupo	1999-2000
Polo	1997-2000
Sharan	1997-2000
Volvo	
S70/V70/C70	1999-2000

Overview of system operation

1 Basic principles and system identification

The Teves 20IE Antilock Brake System is a development of the earlier Teves (ITT) 20GI series ABS, with further improvements to the hydraulic control unit, and the Electronic Control Module (ECM) software. Teves 20GI has been fitted to various passenger vehicles since its introduction in the early 1990s. The system is of the additional or 'add-on' type operating in conjunction with the conventional braking system components. The system may carry the designation Teves 20IE or ITT 20IE although they are both identical.

There are certain discrepancies in vehicle manufacturers technical documentation relating to the identification of Teves 20IE and the earlier Teves (ITT) 20GI system which preceded it. The two systems are visually and operationally identical apart from different ECM pin connections. In most instances it is only possible to differentiate between the two systems by interrogation of the ECM using a suitable system tester.

Depending on application, Teves 20IE may be installed as an antilock braking system only, or as an antilock braking system incorporating traction control.

In ABS mode, the purpose of the system is to apply the vehicle brakes at maximum efficiency without wheel lock or loss of directional stability. Inductive, or 'active' sensors (wheel speed sensors) monitor the speed of the wheels by generating an electrical signal as the wheel is rotated. This information is passed to the ABS-ECM which compares the signals received from each wheel and uses the speed of the fastest wheel as a reference value. The ECM continually monitors the speed of each wheel and if the onset of lock at any wheel is detected (a received speed signal being less than the reference value) a signal is sent to the ABS hydraulic control unit which regulates the brake pressure for the relevant wheel(s).

Where the system incorporates traction control, essentially the reverse principle is applied. When the ECM detects that one or more wheels are rotating faster than the reference value, the brake is actually applied on the relevant wheel(s) to reduce the rotational speed. Additionally, on some installations, when wheel spin is detected, various signals are sent to the engine management ECM to control engine torque and rpm.

Typically, Teves 20IE ABS is comprised of the following components (see illustration 22.1):

a) Hydraulic control unit with integral ABS-ECM.
b) Four inductive wheel speed sensors and associated sensor rings.
c) Brake light switch.
d) ABS warning lamp.
e) Diagnostic connector.

In addition, the conventional brake system is comprised of the following components:

a) Tandem brake master cylinder.
b) Vacuum servo unit.
c) Brake calipers/wheel cylinders and hydraulic hoses and pipes.
d) Pressure regulating/load sensing valve(s) depending on application.

2 Component description and operation

ABS ECM

General

The Teves 20IE Electronic Control Module (ECM) continually monitors wheel speed from the signals provided by the wheel speed sensors, and brake application from the brake light switch signal. If the ECM detects the incidence of wheel lock (or wheel spin, if traction control is incorporated) on one or more wheels, a signal is sent to the hydraulic control unit to modulate the hydraulic pressure to the brake of the locking or spinning wheel(s). The ECM contains two microprocessors and uses digital technology to complete this function and other functions such as, fault code memory and power modules for valve and pump activation (see illustration 22.2).

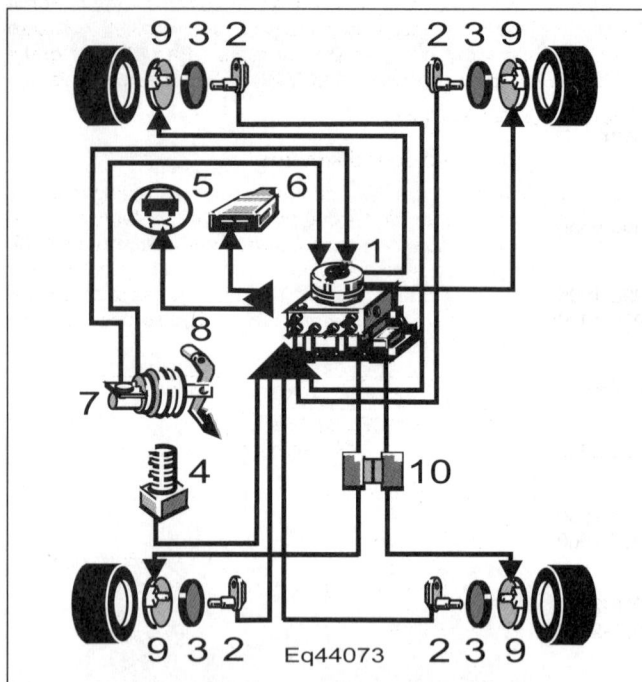

22.1 Typical Teves 20IE main components

1 Hydraulic control unit with integral ABS-ECM	7 Tandem brake master cylinder
2 Inductive wheel speed sensors	8 Vacuum servo unit
3 Sensor rings	9 Brake calipers/wheel cylinders and hydraulic hoses and pipes
4 Brake light switch	
5 ABS warning lamp	10 Pressure regulating/load sensing valve(s) depending on application
6 Diagnostic connector	

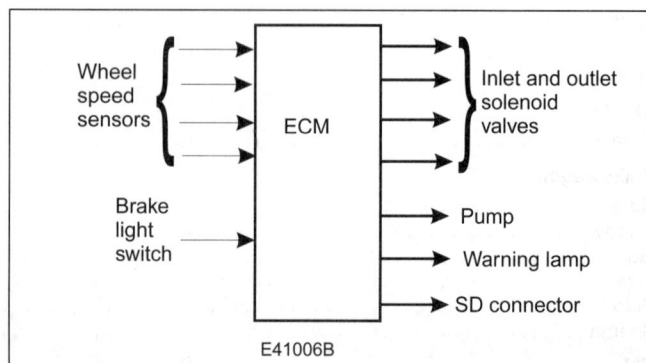

22.2 ECM sensor inputs and control signal outputs

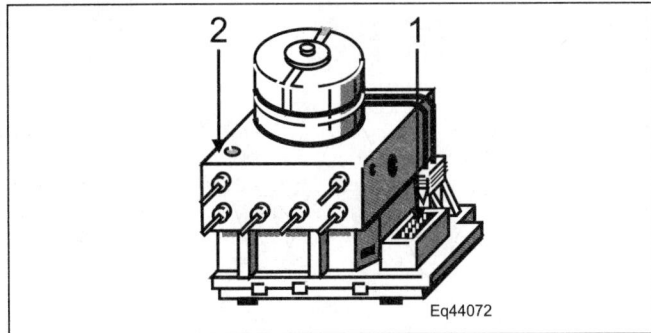

22.3 Teves 20IE ECM (1) and hydraulic control unit (2)

To reduce external electrical connections to a minimum and improve reliability, the ECM is integral with the hydraulic control unit (see illustration 22.3).

Self-test

The Teves 20IE ECM is equipped with a self-test capability that initially examines the ABS system when the ignition is switched on, and then examines the wheel speed sensor signals after a wheel speed of approximately 4 mph is reached from all wheels (engine running). The ABS self-test program continues to examine the signals from the various components as long as the ignition is switched on. If self-test determines that faults are not present, the ABS is ready for operation once a specified vehicle speed has been achieved.

If the ECM detects that a fault is present, all ABS functions are switched off and the warning lamp is turned on. The conventional braking system continues to operate as normal without ABS assistance.

Self-diagnostics

If the ECM detects a fault during the self-test routine, an internal fault code is stored in the ECM memory. Stored fault codes can be retrieved from the SD connector with the aid of a suitable fault code reader. If the fault clears, the code will remain stored until cleared with the FCR.

Hydraulic control unit

Teves 20IE is typically a four-channel system with a separate hydraulic circuit for each brake. On certain models, however, the system is installed in three-channel configuration with a separate hydraulic circuit for each front brake, but with the rear brakes controlled as a pair. The hydraulic control unit consists of an electric motor operating a return pump with eccentric drive and twin radial pistons, inlet and outlet solenoid valves, pressure accumulators and pulsation dampers (see illustration 22.4). The unit controls the hydraulic pressure applied to the brake for each individual front wheel and each individual rear wheel, or pair of wheels. The return pump is switched on when the ABS is activated and returns hydraulic fluid,

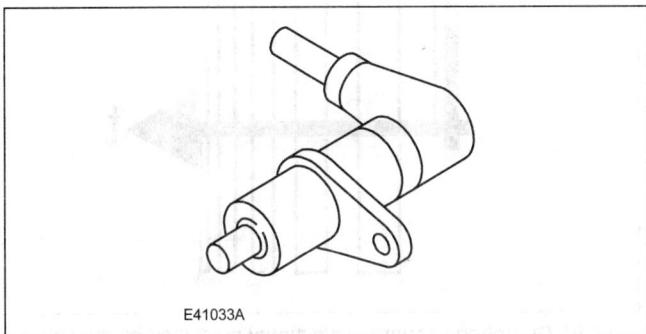

22.5 Typical inductive wheel speed sensor

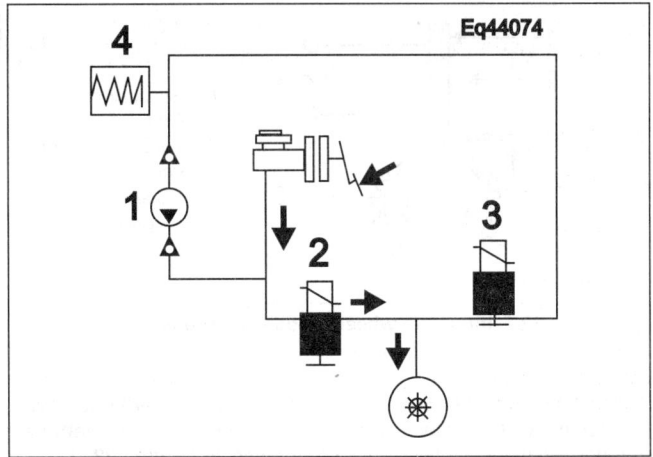

22.4 Hydraulic control unit components

1	Electric motor and return pump	3	Outlet solenoid valve
2	Inlet solenoid valve	4	Pressure accumulator

drained off during the pressure reduction phase, back into the brake circuit.

On some applications, the 'select-low' principle is employed for control of the rear brakes. With the 'select-low' principle, the wheel with the lowest adhesion determines the amount of hydraulic pressure to be supplied to both rear brakes during ABS operation.

In certain installations, the ABS-ECM contains additional software for rear brake hydraulic fluid pressure regulation when ABS is not in operation. This is generally known as 'Electronic Brake Force Distribution' and in these applications, mechanical pressure regulating/load sensing valves are not required as the rear brake hydraulic pressure is controlled by the ABS hydraulic control unit.

Wheel speed sensors

Inductive type wheel speed sensors

The rotational speed of the roadwheels and any changes in the rotational speed are recorded either by inductive, or 'active' wheel speed sensors, one located at each roadwheel (see illustration 22.5).

Where inductive sensors are used, each wheel speed sensor assembly comprises a toothed sensor ring which rotates at roadwheel speed, and an adjacent sensor mounted a set distance from the sensor ring (see illustration 22.6).

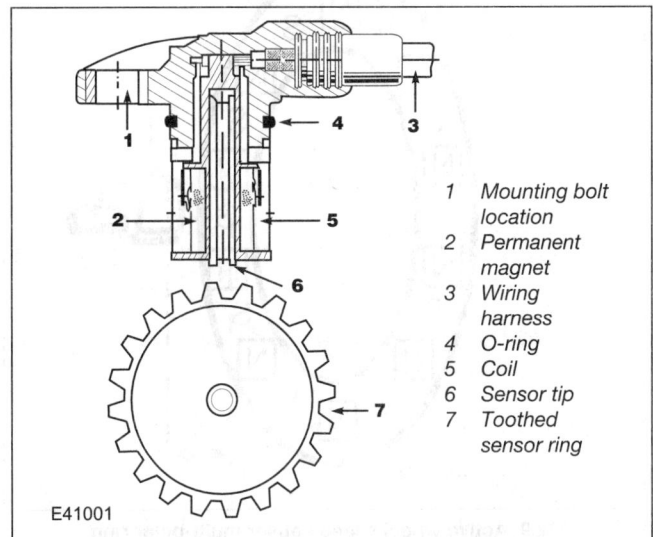

1	Mounting bolt location
2	Permanent magnet
3	Wiring harness
4	O-ring
5	Coil
6	Sensor tip
7	Toothed sensor ring

22.6 Sectional view of an inductive wheel speed sensor

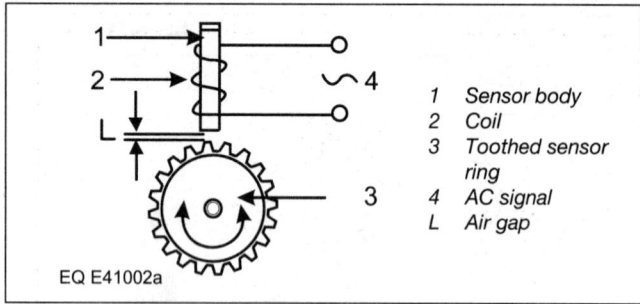

22.7 Inductive wheel speed sensor operation

1 Sensor body
2 Coil
3 Toothed sensor ring
4 AC signal
L Air gap

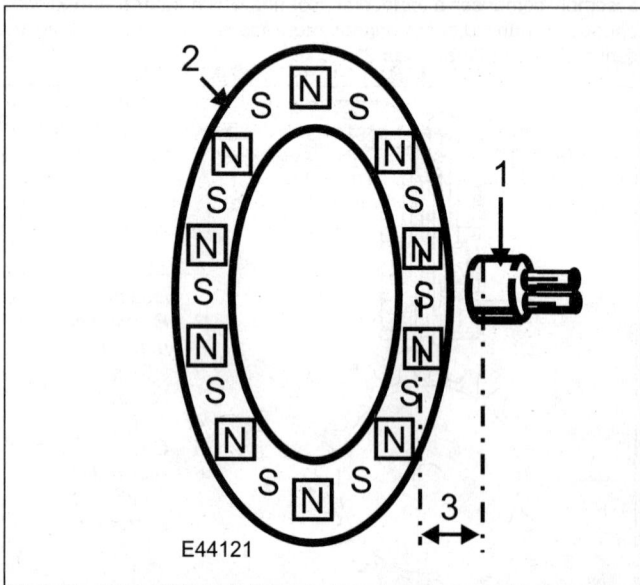

22.8 Inductive wheel speed sensor waveform as viewed on an oscilloscope

The sensors are permanent magnet pulse generator types producing an AC voltage sine wave as the sensor ring teeth pass through the magnetic field of the sensor (see illustration 22.7).

The frequency of the waveform produced by the wheel speed sensor is proportional to the road speed. This AC voltage signal is continually being delivered to the ABS-ECM for processing.

The peak to peak voltage of the speed signal (when viewed upon an oscilloscope) can vary considerably according to wheel speed. An analogue to digital converter (ADC) in the ECM transforms the AC pulse into a digital signal (see illustration 22.8).

'Active' type wheel speed sensors

On the Teves 20IE system fitted to Renault models, the rotational speed of the roadwheels and any changes in the rotational speed are recorded by 'active' wheel speed sensors.

On the majority of antilock brake systems, the wheel speed sensors are of the permanent magnet pulse generator type, as described previously. On the Teves 20IE system with 'active' sensors, the internal electrical resistance of the sensor is altered by changes in intensity and direction of the lines of force of an external magnetic field.

This external magnetic field is created by a unique type of sensor ring known as a 'multi-polar ring'. The multi-polar ring consists of a series of magnetic elements with alternating north/south polarities, located around the circumference of the ring (see illustration 22.9).

The multi-polar ring is attached to the driveshaft or wheel hub, with the adjacent sensor mounted a set distances from the ring. As the ring rotates at roadwheel speed, a DC voltage square wave, constant amplitude signal is produced, the frequency of which is proportional to the road speed. This signal is continually being delivered by the sensor to the ABS-ECM for processing (see illustration 22.10).

ABS electrical wiring harness and relays

An integrated 25-pin main wiring harness is used to connect power and earth to various electrical components and to enable sensor signals to reach the ECM. The ECM, in turn, sends command signals to the hydraulic control unit via internal connections between the two units. The vehicle relays and ABS fuses are typically located in the engine compartment fuse/relay box .

Nominal battery voltage is supplied to the ECM via the ABS main fuses. The ECM then supplies nominal battery voltage to the solenoid valves and return pump and controls the earth for activation of the solenoid valves.

Brake light switch

The brake light switch comprises a switch body and

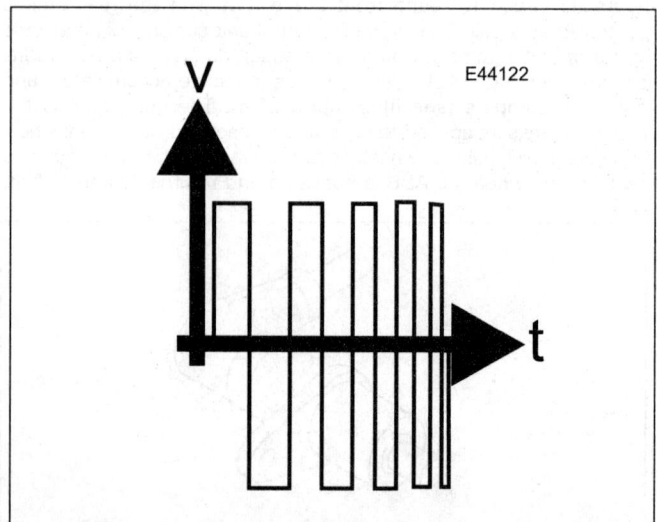

22.9 Active wheel speed sensor multi-polar ring

1 Wheel speed sensor   2 Multi-Polar ring   3 air gap

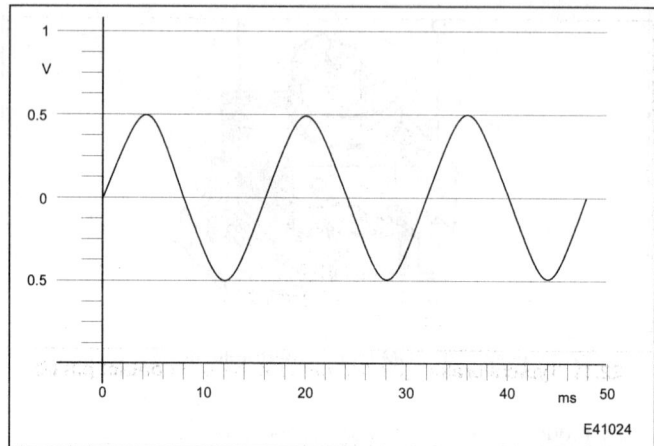

22.10 DC voltage square wave signal produced by the active wheel speed sensors

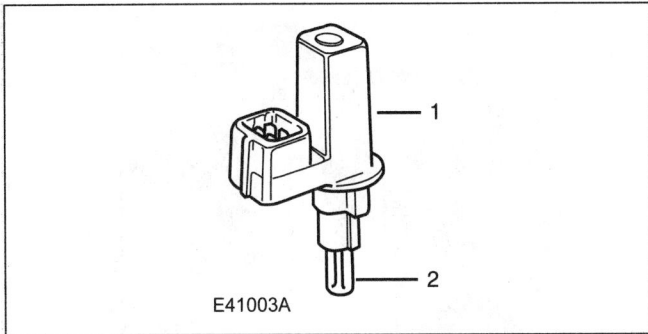

22.11 Typical brake light switch body (1) and contact pin (2)

contact pin and is located above the brake pedal (see illustration 22.11).

When the brake pedal is depressed, closing the brake light switch, a signal is sent to the ECM indicating that the brakes are being applied. Once this signal is received, the ECM will begin monitoring the wheel speed via the wheel speed sensors and activate the ABS if necessary.

ABS warning lamp

After the ignition is switched on, the ABS warning lamp on the instrument panel is illuminated for approximately 2 to 4 seconds as the system executes a self-test routine. If satisfactory operation of the system is detected by the ECM, the light is extinguished. During vehicle operation above a pre-determined wheel speed, the ABS-ECM implements a further self-test cycle whereby ABS operation and wheel speed sensor signals are continually monitored. If a fault is detected, the relevant ECM pin is earthed to illuminate the warning lamp on the instrument panel, and the ABS function is disabled. The warning lamp will remain illuminated until the fault is no longer present (see illustration 22.12).

When the ABS-ECM detects a fault, the fault code is stored and the ABS warning lamp activated. If the fault no longer exists after the next system start (ignition on/off) the ABS warning lamp is extinguished after the self-test cycle, however the fault code remains stored in the ECM memory.

Tandem master cylinder

Typically, the tandem master cylinder comprises a body casting incorporating primary and secondary pressure chambers, primary piston, intermediate piston, floating piston, slotted pin and central valve. The cylinder operates as a conventional master cylinder using vacuum assistance from the vacuum servo unit (see illustration 22.13).

When the brake system is at rest, the central valve in the floating piston rests against the slotted pin. In this condition the central valve is open and brake fluid can discharge out of the pressure chamber

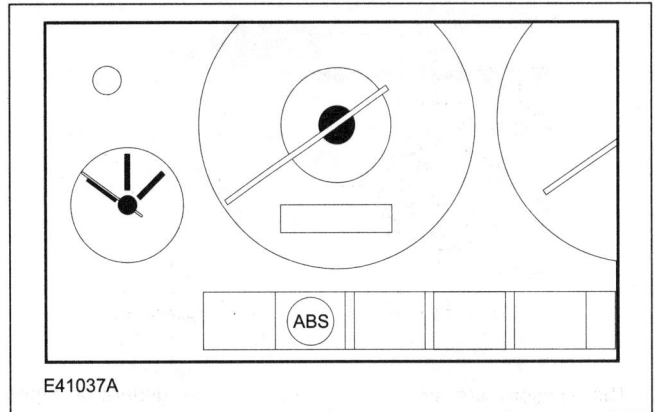

22.12 ABS warning lamp

back into the brake fluid reservoir. When the brake pedal is depressed, the build-up of hydraulic pressure in the primary pressure chamber acts on the intermediate piston and floating piston, moving them down the cylinder bore. The floating piston contacts the seal on the central valve, closing the connection between the intermediate and secondary pressure chambers. Brake hydraulic pressure can now also increase in the secondary pressure chamber.

Vacuum servo unit

The vacuum servo unit is located between the brake pedal and tandem master cylinder. When the brake pedal is depressed, the servo unit increases the force applied by the pedal, reducing the effort required to operate the brakes (see illustration 22.14).

The unit is operated by vacuum created in the engine inlet manifold (or from a separate vacuum pump on diesel engines) which is applied to a diaphragm contained within the unit casing. A pushrod connected to the centre of the diaphragm acts directly on the primary piston in the master cylinder.

When the brake pedal is released, vacuum is applied to both sides of the diaphragm. When the pedal is depressed, one side of the diaphragm is opened to atmosphere and the vacuum acting on the other side deflects the diaphragm which in turn operates the master cylinder primary piston. The resulting force applied to the master cylinder piston is therefore significantly greater than the initial force applied to the brake pedal by the driver.

Pressure regulating/load sensing valve(s)

Depending on vehicle application, pressure regulating valves or load sensing valves may be incorporated to restrict the hydraulic fluid pressure to the rear brakes. The valves may be pressure conscious whereby the hydraulic fluid supply is restricted once a pre-determined pressure is reached, or load conscious whereby the hydraulic pressure is reduced according to vehicle loading.

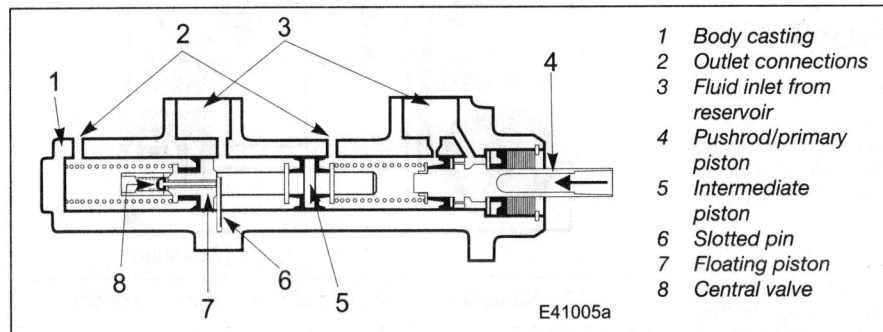

1 Body casting
2 Outlet connections
3 Fluid inlet from reservoir
4 Pushrod/primary piston
5 Intermediate piston
6 Slotted pin
7 Floating piston
8 Central valve

22.13 Sectional view of a typical tandem master cylinder

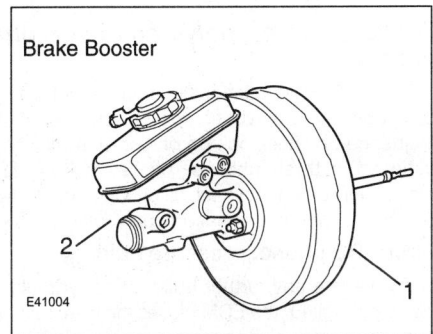

Brake Booster

22.14 Vacuum servo unit (1) and tandem master cylinder (2)

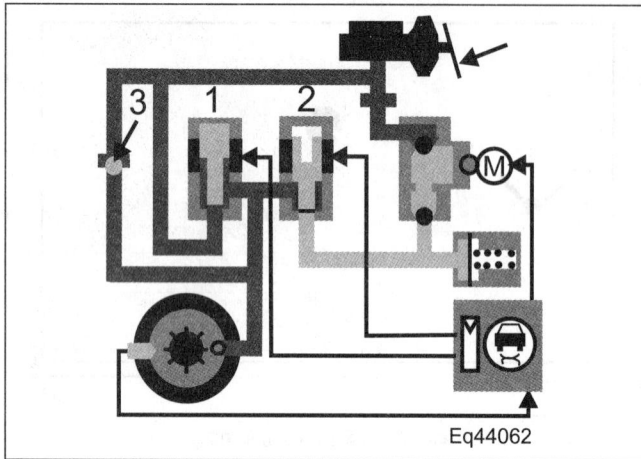

22.15 Brake system operating under conventional control without ABS

1   Inlet solenoid valve   3   One-way valve
2   Outlet solenoid valve

On certain versions, the ABS-ECM contains additional software for rear brake hydraulic fluid pressure regulation during normal braking, generally known as 'Electronic Brake Force Distribution'. In these applications, mechanical pressure regulating/load sensing valves are not required as the rear brake hydraulic pressure is controlled by the ABS hydraulic control unit.

3   System operation

Brake system at rest

When the system is at rest all the brake components are inoperative. Pressure is non-existent in the hydraulic pipes between the tandem master cylinder and the brake calipers. The inlet solenoid valves in the hydraulic control unit are open and the outlet solenoid valves are closed.

Brake system operating under conventional control without ABS

When the brake pedal is activated, the pedal force is applied to the tandem master cylinder by the vacuum servo unit pushrod. The servo unit pushrod acts directly on the pressure piston in the master cylinder which pressurises the hydraulic fluid in the brake pipes to the hydraulic control unit. The inlet solenoid valve and outlet solenoid valve both remain in the 'at rest' position (inlet solenoid valve open and outlet solenoid valve closed). Hydraulic pressure is transmitted to each brake caliper, thus operating the brakes.

When the brake pedal is released, the one-way valve opens allowing the hydraulic pressure in the circuit to rapidly decrease **(see illustration 22.15)**.

Brake system operating in conjunction with ABS control

The ABS-ECM continually monitors wheel speed from the signals provided by the wheel speed sensors. If the ECM detects the incidence of wheel lock on one or more wheels, ABS is automatically initiated in three phases. As Teves 20IE ABS typically operates individually on each wheel, all or any of the wheels could be in any one of the following phases at any particular moment.

First ABS phase, pressure holding

To prevent any further build-up of hydraulic pressure in the circuit being controlled, the ECM closes the inlet solenoid valve and allows the outlet solenoid valve to remain closed. The hydraulic fluid line from the tandem master cylinder to the brake caliper or wheel cylinder is closed, and the hydraulic fluid in the controlled circuit is maintained at a

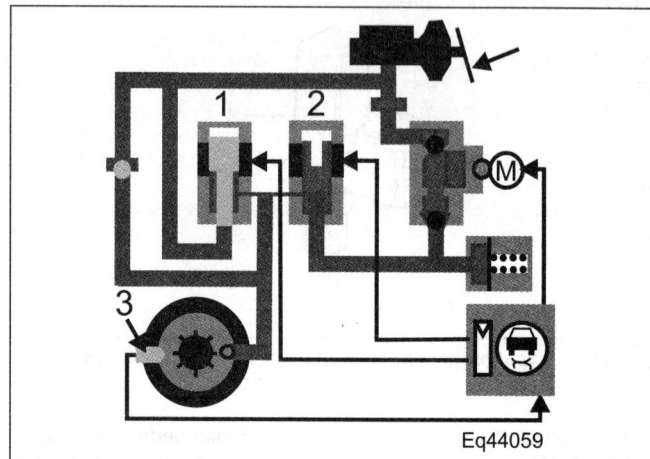

22.16 ABS operation - first phase, pressure holding

1   Inlet solenoid valve   3   Wheel speed sensor
2   Outlet solenoid valve

constant pressure. This effectively removes the braking force from the controlled circuit. The pressure cannot now be increased in that circuit by any further application of the brake pedal **(see illustration 22.16)**.

If the wheel speed sensor signals indicate that wheel rotation has now stabilised, the ECM will instigate the pressure build-up phase, allowing braking to continue. If wheel lock is still detected after the pressure holding phase, the ECM instigates the pressure reduction phase.

Second ABS phase, pressure reduction

If the ECM detects wheel instability, a pressure reduction phase is initiated. The inlet solenoid valve remains closed and the outlet solenoid valve is opened by means of a series of short activation pulses. The pressure in the controlled circuit decreases rapidly as the fluid flows from the brake caliper or wheel cylinder into the pressure accumulator. At the same time, the ECM actuates the electric motor to operate the return pump. The hydraulic fluid is then pumped back into the pressure side of the master cylinder. This process creates a pulsation which can be felt in the brake pedal action, but which is softened by the pulsation damper **(see illustration 22.17)**.

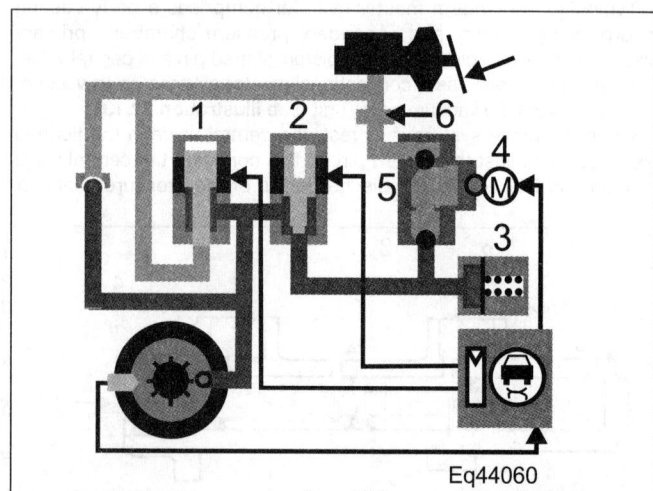

22.17 ABS operation - second phase, pressure reduction

1   Inlet solenoid valve   4   Pump motor
2   Outlet solenoid valve   5   Return pump
3   Pressure accumulator   6   Pulsation damper

Third ABS phase, pressure build-up

The pressure build-up phase is instigated after the wheel rotation has stabilised. The inlet and outlet solenoid valves are returned to the at rest position (inlet solenoid valve open and exhaust solenoid valve closed) which re-opens the hydraulic fluid line from the tandem master cylinder to the brake caliper or wheel cylinder. Hydraulic pressure is reinstated, thus re-introducing operation of the brake. After a brief period, a short pressure holding phase is re-introduced and the ECM continually shifts between pressure build-up and pressure holding until the wheel has decelerated to a sufficient degree where pressure reduction is once more required **(see illustration 22.18)**.

The whole ABS control cycle takes place 4 to 10 times per second for each affected wheel and this ensures maximum braking effect and control during ABS operation.

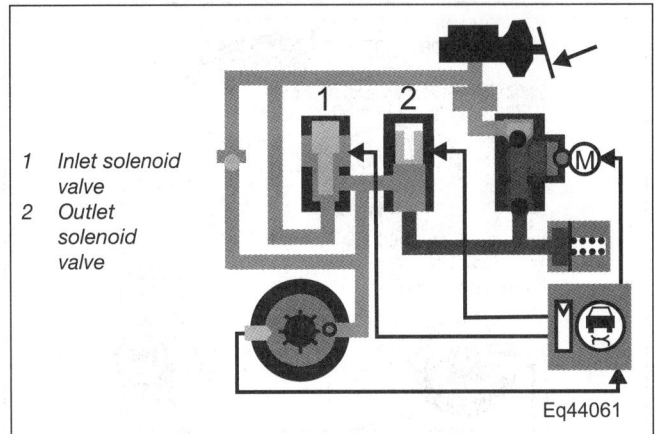

1 Inlet solenoid valve
2 Outlet solenoid valve

Eq44061

22.18 ABS operation - third phase, pressure build-up

Test procedures

Important note: *The test procedures, pin-tables and wiring diagrams contained in this Chapter are necessarily representative of the system depicted. Because of the variations in wiring and other data that often occurs, even between similar vehicles in any particular VM's range, the reader should take great care in identification of ECM pins, and satisfy himself that he has gathered the correct data before failing a particular component.*

4 Wheel speed sensors

Inductive type wheel speed sensors

Checking the wheel speed sensor (general)

1 Inspect the wheel speed sensor for corrosion or damage and check that the sensor is tightly mounted.
2 Check the toothed sensor ring for damage, eccentricity and for broken or missing teeth.
3 Inspect the wheel speed sensor wiring plug for corrosion and damage. One plug for each sensor.
4 Check that the connector terminal pins are pushed fully home and making good contact with the sensor wiring plug.

5 Check the clearance between the sensor and the toothed sensor ring. The clearance is not normally adjustable but is nominally 0.2 to 1.2 mm. If the clearance is excessive, expect a worn sensor tip or problems with the wheel bearings/hub or sensor ring.
6 When carrying out voltage checks with an oscilloscope or voltmeter, the voltage obtained will be proportional to the speed at which the wheel is rotating. In addition to determining that the wheel speed sensors are actually producing a voltage output, it is essential that the output from the sensors on a particular axle is the same for any given wheel speed.

Checking wheel speed sensor output with an oscilloscope

Note: *Refer to the wiring diagrams for specific ECM pin identification according to model.*
7 Switch the ignition off and disconnect the ECM multi-plug or the relevant wheel speed sensor wiring plug.
8 Connect an oscilloscope between the terminal pins for the sensor under test **(see illustration 22.19)**.
9 Select a range to cover 80 Hz on the oscilloscope and a free run time base.
10 Raise the wheel and rotate it by hand at approximately one revolution per second.
11 A sinusoidal wave form should be obtained, with amplitude and duration changing with rotational speed **(see illustration 22.20)**.

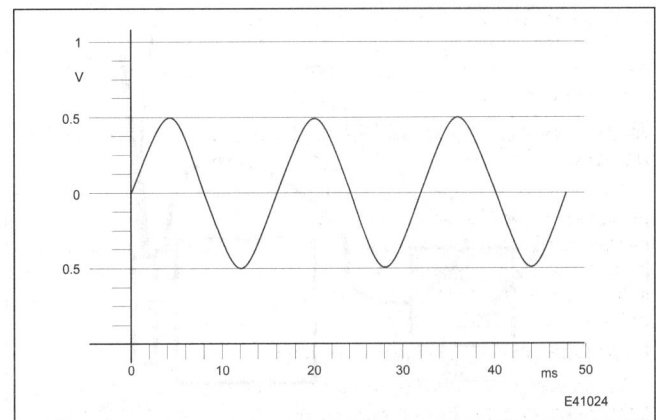

E41040

22.19 Checking inductive wheel speed sensor output with an oscilloscope connected to the sensor wiring plug

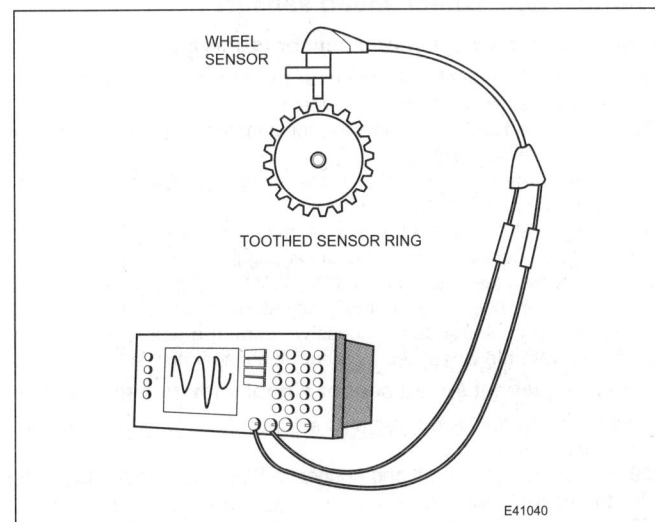

E41024

22.20 Typical inductive wheel speed sensor sine wave as displayed on an oscilloscope

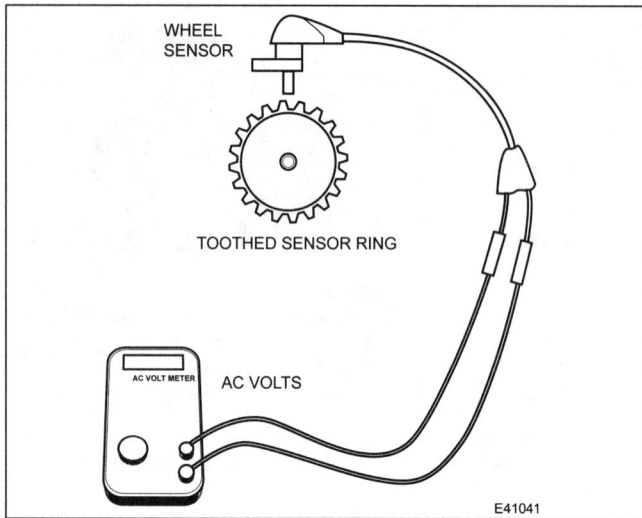

22.21 Checking inductive wheel speed sensor output with a voltmeter connected to the sensor wiring plug

12 If there is no signal, or a very weak or intermittent signal at the ECM, repeat the test at the sensor wiring plug. If there is no change in signal status, the sensor is suspect.

13 If the signal is now satisfactory this indicates a fault in the wiring harness which should be checked for continuity.

Checking wheel speed sensor output with an AC voltmeter

Note: *Refer to the wiring diagrams for specific ECM pin identification according to model.*

14 Switch the ignition off and disconnect the ECM multi-plug or the relevant wheel speed sensor wiring plug.

15 Connect an AC voltmeter between the terminal pins for the sensor under test **(see illustration 22.21)**.

16 Raise the wheel and rotate it by hand at approximately one revolution per second.

17 A voltage of approximately 0.5 to 1.5 volts (AC RMS) should be obtained. If there is no signal, or a very weak or intermittent signal at the ECM, repeat the test at the sensor wiring plug. If there is no change in the signal, the sensor is suspect.

18 If the signal is now satisfactory, this indicates a fault in the wiring

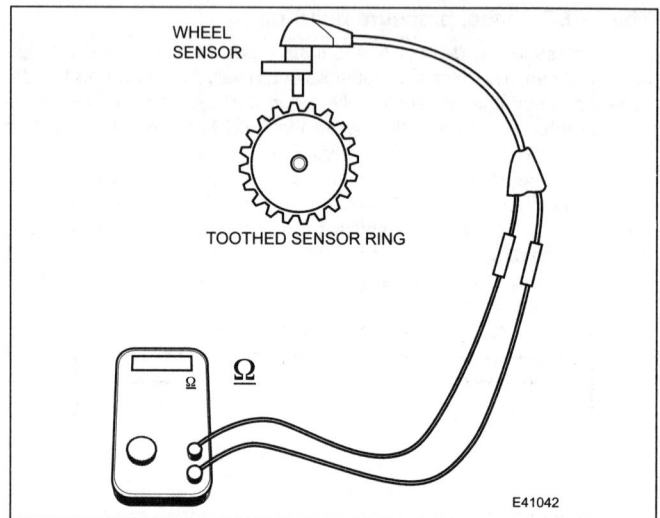

22.23 Checking active wheel speed sensor output with a break-out-box connected between the oscilloscope and ECM multi-plug

harness which should be checked for continuity. **Note:** *This test at least proves that a signal is being generated by the sensor. However, the voltage produced is an average voltage and does not clearly indicate damage to the sensor ring or that the sinewave is regular in formation.*

Checking wheel speed sensor resistance

Note: *Refer to the wiring diagrams for specific ECM pin identification according to model.*

19 Switch the ignition off and disconnect the ECM multi-plug or the relevant wheel speed sensor wiring plug.

20 Connect an ohmmeter between the terminal pins for the sensor under test **(see illustration 22.22)**.

21 The readings obtained should be between 0.7 and 2.2 kohms approximately.

22 If the resistance is excessively high, or open circuit at the ECM, repeat the test at the sensor multi-plug. If there is no change in resistance, the sensor is suspect.

23 If the resistance is now satisfactory, this indicates a fault in the wiring harness which should be checked for continuity. **Note:** *Even if the resistance is within the quoted specifications, this does not prove that the speed sensor can generate an acceptable signal.*

'Active' type wheel speed sensors

Checking the wheel speed sensor (general)

24 Inspect the wheel speed sensor for corrosion or damage and check that the sensor is tightly mounted.

25 Check the toothed sensor ring for damage, eccentricity and for broken or missing teeth.

26 Inspect the wheel speed sensor wiring plug for corrosion and damage. One plug for each sensor.

27 Check that the connector terminal pins are pushed fully home and making good contact with the sensor wiring plug.

28 Check the clearance between the sensor and the toothed sensor ring. The clearance is not normally adjustable but is nominally 0.2 to 2.0 mm. If the clearance is excessive, expect a worn sensor tip or problems with the wheel bearings/hub or sensor ring.

Checking wheel speed sensor signal with an oscilloscope

Note: *Refer to the wiring diagrams for specific ECM pin identification according to model.*

29 Switch the ignition off and connect a BOB between the ECM and the harness multi-plug.

30 Connect an oscilloscope to the relevant BOB terminals for the sensor under test, and set the oscilloscope to measure DC voltage **(see illustration 22.23)**.

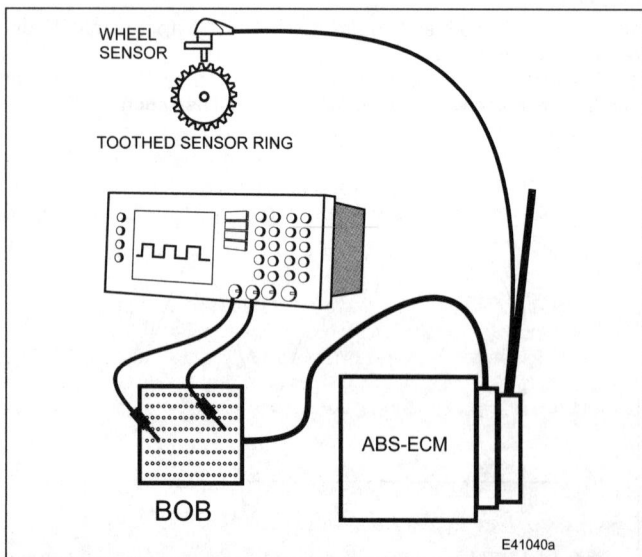

22.22 Checking inductive wheel speed sensor resistance with an ohmmeter connected to the sensor wiring plug

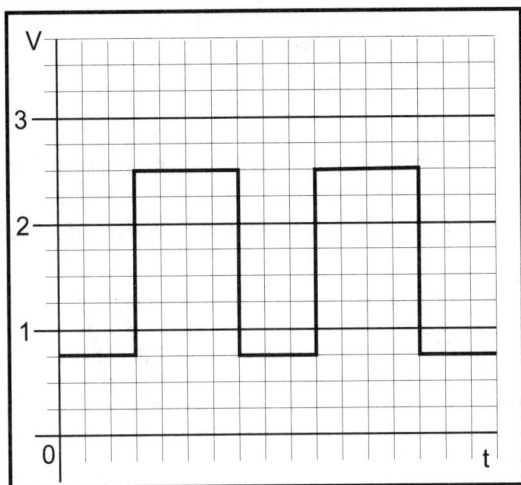

22.24 Active wheel speed sensor square wave output as viewed on an oscilloscope

31 Raise and support the vehicle.

32 Switch the ignition on and rotate the wheel by hand at approximately one revolution per second.

33 A square wave form should be obtained, switching between approximately 0.8 and 1.6 volts as the wheel is rotated **(see illustration 22.24)**.

34 If the output is not as specified, check the supply voltage to the sensor and the wiring harness for continuity.

35 If the supply voltage and wiring are satisfactory, the sensor is suspect.

Checking wheel speed sensor signal with a DC voltmeter

Note: *Refer to the wiring diagrams for specific ECM pin identification according to model.*

36 Switch the ignition off and connect a BOB between the ECM and the harness multi-plug.

37 Connect a DC voltmeter between the relevant BOB terminal for the sensor under test, and earth.

38 Raise and support the vehicle.

39 Switch the ignition on and rotate the wheel very slowly.

40 A voltage switching between approximately 0.8 and 1.6 volts should be obtained as the wheel is rotated.

41 If the voltage is not as specified, check the supply voltage to the sensor and the wiring harness for continuity.

42 If the supply voltage and wiring are satisfactory, the sensor is suspect.

Checking wheel speed sensor supply voltage

43 Switch the ignition off and disconnect the wheel speed sensor wiring harness 2-pin multi-plug, of the sensor under test.

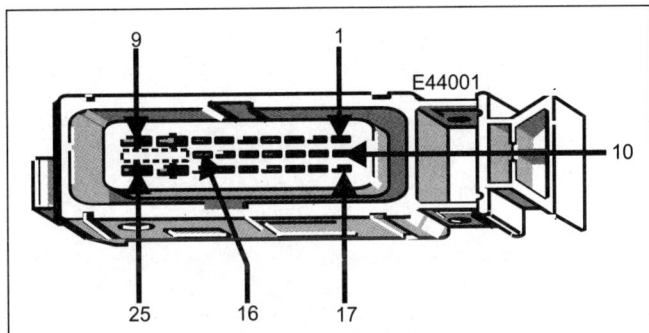

22.26 ECM 25-pin multi-plug

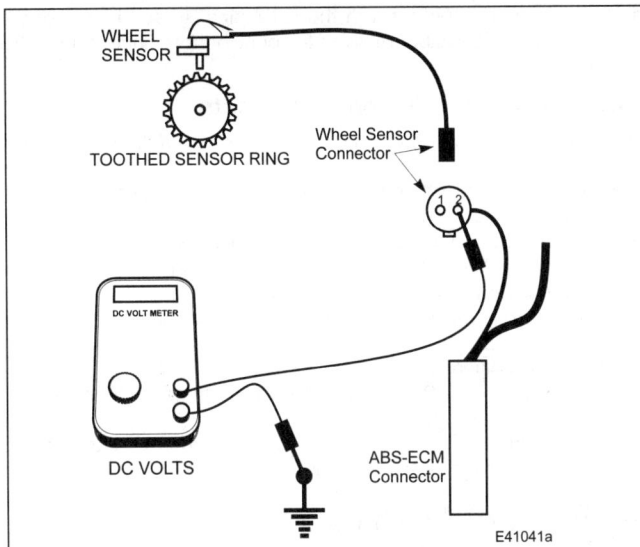

22.25 Checking active wheel speed sensor output using a voltmeter

44 Connect a DC voltmeter between a vehicle earth and pin 2 of the multi-plug **(see illustration 22.25)**.

45 Switch the ignition on. A voltage of approximately 11.0 volts should be obtained.

46 If the voltage is not as specified, repeat the test at the ECM multi-plug using the BOB as described previously. If there is no change in the reading, the ABS-ECM is suspect. If the voltage is now satisfactory, this indicates a fault in the wiring harness which should be checked for continuity. **Note:** *This test at least proves that a supply voltage is available for the sensor. However, the integral electronics in the sensor must be active to produce an output.*

5 System relays

Relay operation tests

1 The relays are integral with the ECM and can only be checked using suitable ABS diagnostic test equipment.

Relay power supply tests

Note: *Refer to the wiring diagrams for specific ECM pin identification according to model.*

Main (solenoid valve) relay

2 Switch the ignition off and disconnect the ECM multi-plug.

3 Attach a negative voltmeter probe to a vehicle earth.

4 Attach a positive voltmeter probe to ECM multi-plug pin 9 **(see illustration 22.26)**. The voltmeter should indicate nbv. If no voltage is found, check the relevant fuse and the supply wiring back to the battery positive terminal.

Pump relay

5 Switch the ignition off and disconnect the ECM multi-plug.

6 Attach a negative voltmeter probe to a vehicle earth.

7 Attach a positive voltmeter probe to ECM multi-plug pin 25. The voltmeter should indicate nbv. If no voltage is found, check the relevant fuse and the supply wiring back to the battery positive terminal.

6 Electronic Control Module

Checking the ECM (general)

1 Inspect the ECM for corrosion or damage and ensure that the unit is securely attached to the hydraulic control unit.

2 Check that the ECM multi-plug terminals are pushed fully home

22

and making good contact with the ECM pins. A fault in any of the above areas are possible reasons for poor performance in the ABS system.

ECM power supply and earth tests

Note: *Refer to the wiring diagrams for specific ECM pin identification according to model.*

3 Switch the ignition off and disconnect the ECM multi-plug.

4 Switch the ignition on.

5 Attach a negative voltmeter probe to a vehicle earth.

6 Attach a positive voltmeter probe to ECM multi-plug pin 4 **(see illustration 22.26)**. The voltmeter should indicate nbv.

7 If no voltage is found, check the supply wiring back to the ignition switch.

8 Switch the ignition off.

9 Connect an ohmmeter between a vehicle earth and ECM multi-plug earth pins 8 and 24 in turn.

10 The ohmmeter should indicate continuity in each case. If not, check the ECM main earth connection and wiring.

7 Solenoid valves

Solenoid valve operation tests

1 The solenoid valves are integral with the ECM and can only be checked using suitable ABS diagnostic test equipment.

8 Hydraulic pump motor

Pump operation test

1 Switch the ignition off and disconnect the pump motor multi-plug.

2 Connect the positive terminal of a 12 volt supply to terminal 2 of the pump motor multi-plug, and the negative terminal to terminal 1 of the multi-plug **(see illustration 22.27)**. The pump motor should now run.

⚠️ *Warning: The test should be made as quickly as possible to avoid damaging the pump.*

3 If the pump does not operate as described, renew the hydraulic control unit.

9 Brake light switch

Checking the brake light switch (general)

1 Check that the brake light switch is correctly and securely mounted and that the plunger moves smoothly with no trace of binding.

2 Check that the wiring multi-plug is pushed fully home and making good contact.

3 Check that no wires have been disconnected.

4 A fault in any of the above areas are possible reasons for failure or malfunctioning of the switch.

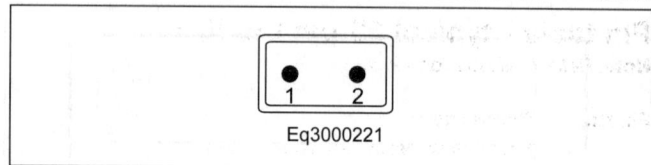

22.27 Pump motor multi-plug terminal identification

Brake light switch voltage and continuity tests

Voltage test

Note: *Refer to the wiring diagrams for specific ECM pin identification according to model.*

5 Switch the ignition off and disconnect the ECM multi-plug from the ECM.

6 Connect a voltmeter between a vehicle earth and the ECM multi-plug brake light switch input pin (typically pin 18) **(see illustration 22.26)**.

7 Switch the ignition on and depress the brake pedal. The voltmeter should indicate nbv.

8 If no voltage is found, the fuse, the brake light switch and the wiring are suspect.

9 Release the brake pedal. The voltage should drop to zero as the switch opens.

Continuity test

10 Switch the ignition off and disconnect the brake light switch multi-plug.

11 Connect an ohmmeter between the terminal pins of the brake light switch.

12 Operate the brake light switch and check for continuity. If the test fails, renew the switch.

10 Warning lamp

Checking the warning lamp (general)

1 Inspect the warning lamp bulb holder contacts in the instrument panel.

2 Check that the instrument panel multi-plug terminal pins are pushed fully home and making good contact.

3 Check that no wires have been disconnected.

4 A fault in any of the above areas are possible reasons for failure or malfunctioning of the warning lamp.

Warning lamp operation test

5 With the ignition switched off, the warning lamp should remain off.

6 Switch the ignition on and the warning lamp should illuminate then extinguish after a few seconds. The lamp should then remain off.

7 If the warning lamp comes on and remains on at any time during vehicle operation, carry out the previously described test procedures on the system components.

Pin table - typical 25-pin (Audi)

Note: *Refer to illustration 22.26*

Pin No.	Connection	Test condition	Voltage
1	Front left wheel speed sensor earth	Roadwheel rotating	0.25 volts (max)
2	Front left wheel speed sensor signal	Roadwheel rotating	0.5 to 1.5 volts AC (approx)
3	Encoding link		
4	Supply from ignition switch	Ignition on	Nbv
5	Rear left wheel speed sensor earth	Roadwheel rotating	0.25 volts (max)
6	Rear left wheel speed sensor signal	Roadwheel rotating	0.5 to 1.5 volts AC (approx)
7	SD connector/ABS warning lamp		
8	ECM earth	Ignition on	0.25 volts (max)
9	Supply from battery	Ignition off	Nbv
10	-	-	-
11	-	-	-
12	-	-	-
13	-	-	-
14	Encoding link		
15	-	-	-
16	ABS warning lamp	Ignition on:	
		Lamp on	2.0 volts (max)
		Lamp off	Nbv
17	-	-	-
18	Brake light switch input	Ignition on:	
		Brake pedal released	0 volts
		Brake pedal depressed	Nbv
19	Front right wheel speed sensor earth	Roadwheel rotating	0.25 volts (max)
20	Front right wheel speed sensor signal	Roadwheel rotating	0.5 to 1.5 volts AC (approx)
21	-	-	-
22	Rear right wheel speed sensor signal	Roadwheel rotating	0.5 to 1.5 volts AC (approx)
23	Rear right wheel speed sensor earth	Roadwheel rotating	0.25 volts (max)
24	ECM earth	Ignition on	0.25 volts (max)
25	Supply from battery	Ignition off	Nbv

Pin table - typical 25-pin (Citroën, Peugeot, Skoda, Volkswagen, Volvo)

Note: *Refer to illustration 22.26*

Pin No.	Connection	Test condition	Voltage
1	Front left wheel speed sensor earth	Roadwheel rotating	0.25 volts (max)
2	Front left wheel speed sensor signal	Roadwheel rotating	0.5 to 1.5 volts AC (approx)
3*	Link to pin 14	-	-
4	Supply from ignition switch	Ignition on	Nbv
5	Rear left wheel speed sensor earth	Roadwheel rotating	0.25 volts (max)
6	Rear left wheel speed sensor signal	Roadwheel rotating	0.5 to 1.5 volts AC (approx)
7	SD connector		
8	ECM earth	Ignition on	0.25 volts (max)
9	Supply from battery	Ignition off	Nbv
10*	Data BUS connection		
11*	Data BUS connection		
12	-	-	-
13	-	-	-
14*	Link to pin 3	-	-
15	-	-	-
16**	ABS warning lamp	Ignition on:	
		Lamp on	2.0 volts (max)
		Lamp off	Nbv
17	-	-	-

Pin table - typical 25-pin (Citroën, Peugeot, Skoda, Volkswagen, Volvo) (continued)

Note: *Refer to illustration 22.26*

Pin No.	Connection	Test condition	Voltage
18	Brake light switch input	Ignition on:	
		Brake pedal released	0 volts
		Brake pedal depressed	Nbv
19	Front right wheel speed sensor earth	Roadwheel rotating	0.25 volts (max)
20	Front right wheel speed sensor signal	Roadwheel rotating	0.5 to 1.5 volts AC (approx)
21	-	-	-
22	Rear right wheel speed sensor signal	Roadwheel rotating	0.5 to 1.5 volts AC (approx)
23	Rear right wheel speed sensor earth	Roadwheel rotating	0.25 volts (max)
24	ECM earth	Ignition on	0.25 volts (max)
25	Supply from battery	Ignition off	Nbv

* Certain models only   ** Not Volvo

Pin table - typical 25-pin (Ford)

Note: *Refer to illustration 22.26*

Pin No.	Connection	Test condition	Voltage
1	Front left wheel speed sensor signal	Roadwheel rotating	0.5 to 1.5 volts AC (approx)
2	Front left wheel speed sensor earth	Roadwheel rotating	0.25 volts (max)
3	-	-	-
4	Supply from ignition switch	Ignition on	Nbv
5	Rear left wheel speed sensor signal	Roadwheel rotating	0.5 to 1.5 volts AC (approx)
6	Rear left wheel speed sensor earth	Roadwheel rotating	0.25 volts (max)
7	SD connector		
8	ECM earth	Ignition on	0.25 volts (max)
9	Supply from battery	Ignition off	Nbv
10	-		
11	-		
12	Navigation control unit	-	-
13	-	-	-
14	Link to 17	-	-
15	-	-	-
16	ABS warning lamp	Ignition on:	
		Lamp on	2.0 volts (max)
		Lamp off	Nbv
17	Link to 14	-	-
18	Brake light switch input	Ignition off/on:	
		Brake pedal released	0 volts
		Brake pedal depressed	Nbv
19	Front right wheel speed sensor signal	Roadwheel rotating	0.5 to 1.5 volts AC (approx)
20	Front right wheel speed sensor earth	Roadwheel rotating	0.25 volts (max)
21	Navigation control unit	-	-
22	Rear right wheel speed sensor earth	Roadwheel rotating	0.25 volts (max)
23	Rear right wheel speed sensor signal	Roadwheel rotating	0.5 to 1.5 volts AC (approx)
24	ECM earth	Ignition on	0.25 volts (max)
25	Supply from battery	Ignition off	Nbv

Pin table - typical 25-pin (Renault)

Note: *Refer to illustration 22.26*

Pin No.	Connection	Test condition	Voltage
1	Front left wheel speed sensor signal	Ignition on/roadwheel rotating	0.8 or 1.6 volts (switching)
2	Front left wheel speed sensor supply	Ignition on	11.0 volts (approx)
3	Supply from ignition switch	Ignition on	Nbv
4	Supply from ignition switch	Ignition on	Nbv
5	Rear left wheel speed sensor signal	Ignition on/roadwheel rotating	0.8 or 1.6 volts (switching)

Pin No.	Connection	Test condition	Voltage
6	Rear left wheel speed sensor supply	Ignition on	11.0 volts (approx)
7	SD connector	-	-
8	ECM earth	Ignition on	0.25 volts (max)
9	Supply from battery	Ignition off	Nbv
10	-	-	-
11	Data BUS connection		
12	-	-	-
13	-	-	-
14	SD connector	-	-
15	ABS warning lamp	Ignition on: Lamp on Lamp off	 2.0 volts (max) Nbv
16	Engine check warning lamp		
17	Instrument panel connection		
18	Brake light switch input	Ignition on: Brake pedal released Brake pedal depressed	 0 volts Nbv
19	Front right wheel speed sensor signal	Ignition on/roadwheel rotating	0.8 or 1.6 volts (switching)
20	Front right wheel speed sensor supply	Ignition on	11.0 volts (approx)
21	-	-	-
22	Rear right wheel speed sensor supply	Ignition on	11.0 volts (approx)
23	Rear right wheel speed sensor signal	Ignition on/roadwheel rotating	0.8 or 1.6 volts (switching)
24	ECM earth	Ignition on	0.25 volts (max)
25	Supply from battery	Ignition off	Nbv

Pin table - typical 25-pin (Seat)

Note: *Refer to illustration 22.26*

Pin No.	Connection	Test condition	Voltage
1*	Front left wheel speed sensor earth	Roadwheel rotating	0.25 volts (max)
2*	Front left wheel speed sensor signal	Roadwheel rotating	0.5 to 1.5 volts AC (approx)
3	-	-	-
4	Supply from ignition switch	Ignition on	Nbv
5*	Rear left wheel speed sensor earth	Roadwheel rotating	0.25 volts (max)
6*	Rear left wheel speed sensor signal	Roadwheel rotating	0.5 to 1.5 volts AC (approx)
7	SD connector		
8	ECM earth	Ignition on	0.25 volts (max)
9	Supply from battery	Ignition off	Nbv
10	-	-	-
11	-	-	-
12	-	-	-
13*	Link to pin 17	-	-
14*	Link to pin 21	-	-
15	-	-	-
16	ABS warning lamp	Ignition on: Lamp on Lamp off	 2.0 volts (max) Nbv
17*	Link to pin 13	-	-
18	Brake light switch input	Ignition off: Brake pedal released Brake pedal depressed	 0 volts Nbv
19	Front right wheel speed sensor signal	Roadwheel rotating	0.5 to 1.5 volts AC (approx)
20	Front right wheel speed sensor earth	Roadwheel rotating	0.25 volts (max)
21*	Link to pin 14	-	-
22*	Rear right wheel speed sensor signal	Roadwheel rotating	0.5 to 1.5 volts AC (approx)
23*	Rear right wheel speed sensor earth	Roadwheel rotating	0.25 volts (max)
24	ECM earth	Ignition on	0.25 volts (max)
25	Supply from battery	Ignition off	Nbv

*Wheel speed sensor signal and earth pins transposed on some models

22

Fault codes

11 General fault codes

1 The Teves 20IE system requires the use of a FCR for obtaining fault codes. Flash codes are not available for output from this system.
2 If a FCR is available, it should be connected to the SD serial connector and used in accordance with the maker's instructions.
3 The FCR can be used for the following purposes:
a) Obtaining fault codes.
b) Clearing fault codes.
c) Obtaining datastream information.
d) Testing the system actuators (solenoid valve relay, pump relay and solenoid valves).

12 Ford and Renault fault codes

1 On Ford and Renault models, internal fault codes are used by the ECM to designate faults in the system components and circuits. A proprietary fault code reader (FCR) or system tester (such as the Ford FDS 2000 or Renault XR25) is required to interrogate the system. No actual fault code numbers are available although the component circuits checked by the ECM are similar to those shown for the other vehicles listed.

Fault code table (Audi, Seat, Skoda, Volkswagen)

Code	Item	Fault
00283	Front left wheel speed sensor	Poor wiring/connections; incorrect air gap; sensor or sensor ring dirty or damaged
00285	Front right wheel speed sensor	Poor wiring/connections; incorrect air gap; sensor or sensor ring dirty or damaged
00287	Rear right wheel speed sensor	Poor wiring/connections; incorrect air gap; sensor or sensor ring dirty or damaged
00290	Rear left wheel speed sensor	Poor wiring/connections; incorrect air gap; sensor or sensor ring dirty or damaged
00668	Supply voltage	Outside expected values
01044	ECM coding	ECM incorrectly coded, poor wiring connection
01130	ABS operation	Signal external interference, poor wiring connections
01276	Hydraulic pump	Faulty operation
65535	ECM	ECM defective

Fault code table (Citroën, Peugeot)

Code	Item	Fault
1	ECM	ECM defective
2	ECM supply voltage	Too high
3	ECM supply voltage	Too low
4	Relay supply voltage	No voltage, voltage too high or too low
5	Front left wheel speed sensor	Incorrect air gap; sensor or sensor ring dirty or damaged
6	Front left wheel speed sensor	Incorrect resistance
7	Front left wheel speed sensor	No signal / incorrect signal
8	Front left roadwheel	Brake or hydraulic system fault
9	Front right wheel speed sensor	Incorrect air gap; sensor or sensor ring dirty or damaged
10	Front right wheel speed sensor	Incorrect resistance
11	Front right wheel speed sensor	No signal / incorrect signal
12	Front right roadwheel	Brake or hydraulic system fault
13	Rear left wheel speed sensor	Incorrect air gap; sensor or sensor ring dirty or damaged
14	Rear left wheel speed sensor	Incorrect resistance
15	Rear left wheel speed sensor	No signal / incorrect signal
16	Rear left roadwheel	Brake or hydraulic system fault
17	Rear right wheel speed sensor	Incorrect air gap; sensor or sensor ring dirty or damaged
18	Rear right wheel speed sensor	Incorrect resistance
19	Rear right wheel speed sensor	No signal / incorrect signal
20	Rear right roadwheel	Brake or hydraulic system fault

Fault code table (Volvo)

Code	Item	Fault
141	Brake fluid pressure switch	Circuit fault
142	Brake light switch	No signal, poor wiring connections
143	Vehicle speed signal	Circuit fault
144	Front brake discs	Overheating (models with TRACS only)
211	Front left wheel speed sensor	No signal or signal outside expected values
212	Front right wheel speed sensor	No signal or signal outside expected values
213	Rear left wheel speed sensor	No signal or signal outside expected values
214	Rear right wheel speed sensor	No signal or signal outside expected values
221	Front left wheel speed sensor	ABS control phase too long - sensor/wiring fault, hydraulic control unit fault
222	Front right wheel speed sensor	ABS control phase too long - sensor/wiring fault, hydraulic control unit fault
223	Rear left wheel speed sensor	ABS control phase too long - sensor/wiring fault, hydraulic control unit fault
224	Rear right wheel speed sensor	ABS control phase too long - sensor/wiring fault, hydraulic control unit fault
311	Front left wheel speed sensor	No signal, wiring short circuit
312	Front right wheel speed sensor	No signal, wiring short circuit
313	Rear left wheel speed sensor	No signal, wiring short circuit
314	Rear right wheel speed sensor	No signal, wiring short circuit
321	Front left wheel speed sensor	Intermittent signal
322	Front right wheel speed sensor	Intermittent signal
323	Rear left wheel speed sensor	Intermittent signal
324	Rear right wheel speed sensor	Intermittent signal
411	Front left inlet solenoid valve	Circuit fault
412	Front left outlet solenoid valve	Circuit fault
413	Front right inlet solenoid valve	Circuit fault
414	Front right outlet solenoid valve	Circuit fault
421	Rear inlet solenoid valve	Circuit fault
422	Rear outlet solenoid valve	Circuit fault
423	Traction control solenoid valve	Circuit fault (models with TRACS only)
431	ECM	Internal circuit fault/ECM defective
432	ECM	Signal interference
433	Supply voltage	Voltage too high
441	Main relay	No voltage supply to solenoid valves
442	ECM	Internal circuit fault/ECM defective
443	Hydraulic pump motor	Circuit fault/pump defective
444	ECM	Internal circuit fault/ECM defective
445	ECM	Internal circuit fault/ECM defective

Wiring diagrams

22.28 25-pin wiring diagram, Audi A3

22.29 25-pin wiring diagram, Citroën Saxo

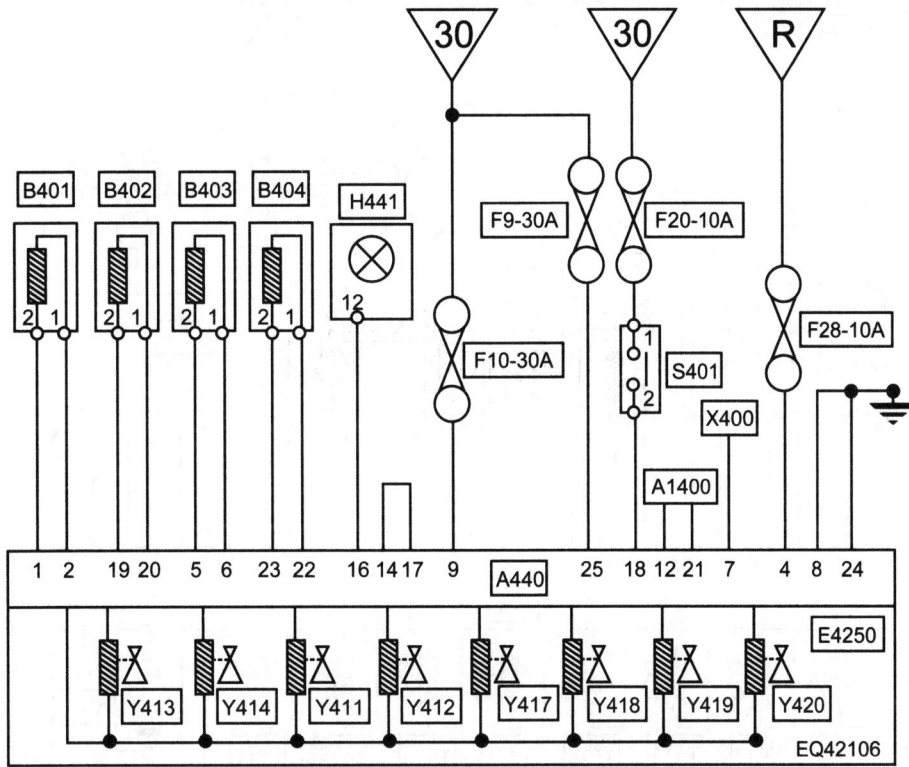

22.30 25-pin wiring diagram, Ford Galaxy

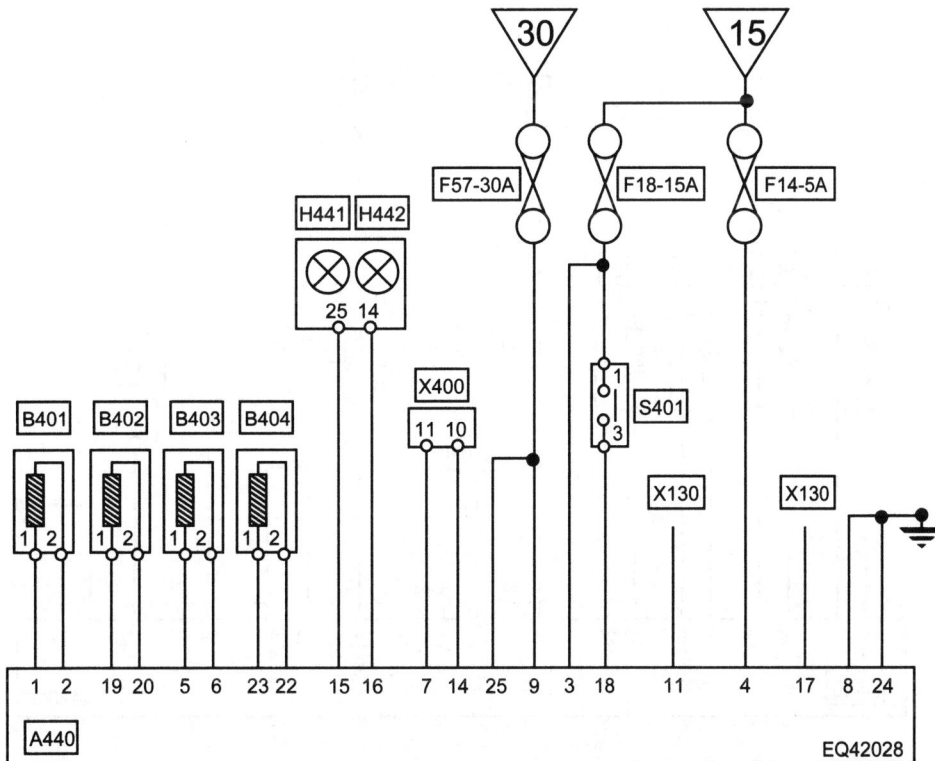

22.31 25-pin wiring diagram, Renault Laguna

22

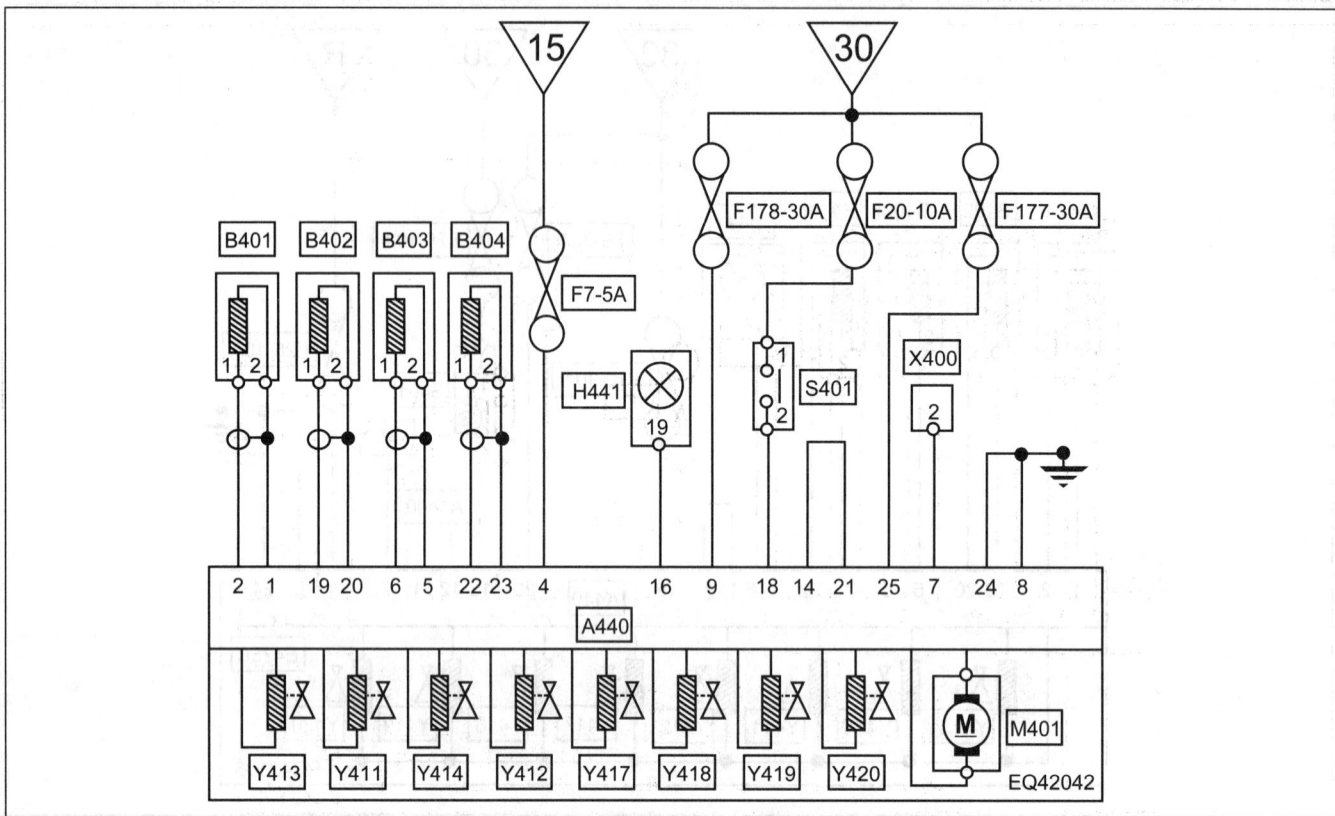

22.32 25-pin wiring diagram, Seat Arosa

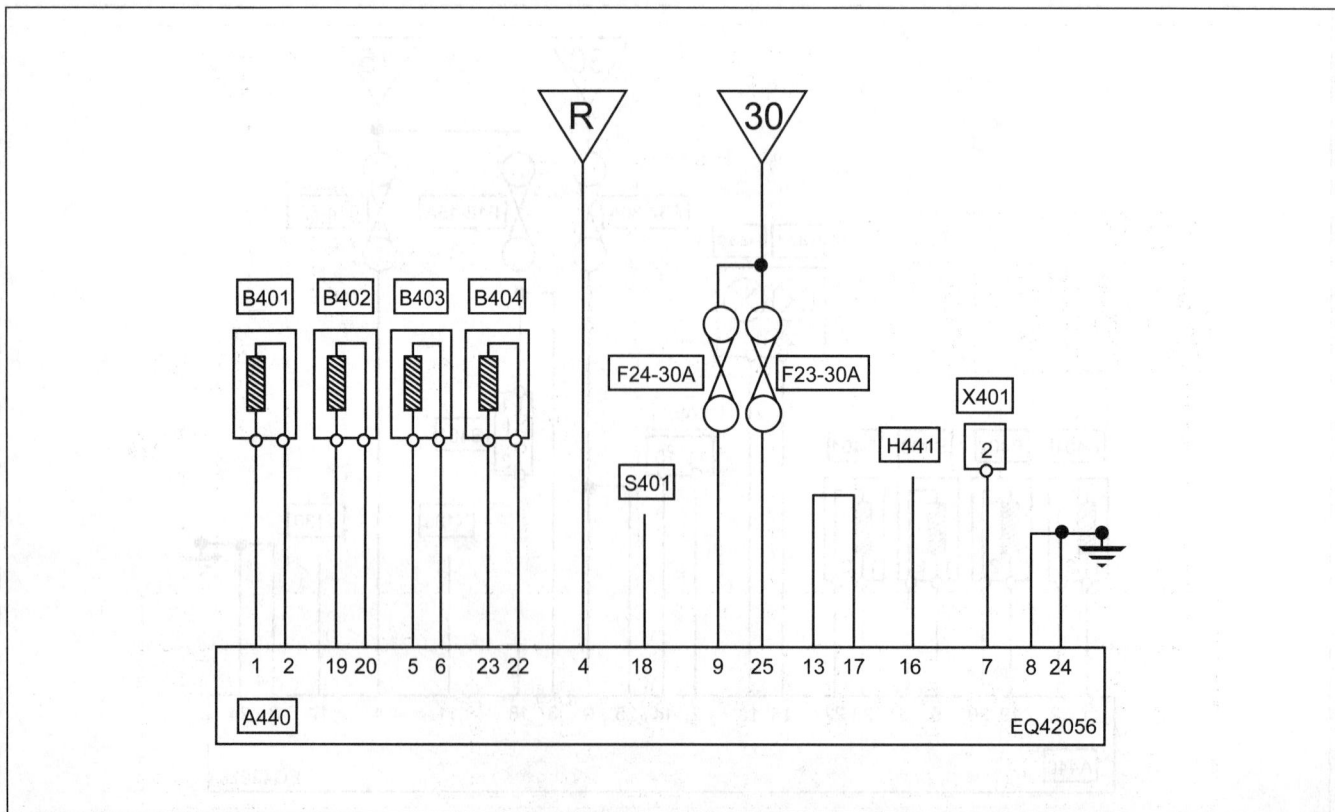

22.33 25-pin wiring diagram, Seat Inca

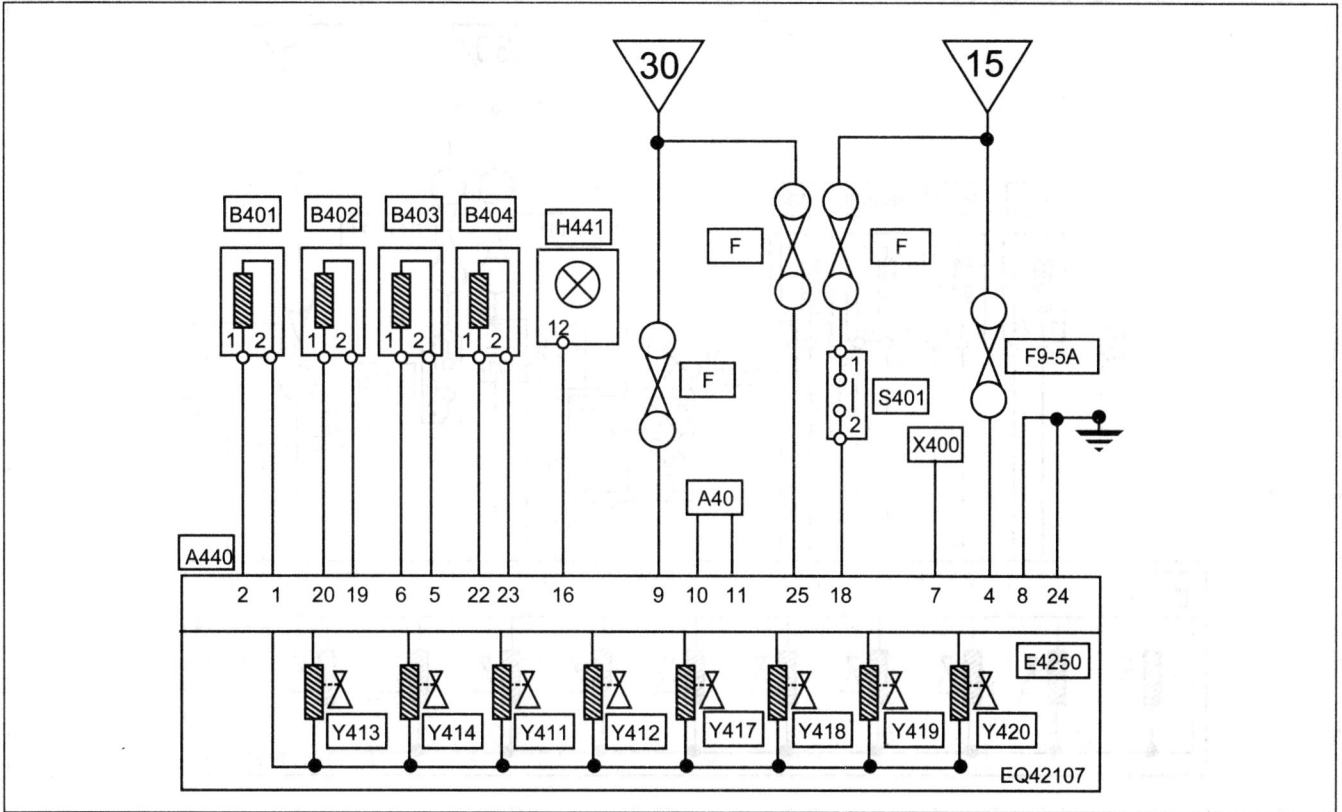

22.34 25-pin wiring diagram, Skoda Octavia

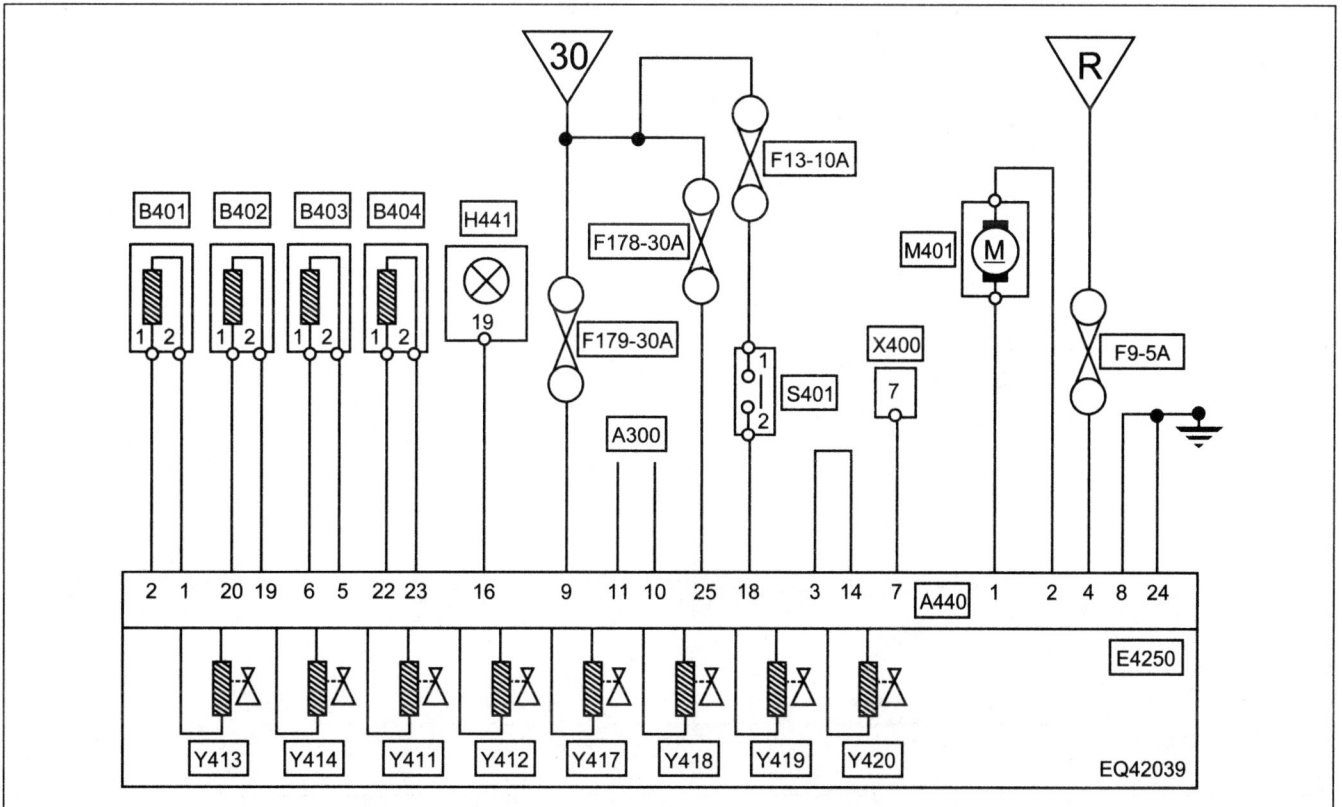

22.35 25-pin wiring diagram, Volkswagen Golf

22

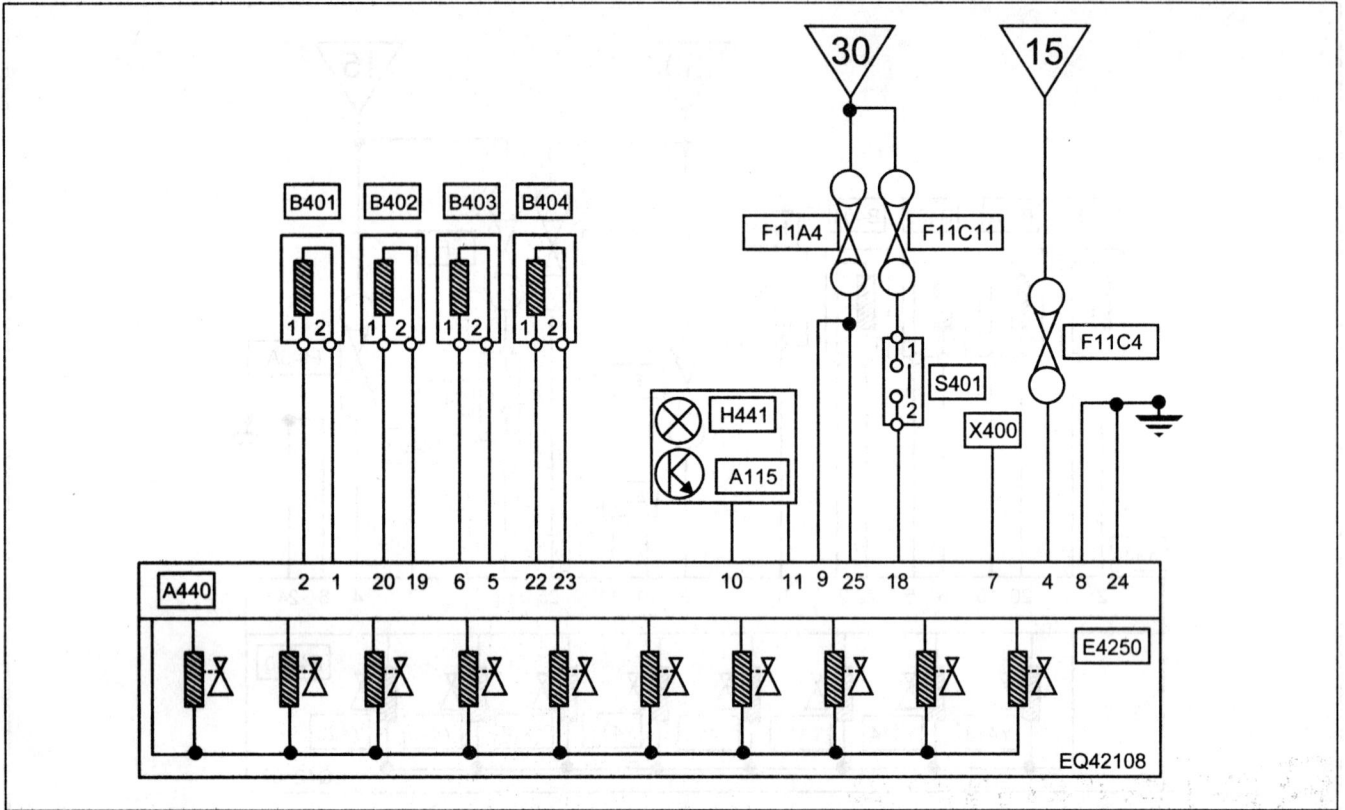

22.36 25-pin wiring diagram, Volvo S70/V70/C70

Reference

Abbreviations and glossary of technical terms

A

A (amperes)

ABS (Antilock Brake System)

A/C (Air Conditioning)

AC (Alternating Current)
An electric current that first flows in one direction and then the opposite. AC voltage is produced by an alternator or by a pulse generator. AC voltage must be rectified to DC before it can be used in the vehicle charging system. AC voltage from a pulse generator is converted to DC by an Analogue to Digital converter.

Actuator
A device such as a relay or solenoid controlled by the ECM.

Actuator driver
See driver (actuator).

ACW (Anti-Clock Wise)
Direction of rotation.

ADC (Analogue to Digital Converter)

ALDL (Assembly Line Diagnostic Link)
The name given to the serial data port used on GM vehicles.

Ammeter
An instrument for measuring current in amperes.

Amp (abbreviation for ampere)
A unit measurement of current flow.

Amplitude
Square waveform: Difference between the maximum and minimum voltage.
AC waveform: Difference between zero and either the maximum or minimum peak.

Analogue signal
An analogue signal is defined as a continuous signal that can change by an infinitely small amount. Any sensor that meets these conditions can also be called an analogue sensor. Typically, an analogue signal is measured by an instrument that uses a needle to progressively sweep across a fixed scale. Any change in the signal will cause the needle to move by a similar amount.

Arcing
Unwanted electrical bridging in a circuit.

AT (Automatic Transmission)

ATA (Automatic Transmission Actuator - Ford term)

ATR (Automatic Transmission Relay - Ford term)

B

Backprobe
A method of obtaining voltage from the multi-plug pin of an electronic component or sensor. The multi-plug must remain connected to the component. The multi-plug insulating boot should be peeled back and the voltmeter positive probe attached to the relevant pin - ignition key on. Note: In this book, the multi-plug diagrams show the terminals of the harness connector. When backprobing the multi-plug (or viewing the sensor connector terminals), the terminal positions will be reversed.

Bar
A unit of pressure. One bar is almost equal to atmospheric pressure.

Barometric pressure
Equal to atmospheric pressure. At sea-level, this is 100 kPa.

Battery
A storage device for electrical energy in chemical form. The primary function of the battery is to provide ignition current during the engine starting period and power to operate the starter motor. This must be accomplished irrespective of adverse temperature conditions. The battery also serves, for a limited time, as a current source to satisfy the electrical demands of the vehicle that are in excess of the alternator output.

BOB (Break Out Box)
A box containing a number of connectors that allows easy access to the ECM input and output signals, without directly probing the ECM pins. The BOB loom terminates in a universal connector. A multi-plug harness of similar construction to the ECM harness is interfaced between the ECM and its multi-plug, and the other end is connected to the BOB loom. The BOB will now intercept all signals that go to and from the ECM.

BOO (Brake On/Off switch - Ford term)

Bra (Brake signal)

C

°C (Celsius or Centigrade)
Measurement of temperature. Centigrade is more accurately used as a measure of angle (hundredths of a grade).

Cable
Heavy electrical wire used to conduct high voltage or high current. ie battery cables.

Cadence braking
An advanced driving technique whereby the brakes are applied and released by rapidly depressing and releasing the brake pedal many times during the braking manoeuvre.

Celsius
See °C

Circuit
An electrical path through which current can flow and that begins and ends at the current source. A circuit is NOT complete unless the current can return to the source. In modern systems the current flows from the positive terminal of the battery, through wires or cables and switches to the load (ie a starter motor). The return is through earth to the negative terminal of the battery.

Conductor
A material that will pass electrical current efficiently. A good conductor depends on material used, length, cross sectional area and temperature.

Control signal
See driver (actuator), relay driver, final stage.

Corrosion
Deterioration and crumbling of a component by chemical action. Sensor terminals and multi-plugs are particularly susceptible to this complaint.

CPU (Central Processing Unit)

Cranking
Rotating the engine by use of the starter motor.

Current
The flow of electrons through a conductor and measured in amps.

CW (clockwise)
Direction of rotation.

D

Datastream
Once the FCR has decoded a fault, a datastream enquiry (some systems only) is a quick method of determining where the fault might lie. This data may take various forms but is essentially electrical data on voltage, frequency, dwell or pulse duration, temperature etc, provided by the various sensors and actuators. Unfortunately, such data is not available from all vehicle systems. Since the data is in real time, various tests can be made and the response of the sensor or actuator evaluated.

DC (Direct Current)
An electrical current source which flows in only one direction.

Degree
1/360 part of a circle.

Digital signal
A digital signal is represented by a code that has two states, on and off. In simple terms, the signal consists of a series of digital pulses when the frequency, pulse width or number of pulses is used to indicate a specific value.

Because the ECM works in a digital fashion, all analogue signals must pass through an analogue to digital converter when the signal will be stored by the ECM in digital format. A digital signal from a digital sensor does not need converting, and processing by the ECM is therefore much faster.

DIN
International standard used in the automotive industry.

Diesel
A fuel injected engine that uses the high temperature generated in compression to ignite the charge.

Diode
A transistor that allows current flow in one direction alone.

DMM (Digital Multi-Meter)
An instrument designed for automotive use that can measure voltage, current, resistance and sometimes covers other functions such as dwell, duty cycle, frequency, rpm etc.

Driver (actuator) - refer also to relay driver, control signal and final stage

The system actuators are supplied with a voltage feed from either the ignition switch or from one of the ABS or fuel injection system relays. The earth connection is then connected to an ECM earth pin. When the ECM actuates the component, it drives the appropriate ECM pin to earth by completing the circuit internally for as long as the actuation is required.

This signal could be termed a 'driver' or a 'final stage' or a 'control' signal. In this book we have settled for the 'driver' term. Examples of an actuator are, fuel injector, relay, solenoid valve etc.

DVM (Digital VoltMeter)

Dynamic testing

Testing a device whilst it is running as opposed to static testing.

E

Earth

A path for current to return to the power source.

EC (European Community)

ECM (Electronic control module)

A computer control unit that assimilates information from various sensors and computes an output. Used to control the various vehicle systems ie; engine management, antilock brakes, automatic transmission, air bags, etc.

ECU (Electronic Control Unit - see ECM)

EGS (Electronic Transmission Control)

The common abbreviation 'EGS' is the German abbreviation for this component.

EML (Electronic throttle)

The common abbreviation 'EML' is the German abbreviation for this component.

EMS (Engine Management System)

EPROM (Electronic Programmable Read Only Memory)

F

Fahrenheit Temperature scale.

Fast codes

Digital fault codes emitted by an ECM that are too fast to be displayed on an LED lamp, or on an instrument panel warning lamp. A digital FCR instrument is required for capturing fast codes.

Fault codes

Electronics are now extensively used throughout the modern vehicle and often control functions such as the, suspension, automatic transmission, air conditioning, suspension, antilock brakes, air bags and myriad others.

Most modern vehicle ECMs have the facility of making self-diagnostic checks upon the sensors and actuators used in a particular system. A fault in one of the components or circuits causes a flag or code to be set in the ECM memory.

If a suitable code reading device is attached to the serial port on the vehicle harness, these faults can then be read out from the ECM and displayed in the form of a two or three digit output code.

FCR (Fault Code Reader)

A device that can be connected to the vehicle serial (diagnostic port) to interrogate the vehicle ECM(s). Fault codes and datastream information can then be read from the ECM. In some instances, vehicle actuators can be actuated from the controls on the FCR.

The codes may be described as slow or fast and some ECMs are capable of emitting both types. Slow codes can be captured by an LED tool, whereas fast codes must be captured by a digital FCR.

Final stage

See driver (actuator), relay driver and control signal.

Flash codes

Fault codes of the slow variety that are output on an instrument panel warning lamp, or via an LED lamp.

Frequency

Frequency of a pulse. Usually measured in Hz.

Fuse

A small component containing a sliver of metal that is inserted into a circuit. The fuse will blow at a specified current rating, in order to protect the circuit from voltage overload.

Fuselink (also known as a fusible link)

A heavy duty circuit protection component that can burnout if the circuit becomes overloaded.

G

GM (General Motors)

Manufacturer of Opel and Vauxhall in Europe. The parent company is based in the USA.

GND (ground)

USA term for earth. See also earth.

H

Hard faults

Generally refers to faults logged by an ECM self-diagnosis routine. The faults are usually present at the moment of testing.

Hydraulic control unit

The ABS hydraulic control unit consists basically of a series of solenoid valves and an electrically driven fluid return pump. Under ECM control, the unit regulates the hydraulic pressure applied to each brake, or pair of brakes, during a controlled ABS braking cycle. Often referred to as a hydraulic modulator.

Hz (Hertz)

Frequency in cycles per second.

I

ID (Identification)

Impedance

Resistance to the flow of current and often used to describe the resistance of a voltmeter. A minimum 10 megohm impedance is recommended for instruments used to measure values in electronic circuits.

Insulator A material that will not pass current readily, and therefore used to prevent electrical leakage.

inst panel

Instrument panel on the vehicle dashboard or facia.

ISO (International Standards Organisation)

J

Jumper lead
A small electrical cable that is used to bridge a component or wire on a temporary basis.

J1930
SAE standard for acronyms describing electrical and electronic components.

K

kohms (kilohms)
A resistance measurement equal to 1000 ohms.

kPa (kilopascals)
International standard for the measurement of pressure and vacuum.

kV (kilovolt)
A unit of secondary voltage measurement equal to 1000 volts.

L

LED (Light Emitting Diode)

Lb/in² (Pounds per Square Inch)
An Imperial measurement of pressure.

LHS (Left-Hand Side)
Viewed from the drivers seat.

Limp home see LOS.

LOS (Limited Operating Strategy)
Often called LIMP HOME, this is a safety system incorporated into current engine management ECMs and also many of the ECMs controlling other vehicle systems (automatic transmission for example). LOS allows the vehicle to be driven to a service area if a fault occurs. Some LOS systems are so sophisticated that the driver may be unaware from the way that the vehicle operates, that a fault has occurred.

When the system perceives that a sensor is operating outside of its design parameters, a substitute value is used which allows the engine, or other system, to operate. In an engine management system, the substitute value is usually that for a hot or semi-hot engine and this means that the engine may be difficult to start and run badly when it is cold. The instrument panel warning light (where fitted) is switched on to indicate that a fault has occurred.

M

mA (milliamperes)

Magnet
A substance that has the ability to attract iron.

Magnetic field
The space around a magnet that is filled by invisible lines of magnetic force.

Master cylinder
A mechanical device which converts driver pressure applied to the brake pedal into hydraulic pressure to apply the brakes. Current units consist of dual hydraulic chambers and pistons (one for the primary braking circuit and one for the secondary circuit) and are referred to as 'tandem' master cylinders.

Max
Abbreviation for maximum.

Molecule
The smallest particle into which a chemical compound may be divided.

ms (millisecond)
1/1000 second (0.001 s).

MT (Manual Transmission)

Multimeter See DMM.

Multi-plug
A multiple-terminal connecting plug in the wiring harness. Often used to connect the harness to a sensor or actuator. In this book, the multi-plug diagrams show the terminals of the harness connector. When backprobing the multi-plug (or viewing the sensor connector terminals), the terminal positions will be reversed.

mV (millivolt)
one millivolt = 1/1000 of a volt (0.001 V).

MY (Model Year)
Most VMs start manufacturing their latest models in the months leading up to the end of a particular year. The actual date when manufacturing commences is usually termed the 'model year' date, and the year used is usually that of the following year. For example, September 1995 could mark the commencement of manufacture of 1996 'model year' vehicles.

N

Nbv (nominal battery voltage)
Nominally 12 volts, the voltage will vary under engine operating conditions:
Engine stopped: 12 - 13 volts.
Engine cranking: 9.0 to 12.0 volts.
Engine running: 13.8 to 14.8 volts.

Nearside
Side nearest to the kerb on any vehicle - irrespective of whether it is left-hand or right-hand drive.

NEEC (New European Economic Community)

Newton (N)
An international unit of force that is independent of gravity. This unit was introduced because gravity varies in different parts of the world. The Newton is defined as the force required to accelerate a mass of 1kg at 1 metre per second per second. Newton units of force are measured as N/m² and called Pascal units. This unit is very small, and measured in MPa (1 000 000 Pascals) or kPa (1000 Pascals). See also Pascal.

Non-sinusoidal
Waveforms such as sawtooth, square, ripple, etc.

Non-volatile memory
ECM memory that is able to retain information - even when the vehicle battery is disconnected.

NTC (Negative Temperature Co-efficient)
A thermister in which the resistance falls as the temperature rises. An NTC resistor decreases (negatively) in resistance as the temperature rises.

O

OBD (On-Board Diagnostics)

Ohm
A unit of resistance that opposes current flow in a circuit.

Ohmmeter
An instrument used to measure resistance in ohms.

Ohms law
Volts = Amps x Ohms (V = I x R)
Amps = Volts / Ohms (I = V / R)
Ohms = Volts / Amps (R = V / I)
Also:
Power (Watts) = Volts x Amps

OOR (Outside Operating Range)

Open circuit
A break in an electrical circuit which prevents the flow of current.

Oscilloscope
A high speed voltmeter that visually displays a change in voltage against time. Used to display sensor or actuator waveforms.

OVP (Over Voltage Protection)

P

Pascal
International standard for the measurement of pressure and vacuum. Refer also to Newton.

Percent
Parts of a hundred.

Permanent magnet
A magnet that has a magnetic field at all times.

Polarity
A positive or negative state with reference to two electrical poles.

Pot (potentiometer)
A variable resistance.

Probe
A method of obtaining voltage from the multi-plug pin of an electronic component or sensor. The multi-plug should be disconnected from the component and the voltmeter positive test lead used to probe the relevant pin.

PROM (Programmable Read Only Memory)

PSI (Pounds per Square Inch)
An Imperial measurement of pressure.

PTC (Positive Temperature Co-efficient)
A thermister in which the resistance rises as the temperature rises. A PTC resistor increases (positively) in resistance as the temperature rises.

Pulse
A digital signal sent by the ECM.

Pulse width
The time period during which an electronic component is energised. It is usually measure in milliseconds.

PWM (Pulse Width Modulated)

R

RAM (Random Access Memory - computer term)

Reference voltage
During normal engine operation, battery voltage could vary between 9.5 (cranking) and 14.5 (running). To minimise the effect on engine sensors (for which the ECM would need to compensate), many ECM voltage supplies to the sensors are made at a constant value (known as a reference voltage) of 5.0 volts.

Relay
An electro-magnetic switching solenoid controlled by a fine shunt coil. A small current activates the shunt winding, which then exerts magnetic force to close the relay switching contacts. The relay is often used when a low current circuit is required to connect one or more circuits that operate at high current levels. The relay terminal numbers are usually annotated to the DIN standard, to which most (but not all) European VMs subscribe. Typical relay annotation to DIN standard:

30   Supply voltage direct from the battery positive terminal.
31   Earth return direct to battery.
85   Relay earth for energising system. May be connected direct to earth, or 'driven' to earth through the ECM.
85b Relay earth for output. May be connected direct to earth, or 'driven' to earth through the ECM.
86   Energising system supply. May arrive from battery positive or through the ignition switch.
87   Output from first relay or first relay winding. This terminal will often provide power to the second relay terminal 86 and provide voltage to the ECM, solenoids etc.
87b Output from second relay or second relay winding. Often provides power to the ABS hydraulic pump.

Relay control/relay driver
The system relays are supplied with a voltage feed from either the battery, ignition switch or from one of the system relays. The earth connection is then connected to an ECM earth pin. When the ECM actuates the relay, it drives the appropriate ECM pin to earth by completing the circuit internally for as long as the actuation is required.

Depending upon the relay, the input signal may be instigated by switching on the ignition or cranking the engine. Once the ECM has received the signal, the ECM will 'drive' the relay to earth by completing the circuit internally. This signal could be termed a 'driver' or a 'final stage' or a 'control' signal. In this book we have settled for the 'driver' term.

Res
Abbreviation of resistance.

Resistance
Opposition to the flow of current.

Return
Term used to describe the earth return path to an ECM or module of typically a sensor, or relay when the return is not directly connected to earth. The ECM or module will internally connect the return to one of its own earth connections. By this method the number of earth connections is much reduced.

RFI (Radio Frequency Interference)
The EMS is susceptible to outside interference. Radiated RFI can be a problem if the levels are high enough and this can emanate from items such as a faulty ignition secondary HT circuit or a faulty alternator. Excess RFI can disrupt and affect ECM operation.

RHS (Right-Hand Side)
Viewed from the drivers seat.

RMS (Root mean square)
AC equivalent to DC voltage. Can be calculated from AC amplitude by the formula:
 AC amplitude x 0.707.

ROM (Read Only Memory - computer term)

RPM (revolutions per minute)
A measure of engine speed.

S

SAE (Society of Automotive Engineers)
The Society sets standards for automotive engineering.

Scanner See FCR.
US term for a FCR.

Scope
See oscilloscope. Abbreviation for an oscilloscope.

SD (Self-Diagnostics)

Self-diagnosis of serial data See fault codes.

Sensor
A device that can measure one or more of the following parameters: wheel speed, temperature, position, airflow, pressure etc., and returns this information to the ECM, in the form of a voltage or current signal, for processing by the ECM.

Serial data port
The serial port is an output terminal from the ECM. Signals have therefore been processed and faults or values are output to the terminal as a coded digital signal.

Short, short to earth, or short circuit
When electricity goes to earth and takes a shorter path back to the power source. Because extremely high current values are present, the condition can cause an electrical fire.

Signal voltage
A varying voltage returned to the ECM by a sensor so that the ECM can detect load, or temperature.

Sinusoidal
A sine wave (ie a wheel speed sensor waveform) where the amplitude of the positive part of the waveform is roughly equal to the amplitude of the negative part of the waveform.

Slow codes
Fault codes emitted by an ECM that are slow enough to be displayed on an LED lamp or instrument panel warning lamp.

Soft faults
Generally refers to intermittent faults logged by an ECM self-diagnosis routine. The faults are often not present at the moment of testing, but have been logged at some period in the past.

Solenoid
An electrical device that produces a mechanical effort when energised.

STAR (Self Test Automatic Readout - Ford term)
Electronic FCR test

STC (Self-Test Connector - Ford term)
Refer to Self Diagnosis

STI (Self-test Input - Ford term)

STO (Self-test Output - Ford term)

Suppression Reduction of radio or television interference generated by the high voltage ignition system. Typical means used are radio capacitors or resistive components in the secondary ignition circuit.

Suppressor Used to prevent radio interference. See capacitor.

SYOV (System Overview)
A term to describe the technical description of how the system operates.

T

Tandem master cylinder See Master cylinder.

TxD (Diagnostic Link signal - BMW term)

TDCL (Total Diagnostic Communication Link - Toyota term)

Temp
Abbreviation for temperature.

Terminal
An electrical connecting point.

Thermister
A potentiometer controlled by temperature.

Transistor
An electronic switching component.

Trouble codes
US term for fault codes.

U

U-Batt
Battery voltage - BMW term

U-Verst
Supply voltage - BMW term

V

Vacuum
A negative pressure or a pressure less than atmospheric. Measured in millibars or inches of mercury. A perfect vacuum exists in a space which is totally empty. It contains no atoms or molecules and therefore has no pressure. In practice a perfect vacuum cannot be achieved.

Vacuum gauge
A gauge used to measure the amount of vacuum in the engine intake system.

Vacuum servo unit
The vacuum servo unit utilises engine inlet manifold vacuum to increase the force applied by the driver on the brake pedal, thus reducing the effort required to operate the brakes.

VIN (Vehicle Identification Number)
A serial number to identify the vehicle. The number often contains coded letters to identify model & year.

VM
Vehicle manufacturer

Volt
A unit of electrical pressure.

Voltage
Electrical pressure.

Voltage drop
Voltage drop is voltage expended when a current flows through a resistance. The greater the resistance then the higher the voltage drop. The total voltage drop in any automotive circuit should be no more than 10%.

Voltmeter
An instrument used to measure voltage in a circuit in volts.

VSS (Vehicle Speed Sensor)
A sensor to measure the road speed of the vehicle.

W

Watt
A unit of electrical power. 746 watts are equal to one mechanical horsepower.

Wheel speed sensor
In the ABS system, the rotational speed of the roadwheels and any changes in the rotational speed are recorded by the wheel speed sensors. Typically, each assembly comprises a toothed sensor ring which rotates at roadwheel speed and an adjacent sensor mounted a set distance from the sensor ring. The sensors are typically permanent magnet pulse generators producing an AC voltage sine wave as the sensor ring teeth pass through the magnetic field of the sensor. On later systems, 'active' wheel speed sensors may be used, which receive a DC voltage from the ECM. The voltage is switched on and off as the wheel rotates, generating a DC square wave signal.

Wiggle test
A test in which a suspect connection is wiggled or gently tapped, or a gentle heat or a cooling effect is applied. If the voltage or resistance of the circuit alters during the test, that connection may be suspect.

Warnings: Precautions to be taken with automotive electronic circuits

1 VERY IMPORTANT: Avoid severe damage to the ECM by switching the ignition OFF before disconnecting the multi-plug to this component. The ignition must also be switched off before disconnecting any other ABS wiring connector, multi-plug or component.

2 Safety is extremely important when carrying out any work on the braking system. All connections and components must be scrupulously clean and all repairs must be of a permanent nature. **DO NOT** make temporary repairs or attempt to modify the system in any way.

3 DO NOT attempt to dismantle or repair the hydraulic control unit. This is generally a sealed unit and replacement internal components are not available separately.

4 DO NOT drive the car with the ABS ECM disconnected, or with test equipment connected to the ECM or multi-plug. The test equipment may inadvertently operate the solenoid valves in the hydraulic control unit, causing individual wheel lock and loss of vehicle control.

5 When taking voltage readings at a multi-plug or terminal block, the use of meter leads with thin probes is strongly recommended. However, it is useful to attach a paper clip or split pin to the terminal, and attach the voltmeter to the clip. Be very careful not to short out these clips. A number of ECMs employ gold plated pins at the ECM multi-plug. Particular care should be taken that this plating is not removed by insensitive probing. To be entirely safe when probing the ECM multi-plug, the use of a break-out box is strongly recommended.

6 DO NOT use an analogue voltmeter, or a digital voltmeter with an electrical impedance of less than 10 megohms, to take voltage readings at an ECM with the ECM in circuit. (Unless the manufacturer of the equipment warranties that no damage will ensue).

7 To prevent damage to a DMM or to the vehicle electronic system, the appropriate measuring range should be selected before the instrument probes are connected to the vehicle.

8 During resistance tests with an ohmmeter, always ensure that the ignition is OFF and that the circuit is isolated from a voltage supply. Resistance tests should **NOT** be made at the ECM pins. Damage could be caused to sensitive components, and in any case results would be meaningless.

9 When removing battery cables, good electrical procedure dictates that the earth (negative) cable is disconnected before the live (positive) cable. This will prevent spurious voltage spikes that can cause damage to electronic components.

10 Most modern radios are coded as a security measure and the radio will lose its coding and its pre-selected stations when the battery is disconnected. The code should be obtained from the vehicle owner before disconnecting the battery for component renewal, or to make other repairs.

11 Use protected jumper cables when jump starting a vehicle equipped with an ECM. If unprotected cables are used, and the vehicle earth cables are in poor condition, a voltage spike may destroy the ECM.

12 When a battery is discharged, by far the best course of action is to recharge the battery (or renew if faulty), before attempting to start the vehicle. The ECM is put at risk from defective components such as battery, starter, battery cables and earth cables.

13 DO NOT use a boost charger or allow a voltage higher than 16.0 volt when attempting to start an engine. The battery leads should be disconnected before a boost charger is used to quick charge the battery.

14 The ECM must not be exposed to a temperature exceeding 80° C. If the vehicle is to be placed in a vehicle spray booth, the ECM must be disconnected and removed from the car to a place of safety.

15 Disconnect all ECMs when welding repairs are to be made upon a vehicle.

16 Many modern vehicles are now equipped with SRS (Supplemental Restraint System) which is an airbag assembly which may be installed in the steering wheel, passenger compartment facia and, in many instances, in the sides of the front seats. Extreme caution must be exercised when repairing components situated close to the wiring or components of the SRS. In some vehicles, the SRS wiring runs under the facia, and related SRS components are situated in the steering wheel, in and around the under facia area, and adjacent to some components used in the ABS. Any damage to the SRS wiring must be repaired by renewing the whole harness. Improper removal or disturbance of SRS components or wiring could lead to SRS failure or accidental deployment. Failure to observe these precautions can lead to unexpected deployment of the SRS and severe personal injury. In addition, the SRS must be repaired and serviced according to the procedures laid down by the manufacturer. Any impairment of the SRS could lead to its failure to deploy in an emergency and leave the vehicle occupants unprotected.

Master component key

15	Ignition switch supply	M401	Hydraulic pump motor
30	Battery supply	P401	Tachometer
50	Starter circuit	R	Ignition switch accessory position
A40	Engine management control module	R2	Resistor
A115	Instrument cluster	R430	Brake pedal position sensor
A300	Automatic transmission control unit	S55	Oil Pressure Switch
A440	ABS control module (ABS-ECM)	S400	ABS switch
A1400	Navigation control module (Nav-ECM)	S401	Brake light switch
B101	Vehicle Speed Signal	S403	Hand brake warning switch
B401	Inductive wheel speed sensor, front left	S410	ABS High pressure switch
B402	Inductive wheel speed sensor, front right	S411	ABS Warning pressure switch
B403	Inductive wheel speed sensor, rear left	S412	ABS pressure control switch
B404	Inductive wheel speed sensor, rear right	S413	Brake Fluid level sensor
B407	Inductive wheel speed sensor, rear axle	S414	ABS pressure switch
B410	Pump motor sensor	V1	Diode
B411	Active wheel speed sensor, front left	X10	Self Diagnosis connector, 10-pin
B412	Active wheel speed sensor, front right	X16	Self Diagnosis connector, 16-pin
B413	Active wheel speed sensor, rear left	X20	Self Diagnosis connector, 20-pin
B414	Active wheel speed sensor, rear right	X130	Instrument Panel Connector
B465	Acceleration sensor	X400	Self Diagnosis connector, ABS
B495	Suspension G-Force sensor	X401	Self Diagnosis connector, ABS, 1-pin
E167	Brake light	X402	Self Diagnosis connector, ABS, 2-pin
E700	Electronic throttle	X403	Self Diagnosis connector, ABS, 3-pin
E4250	Hydraulic control unit	X404	Self Diagnosis connector, ABS, 4-pin
F	Fuse	X406	Self Diagnosis connector, ABS, 6-pin
F1-46	Fuses	X435	Instrument cluster connector
FM5	Fuse	Y401	Solenoid valve, front left
FMF3	Fuse	Y402	Solenoid valve, front right
G1	Battery	Y403	Solenoid valve, rear left
G50	Alternator	Y404	Solenoid valve, rear right
G450	Alternator	Y405	Solenoid valve, rear combined
H41	Engine check light	Y406	Solenoid valve, left front restricting
H440	Handbrake warning lamp	Y407	Solenoid valve, right front restricting
H441	ABS warning lamp	Y411	Solenoid valve, front left inlet
H442	Hand brake/fluid level Warning lamp	Y412	Solenoid valve, front left outlet
H450	Traction control Warning lamp	Y413	Solenoid valve, front right inlet
H460	Brake light	Y414	Solenoid valve, front right outlet
K1	Ignition relay	Y415	Solenoid valve, rear axle inlet
K30	Injection relay	Y416	Solenoid valve, rear axle outlet
K40	Reverse polarity protection relay	Y417	Solenoid valve, rear left inlet
K430	Main ABS relay	Y418	Solenoid valve, rear left outlet
K431	Secondary relay	Y419	Solenoid valve, rear right inlet
K435	Hydraulic pump relay	Y420	Solenoid valve, rear right outlet
K440	Solenoid valve relay	Y430	Solenoid valve Inlet Isolating
K445	Overvoltage (surge) protection relay	Y431	Solenoid valve Outlet Isolating
K450	ABS warning Lamp Relay	Y450	Solenoid valve, water
K530	Main relay		